Introductory Algebra
Concepts with Applications

Charles P. McKeague

*xyz*textbooks

2ND PRINTING — December 2012
3RD PRINTING — February 2013
4TH PRINTING — March 2014
5TH PRINTING — August 2014

Introductory Algebra:
Concepts with Applications
Charles P. McKeague

Project Manager: Staci Truelson

Production Manager: Matthew Hoy

Developmental Editor: Katherine Heistand Shields

Illustrator and Cover Design: Kaela Soohoo

Illustrator: Joyce Lin

Cover Image: Katherine Heistand Shields

Text Designer: Devin Christ

Production Manager: Matthew Hoy

Reviewer: Jeffrey Saikali

AIE Edition: ISBN-13: 978-1-936368-09-9 / ISBN-10: 1-936368-09-9
Student Edition: ISBN-13: 978-1-936368-08-2 / ISBN-10: 1-936368-08-0

For product information and technology assistance, contact us at
XYZ Textbooks, 1-877-745-3499

For permission to use material from this text or product,
e-mailed: **info@mathtv.com**

XYZ Textbooks
1339 Marsh St.
San Luis Obispo, CA 93401
USA

For your course and learning solutions, visit **www.xyztextbooks.com**

Printed in the United States of America

Brief Contents

Contents

5 Factoring Polynomials 383

6 Rational Expressions and Equations 451

7 Systems of Linear Equations 523

8 Radical Expressions and Equations 569

9 Quadratic Equations and Functions 629

Preface

XYZ Textbooks is on a mission to improve the quality and affordability of course materials for mathematics. Our 2013 Concepts with Applications Series brings a new level of technology innovation, which is certain to improve student proficiency.

As you can see, these books are presented in a format in which each section is organized by easy to follow learning objectives. Each objective includes helpful example problems, as well as additional practice problems in the margins, and are referenced in each exercise set. To help students relate what they are studying to the real world, each chapter and section begin with a real-life application of key concepts. You and your students will encounter extensions of these opening applications in the exercise sets.

We have put features in place that help your students stay on the trail to success. With that perspective in mind, we have incorporated this navigational theme into the following new and exciting features.

QR Codes

Unique technology embedded in each exercise set allows students with a smart phone or other internet-connected mobile device to view our video tutorials without being tied to a computer.

SCAN TO ACCESS

SAMPLE

10. Noelle recieves $17,000 in two loans. One loan charges 5% interest per year and the other 6.5%. If her total interest after 1 year is $970, how much was each loan?
$9,000 at 5%, $8,000 at 6.5%

Vocabulary Review

A list of fill-in-the-blank sentences appears at the beginning of each exercise set to help students better comprehend and verbalize concepts.

Vocabulary Review

Choose the correct words to fill in the blanks below.

 prime composite remainder lowest terms

1. A _____ number is any whole number greater than 1 that has exactly two divisors: 1 and the number itself.
2. A number is a divisor of another number if it divides it without a _____.
3. Any whole number greater than 1 that is not a prime number is called a _____ number.
4. A fraction is said to be in _____ if the numerator and the denominator have no factors in common other than the number 1.

Getting Ready For Class

Simple writing exercises for students to answer after reading each section increases their understanding of key concepts.

GETTING READY FOR CLASS

After reading through the preceding section, respond in your own words and in complete sentences.

A. What is a prime number?
B. Why is the number 22 a composite number?
C. Factor 120 into a product of prime factors.
D. How would you reduce a fraction to lowest terms?

Key words

A list of important vocabulary appears at the beginning of each section to help students prepare for new concepts. These words can also be found in blue italics within the sections.

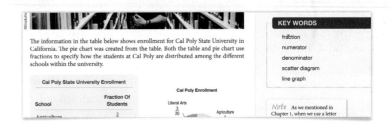

Navigation Skills

A discussion of important study skills appearing in each chapter anticipates students' needs as they progress through the course.

Navigation Skills: Prepare, Study, Achieve

Your instructor is a vital resource for your success in this class. Make note of your instructor's office hours and utilize them regularly. Compile a resource list that you keep with your class materials. This list should contain your instructor's office hours and contact information (e.g., office phone number or e-mail), as well as classmates' contact information that you can utilize outside of class. Communicate often with your classmates about how the course is going for you and any questions you may have. Odds are that someone else has the same question and you may be able to work together to find the answer.

Find the Mistake

Complete sentences that include common mistakes appear in each exercise set to help students identify errors and aid their comprehension of the concepts presented in that section.

Find the Mistake

Each sentence below contains a mistake. Circle the mistake and write the correct word or phrase on the line provided.

1. The set of all inputs for a function is called the range. _____
2. For the relation (1, 2), (3, 4), (5, 6), (7, 8), the domain is {2, 4, 6, 8}. _____
3. Graphing the relation $y = x^2 - 3$ will show that the domain is $\{x \mid x \geq -3\}$. _____
4. If a horizontal line crosses the graph of a relation in more than one place, the relation cannot be a function. _____

Landmark Review

A review appears in the middle of each chapter for students to practice key skills, check progress, and address difficulties.

Landmark Review: Checking Your Progress [7.1– 7.3]

Solve the following systems.

1. $y = 2x + 5$
 $y = -3x - 5$

2. $2x - y = -2$
 $2x - 3y$

3. $2x - y = 1$
 $y = -2x + 5$

4. $2x + 5y = 0$
 $x - 5y = -9$

5. $3x - 3y = 4$
 $x = 3y - 1$

6. $-6x + 2y = -9$
 $y = 3x + 4$

7. $x - y = 11$
 $x + y = 1$

8. $-3x + y = -15$
 $-2x + 3y = -17$

9. $2x + 5y = 17$
 $x - 4y = -24$

10. $x + 4y = 5$
 $3x + 8y = 14$

End of Chapter Reviews, Cumulative Reviews, and Tests

Problems at the end of each chapter provide students with a comprehensive review of each chapter's concepts.

CHAPTER 5 Test

Write each ratio as a fraction in lowest terms. [5.1]

1. 48 to 18
2. $\frac{5}{8}$ to $\frac{3}{4}$
3. 6 to $2\frac{4}{7}$
4. 0.14 to 0.4

Find the unknown term in each proportion. [5.3, 5.4]

10. $\frac{4}{7} = \frac{24}{x}$
11. $\frac{2.5}{5} = \frac{1.5}{x}$
12. **Baseball** A baseball player gets 13 hits in his first 20 games of the season. If he continues at the same rate, how many hits will he get in 60 games? [5.5]

Trail Guide Projects

Individual or group projects appear at the end of each chapter for students to apply concepts learned to real life.

CHAPTER 2 Trail Guide Project

Supplies Needed

A Piece of graph paper
B 4 colored pens or pencils

Color by Fractions

This project will provide a visual approach for working with the multiplication and division of fractions. You will need a piece of graph paper and 4 colored pens or pencils.

1. On your graph paper, draw two squares each with a side length of 20 boxes. Fill in the boxes of your first square using your pens or pencils. You may choose any pattern as long as you use all four colors and fill in every box in your square.

XYZ Textbooks is committed to helping students achieve their goals of success. This new series of highly-developed, innovative books will help students prepare for the course ahead and navigate their way to a successful course completion.

Supplements

Instructor

Name: _____

Phone: _____

Email: _____

Office Hours: _____

Teacher Assistant

Name: _____

Phone: _____

Email: _____

Office Hours: _____

On-Campus Tutoring

Location: _____

Hours: _____

Study Group

Name: _____

Phone: _____

Email: _____

Name: _____

Phone: _____

Email: _____

Name: _____

Phone: _____

Email: _____

For the Instructor

Online Homework XYZ Homework provides powerful online instructional tools for faculty and students. Randomized questions provide unlimited practice and instant feedback with all the benefits of automatic grading. Tools for instructors can be found at *www.xyzhomework.com* and include the following:

- Quick setup of your online class
- More than 3,000 randomized questions, similar to those in the textbook, for use in a variety of assessments, including online homework, quizzes, and tests
- Text and videos designed to supplement your instruction
- Automated grading of online assignments
- Flexible gradebook
- Message boards and other communication tools, enhanced with calculator-style input for proper mathematics notation

MathTV.com With more than 8,000 videos, MathTV.com provides the instructor with a useful resource to help students learn the material. MathTV.com features videos of most examples in the book, explained by the author and a variety of peer instructors. If a problem can be completed more than one way, the peer instructors often solve it by different methods. Instructors can also use the site's Build a Playlist feature to create a custom list of videos for posting on their class blog or website.

For the Student

Online Homework XYZ Homework provides powerful online instruction and homework practice for students. Benefits for the student can be found at *www.xyzhomework.com* and include the following:

- Unlimited practice with problems similar to those in the text
- Online quizzes and tests for instant feedback on performance
- Online video examples
- Convenient tracking of class programs

MathTV.com Students have access to math instruction 24 hours a day, seven days a week, on MathTV.com. Assistance with any problem or subject is never more than a few clicks away.

Online Book This text is available online for both instructors and students. Tightly integrated with MathTV.com, students can read the book and watch videos of the author and peer instructors explaining most examples. Access to the online book is available free with the purchase of a new book.

Additional Chapter Features

Learning Objectives

We have provided a list of the section's important concepts at the beginning of each section near the section title, noted with an orange capital letter, which is reiterated beside each learning objective header in the section.

Example and Practice Problems

Example problems, with work, explanations, and solutions are provided for every learning objective. Beside each example problem, we have also provided a corresponding practice problem and its answer in the margin.

Margin Notes

In many sections, you will find yellow notes in the margin that correspond to the section's current discussion. These notes contain important reminders, fun facts, and helpful advice for working through problems.

Colored Boxes

Throughout each section, we have highlighted important definitions, properties, rules, how to's, and strategies in colored boxes for easy reference.

Blueprint for Problem Solving

We provide students with a step-by-step guide for solving application problems in a clear and efficient manner.

Facts from Geometry

Throughout the book, we highlight specific geometric topics and show students how they apply to the concepts they are learning in the course.

Using Technology

Throughout the book, we provide students with step-by-step instructions for how to work some problems on a scientific or graphing calculator.

Descriptive Statistics

In some sections, we call out to the use of descriptive statistics as they are applied to the concepts students are learning, and thus, applied to real life.

Vocabulary Review

At the beginning of every exercise set, these fill-in-the-blank sentences help students comprehend and verbalize concepts from the corresponding section.

Problems

Each section contains problems ranging in difficulty that allow students to practice each learning objective presented in the section. The majority of the sections will also include problems in one or more of the following categories:

- Applying the Concepts
- Extending the Concepts or One Step Further
- Improving your Quantitative Literacy
- Getting Ready for the Next Section
- Extending the Concepts

Chapter Summary

At the end of each chapter, we provide a comprehensive summary of the important concepts from the chapter. Beside each concept, we also provide additional example problems for review.

Answers to Odd-Numbered Questions

At the end of the book, we have supplied answers for all odd-numbered problems that appear in the following features:

- Landmark Reviews
- Vocabulary Reviews
- Exercise Sets
- Find the Mistake Exercises

Navigation Skills at a Glance

- Compile a resource list for assistance.
- Set short- and long-term goals for the course.
- Complete homework completely and efficiently.
- Pay attention to instructions, and work ahead.
- Be an active listener while in class.
- Engage your senses when studying and memorizing concepts.
- Study with a partner or group.
- Maintain a positive academic self-image, and avoid burnout.
- Prepare early for the final exam, and reward yourself for a job well done!

Prealgebra Review

One of the most beautiful views in the world can be found from Skydeck 88 on the eighty-eighth floor of the Eureka Tower in Melbourne, Australia. Once you enter the building from the ground floor, it won't take long to get there. The elevators in Eureka Tower are the the fastest in the southern hemisphere, traveling at over twenty miles per hour! When you arrive at Skydeck 88, you can stand in a glass cube that extends almost 10 feet out of the building, nearly 1,000 feet above the ground. The glass on the upper ten floors of the tower is plated in twenty-four carat gold, making the building also beautiful to look at from the ground.

Suppose you wanted to visit Skydeck 88 with some friends. The admission prices are listed in the table in Australian dollars.

	Skydeck Entry	Sun & Stars Entry (2 entries per day)
Adult	$17.50	$21.00
Child	$9.00	$12.50
Concession	$11.00	$14.50
Family of 4	$39.90	$49.50
Family of 3	$29.50	$39.50
Extra child	$7.00	$10.50

How much would it cost for you and two adult friends to purchase the Skydeck Entry and to pay the concession fee? How would you split this cost evenly between the three of you? How much would you pay if you treated your niece and nephew to the Sun & Stars Entry? Answering all of these questions requires that you know how to work with decimal numbers, which is one of the concepts we will review in this chapter.

1

0.1 Factors and Simplifying Fractions

©iStockphoto.com/kentarus

KEY WORDS

prime number

composite number

divisor

factor

prime factor

fraction

numerator

denominator

lowest terms

equivalent fraction

improper fraction

proper fraction

Note You may have already noticed that the word *divisor* as we are using it here means the same as the word *factor*. A divisor and a factor of a number are the same thing. A number can't be a divisor of another number without also being a factor of it.

Practice Problems

1. Which of the numbers below are prime numbers, and which are composite? For those that are composite, give two divisors other than the number itself and 1.

37, 39, 51, 59

Answers

1. Prime: 37, 59
Composite: 39, 3 and 13; 51, 3 and 17

A Prime Numbers

Suppose you and a friend decide to split a medium-sized pizza for lunch. When the pizza is delivered, you find that it has been cut into eight equal pieces. If you eat four pieces, you have eaten $\frac{4}{8}$ of the pizza, but you also know that you have eaten $\frac{1}{2}$ of the pizza. The fraction $\frac{4}{8}$ is equivalent to the fraction $\frac{1}{2}$; that is, they both have the same value. The mathematical process we use to rewrite $\frac{4}{8}$ as $\frac{1}{2}$ is called *reducing to lowest terms*. Before we look at that process, we need to review some terms. Here is our first one.

> **DEFINITION** prime number
>
> A *prime number* is any whole number greater than 1 that has exactly two divisors—itself and 1. (A number is a divisor of another number if it divides it without a remainder.)
>
> $$\text{Prime numbers} = \{2, 3, 5, 7, 11, 13, 17, 19, 23, 29, 31, 37, \ldots\}$$

The list goes on indefinitely. Each number in the list has exactly two distinct divisors—itself and 1.

> **DEFINITION** composite number
>
> Any whole number greater than 1 that is not a prime number is called a *composite number*. A composite number always has at least one divisor other than itself and 1.

EXAMPLE 1 Identify each of the numbers below as either a prime number or a composite number. For those that are composite, give two divisors other than the number itself or 1.

a. 43 **b.** 12

Solution **a.** 43 is a prime number, because the only numbers that divide it without a remainder are 43 and 1.

b. 12 is a composite number, because it can be written as $12 = 4 \cdot 3$, which means that 4 and 3 are divisors of 12. (These are not the only divisors of 12; other divisors are 1, 2, 6, and 12.)

Every composite number can be written as the product of prime factors. Let's look at the composite number 108. One way we can write 108 is $2 \cdot 54$. The number 2 is a prime number, but 54 is not prime. Because 54 can be written as $2 \cdot 27$, we have

$$108 = 2 \cdot 54$$
$$= 2 \cdot 2 \cdot 27$$

The number 27 can be written as $3 \cdot 9$ or $3 \cdot 3 \cdot 3$ (because $9 = 3 \cdot 3$). Overall,

$$108 = 2 \cdot 54$$
$$108 = 2 \cdot 2 \cdot 27$$
$$108 = 2 \cdot 2 \cdot 3 \cdot 9$$
$$108 = 2 \cdot 2 \cdot 3 \cdot 3 \cdot 3$$

This last line is the number 108 written as the product of prime factors. We can use exponents to rewrite the last line.

$$108 = 2^2 \cdot 3^3$$

EXAMPLE 2 Factor 60 into a product of prime factors.

Solution We begin by writing 60 as $6 \cdot 10$ and continue factoring until all factors are prime numbers.

$$60 = 6 \cdot 10$$
$$= 2 \cdot 3 \cdot 2 \cdot 5$$
$$= 2^2 \cdot 3 \cdot 5$$

Notice that if we had started by writing 60 as $3 \cdot 20$, we would have achieved the same result. In other words, it doesn't matter which two factors you choose to begin, as long as you continue factoring until you have a product of primes.

$$60 = 3 \cdot 20$$
$$= 3 \cdot 2 \cdot 10$$
$$= 3 \cdot 2 \cdot 2 \cdot 5$$
$$= 2^2 \cdot 3 \cdot 5$$

B Simplifying Fractions

As we proceed through Chapter 0, from time to time we will review some of the major concepts associated with fractions. To begin, here is the formal definition of a *fraction*:

> **DEFINITION** fraction
>
> If a and b are integers, then the expression
> $$\frac{a}{b} \quad b \neq 0$$
> is called a *fraction* or a rational number. The top number a is called the **numerator**, and the bottom number b is called the **denominator**. The restriction $b \neq 0$ keeps us from writing an expression that is undefined. (As you will see, division by zero is not allowed.)

We can use the method of factoring numbers into prime factors to help reduce fractions to *lowest terms*. Here is the definition for lowest terms:

> **DEFINITION** fraction in lowest terms
>
> A *fraction* is said to be in *lowest terms* if the numerator and the denominator have no factors in common other than the number 1.

EXAMPLE 3 The fractions $\frac{1}{2}$, $\frac{1}{3}$, $\frac{2}{3}$, $\frac{1}{4}$, $\frac{3}{4}$, $\frac{1}{5}$, $\frac{2}{5}$, $\frac{3}{5}$, and $\frac{4}{5}$ are all in lowest terms, because in each case the numerator and the denominator have no factors other than 1 in common. That is, in each fraction, no number other than 1 divides both the numerator and the denominator exactly (without a remainder).

Note The numbers 2 and 54 are not the only divisors of 108. We could have begun with $3 \cdot 36$, $4 \cdot 27$, or even $6 \cdot 18$ and still end up with the same product of primes.

2. Factor 90 into a product of prime factors.

Note There are some shortcuts to finding the divisors of a number. For instance, if a number ends in 0 or 5, then it is divisible by 5. If a number ends in an even digit (0, 2, 4, 6, or 8), then it is divisible by 2. A number is divisible by 3 if the sum of its digits is divisible by 3. For example, 921 is divisible by 3 because the sum of its digits is $9 + 2 + 1 = 12$, which is divisible by 3.

3. Which of the following fractions are in lowest terms?
$$\frac{1}{6}, \frac{2}{8}, \frac{15}{25}, \frac{9}{13}$$

Answers

2. $2 \cdot 3^2 \cdot 5$

3. $\frac{1}{6}, \frac{9}{13}$

4. Reduce $\frac{12}{18}$ to lowest terms by dividing the numerator and the denominator by 6.

5. Reduce the fraction $\frac{15}{30}$ to lowest terms by first factoring the numerator and the denominator into prime factors and then dividing out the factors they have in common.

Note A common mistake when reducing to lowest terms is to write a fraction, such as

$\frac{12}{15}$ as $\frac{3+3+3+3}{3+3+3+3+3} \neq \frac{1}{3}$.

When finding an equivalent fraction, we can only divide out common factors, not common terms.

6. Reduce $\frac{30}{35}$ to lowest terms.

EXAMPLE 4 The fraction $\frac{6}{8}$ is not written in lowest terms, because the numerator and the denominator are both divisible by 2. To write $\frac{6}{8}$ in lowest terms, we divide both the numerator and the denominator by 2.

$$\frac{6}{8} = \frac{6 \div 2}{8 \div 2} = \frac{3}{4}$$

The fraction $\frac{3}{4}$ is in lowest terms, because 3 and 4 have no factors in common except the number 1.

Reducing a fraction to lowest terms is simply a matter of dividing the numerator and the denominator by all the factors they have in common to produce an *equivalent fraction* of the same value.

EXAMPLE 5 Reduce the fraction $\frac{12}{15}$ to lowest terms by first factoring the numerator and the denominator into prime factors and then dividing both the numerator and the denominator by the factor they have in common.

Solution The numerator and the denominator factor as follows:

$$12 = 2 \cdot 2 \cdot 3 \quad \text{and} \quad 15 = 3 \cdot 5$$

The factor they have in common is 3. We can divide both terms of the fraction by 3 to produce an equivalent fraction.

$$\frac{12}{15} = \frac{2 \cdot 2 \cdot 3}{3 \cdot 5} \qquad \text{Factor the numerator and the denominator completely.}$$

$$= \frac{2 \cdot 2 \cdot 3 \div 3}{3 \cdot 5 \div 3} \qquad \text{Divide by 3.}$$

$$= \frac{4}{5}$$

The fraction $\frac{4}{5}$ is equivalent to $\frac{12}{15}$ and is in lowest terms, because the numerator and the denominator have no factors other than 1 in common.

We can shorten the work involved in reducing fractions to lowest terms by using the slash to indicate division. For example, we can write the above problem this way:

$$\frac{12}{15} = \frac{2 \cdot 2 \cdot \cancel{3}}{\cancel{3} \cdot 5} = \frac{4}{5}$$

So long as we understand that the slashes through the 3s indicate that we have divided both the numerator and the denominator by 3, we can use this notation.

EXAMPLE 6 Laura is having a party. She puts 4 six-packs of diet soda in a cooler for her guests. At the end of the party, she finds that only 4 sodas have been consumed. What fraction of the sodas are left? Write your answer in lowest terms.

Solution She had 4 six-packs of soda, which is $4(6) = 24$ sodas. Only 4 were consumed at the party, so 20 are left. The fraction of sodas left is

$$\frac{20}{24}$$

Factoring 20 and 24 completely and then dividing out the factors they have in common gives us

$$\frac{20}{24} = \frac{\cancel{2} \cdot \cancel{2} \cdot 5}{\cancel{2} \cdot \cancel{2} \cdot 2 \cdot 3} = \frac{5}{6}$$

Note The slashes in Example 6 indicate that we have divided both the numerator and the denominator by $2 \cdot 2$, which is equal to 4. With some fractions it is apparent at the start what number divides the numerator and the denominator. For instance, you may have recognized that both 20 and 24 in Example 6 are divisible by 4. We can divide both terms by 4 without factoring first. This still guarantees that dividing both terms of the fraction by 4 will produce an equivalent fraction.

$$\frac{20}{24} = \frac{20 \div 4}{24 \div 4} = \frac{5}{6}$$

Answers

4. $\frac{2}{3}$

5. $\frac{1}{2}$

6. $\frac{6}{7}$

EXAMPLE 7 Reduce $\frac{6}{42}$ to lowest terms.

Solution We begin by factoring both terms. We then divide through by any factors common to both terms.

$$\frac{6}{42} = \frac{2 \cdot 3}{2 \cdot 3 \cdot 7} = \frac{1}{7}$$

We must be careful in a problem like this to remember that the slashes indicate division. They are used to indicate that we have divided both the numerator and the denominator by $2 \cdot 3 = 6$. The result of dividing the numerator 6 by $2 \cdot 3$ is 1. It is a very common mistake to call the numerator 0 instead of 1 or to leave the numerator out of the answer. Be careful not to make this mistake.

EXAMPLE 8 Reduce $\frac{4}{40}$ to lowest terms.

Solution
$$\frac{4}{40} = \frac{2 \cdot 2 \cdot 1}{2 \cdot 2 \cdot 2 \cdot 5}$$
$$= \frac{1}{10}$$

All the fractions we have worked with so far have been *proper fractions*. The following example involves an *improper fraction*. Let's review a formal definition before working the problem.

> **DEFINITION** improper fraction, proper fraction
>
> A fraction in which the denominator is less than or equal to the numerator is called an **improper fraction**; whereas, a fraction in which the denominator is greater than the numerator is called a **proper fraction**.

EXAMPLE 9 Reduce $\frac{105}{30}$ to lowest terms.

Solution
$$\frac{105}{30} = \frac{3 \cdot 5 \cdot 7}{2 \cdot 3 \cdot 5}$$
$$= \frac{7}{2}$$

GETTING READY FOR CLASS

Each section of the book will end with some problems and questions like the ones below. They are for you to answer after you have read through the section but before you go to class. All of them require that you give written responses in complete sentences. Writing about mathematics is a valuable exercise. As with all problems in this course, approach these writing exercises with a positive point of view. You will get better at giving written responses to questions as the course progresses. Even if you never feel comfortable writing about mathematics, just attempting the process will increase your understanding and ability in this course.

After reading through the preceding section, respond in your own words and in complete sentences.

A. What is a prime number?

B. Why is the number 22 a composite number?

C. Factor 120 into a product of prime factors.

D. What is meant by the phrase "a fraction in lowest terms"?

7. Reduce $\frac{8}{72}$ to lowest terms.

8. Reduce $\frac{5}{50}$ to lowest terms.

9. Reduce $\frac{120}{25}$ to lowest terms.

Note Unless instructed otherwise, it is perfectly acceptable to leave a final answer as an improper fraction as we have done in Example 9. However, for application problems, mixed numbers or decimals are preferred.

Answers

7. $\frac{1}{9}$

8. $\frac{1}{10}$

9. $\frac{24}{5}$

Vocabulary Review

Choose the correct words to fill in the blanks below.

numerator denominator composite number lowest terms prime number equivalent

1. Any whole number greater than 1 that has exactly two divisors, 1 and the number itself, is called a
_____ .

2. A _____ is any whole number greater than 1 that is not a prime number.

3. The top number of a fraction is called the _____ .

4. The restriction $b \neq 0$ for the _____ of a fraction keeps us from writing an expression that is undefined.

5. Fractions with the same value are said to be _____ .

6. A fraction is in _____ if the numerator and the denominator have no factors in common other than the number 1.

Problems

A Identify each of the numbers below as either a prime number or a composite number. For those that are composite, give at least one divisor (factor) other than the number itself or the number 1.

1. 11

2. 23

3. 105

4. 41

5. 81

6. 50

7. 13

8. 219

Factor each of the following into a product of prime factors.

9. 12

10. 18

11. 81

12. 210

13. 215

14. 75

15. 15

16. 42

B Reduce each fraction to lowest terms.

17. $\dfrac{5}{10}$

18. $\dfrac{3}{6}$

19. $\dfrac{4}{6}$

20. $\dfrac{4}{10}$

21. $\dfrac{8}{10}$

22. $\dfrac{6}{10}$

23. $\dfrac{36}{20}$

24. $\dfrac{32}{12}$

25. $\dfrac{42}{66}$

26. $\dfrac{36}{60}$

27. $\dfrac{24}{40}$

28. $\dfrac{50}{75}$

29. $\dfrac{14}{98}$

30. $\dfrac{12}{84}$

31. $\dfrac{70}{90}$

32. $\dfrac{80}{90}$

SCAN TO ACCESS

33. $\dfrac{42}{30}$ **34.** $\dfrac{60}{36}$ **35.** $\dfrac{18}{90}$ **36.** $\dfrac{150}{210}$ **37.** $\dfrac{110}{70}$ **38.** $\dfrac{45}{75}$

39. $\dfrac{180}{108}$ **40.** $\dfrac{105}{30}$ **41.** $\dfrac{96}{108}$ **42.** $\dfrac{66}{84}$ **43.** $\dfrac{126}{165}$ **44.** $\dfrac{210}{462}$

45. $\dfrac{102}{114}$ **46.** $\dfrac{255}{285}$ **47.** $\dfrac{294}{693}$ **48.** $\dfrac{273}{385}$

49. Reduce each fraction to lowest terms.

a. $\dfrac{6}{51}$ **b.** $\dfrac{6}{52}$ **c.** $\dfrac{6}{54}$ **d.** $\dfrac{6}{56}$ **e.** $\dfrac{6}{57}$

50. Reduce each fraction to lowest terms.

a. $\dfrac{6}{42}$ **b.** $\dfrac{6}{44}$ **c.** $\dfrac{6}{45}$ **d.** $\dfrac{6}{46}$ **e.** $\dfrac{6}{48}$

51. Reduce each fraction to lowest terms.

a. $\dfrac{2}{90}$ **b.** $\dfrac{3}{90}$ **c.** $\dfrac{5}{90}$ **d.** $\dfrac{6}{90}$ **e.** $\dfrac{9}{90}$

52. Reduce each fraction to lowest terms.

a. $\dfrac{3}{105}$ **b.** $\dfrac{5}{105}$ **c.** $\dfrac{7}{105}$ **d.** $\dfrac{15}{105}$ **e.** $\dfrac{21}{105}$

53. The answer to each problem below is wrong. Give the correct answer.

 a. $\dfrac{5}{15} = \dfrac{\cancel{5}}{3 \cdot \cancel{5}} = \dfrac{0}{3}$

 b. $\dfrac{5}{6} = \dfrac{3 + \cancel{2}}{4 + \cancel{2}} = \dfrac{3}{4}$

 c. $\dfrac{6}{30} = \dfrac{\cancel{2} \cdot 3}{\cancel{2} \cdot 3 \cdot 5} = 5$

54. The answer to each problem below is wrong. Give the correct answer.

 a. $\dfrac{10}{20} = \dfrac{7 + \cancel{3}}{17 + \cancel{3}} = \dfrac{7}{17}$

 b. $\dfrac{9}{36} = \dfrac{\cancel{3} \cdot \cancel{3}}{2 \cdot 2 \cdot \cancel{3} \cdot \cancel{3}} = \dfrac{0}{4}$

 c. $\dfrac{4}{12} = \dfrac{\cancel{2} \cdot \cancel{2}}{\cancel{2} \cdot \cancel{2} \cdot 3} = 3$

55. Which of the fractions $\frac{6}{8}, \frac{15}{20}, \frac{9}{16}, \frac{21}{28}$ does not reduce to $\frac{3}{4}$? **56.** Which of the fractions $\frac{4}{9}, \frac{10}{15}, \frac{8}{12}, \frac{6}{12}$ do not reduce to $\frac{2}{3}$?

The number line below extends from 0 to 2, with the segment from 0 to 1 and the segment from 1 to 2 each divided into 8 equal parts. Locate each of the following numbers on this number line.

57. $\dfrac{1}{2}, \dfrac{2}{4}, \dfrac{4}{8}$, and $\dfrac{8}{16}$ **58.** $\dfrac{3}{2}, \dfrac{6}{4}, \dfrac{12}{8}$, and $\dfrac{24}{16}$ **59.** $\dfrac{5}{4}, \dfrac{10}{8}$, and $\dfrac{20}{16}$ **60.** $\dfrac{1}{4}, \dfrac{2}{8}$, and $\dfrac{4}{16}$

Applying the Concepts

61. Income A family's monthly income is $2,400, and they spend $600 each month on food. Write the amount they spend on food as a fraction of their monthly income in lowest terms.

62. Hours and Minutes There are 60 minutes in 1 hour. What fraction of an hour is 20 minutes? Write your answer in lowest terms.

63. Computers Use the chart to answer the questions about computers used by students.

 a. Is the percentage of Apple and Dell computers less than or more than half of the computers used by students?

 b. Is the percentage of Toshiba and HP computers used less than or more than one-third of computers used by students?

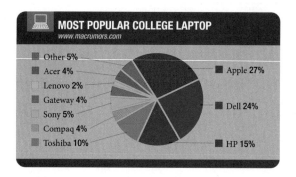

MOST POPULAR COLLEGE LAPTOP
www.macrumors.com

- Other 5%
- Acer 4%
- Lenovo 2%
- Gateway 4%
- Sony 5%
- Compaq 4%
- Toshiba 10%
- Apple 27%
- Dell 24%
- HP 15%

64. Students The table shows the number of male and female students enrolled at a small community college. What fraction of the students is male? Write your answer in lowest terms.

Community College Enrollment	
Men	1,400
Women	1,800

65. Final Exams Suppose 33 people took the final exam in a math class. If 11 people got an A on the final exam, what fraction of the students did not get an A on the exam? Write your answer in lowest terms.

66. Income Tax A person making $21,000 a year pays $3,000 in income tax. What fraction of the person's income is paid as income tax? Write your answer in lowest terms.

Nutrition The nutrition labels below are from two different kinds of cheese crackers. Use them to work Problems 67–72.

67. What fraction of the calories in Cheez-It crackers comes from fat?

68. What fraction of the calories in Goldfish crackers comes from fat?

69. For Cheez-It crackers, what fraction of the total fat is from saturated fat?

70. For Goldfish crackers, what fraction of the total fat is from saturated fat?

CHEEZ-IT CRACKERS

Nutrition Facts
Serving Size 30 g. (About 27 crackers)
Servings Per Container: 9

Amount Per Serving

Calories 150 Calories from fat 70

	% Daily Value*
Total Fat 8g	12%
Saturated Fat 2g	10%
Trans Fat 0g	
Polysaturated Fat 4g	
Monounsaturated Fat 2g	
Cholesterol 0mg	0%
Sodium 230mg	10%
Total Carbohydrate 17g	6%
Dietary Fiber less than 1g	3%
Sugars 0g	
Protein 3g	

Vitamin A 2%	•	Vitamin C 0%
Calcium 4%	•	Iron 6%

*Percent Daily Values are based on a 2,000 calorie diet

GOLDFISH CRACKERS

Nutrition Facts
Serving Size 55 pieces
Servings Per Container About 4

Amount Per Serving

Calories 140 Calories from fat 45

	% Daily Value*
Total Fat 5g	8%
Saturated Fat 1g	5%
Cholesterol 5mg	2%
Sodium 250mg	10%
Total Carbohydrate 20g	7%
Dietary Fiber 1g	4%
Sugars 1g	
Protein 4g	

Vitamin A 0%	•	Vitamin C 0%
Calcium 4%	•	Iron 2%

*Percent Daily Values are based on a 2,000 calorie diet

71. What fraction of the total carbohydrates in Cheez-It crackers is from sugar?

72. What fraction of the total carbohydrates in Goldfish crackers is from sugar?

73. Below are the 2011 prices for adult tickets at Disneyland based on the number of days planned for the visit. Complete the table to find the average cost per day for each type of ticket.

Duration	Total Cost	Cost per Day
1 day	$80	
2 days	$158	
3 days	$199	
4 days	$219	
5 days	$231	

Source: disneyland.disney.go.com

74. Suppose these are the average adult ticket prices for movies in the United States over the past 30 years.

a. During which 5 year period did the prices go up the most? Which 5 year period did they go up the least?

b. Approximately what fraction represents the price in 1980 to the price in 2010?

75. In Universal City, Los Angeles, the adult price to see Harry Potter and the Deathly Hollows Part 2: IMAX 3D was $19.50. Refer to the chart in problem 74 to answer the following questions.

 a. How many times more expensive is this than the average 2010 adult ticket price?

 b. How many times more expensive is this than the average 1980 adult ticket price?

Estimating

76. Which of the following fractions is closest to the number $\frac{1}{2}$?

 a. $\dfrac{5}{4}$ **b.** $\dfrac{5}{8}$ **c.** $\dfrac{5}{16}$

77. Which of the following fractions is closest to the number $\frac{1}{2}$?

 a. $\dfrac{3}{8}$ **b.** $\dfrac{5}{8}$ **c.** $\dfrac{7}{8}$

78. Which of the following fractions is closest to the number $\frac{1}{2}$?

 a. $\dfrac{3}{4}$ **b.** $\dfrac{5}{8}$ **c.** $\dfrac{9}{16}$

79. Which of the following fractions is closest to the number 1?

 a. $\dfrac{3}{4}$ **b.** $\dfrac{7}{8}$ **c.** $\dfrac{15}{16}$

Calculator Problems

Use your calculator to help fill in the blanks for the following problems.

80. $\dfrac{5}{7} = \dfrac{\square}{574}$

81. $\dfrac{2}{3} = \dfrac{\square}{2,223}$

82. $\dfrac{8}{9} = \dfrac{728}{\square}$

83. $\dfrac{7}{8} = \dfrac{34,972}{\square}$

84. $\dfrac{3,787}{4,869} = \dfrac{\square}{9}$

85. $\dfrac{684}{836} = \dfrac{\square}{11}$

86. $\dfrac{262}{1,965} = \dfrac{\square}{15}$

87. $\dfrac{616}{1,672} = \dfrac{\square}{19}$

88. Ticket Prices The graphic shows the rise in Disneyland ticket prices since 1981. Which of the following is the closest fraction of the ticket price in 1981 to the ticket price in 2011?

a. $\dfrac{1}{5}$ b. $\dfrac{1}{7}$ c. $\dfrac{1}{10}$ d. $\dfrac{1}{12}$

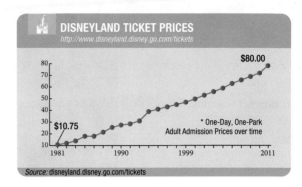

Find the Mistake

Each sentence below contains a mistake. Circle the mistake and write the correct word(s) or numbers(s) on the line provided.

1. A number is a divisor of another number if it divides it with a remainder. _____

2. 31 is a composite number because the only numbers that divide it without a remainder are 31 and 1. _____

3. Factoring 112 into the product of primes gives us $2 \cdot 7$. _____

4. To get the fraction $\dfrac{24}{36}$ in lowest terms, we divide the numerator and the denominator by 2. _____

Navigation Skills: Prepare, Study, Achieve

At the end of the first exercise set of each chapter, we provide an important discussion of study skills that will help you succeed in this course. Pay special attention to these skills. Ponder each one and apply it to your life. Your success is in your hands.

Studying is the key to success in this course. However, many students have never learned effective skills for studying. Study skills include but are not limited to the following:

- Work done on problems for practice and homework
- Amount of time spent studying
- Time of day and location for studying
- Management of distractions during study sessions
- Material chosen to review
- Order and process of review

Let's begin our discussion with the topic of homework. From the first day of class, we recommend you spend two hours on homework for every hour you are scheduled to attend class. Any less may drastically impact your success in this course. To help visualize this commitment, map out a weekly schedule that includes your classes, work shifts, extracurriculars, and any additional obligations. Fill in the hours you intend to devote to completing assignments and studying for this class. Post this schedule at home and keep a copy with your study materials to remind you of your commitment to success.

OBJECTIVES

A Multiply fractions.

B Divide fractions.

C Add and subtract fractions with a common denominator.

D Find LCDs and solve problems involving fractions with different denominators.

KEY WORDS

multiplication

product

division

reciprocal

least common denominator

©iStockphoto.com/wojciech_gajda

Dock diving is a sport where dogs compete to see who can jump the highest or the farthest off the end of a dock into a body of water. Many of the dogs' handlers use a chase object to lure the dog as far as possible off the dock. The handler commands the dog to begin a running start and then throws the chase object off over the water. The dog leaps after the object, and once he splashes into the water, an event official measures his jump distance. The official measures the dog's final distance from the end of the dock to the point where the base of the dog's tail first hit the water.

Let's say that the top two dogs jumped 21 feet $10\frac{1}{4}$ inches and 21 feet $10\frac{7}{8}$ inches. How much farther did the second dog jump? To solve this problem, we need to know how to subtract $\frac{1}{4}$ from $\frac{7}{8}$. After reading this section, you will be more comfortable working problems involving fractions such as this. First, let's discuss a bit about the multiplication of fractions.

A Multiplication of Fractions

A recipe calls for $\frac{3}{4}$ cup of flour. If you are making only $\frac{1}{2}$ the recipe, how much flour do you use? This question can be answered by multiplying $\frac{1}{2}$ and $\frac{3}{4}$ because we need to find out how much is $\frac{1}{2}$ of $\frac{3}{4}$. Here is the problem written in symbols:

$$\frac{1}{2} \cdot \frac{3}{4} = \frac{3}{8}$$

As you can see from this example, to multiply two fractions, we multiply the numerators and then multiply the denominators. We begin this section with the rule for multiplication of fractions.

Note: As you may have covered in a previous math class, many operations in mathematics are represented in word problems as key words. In this case, the explanation used the word *of* to represent multiplication. We will revisit this and additional key words later in the book.

> **RULE** Multiplication of Fractions
>
> The product of two fractions is the fraction whose numerator is the product of the two numerators and whose denominator is the product of the two denominators. We can write this rule in symbols as follows:
>
> If a, b, c, and d represent any numbers and b and d are not zero, then
>
> $$\frac{a}{b} \cdot \frac{c}{d} = \frac{a \cdot c}{b \cdot d}$$

EXAMPLE 1 Multiply: $\dfrac{3}{5} \cdot \dfrac{2}{7}$.

Solution Using our rule for multiplication, we multiply the numerators and multiply the denominators.

$$\frac{3}{5} \cdot \frac{2}{7} = \frac{3 \cdot 2}{5 \cdot 7} = \frac{6}{35}$$

The product of $\frac{3}{5}$ and $\frac{2}{7}$ is the fraction $\frac{6}{35}$. The numerator 6 is the product of 3 and 2, and the denominator 35 is the product of 5 and 7.

EXAMPLE 2 Multiply: $\dfrac{3}{8} \cdot 5$.

Solution The number 5 can be written as $\frac{5}{1}$. That is, 5 can be considered a fraction with numerator 5 and denominator 1. Writing 5 this way enables us to apply the rule for multiplying fractions.

$$\frac{3}{8} \cdot 5 = \frac{3}{8} \cdot \frac{5}{1}$$

$$= \frac{3 \cdot 5}{8 \cdot 1} = \frac{15}{8}$$

EXAMPLE 3 Multiply: $\dfrac{1}{2} \cdot \dfrac{3}{4} \cdot \dfrac{1}{5}$.

Solution We find the product inside the parentheses first and then multiply the result by $\frac{1}{2}$.

$$\frac{1}{2} \cdot \frac{3}{4} \cdot \frac{1}{5} = \frac{1}{2} \cdot \frac{3 \cdot 1}{4 \cdot 5}$$

$$= \frac{1}{2} \cdot \frac{3}{20}$$

$$= \frac{1 \cdot 3}{2 \cdot 20} = \frac{3}{40}$$

The answers to all the examples so far in this section have been in lowest terms. Let's see what happens when we multiply two fractions to get a product that is not in lowest terms.

EXAMPLE 4 Multiply: $\dfrac{15}{8} \cdot \dfrac{4}{9}$.

Solution Multiplying the numerators and multiplying the denominators, we have

$$\frac{15}{8} \cdot \frac{4}{9} = \frac{15 \cdot 4}{8 \cdot 9} = \frac{60}{72}$$

The product is $\frac{60}{72}$, which can be reduced to lowest terms by factoring 60 and 72 and then dividing out any factors they have in common.

$$\frac{60}{72} = \frac{\cancel{2} \cdot \cancel{2} \cdot \cancel{3} \cdot 5}{\cancel{2} \cdot \cancel{2} \cdot 2 \cdot \cancel{3} \cdot 3} = \frac{5}{6}$$

We can actually save ourselves some time by factoring before we multiply. Here's how it is done:

$$\frac{15}{8} \cdot \frac{4}{9} = \frac{15 \cdot 4}{8 \cdot 9}$$

$$= \frac{(3 \cdot 5) \cdot (2 \cdot 2)}{(2 \cdot 2 \cdot 2) \cdot (3 \cdot 3)}$$

$$= \frac{\cancel{2} \cdot \cancel{2} \cdot \cancel{3} \cdot 5}{\cancel{2} \cdot \cancel{2} \cdot 2 \cdot \cancel{3} \cdot 3} = \frac{5}{6}$$

The result is the same in both cases. Reducing to lowest terms before we actually multiply takes less time.

Practice Problems

1. Multiply: $\dfrac{2}{3} \cdot \dfrac{5}{9}$.

2. Multiply: $\dfrac{2}{5} \cdot 7$.

3. Multiply: $\dfrac{1}{3} \cdot \dfrac{4}{5} \cdot \dfrac{1}{3}$.

4. Multiply: $\dfrac{12}{25} \cdot \dfrac{5}{6}$.

Answers

1. $\frac{10}{27}$

2. $\frac{14}{5}$

3. $\frac{4}{45}$

4. $\frac{2}{5}$

B Division of Fractions

Before we define division with fractions, we must first introduce the idea of *reciprocals*. Look at the following multiplication problems:

$$\frac{3}{4} \cdot \frac{4}{3} = \frac{12}{12} = 1 \qquad \frac{7}{8} \cdot \frac{8}{7} = \frac{56}{56} = 1$$

In each case the product is 1. Whenever the product of two numbers is 1, we say the two numbers are reciprocals.

> **DEFINITION** reciprocal
>
> Two numbers whose product is 1 are said to be **reciprocals**. In symbols, the reciprocal of $\frac{a}{b}$ is $\frac{b}{a}$, because
>
> $$\frac{a}{b} \cdot \frac{b}{a} = \frac{a \cdot b}{a \cdot b} = 1 \qquad (a \neq 0, b \neq 0)$$

Every number has a reciprocal except 0. The reason 0 does not have a reciprocal is because the product of any number with 0 is 0. It can never be 1. Reciprocals of whole numbers are fractions with 1 as the numerator. For example, the reciprocal of 5 is $\frac{1}{5}$, because

$$5 = \frac{1}{5} = \frac{5}{1} \cdot \frac{1}{5} = \frac{5}{5} = 1$$

Table 1 lists some numbers and their reciprocals.

Table 1		
Number	**Reciprocal**	**Reason**
$\frac{3}{4}$	$\frac{4}{3}$	Because $\frac{3}{4} \cdot \frac{4}{3} = \frac{12}{12} = 1$
$\frac{9}{5}$	$\frac{5}{9}$	Because $\frac{9}{5} \cdot \frac{5}{9} = \frac{45}{45} = 1$
$\frac{1}{3}$	3	Because $\frac{1}{3} \cdot 3 = \frac{1}{3} \cdot \frac{3}{1} = \frac{3}{3} = 1$
7	$\frac{1}{7}$	Because $7 \cdot \frac{1}{7} = \frac{7}{1} \cdot \frac{1}{7} = \frac{7}{7} = 1$

Division with fractions is accomplished by using reciprocals. More specifically, we can define division by a fraction to be the same as multiplication by its reciprocal.

> **RULE** Division of Fractions
>
> Division of fractions is the equivalent of multiplication by the reciprocal. We can write this rule in symbols as follows:
>
> If a, b, c, and d represent any numbers and b and d are not zero, then
>
> $$\frac{a}{b} \div \frac{c}{d} = \frac{a}{b} \cdot \frac{d}{c}$$

This definition states that dividing by the fraction $\frac{c}{d}$ is exactly the same as multiplying by its reciprocal $\frac{d}{c}$.

In some math classes, multiplication by the reciprocal is referred to as "inverting the divisor and multiplying" or "flip that guy and multiply." No matter how you say it, division by any number (except zero) is always equivalent to multiplication by its reciprocal. Here are additional examples that involve division by fractions.

Note Defining division to be the same as multiplication by the reciprocal does make sense. If we divide 6 by 2, we get 3. On the other hand, if we multiply 6 by $\frac{1}{2}$ (the reciprocal of 2), we also get 3. Whether we divide by 2 or multiply by $\frac{1}{2}$, we get the same result.

EXAMPLE 5 Divide: $\dfrac{3}{8} \div \dfrac{9}{4}$.

Solution Dividing by $\dfrac{9}{4}$ is the same as multiplying by its reciprocal, which is $\dfrac{4}{9}$.

$$\dfrac{3}{8} \div \dfrac{9}{4} = \dfrac{3}{8} \cdot \dfrac{4}{9}$$

$$= \dfrac{3 \cdot 2 \cdot 2}{2 \cdot 2 \cdot 2 \cdot 3 \cdot 3}$$

$$= \dfrac{1}{6}$$

The quotient of $\dfrac{3}{8}$ and $\dfrac{9}{4}$ is $\dfrac{1}{6}$.

EXAMPLE 6 Divide: $\dfrac{2}{3} \div 2$.

Solution The reciprocal of 2 is $\dfrac{1}{2}$. Applying the definition for division of fractions, we have

$$\dfrac{2}{3} \div 2 = \dfrac{2}{3} \cdot \dfrac{1}{2}$$

$$= \dfrac{2 \cdot 1}{3 \cdot 2}$$

$$= \dfrac{1}{3}$$

C Addition and Subtraction of Fractions (Common Denominator)

The solution to a problem that involves the addition of fractions is simply the sum of the numerators placed over the denominator both fractions share, also called the *common denominator*. In symbols, we have the following rule:

> **RULE** Addition and Subtraction of Fractions
>
> If a, b, and c are integers, and c is not equal to 0, then
> $$\dfrac{a}{c} + \dfrac{b}{c} = \dfrac{a+b}{c}$$
> This rule holds for subtraction as well. That is,
> $$\dfrac{a}{c} - \dfrac{b}{c} = \dfrac{a-b}{c}$$

In Examples 7–10, find the sum or difference of the two fractions. (Add or subtract as indicated.) Reduce all answers to lowest terms. (Assume all variables represent nonzero numbers.)

EXAMPLE 7 $\dfrac{3}{8} + \dfrac{1}{8} = \dfrac{3+1}{8}$ Add numerators; keep the same denominator.

$$-\dfrac{4}{8}$$ The sum of 3 and 1 is 4.

$$= \dfrac{1}{2}$$ Reduce to lowest terms.

EXAMPLE 8 $\dfrac{a+5}{8} - \dfrac{3}{8} = \dfrac{a+5-3}{8}$ Combine numerators; keep the same denominator.

$$= \dfrac{a+2}{8}$$ The difference of 5 and 3 is 2.

Note This answer is in lowest terms because you cannot combine the variable a and the number 2 as they are not similar terms. We will discuss the concept of similar terms in the next chapter.

5. Divide: $\dfrac{5}{9} \div \dfrac{10}{3}$.

6. Divide: $\dfrac{3}{4} \div 3$.

Find the sum or difference. Reduce all answers to lowest terms.

7. $\dfrac{3}{10} + \dfrac{1}{10}$

8. $\dfrac{a+5}{12} + \dfrac{3}{12}$

Answers

5. $\dfrac{1}{6}$

6. $\dfrac{1}{4}$

7. $\dfrac{2}{5}$

8. $\dfrac{a+8}{12}$

9. $\dfrac{8}{7} - \dfrac{5}{7}$

EXAMPLE 9 $\dfrac{9}{5} - \dfrac{3}{5} = \dfrac{9-3}{5}$ *Subtract numerators; keep the same denominator.*

$$= \dfrac{6}{5}$$ *The difference of 9 and 3 is 6.*

10. $\dfrac{5}{9} + \dfrac{8}{9} + \dfrac{5}{9}$

EXAMPLE 10 $\dfrac{3}{7} + \dfrac{2}{7} + \dfrac{9}{7} = \dfrac{3+2+9}{7}$

$$= \dfrac{14}{7}$$

$$= 2$$

As Examples 7–10 indicate, addition and subtraction are simple straightforward processes when all the fractions have the same denominator. We will now turn our attention to the process of adding fractions that have different denominators.

D Addition and Subtraction of Fractions (Different Denominators)

In order to get started, we need the following definition:

> **DEFINITION** least common denominator
>
> The *least common denominator* (LCD) for a set of denominators is the smallest number that is exactly divisible by each denominator. (Note that, in some books, the least common denominator is also called the least common multiple.)

In other words, all the denominators of the fractions involved in a problem must divide into the least common denominator exactly. That is, they divide it without leaving a remainder.

11. Find the LCD for the fractions $\frac{5}{18}$ and $\frac{3}{14}$.

EXAMPLE 11 Find the LCD for the fractions $\dfrac{5}{12}$ and $\dfrac{7}{18}$.

Solution The least common denominator for the denominators 12 and 18 must be the smallest number divisible by both 12 and 18. We can factor 12 and 18 completely and then build the LCD from these factors. Factoring 12 and 18 completely gives us

$$12 = 2 \cdot 2 \cdot 3 \qquad 18 = 2 \cdot 3 \cdot 3$$

Now, if 12 is going to divide the LCD exactly, then the LCD must have factors of $2 \cdot 2 \cdot 3$. If 18 is to divide it exactly, it must have factors of $2 \cdot 3 \cdot 3$. We don't need to repeat the factors that 12 and 18 have in common.

Note The ability to find least common denominators is very important in mathematics. The discussion here is a detailed explanation of how to find an LCD.

$$12 = 2 \cdot 2 \cdot 3$$

12 divides the LCD.

$$\text{LCD} = 2 \cdot 2 \cdot 3 \cdot 3 = 36$$

$$18 = 2 \cdot 3 \cdot 3$$

18 divides the LCD.

The LCD for 12 and 18 is 36. It is the smallest number that is divisible by both 12 and 18; 12 divides it exactly three times, and 18 divides it exactly two times.

We can visualize the results in Example 11 with the diagram below. It shows that 36 is the smallest number that both 12 and 18 divide evenly. As you can see, 12 divides 36 exactly 3 times, and 18 divides 36 exactly 2 times.

EXAMPLE 12 Add: $\frac{5}{12} + \frac{7}{18}$.

Solution We can add fractions only when they have the same denominators. In Example 11, we found the LCD for $\frac{5}{12}$ and $\frac{7}{18}$ to be 36. We change $\frac{5}{12}$ and $\frac{7}{18}$ to equivalent fractions that have 36 for a denominator. Recall from a previous math class that if the numerator and the denominator are multiplied by the same nonzero number, the resulting fraction is equivalent to the original fraction.

$$\frac{5}{12} = \frac{5 \cdot 3}{12 \cdot 3} = \frac{15}{36}$$

$$\frac{7}{18} = \frac{7 \cdot 2}{18 \cdot 2} = \frac{14}{36}$$

The fraction $\frac{15}{36}$ is equivalent to $\frac{5}{12}$, because it was obtained by multiplying both the numerator and the denominator by 3. Likewise, $\frac{14}{36}$ is equivalent to $\frac{7}{8}$, because it was obtained by multiplying the numerator and the denominator by 2. All we have left to do is to add numerators.

$$\frac{15}{36} + \frac{14}{36} = \frac{29}{36}$$

The sum of $\frac{5}{12}$ and $\frac{7}{18}$ is the fraction $\frac{29}{36}$. Let's write the complete problem again step by step.

$$\frac{5}{12} + \frac{7}{18} = \frac{5 \cdot 3}{12 \cdot 3} + \frac{7 \cdot 2}{18 \cdot 2} \qquad \text{Rewrite each fraction as an equivalent fraction with denominator 36.}$$

$$= \frac{15}{36} + \frac{14}{36}$$

$$= \frac{29}{36} \qquad \text{Add numerators; keep the common denominator.}$$

EXAMPLE 13 Find the LCD for $\frac{3}{4}$ and $\frac{1}{6}$.

Solution We factor 4 and 6 into products of prime factors and build the LCD from these factors.

$$4 = 2 \cdot 2$$
$$\text{LCD} = 2 \cdot 2 \cdot 3 = 12$$
$$6 = 2 \cdot 3$$

The LCD is 12. Both denominators divide it exactly; 4 divides 12 exactly 3 times, and 6 divides 12 exactly 2 times.

EXAMPLE 14 Add: $\frac{3}{4} + \frac{1}{6}$.

Solution In Example 13, we found that the LCD for these two fractions is 12. We begin by changing $\frac{3}{4}$ and $\frac{1}{6}$ to equivalent fractions with denominator 12.

$$\frac{3}{4} = \frac{3 \cdot 3}{4 \cdot 3} = \frac{9}{12}$$

$$\frac{1}{6} = \frac{1 \cdot 2}{6 \cdot 2} = \frac{2}{12}$$

The fraction $\frac{9}{12}$ is equal to the fraction $\frac{3}{4}$, because it was obtained by multiplying the numerator and the denominator of $\frac{3}{4}$ by 3. Likewise, $\frac{2}{12}$ is equivalent to $\frac{1}{6}$, because it was obtained by multiplying the numerator and the denominator of $\frac{1}{6}$ by 2. To complete the problem, we add numerators.

$$\frac{9}{12} + \frac{2}{12} = \frac{11}{12}$$

12. Add: $\frac{5}{18} + \frac{3}{14}$.

13. Find the LCD for $\frac{2}{9}$ and $\frac{4}{15}$.

14. Add: $\frac{2}{9} + \frac{4}{15}$.

Note We can visualize the work in Example 14 using circles and shading:

Answers

12. $\frac{31}{63}$

13. 45

14. $\frac{22}{45}$

The sum of $\frac{3}{4}$ and $\frac{1}{6}$ is $\frac{11}{12}$. Here is how the complete problem looks:

$$\frac{3}{4} + \frac{1}{6} = \frac{3 \cdot 3}{4 \cdot 3} + \frac{1 \cdot 2}{6 \cdot 2}$$

Rewrite each fraction as an equivalent fraction with denominator 12.

$$= \frac{9}{12} + \frac{2}{12}$$

$$= \frac{11}{12}$$

Add numerators; keep the same denominator.

15. Subtract: $\frac{8}{25} - \frac{3}{20}$.

EXAMPLE 15 Subtract: $\dfrac{7}{15} - \dfrac{3}{10}$.

Solution Let's factor 15 and 10 completely and use these factors to build the LCD.

$$15 = 3 \cdot 5$$

15 divides the LCD.

$$\text{LCD} = 2 \cdot 3 \cdot 5 = 30$$

$$10 = 2 \cdot 5$$

10 divides the LCD.

Changing to equivalent fractions and subtracting, we have

$$\frac{7}{15} - \frac{3}{10} = \frac{7 \cdot 2}{15 \cdot 2} - \frac{3 \cdot 3}{10 \cdot 3}$$

Rewrite as equivalent fractions with the LCD for the denominator.

$$= \frac{14}{30} - \frac{9}{30}$$

$$= \frac{5}{30}$$

Subtract numerators; keep the LCD.

$$= \frac{1}{6}$$

Reduce to lowest terms.

As a summary of what we have done so far, and as a guide to working other problems, we now list the steps involved in adding and subtracting fractions with different denominators.

> **HOW TO** **Add or Subtract Any Two Fractions**
>
> **Step 1:** Factor each denominator completely, and use the factors to build the LCD. (Remember, the LCD is the smallest number divisible by each of the denominators in the problem.)
>
> **Step 2:** Rewrite each fraction as an equivalent fraction that has the LCD for its denominator. This is done by multiplying both the numerator and the denominator of the fraction in question by the appropriate whole number.
>
> **Step 3:** Add or subtract the numerators of the fractions produced in Step 2. This is the numerator of the sum or difference. The common denominator of the sum or difference is the LCD.
>
> **Step 4:** Reduce the fraction produced in Step 3 to lowest terms if it is not already in lowest terms.

The idea behind adding or subtracting fractions is really very simple. We can only add or subtract fractions that have the same denominators. If the fractions we are trying to add or subtract do not have the same denominators, we rewrite each of them as an equivalent fraction with the LCD for a denominator.

Here are some additional examples of sums and differences of fractions.

EXAMPLE 16 Subtract: $\frac{3}{5} - \frac{1}{6}$.

Solution The LCD for 5 and 6 is their product, 30. We begin by rewriting each fraction with this common denominator.

$$\frac{3}{5} - \frac{1}{6} = \frac{3 \cdot 6}{5 \cdot 6} - \frac{1 \cdot 5}{6 \cdot 5}$$

$$= \frac{18}{30} - \frac{5}{30}$$

$$= \frac{13}{30}$$

EXAMPLE 17 Add: $\frac{1}{6} + \frac{1}{8} + \frac{1}{4}$.

Solution We begin by factoring the denominators completely and building the LCD from the factors that result.

$$6 = 2 \cdot 3$$

8 divides the LCD.

$$8 = 2 \cdot 2 \cdot 2 \qquad \text{LCD} = 2 \cdot 2 \cdot 2 \cdot 3 = 24$$

$$4 = 2 \cdot 2$$

4 divides the LCD. 6 divides the LCD.

We then change to equivalent fractions and add as usual.

$$\frac{1}{6} + \frac{1}{8} + \frac{1}{4} = \frac{1 \cdot 4}{6 \cdot 4} = \frac{1 \cdot 3}{8 \cdot 3} + \frac{1 \cdot 6}{4 \cdot 6} = \frac{4}{24} + \frac{3}{24} + \frac{6}{24} = \frac{13}{24}$$

EXAMPLE 18 Subtract: $3 - \frac{5}{6}$.

Solution The denominators are 1 (because $3 = \frac{3}{1}$) and 6. The smallest number divisible by both 1 and 6 is 6.

$$3 - \frac{5}{6} = \frac{3}{1} - \frac{5}{6} = \frac{3 \cdot 6}{1 \cdot 6} - \frac{5}{6} = \frac{18}{6} = \frac{5}{6} = \frac{13}{6}$$

GETTING READY FOR CLASS

After reading through the preceding section, respond in your own words and in complete sentences.

A. What is a reciprocal?

B. When adding two fractions with the same denominators, we always add their _____, but we never add their _____. Explain.

C. What does the abbreviation LCD stand for?

D. What is the first step when finding the LCD for the fractions $\frac{5}{12}$ and $\frac{7}{18}$?

16. Subtract: $\frac{3}{4} - \frac{1}{5}$.

17. Add: $\frac{1}{9} + \frac{1}{4} + \frac{1}{6}$.

18. Subtract: $2 - \frac{3}{4}$.

Answers

16. $\frac{11}{20}$

17. $\frac{19}{36}$

18. $\frac{5}{4}$

Vocabulary Review

Choose the correct words to fill in the blanks below.

reciprocal factors least common denominator

multiplication division

1. The rule for the _____ of fractions is given by $\frac{a}{b} \cdot \frac{c}{d} = \frac{a \cdot c}{b \cdot d}$.

2. The rule for the _____ of fractions is given by $\frac{a}{b} \div \frac{c}{d} = \frac{a}{b} \cdot \frac{d}{c}$.

3. The _____ of 4 is $\frac{1}{4}$.

4. The smallest number that is exactly divisible by each denominator in a set of denominators is called the _____.

5. The first step to add any two fractions is to factor each denominator completely, and use the _____ to build the least common denominator.

Problems

A Find each of the following products. (Multiply.)

1. $\frac{2}{3} \cdot \frac{4}{5}$ **2.** $\frac{5}{6} \cdot \frac{7}{4}$ **3.** $\frac{1}{2} \cdot \frac{7}{4}$ **4.** $\frac{3}{5} \cdot \frac{4}{7}$ **5.** $\frac{5}{3} \cdot \frac{3}{5}$ **6.** $\frac{4}{7} \cdot \frac{7}{4}$

7. $\frac{3}{4} \cdot 9$ **8.** $\frac{2}{3} \cdot 5$ **9.** $\frac{6}{7} \cdot \frac{7}{6}$ **10.** $\frac{2}{9} \cdot \frac{9}{2}$ **11.** $\frac{1}{2} \cdot \frac{1}{3} \cdot \frac{1}{4}$ **12.** $\frac{2}{3} \cdot \frac{4}{5} \cdot \frac{1}{3}$

13. $\frac{2}{5} \cdot \frac{3}{5} \cdot \frac{4}{5}$ **14.** $\frac{1}{4} \cdot \frac{3}{4} \cdot \frac{3}{4}$ **15.** $\frac{3}{2} \cdot \frac{5}{2} \cdot \frac{7}{2}$ **16.** $\frac{4}{3} \cdot \frac{5}{3} \cdot \frac{7}{3}$ **17.** $\frac{2}{3} \cdot \frac{2}{3} \cdot \frac{2}{3}$ **18.** $\frac{3}{4} \cdot \frac{3}{4} \cdot \frac{3}{4}$

Multiply each of the following. Be sure all answers are written in lowest terms.

19. $\frac{9}{20} \cdot \frac{4}{3}$ **20.** $\frac{135}{16} \cdot \frac{2}{45}$ **21.** $\frac{3}{4} \cdot 12$ **22.** $\frac{3}{4} \cdot 20$

23. $\frac{1}{3}$ (3) **24.** $\frac{1}{5}$ (5) **25.** $\frac{2}{5} \cdot 20$ **26.** $\frac{3}{5} \cdot 15$

27. $\frac{72}{35} \cdot \frac{55}{108} \cdot \frac{7}{110}$ **28.** $\frac{32}{27} \cdot \frac{72}{49} \cdot \frac{1}{40}$

B Find the quotient in each case by replacing the divisor by its reciprocal and multiplying.

29. $\dfrac{3}{4} \div \dfrac{1}{5}$ **30.** $\dfrac{1}{3} \div \dfrac{1}{2}$ **31.** $\dfrac{2}{3} \div \dfrac{1}{2}$ **32.** $\dfrac{5}{8} \div \dfrac{1}{4}$

33. $6 \div \dfrac{2}{3}$ **34.** $8 \div \dfrac{3}{4}$ **35.** $20 \div \dfrac{1}{10}$ **36.** $16 \div \dfrac{1}{8}$

37. $\dfrac{3}{4} \div 2$ **38.** $\dfrac{3}{5} \div 2$ **39.** $\dfrac{7}{8} \div \dfrac{7}{8}$ **40.** $\dfrac{4}{3} \div \dfrac{4}{3}$

41. $\dfrac{7}{8} \div \dfrac{8}{7}$ **42.** $\dfrac{4}{3} \div \dfrac{3}{4}$ **43.** $\dfrac{9}{16} \div \dfrac{3}{4}$ **44.** $\dfrac{25}{36} \div \dfrac{5}{6}$

45. $\dfrac{25}{46} \div \dfrac{40}{69}$ **46.** $\dfrac{25}{24} \div \dfrac{15}{36}$ **47.** $\dfrac{13}{28} \div \dfrac{39}{14}$ **48.** $\dfrac{28}{125} \div \dfrac{5}{2}$

49. $\dfrac{27}{196} \div \dfrac{9}{392}$ **50.** $\dfrac{16}{135} \div \dfrac{2}{45}$ **51.** $\dfrac{25}{18} \div 5$ **52.** $\dfrac{30}{27} \div 6$

53. $6 \div \dfrac{4}{3}$ **54.** $12 \div \dfrac{4}{3}$ **55.** $\dfrac{4}{3} \div 6$ **56.** $\dfrac{4}{3} \div 12$

C Find the following sums and differences, and reduce to lowest terms. (Add or subtract as indicated.)

57. $\dfrac{3}{6} + \dfrac{1}{6}$ **58.** $\dfrac{2}{5} + \dfrac{3}{5}$ **59.** $\dfrac{5}{8} - \dfrac{3}{8}$ **60.** $\dfrac{6}{7} - \dfrac{1}{7}$

61. $\dfrac{3}{4} - \dfrac{1}{4}$ **62.** $\dfrac{7}{9} - \dfrac{4}{9}$ **63.** $\dfrac{2}{3} - \dfrac{1}{3}$ **64.** $\dfrac{9}{8} - \dfrac{1}{8}$

65. $\dfrac{1}{4} + \dfrac{2}{4} + \dfrac{3}{4}$

66. $\dfrac{2}{5} + \dfrac{3}{5} + \dfrac{4}{5}$

67. $\dfrac{x+7}{2} - \dfrac{1}{2}$

68. $\dfrac{x+5}{4} - \dfrac{3}{4}$

69. $\dfrac{1}{10} + \dfrac{3}{10} + \dfrac{4}{10}$

70. $\dfrac{3}{20} + \dfrac{1}{20} + \dfrac{4}{20}$

71. $\dfrac{1}{3} + \dfrac{4}{3} + \dfrac{5}{3}$

72. $\dfrac{5}{4} + \dfrac{4}{4} + \dfrac{3}{4}$

D Find the LCD for each of the following; then use the methods developed in this section to add or subtract as indicated.

73. $\dfrac{4}{9} + \dfrac{1}{3}$

74. $\dfrac{1}{2} + \dfrac{1}{4}$

75. $2 + \dfrac{1}{3}$

76. $3 + \dfrac{1}{2}$

77. $\dfrac{3}{4} + 1$

78. $\dfrac{3}{4} + 2$

79. $\dfrac{1}{2} + \dfrac{2}{3}$

80. $\dfrac{1}{8} + \dfrac{3}{4}$

81. $\dfrac{1}{4} + \dfrac{1}{5}$

82. $\dfrac{1}{3} + \dfrac{1}{5}$

83. $\dfrac{1}{2} + \dfrac{1}{5}$

84. $\dfrac{1}{2} - \dfrac{1}{5}$

85. $\dfrac{5}{12} + \dfrac{3}{8}$

86. $\dfrac{9}{16} + \dfrac{7}{12}$

87. $\dfrac{8}{30} - \dfrac{1}{20}$

88. $\dfrac{9}{40} - \dfrac{1}{30}$

89. $\dfrac{3}{10} + \dfrac{1}{100}$

90. $\dfrac{9}{100} + \dfrac{7}{10}$

91. $\dfrac{10}{36} + \dfrac{9}{48}$

92. $\dfrac{12}{28} + \dfrac{9}{20}$

Find the Mistake

Each sentence below contains a mistake. Circle the mistake and write the correct word(s) or numbers(s) on the line provided.

1. To multiply the fractions $\dfrac{2}{8}$ and $\dfrac{3}{4}$, we put the product of 2 and 3 in the numerator over 8 in the denominator.

2. The reciprocal of $\dfrac{5}{20}$ is $\dfrac{1}{5}$.

3. To divide $\dfrac{7}{8}$ by $\dfrac{2}{5}$, multiply $\dfrac{2}{5}$ by $\dfrac{8}{7}$.

4. The least common denominator for the fractions $\dfrac{3}{7}$ and $\dfrac{5}{6}$ is 13.

0.3 Decimal Notation

KEY WORDS

place value

expanded form

decimal point

decimal fraction

round

April 3rd is International Pillow Fight Day. In 2010, the 5th Pillow Fight NYC took place in Union Square, New York City. People from all over the city filled the square and wielded pillows. When the fight began, people lightly swung their pillows from side to side, hoping it would land on the nearest victim before getting hit themselves. The city's inaugural pillow fight took place in 2006 and had 500 participants. By 2010, the event garnered at least 5000 people.

In this section, we will learn about *place value* and its relation to decimals. Comparing the number of participants in the 2006 pillow fight to the 2010 pillow fight, we can see that the 5 moves from the hundreds place value to the thousands place value.

A Expanded Form

What is interesting and useful about decimals is their relationship to fractions and to powers of ten. The work we have done up to now—especially our work with fractions—can be used to develop the properties of decimal numbers.

Recall from a previous math class the idea of place value for the digits in a whole number. The name and the place value of each of the first seven columns in our number system are as follows:

Millions Column	Hundred Thousands Column	Ten Thousands Column	Thousands Column	Hundreds Column	Tens Column	Ones Column
1,000,000	100,000	10,000	1,000	100	10	1

As we move from right to left, we multiply by 10 each time. The value of each column is 10 times the value of the column on its right, with the rightmost column being 1. Up until now we have always looked at place value as increasing by a factor of 10 each time we move one column to the left.

Ten Thousands		Thousands		Hundreds		Tens		Ones
10,000	←	1,000	←	100	←	10	←	1
	Multiply by 10		Multiply by 10		Multiply by 10		Multiply by 10	

To understand the idea behind decimal numbers, we notice that moving in the opposite direction, from left to right, we *divide* by 10 each time.

Ten Thousands	Thousands	Hundreds	Tens	Ones
10,000 \longrightarrow	1,000 \longrightarrow	100 \longrightarrow	10 \longrightarrow	1
Divide by 10	Divide by 10	Divide by 10	Divide by 10	

If we keep going to the right, the next column will have to be

$$1 \div 10 = \frac{1}{10} \qquad \text{Tenths}$$

The next one after that will be

$$\frac{1}{10} \div 10 = \frac{1}{10} \cdot \frac{1}{10} = \frac{1}{100} \qquad \text{Hundredths}$$

After that, we have

$$\frac{1}{100} \div 10 = \frac{1}{100} \cdot \frac{1}{10} = \frac{1}{1,000} \qquad \text{Thousandths}$$

We could continue this pattern as long as we wanted. We simply divide by 10 to move one column to the right. (And remember, dividing by 10 gives the same result as multiplying by $\frac{1}{10}$.)

To show where the ones column is, we use a *decimal point* between the ones column and the tenths column.

Thousands	Hundreds	Tens	Ones		Tenths	Hundredths	Thousandths	Ten-Thousandths	Hundred-Thousandths
1,000	100	10	1	.	$\frac{1}{10}$	$\frac{1}{100}$	$\frac{1}{1,000}$	$\frac{1}{10,000}$	$\frac{1}{100,000}$

↑ Decimal point

It is important to recognize the ones column, not the decimal point, can be thought of as the middle column, with columns larger than 1 to the left and columns smaller than 1 to the right. The first column to the right of the ones column is the tenths column, not the "oneths" column. The next column to the right is the hundredths column, the next is the thousandths column, and so on. The decimal point is always written between the ones column and the tenths column.

We can use the place value of decimal fractions to write them in *expanded form*.

EXAMPLE 1 Write 423.576 in expanded form.

Solution $423.576 = 400 + 20 + 3 + \dfrac{5}{10} + \dfrac{7}{100} + \dfrac{6}{1,000}$

EXAMPLE 2 Write each number in words.
 a. 0.4 **b.** 0.04 **c.** 0.004

Solution
 a. 0.4 is "four tenths."
 b. 0.04 is "four hundredths."
 c. 0.004 is "four thousandths."

When a decimal fraction contains digits to the left of the decimal point, we use the word "and" to indicate where the decimal point is when writing the number in words.

Note Because the digits to the right of the decimal point have fractional place values, numbers with digits to the right of the decimal point are called *decimal fractions*. In this book, we will also call them *decimal numbers*, or simply *decimals* for short.

Practice Problems

1. Write 785.462 in expanded form.

2. Write in words.
 a. 0.06
 b. 0.7
 c. 0.008

Answers

1. $700 + 80 + 5 + \frac{4}{10} + \frac{6}{100} + \frac{2}{1,000}$

2. a. Six hundredths **b.** Seven tenths
 c. Eight thousandths

EXAMPLE 3 Write each number in words.

 a. 5.4 **b.** 5.04 **c.** 5.004

Solution

 a. 5.4 is "five and four tenths."

 b. 5.04 is "five and four hundredths."

 c. 5.004 is "five and four thousandths."

EXAMPLE 4 Write 3.64 in words.

Solution The number 3.64 is read "three and sixty-four hundredths." The place values of the digits are as follows:

$$\underset{\substack{\uparrow\qquad\uparrow\qquad\uparrow\\ 3\text{ ones }\;6\text{ tenths }\;4\text{ hundredths}}}{3.64}$$

We read the decimal part as "sixty-four hundredths" because

$$6\text{ tenths} + 4\text{ hundredths} = \frac{6}{10} + \frac{4}{100} = \frac{60}{100} + \frac{4}{100} = \frac{64}{100}$$

EXAMPLE 5 Write 25.4936 in words.

Solution Using the idea given in Example 4, we write 25.4936 in words as "twenty-five and four thousand, nine hundred thirty-six ten-thousandths."

EXAMPLE 6 Write the following numbers in order from largest to smallest.

 0.03 0.3 0.032 0.006 0.003 0.2

Solution We know that the numbers with a digit greater than zero in the tenths place are greater than those with a zero in the tenths place, so our greatest numbers are 0.3 and 0.2.

 $0.3 > 0.2$

Next, we look to the numbers with first digits greater than zero in the hundredths place.

 $0.032 > 0.030$ so $0.032 > 0.03$

Finally, we look at numbers with the first digits greater than zero in the thousandths place.

 $0.006 > 0.003$

Our complete list looks like this: 0.3 0.2 0.032 0.03 0.006 0.003

In order to understand addition and subtraction of decimals in the next section, we need to be able to convert decimal numbers to fractions or mixed numbers.

EXAMPLE 7 Write each number as a fraction or a mixed number. Do not reduce to lowest terms.

 a. 0.004 **b.** 3.64 **c.** 25.4936

Solution

 a. Because 0.004 is 4 thousandths, we write

$$\text{Three digits after the decimal point} \rightarrow 0.004 = \frac{4}{1{,}000} \leftarrow \text{Three zeros}$$

 b. Looking over the work in Example 4, we can write

$$\text{Two digits after the decimal point} \rightarrow 3.64 = 3\,\frac{64}{100} \leftarrow \text{Two zeros}$$

 c. From the way in which we wrote 25.4936 in words in Example 5, we have

$$\text{Four digits after the decimal point} \rightarrow 25.4936 = 25\,\frac{4936}{10{,}000} \leftarrow \text{Four zeros}$$

3. Write each number in words.

 a. 5.06

 b. 4.7

 c. 3.008

Note Sometimes we name decimal fractions by simply reading the digits from left to right and using the word "point" to indicate where the decimal point is. For example, using this method the number 5.04 is read "five point zero four." But for the purposes of this section, we will continue to use place values when naming decimal fractions.

4. Write 5.98 in words.

5. Write 305.406 in words.

6. Write the following numbers in order from largest to smallest.

 0.01 0.001 0.012 0.2 0.1 0.12

7. Write each number as a fraction or a mixed number. Do not reduce to lowest terms.

 a. 0.06

 b. 0.7

 c. 0.008

Note Recall from a previous math class that a *mixed number* is the sum of a whole number and a proper fraction. In Example 6b, $3\frac{64}{100}$ is a mixed number.

Answers

3. a. Five and six hundredths
 b. Four and seven tenths
 c. Three and eight thousandths
4. Five and ninety eight hundredths
5. Three hundred five and four hundred six thousandths
6. 0.2 0.12 0.1 0.012 0.01 0.001
7. a. $\frac{6}{100}$ **b.** $\frac{98}{100}$ **c.** $\frac{8}{1{,}000}$

B Rounding Decimal Numbers

The rule for rounding decimal numbers is similar to the rule for rounding whole numbers. If the digit in the column to the right of the one we are rounding to is 5 or more, we add 1 to the digit in the column we are rounding to; otherwise, we leave it alone. We then replace all digits to the right of the column we are rounding to with zeros if they are to the left of the decimal point; otherwise, we simply delete them. Table 1 illustrates the procedure.

Table 1	Round to the Nearest		
	Whole Number	Tenth	Hundredth
24.785	25	24.8	24.79
2.3914	2	2.4	2.39
0.98243	1	1.0	0.98
14.0942	14	14.1	14.09
0.545	1	0.5	0.55

8. Round 8,935.042 to the nearest hundred.

EXAMPLE 8 Round 9,235.492 to the nearest hundred.

Solution The number next to the hundreds column is 3, which is less than 5. We change all digits to the right to 0, and we can drop all digits to the right of the decimal point, so we write

$$9,200$$

9. Round 0.05067 to the nearest ten-thousandth.

EXAMPLE 9 Round 0.00346 to the nearest ten-thousandth.

Solution Because the number to the right of the ten-thousandths column is more than 5, we add 1 to the 4 and get

$$0.0035$$

10. Round each number in the bar chart to the nearest tenth of a dollar.

EXAMPLE 10 The bar chart below shows some ticket prices for a recent major league baseball season. Round each ticket price to the nearest dollar.

Solution Using our rule for rounding decimal numbers, we have the following results:

Least expensive: $14.31 rounds to $14.
League average: $26.74 rounds to $27.
Most expensive: $52.32 rounds to $52.

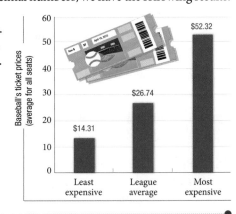

GETTING READY FOR CLASS

After reading through the preceding section, respond in your own words and in complete sentences.

A. Write 754.326 in expanded form.

B. Write $400 + 70 + 5 + \frac{1}{10} + \frac{3}{100} + \frac{7}{1,000}$ in decimal form.

C. Write seventy-two and three tenths in decimal form.

D. What is the rule for rounding decimal numbers?

Answers

8. 8,900

9. 0.0507

10. $14.30, $26.70, $52.60

Vocabulary Review

Choose the correct words to fill in the blanks below.

hundredths tenths decimal point multiply divide

1. Moving from right to left in the place value columns, we _____ by 10 each time.

2. Moving from left to right in the place value columns, we _____ by 10 each time.

3. The symbol used between the ones column and the tenths column is called a _____.

4. If we write 6.3 in words, we get "six and three _____."

5. If we round 0.0456 to the nearest _____ column, we get the answer 0.05.

Problems

A Write out the name of each number in words.

1. 0.3 **2.** 0.03

3. 0.015 **4.** 0.0015

5. 3.4 **6.** 2.04

7. 52.7 **8.** 46.8

Write each number as a fraction or a mixed number. Do not reduce your answers.

9. 405.36 **10.** 362.78

11. 9.009 **12.** 60.06

13. 1.234 **14.** 12.045

15. 0.00305 **16.** 2.00106

SCAN TO ACCESS

Give the place value of the 5 in each of the following numbers.

17. 458.327 **18.** 327.458 **19.** 29.52 **20.** 25.92

21. 0.00375 **22.** 0.00532 **23.** 275.01 **24.** 0.356

25. 539.76 **26.** 0.123456 **27.** 0.00045 **28.** 0.0000356

Write each of the following as a decimal number.

29. Fifty-five hundredths

30. Two hundred thirty-five ten-thousandths

31. Six and nine tenths

32. Forty-five thousand and six hundred twenty-one thousandths

33. Eleven and eleven hundredths

34. Twenty-six thousand, two hundred forty-five and sixteen hundredths

35. One hundred and two hundredths

36. Seventy-five and seventy-five hundred-thousandths

37. Three thousand and three thousandths

38. One thousand, one hundred eleven and one hundred eleven thousandths

B Complete the following table.

		Round the the Nearest			
	Number	Whole Number	Tenth	Hundredth	Thousandth
39.	47.5479	_____	_____	_____	_____
40.	100.9256	_____	_____	_____	_____
41.	0.8175	_____	_____	_____	_____
42.	29.9876	_____	_____	_____	_____
43.	0.1562	_____	_____	_____	_____
44.	128.9115	_____	_____	_____	_____
45.	2,789.3241	_____	_____	_____	_____
46.	0.8743	_____	_____	_____	_____
47.	99.9999	_____	_____	_____	_____
48.	71.7634	_____	_____	_____	_____

49. Write the following numbers in order from smallest to largest.

 0.02 0.05 0.025 0.052 0.005 0.002

50. Write the following numbers in order from smallest to largest.

 0.2 0.02 0.4 0.04 0.42 0.24

51. Which of the following numbers will round to 7.5?

 7.451 7.449 7.54 7.56

52. Which of the following numbers will round to 3.2?

 3.14999 3.24999 3.279 3.16111

Change each decimal to a fraction, and then reduce to lowest terms.

53. 0.25 **54.** 0.75 **55.** 0.35 **56.** 0.65

57. 0.125 **58.** 0.375 **59.** 0.625 **60.** 0.0625

61. 0.875 **62.** 0.1875 **63.** 0.012 **64.** 0.045

65. The following table lists the five tallest buildings in the world. Complete the chart by filing in each height in words.

Building	City, Country	Height (ft)	Height in Words
Burj Khalifa	Dubai, The United Arab Emirates	2,716	
Taipei 101	Taipei, Taiwan	1,667	
World Financial Center	Shangai, China	1,614	
Petronus Tower 1	Kuala Lunpur, Malaysia	1,483	
Petronus Tower 2	Kuala Lunpur, Malaysia	1,483	

66. The following table lists the famous "seven summits," the highest mountains on each of the seven continents. Complete the list by filling in the height in words of each mountain.

Summit	Continent	Height (ft)	Height in Words
Everest	Asia	29,035	
Aconcagua	South America	23,841	
McKinley	North America	20,320	
Kilimanjaro	Africa	19,340	
Elbrus	Europe	18,320	
Vinson Massif	Antartica	16,050	
Carstensz Pyramid	Australia	16,024	

Source: 7summits.com

Estimating

For each pair of numbers, choose the number that is closest to 10.

67. 9.9 and 9.99 **68.** 8.5 and 8.05 **69.** 10.5 and 10.05 **70.** 10.9 and 10.99

For each pair of numbers, choose the number that is closest to 0.

71. 0.5 and 0.05 **72.** 0.10 and 0.05 **73.** 0.01 and 0.02 **74.** 0.1 and 0.01

Find the Mistake

Each sentence below contains a mistake. Circle the mistake and write the correct word(s) or numbers(s) on the line provided.

1. The value of each column to the left of the decimal point is 10 times the value of the column on its left.

2. The fourth column to the right of the decimal point is called the thousandths column.

3. 2,568 rounded to the nearest hundred is 2,560.

4. 2.568 rounded to the nearest tenth is 2.5.

0.4 Operations with Decimals

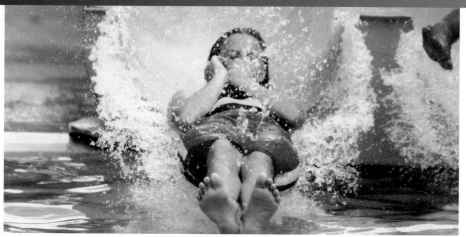

KEY WORDS

column

fractional part

estimate

operation

remainder

At Noah's Ark Waterpark in Wisconsin, thrill seekers brave the country's first nearly-vertical loop waterslide. From a ten-story platform, a person enters the top of the waterslide's tube. Once the door of the tube closes, a countdown from ten begins. Then the floor drops from beneath the rider who plummets to the base of the loop, gaining as much speed as possible. Water sprays the inside of the tube to keep the ride smooth around the loop, and the rider comes shooting out happily on the other side.

The waterpark ran a summer special charging $34.50 for an adult ticket and $30.66 for a child's ticket. Suppose you bought tickets for two adults and two children. You would need to know how to perform operations with decimals in order to calculate how much you paid for the four tickets. In this section, we will do just that.

A Addition and Subtraction with Decimals

Let's begin with an example involving wages. Suppose you are earning $8.50 an hour and you receive a raise of $1.25 an hour. Your new hourly rate of pay is

$$
\begin{array}{r}
\$8.50 \\
+ \ \$1.25 \\
\hline
\$9.75
\end{array}
$$

To add the two rates of pay, we align the decimal points, and then add in columns.

> **RULE** Addition (or Subtraction) of Decimals
>
> To add (or subtract) decimal numbers, we line up the decimal points and add (or subtract) as usual. The decimal point in the result is written directly below the decimal points in the problem.

We will use this rule for Examples 1–4.

EXAMPLE 1 Subtract: 39.812 − 14.236.

Solution We write the numbers vertically, with the decimal points lined up, and subtract as usual.

$$
\begin{array}{r}
39.812 \\
- \ 14.236 \\
\hline
25.576
\end{array}
$$

Practice Problems

1. Subtract: 78.674 − 23.431.

Answer

1. 55.243

2. Add: $16 + 0.033 + 4.6 + 0.08$.

Note Writing the extra zeros here is really equivalent to finding a common denominator for the fractional parts of the original four numbers—now we have a thousandths column in all the numbers.

3. Subtract: $6.7 - 2.0563$.

4. Subtract 5.89 from the sum of 7 and 3.567.

5. Change to a fractions and multiply.

$$0.4 \times 0.6$$

Note To indicate multiplication, we are using a \times sign here instead of a dot so we won't confuse the decimal points with the multiplication symbol.

6. Change to fractions and multiply.

$$3.5 \times 0.04$$

Answers

2. 20.713
3. 4.6437
4. 4.677
5. 0.24
6. 0.14

EXAMPLE 2 Add: $8 + 0.002 + 3.1 + 0.04$.

Solution To make sure we keep the digits in the correct columns, we can write zeros to the right of the rightmost digits.

$$8 = 8.000$$
$$3.1 = 3.100$$
$$0.04 = 0.040$$

This doesn't change the value of any of the numbers, and it makes our task easier. Now we have

$$
\begin{array}{r}
8.000 \\
0.002 \\
3.100 \\
+ \ 0.040 \\
\hline
11.142
\end{array}
$$

EXAMPLE 3 Subtract: $5.9 - 3.0814$.

Solution In this case, it is very helpful to write 5.9 as 5.9000, since we will have to borrow in order to subtract.

$$
\begin{array}{r}
5.9000 \\
- \ 3.0814 \\
\hline
2.8186
\end{array}
$$

EXAMPLE 4 Subtract 3.09 from the sum of 9 and 5.472.

Solution Writing the problem in symbols, we have

$$(9 + 5.472) - 3.09 = 14.472 - 3.09$$
$$= 11.382$$

B Multiplication with Decimals

EXAMPLE 5 Change each decimal to a fraction and multiply.

$$0.5 \times 0.3$$

Solution Changing each decimal to a fraction and multiplying, we have

$$0.5 \times 0.3 = \frac{5}{10} \times \frac{3}{10} \qquad \textit{Change to fractions.}$$

$$= \frac{15}{100} \qquad \textit{Multiply numerators and multiply denominators.}$$

$$= 0.15 \qquad \textit{Write the answer in decimal form.}$$

The result is 0.15, which has two digits to the right of the decimal point.

EXAMPLE 6 Multiply: 2.1×0.07.

Solution $2.1 \times 0.07 = 2\frac{1}{10} \times \frac{7}{100} \qquad \textit{Change to fractions.}$

$$= \frac{21}{10} \times \frac{7}{100}$$

$$= \frac{147}{1,000} \qquad \textit{Multiply numerators and denominators.}$$

$$= 0.147 \qquad \textit{Write the answer as a decimal.}$$

Again, the digits in the answer come from multiplying $21 \times 7 = 147$. The decimal point is placed so that there are three digits to its right, because that is the total number of digits to the right of the decimal points in 2.1 and 0.07.

EXAMPLE 7 How many digits will be to the right of the decimal point in the following product?

$$2.987 \times 24.82$$

Solution There are three digits to the right of the decimal point in 2.987 and two digits to the right in 24.82. Therefore, there will be $3 + 2 = 5$ digits to the right of the decimal point in their product.

EXAMPLE 8 Multiply: 3.05×4.36.

Solution We can set this up as if it were a multiplication problem with whole numbers. We multiply and then place the decimal point in the correct position in the answer.

$$
\begin{array}{r}
3.05 \\
\times\ 4.36 \\
\hline
1830 \\
915 \\
1220 \\
\hline
13.2980
\end{array}
$$

←— Move decimal point 2 digits to the right.
←— Move decimal point 2 digits to the right, then multiply.

The decimal point is placed so that there are $2 + 2 = 4$ digits to its right.

We summarize this discussion with a rule.

> **RULE** Multiplication with Decimals
>
> To multiply two decimal numbers:
>
> 1. Multiply as you would if the decimal points were not there.
>
> 2. Place the decimal point in the answer so that the number of digits to its right is equal to the total number of digits to the right of the decimal points in the original two numbers in the problem.

As you can see, multiplying decimal numbers is just like multiplying whole numbers, except that we must place the decimal point in the result in the correct position.

Estimating

Look back to Example 8. We could have placed the decimal point in the answer by rounding the two numbers to the nearest whole number and then multiplying them. Because 3.05 rounds to 3 and 4.36 rounds to 4, and the product of 3 and 4 is 12, we estimate that the answer to 3.05×4.36 will be close to 12. We then place the decimal point in the product 132980 between the 3 and the 2 in order to make it into a number close to 12.

EXAMPLE 9 Estimate the answer to each of the following products.

 a. 29.4×8.2 **b.** 68.5×172 **c.** $(6.32)^2$

Solution **a.** Because 29.4 is approximately 30 and 8.2 is approximately 8, we estimate this product to be about $30 \times 8 = 240$. (If we were to multiply 29.4 and 8.2, we would find the product to be exactly 241.08.)

 b. Rounding 68.5 to 70 and 172 to 170, we estimate this product to be $70 \times 170 = 11,900$. (The exact answer is 11,782.) Note here that we do not always round the numbers to the nearest whole number when making estimates. The idea is to round to numbers that will be easy to multiply.

 c. Because 6.32 is approximately 6 and $6^2 = 36$, we estimate our answer to be close to 36. (The actual answer is 39.9424.)

7. How many digits will be to the right of the decimal point in the following product?

$$3.706 \times 55.88$$

8. Multiply: 4.03×5.22.

9. Estimate the answer to each product.
 a. 82.3×5.8
 b. 37.5×178
 c. $(8.21)^2$

Answers
7. 5
8. 21.0366
9. a. 480 **b.** 7,200 **c.** 64

C Division with Decimals

Suppose three friends go out for lunch and their total bill comes to $18.75. If they decide to split the bill equally, how much does each person owe? To find out, they will have to divide 18.75 by 3. If you think about the dollars and cents separately, you will see that each person owes $6.25. Therefore,

$$\$18.75 \div 3 = \$6.25$$

However, many division problems that involve decimals are not as intuitive as this one. Here are some examples that will help you find out how to do division with any combination of decimal numbers.

EXAMPLE 10 Divide: 5,974 ÷ 20.

Solution
```
       298
  20)5974
      40 ↓
      197
      180 ↓
       174
       160
        14    Remainder of 14
```

In the past, we have written this answer as $298\frac{14}{20}$ or, after reducing the fraction, $298\frac{7}{10}$. Because $\frac{7}{10}$ can be written as 0.7, we could also write our answer as 298.7. This last form of our answer is exactly the same result we obtain if we write 5,974 as 5,974.0 and continue the division until we have no remainder. Here is how it looks:

```
        298.7
  20)5974.0
      40 ↓
      197
      180 ↓
       174
       160 ↓
       140
       140
         0
```

Let's try another division problem. This time one of the numbers in the original problem will be a decimal.

EXAMPLE 11 Divide: 34.8 ÷ 4.

Solution We can use the ideas from Example 13 and divide as usual. The decimal point in the answer will be placed directly above the decimal point in the problem.

```
        8.7
   4)34.8
      32 ↓
       28
       28
        0
```

The answer is 8.7.

Check:
```
     8.7
   × 4
    34.8
```

10. Divide: 4,626 ÷ 30.

Note We can estimate the answer to Example 10 by rounding 5,974 to 6,000 and dividing by 20.

$$\frac{6,000}{20} = 300$$

11. Divide: 33.5 ÷ 5.

Note We never need to make a mistake with division, because we can always check our results with multiplication.

RULE Division with Decimals

To divide a decimal by a whole number, we do the usual long division as if there were no decimal point involved. The decimal point in the answer is placed directly above the decimal point in the problem.

Here are some more examples to illustrate the procedure:

EXAMPLE 12 Divide: $1{,}138.9 \div 35$.

Solution

$$
\begin{array}{r}
32.54 \\
35\overline{)1138.90} \\
105 \\
\hline
88 \\
70 \\
\hline
18\,9 \\
17\,5 \\
\hline
1\,40 \\
1\,40 \\
\hline
0
\end{array}
$$

Hint: write 0 after the 9. It doesn't change the original number, but it gives us another digit to bring down.

The answer is 32.54.

Check:

$$
\begin{array}{r}
32.54 \\
\times\ \ 35 \\
\hline
16270 \\
97620 \\
\hline
1{,}138.90
\end{array}
$$

Until now we have considered only division by whole numbers. Extending division to include division by decimal numbers is a matter of knowing what to do about the decimal point in the divisor.

EXAMPLE 13 Divide: $31.35 \div 3.8$.

Solution In fraction form, this problem is equivalent to

$$\frac{31.35}{3.8}$$

If we want to write the divisor as a whole number, we can multiply the numerator and the denominator of this fraction by 10.

$$\frac{31.35 \times 10}{3.8 \times 10} = \frac{313.5}{38}$$

Since this fraction is equivalent to the original fraction, our original division problem is equivalent to

$$
\begin{array}{r}
8.25 \\
38\overline{)313.50} \\
304 \\
\hline
95 \\
76 \\
\hline
190 \\
190 \\
\hline
0
\end{array}
$$

Put 0 after the last digit

12. Divide: $1{,}138.5 \div 25$.

Note We do not always use the rules for rounding numbers to make estimates. For example, to estimate the answer to Example 13, $31.35 \div 3.8$, we can get a rough estimate of the answer by reasoning that 3.8 is close to 4 and 31.35 is close to 32. Therefore, our answer will be approximately $32 \div 4 = 8$.

13. Divide: $13.23 \div 4.2$.

Answers

12. 45.54

13. 3.15

We can summarize division with decimal numbers by listing the following points, as illustrated by the previous examples.

> **Summary of Division with Decimals**
>
> 1. We divide decimal numbers by the same process used to divide whole numbers. The decimal point in the answer is placed directly above the decimal point in the dividend.
>
> 2. We are free to write as many zeros after the last digit in a decimal number as we need.
>
> 3. If the divisor is a decimal, we can change it to a whole number by moving the decimal point to the right as many places as necessary, so long as we move the decimal point in the dividend the same number of places.

14. Divide, and round your answer to the nearest hundredth.
$0.4553 \div 0.32$

EXAMPLE 14 Divide, and round the answer to the nearest hundredth.

$$0.3778 \div 0.25$$

Solution First, we move the decimal point two places to the right.

$$0.25. \qquad 0.37.78$$

Then we divide, using long division.

$$
\begin{array}{r}
1.5112 \\
25\overline{)37.7800} \\
\underline{25} \\
12\,7 \\
\underline{12\,5} \\
28 \\
\underline{25} \\
30 \\
\underline{25} \\
50 \\
\underline{50} \\
0
\end{array}
$$

Rounding to the nearest hundredth, we have 1.51. We actually did not need to have this many digits to round to the hundredths column. We could have stopped at the thousandths column and rounded off.

15. Divide, and round to the nearest tenth.
$19 \div 0.06$

EXAMPLE 15 Divide, and round the answer to the nearest tenth.

$$17 \div 0.03$$

Solution Because we are rounding to the nearest tenth, we will continue dividing until we have a digit in the hundredths column. We don't have to go any further to round to the tenths column.

$$
\begin{array}{r}
566.66 \\
0.03.\overline{)17.00.00} \\
\underline{15} \\
20 \\
\underline{18} \\
20 \\
\underline{18} \\
20 \\
\underline{18} \\
20 \\
\underline{18} \\
2
\end{array}
$$

Rounding to the nearest tenth, we have 566.7.

Answers

14. 1.423

15. 316.7

EXAMPLE 16 A telephone company charges $0.43 for the first minute and then $0.33 for each additional minute for a long-distance call. If a long-distance call costs $3.07, how many minutes was the call?

Solution To solve this problem we need to find the number of additional minutes for the call. To do so, we first subtract the cost of the first minute from the total cost, and then we divide the result by the cost of each additional minute. Without showing the actual arithmetic involved, the solution looks like this:

$$\underset{\substack{\uparrow \\ \text{Cost of each additional minute}}}{\text{The number of additional minutes}} = \dfrac{\overset{\substack{\text{Total cost} \\ \text{of the call} \\ \downarrow}}{3.07} - \overset{\substack{\text{Cost of the} \\ \text{first minute} \\ \downarrow}}{0.43}}{0.33} = \dfrac{2.64}{0.33} = 8$$

The call was 9 minutes long. (The number 8 is the number of additional minutes past the first minute.)

GETTING READY FOR CLASS

After reading through the preceding section, respond in your own words and in complete sentences.

A. Briefly explain the rule for adding and subtracting decimals.

B. Where do you place the decimal point in an answer to a multiplication problem that involves decimals? Explain.

C. The answer to the division problem in Example 13 is $298\frac{14}{20}$. In your own words, explain why we can also write this number as 298.7.

D. In Example 15, we place a 0 at the end of a number without changing the value of the number. Why is the placement of this 0 helpful?

16. If the phone company in Example 19 charged $4.39 for a call, how long was the call?

EXERCISE SET 0.4

Vocabulary Review

Choose the correct words to fill in the blanks below.

divide multiplying add right above

1. To _____ decimal numbers, we first line up the decimal points and then add as usual.

2. When _____ decimal numbers, make sure to place the decimal point in the answer so that the number of digits to its right is equal to the total number of digits to the _____ of the decimal points in the original two numbers.

3. To _____ a decimal by a whole number, do the usual long division and then place the decimal point in the answer directly _____ the decimal point in the problem.

Problems

A Find each of the following sums. (Add.)

1. 2.91 + 3.28

2. 8.97 + 2.04

3. 0.04 + 0.31 + 0.78

4. 0.06 + 0.92 + 0.65

5. 3.89 + 2.4

6. 7.65 + 3.8

7. 4.532 + 1.81 + 2.7

8. 9.679 + 3.49 + 6.5

9. 0.081 + 5 + 2.94

10. 0.396 + 7 + 3.96

11. 5.0003 + 6.78 + 0.004

12. 27.0179 + 7.89 + 0.009

13.
```
  7.123
  8.12
+ 9.1
```

14.
```
  5.432
  4.32
+ 3.2
```

15.
```
  9.001
  8.01
+ 7.1
```

16.
```
  6.003
  5.02
+ 4.1
```

17.
```
   89.7854
    3.4
   65.65
+ 100.006
```

18.
```
   57.4698
    9.89
   32.032
+ 572.0079
```

19.
```
  543.21
+ 123.45
```

20.
```
  987.654
+ 456.789
```

21.
```
  14.2586
   5.34
   6.2
+ 12.293
```

22.
```
  243.75
+ 582.346
```

Find each of the following differences. (Subtract.)

23. $99.34 - 88.23$ **24.** $47.69 - 36.58$ **25.** $5.97 - 2.4$ **26.** $9.87 - 1.04$

27. $6.3 - 2.08$ **28.** $7.5 - 3.04$ **29.** $149.37 - 28.96$ **30.** $796.45 - 32.68$

31. $45 - 0.067$ **32.** $48 - 0.075$ **33.** $8 - 0.327$ **34.** $12 - 0.962$

35. $765.432 - 234.567$ **36.** $654.321 - 123.456$ **37.** $10 - 1.942$ **38.** $10 - 5.0874$

Subtract each decimal.

39. $\begin{array}{r} 34.07 \\ -\ 6.18 \\ \hline \end{array}$ **40.** $\begin{array}{r} 25.008 \\ -\ 3.119 \\ \hline \end{array}$ **41.** $\begin{array}{r} 40.04 \\ -\ 4.40 \\ \hline \end{array}$ **42.** $\begin{array}{r} 50.05 \\ -\ 5.50 \\ \hline \end{array}$ **43.** $\begin{array}{r} 768.436 \\ -\ 356.998 \\ \hline \end{array}$ **44.** $\begin{array}{r} 495.237 \\ -\ 247.668 \\ \hline \end{array}$

Add and subtract as indicated.

45. $(7.8 - 4.3) + 2.5$ **46.** $(8.3 - 1.2) + 3.4$ **47.** $7.8 - (4.3 + 2.5)$ **48.** $8.3 - (1.2 + 3.4)$

49. $(9.7 - 5.2) - 1.4$ **50.** $(7.8 - 3.2) - 1.5$ **51.** $9.7 - (5.2 - 1.4)$ **52.** $7.8 - (3.2 - 1.5)$

53. Subtract 5 from the sum of 8.2 and 0.072. **54.** Subtract 8 from the sum of 9.37 and 2.5.

55. What number is added to 0.035 to obtain 4.036? **56.** What number is added to 0.043 to obtain 6.054?

B Find each of the following products. (Multiply.)

57. 0.7
\times 0.4

58. 0.8
\times 0.3

59. 0.07
\times 0.4

60. 0.8
\times 0.03

61. 0.03
\times 0.09

62. 0.07
\times 0.002

63. 2.6(0.3)

64. 8.9(0.2)

65. 0.9
\times 0.88

66. 0.8
\times 0.99

67. 3.12
\times 0.005

68. 4.69
\times 0.006

69. 4.003
\times 6.07

70. 7.0001
\times 3.04

71. 5(0.006)

72. 7(0.005)

73. 75.14
\times 2.5

74. 963.8
\times 0.24

75. 0.1
\times 0.02

76. 0.3
\times 0.02

77. 2.796(10)

78. 97.531(100)

79. 0.0043
\times 100

80. 12.345
\times 1,000

81. 49.94
\times 1,000

82. 157.02
\times 10,000

83. 987.654
\times 10,000

84. 1.23
\times 100,000

85. 1,000
\times 0.001

86. 0.0001
\times 10,000

A B Perform the following operations.

87. 2.1(3.5 − 2.6)

88. 5.4(9.9 − 6.6)

89. 0.05(0.02 + 0.03)

90. 0.04(0.07 + 0.09)

91. 2.02(0.03 + 2.5)

92. 4.04(0.05 + 6.6)

93. (2.1 + 0.03)(3.4 + 0.05)

94. (9.2 + 0.01)(3.5 + 0.03)

95. $(2.1 - 0.1)(2.1 + 0.1)$ **96.** $(9.6 - 0.5)(9.6 + 0.5)$ **97.** $3.08 - 0.2(5 + 0.03)$ **98.** $4.09 + 0.5(6 + 0.02)$

99. $4.23 - 5(0.04 + 0.09)$ **100.** $7.89 - 2(0.31 + 0.76)$ **101.** $2.5 + 10(4.3)^2$ **102.** $3.6 + 15(2.1)^2$

103. $100(1 + 0.08)^2$ **104.** $500(1 + 0.12)^2$ **105.** $(1.5)^2 + (2.5)^2 + (3.5)^2$ **106.** $(1.1)^2 + (2.1)^2 + (3.1)^2$

C Perform each of the following divisions.

107. $394 \div 20$ **108.** $486 \div 30$ **109.** $248 \div 40$ **110.** $372 \div 80$

111. $5\overline{)26}$ **112.** $8\overline{)36}$ **113.** $25\overline{)276}$ **114.** $50\overline{)276}$

115. $28.8 \div 6$ **116.** $15.5 \div 5$ **117.** $77.6 \div 8$ **118.** $31.48 \div 4$

119. $35\overline{)92.05}$ **120.** $26\overline{)146.38}$ **121.** $45\overline{)190.8}$ **122.** $55\overline{)342.1}$

123. $86.7 \div 34$ **124.** $411.4 \div 44$ **125.** $29.7 \div 22$ **126.** $488.4 \div 88$

127. $4.5\overline{)29.25}$ **128.** $3.3\overline{)21.978}$ **129.** $0.11\overline{)1.089}$ **130.** $0.75\overline{)2.40}$

131. $2.3\overline{)0.115}$ **132.** $6.6\overline{)0.198}$ **133.** $0.012\overline{)1.068}$ **134.** $0.052\overline{)0.23712}$

135. $1.1\overline{)2.42}$ **136.** $2.2\overline{)7.26}$ **137.** $0.15\overline{)1.8}$ **138.** $0.04\overline{)0.0064}$

Carry out each of the following divisions only so far as needed to round the results to the nearest hundredth.

139. $26\overline{)35}$ **140.** $18\overline{)47}$ **141.** $3.3\overline{)56}$ **142.** $4.4\overline{)75}$

143. $0.1234 \div 0.5$ **144.** $0.543 \div 2.1$ **145.** $19 \div 7$ **146.** $16 \div 6$

Applying the Concepts

147. Hot Air Balloon Since the pilot of a hot air balloon can only control the balloon's altitude, he relies on the winds for travel. To ride on the jet streams, a hot air balloon must rise as high as 12 kilometers. Convert this to miles by dividing by 1.61. Round your answer to the nearest tenth of a mile.

148. Hot Air Balloon December and January are the best times for traveling in a hot-air balloon because the jet streams in the Northern Hemisphere are the strongest. They reach speeds of 400 kilometers per hour. Convert this to miles per hour by dividing by 1.61. Round to the nearest whole number.

149. Wages If a woman earns $45.30 for working 6 hours, how much does she earn per hour?

150. Wages How many hours does a person making $7.65 per hour have to work in order to earn $191.25?

151. Gas Mileage If a car travels 336 miles on 15 gallons of gas, how far will the car travel on 1 gallon of gas?

152. Gas Mileage If a car travels 392 miles on 16 gallons of gas, how far will the car travel on 1 gallon of gas?

153. Wages Suppose a woman earns $8.34 an hour for the first 40 hours she works in a week and then $12.51 an hour in overtime pay for each additional hour she works in the same week. If she makes $371.13 in one week, how many hours did she work overtime?

154. Wages Suppose a woman makes $386.10 in one week. If she is paid $7.80 an hour for the first 36 hours she works and then $11.70 an hour in overtime pay for each additional hour she works in the same week, how many hours did she work overtime that week?

155. Phone Bill Suppose a telephone company charges $0.41 for the first minute and then $0.32 for each additional minute for a long-distance call. If a long-distance call costs $2.33, how many minutes was the call?

156. Phone Bill Suppose a telephone company charges $0.45 total for the first three minutes and then $0.29 for each additional minute for a long-distance call. If a long-distance call costs $2.77, how many minutes was the call?

157. The 2012 proposed federal budget contains $2.627 trillion in revenue and $3.729 trillion in spending. An alternative congressional plan contains $2.533 trillion in revenue and $3.529 trillion in spending. Calculate the deficits (revenue − spending) for each budget and compare.

158. The 2012 proposed federal budget contains $241.6 billion in interest and $925.2 billion in defense. Approximately how many times greater is the defense budget?

Find the Mistake

Each sentence below contains a mistake. Circle the mistake and write the correct word(s) or numbers(s) on the line provided.

1. To subtract 2.35 from 16.854, line up the right most digit of each number and subtract as usual.

2. The product of 7.19 and 189.4 will have 4 digits to the right of the decimal. _____

3. To estimate the answer to the problem 6.79 × 28.6, we align the decimals and multiply as usual.

4. When dividing with decimal numbers, the decimal point in the answer is placed one digit to the right of the decimal in the dividend. _____

Landmark Review: Checking Your Progress

This feature is intended as a review of key skills from the preceding sections in the chapter. Each review will give you the opportunity to utilize important concepts and check your progress as you practice problems of different types. Take note of any problems in this review that you find difficult. There is no better time than now to revisit and master those difficult concepts in preparation for any upcoming exams or subsequent chapters.

Identify each number as a prime or composite number. For those that are composite, factor into the product of prime factors.

1. 25 **2.** 19 **3.** 31 **4.** 21

Simplify each of the following expressions. Make sure each answer is written in lowest terms.

5. $\dfrac{2}{3} \cdot \dfrac{1}{2}$ **6.** $\dfrac{4}{5} \div \dfrac{3}{4}$ **7.** $\dfrac{3}{8} + \dfrac{5}{8}$ **8.** $\dfrac{5}{6} - \dfrac{1}{3}$

Write each number as a fraction or a mixed number. Do not reduce your answers.

9. 15.7 **10.** 0.0046

Write each number as a decimal.

11. $1\dfrac{4}{10}$ **12.** $11\dfrac{11}{100}$

Add or subtract each of the following.

13. $1.08 + 0.004 + 5$ **14.** $143.25 - 23.13$

Multiply or divide each of the following. Round to the nearest thousandth if necessary.

15. $2.3(0.3)$ **16.** $\begin{array}{r} 3.004 \\ \times\ 2.03 \end{array}$ **17.** $428 \div 12$ **18.** $532 \div 6$

0.5 Percents, Decimals, and Fractions

Image © Richard Ling, 2005

OBJECTIVES

A Write percents as fractions.

B Write percents as decimals.

C Write decimals as percents.

D Write percents as fractions.

E Write fractions as percents.

KEY WORDS

percent

pie chart

Scientists have discovered a toxin in the spit of a sea snail that works with greater effectiveness but in smaller dosages and without the addictive risk of the painkiller morphine. The marine cone snail dwells on the ocean floor. The snail shoots its harpoon-like teeth coated in toxic saliva into its prey to paralyze it. Researchers have discovered how to isolate the saliva's toxin and put it into a pill for humans in pain to ingest. A patient will feel the same pain-reducing effects with 1% of a dose of a popular neuropathic painkiller prescribed in hospitals. If the prescription for an adult of the popular painkiller starts at 300 milligrams, how many milligrams of the sea snail drug would be dosed? To answer this question, we need to review percents and how they interact with decimals and fractions.

A The Meaning of Percent

Percent means "per hundred." Writing a number as a percent is a way of comparing the number with 100. For example, the number 42% (the % symbol is read "percent") is the same as 42 one-hundredths; that is,

$$42\% = \frac{42}{100}$$

Percents are really fractions (or ratios) with denominator 100. Here are some examples that show the meaning of percent.

EXAMPLE 1

a. $50\% = \dfrac{50}{100}$

b. $75\% = \dfrac{75}{100}$

c. $25\% = \dfrac{25}{100}$

d. $33\% = \dfrac{33}{100}$

e. $6\% = \dfrac{6}{100}$

f. $160\% = \dfrac{160}{100}$

Practice Problems

1. Write each number as an equivalent fraction without the % symbol.

a. 40%

b. 80%

c. 15%

d. 37%

e. 8%

f. 150%

Answers

1. a. $\frac{40}{100}$ **b.** $\frac{80}{100}$ **c.** $\frac{15}{100}$ **d.** $\frac{37}{100}$
e. $\frac{8}{100}$ **f.** $\frac{150}{100}$

B Changing Percents to Decimals

To change a percent to a decimal number, we simply use the meaning of percent.

2. Change 25.2% to a decimal.

EXAMPLE 2 Change 35.2% to a decimal.

Solution We drop the % symbol and write 35.2 over 100.

$$35.2\% = \frac{35.2}{100}$$ *Use the meaning of % to convert to a fraction with denominator 100.*

$$= 0.352$$ *Divide 35.2 by 100.*

We see from Example 2 that 35.2% is the same as the decimal 0.352. The result is that the % symbol has been dropped and the decimal point has been moved two places to the *left*. Because % means "per hundred," we will always end up moving the decimal point two places to the left when we change percents to decimals. Because of this, we can write the following rule:

> **RULE** Changing a Percent to a Decimal
>
> To change a percent to a decimal, drop the % symbol and move the decimal point two places to the left.

Here are some examples to illustrate how to use this rule:

3. Change each percent to a decimal.

 a. 40%

 b. 80%

 c. 15%

 d. 5.6%

 e. 4.86%

 f. 0.6%

EXAMPLE 3

 a. 25% = 0.25

 b. 75% = 0.75

 c. 50% = 0.50

 d. 6.80% = 0.068

 e. 3.62% = 0.0362

 f. 0.4% = 0.004

Notice that the results in Example 3a, b, and c are consistent with the results in Example 1a, b, and c.

Notice in 3d and e that we put a 0 in front of the 6 so we can move the decimal point two places to the left.

In 3f, we put two 0's in front of the 4 in order to be able to move the decimal point two places to the left.

4. Write 0.58% as a decimal number.

EXAMPLE 4 Suppose a cortisone cream is 0.5% hydrocortisone. Writing this number as a decimal, we have

$$0.5\% = 0.005$$

C Changing Decimals to Percents

Now we want to do the opposite of what we just did in Examples 2, 3 and 4. We want to change decimals to percents. We know that 42% written as a decimal is 0.42, which means that in order to change 0.42 back to a percent, we must move the decimal point two places to the *right* and use the % symbol.

$$0.42 = 42\%$$ *Notice that we don't show the new decimal point if it is at the end of the number.*

Answers

2. 0.252

3. a. 0.40 **b.** 0.80 **c.** 0.15

 d. 0.056 **e.** 0.0486 **f.** 0.006

4. 0.0058

RULE Changing a Decimal to a Percent

To change a decimal to a percent, we move the decimal point two places to the right and use the % symbol.

Example 5 shows how we use this rule.

EXAMPLE 5

a. $0.27 = 27\%$

b. $4.89 = 489\%$

c. $0.2 = 0.20 = 20\%$ *Notice here that we put a 0 after the 2 so we can move the decimal point two places to the right.*

d. $0.09 = 09\% = 9\%$ *Notice that we can drop the 0 at the left without changing the value of the number.*

e. $25 = 25.00 = 2,500\%$ *Here, we put two 0's after the 5 so we can move the decimal point two places to the right.*

EXAMPLE 6 A softball player has a batting average of 0.650. As a percent, this number is $0.650 = 65.0\%$.

As you can see from the examples above, percent is just a way of comparing numbers to 100. To multiply decimals by 100, we move the decimal point two places to the right. To divide by 100, we move the decimal point two places to the left. Because of this, it is a fairly simple procedure to change percents to decimals and decimals to percents.

D Changing Percents to Fractions

Recall from the beginning of this section our discussion of percents and how they are really fractions (or ratios) with denominator 100. We will now explore further how to change a percent to a fraction. To do so, drop the % symbol and write the original number over 100.

EXAMPLE 7 The pie chart shows who pays for college expenses. Change each percent to a fraction.

Solution In each case, we drop the percent symbol and write the number over 100. Then we reduce to lowest terms if possible.

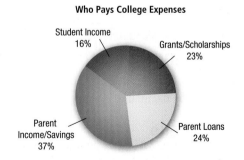

Who Pays College Expenses

Student Income 16%

Grants/Scholarships 23%

Parent Loans 24%

Parent Income/Savings 37%

$16\% = \dfrac{16}{100} = \dfrac{4}{25}$ $23\% = \dfrac{23}{100}$ $24\% = \dfrac{24}{100} = \dfrac{6}{25}$ $37\% = \dfrac{37}{100}$

5. Change each decimal as a percent.
 a. 0.35
 b. 5.77
 c. 0.4
 d. 0.03
 e. 45

6. Write 0.69 as a percent.

7. Change 82% to a fraction in lowest terms.

Answers

5. **a.** 35% **b.** 577% **c.** 40%
 d. 3% **e.** 4,500%
6. 69%
7. $\frac{41}{50}$

8. Change 6.5% to a fraction in lowest terms.

EXAMPLE 8 Change 4.5% to a fraction in lowest terms.

Solution We begin by writing 4.5 over 100.

$$4.5\% = \frac{4.5}{100}$$

We now multiply the numerator and the denominator by 10 so the numerator will be a whole number.

$$\frac{4.5}{100} = \frac{4.5 \times 10}{100 \times 10}$$ *Multiply the numerator and the denominator by 10.*

$$= \frac{45}{1,000}$$

$$= \frac{9}{200}$$ *Reduce to lowest terms.*

The next example involves changing a mixed number into a complex fraction in order to reduce to lowest terms. A *complex fraction* contains fractions in the numerator and the denominator.

9. Change $42\frac{1}{2}$% to a fraction in lowest terms.

EXAMPLE 9 Change $32\frac{1}{2}$% to a fraction in lowest terms.

Solution Writing $32\frac{1}{2}$% over 100 produces a complex fraction. We change $32\frac{1}{2}$ to an improper fraction and simplify.

$$32\frac{1}{2}\% = \frac{32\frac{1}{2}}{100}$$

$$= \frac{\frac{65}{2}}{100}$$ *Change $32\frac{1}{2}$ to the improper fraction $\frac{65}{2}$.*

$$= \frac{65}{2} \times = \frac{1}{100}$$ *Divide by 100 is the same as multiplying by $\frac{1}{100}$.*

$$= \frac{5 \cdot 13 \cdot 1}{2 \cdot 5 \cdot 20}$$ *Multiply.*

$$= \frac{13}{40}$$ *Reduce to lowest terms.*

Note that we could have changed our original mixed number to a decimal first and then changed to a fraction.

$$32\frac{1}{2}\% = 32.5\% = \frac{32.5}{100} = \frac{32.5 \times 10}{100 \times 10} = \frac{325}{1000} = \frac{5 \cdot 5 \cdot 13}{5 \cdot 5 \cdot 40} = \frac{13}{40}$$

The result is the same in both cases.

E Changing Fractions to Percents

To change a fraction to a percent, we can change the fraction to a decimal and then change the decimal to a percent.

10. Change $\frac{9}{10}$ to a percent.

EXAMPLE 10 Suppose the price your bookstore pays for your textbook is $\frac{7}{10}$ of the price you pay for your textbook. Write $\frac{7}{10}$ as a percent.

Solution We can change $\frac{7}{10}$ to a decimal by dividing 7 by 10.

$$\begin{array}{r} 0.7 \\ 10\overline{)7.0} \\ \underline{7\,0} \\ 0 \end{array}$$

We then change the decimal 0.7 to a percent by moving the decimal point two places to the right and using the % symbol.

$$0.7 = 70\%$$

Answers

8. $\frac{13}{200}$

9. $\frac{17}{40}$

10. 90%

You may have noticed that we could have saved some time by simply writing $\frac{7}{10}$ as an equivalent fraction with denominator 100; that is,

$$\frac{7}{10} = \frac{7 \cdot 10}{10 \cdot 10} = \frac{70}{100} = 70\%$$

This is a good way to convert fractions like $\frac{7}{10}$ to percents. It works well for fractions with denominators of 2, 4, 5, 10, 20, 25, and 50, because they are easy to change to fractions with denominators of 100.

EXAMPLE 11 Change $\frac{3}{8}$ to a percent.

Solution We write $\frac{3}{8}$ as a decimal by dividing 3 by 8. We then change the decimal to a percent by moving the decimal point two places to the right and using the % symbol.

$$\frac{3}{8} = 0.375 = 37.5\%$$

$$
\begin{array}{r}
.375 \\
8{\overline{\smash{\big)}\,3.000}} \\
\underline{2\,4} \\
6\,0 \\
\underline{5\,6} \\
4\,0 \\
\underline{4\,0} \\
0
\end{array}
$$

11. Change $\frac{5}{8}$ to a percent.

EXAMPLE 12 Change $\frac{5}{12}$ to a percent.

Solution We begin by dividing 5 by 12.

$$
\begin{array}{r}
.4166 \\
12{\overline{\smash{\big)}\,5.0000}} \\
\underline{4\,8} \\
2\,0 \\
\underline{1\,2} \\
8\,0 \\
\underline{7\,2} \\
8\,0 \\
\underline{7\,2}
\end{array}
$$

Because the 6s repeat indefinitely, we can use mixed number notation to write

$$\frac{5}{12} = 0.41\overline{6} = 41\frac{2}{3}\%$$

Or rounding, we can write

$$\frac{5}{12} \approx 41.7\% \qquad \textit{Round to the nearest tenth of a percent.}$$

12. Change $\frac{7}{12}$ to a percent.

Note When rounding off, let's agree to round off to the nearest thousandth and then move the decimal point. Our answers in percent form will then be accurate to the nearest tenth of a percent, as in Example 12.

EXAMPLE 13 Change $2\frac{1}{2}$ to a percent.

Solution We first change to a decimal and then to a percent.

$$2\frac{1}{2} = 2.5$$

$$= 250\%$$

13. Change $3\frac{3}{4}$ to a percent.

Answers

11. 62.5%

12. $58\frac{1}{3}\% = 58.3\%$

13. 375%

Table 1 lists some of the most commonly used fractions and decimals and their equivalent percents.

Table 1

Fraction	Decimal	Percent
$\frac{1}{2}$	0.5	50%
$\frac{1}{4}$	0.25	25%
$\frac{3}{4}$	0.75	75%
$\frac{1}{3}$	$0.\overline{3}$	33.33%
$\frac{2}{3}$	$0.\overline{6}$	66.67%
$\frac{1}{5}$	0.2	20%
$\frac{2}{5}$	0.4	40%
$\frac{3}{5}$	0.6	60%
$\frac{4}{5}$	0.8	80%

GETTING READY FOR CLASS

After reading through the preceding section, respond in your own words and in complete sentences.

A. What is the relationship between the word *percent* and the number 100?

B. Explain in words how you would change 25% to a decimal.

C. Explain in words how you would change 25% to a fraction.

D. After reading this section, you know that $\frac{1}{2}$, 0.5, and 50% are equivalent. Show mathematically why this is true.

Vocabulary Review

Choose the correct words to fill in the blanks below.

right left percent fraction decimal denominator

1. Writing a number as a _____ is a way of comparing the number with 100.

2. A percent is really a fraction with a _____ of 100.

3. To change a percent to a decimal, drop the percent symbol and move the decimal point two places to the _____.

4. To change a decimal to a percent, move the decimal point two places to the _____ and use the percent symbol.

5. To change a percent to a _____, drop the percent symbol and write the original number over 100.

6. To change a fraction to a percent, change the fraction to a decimal and then change the _____ to a percent.

A Write each percent as a fraction with denominator 100.

1. 20% **2.** 40% **3.** 60% **4.** 80% **5.** 24% **6.** 48%

7. 65% **8.** 35% **9.** 48% **10.** 75% **11.** 10% **12.** 2%

B Change each percent to a decimal.

13. 23% **14.** 34% **15.** 92% **16.** 87% **17.** 9% **18.** 7%

19. 3.4% **20.** 5.8% **21.** 6.34% **22.** 7.25% **23.** 0.9% **24.** 0.6%

C Change each decimal to a percent.

25. 0.23 **26.** 0.34 **27.** 0.92 **28.** 0.87 **29.** 0.45

30. 0.54 **31.** 0.03 **32.** 0.04 **33.** 0.6 **34.** 0.9

SCAN TO ACCESS

35. 0.8 **36.** 0.5 **37.** 0.27 **38.** 0.62 **39.** 1.23 **40.** 2.34

41. 0.7 **42.** 0.07

D Change each percent to a fraction in lowest terms.

43. 60% **44.** 40% **45.** 75% **46.** 25% **47.** 4% **48.** 2%

49. 26.5% **50.** 34.2% **51.** 71.87% **52.** 63.6% **53.** 0.75% **54.** 0.45%

55. $6\frac{1}{4}\%$ **56.** $5\frac{1}{4}\%$ **57.** $33\frac{1}{3}\%$ **58.** $66\frac{2}{3}\%$ **59.** $2\frac{1}{2}\%$ **60.** $5\frac{1}{3}\%$

E Change each fraction or mixed number to a percent.

61. $\frac{1}{2}$ **62.** $\frac{1}{4}$ **63.** $\frac{3}{4}$ **64.** $\frac{2}{3}$ **65.** $\frac{1}{3}$ **66.** $\frac{1}{5}$

67. $\frac{4}{5}$ **68.** $\frac{1}{6}$ **69.** $\frac{7}{8}$ **70.** $\frac{1}{8}$ **71.** $\frac{7}{50}$ **72.** $\frac{9}{25}$

73. $3\frac{1}{4}$ **74.** $2\frac{1}{8}$ **75.** $1\frac{1}{2}$ **76.** $1\frac{3}{4}$ **77.** $2\frac{1}{2}$ **78.** $3\frac{3}{4}$

79. $\frac{21}{43}$ to the nearest tenth of a percent **80.** $\frac{36}{49}$ to the nearest tenth of a percent

Calculator Problems

Use a calculator to write each fraction as a decimal, and then change the decimal to a percent. Round all answers to the nearest tenth of a percent.

81. $\frac{29}{37}$ **82.** $\frac{18}{83}$ **83.** $\frac{6}{51}$ **84.** $\frac{8}{95}$ **85.** $\frac{236}{327}$ **86.** $\frac{568}{732}$ **87.** $\frac{525}{739}$ **88.** $\frac{1024}{2575}$

Applying the Concepts

89. Physiology The human body is between 50% and 75% water. Write each of these percents as a decimal.

90. Alcohol Consumption In the United States, 2.7% of those over 15 years of age drink more than 6.3 ounces of alcohol per day. In France, the same figure is 9%. Write each of these percents as a decimal.

91. iPhones Use the chart to work the following problem.

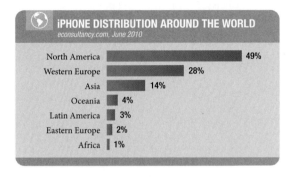

iPHONE DISTRIBUTION AROUND THE WORLD
econsultancy.com, June 2010

North America	49%
Western Europe	28%
Asia	14%
Oceania	4%
Latin America	3%
Eastern Europe	2%
Africa	1%

a. Convert each percent to a fraction.

b. Convert each percent to a decimal.

92. Commuting The pie chart below shows how we, as a nation, commute to work. Change each percent to a fraction in lowest terms.

How We Commute

Car pool 11%
Public transit 5%
Drive alone 80%
Other 4%

93. Nutrition Although, nutritionally, breakfast is the most important meal of the day, only $\frac{1}{5}$ of the people in the United States consistently eat breakfast. What percent of the population is this?

94. Children In School In Belgium, 96% of all children between 3 and 6 years of age go to school. In Sweden, the same figure is only 25%. In the United States, the figure is 60%. Write each of these percents as a fraction in lowest terms.

95. Student Enrollment Using the pie chart below, change each fraction to a percent.

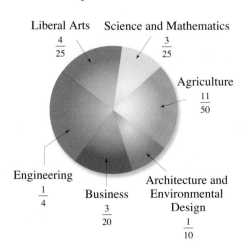

Liberal Arts $\frac{4}{25}$ Science and Mathematics $\frac{3}{25}$

Agriculture $\frac{11}{50}$

Engineering $\frac{1}{4}$

Business $\frac{3}{20}$

Architecture and Environmental Design $\frac{1}{10}$

96. Mac Users The chart shows the increase in Mac users since 2003. Use the chart to answer the following questions.

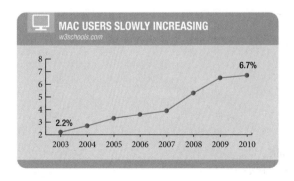

MAC USERS SLOWLY INCREASING
w3schools.com

2.2% — 6.7%

2003 2004 2005 2006 2007 2008 2009 2010

a. In 2006, estimate what fraction of people were Mac users.

b. During what year did the fraction of Mac users equal about $\frac{1}{20}$?

97. Number of Teachers The ratio of the number of teachers to the number of students in secondary schools in Japan is 1 to 17. In the United States, the ratio is 1 to 19. Write each of these ratios as a fraction and then as a percent. Round your answers to the nearest tenth.

98. Mountain Climbing Mountain climbing is generally known as a dangerous hobby or occupation. One way mountains are compared in terms of danger is their death to summit ratio, expressed as a percent. The famed Mount Everest, from 1922 to 2010, has had 5,101 summits and 219 deaths. Find the death to summit ratio as a percent. Round your answer to the nearest tenth.

99. Summit Deaths While climbing Mount Everest is certainly dangerous, there are far more dangerous climbs. Widely recognized by mountain climbers as the most dangerous summits are Ammapurna I in Nepal, with 130 summits and 63 deaths, and Gongjashan in China, with 22 summits and 16 deaths. Compare these two deadly mountains by finding their death to summit ratios as percents. Round your answer to the nearest percent.

100. National Defense In the proposed 2012 federal budget of $3,728.7 billion, $925.2 billion is proposed for defense. What percent of the federal budget is proposed for defense? Round your answer to the nearest percent.

101. Federal Budget In the proposed 2010 federal budget of $3,728.7 billion, $866.1 billion is proposed for health care, $121.1 billion for education, and $431.5 billion for welfare. What is the combined percent proposed for health, education, and welfare? Round your answer to the nearest percent.

102. Oxygen Ang Rita, a famous Mountain Climber from Nepal earned the nickname "The Snow Leopard" when he climbed Mount Everest 10 times without using supplemental oxygen. This has only been accomplished a total of 173 times. What percent of ascents without supplemental oxygen has Ang Rita participated in? Round your answer to the nearest tenth.

Find the Mistake

Each sentence below contains a mistake. Circle the mistake and write the correct word(s) or numbers(s) on the line provided.

1. The fraction $\frac{36}{100}$ given as a percent is 3.6%. _____

2. Changing the percent 57.4% to a decimal gives us 5.74. _____

3. Changing the decimal 0.1465 to a percent gives us 0.001465%. _____

4. To change a fraction to a percent, first change the fraction to its reciprocal and then change to a percent.

0.6 Exponents and Order of Operations

OBJECTIVES

A Evaluate exponential expressions.

B Use order of operations to evaluate expressions.

KEY WORDS

base

exponent

order of operations

Each year, the coastal waters of Qingdao, China are infiltrated with a vibrant green algae bloom that covers more than 150 square miles. The algae thrive in high temperatures and in polluted waters caused by runoff from agriculture land and fish farms. The bloom sucks oxygen out of the water, threatening the local marine life. As it washes onshore and dries in the hot sun, it gives off a noxious odor that smells like rotten eggs. Soldiers and other volunteers work tirelessly to clear the water and beaches of the algae before the bloom drastically damages the local ecosystem.

The measurement used to describe the area covered by the algae was *square miles*. This term comes from the surface measurement made up of multiple squares (150 in the above example), each with a side length of one mile. Another notation used for square miles is mi^2, which is read "miles to the second power" or "miles squared." This second power is an exponent and is a shorthand way of writing repeated multiplication.

A Exponential Expressions

In the expression 2^3, the 2 is called the *base* and the 3 is called the *exponent*. The expression 2^3 is read "2 to the third power" or "2 cubed." The exponent 3 tells us to use the base 2 as a multiplication factor three times.

$$2^3 = 2 \cdot 2 \cdot 2 \qquad \text{2 is used as a factor three times.}$$

We can simplify the expression by multiplication.

$$2^3 = 2 \cdot 2 \cdot 2$$
$$= 4 \cdot 2$$
$$= 8$$

The expression 2^3 is equal to the number 8. We can summarize this discussion with the following definition:

> **DEFINITION** exponent
>
> An *exponent* is a number that indicates how many times the base is to be used as a factor. Exponents indicate repeated multiplication.

For example, in the expression 5^2, 5 is the base and 2 is the exponent. The meaning of the expression is

$$5^2 = 5 \cdot 5 \qquad \text{5 is used as a factor two times.}$$
$$= 25$$

The expression 5^2 is read "5 to the second power" or "5 squared."

Table 1 shows some more examples:

Table 1 Expression	Explanation
3^2	The base is 3, and the exponent is 2. The expression is read "3 to the second power" or "3 squared."
4^3	The base is 4, and the exponent is 3. The expression is read "4 to the third power" or "4 cubed."
2^4	The base is 2, and the exponent is 4. The expression is read "2 to the fourth power."

As you can see from these examples, a base raised to the second power is also said to be *squared*, and a base raised to the third power is also said to be *cubed*. These are the only two exponents (2 and 3) that have special names. All other exponents are referred to only as "fourth powers," "fifth powers," "sixth powers," and so on.

The next examples show how we can simplify expressions involving exponents by using repeated multiplication.

EXAMPLE 1 Simplify each of the following by using repeated multiplication.

a. $3^2 = 3 \cdot 3 = 9$

b. $4^2 = 4 \cdot 4 = 16$

c. $3^3 = 3 \cdot 3 \cdot 3 = 9 \cdot 3 = 27$

d. $3^4 = 3 \cdot 3 \cdot 3 \cdot 3 = 9 \cdot 9 = 81$

e. $2^4 = 2 \cdot 2 \cdot 2 \cdot 2 = 4 \cdot 4 = 16$

USING TECHNOLOGY Calculators

Here is how we use a calculator to evaluate exponents, as we did in Example 1e:

Scientific Calculator: 2 $\boxed{x^y}$ 4 $\boxed{=}$

Graphing Calculator: 2 $\boxed{\wedge}$ 4 $\boxed{\text{ENT}}$ or 2 $\boxed{x^y}$ 4 $\boxed{\text{ENT}}$ (depending on the calculator)

Finally, we should consider what happens when the numbers 0 and 1 are used as exponents.

DEFINITION exponent 1

Any number with an *exponent 1*(raised to the first power) is itself. That is, if we let the letter a represent any number, then

$$a^1 = a$$

To take care of the cases when 0 is used as an exponent, we must use the following definition:

DEFINITION exponent 0

Any number other than 0 that has an *exponent 0* (raised to the 0 power) is 1. That is, if a represents any nonzero number, then it is always true that

$$a^0 = 1$$

Practice Problems

1. Simplify each of the following by using repeated multiplication.

 a. 5^2

 b. 9^2

 c. 2^3

 d. 1^3

 e. 2^5

Answers

1. a. 25 b. 81 c. 8
 d. 1 e. 32

Any number raised to the first power is itself.

EXAMPLE 2

 a. $5^1 = 5$

 b. $9^1 = 9$

2. Simplify each of the following expressions.

 a. 7^1

 b. 4^1

Any nonzero number raised to the 0 power is 1.

EXAMPLE 3

 a. $4^0 = 1$

 b. $8^0 = 1$

3. Simplify each of the following expressions.

 a. 9^0

 b. 1^0

B Order of Operations

The symbols we use to specify operations, along with the symbols we use for grouping, () and [], serve the same purpose in mathematics as punctuation marks in English. They may be called the punctuation marks of mathematics. Consider the following sentence:

> Kendra said Tanner is tall.

It can have two different meanings, depending on how we punctuate it:

1. "Kendra," said Tanner, "is tall."
2. Kendra said, "Tanner is tall."

Without the punctuation marks, we don't know which meaning the sentence has. Now consider the following mathematical expression:

> $4 + 5 \cdot 2$

It can have two different meanings, depending on the order in which we perform the operations. If we add 4 and 5 first, and then multiply 2, we get 18. On the other hand, if we multiply 5 and 2 first, and then add 4, we get 14. There seem to be two different answers. In mathematics, we want to avoid situations in which two different results are possible. Therefore, we follow the rule for order of operations.

RULE Order of Operations

When evaluating mathematical expressions, we will perform the operations in the following order:

1. If the expression contains grouping symbols, such as parentheses (), brackets [], or a fraction bar, then we perform the operations inside the grouping symbols, or above and below the fraction bar, first.

2. Then we evaluate, or simplify, any numbers with exponents.

3. Then we do all multiplications and divisions in order, starting at the left and moving right.

4. Finally, we do all additions and subtractions, from left to right.

According to our rule, the expression $4 + 5 \cdot 2$ would have to be evaluated by multiplying 5 and 2 first, and then adding 4. The correct answer—and the only answer—to this problem is 14.

$$4 + 5 \cdot 2 = 4 + 10 \qquad \text{Multiply first.}$$
$$= 14 \qquad \text{Then add.}$$

Answers

2. a. 7 **b.** 4

3. a. 1 **b.** 1

Here are some more examples that illustrate how we apply the rule for order of operations to simplify (or evaluate) expressions.

4. Simplify: $5 \cdot 7 - 3 \cdot 6$.

EXAMPLE 4 Simplify: $4 \cdot 8 - 2 \cdot 6$.

Solution We multiply first and then subtract.

$$4 \cdot 8 - 2 \cdot 6 = 32 - 12 \qquad \text{Multiply first.}$$
$$= 20 \qquad \text{Then subtract.}$$

5. Simplify: $7 + 3(6 + 4)$.

EXAMPLE 5 Simplify: $5 + 2(7 - 1)$.

Solution According to the rule for the order of operations, we must do what is inside the parentheses first.

$$5 + 2(7 - 1) = 5 + 2(6) \qquad \text{Work inside parentheses first.}$$
$$= 5 + 12 \qquad \text{Then multiply.}$$
$$= 17 \qquad \text{Then add.}$$

6. Simplify: $6 \cdot 3^2 + 64 \div 2^4 - 2$.

EXAMPLE 6 Simplify: $9 \cdot 2^3 + 36 \div 3^2 - 8$.

Solution $9 \cdot 2^3 + 36 \div 3^2 - 8 = 9 \cdot 8 + 36 \div 9 - 8$ Work exponents first.

$$= 72 + 4 - 8 \qquad \text{Then multiply and divide, left to right.}$$
$$= 76 - 8 \qquad \text{Add and subtract, left to right.}$$
$$= 68$$

7. Simplify: $5 + 3[24 - 5(6 - 2)]$.

EXAMPLE 7 Simplify: $3 + 2[10 - 3(5 - 2)]$.

Solution The brackets, [], are used in the same way as parentheses. In a case like this, we move to the innermost grouping symbols first and begin simplifying.

$$3 + 2[10 - 3(5 - 2)] = 3 + 2[10 - 3(3)]$$
$$= 3 + 2[10 - 9]$$
$$= 3 + 2[1]$$
$$= 3 + 2$$
$$= 5$$

We can use the rule for order of operations to simplify expressions involving decimal numbers and addition, subtraction, and multiplication.

8. Perform the indicated operations: $0.03(5.5 + 0.02)$

EXAMPLE 8 Perform the indicated operations: $0.05(4.2 + 0.03)$.

Solution We begin by adding inside the parentheses.

$$0.05(4.2 + 0.03) = 0.05(4.23) \qquad \text{Add.}$$
$$= 0.2115 \qquad \text{Multiply.}$$

Answers

4. 17

5. 37

6. 56

7. 17

8. 0.1656

We could also have used the distributive property first, and the result would be unchanged. We will learn more about the distributive property in the next chapter, but for now, here is what the use of the property would look like:

$$0.05(4.2 + 0.03) = 0.05(4.2) + 0.05(0.03) \quad \text{Distributive property}$$
$$= 0.210 + 0.0015 \quad \text{Multiply.}$$
$$= 0.2115 \quad \text{Add.}$$

EXAMPLE 9 Simplify: $4.8 + 12(3.2)^2$.

Solution According to the rule for order of operations, we must first evaluate the number with an exponent, then multiply, and finally add.

$$4.8 + 12(3.2)^2 = 4.8 + 12(10.24) \quad (3.2)^2 = 10.24$$
$$= 4.8 + 122.88 \quad \text{Multiply.}$$
$$= 127.68 \quad \text{Add.}$$

EXAMPLE 10 Sally earns $7.25 for each of the first 36 hours she works in one week and $10.88 in overtime pay for each additional hour she works in the same week. How much money will she make if she works 42 hours in one week?

Solution The difference between 42 and 36 is 6 hours of overtime pay. The total amount of money she will make is

$$\underbrace{7.25(36)}_{\substack{\text{Pay for the} \\ \text{first 36 hours}}} + \underbrace{10.88(6)}_{\substack{\text{Pay for the} \\ \text{next 6 hours}}} = 261.00 + 65.28$$
$$= 326.28$$

She will make $326.28 for working 42 hours in one week.

GETTING READY FOR CLASS

After reading through the preceding section, respond in your own words and in complete sentences.

A. What is an exponent?

B. In symbols, show that any number other than zero raised to the 0 power is 1.

C. Why is using the order of operations important to finding the correct solution to a problem?

D. Briefly list the steps of the order of operations.

9. Simplify: $5.7 + 14(2.4)^2$.

10. How much will Sally make if she works 50 hours in one week?

Answers

9. 86.34
10. $413.32

Vocabulary Review

Choose the correct words to fill in the blanks below.

zero	first	base	product	difference
factor	sum	exponent	quotient	

1. For the exponential expression 5^3, the 5 is called the _____ and the 3 is called the

_____.

2. An exponent is a number that indicates how many times the base is used as a _____.

3. Any number raised to the _____ power is the number itself.

4. Any non-zero number raised to the _____ power is 1.

5. Twice the _____ of 6 and 1 is written in symbols as $2(6 + 1)$.

6. 4 added to 5 times the _____ of 3 and 2 is written in symbols as $5(3 - 2) + 4$.

7. 7 subtracted from the _____ of 4 and 3 is written in symbols as $4 \cdot 3 - 7$.

8. The _____ of 8 and 2 plus 9 is written in symbols as $8 \div 2 + 9$.

The following is a list of steps to perform operations when evaluating mathematical expressions. Number the steps in the correct order.

_____ Do all multiplications and divisions in order, left to right.

_____ Simplify any numbers with exponents.

_____ Perform operations inside the grouping symbols, or above and below the fraction bar.

_____ Do all additions and subtractions, in order left to right.

Problems

A For each of the following expressions, name the base and the exponent.

1. 4^5 **2.** 5^4 **3.** 3^6 **4.** 6^3 **5.** 8^2 **6.** 2^8

7. 9^1 **8.** 1^9 **9.** 4^0 **10.** 0^4 **11.** 4^8 **12.** 12^2

Use the definition of exponents to simplify each of the following expressions.

13. 6^2 **14.** 7^2 **15.** 2^3 **16.** 2^4 **17.** 1^4 **18.** 5^1

19. 9^0 **20.** 27^0 **21.** 9^2 **22.** 8^2 **23.** 10^1 **24.** 8^1

25. 12^1 **26.** 16^0 **27.** 45^0 **28.** 3^4 **29.** 3^3 **30.** 3^2

B Use the rule for the order of operations to simplify each expression.

31. $16 - 8 + 4$ **32.** $16 - 4 + 8$ **33.** $20 \div 2 \cdot 10$ **34.** $40 \div 4 \cdot 5$

35. $20 - 4 \cdot 4$ **36.** $30 - 10 \cdot 2$ **37.** $3 + 5 \cdot 8$ **38.** $7 + 4 \cdot 9$

39. $3 \cdot 6 - 2$ **40.** $5 \cdot 1 + 6$ **41.** $6 \cdot 2 + 9 \cdot 8$ **42.** $4 \cdot 5 + 9 \cdot 7$

43. $4 \cdot 5 - 3 \cdot 2$ **44.** $5 \cdot 6 - 4 \cdot 3$ **45.** $5^2 + 7^2$ **46.** $4^2 + 9^2$

47. $480 + 12(32)^2$ **48.** $360 + 14(27)^2$ **49.** $3 \cdot 2^3 + 5 \cdot 4^2$ **50.** $4 \cdot 3^2 + 5 \cdot 2^3$

51. $8 \cdot 10^2 - 6 \cdot 4^3$ **52.** $5 \cdot 11^2 - 3 \cdot 2^3$ **53.** $2(3 + 6 \cdot 5)$ **54.** $8(1 + 4 \cdot 2)$

55. $19 + 50 \div 5^2$ **56.** $9 + 8 \div 2^2$ **57.** $9 - 2(4 - 3)$ **58.** $15 - 6(9 - 7)$

59. $4 \cdot 3 + 2(5 - 3)$ **60.** $6 \cdot 8 + 3(4 - 1)$ **61.** $4[2(3) + 3(5)]$ **62.** $3[2(5) + 3(4)]$

63. $(7 - 3)(8 + 2)$ **64.** $(9 - 5)(9 + 5)$ **65.** $3(9 - 2) + 4(7 - 2)$ **66.** $7(4 - 2) - 2(5 - 3)$

67. $18 + 12 \div 4 - 3$ **68.** $20 + 16 \div 2 - 5$ **69.** $4(10^2) + 20 \div 4$ **70.** $3(4^2) + 10 \div 5$

71. $8 \cdot 2^4 + 25 \div 5 - 3^2$ **72.** $5 \cdot 3^4 + 16 \div 8 - 2^2$ **73.** $5 + 2[9 - 2(4 - 1)]$ **74.** $6 + 3[8 - 3(1 + 1)]$

75. $3 + 4[6 + 8(2 - 0)]$ **76.** $2 + 5[9 + 3(4 - 1)]$ **77.** $\dfrac{15 + 5(4)}{17 - 12}$ **78.** $\dfrac{20 + 6(2)}{11 - 7}$

Multiply each of the following. Be sure all answers are written in lowest terms.

79. $\left(\dfrac{3}{4}\right)^2 \cdot \dfrac{8}{9}$ **80.** $\left(\dfrac{5}{6}\right)^2 \cdot \dfrac{12}{15}$ **81.** $\left(\dfrac{1}{2}\right)^2 \cdot \left(\dfrac{3}{5}\right)^2$

82. $\left(\dfrac{3}{8}\right)^2 \cdot \left(\dfrac{4}{3}\right)^2$ **83.** $\left(\dfrac{1}{2}\right)^2 \cdot 8 + \left(\dfrac{1}{3}\right)^2 \cdot 9$ **84.** $\left(\dfrac{2}{3}\right)^2 \cdot 9 + \left(\dfrac{1}{2}\right)^2 \cdot 4$

Simplify each expression as much as possible.

85. $10 \div \left(\dfrac{1}{2}\right)^2$ **86.** $12 \div \left(\dfrac{1}{4}\right)^2$

87. $\dfrac{18}{35} \div \left(\dfrac{6}{7}\right)^2$ **88.** $\dfrac{48}{55} \div \left(\dfrac{8}{11}\right)^2$

89. $\dfrac{4}{5} \div \dfrac{1}{10} + 5$ **90.** $\dfrac{3}{8} \div \dfrac{1}{16} + 4$

91. $10 + \dfrac{11}{12} \div \dfrac{11}{24}$ **92.** $15 + \dfrac{13}{14} \div \dfrac{13}{42}$

93. $24 \div \left(\dfrac{2}{5}\right)^2 + 25 \div \left(\dfrac{5}{6}\right)^2$

94. $18 \div \left(\dfrac{3}{4}\right)^2 + 49 \div \left(\dfrac{7}{9}\right)^2$

95. $100 \div \left(\dfrac{5}{7}\right)^2 + 200 \div \left(\dfrac{2}{3}\right)^2$

96. $64 \div \left(\dfrac{8}{11}\right)^2 + 81 \div \left(\dfrac{9}{11}\right)^2$

Translate each English expression into an equivalent mathematical expression written in symbols. Then simplify.

97. 8 times the sum of 4 and 2

98. 3 times the difference of 6 and 1

99. Twice the sum of 10 and 3

100. 5 times the difference of 12 and 6

101. 4 added to 3 times the sum of 3 and 4

102. 25 added to 4 times the difference of 7 and 5

103. 9 subtracted from the quotient of 20 and 2

104. 7 added to the quotient of 6 and 2

105. The sum of 8 times 5 and 5 times 4

106. The difference of 10 times 5 and 6 times 2

Find the Mistake

Each sentence below contains a mistake. Circle the mistake and write the correct word(s) or numbers(s) on the line provided.

1. The expression 8^5 is read "5 to the 8th power." _____

2. In the expression 16^3, the 16 is used as a factor two times. _____

3. To simplify $9^2 + 4 - 6 \cdot 7$ using the order of operations, we first find the product of 6 and 7. _____

4. To work the problem $5 + 6 - 2 + 18$ using the order of operations, we add and subtract right to left. _____

Smart Cyber Shopping

When shopping for an item online, it is very important to research and compare prices before clicking the "Place Your Order" button. Doing so will help you save a lot of money. During your research, you may choose to go with a sale item, or even a used or refurbished version. For example, suppose you are a photographer looking to purchase a Canon EOS 5D Mark ii DSLR camera. A quick online search yields multiple vendors offering a price of just under $2,000 to over $4,000 depending on condition, location, etc. For this project, you will research the price of an item of your choosing and use the math concepts we have reviewed in this chapter to decide on the most worthwhile purchase.

1. Choose an item that is offered by multiple online sellers.

2. Choose five sellers and collect the following information:
 a. Seller's name
 b. Seller's location
 c. Item description (including sale item, or used/refurbished)
 d. Retail price
 e. Sales price (if applicable)
 f. Shipping cost
 g. Sales tax rate and sale tax (if applicable)
 h. Total purchase price

Note: You may choose to include a used or refurbished item, but this is not required to complete the project.

3. Make a table that clearly presents the information from step 2.

4. Based on the data you have gathered, calculate the following:
 a. Difference in dollars between the least expensive and most expensive price before tax and shipping cost
 b. Difference in dollars between the least expensive and most expensive total purchase price
 c. Percent difference between the least expensive and most expensive price before tax and shipping cost
 d. Percent difference between the least expensive and most expensive total purchase price

5. Are there any other factors or calculations you would consider before making a purchase?

6. Which item would you purchase and why?

The number(s) in brackets next to each heading indicates the section(s) in which that topic is discussed.

Factors and Simplifying Fractions [0.1]

A *prime number* is any whole number greater than 1 that has exactly two divisors—itself and 1. (A number is a divisor of another number if it divides it without a remainder.)

A *fraction in lowest terms* is one that has a numerator and a denominator with no factors in common other than the number 1.

EXAMPLES

1. Reduce the fraction $\frac{8}{12}$ to lowest terms.

$$\frac{8}{12} = \frac{2 \cdot 2 \cdot 2}{2 \cdot 2 \cdot 3}$$

$$= \frac{2}{3}$$

Operations with Fractions [0.2]

$$\frac{a}{b} \cdot \frac{c}{d} = \frac{(a \cdot c)}{(b \cdot d)}$$
Multiplication of fractions

$$\frac{a}{b} \div \frac{c}{d} = \frac{a}{b} \cdot \frac{d}{c}$$
Division by a fraction

$$\frac{a}{c} + \frac{b}{c} = \frac{(a + b)}{c}$$
Addition of fractions with same denominator

$$\frac{a}{c} - \frac{b}{c} = \frac{(a - b)}{c}$$
Subtraction of fractions with same denominator

2. Subtract: $\frac{8}{15} - \frac{5}{10}$.

$$\frac{8}{15} - \frac{5}{10} = \frac{8 \cdot 2}{15 \cdot 2} - \frac{5 \cdot 3}{10 \cdot 3}$$

$$= \frac{16}{30} - \frac{15}{30}$$

$$= \frac{16 - 15}{30}$$

$$= \frac{1}{30}$$

To add or subtract any two fractions,

1. Factor each denominator completely, and use the factors to build the LCD. (Remember, the LCD is the smallest number divisible by each of the denominators in the problem.)

2. Rewrite each fraction as an equivalent fraction that has the LCD for its denominator. This is done by multiplying both the numerator and the denominator of the fraction in question by the appropriate whole number.

3. Add or subtract the numerators of the fractions produced in Step 2. This is the numerator of the sum or difference. The denominator of the sum or difference is the LCD.

4. Reduce the fraction produced in Step 3 to lowest terms if it is not already in lowest terms.

Decimal Notation [0.3]

3. Write thirty-five thousandths as a decimal number.

0.035

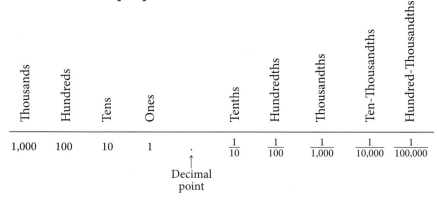

Addition (or Subtraction) of Decimals [0.4]

4. Add: $42.38 + 37.425$.

$$\begin{array}{r} 42.38 \\ +37.425 \\ \hline 79.805 \end{array}$$

To add (or subtract) decimal numbers, we line up the decimal points and add (or subtract) as usual. The decimal point in the result is written directly below the decimal points in the problem.

Multiplication with Decimals [0.4]

5. Multiply: 0.5×0.04.

$$\begin{array}{r} 0.5 \\ \times 0.04 \\ \hline 0.020 \end{array}$$

To multiply with decimal numbers:

1. Multiply as you would if the decimal points were not there.

2. Place the decimal point in the answer so that the number of digits to its right is equal to the total number of digits to the right of the decimal points in the original two numbers in the problem.

Division with Decimals [0.4]

6. Divide: $32.8 \div 4$.

$$\begin{array}{r} 8.2 \\ 4\overline{)32.8} \\ \underline{32} \\ 0\,8 \\ \underline{8} \\ 0 \end{array}$$

To divide a decimal by a whole number, we do the usual long division as if there were no decimal point involved. The decimal point in the answer is placed directly above the decimal point in the problem.

Percents, Decimals, and Fractions [0.5]

7. Change 48.3% to a decimal.

$$48.3\% = 0.483$$

Percent means "per hundred." Writing a number as a percent is a way of comparing the number with 100. Percents are really fractions (or ratios) with denominator 100.

Exponential Expressions [0.6]

8. Simplify.

 a. $3^3 = 27$

 b. $4^1 = 4$

 c. $7^0 = 1$

An *exponent* is a number that indicates how many times the base is to be used as a factor. Exponents indicate repeated multiplication.

Order of Operations [0.6]

9. Simplify: $4 \cdot 10^2 - 3 \cdot 5^2$

$$4 \cdot 10^2 - 3 \cdot 5^2 = 4 \cdot 100 - 3 \cdot 25$$
$$= 400 - 75$$
$$= 325$$

1. If the expression contains grouping symbols, such as parentheses (), brackets [], or a fraction bar, then we perform the operations inside the grouping symbols, or above and below the fraction bar, first.

2. Then we evaluate, or simplify, any numbers with exponents.

3. Then we do all multiplications and divisions in order, starting at the left and moving right.

4. Finally, we do all additions and subtractions, from left to right.

At the end of each chapter, we will provide a review of problems similar to those you have seen in the chapter's previous sections. These problems are meant to supplement, not replace, a thorough review of the concepts, examples, and exercise sets presented in the chapter.

Factor each number into a product of prime factors. [0.1]

1. 225 **2.** 270

3. 308 **4.** 189

Reduce to lowest terms. [0.1]

5. $\dfrac{84}{108}$ **6.** $\dfrac{36}{117}$

7. $\dfrac{40}{96}$ **8.** $\dfrac{96}{204}$

Perform the indicated operation. [0.2]

9. $\dfrac{18}{15} \cdot \dfrac{10}{14}$ **10.** $\dfrac{20}{27} \cdot \dfrac{6}{8}$

11. $\dfrac{6}{35} \div \dfrac{12}{30}$ **12.** $\dfrac{24}{35} \div \dfrac{32}{40}$

13. $\dfrac{5}{6} + \dfrac{3}{8}$ **14.** $\dfrac{7}{12} - \dfrac{3}{8}$

15. $\dfrac{7}{18} - \dfrac{1}{3}$ **16.** $\dfrac{17}{32} + \dfrac{5}{12}$

17. $\left(\dfrac{4}{5}\right)^2 \left(\dfrac{15}{32}\right)$ **18.** $\dfrac{10}{21} \cdot \dfrac{27}{55} \cdot \dfrac{77}{6}$

Change each decimal to a fraction, and then reduce to lowest terms. [0.3]

19. 0.6 **20.** 0.245 **21.** 0.704 **22.** 0.34

Perform the indicated operations. [0.4]

23. $6.21 + 9.3 + 6.491$ **24.** $13.075 + 3.2 + 0.76$

25. $48.2 - 25.071$ **26.** $56.03 - 7.9$

27. $32.9 - (6.41 + 2.1)$ **28.** $16 - (3.21 - 0.0215)$

29. 5.7×3.5 **30.** 5.007×3.4

31. $8\overline{)35}$ **32.** $0.91 \div 1.3$

Change each percentage to a decimal and a fraction in lowest terms. [0.5]

33. 78% **34.** 90%

Change each decimal to a percentage and a fraction in lowest terms. [0.5]

35. 0.88 **36.** 1.25

Change each fraction into a percentage and a decimal. [0.5]

37. $\dfrac{13}{25}$ **38.** $\dfrac{7}{20}$

Simplify. [0.6]

39. 18^0 **40.** 4^4 **41.** 5^3 **42.** 9^1

Simplify each expression. [0.6]

43. $12 - 25 \div 5$ **44.** $(6 \cdot 5) - (2 \cdot 7)$

45. $\left(\dfrac{5}{9}\right)\left(\dfrac{3}{5}\right) - \dfrac{1}{6}$ **46.** $56 - 4[17 - (21 - 9)]$

47. $4(3)^2 \div 6(3)$ **48.** $5(10 - 3) + 2(3 - 2)$

49. $15 - 3[17 - 3(6 - 2)]$ **50.** $\dfrac{48 - 5(6)}{6 - 3}$

The chart shows the percent of iPads sold per state. Use it to answer the following questions.

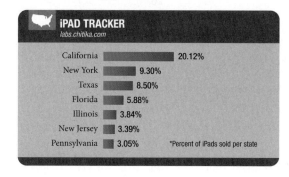

51. Change the percent of iPads sold in California to a decimal and a fraction in lowest terms.

52. What is the difference between the percent of iPads sold in Texas compared to New Jersey?

At the end of each chapter, we will provide a test that you should use as a helpful study tool for a chapter exam. Remember, it is still important to study every aspect of the material presented in the chapter, and not rely solely on the Chapter Test to study for an exam.

Factor each number into a product of prime factors. [0.1]

1. 90

2. 84

3. 630

4. 168

Reduce to lowest terms. [0.1]

5. $\dfrac{36}{64}$

6. $\dfrac{48}{80}$

7. $\dfrac{24}{42}$

8. $\dfrac{88}{104}$

Perform the indicated operation. [0.2]

9. $\dfrac{15}{18} \cdot \dfrac{2}{3}$

10. $\dfrac{12}{15} \cdot \dfrac{20}{28}$

11. $\dfrac{7}{9} \div \dfrac{1}{18}$

12. $\dfrac{16}{21} \div \dfrac{4}{9}$

13. $\dfrac{5}{6} - \dfrac{1}{3}$

14. $\dfrac{7}{12} + \dfrac{8}{9}$

15. $\dfrac{6}{16} - \dfrac{1}{8}$

16. $\dfrac{13}{24} + \dfrac{7}{16}$

17. $\left(\dfrac{3}{5}\right)^2\left(\dfrac{50}{63}\right)$

18. $\dfrac{10}{21} \cdot \dfrac{35}{36} \cdot \dfrac{18}{30}$

Change each decimal to a fraction, and then reduce to lowest terms. [0.3]

19. 0.48

20. 0.3125

21. 0.4375

22. 0.455

Perform the indicated operations. [0.4]

23. $5.3 + 6.02 + 0.7$

24. $14.326 + 9 + 4.75$

25. $36.79 - 14.517$

26. $28.3 - 6.517$

27. $19.7 - (3.61 + 2.5)$

28. $15 - (7.31 - 4.9)$

29. 3.2×7.9

30. 5.003×10.4

31. $8\overline{)42}$

32. $1.4 \div 0.4$

Change each percentage to a decimal and a fraction in lowest terms. [0.5]

33. 48%

34. 36%

Change each decimal to a percentage and a fraction in lowest terms. [0.5]

35. 0.55

36. 1.80

Change each fraction into a percentage and a decimal. [0.5]

37. $\dfrac{5}{8}$

38. $\dfrac{3}{16}$

Simplify. [0.6]

39. 4^3

40. 64^0

41. 5^2

42. 81^1

Simplify each expression. [0.6]

43. $9 - 16 \div 2$

44. $6 \cdot 3 - 2 \cdot 4$

45. $\left(\dfrac{2}{3}\right)^3\left(\dfrac{9}{4}\right) + \dfrac{5}{2}$

46. $17 - 3[4 - (6 - 5)]$

47. $3(5)^2 \div (9 - 2^2)$

48. $3(9 - 2) - 4(6 - 4)$

49. $3 + 7[9 - 2(7 - 5)]$

50. $\dfrac{25 - 3(5)}{10 - 8}$

The chart shows the extent to which Americans say they know a foreign language. Use it to answer the following questions.

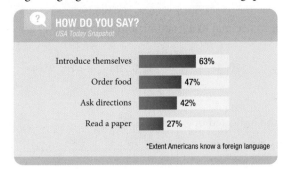

51. Change the percent of those who say they can ask directions to a decimal and a fraction in lowest terms.

52. What is the difference between the percent of those who say they can introduce themselves and those who say they can read a paper?

Real Numbers and Algebraic Expressions

The Three Gorges Dam
Sandouping, China
Image © 2010 DigitalGlobe

A dam built on a large river can produce hydroelectric power. Gravity forces water through the propellers of a turbine near the bottom of the dam wall. The turbine rotates a metal shaft in a generator, producing electricity. The Three Gorges Dam on the Yangtze River in China is the world's largest hydroelectric power project. The total electricity generating capacity of the dam is estimated to be 18,000 megawatts of power. However, the environmental and human impact of the project is still a source of debate.

When the dam was completed in 2006, the 660-kilometer-long reservoir displaced approximately 1,300,000 people.

If 1,503 villages were submerged, how many people lived in each village before being relocated? (For the purpose of this problem, let's assume there were an equal number of people in each village.) We can find the answer to this question by solving this division problem:

$$\text{People relocated from each village} = \frac{\text{Total people relocated}}{\text{\# of villages}} = \frac{1,300,000}{1,503}$$

This type of division is one of the skills you will use as you work through this chapter.

OBJECTIVES

A Locate and label points on the real number line.

B Change a fraction to an equivalent fraction with a new denominator.

C Simplify expressions containing absolute value.

D Identify the opposite of a number.

E Translate between phrases written in English and expressions written in symbols.

F Find the perimeter and area of squares, rectangles, and triangles.

KEY WORDS

real number line

coordinate

equivalent fractions

absolute value

opposites

variable

algebraic expression

algebraic equation

algebraic inequality

area

perimeter

Note If there is no sign (+ or −) in front of a non-zero number, the number is assumed to be positive (+). The number zero has no sign.

©iStockphoto.com/KKIDD

Each winter, thousands of spectators come to a mountain resort in Wyoming to watch the World Championship Snow King Hill Climb. One by one, snowmobilers attempt to race up a 1,500-foot snow-covered slope on Snow King Mountain. At its steepest point, the slope measures 45 degrees! The steep slope often overpowers the snowmobile, which comes tumbling back down the hill, sometimes taking the rider with it.

The weather and the conditions of the snow may also play a part in the success of the snowmobilers. The table and bar chart shown here give the record low temperature, in degrees Fahrenheit, for each month of the year in the nearby city of Jackson, Wyoming. Notice that some of these temperatures are represented by negative numbers which we will begin working with in this section.

FIGURE 1

A The Real Number Line

To represent negative numbers in algebra, we use what is called the *real number line*. Here is how we construct a real number line: We first draw a straight line and label a convenient point on the line with 0. Then we mark off equally spaced distances in both directions from 0. Label the points to the right of 0 with the numbers 1, 2, 3, . . .(the dots mean "and so on"). The points to the left of 0 we label in order, −1, −2, −3, Here is what it looks like:

The numbers increase in value going from left to right. If we "move" to the right, we are moving in the positive direction. If we move to the left, we are moving in the negative direction. When we compare two numbers on the number line, the number on the left is always smaller than the number on the right. For instance, -3 is smaller than -1 because it is to the left of -1 on the number line.

EXAMPLE 1 Locate and label the points on the real number line associated with the numbers $-3.5, -1\frac{1}{4}, \frac{1}{2}, \frac{3}{4}, 2.5$.

Solution We draw a real number line from -4 to 4 and label the points in question.

Practice Problems

1. Locate and label the points associated with $-2, -\frac{1}{2}, 0, 1.5, 2.75$.

> **DEFINITION** coordinate
>
> The number associated with a point on the real number line is called the **coordinate** of that point.

In the preceding example, the numbers $\frac{1}{2}, \frac{3}{4}, 2.5, -3.5$, and $-1\frac{1}{4}$ are the coordinates of the points they represent.

> **DEFINITION** real numbers
>
> The numbers that can be represented with points on the real number line are called **real numbers**.

Note There are other numbers on the number line that you may not be as familiar with. They are irrational numbers such as $\pi, \sqrt{2}, \sqrt{3}$. We will introduce these numbers later in the book.

Real numbers include whole numbers, fractions, decimals, and other numbers that are not as familiar to us as these.

B Equivalent Fractions on the Number Line

The number line can be used to visualize fractions. Recall that for the fraction $\frac{a}{b}$, a is called the numerator and b is called the denominator. As long as $\frac{a}{b}$ is a proper fraction where $a < b$, the denominator b indicates the number of equal parts in the interval from 0 to 1 on the number line. The numerator a indicates how many of those parts we have. If we take that part of the number line from 0 to 1 and divide it into *three equal parts,* we say that we have divided it into *thirds* (Figure 2). Each of the three segments is $\frac{1}{3}$ (one third) of the whole segment from 0 to 1.

FIGURE 2

Two of these smaller segments together are $\frac{2}{3}$ (two thirds) of the whole segment. And three of them would be $\frac{3}{3}$ (three thirds), or the whole segment.

Let's do the same thing again with six equal divisions of the segment from 0 to 1 (Figure 3). In this case we say each of the smaller segments has a length of $\frac{1}{6}$ (one sixth).

FIGURE 3

The same point we labeled with $\frac{1}{3}$ in Figure 2 is now labeled with $\frac{2}{6}$. Likewise, the point we labeled earlier with $\frac{2}{3}$ is now labeled $\frac{4}{6}$. It must be true then that

$$\frac{2}{6} = \frac{1}{3} \quad \text{and} \quad \frac{4}{6} = \frac{2}{3}$$

Actually, there are many fractions that name the same point as $\frac{1}{3}$. If we were to divide the segment between 0 and 1 into 12 equal parts, 4 of these 12 equal parts $\left(\frac{4}{12}\right)$ would be the same as $\frac{2}{6}$ or $\frac{1}{3}$; that is,

$$\frac{4}{12} = \frac{2}{6} = \frac{1}{3}$$

Even though these three fractions look different, each names the same point on the number line, as shown in Figure 4. All three fractions have the same value because they all represent the same number.

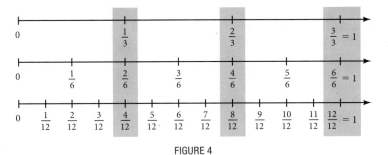

FIGURE 4

Fractions that represent the same number are said to be **equivalent**. Equivalent fractions may look different, but they must have the same value.

It is apparent that every fraction has many different representations, each of which is equivalent to the original fraction. The next two properties review a way of changing the terms of a fraction without changing its value.

PROPERTY Equivalent Fractions Using Multiplication

Multiplying the numerator and denominator of a fraction by the same nonzero number never changes the value of the fraction.

PROPERTY Equivalent Fractions Using Division

Dividing the numerator and denominator of a fraction by the same nonzero number never changes the value of the fraction.

EXAMPLE 2 Write $\frac{3}{4}$ as an equivalent fraction with denominator 20.

Solution The denominator of the original fraction is 4. The fraction we are trying to find must have a denominator of 20. We know that if we multiply 4 by 5, we get 20. The property for equivalent fractions using multiplication indicates that we are free to multiply the denominator by 5 as long as we do the same to the numerator.

$$\frac{3}{4} = \frac{3 \cdot 5}{4 \cdot 5} = \frac{15}{20}$$

The fraction $\frac{15}{20}$ is equivalent to the fraction $\frac{3}{4}$.

C Absolute Values

Representing numbers on the number line lets us give each number two important properties: a direction from zero and a distance from zero. The direction from zero is represented by the sign in front of the number. (A number without a sign is understood to be positive.) The distance from zero is called the *absolute value* of the number, as the following definition indicates:

> **DEFINITION** absolute value
>
> The *absolute value* of a real number is its distance from zero on the number line. If x represents a real number, then the absolute value of x is written $|x|$.

EXAMPLE 3 Write each expression without absolute value symbols.

a. $|5|$ **b.** $|-5|$ **c.** $\left|-\frac{1}{2}\right|$ **d.** $-|-8|$

Solution

a. $|5| = 5$ The number 5 is 5 units from zero.

b. $|-5| = 5$ The number -5 is 5 units from zero.

c. $\left|-\frac{1}{2}\right| = \frac{1}{2}$ The number $-\frac{1}{2}$ is $\frac{1}{2}$ units from zero.

d. $-|-8| = -8$ The number -8 is 8 units from zero, but the preceding negative sign gives the value a direction of 8 units to the left of zero.

The absolute value of a number is *never* negative. It is the distance the number is from zero without regard to which direction it is from zero. When working with the absolute value of sums and differences, we must simplify the expression inside the absolute value symbols first and then find the absolute value of the simplified expression.

EXAMPLE 4 Simplify each expression.

a. $|8 - 3|$ **b.** $|3 \cdot 2^3 + 2 \cdot 3^2|$ **c.** $|9 - 2| - |8 - 6|$

Solution

a. $|8 - 3| = |5| = 5$

b. $|3 \cdot 2^3 + 2 \cdot 3^2| = |3 \cdot 8 + 2 \cdot 9| = |24 + 18| = |42| = 42$

c. $|9 - 2| - |8 - 6| = |7| - |2| = 7 - 2 = 5$

2. Write $\frac{5}{8}$ as an equivalent fraction with denominator 48.

> *Note* To help you understand absolute value, consider the distance between two cities. If you were to travel from city A to city B, the distance would be a positive measurement. The same goes for traveling from city B back to city A. Just as with absolute value, the distance from one point to another is always positive, regardless of the direction.

3. Write each expression without absolute value symbols.

 a. $|7|$

 b. $|-7|$

 c. $\left|-\frac{3}{4}\right|$

 d. $-|-12|$

> *Note* The value of the expression in part d of Example 3 is much different than the value of the expression $-(-8)$, where we use the parentheses instead of absolute value symbols.
> $-|-8| = -8$, whereas $-(-8) = 8$

4. Simplify each expression.

 a. $|7 - 2|$

 b. $|2 \cdot 3^2 + 5 \cdot 2^2|$

 c. $|10 - 4| - |11 - 9|$

Answers

2. $\frac{30}{48}$

3. a. 7 **b.** 7 **c.** $\frac{3}{4}$ **d.** -12

4. a. 5 **b.** 38 **c.** 4

D Opposites

Another important concept associated with numbers on the number line is that of *opposites*. Here is the definition:

> **DEFINITION** opposites
>
> Numbers the same distance from zero but in opposite directions from zero are called *opposites*.

5. Give the opposite of each number.
 a. 8
 b. −5
 c. $-\dfrac{2}{3}$
 d. −4.2

EXAMPLE 5 Give the opposite of each number.

a. 5 **b.** −3 **c.** $\dfrac{1}{4}$ **d.** −2.3

Solution

	Number	Opposite	
a.	5	−5	5 and −5 are opposites.
b.	−3	3	−3 and 3 are opposites.
c.	$\dfrac{1}{4}$	$-\dfrac{1}{4}$	$\frac{1}{4}$ and $-\frac{1}{4}$ are opposites.
d.	2.3	2.3	−2.3 and 2.3 are opposites.

Each negative number is the opposite of some positive number, and each positive number is the opposite of some negative number. The opposite of a negative number is a positive number. In symbols, if a represents a positive number, then

$$-(-a) = a$$

Opposites always have the same absolute value. And, when you add any two opposites, the result is always zero.

$$a + (-a) = 0$$

Also, you will see that understanding opposites is important to graphing subtraction of real numbers in the next section.

E Variables: An Intuitive Look

In the previous chapter, you probably noticed the use of letters instead of numbers in certain expressions. These letters are called *variables*, which we will now discuss in greater detail.

When you filled out the application for the school you are attending, there was a space to fill in your first name. "First name" is a variable quantity because the value it takes depends on who is filling out the application. For example, if your first name is Manuel, then the value of "First Name" is Manuel. However, if your first name is Christa, then the value of "First Name" is Christa.

If we denote "First Name" as F, "Last Name" as L, and "Whole Name" as W, then we take the concept of a variable further and write the relationship between the names this way:

$$F + L = W$$

We use the $+$ symbol loosely here to represent writing the names together with a space between them. This relationship we have written holds for all people who have only a first name and a last name. For those people who have a middle name, the relationship between the names is

$$F + M + L = W$$

where M represents a middle name.

A similar situation exists in algebra when we let a letter stand for a number or a group of numbers. For instance, if we say "let a and b represent numbers," then a and b are called variables because the values they take on vary. Here is the formal definition:

DEFINITION variable

A *variable* is a letter that stands for (represents) a mathematical quantity.

Furthermore, much of what we do in algebra involves comparison of quantities. We will begin by listing some symbols used to compare mathematical quantities. The comparison symbols fall into two major groups: equality symbols and inequality symbols. We use the variables a and b in the following lists so that the relationships shown there are true for all numbers that we will encounter in this book. By using variables, the following statements are general statements about all numbers, rather than specific statements about only a few numbers.

PROPERTY Comparison Symbols

Equality:	$a = b$	a is equal to b
	$a \neq b$	a is not equal to b
Inequality:	$a < b$	a is less than b
	$a > b$	a is greater than b
	$a \geq b$	a is greater than or equal to b
	$a \leq b$	a is less than or equal to b

The symbols for inequality, $<$ and $>$, always point to the smaller of the two quantities being compared. For example, $3 < x$ means 3 is smaller than x. In this case, we can say "3 is less than x" or "x is greater than 3"; both statements are correct. Similarly, the expression $5 > y$ can be read as "5 is greater than y" or as "y is less than 5" because the inequality symbol is pointing to y, meaning y is the smaller of the two quantities.

Next, we consider the symbols used to represent the four basic operations: addition, subtraction, multiplication, and division.

PROPERTY Operation Symbols

Addition:	$a + b$	The *sum* of a and b
Subtraction:	$a - b$	The *difference* of a and b
Multiplication:	$a \cdot b, (a)(b), a(b), (a)b, ab$	The *product* of a and b
Division:	$a \div b, a/b, \dfrac{a}{b}, b\overline{)a}$	The *quotient* of a and b

Note In the past, you may have used the notation 3×5 to denote multiplication. In algebra, it is best to avoid this notation if possible, because the multiplication symbol \times can be confused with the variable x when written by hand.

When we encounter the word *sum*, the implied operation is addition. To find the sum of two numbers, we simply add them. *Difference* implies subtraction, *product* implies multiplication, and *quotient* implies division. Notice also that there is more than one way to write the product or quotient of two numbers.

PROPERTY Grouping Symbols

Parentheses () and brackets [] are the symbols used for grouping numbers together. Occasionally, braces { } are also used for grouping, although they are usually reserved for solution set notation.

The following table illustrates the relationship between the symbols for comparing, operating, and grouping and the English language.

Mathematical Expression	Written Equivalent
$4 + 1 = 5$	The sum of 4 and 1 is 5.
$8 - 1 < 10$	The difference of 8 and 1 is less than 10.
$2(3 + 4) = 14$	Twice the sum of 3 and 4 is 14.
$3x \geq 15$	The product of 3 and x is greater than or equal to 15.
$\dfrac{y}{2} = y - 2$	The quotient of y and 2 is equal to the difference of y and 2.

Lastly, remember that a variable is a letter used to represent a number or a group of numbers. An expression that contains variables with other numbers and symbols is called an *algebraic expression*. Here is a formal definition:

DEFINITION algebraic expression

An *algebraic expression* is an expression that contains any combination of numbers, variables, operation symbols, and grouping symbols.

Each of the following is an algebraic expression:

$$3x + 5 \qquad 4t^2 - 9 \qquad x^2 - 6xy + y^2$$

$$-15x^2y^4z^5 \qquad \frac{a^2 - 9}{a - 3} \qquad \frac{(x - 3)(x + 2)}{4x}$$

You will encounter a variety of algebraic expressions in this book.

Furthermore, when we set an algebraic expression equal to a value or another expression, the result is an *algebraic equation*. Here are some examples of algebraic equations:

$$4x + 2 = 10 \qquad 2a^2 + 3 = -1 \qquad \frac{(y + 2)(y - 3)}{6y}$$

If an inequality symbol is used in place of an equals sign, then the result is called an *algebraic inequality*.

F Formulas for Area and Perimeter

FACTS FROM GEOMETRY Formulas for Area and Perimeter

A square, rectangle, and triangle are shown in the following figures. Note that we have labeled the dimensions of each with variables. The formulas for the perimeter and area of each object are given in terms of its dimensions.

Square

Perimeter = $4s$

Area = s^2

Rectangle

Perimeter = $2l + 2w$

Area = lw

Triangle

Perimeter = $a + b + c$

Area = $\frac{1}{2}bh$

The formula for perimeter gives us the distance around the outside of the object along its sides, whereas the formula for area gives us a measure of the amount of surface the object has.

> *Note* The vertical line labeled h in the triangle is its height, or altitude. It extends from the top of the triangle down to the base, meeting the base at an angle of 90°. The altitude of a triangle is always perpendicular to the base. The small square shown where the altitude meets the base is used to indicate that the angle formed is 90°.

EXAMPLE 6 Find the perimeter and area of each figure.

a.

5 ft

b.

6 in.

8 in.

c.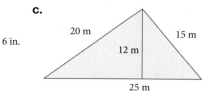

20 m 15 m

12 m

25 m

Solution We use the preceding formulas to find the perimeter and the area. In each case, the units for perimeter are linear units, whereas the units for area are square units.

a. Perimeter = $4s = 4 \cdot 5$ feet = 20 feet

Area = $s^2 = (5 \text{ feet})^2 = 25$ square feet

b. Perimeter = $2l + 2w = 2(8 \text{ inches}) + 2(6 \text{ inches}) = 28$ inches

Area = $lw = (8 \text{ inches})(6 \text{ inches}) = 48$ square inches

c. Perimeter = $a + b + c = (20 \text{ meters}) + (25 \text{ meters}) + (15 \text{ meters}) = 60$ meters

Area = $\frac{1}{2}bh = \frac{1}{2}(25 \text{ meters})(12 \text{ meters}) = 150$ square meters

6. Find the perimeter and area of the figures in Example 6, with the following changes:

a. $s = 4$ feet

b. $l = 7$ inches
 $w = 5$ inches

c. $a = 40$ meters
 $b = 50$ meters (base)
 $c = 30$ meters
 $h = 24$ meters

GETTING READY FOR CLASS

After reading through the preceding section, respond in your own words and in complete sentences.

A. What is a real number?

B. What is an equivalent fraction of $\frac{2}{6}$? Explain.

C. What is a variable?

D. Explain how you find the perimeter and the area of a rectangle.

Answer

6. a. Perimeter = 16 feet
 Area = 16 square feet

b. Perimeter = 24 inches
 Area = 35 square inches

c. Perimeter = 120 meters
 Area = 600 square meters

Vocabulary Review

Choose the correct words to fill in the blanks below.

absolute value nonzero real algebraic expression coordinate

equivalent variables opposites real number line

1. The _____ is a straight line with points that represent positive and negative numbers and zero.
2. The number associated with a point on the number line is called a _____.
3. Numbers that can be represented with points on a number line are called _____ numbers.
4. Although _____ fractions look different, they have the same value.
5. Multiplying (or dividing) the numerator and denominator of a fraction by the same _____ number never changes the value of the fraction.
6. The _____ of a real number is its distance from zero on the number line.
7. Numbers with the same absolute value but have points in opposite directions from zero on the number line are called _____.
8. An _____ contains any combination of numbers, _____, operation symbols, and grouping symbols.

Problems

A Draw a number line that extends from −5 to 5. Label the points with the following coordinates.

1. 5 **2.** −2 **3.** −4 **4.** −3

5. 1.5 **6.** −1.5 **7.** $\dfrac{9}{4}$ **8.** $\dfrac{8}{3}$

B Write each of the following fractions as an equivalent fraction with denominator 24.

9. $\dfrac{3}{4}$ **10.** $\dfrac{5}{6}$ **11.** $\dfrac{1}{2}$ **12.** $\dfrac{1}{8}$ **13.** $\dfrac{5}{8}$ **14.** $\dfrac{7}{12}$

SCAN TO ACCESS

Write each fraction as an equivalent fraction with denominator 60.

15. $\dfrac{3}{5}$ **16.** $\dfrac{5}{12}$ **17.** $\dfrac{11}{30}$

18. $\dfrac{9}{10}$ **19.** $\dfrac{5}{6}$ **20.** $\dfrac{2}{3}$

C Simplify each expression.

21. $|8 - 2|$

22. $|6 - 1|$

23. $|5 \cdot 2^3 - 2 \cdot 3^2|$

24. $|2 \cdot 10^2 + 3 \cdot 10|$

25. $|7 - 2| - |4 - 2|$

26. $|10 - 3| - |4 - 1|$

27. $10 - |7 - 2(5 - 3)|$

28. $12 - |9 - 3(7 - 5)|$

29. $15 - |8 - 2(3 \cdot 4 - 9)| - 10$

30. $25 - |9 - 3(4 \cdot 5 - 18)| - 20$

31. $25 - 2|8 - 2 \cdot 3| - 4$

32. $30 - 4|6 - 4(3 - 2)| - 7 \cdot 2$

C D For each of the following numbers, give the opposite and the absolute value. (Assume all variables are nonzero.)

33. 10

34. 8

35. $\dfrac{3}{4}$

36. $\dfrac{5}{7}$

37. $\dfrac{11}{2}$

38. $\dfrac{16}{3}$

39. -3

40. -5

41. $-\dfrac{2}{5}$

42. $-\dfrac{3}{8}$

43. x

44. a

E Place one of the symbols $<$ or $>$ between each of the following to make the resulting statement true.

45. $-5 \quad -3$

46. $-8 \quad -1$

47. $-3 \quad -7$

48. $-6 \quad 5$

49. $|-4| \quad -|-4|$

50. $3 \quad -|-3|$

51. $7 \quad -|-7|$

52. $-7 \quad |-7|$

53. $-\dfrac{3}{4} \quad -\dfrac{1}{4}$

54. $-\dfrac{2}{3} \quad -\dfrac{1}{3}$

55. $-\dfrac{3}{2} \quad -\dfrac{3}{4}$

56. $-\dfrac{8}{3} \quad -\dfrac{17}{3}$

For each sentence below, write an equivalent algebraic equation or inequality.

57. The sum of x and 5 is 14.

58. The difference of x and 4 is 8.

59. The product of 5 and y is less than 30.

60. The product of 8 and y is greater than 16.

61. The product of 5 and y is greater than or equal to the difference of y and 16.

62. The product of 3 and y is less than or equal to the sum of y and 6.

63. The quotient of x and 3 is equal to the sum of x and 2.

64. The quotient of x and 2 is equal to the difference of x and 4.

F Find the perimeter and area of each figure.

65.

1 in.

1 in.

66.

15 mm

15 mm

67.

0.75 in.

1.5 in.

68.

1.5 cm

4.5 cm

69.

2.75 cm 3.5 cm

2.5 cm

4 cm

70.

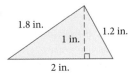

1.8 in. 1.2 in.

1 in.

2 in.

..

Applying the Concepts

71. Football Yardage A football team gains 6 yards on one play and then loses 8 yards on the next play. To what number on the number line does a loss of 8 yards correspond? The total yards gained or lost on the two plays corresponds to what negative number?

72. Checking Account Balance A woman has a balance of $20 in her checking account. If she writes a check for $30, what negative number can be used to represent the new balance in her checking account?

Temperature In the United States, temperature is measured on the Fahrenheit temperature scale. On this scale, water boils at 212 degrees and freezes at 32 degrees. To denote a temperature of 32 degrees on the Fahrenheit scale, we write

$$32°F, \text{ which is read "32 degrees Fahrenheit"}$$

Use this information for Problems 73 and 74.

73. Temperature and Altitude Marilyn is flying from Seattle to San Francisco on a Boeing 737 jet. When the plane reaches an altitude of 35,000 feet, the temperature outside the plane is 64 degrees below zero Fahrenheit. Represent the temperature with a negative number. If the temperature outside the plane gets warmer by 10 degrees, what will the new temperature be?

74. Temperature Change At 10:00 in the morning in White Bear Lake, Minnesota, John notices the temperature outside is 10 degrees below zero Fahrenheit. Write the temperature as a negative number. An hour later it has warmed up by 6 degrees. What is the temperature at 11:00 that morning?

Wind Chill The table to the right gives wind chill temperatures. The left column gives the air temperature, and the first row is wind speed in miles per hour. The numbers within the table indicate how cold the weather will feel. For example, if the thermometer reads 30°F and the wind is blowing at 15 miles per hour, the wind chill temperature is 9°F. Use the table to answer Problems 75 and 76.

75. Find the wind chill temperature if the thermometer reads 20°F and the wind is blowing at 25 miles per hour.

76. Which will feel colder: a day with an air temperature of 10°F with a 25-mile-per-hour wind, or a day with an air temperature of 25° F and a 10-mile-per-hour wind?

Wind Chill Temperatures

Air Temperature (°F)	Wind Speed (mph)				
	10	15	20	25	30
30°	16°	9°	4°	1°	−2°
25°	10°	2°	−3°	−7°	−10°
20°	3°	−5°	−10°	−15°	−18°
15°	−3°	−11°	−17°	−22°	−25°
10°	−9°	−18°	−24°	−29°	−33°
5°	−15°	−25°	−31°	−36°	−41°
0°	−22°	−31°	−39°	−44°	−49°
−5°	−27°	−38°	−46°	−51°	−56°

77. Scuba Diving Steve is scuba diving near his home in Maui. At one point he is 100 feet below the surface. Represent this number with a negative number. If he descends another 5 feet, what negative number will represent his new position?

78. Pizza Sales The chart shows top sellers of frozen pizza in one year. Write a mathematical statement using one of the symbols < or > to compare the following:

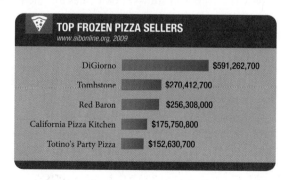

TOP FROZEN PIZZA SELLERS
www.aibonline.org, 2009

DiGiorno	$591,262,700
Tombstone	$270,412,700
Red Baron	$256,308,000
California Pizza Kitchen	$175,750,800
Totino's Party Pizza	$152,630,700

a. Digiorno to California Pizza Kitchen

b. Totino's Party Pizza to Tombstone

c. Red Baron to California Pizza Kitchen

79. Geometry Find the area and perimeter of an $8\frac{1}{2}$-by-11-inch piece of notebook paper.

80. Geometry Find the area and perimeter of an $8\frac{1}{2}$-by-$5\frac{1}{2}$-inch piece of paper.

Calories and Exercise The table here gives the amount of energy expended per hour for various activities for a person weighing 120, 150, or 180 pounds. Use the table to answer questions 81–84.

81. Suppose you weigh 120 pounds. How many calories will you burn if you play handball for 2 hours and then ride your bicycle for an hour?

82. How many calories are burned by a person weighing 150 pounds who jogs for $\frac{1}{2}$ hour and then goes bicycling for 2 hours?

83. Two people go skiing. One weighs 180 pounds and the other weighs 120 pounds. If they ski for 3 hours, how many more calories are burned by the person weighing 180 pounds?

Energy Expended from Exercising

Activity	Calories per Hour		
	120 lb	150 lb	180 lb
Bicycling	299	374	449
Bowling	212	265	318
Handball	544	680	816
Horseback trotting	278	347	416
Jazzercise	272	340	408
Jogging	544	680	816
Skiing (downhill)	435	544	653

84. Two people spend an evening bowling for 3 hours. If one weighs 120 pounds and the other weighs 150 pounds, how many more calories are burned during the evening by the person weighing 150 pounds?

85. Cookie Sales The chart shows cookie sales for one year. Use it to answer the following questions.

a. What was the revenue for Oreo cookie sales during the year?

b. True or false? The chart shows sales of private label cookies to be less than $400,000,000.

c. True or false? The chart shows sales of Lofthouse cookies to be more than $100,000,000.

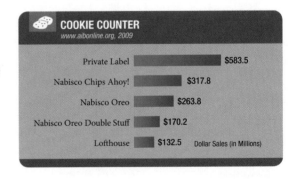

Improving Your Quantitative Literacy

86. Quantitative literacy is a subject discussed by many people involved in teaching mathematics. The person they are concerned with when they discuss it is you. We are going to work at improving your quantitative literacy, but before we do that we should answer the question, What is quantitative literacy? Lynn Arthur Steen, a noted mathematics educator, has stated that quantitative literacy is "the capacity to deal effectively with the quantitative aspects of life."

 a. Give a definition for the word *quantitative*.

 b. Give a definition for the word *literacy*.

 c. Are there situations that occur in your life that you find distasteful, or that you try to avoid, because they involve numbers and mathematics? If so, list some of them here. (For example, some people find the process of buying a car particularly difficult because they feel that the numbers and details of the financing are beyond them.)

Find the Mistake

Each sentence below contains a mistake. Circle the mistake and write the correct word(s) or numbers(s) on the line provided.

1. Writing the fraction $\frac{4}{11}$ as an equivalent fraction with a denominator of 33 gives us $\frac{4}{33}$. _____

2. A real number and its opposite will have different absolute values. _____

3. In the algebraic expression $2x - y$, the 2 is the variable. _____

4. The perimeter of a rectangle with a length of 32 and a width of 8 is 256. _____

Navigation Skills: Prepare, Study, Achieve

Your instructor is a vital resource for your success in this class. Make note of your instructor's office hours and utilize them regularly. Compile a resource list that you keep with your class materials. This list should contain your instructor's office hours and contact information (e.g., office phone number or e-mail), as well as classmates' contact information that you can utilize outside of class. Communicate often with your classmates about how the course is going for you and any questions you may have. Odds are that someone else has the same question and you may be able to work together to find the answer.

OBJECTIVES

A Add any combination of positive and negative numbers.

B Subtract any combination of positive and negative numbers.

C Simplify expressions using the rule for order of operations.

D Translate sentences from English into symbols and then simplify.

E Find the complement and the supplement of an angle.

KEY WORDS

negative numbers

angles

complementary

supplementary

Image © sxc.hu, palm3559, 2007

If you have an extra $20,000 and a knack for poker, you may receive a coveted invitation to the National Heads-Up Poker Championship held annually in the United States. The money would be your buy-in to the single-elimination tournament, which hosts one-on-one matches in bracket form. The last player at the table is crowned champion and receives a prize of $500,000.

No invitation? Instead, suppose that you are playing a friendly game of poker with some friends, and you lose $3 on the first hand and $4 on the second hand. If you represent winning with positive numbers and losing with negative numbers, how can you translate this situation into symbols? Because you lost $3 and $4 for a total of $7, one way to represent this situation is with addition of negative numbers:

$$(-\$3) + (-\$4) = -\$7$$

From this equation, we see that the sum of two negative numbers is a negative number. To generalize addition with positive and negative numbers, we use the number line.

Because real numbers have both a distance from zero (absolute value) and a direction from zero ($+$ or $-$ sign), we can think of addition of two numbers in terms of distance and direction from zero.

A Adding Positive and Negative Numbers

Let's look at a problem for which we know the answer. Suppose we want to add the numbers 3 and 4. The problem is written $3 + 4$. To put it on the number line, we read the problem as follows:

1. The 3 tells us to "start at the origin and move 3 units in the positive direction."
2. The $+$ sign is read "and then move."
3. The 4 means "4 units in the positive direction."

To summarize, $3 + 4$ means to start at the origin, move 3 units in the positive direction, and then move an additional 4 units also in the positive direction.

We end up at 7, which is the answer to our problem: $3 + 4 = 7$.

Let's try other combinations of positive and negative 3 and 4 on the number line.

EXAMPLE 1 Add $3 + (-4)$.

Solution Starting at the origin, move 3 units in the positive direction and then 4 units in the negative direction.

We end up at -1; therefore, $3 + (-4) = -1$.

EXAMPLE 2 Add $-3 + 4$.

Solution Starting at the origin, move 3 units in the negative direction and then 4 units in the positive direction.

We end up at 1; therefore, $-3 + 4 = 1$.

EXAMPLE 3 Add $-3 + (-4)$.

Solution Starting at the origin, move 3 units in the negative direction and then 4 units in the negative direction.

We end up at -7; therefore, $-3 + (-4) = -7$. Here is a summary of what we have just completed:

$$3 + 4 = 7$$
$$3 + (-4) = -1$$
$$-3 + 4 = 1$$
$$-3 + (-4) = -7$$

If we look closely at the preceding addition problems, we can summarize our results with the following rule:

> **RULE** Adding Real Numbers
>
> To add two real numbers with
>
> 1. the *same* sign, simply add their absolute values and use the common sign. (If both numbers are positive, then the answer is positive. If both numbers are negative, then the answer is negative.)
>
> 2. *different* signs, subtract the smaller absolute value from the larger. The answer will have the sign of the number with the larger absolute value.

Practice Problems

1. Add $2 + (-5)$. You can use the number line in Example 1 if you like.

2. Add $-2 + 5$.

3. Add $-2 + (-5)$.

Note This rule is what we have been working towards. It is very important. Be sure that you understand it and can use it. The problems we have done up to this point have been done simply to justify this rule. Now that we have the rule, we no longer need to do our addition problems on the number line.

Answers

1. -3
2. 3
3. -7

This rule covers all possible combinations of addition with real numbers. You must memorize it. After you have worked a number of problems, it will seem almost automatic.

EXAMPLE 4 Add all combinations of positive and negative 10 and 13.

Solution Rather than work these problems on the number line, we use the rule for adding positive and negative numbers to obtain our answers.

$$10 + 13 = 23$$
$$10 + (-13) = -3$$
$$-10 + 13 = 3$$
$$-10 + (-13) = -23$$

4. Add all possible combinations of positive and negative 9 and 12.
 a. 9 + 12
 b. 9 + (12)
 c. −9 + 12
 d. −9 + (12)

B Subtracting Positive and Negative Numbers

Because we want to make as few rules as possible, we can define subtraction in terms of addition. By doing so, we can then use the rule for addition to solve our subtraction problems.

> **RULE** Subtracting Real Numbers
>
> To subtract one real number from another, simply add its opposite.
> Algebraically, the rule is written like this: If a and b represent two real numbers, then it is always true that
>
> $$\underbrace{a - b}_{\text{To subtract } b,} = \underbrace{a + (-b)}_{\text{add the opposite of } b.}$$

This is how subtraction is defined in algebra. This definition of subtraction will not conflict with what you already know about subtraction, but it will allow you to do subtraction using negative numbers.

EXAMPLE 5 Subtract all possible combinations of positive and negative 7 and 2.
Solution

$$\left.\begin{array}{l} 7 - 2 = 7 + \ \ (-2) = 5 \\ -7 - 2 = -7 + (-2) = -9 \end{array}\right\}$$ Subtracting 2 is the same as adding −2.

$$\left.\begin{array}{l} 7 - (-2) = 7 + 2 = 9 \\ -7 - (-2) = -7 + 2 = -5 \end{array}\right\}$$ Subtracting −2 is the same as adding 2.

5. Subtract all possible combinations of positive and negative 4 and 7.
 a. 7 − 4
 b. −7 − 4
 c. 7 − (−4)
 d. −7 − (−4)

Notice that each subtraction problem is first changed to an addition problem. The rule for addition is then used to arrive at the answer.

EXAMPLE 6 Subtract all combinations of positive and negative 8 and 13.
Solution

$$\left.\begin{array}{l} 8 - 13 = 8 + (-13) = -5 \\ -8 - 13 = -8 + (-13) = -21 \end{array}\right\}$$ Subtracting 13 is the same as adding −13.

$$\left.\begin{array}{l} 8 - (-13) = 8 + 13 = 21 \\ -8 - (-13) = -8 + 13 = 5 \end{array}\right\}$$ Subtracting −13 is the same as adding 13.

6. Subtract all combinations of positive and negative 4 and 6.
 a. 6 − 4
 b. −6 − 4
 c. 6 − (−4)
 d. −6 − (−4)

Answers
 4. a. 21 **b.** −3 **c.** 3 **d.** −21
 5. a. 3 **b.** −11 **c.** 11 **d.** −3
 6. a. 2 **b.** −10 **c.** 10 **d.** −2

We have defined subtraction in terms of addition, and we still obtain answers consistent with the answers we are used to getting with subtraction. Moreover, we now can do subtraction problems involving both positive and negative numbers.

As you proceed through the following examples and the problem set, you will begin to notice shortcuts you can use in working the problems. You will not always have to change subtraction to addition of the opposite to be able to get answers quickly. Use all the shortcuts you wish as long as you consistently get the correct answers.

C Order of Operations

Recall our discussion about the order of operations in the previous chapter. Let's apply this rule to some problems involving real numbers.

EXAMPLE 7 Add $-3 + 2 + (-4)$.

Solution Applying the rule for order of operations, we add left to right.

$$-3 + 2 + (-4) = -1 + (-4)$$
$$= -5$$

7. Add $-5 + 2 + (-7)$.

EXAMPLE 8 Add $-8 + [2 + (-5)] + (-1)$.

Solution Adding inside the brackets first and then left to right, we have

$$-8 + [2 + (-5)] + (-1) = -8 + (-3) + (-1)$$
$$= -11 + (-1)$$
$$= -12$$

8. Add $-4 + [2 + (-3)] + (-1)$.

EXAMPLE 9 Simplify each expression as much as possible.

a. $7 + (-3) - 5$ **b.** $8 - (-2) - 6$ **c.** $-2 - (-3 + 1) - 5$

Solution

a. $7 + (-3) - 5 = 7 + (-3) + (-5)$ *Begin by changing all subtractions to*
$\qquad\qquad\quad\; = 4 + (-5)$ *additions. Then add left to right.*
$\qquad\qquad\quad\; = -1$

b. $8 - (-2) - 6 = 8 + 2 + (-6)$ *Begin by changing all subtractions to*
$\qquad\qquad\quad\; = 10 + (-6)$ *additions. Then add left to right.*
$\qquad\qquad\quad\; = 4$

c. $-2 - (-3 + 1) - 5 = -2 - (-2) - 5$ *Do what is in the parentheses first.*
$\qquad\qquad\qquad\qquad = -2 + 2 + (-5)$
$\qquad\qquad\qquad\qquad = -5$

9. Simplify each expression as much as possible.

a. $4 + (-2) - 3$

b. $9 - 4 - 5$

c. $-3 - (-5 + 1) - 6$

Note When writing lines of simplified calculations or algebraic expression (as we will encounter later in this chapter), it is important to remember to begin each line with an equals sign, as we have done here.

The next two examples involve multiplication and exponents as well as addition and subtraction. Remember, according to the rule for order of operations, we evaluate the numbers containing exponents and multiply before we subtract.

EXAMPLE 10 Simplify $2 \cdot 5 - 3 \cdot 8 - 4 \cdot 9$.

Solution First, we multiply left to right, and then we subtract.

$$2 \cdot 5 - 3 \cdot 8 - 4 \cdot 9 = 10 - 24 - 36$$
$$= -14 - 36$$
$$= -50$$

10. Simplify $3 \cdot 9 - 4 \cdot 10 - 5 \cdot 11$.

Answers
7. -10
8. -6
9. a. 1 **b.** 0 **c.** 5
10. -68

11. Simplify $4 \cdot 2^5 - 3 \cdot 5^2$.

EXAMPLE 11 Simplify $3 \cdot 2^3 + 2 \cdot 4^2$.

Solution We begin by evaluating each number that contains an exponent. Then we multiply before we subtract.

$$3 \cdot 2^3 - 2 \cdot 4^2 = 3 \cdot 8 + 2 \cdot 16$$
$$= 24 + 32$$
$$= 56$$

D Writing Words in Symbols

Just as we discussed in the previous section, it is important to be able to translate the English language into symbols for operations with real numbers. The next four examples will let you practice doing so.

12. Add 8 to -5.

EXAMPLE 12 Add 7 to -3.

Solution First, we write the problem in terms of addition, then we simplify.

$$-3 + 7 = 4$$

13. Add -3 to 8.

EXAMPLE 13 Add -5 to 2.

Solution $2 + (-5) = -3$

14. Find the difference of 8 and 6.

EXAMPLE 14 Find the difference of 9 and 2.

Solution Written in symbols, the problem looks like this:

$$9 - 2 = 7$$

The difference of 9 and 2 is 7.

15. Find the difference of 4 and -7.

EXAMPLE 15 Find the difference of 3 and -5.

Solution Subtracting -5 from 3 we have

$$3 - (-5) = 3 + 5$$
$$= 8$$

E Complementary and Supplementary Angles

FACTS FROM GEOMETRY Complementary and Supplementary Angles

If you have studied geometry at all, you know that there are 360° in a full rotation—the number of degrees swept out by the radius of a circle as it rotates once around the circle.

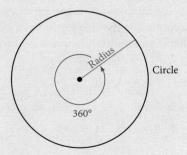

Answers
11. 53
12. 3
13. 5
14. 2
15. 11

We can apply our knowledge of algebra to help solve some simple geometry problems. Before we do, however, we need to review some of the vocabulary associated with angles.

> **DEFINITION** complementary and supplementary angles
>
> In geometry, two angles that add to 90° are called *complementary angles*. In a similar manner, two angles that add to 180° are called *supplementary angles*.
>
>
>
> Complementary angles: $x + y = 90°$ Supplementary angles: $x + y = 180°$

EXAMPLE 16 Find x in each of the following diagrams.

a.

b.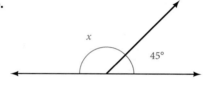

Solution We use subtraction to find each angle.

a. Because the two angles are complementary, we can find x by subtracting 30° from 90°.
$$x = 90° - 30° = 60°$$
We say 30° and 60° are complementary angles. The complement of 30° is 60°.

b. The two angles in the diagram are supplementary. To find x, we subtract 45° from 180°.
$$x = 180° - 45° = 135°$$
We say 45° and 135° are supplementary angles. The supplement of 45° is 135°.

Subtracting and Taking Away

For some people taking algebra for the first time, subtraction of positive and negative numbers can be a problem. These people may believe that $-5 - 9$ should be -4 or 4, not -14. If this is happening to you, you probably are thinking of subtraction in terms of taking one number away from another. Thinking of subtraction in this way works well with positive numbers if you always subtract the smaller number from the larger. In algebra, however, we encounter many situations other than this. The definition of subtraction, that $a - b = a + (-b)$, clearly indicates the correct way to use subtraction; that is, when working subtraction problems, you should think "addition of the opposite," not "take one number away from another." To be successful in algebra, you need to apply properties and definitions exactly as they are presented here.

> **GETTING READY FOR CLASS**
>
> *After reading through the preceding section, respond in your own words and in complete sentences.*
>
> **A.** Explain how you would add 3 and -5 on the number line.
>
> **B.** Define subtraction in terms of addition.
>
> **C.** Why is it important to understand absolute value in order to know how to add positive and negative numbers?
>
> **D.** What are complementary and supplementary angles?

16. Find x in each of the following diagrams.

a.

b.

Answers

16. a. 45° **b.** 120°

EXERCISE SET 1.2

Vocabulary Review

Choose the correct words to fill in the blanks below.

opposite different same complementary supplementary

1. To add two real numbers with the _____ sign, add their absolute values and use the common sign.
2. To add two real numbers with _____ signs, subtract the smaller absolute value from the larger and add the sign of the number with the larger absolute value.
3. To subtract one real number from another, add its _____.
4. Two angles that add to 90° are called _____ angles.
5. Two angles that add to 180° are called _____ angles.

Problems

A Work the following addition problems. You may want to begin by doing a few on the number line.

1. $6 + (-3)$

2. $7 + (-8)$

3. $13 + (-20)$

4. $15 + (-25)$

5. $18 + (-32)$

6. $6 + (-9)$

7. $-6 + 3$

8. $-8 + 7$

9. $-30 + 5$

10. $-18 + 6$

11. $-6 + (-6)$

12. $-5 + (-5)$

13. $-9 + (-10)$

14. $-8 + (-6)$

15. $-10 + (-15)$

16. $-18 + (-30)$

B Subtract.

17. $5 - 8$

18. $6 - 7$

19. $3 - 9$

20. $2 - 7$

SCAN TO ACCESS

21. $5 - 5$

22. $8 - 8$

23. $-8 - 2$

24. $-6 - 3$

25. $-4 - 12$

26. $-3 - 15$

27. $-6 - 6$

28. $-3 - 3$

29. $-8 - (-1)$

30. $-6 - (-2)$

31. $15 - (-20)$

32. $20 - (-5)$

33. $-4 - (-4)$

34. $-5 - (-5)$

35. $-3 - (-12)$

36. $-7 - (-20)$

C Work the following problems using the rule for addition of real numbers. You may want to refer back to the rule for order of operations.

37. $5 + (-6) + (-7)$

38. $6 + (-8) + (-10)$

39. $3 - 2 - 5$

40. $4 - 8 - 6$

41. $-22 + 4 - 10$

42. $-13 + 6 - 5$

43. $5 + [6 + (-2)] + (-3)$

44. $10 + [8 + (-5)] + (-20)$

45. $[6 + (-2)] + [3 + (-1)]$

46. $[18 + (-5)] + [9 + (-10)]$

47. $20 + (-6) + [3 + (-9)]$

48. $18 + (-2) + [9 + (-13)]$

49. $16 - [(4 - 5) - 1]$

50. $15 - [(4 - 2) - 3]$

51. $5 - [(2 - 3) - 4]$

52. $6 - [(4 - 1) - 9]$

53. $21 - [-(3 - 4) - 2] - 5$

54. $30 - [-(10 - 5) - 15] - 25$

55. $12 - [-(7 - 2) - 5] - 9$

56. $13 - [-(5 - 12) - 3] - 7$

The problems that follow involve some multiplication. Be sure that you work inside the parentheses first, then multiply, and finally, add left to right. Use the rule for order of operations to simplify each expression as much as possible.

57. $-5 + 2(-3 + 7)$

58. $-3 + 4(-2 + 7)$

59. $9 + 3(-8 + 10)$

60. $4 + 5(-2 + 6)$

61. $2 \cdot 8 - 3 \cdot 5$

62. $3 \cdot 4 - 6 \cdot 7$

63. $3 \cdot 5 - 2 \cdot 7$

64. $6 \cdot 10 - 5 \cdot 20$

65. $2 \cdot 3^2 + 5 \cdot 2^2$

66. $3 \cdot 7^2 + 2 \cdot 8^2$

67. $4 \cdot 3^3 - 5 \cdot 2^3$

68. $3 \cdot 6^2 - 2 \cdot 3^2 - 8 \cdot 6^2$

D Write the numerical expression that is equivalent to each of the following phrases and then simplify.

69. The sum of 5 and 9

70. The sum of 6 and -3

71. Four added to the sum of -7 and -5

72. Six added to the sum of -9 and 1

73. Subtract 4 from -7.

74. Subtract 5 from -19.

75. Subtract 17 from the sum of 4 and -5.

76. Subtract -6 from the sum of 6 and -3.

77. The difference of 8 and 5

78. The difference of 5 and 8

79. The difference of 8 and -5

80. The difference of 5 and -8

Answer the following questions.

81. What number do you add to -8 to get -5?

82. The sum of what number and -12 is 8?

83. What number do you subtract from 8 to get -2?

84. What number do you subtract from 1 to get 5?

E Find x in each of the following diagrams.

85.

86.

87.

88.

..

Applying the Concepts

89. Temperature Change The temperature inside a space shuttle is 73°F before reentry. During reentry the temperature inside the craft increases 10°. On landing it drops 8°F. Write an expression using the numbers 73, 10, and 8 to describe this situation. What is the temperature inside the shuttle on landing?

90. Temperature Change The temperature at noon is 23°F. Six hours later it has dropped 19°F, and by midnight it has dropped another 10°F. Write a subtraction problem that gives the temperature at midnight. What is the temperature at midnight?

91. Temperature Change The coldest temperature ever recorded in Antarctica was −129°F, while the highest recorded was 59°F. What is the absolute value of the difference between these two temperatures?

92. Temperature Change The annual mean temperature at the South Pole is −58°F, while that of the North Pole is 0°F. What is the absolute value of the difference between these two temperatures?

93. Temperature Change The highest temperature ever recorded on Earth was 136°F in Libya, while the coldest temperature ever recorded on Earth was −129°F in Antarctica. What is the absolute value of the difference between these two temperatures?

94. Temperature Change The highest North American temperature ever recorded was 134°F in Death Valley, CA, while the coldest North American temperature ever recorded was −87° in north Ice, Greenland. What is the absolute value of the difference between these two temperatures?

Profit Revenue and Costs In business, the difference of revenue and cost is profit, or $P = R - C$, where P is profit, R is revenue, and C is costs. Let's suppose the bar charts below show the cost and revenue for a children's shoe company called First Steps for a recent 5-year period. Use this information to answer the questions below.

95. What was the profit for the year 2007?

96. In what year was the largest increase in costs from the previous year? How much was the increase?

97. What was the only year the company had a loss? (a negative profit.) How much was the loss?

98. In which year was the profit the greatest? What was the profit in that year?

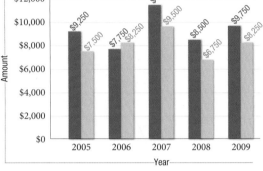

First Steps Shoe Company: Revenue and Cost

99. **Health Clubs** The chart shows the membership in US Health Clubs over a five-year period. Use it to answer the following questions.

a. How many more members were there in 2003 than 2001?

b. How many more members were there in 2004 than 2000?

c. If the trend from 2004 to 2005 continued, how many members were there be in 2006?

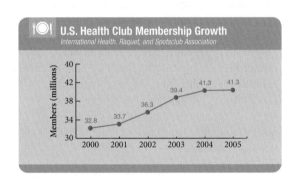

Find the Mistake

Each sentence below contains a mistake. Circle the mistake and write the correct word(s) or numbers(s) on the line provided.

1. The difference of −22 and −17 is −39. _____

2. To subtract one real number from another, subtract its opposite. _____

3. To work the problem $4^3 - (-8 + 4) - 2$ using the order of operations, we subtract the 2 before working what is inside the parentheses. _____

4. The supplement of an angle 88° is 2. _____

1.3 Multiplication of Real Numbers

OBJECTIVES

A Multiply any combination of positive and negative numbers.

B Simplify expressions using the rule for order of operations.

C Multiply positive and negative fractions.

D Multiply using the multiplication property of zero.

Facebook made its Wall Street debut as a publicly traded company in May 2012. The stock opened at $42.05 per share, up 13% from its IPO pricing of $38. Then it hit $42.99 before falling back to $40 by the end of the first ten minutes of trading. Suppose you own 5 shares of Facebook stock during the time when the stock fell from $42.99 to $40, approximately a $3 drop per share. How much money have you lost? Intuitively, we know the loss is $15. Because it is a loss, we can express it as $-$15. To describe this situation with numbers, we would write

$$5 \text{ shares each lose } \$3. \rightarrow \quad 5(-3) = -15 \quad \leftarrow \text{ Total of } \$15$$

Reasoning in this manner, we conclude that the product of a positive number with a negative number is a negative number. Let's look at multiplication in more detail.

A Multiplication of Real Numbers

From our experience with counting numbers, we know that multiplication is simply repeated addition; that is, $3(5) = 5 + 5 + 5$. We will use this fact, along with our knowledge of negative numbers, to develop the rule for multiplication of any two real numbers. The following example illustrates multiplication with all of the possible combinations of positive and negative numbers.

EXAMPLE 1 Multiply.

a. $3(5)$ **b.** $3(-5)$ **c.** $-3(5)$ **d.** $-3(-5)$

Solution

a. Two positives: $3(5) = 5 + 5 + 5$

 $= 15$ *Positive answer*

b. One positive: $3(-5) = -5 + (-5) + (-5)$

 $= -15$ *Negative answer*

c. One negative: $-3(5) = -15$ *Negative answer*

To understand why multiplying a negative by a positive (or a positive by a negative) results in a negative answer, we consider the number line. Recall that, on the number line, the negative sign means to move to the left of the zero. If we think of -3 in the problem $-3(5)$ as moving 5 units to the left of the zero 3 times, then we have

Practice Problems

1. Multiply.

 a. $4(2)$

 b. $4(-2)$

 c. $-4(2)$

 d. $-4(-2)$

d. Two negatives: $-3(-5) = ?$

With two negatives, $-3(-5)$, it is not possible to work the problem in terms of repeated addition. (It doesn't "make sense" to write -5 down a -3 number of times.) The answer is probably 15 (that's just a guess), but we need some justification for saying so. We will solve a different problem and in doing so get the answer to the problem $(-3)(-5)$.

Here is a problem to which we know the answer. We will work it two different ways.

$$-3[5 + (-5)] = -3(0) = 0$$

The answer is zero. We also can work the problem by distributing the -3 over both terms inside the parentheses. This distributive property is one of the properties we will cover more in depth later in the chapter.

$$-3[5 + (-5)] = -3(5) + (-3)(-5)$$
$$= -15 + ?$$

Because the answer to the problem is 0, our ? must be 15. (What else could we add to -15 to get 0? Only 15.)

Note You may have to read the explanation for Example 1d several times before you understand it completely. The purpose of the explanation in Example 1d is simply to justify the fact that the product of two negative numbers is a positive number.

Here is a summary of the results we have obtained:

Original numbers have		The answer is
the same sign	$3(5) = 15$	positive
different signs	$3(-5) = -15$	negative
different signs	$-3(5) = -15$	negative
the same sign	$-3(-5) = 15$	positive

By examining Example 1 and the preceding table, we can use the information there to write the following rule. This rule tells us how to multiply any two real numbers.

> **RULE** **Multiplying Real Numbers**
>
> To multiply any two real numbers, simply multiply their absolute values. The sign of the answer is
>
> **1.** *positive* if both numbers have the same sign (both + or both −).
>
> **2.** *negative* if the numbers have opposite signs (one +, the other −).

Note Students have trouble with the expression $-3(-5)$ because they want to subtract rather than multiply. Because we are very precise with the notation we use in algebra, the expression $-3(-5)$ has only one meaning: multiplication. A subtraction problem that uses the same numbers is $-3 - 5$. Compare the two following lists:

All Multiplication	No Multiplication
$5(4)$	$5 + 4$
$-5(4)$	$-5 + 4$
$5(-4)$	$5 - 4$
$-5(-4)$	$-5 - 4$

The following example illustrates how we use the preceding rule to multiply real numbers.

EXAMPLE 2 Multiply.

a. $-8(-3) = 24$

b. $-10(-5) = 50$ *If the two numbers in the product have the same sign, the answer is positive.*

c. $-4(-7) = 28$

d. $5(-7) = -35$

e. $-4(8) = -32$ *If the two numbers in the product have different signs, the answer is negative.*

f. $-6(10) = -60$

2. Multiply.

a. $-4(-9)$

b. $-11(-12)$

c. $2(-7)$

d. $4(-3)$

e. $-2(10)$

f. $-3(13)$

Answers

2. a. 36 **b.** 132 **c.** 14 **d.** -12
 e. -20 **f.** -39

B Using Order of Operations

In the following example, we combine the rule for order of operations with the rule for multiplication to simplify expressions. Remember, the rule for order of operations specifies that we are to work inside the parentheses first and then simplify numbers containing exponents. After this, we multiply and divide, left to right. The last step is to add and subtract, left to right.

EXAMPLE 3 Simplify as much as possible.

a. $-5(-3)(-4)$ **b.** $4(-3) + 6(-5) - 10$ **c.** $(-2)^3$
d. $-3(-2)^3 - 5(-4)^2$ **e.** $6 - 4(7 - 2)$

Solution

a. $-5(-3)(-4) = 15(-4)$
$$= -60$$

b. $4(-3) + 6(-5) - 10 = -12 + (-30) - 10$ Multiply.
$$= -42 - 10$$ Add.
$$= -52$$ Subtract.

c. $(-2)^3 = (-2)(-2)(-2)$ Definition of exponents
$$= -8$$ Multiply, left to right.

d. $-3(-2)^3 - 5(-4)^2 = -3(-8) - 5(16)$ Exponents first
$$= 24 - 80$$ Multiply.
$$= -56$$ Subtract.

e. $6 - 4(7 - 2) = 6 - 4(5)$ Inside parentheses first
$$= 6 - 20$$ Multiply.
$$= -14$$ Subtract.

EXAMPLE 4 Figure 1 gives the calories that are burned in 1 hour for a variety of exercise by a person weighing 150 pounds. Figure 2 gives the calories that are consumed by eating some popular fast foods. Find the net change in calories for a 150-pound person playing handball for 2 hours and then eating a Whopper.

FIGURE 1

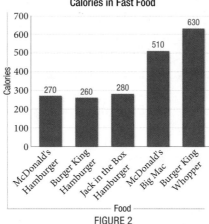

FIGURE 2

Solution The net change in calories will be the difference of the calories gained from eating and the calories lost from exercise.

$$\text{Net change in calories} = 630 - 2(680) = -730 \text{ calories}$$

3. Simplify as much as possible.

 a. $-7(-6)(-1)$

 b. $2(-8) + 3(-7) - 4$

 c. $(-3)^5$

 d. $-5(-2)^3 - 7(-3)^2$

 e. $7 - 3(4 - 9)$

Note It is important to pay close attention to the order of operations when simplifying problems involving even and odd exponents, and negative signs. Compare the following lists:

$2^3 = 8$	$2^4 = 16$
$(-2)^3 = -8$	$(-2)^4 = 16$
$-2^3 = -8$	$-2^4 = -16$
$-(-2)^3 = 8$	$-(-2)^4 = -16$

4. Find the net change in calories for a 150-pound person jogging for 2 hours then eating a Big Mac.

C Multiplying Fractions

Previously, we mentioned that to multiply two fractions we multiply numerators and multiply denominators. We can apply the rule for multiplication of positive and negative numbers to fractions in the same way we apply it to other numbers. We multiply absolute values: The product is positive if both fractions have the same sign and negative if they have different signs. Here is an example:

5. Multiply.

a. $-\dfrac{2}{3}\left(\dfrac{5}{9}\right)$

b. $-8\left(\dfrac{1}{2}\right)$

c. $-\dfrac{3}{4}\left(-\dfrac{4}{3}\right)$

EXAMPLE 5 Multiply.

a. $-\dfrac{3}{4}\left(\dfrac{5}{7}\right)$ b. $-6\left(\dfrac{1}{2}\right)$ c. $-\dfrac{2}{3}\left(-\dfrac{3}{2}\right)$

Solution

a. $-\dfrac{3}{4}\left(\dfrac{5}{7}\right) = -\dfrac{3 \cdot 5}{4 \cdot 7}$ *Different signs give a negative answer.*

$= -\dfrac{15}{28}$

b. $-6\left(\dfrac{1}{2}\right) = -\dfrac{6}{1}\left(\dfrac{1}{2}\right)$ *Different signs give a negative answer.*

$= -\dfrac{6}{2}$

$= -3$

c. $-\dfrac{2}{3}\left(-\dfrac{3}{2}\right) = \dfrac{2 \cdot 3}{3 \cdot 2}$ *Same signs give a positive answer.*

$= \dfrac{6}{6}$

$= 1$

D The Multiplication Property of Zero

As a reminder, we will cover properties of real numbers later in the chapter. For now we will consider the multiplication property of zero. It is of great importance, as you will see, throughout our study of algebra and deserves special attention.

> **RULE** Multiplication Property of Zero
>
> The product of any real number a and zero is 0.
>
> $$a(0) = 0 \qquad a \text{ is a real number.}$$
>
> $$-a(0) = 0 \qquad a \text{ is a real number.}$$
>
> $$0(0) = 0$$

6. Multiply.
a. $8(0)$
b. $0(12)$
c. $8{,}4320(0)$
d. $(-4)(-3)(0)$

EXAMPLE 6 Multiply.

a. $5(0)$ b. $0(6)$ c. $5{,}398{,}423(0)$ d. $(-1)(-6)(0)$

Solution

a. $5(0) = 0$

b. $0(6) = 0$

c. $5{,}398{,}423(0) = 0$

d. $(-1)(-6)(0) = 0$

Answers

5. a. $-\dfrac{10}{27}$ b. -4 c. 1
6. a. 0 b. 0 c. 0 d. 0

Whenever you see a 0 as one of the numbers to be multiplied, you can simply write the answer down without any calculations. It doesn't matter what the other numbers multiply to be or if the numbers are positive, negative, fractions, very large, or very small.

EXAMPLE 7 Multiply.

a. $\frac{1}{12}(0)$ **b.** $25^3(0)$ **c.** $(-3)\left(\frac{9}{13}\right)(0)$ **d.** $(465,832)(0.00026)(0)$

Solution

a. $\frac{1}{12}(0) = 0$

b. $25^3(0) = 0$

c. $(-3)\left(\frac{9}{13}\right)(0) = 0$

d. $(465,832)(0.00026)(0) = 0$

GETTING READY FOR CLASS

After reading through the preceding section, respond in your own words and in complete sentences.

A. Explain how a multiplication problem is repeated addition.

B. How do you multiply two numbers with different signs?

C. How do you multiply two negative numbers?

D. Explain the multiplication property of zero.

7. Multiply.

 a. $\frac{1}{2}(0)$

 b. $4^3(0)$

 c. $-14\left(\frac{8}{17}\right)(0)$

 d. $(543,287)(0.42987)(0)$

Answers

7. a. 0 **b.** 0 **c.** 0 **d.** 0

EXERCISE SET 1.3

Vocabulary Review

Choose the correct words to fill in the blanks below.

negative zero positive absolute values

1. To multiply two real numbers, first multiply their _____.

2. The sign of an answer to a multiplication problem involving two numbers will be _____ if both numbers have the same sign.

3. The sign of an answer to a multiplication problem involving two numbers will be _____ if both numbers have opposite signs.

4. The product of any real number and _____ is 0.

Problems

A Use the rule for multiplying two real numbers to find each of the following products.

1. $7(-6)$ **2.** $8(-4)$ **3.** $-8(2)$ **4.** $-16(3)$

5. $-3(-1)$ **6.** $-7(-1)$ **7.** $-11(-11)$ **8.** $-12(-12)$

B Use the rule for order of operations to simplify each expression as much as possible.

9. $-3(2)(-1)$ **10.** $-2(3)(-4)$ **11.** $-3(-4)(-5)$

12. $-5(-6)(-7)$ **13.** $-2(-4)(-3)(-1)$ **14.** $-1(-3)(-2)(-1)$

15. $(-7)^2$ **16.** $(-8)^2$

17. $(-3)^3$ **18.** $(-2)^4$

19. $-2(2-5)$ **20.** $-3(3-7)$

SCAN TO ACCESS

21. $-5(8 - 10)$

22. $-4(6 - 12)$

23. $(4 - 7)(6 - 9)$

24. $(3 - 10)(2 - 6)$

25. $(-3 - 2)(-5 - 4)$

26. $(-3 - 6)(-2 - 8)$

27. $-3(-6) + 4(-1)$

28. $-4(-5) + 8(-2)$

29. $2(3) - 3(-4) + 4(-5)$

30. $5(4) - 2(-1) + 5(6)$

31. $4(-3)^2 + 5(-6)^2$

32. $2(-5)^2 + 4(-3)^2$

33. $7(-2)^3 - 2(-3)^3$

34. $10(-2)^3 - 5(-2)^4$

35. $6 - 4(8 - 2)$

36. $7 - 2(6 - 3)$

37. $9 - 4(3 - 8)$

38. $8 - 5(2 - 7)$

39. $-4(3 - 8) - 6(2 - 5)$

40. $-8(2 - 7) - 9(3 - 5)$

41. $7 - 2[-6 - 4(-3)]$

42. $6 - 3[-5 - 3(-1)]$

43. $7 - 3[2(-4 - 4) - 3(-1 - 1)]$

44. $5 - 3[7(-2 - 2) - 3(-3 + 1)]$

45. $8 - 6[-2(-3 - 1) + 4(-2 - 3)]$

46. $4 - 2[-3(-1 + 8) + 5(-5 + 7)]$

47. $-5 - 2(-3 - 8)$

48. $-13 - 4(-2 - 6)$

49. $-5(2 - 3) - 7(4 - 9)$

50. $-8(-1 - 6) - 5(-2 - 3)$

C Multiply the following fractions.

51. $-\dfrac{2}{3} \cdot \dfrac{5}{7}$

52. $-\dfrac{6}{5} \cdot \dfrac{2}{7}$

53. $-8\left(\dfrac{1}{2}\right)$

54. $-12\left(\dfrac{1}{3}\right)$

55. $-\dfrac{3}{4}\left(-\dfrac{4}{3}\right)$

56. $-\dfrac{5}{8}\left(-\dfrac{8}{5}\right)$

57. $\left(-\dfrac{3}{4}\right)^2$

58. $\left(-\dfrac{2}{5}\right)^2$

D Find the following products.

59. $-2(0)$

60. $2x(0)$

61. $-7x(0)$

62. $\dfrac{1}{5}(0)$

63. $3x(-5)(0)$

64. $(16^5)\left(\dfrac{4}{19}\right)(0)(x)$

65. $(4{,}285)(x)(y)(0)$

66. $\left(\dfrac{1}{151}\right)(3)(z)(0)$

Simplify.

67. $2(5 - 5) + 4$

68. $5x(2) + 3x(0) + 4$

69. $6(3 + 1) - 5(0)$

70. $2(3) + 4(5) - 5(2)$

71. $\left(\dfrac{1}{2} \cdot 18\right)^2$

72. $\left[\dfrac{1}{2}(-10)\right]^2$

73. $\left(\dfrac{1}{2} \cdot 3\right)^2$

74. $\left(\dfrac{1}{2} \cdot 5\right)^2$

Applying the Concepts

75. Temperature Change The temperature is 25°F at 5:00 in the afternoon. If the temperature drops 6°F every hour after that, what is the temperature at 9:00 in the evening?

76. Investment Value Suppose you purchase $500 worth of a mutual fund and find that the value of your purchase doubles every 2 years. Write a sequence of numbers that gives the value of your purchase every 2 years for the first 10 years you own it. Is this sequence a geometric sequence?

77. Reading Charts Refer to the bar charts below to find the net change in calories for a 150-pound person who bowls for 3 hours and then eats a cheeseburger with onion and a strawberry shake.

78. Reading Charts Refer to the bar charts below to find the net change in calories for a 150-pound person who plays handball for 1.5 hours and then eats a Double-Double with onion and fries.

Source: www.in-n-out.com/nutritional info.asp

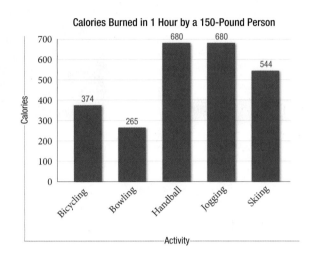

79. Using a Calculator You own 245 shares of a stock that rises $1.08 per share. How much did your investment gain in value?

80. Using a Calculator You own 548 shares of a stock that loses $2.54 per share. How much did your investment lose in value? (Hint: Answer should be a positive number.)

81. Using a Calculator You own 485 shares of stock *x* and 795 shares of stock *y*. During one week, *x* increases by $3.47 per share, while *y* loses $1.05 per share. How much did the value of your investment change over the week?

82. Using a Calculator You own 1,150 shares of stock A and 575 shares of stock B. During one week stock A increases by $2.57 per share, while stock B loses $6.40 per share. How much did the value of your investment change over the week?

Find the Mistake

Each sentence below contains a mistake. Circle the mistake and write the correct word(s) or numbers(s) on the line provided.

1. Multiplying a positive number by a negative number will give a positive answer. _____

2. Multiplying two numbers with the same sign will give a negative answer. _____

3. Simplifying the expression $(-15)(-3)(6)(-2)$ gives us 540. _____

4. The number -12 raised to the third power is equal to the number 12 raised to the third power. _____

OBJECTIVES

A Divide any combination of positive and negative numbers.

B Divide positive and negative fractions.

C Simplify expressions using the rule for order of operations.

KEY WORDS

dividing with zero

Image © Ben Barczi, 2010

In 2010, a Peruvian inventor won the $200,000 grand prize for a competition that requested its entrants to suggest a new and ingenious way to save the earth. To battle global warming, the inventor proposed covering 173 acres of dry rocky land in the Andes Mountains with a mixture made from egg whites, lime, and water. The mixture creates a whitewash that when spread over the rocks would reflect sunlight and ideally reduce temperatures, thus slowing the melting of the area's glaciers.

Suppose it took 50 gallons to cover an acre of the mountaintop. Let's say each of a five-member team used a 2-gallon bucket to transport the whitewash mixture up the slope and pour it onto the rocks. How many trips up the slope would each team member have to make for the team to cover the acre? We can represent the answer to this question with division:

$$\frac{50}{2(5)} = \frac{50}{10} = 5$$

Here is a more detailed discussion of division with positive and negative numbers.

A Dividing Positive and Negative Numbers

The last of the four basic operations is division. We will use the same approach to define division as we used for subtraction; that is, we will define division in terms of rules we already know.

Recall that we developed the rule for subtraction of real numbers by defining subtraction in terms of addition. We changed our subtraction problems to addition problems and then added to get our answers. Because we already have a rule for multiplication of real numbers, and division is the inverse operation of multiplication, we will simply define division in terms of multiplication.

We know that division by the number 2 is the same as multiplication by its reciprocal $\frac{1}{2}$; that is, 6 divided by 2 is 3, which is the same as 6 times $\frac{1}{2}$. Similarly, dividing a number by 5 gives the same result as multiplying by $\frac{1}{5}$. We can extend this idea to all real numbers with the following rule:

> **RULE** Dividing Real Numbers
>
> If a and b represent any two real numbers (b cannot be 0), then it is always true that
>
> $$a \div b = \frac{a}{b} = a\left(\frac{1}{b}\right)$$

Notice that we are using parentheses in the above rule to imply multiplication. The expression $a\left(\frac{1}{b}\right)$ can also be written as $a \cdot \frac{1}{b}$. A common mistake is to forget the parentheses and write the expression as $a\frac{1}{b}$. This would instead imply a mixed number, written as $a + \frac{1}{b}$, and would be an error.

Note We are defining division this way simply so that we can use what we already know about multiplication to do division problems. We actually want as few rules as possible. Defining division in terms of multiplication allows us to avoid writing a separate rule for division.

EXAMPLE 1 Write each division problem as an equivalent multiplication problem, and then multiply.

a. $\dfrac{6}{2}$ **b.** $\dfrac{6}{-2}$ **c.** $\dfrac{-6}{2}$ **d.** $\dfrac{-6}{-2}$

Solution

a. $\dfrac{6}{2} = 6\left(\dfrac{1}{2}\right) = 3$ *The product of two positives is positive.*

b. $\dfrac{6}{-2} = 6\left(-\dfrac{1}{2}\right) = -3$

 The product of a positive and a negative is a negative.

c. $\dfrac{-6}{2} = -6\left(\dfrac{1}{2}\right) = -3$

d. $\dfrac{-6}{-2} = -6\left(-\dfrac{1}{2}\right) = 3$ *The product of two negatives is positive.*

The second step in these examples is used only to show that we *can* write division in terms of multiplication. [In actual practice we wouldn't write $\frac{6}{2}$ as $6\left(\frac{1}{2}\right)$.] The answers, therefore, follow from the rule for multiplication; that is, like signs produce a positive answer, and unlike signs produce a negative answer.

Here are some examples. This time we will not show division as multiplication by the reciprocal. We will simply divide. If the original numbers have the same signs, the answer will be positive. If the original numbers have different signs, the answer will be negative.

EXAMPLE 2 Divide.

a. $\dfrac{12}{6} = 2$ *Like signs give a positive answer.*

b. $\dfrac{12}{-6} = -2$ *Unlike signs give a negative answer.*

c. $\dfrac{-12}{6} = -2$ *Unlike signs give a negative answer.*

d. $\dfrac{-12}{-6} = 2$ *Like signs give a positive answer.*

e. $\dfrac{15}{-3} = -5$ *Unlike signs give a negative answer.*

f. $\dfrac{-40}{-5} = 8$ *Like signs give a positive answer.*

g. $\dfrac{-14}{2} = -7$ *Unlike signs give a negative answer.*

B Division with Fractions

We can apply the definition of division with real numbers to fractions. Remember that dividing by a fraction is equivalent to multiplying by its reciprocal. For example, we can divide a number, even if it is another fraction, by the fraction $\frac{3}{4}$, by multiplying the number by the reciprocal of $\frac{3}{4}$, which is $\frac{4}{3}$.

$$\frac{2}{5} \div \frac{3}{4} = \frac{2}{5} \cdot \frac{4}{3} = \frac{8}{15}$$

Practice Problems

1. Change each problem to an equivalent multiplication problem, and then multiply to get your answer.

 a. $\dfrac{12}{4}$

 b. $\dfrac{12}{-4}$

 c. $\dfrac{-12}{4}$

 d. $\dfrac{-12}{-4}$

Note What we are saying here is that the work shown in Example 1 is shown simply to justify the answers we obtain. In the future, we won't show the middle step in these kinds of problems. Even so, we need to know that division is *defined* to be multiplication by the reciprocal.

2. Divide.

 a. $\dfrac{18}{3}$

 b. $\dfrac{18}{-3}$

 c. $\dfrac{-18}{3}$

 d. $\dfrac{-18}{-3}$

 e. $\dfrac{30}{-10}$

 f. $\dfrac{-50}{-25}$

 g. $\dfrac{-21}{3}$

Answers

1. a. 3 **b.** -3 **c.** -3 **d.** 3
2. a. 6 **b.** -6 **c.** -6 **d.** 6
 e. -3 **f.** 2 **g.** -7

3. Divide.

a. $\dfrac{3}{4} \div \dfrac{5}{7}$

b. $-\dfrac{7}{8} \div \dfrac{3}{5}$

c. $10 \div \left(-\dfrac{5}{6}\right)$

EXAMPLE 3 Divide.

a. $\dfrac{2}{3} \div \dfrac{5}{7}$ **b.** $-\dfrac{3}{4} \div \dfrac{7}{9}$ **c.** $8 \div \left(-\dfrac{4}{5}\right)$

Solution

a. $\dfrac{2}{3} \div \dfrac{5}{7} = \dfrac{2}{3} \cdot \dfrac{7}{5}$ Rewrite as multiplication by the reciprocal.

$\phantom{\dfrac{2}{3} \div \dfrac{5}{7}} = \dfrac{14}{15}$ Multiply.

b. $-\dfrac{3}{4} \div \dfrac{7}{9} = -\dfrac{3}{4} \cdot \dfrac{9}{7}$ Rewrite as multiplication by the reciprocal.

$\phantom{-\dfrac{3}{4} \div \dfrac{7}{9}} = -\dfrac{27}{28}$ Multiply.

c. $8 \div \left(-\dfrac{4}{5}\right) = \dfrac{8}{1}\left(-\dfrac{5}{4}\right)$ Rewrite as multiplication by the reciprocal.

$\phantom{8 \div \left(-\dfrac{4}{5}\right)} = -\dfrac{40}{4}$ Multiply.

$\phantom{8 \div \left(-\dfrac{4}{5}\right)} = -10$ Divide 40 by 4 to simplify.

C Using Order of Operations

The last step in each of the following examples involves reducing a fraction to lowest terms. Recall that to reduce a fraction to lowest terms, we divide the numerator and denominator by the largest number that divides each of them exactly. For example, to reduce $\frac{15}{20}$ to lowest terms, we divide 15 and 20 by 5 to get $\frac{3}{4}$.

4. Simplify.

a. $\dfrac{-6(5)}{9}$

b. $\dfrac{10}{-7-1}$

EXAMPLE 4 Simplify as much as possible.

a. $\dfrac{-4(5)}{6}$ **b.** $\dfrac{30}{-4-5}$

Solution

a. $\dfrac{-4(5)}{6} = \dfrac{-20}{6}$ Simplify numerator.

$\phantom{\dfrac{-4(5)}{6}} = -\dfrac{10}{3}$ Reduce to lowest terms by dividing numerator and denominator by 2.

b. $\dfrac{30}{-4-5} = \dfrac{30}{-9}$ Simplify denominator.

$\phantom{\dfrac{30}{-4-5}} = -\dfrac{10}{3}$ Reduce to lowest terms by dividing numerator and denominator by 3.

In the examples that follow, the numerators and denominators contain expressions that are somewhat more complicated than those we have seen thus far. To apply the rule for order of operations to these examples, we treat fraction bars the same way we treat grouping symbols; that is, fraction bars separate numerators and denominators so that each will be simplified separately.

Answers

3. a. $\dfrac{21}{20}$ **b.** $-\dfrac{35}{24}$ **c.** -12

4. a. $-\dfrac{10}{3}$ **b.** $-\dfrac{5}{4}$

EXAMPLE 5 Simplify.

a. $\dfrac{2(-3) + 4}{12}$ **b.** $\dfrac{5(-4) + 6(-1)}{2(3) - 4(1)}$

Solution

a. $\dfrac{2(-3) + 4}{12} = \dfrac{-6 + 4}{12}$ In the numerator, we multiply before we add.

$= \dfrac{-2}{12}$ Add.

$= -\dfrac{1}{6}$ Reduce to lowest terms by dividing the numerator and the denominator by 2.

b. $\dfrac{5(-4) + 6(-1)}{2(3) - 4(1)} = \dfrac{-20 + (-6)}{6 - 4}$ Multiply before adding.

$= \dfrac{-26}{2}$ Simplify numerator and denominator.

$= -13$ Divide -26 by 2.

We must be careful when we are working with expressions such as $(-5)^2$ and -5^2 that we include the negative sign with the base only when parentheses indicate we are to do so.

Unless there are parentheses to indicate otherwise, we consider the base to be only the number directly below and to the left of the exponent. If we want to include a negative sign with the base, we must use parentheses.

To simplify a more complicated expression, we follow the same rule. For example,

$$7^2 - 3^2 = 49 - 9$$ The bases are 7 and 3; the sign between the two terms is a subtraction sign.

For another example,

$$5^3 - 3^4 = 125 - 81$$ We simplify exponents first, then subtract.

EXAMPLE 6 Simplify.

a. $\dfrac{5^2 - 3^2}{-5 + 3}$ **b.** $\dfrac{(3 + 2)^2}{-3^2 - 2^2}$

Solution

a. $\dfrac{5^2 - 3^2}{-5 + 3} = \dfrac{25 - 9}{-2}$ Simplify numerator and denominator separately.

$= \dfrac{16}{-2}$

$= -8$

b. $\dfrac{(3 + 2)^2}{-3^2 - 2^2} = \dfrac{5^2}{-9 - 4}$ Simplify numerator and denominator separately.

$= \dfrac{25}{-13}$

$= -\dfrac{25}{13}$

5. Simplify.

a. $\dfrac{3(-4) + 9}{6}$

b. $\dfrac{6(-2) + 5(-3)}{5(4) - 11}$

6. Simplify.

a. $\dfrac{4^2 - 2^2}{-4 + 2}$

b. $\dfrac{(4 + 3)^2}{-4^2 - 3^2}$

Answers

5. a. $-\dfrac{1}{2}$ **b.** -3

6. a. -6 **b.** $-\dfrac{49}{25}$

Division with the Number 0

For every division problem there is an associated multiplication problem involving the same numbers. For example, the following two problems say the same thing about the numbers 2, 3, and 6:

<p style="text-align:center">Division Multiplication</p>

$$\frac{6}{3} = 2 \qquad 6 = 2(3)$$

We can use this relationship between division and multiplication to clarify division involving the number 0.

First, dividing 0 by a number other than 0 is allowed and always results in 0. To see this, consider dividing 0 by 5. We know the answer is 0 because of the relationship between multiplication and division. This is how we write it:

$$\frac{0}{5} = 0 \qquad \text{because} \qquad 0 = 0(5)$$

However, dividing a nonzero number by 0 is not allowed in the real numbers. Suppose we were attempting to divide 5 by 0. We don't know if there is an answer to this problem, but if there is, let's say the answer is a number that we can represent with the letter n. If 5 divided by 0 is a number n, then

$$\frac{5}{0} = n \qquad \text{and} \qquad 5 = n(0)$$

This is impossible, because no matter what number n is, when we multiply it by 0 the answer must be 0. It can never be 5. In algebra, we say expressions like $\frac{5}{0}$ are undefined because there is no answer to them; that is, division by 0 is not allowed in the real numbers.

The only other possibility for division involving the number 0 is 0 divided by 0. We will treat problems like $\frac{0}{0}$ as if they were undefined also. Here is a rule that summarizes this information:

> **RULE** Dividing with Zero
>
> 1. Division by 0 is not defined.
>
> $$a \div 0 \text{ or } \frac{a}{0} \text{ is not defined, for all real numbers, } a.$$
>
> 2. Zero divided by any number is 0.
>
> $$a \div 0 \text{ or } \frac{0}{a} = 0 \text{ for all real numbers } a \text{ so long as } a \neq 0.$$

> **GETTING READY FOR CLASS**
>
> *After reading through the preceding section, respond in your own words and in complete sentences.*
>
> **A.** Why do you define division in terms of multiplication?
>
> **B.** How do you divide fractions?
>
> **C.** Why is it important to use the order of operations when dividing real numbers?
>
> **D.** Why is division by 0 not allowed with real numbers?

Vocabulary Review

Choose the correct words to fill in the blanks below.

> lowest terms zero negative reciprocal positive

1. We define division as the same as multiplication by the _____.

2. When dividing two real numbers with the same sign, the answer will be _____.

3. When dividing two real numbers with different signs, the answer will be _____.

4. Simplify the numerator and denominator of a fraction separately before reducing it to _____.

5. Dividing by _____ is not defined, whereas zero divided by any non-zero number is 0.

Problems

A Find the following quotients (divide).

1. $\dfrac{8}{-4}$ **2.** $\dfrac{10}{-5}$ **3.** $\dfrac{-48}{16}$ **4.** $\dfrac{-32}{4}$ **5.** $\dfrac{-7}{21}$ **6.** $\dfrac{-25}{100}$

7. $\dfrac{-39}{-13}$ **8.** $\dfrac{-18}{-6}$ **9.** $\dfrac{-6}{-42}$ **10.** $\dfrac{-4}{-28}$ **11.** $\dfrac{0}{-32}$ **12.** $\dfrac{0}{17}$

The following problems review all four operations with positive and negative numbers. Perform the indicated operations.

13. $-3 + 12$ **14.** $5 + (-10)$ **15.** $-3 - 12$ **16.** $5 - (-10)$

17. $-3(12)$ **18.** $5(-10)$ **19.** $-3 \div 12$ **20.** $5 \div (-10)$

B Divide and reduce all answers to lowest terms.

21. $\dfrac{4}{5} \div \dfrac{3}{4}$ **22.** $\dfrac{6}{8} \div \dfrac{3}{4}$ **23.** $-\dfrac{5}{6} \div \left(-\dfrac{5}{8}\right)$ **24.** $-\dfrac{7}{9} \div \left(-\dfrac{1}{6}\right)$

25. $\dfrac{10}{13} \div \left(-\dfrac{5}{4}\right)$ **26.** $\dfrac{5}{12} \div \left(-\dfrac{10}{3}\right)$ **27.** $-\dfrac{5}{6} \div \dfrac{5}{6}$ **28.** $-\dfrac{8}{9} \div \dfrac{8}{9}$

29. $-\dfrac{3}{4} \div \left(-\dfrac{3}{4}\right)$ **30.** $-\dfrac{6}{7} \div \left(-\dfrac{6}{7}\right)$ **31.** $-\dfrac{5}{12} \div \left(-\dfrac{3}{4}\right)$ **32.** $-\dfrac{4}{5} \div \dfrac{36}{25}$

C The following problems involve more than one operation. Simplify as much as possible.

33. $\dfrac{3(-2)}{-10}$

34. $\dfrac{4(-3)}{24}$

35. $\dfrac{-5(-5)}{-15}$

36. $\dfrac{-7(-3)}{-35}$

37. $\dfrac{-8(-7)}{-28}$

38. $\dfrac{-3(-9)}{-6}$

39. $\dfrac{27}{4-13}$

40. $\dfrac{27}{13-4}$

41. $\dfrac{20-6}{5-5}$

42. $\dfrac{10-12}{3-3}$

43. $\dfrac{2\cdot5-10}{-3+9}$

44. $\dfrac{2\cdot4-8}{2+8}$

45. $\dfrac{15(-5)-25}{2(-10)}$

46. $\dfrac{10(-3)-20}{5(-2)}$

47. $\dfrac{27-2(-4)}{-3(5)}$

48. $\dfrac{20-5(-3)}{10(-3)}$

49. $\dfrac{12-6(-2)}{12(-2)}$

50. $\dfrac{3(-4)+5(-6)}{10-6}$

51. $\dfrac{5^2-2^2}{-5+2}$

52. $\dfrac{7^2-4^2}{-7+4}$

53. $\dfrac{8^2-2^2}{8^2+2^2}$

54. $\dfrac{4^2-6^2}{4^2+6^2}$

55. $\dfrac{(5+3)^2}{-5^2-3^2}$

56. $\dfrac{(7+2)^2}{-7^2-2^2}$

57. $\dfrac{(8-4)^2}{8^2-4^2}$

58. $\dfrac{(6-2)^2}{6^2-2^2}$

59. $\dfrac{-4\cdot3^2-5\cdot2^2}{-8(7)}$

60. $\dfrac{-2\cdot5^2+3\cdot2^3}{-3(13)}$

61. $\dfrac{3\cdot10^2+4\cdot10+5}{345}$

62. $\dfrac{5\cdot10^2+6\cdot10+7}{567}$

63. $\dfrac{7-[(2-3)-4]}{-1-2-3}$

64. $\dfrac{2-[(3-5)-8]}{-3-4-5}$

65. $\dfrac{6(-4)-2(5-8)}{-6-3-5}$

66. $\dfrac{3(-4)-5(9-11)}{-9-2-3}$

67. $\dfrac{3(-5-3)+4(7-9)}{5(-2)+3(-4)}$

68. $\dfrac{-2(6-10)-3(8-5)}{6(-3)-6(-2)}$

69. $\dfrac{|3-9|}{3-9}$

70. $\dfrac{|4-7|}{4-7}$

71. $\dfrac{|-2-3|}{-2-3}$

72. $\dfrac{|4-12|}{4-12}$

73. Simplify each expression.

a. $20 \div 4 \cdot 5$

b. $-20 \div 4 \cdot 5$

c. $20 \div (-4) \cdot 5$

d. $20 \div 4(-5)$

e. $-20 \div 4(-5)$

74. Simplify each expression.

a. $32 \div 8 \cdot 4$

b. $-32 \div 8 \cdot 4$

c. $32 \div (-8) \cdot 4$

d. $32 \div 8(-4)$

e. $-32 \div 8(-4)$

75. Simplify each expression.

a. $8 \div \dfrac{4}{5}$

b. $8 \div \dfrac{4}{5} - 10$

c. $8 \div \dfrac{4}{5}(-10)$

d. $8 \div \left(-\dfrac{4}{5}\right) - 10$

76. Simplify each expression.

a. $10 \div \dfrac{5}{6}$

b. $10 \div \dfrac{5}{6} - 12$

c. $10 \div \dfrac{5}{6}(-12)$

d. $10 \div \left(-\dfrac{5}{6}\right) - 12$

77. What is the quotient of -12 and -4?

78. The quotient of -4 and -12 is what number?

79. What number do we divide by -5 to get 2?

80. What number do we divide by -3 to get 4?

81. Twenty-seven divided by what number is -9?

82. Fifteen divided by what number is -3?

83. If the quotient of -20 and 4 is decreased by 3, what number results?

84. If -4 is added to the quotient of 24 and -8, what number results?

Applying the Concepts

85. **Investment** Suppose that you and 3 friends bought equal shares of an investment for a total of $15,000 and then sold it later for only $13,600. How much did each person lose?

86. **Investment** If 8 people invest $500 each in a stamp collection and after a year the collection is worth $3,800, how much did each person lose?

87. **Temperature Change** Suppose that the temperature outside is dropping at a constant rate. If the temperature is 75°F at noon and drops to 61°F by 4:00 in the afternoon, by how much did the temperature change each hour?

88. **Temperature Change** In a chemistry class, a thermometer is placed in a beaker of hot water. The initial temperature of the water is 165°F. After 10 minutes the water has cooled to 72°F. If the water temperature drops at a constant rate, by how much does the water temperature change each minute?

89. **Sewing** If $\frac{6}{7}$ yard of material is needed to make a blanket, how many blankets can be made from 12 yards of material?

90. **Manufacturing** A clothing manufacturer is making scarves that require $\frac{3}{8}$ yard for material each. How many can be made from 27 yards of material?

91. **Capacity** Suppose a bag holds exactly $\frac{1}{4}$ pound of candy. How many of these bags can be filled from 12 pounds of candy?

92. **Capacity** A certain size bottle holds exactly $\frac{4}{5}$ pint of liquid. How many of these bottles can be filled from a 20-pint container?

93. **Cooking** A man is making cookies from a recipe that calls for $\frac{3}{4}$ teaspoon of oil. If the only measuring spoon he can find is a $\frac{1}{8}$ teaspoon, how many of these will he have to fill with oil in order to have a total of $\frac{3}{4}$ teaspoon of oil?

94. **Cooking** A cake recipe calls for $\frac{1}{2}$ cup of sugar. If the only measuring cup available is a $\frac{1}{8}$ cup, how many of these will have to be filled with sugar to make a total of $\frac{1}{2}$ cup of sugar?

95. **Cartons of Milk** If a small carton of milk holds exactly $\frac{1}{2}$ pint, how many of the $\frac{1}{2}$-pint cartons can be filled from a 14-pint container?

96. **Pieces of Pipe** How many pieces of pipe that are $\frac{2}{3}$ foot long must be laid together to make a pipe 16 feet long?

97. Internet Mailing Lists A company sells products on the Internet through an e-mail list. They predict that they sell one $50 product for every 25 people on their mailing list.

 a. What is their projected revenue if their list contains 10,000 e-mail addresses?

 b. What is their projected revenue if their list contains 25,000 e-mail addresses?

 c. They can purchase a list of 5,000 e-mail addresses for $5,000. Is this a wise purchase?

98. Internet Mailing Lists A new band has a following on the Internet. They sell their CDs through an e-mail list. They predict that they sell one $15 CD for every 10 people on their mailing list.

 a. What is their projected revenue if their list contains 5,000 e-mail addresses?

 b. What is their projected revenue if their list contains 20,000 e-mail addresses?

 c. If they need to make $45,000, how many people do they need on their e-mail list?

99. South Dakota holds two weather records: the fastest temperature rise and the fastest temperature drop. This change is caused by Chinook winds, which can warm temperatures rapidly. As the wind dies down, the resulting temperature plummets. The fastest temperature rise occurred in Spearfish, SD when the temperature rose 49° in 2 minutes, while the fastest temperature dropped 47° occurred in Rapid City, SD, when the temperature dropped 47° in 5 minutes. Find the change in temperature per minute for Spearfish and Rapid City.

South Dakota

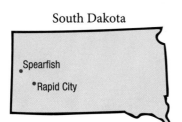

100. Montana holds two weather records for the greatest decrease and the greatest increase in temperature in 24 hours. In Browning, Montana, the temperature dropped from 44°F to −56°F in a 24–hour period. In Loma, Montana, the temperature rose from −54°F to 48°F. Both changes were the result of arctic cold fronts (one entering, one leaving). Find the change in temperature per hour for Browing and Loma.

Montana

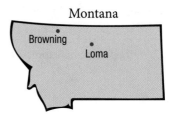

Find the Mistake

Each sentence below contains a mistake. Circle the mistake and write the correct word(s) or numbers(s) on the line provided.

1. The division problem $88 \div 4$ is the same as $4 \cdot \left(\frac{1}{88}\right)$. _____

2. Dividing two numbers with unlike signs gives a positive answer. _____

3. Dividing −55 by −5 will give us −11. _____

4. Simplifying $\frac{-3(6)}{9}$ gives us 2. _____

Landmark Review: Checking Your Progress

Give the opposite and absolute value for each of the following numbers.

1. $\dfrac{3}{4}$

2. -27

Simplify each expression.

3. $|15 - 4|$

4. $15 - 3|5 - 2 \cdot 3| - 3$

Write an equivalent algebraic expression in symbols.

5. The sum of y and 15 is 27.

6. The product of 3 and x is less than 24.

Add or subtract the following.

7. $-42 + 3.$

8. $-13 - 5$

9. $4 - 7$

10. $-52 + 37 - 15$

Which of the following equations matches with each illustration?

a.

b.

11. $x + y = 180$

12. $x + y = 90$

Simplify each of the following expressions.

13. $(-5)^2$

14. $-5(14 - 24)$

15. $\dfrac{15}{25 - 30}$

16. $\dfrac{52 - 13}{5 - 5}$

17. Thirty divided by what number is -10?

18. If the quotient of 8 and -4 is decreased by 2, what number results?

1.5 Properties of Real Numbers

KEY WORDS

- property
- commutative
- associative
- grouping
- distributive
- identity
- inverse
- symmetric

In this section, we will list all the facts (properties) that you know from past experience are true about numbers in general. We will give each property a name so we can refer to it later in this book. Mathematics is very much like a game. The game involves numbers. The rules of the game are the properties we are developing in this chapter. The goal of the game is to extend the basic rules to as many situations as possible.

You know from past experience with numbers that it makes no difference in which order you add two numbers; that is, $3 + 5$ is the same as $5 + 3$. Imagine a surfer catching some waves on a beautiful sunny day. He is able to stand up on the first 3 waves but then immediately wipes out into the water. He has successful rides on the next 5 waves. Using our knowledge of addition, we know that the surfer attempted a total of 8 waves. Even if the surfer had successful rides on the first 5 waves and wiped out on the next three, the total waves he attempted would still be 8. This is an example of the *commutative property of addition*. We say addition is a commutative operation. Changing the order of the numbers does not change the answer. This fact about numbers is one of the properties of real numbers we will cover in this section.

Multiplication is another commutative operation because $3(5)$ is the same as $5(3)$. Changing the order of the two numbers you are multiplying does not change the answer.

A The Commutative and Associative Properties

For all properties listed in this section, a, b, and c represent real numbers.

> **PROPERTY** Commutative Property of Addition
>
> *In symbols:* $a + b = b + a$
>
> *In words:* Changing the *order* of the numbers in a sum will not change the result.

> **PROPERTY** Commutative Property of Multiplication
>
> *In symbols:* $a \cdot b = b \cdot a$
>
> *In words:* Changing the *order* of the numbers in a product will not change the result.

The statement $5 + 8 = 8 + 5$ is an example of the commutative property of addition. The statement $2 \cdot y = y \cdot 2$ is an example of the commutative property of multiplication.

1. Simplify $9 + y + 4$.

> **Note** At this point, some students are confused by the expression $x + 8$; they feel that there is more to do, but they don't know what. At this point, there isn't any more that can be done with $x + 8$ unless we know what x is. So $x + 8$ is as far as we can go with this problem.

EXAMPLE 1 Simplify $5 + x + 3$.

Solution The expression $5 + x + 3$ can be simplified using the commutative property of addition.

$$5 + x + 3 = x + 5 + 3 \qquad \text{Commutative property of addition}$$
$$= x + 8 \qquad \text{Add.}$$

The other two basic operations, subtraction and division, are not commutative. The order in which we subtract or divide two numbers makes a difference in the answer.

Another property of numbers that you have used many times has to do with grouping. You know that when we add three numbers it makes no difference which two we add first. When adding $3 + 5 + 7$, we can add the 3 and 5 first and then the 7, or we can add the 5 and 7 first and then the 3. Mathematically, it looks like this: $(3 + 5) + 7 = 3 + (5 + 7)$. This property is true of multiplication as well. Operations that behave in this manner are called *associative* operations. The answer will not change when we change the association (or grouping) of the numbers.

> **PROPERTY** Associative Property of Addition
>
> *In symbols:* $a + (b + c) = (a + b) + c$
>
> *In words:* Changing the *grouping* of the numbers in a sum will not change the result.

> **PROPERTY** Associative Property of Multiplication
>
> *In symbols:* $a(bc) = (ab)c$
>
> *In words:* Changing the *grouping* of the numbers in a product will not change the result.

The following examples illustrate how the associative properties can be used to simplify expressions that involve both numbers and variables.

2. Simplify.

 a. $3 + (7 + x)$

 b. $3(4x)$

EXAMPLE 2 Simplify.

 a. $4 + (5 + x)$ **b.** $5(2x)$

Solution

 a. $4 + (5 + x) = (4 + 5) + x \qquad \text{Associative property of addition}$
$$= 9 + x \qquad \text{Add.}$$

 b. $5(2x) = (5 \cdot 2)x \qquad \text{Associative property of multiplication}$
$$= 10x \qquad \text{Multiply.}$$

3. Simplify.

 a. $\frac{1}{3}(3x)$

 b. $2\left(\frac{1}{2}x\right)$

 c. $16\left(\frac{3}{4}x\right)$

EXAMPLE 3 Simplify.

 a. $\frac{1}{5}(5x)$ **b.** $3\left(\frac{1}{3}x\right)$ **c.** $12\left(\frac{2}{3}x\right)$

Solution

 a. $\quad \frac{1}{5}(5x) = \left(\frac{1}{5} \cdot 5\right)x \qquad \text{Associative property of multiplication}$
$$= 1x \qquad \text{Multiply.}$$
$$= x$$

 b. $3\left(\frac{1}{3}x\right) = \left(3 \cdot \frac{1}{3}\right)x \qquad \text{Associative property of multiplication}$
$$= 1x \qquad \text{Multiply.}$$
$$= x$$

 c. $12\left(\frac{2}{3}x\right) = \left(12 \cdot \frac{2}{3}\right)x \qquad \text{Associative property of multiplication}$
$$= 8x \qquad \text{Multiply.}$$

B The Distributive Property

The associative and commutative properties apply to problems that are either all multiplication or all addition. There is a third basic property that involves both addition and multiplication. It is called the *distributive property* and looks like this:

> **PROPERTY** Distributive Property
>
> *In symbols:* $a(b + c) = ab + ac$
>
> *In words:* Multiplication *distributes* over addition.

You will see as we progress through the book that the distributive property is used very frequently in algebra. We can give a visual justification to the distributive property by finding the areas of rectangles. Figure 1 shows a large rectangle that is made up of two smaller rectangles. We can find the area of the large rectangle two different ways.

Method 1

We can calculate the area of the large rectangle directly by finding its length and width. The width is 5 inches, and the length is $(3 + 4)$ inches.

$$\text{Area of large rectangle} = 5(3 + 4)$$
$$= 5(7)$$
$$= 35 \text{ square inches}$$

FIGURE 1

Method 2

Because the area of the large rectangle is the sum of the areas of the two smaller rectangles, we can find the area of each small rectangle and then add to find the area of the large rectangle.

$$\begin{aligned}
\text{Area of large rectangle} &= \text{Area of rectangle I} + \text{Area of rectangle II} \\
&= \quad\quad 5(3) \quad\quad + \quad\quad 5(4) \\
&= \quad\quad\; 15 \quad\quad\;\; + \quad\quad\;\; 20 \\
&= \quad\quad 35 \text{ square inches}
\end{aligned}$$

In both cases the result is 35 square inches. Because the results are the same, the two original expressions must be equal. Stated mathematically, $5(3 + 4) = 5(3) + 5(4)$. We can either add the 3 and 4 first and then multiply that sum by 5, or we can multiply the 3 and the 4 separately by 5 and then add the products. The following examples illustrate how we use the distributive property with algebraic expressions.

EXAMPLE 4 Apply the distributive property to each expression, and then simplify.

a. $2(x + 3)$ **b.** $5(2x - 8)$ **c.** $5(2x + 4y)$ **d.** $4(2a + 3) + 8$ **e.** $-(a + 7)$

Solution

a. $2(x + 3) = 2(x) + 2(3)$ Distributive property
$ = 2x + 6$ Multiply.

b. $5(2x - 8) = 5(2x) - 5(8)$ Distributive property
$ = 10x - 40$ Multiply.

Notice in Part b that multiplication distributes over subtraction as well as addition.

c. $5(2x + 4y) = 5(2x) + 5(4y)$ Distributive property
$ = 10x + 20y$ Multiply.

d. $4(2a + 3) + 8 = 4(2a) + 4(3) + 8$ Distributive property
$ = 8a + 12 + 8$ Multiply.
$ = 8a + 20$ Add.

e. $-(a + 7) = (-1)(a) + (-1)(1)$ Distributive property
$ = -a - 7$ Multiply.

Note Because subtraction is defined in terms of addition, it is also true that the distributive property applies to subtraction as well as addition; that is, $a(b - c) = ab - ac$ for any three real numbers a, b, and c.

Note Students sometimes mistake $a(bc)$ as $ab \cdot ac$, which is not true. The expression $ab \cdot ac$ *simplifies to* a^2bc, which is not equal to the original expression. The distributive property states that multiplication distributes over addition, not multiplication.

4. Apply the distributive property to each expression, and then simplify the result.

a. $3(x + 2)$

b. $4(2x - 5)$

c. $3(7x - 6y)$

d. $5(7a - 3) + 2$

e. $-(b - 2)$

Note The negative sign in Example 4 part d is the same as distributing a -1 over the sum.

Answers

4. a. $3x + 6$ **b.** $8x - 20$
 c. $21x - 18y$ **d.** $35a - 13$
 e. $-b + 2$

5. Apply the distributive property to the expression, and then simplify the result.

$$\frac{1}{3}(3x + 6)$$

Next, we have some expressions to simplify that involve fractions.

EXAMPLE 5 Apply the distributive property to each expression, and then simplify the result.

a. $\frac{1}{2}(3x + 6)$ **b.** $3\left(\frac{1}{3}x + 5\right)$ **c.** $a\left(1 + \frac{1}{a}\right)$ **d.** $12\left(\frac{2}{3}x + \frac{1}{2}y\right)$

Solution

a. $\frac{1}{2}(3x + 6) = \frac{1}{2}(3x) + \frac{1}{2}(6)$ Distributive property

$\qquad\qquad = \frac{3}{2}x + 3$ Multiply.

b. $3\left(\frac{1}{3}x + 5\right) = 3 \cdot \frac{1}{3}x + 3 \cdot 5$ Distributive property

$\qquad\qquad = x + 15$ Multiply.

c. $a\left(1 + \frac{1}{a}\right) = a \cdot 1 + a \cdot \frac{1}{a}$ Distributive property

$\qquad\qquad = a + 1$ Multiply.

d. $12\left(\frac{2}{3}x + \frac{1}{2}y\right) = 12 \cdot \frac{2}{3}x + 12 \cdot \frac{1}{2}y$ Distributive property

$\qquad\qquad = 8x + 6y$ Multiply.

Special Numbers

In addition to the three properties mentioned so far, we want to include in our list two special numbers that have unique properties. They are the numbers 0 and 1.

PROPERTY Additive Identity Property

There exists a unique number 0 such that

In symbols: $a + 0 = a$ and $0 + a = a$

In words: Adding zero to any number does not change that value of the number.

PROPERTY Multiplicative Identity Property

There exists a unique number 1 such that

In symbols: $a(1) = a$ and $(1)a = a$

In words: Multiplying any number by 1 does not change the value of the number.

PROPERTY Additive Inverse Property

In symbols: $a + -a = 0$

In words: Opposites add to 0.

PROPERTY Multiplicative Inverse Property

For every real number a, except 0, there exists a unique real number $\frac{1}{a}$ such that

In symbols: $a\left(\frac{1}{a}\right) = 1$

In words: Reciprocals multiply to 1.

Answer

5. $x + 2$

C Identifying Properties

A final property we would like to mention in this chapter is the *symmetric property*. This property allows us to move our variable in an equation from the right-hand side of the equals sign to the left-hand side once we have solved the equation.

> **PROPERTY** Symmetric Property
>
> *In symbols:* $a = b$, then $b = a$
>
> *In words:* If one quantity equals another quantity, then the second quantity also equals the first.

Of all the basic properties listed, the commutative, associative, and distributive properties are the ones we will use most often. They are important because they will be used as justifications or reasons for many of the things we will do.

The following example illustrates how we use the preceding properties. Each sub-example contains an algebraic expression that has been changed in some way. The property that justifies the change is written to the right.

EXAMPLE 6 State the property that justifies the given statement.

a. $x + 5 = 5 + x$ — Commutative property of addition

b. $(2 + x) + y = 2 + (x + y)$ — Associative property of addition

c. $6(x + 3) = 6x + 18$ — Distributive property

d. $2 + (-2) = 0$ — Additive inverse property

e. $3\left(\dfrac{1}{3}\right) = 1$ — Multiplicative inverse property

f. $(2 + 0) + 3 = 2 + 3$ — Additive identity property

g. $(2 + 3) + 4 = 3 + (2 + 4)$ — Commutative and associative properties of addition

h. $(x + 2) + y = (x + y) + 2$ — Commutative and associative properties of addition

i. If $\dfrac{3}{4} = x$, then $x = \dfrac{3}{4}$. — Symmetric property

As a final note on the properties of real numbers, we should mention that although some of the properties are stated for only two or three real numbers, they hold for as many numbers as needed. For example, the distributive property holds for expressions like $3(x + y + z + 5 + 2)$; that is,

$$3(x + y + z + 5 + 2) = 3x + 3y + 3z + 15 + 6$$

It is not important how many numbers are contained in the sum, only that it is a sum. Multiplication, you see, distributes over addition, whether there are two numbers in the sum or 200.

> **GETTING READY FOR CLASS**
>
> *After reading through the preceding section, respond in your own words and in complete sentences.*
>
> **A.** What is the commutative property of addition?
>
> **B.** Write the commutative property of multiplication in symbols and words.
>
> **C.** Use the associative property of multiplication to simplify $4(3x)$.
>
> **D.** How do you use the distributive property to find the area of a rectangle?

6. State the property that justifies the given statement.
a. $5 \cdot 4 = 4 \cdot 5$
b. $x + 7 = 7 + x$
c. $(3 + a) + b = 3 + (a + b)$
d. $6(x + 5) = 6x + 30$
e. $4\left(\dfrac{1}{4}\right) = 1$
f. $8 + (-8) = 0$
g. $(x + 2) + 3 = x + (2 + 3)$
h. $5(1) = 5$
i. If $\dfrac{2}{3} = y$, then $y = \dfrac{2}{3}$.

Answers
6. a. Commutative property of multiplication
b. Commutative property of addition
c. Associative property of addition
d. Distributive property
e. Multiplicative inverse
f. Additive inverse
g. Associative property of addition
h. Multiplicative identity
i. Symmetric property

Vocabulary Review

Match the following definitions on the left with their property names on the right.

1. $a + b = b + a$ — e

2. $a \times b = b \times a$ — b

3. $a + (b + c) = (a + b) + c$ — d

4. $a(bc) = (ab)c$ — h

5. $a(b + c) = ab + ac$ — c

6. $a + 0 = a$ and $0 + a = a$ — f

7. $a(1) = a$ and $1(a) = a$ — j

8. $a + (-a) = 0$ — i

9. $a\left(\frac{1}{a}\right) = 1$ — a

10. $a = b$ then $b = a$

a. Multiplicative inverse property
b. Commutative property of multiplication
c. Distributive property
d. Associative property of addition
e. Commutative property of addition
f. Additive identity property
g. Symmetric property
h. Associative property of multiplication
i. Additive inverse property
j. Multiplicative identity property

Problems

A Use the associative property to rewrite each of the following expressions, and then simplify the result.

1. $4 + (2 + x)$

2. $5 + (6 + x)$

3. $(x + 2) + 7$

4. $(x + 8) + 2$

5. $3(5x)$

6. $5(3x)$

7. $9(6y)$

8. $6(9y)$

9. $\frac{1}{2}(3a)$

10. $\frac{1}{3}(2a)$

11. $\frac{1}{3}(3x)$

12. $\frac{1}{4}(4x)$

13. $\frac{1}{2}(2y)$

14. $\frac{1}{7}(7y)$

15. $\frac{3}{4}\left(\frac{4}{3}x\right)$

16. $\frac{3}{2}\left(\frac{2}{3}x\right)$

17. $\frac{6}{5}\left(\frac{5}{6}a\right)$

18. $\frac{2}{5}\left(\frac{5}{2}a\right)$

19. $3(5xy)$

20. $12(3ab)$

21. $15(2xy)$

22. $5(4cd)$

B Apply the distributive property to each of the following expressions. Simplify when possible.

23. $8(x + 2)$

24. $5(x + 3)$

25. $8(x - 2)$

26. $5(x - 3)$

27. $4(y + 1)$

28. $4(y - 1)$

29. $3(6x + 5)$

30. $3(5x + 6)$

31. $2(3a + 7)$

32. $5(3a + 2)$

33. $9(6y - 8)$

34. $2(7y - 4)$

35. $\frac{1}{3}(3x + 6)$

36. $\frac{1}{2}(2x + 4)$

37. $-3(x - 4y)$

38. $-4(3a - 5b)$

39. $6(2x + 3y)$

40. $8(3x + 2y)$

41. $4(3a - 2b)$

42. $5(4a - 8b)$

43. $\frac{1}{2}(6x + 4y)$

44. $\frac{1}{3}(6x + 9y)$

45. $4(a + 4) + 9$

46. $6(a + 2) + 8$

47. $-2(2a - 4) + 3$

48. $-5(3x - 2) + 5$

49. $2(3x + 5) + 2$

50. $7(2x + 1) + 3$

51. $7(2x + 4) + 10$

52. $3(5x + 6) + 20$

53. $2(4x - 5) - 7$

54. $3(5x - 7) - 9$

Apply the distributive property and simplify, if possible.

55. $\frac{1}{2}(4x + 2)$

56. $\frac{1}{3}(6x + 3)$

57. $\frac{3}{4}(8x - 4)$

58. $\frac{2}{5}(5x + 10)$

59. $-\frac{1}{4}(8x - 16)$

60. $-\frac{3}{5}(15x + 5)$

61. $\frac{5}{6}(6x + 12)$

62. $\frac{2}{3}(9x - 3)$

63. $10\left(\dfrac{3}{5}x + \dfrac{1}{2}\right)$

64. $8\left(\dfrac{1}{4}x - \dfrac{5}{8}\right)$

65. $-12\left(\dfrac{1}{4}x + \dfrac{2}{3}\right)$

66. $-15\left(\dfrac{2}{3}a - \dfrac{3}{5}\right)$

67. $15\left(\dfrac{1}{3}x + \dfrac{2}{5}\right)$

68. $12\left(\dfrac{1}{12}m + \dfrac{1}{6}\right)$

69. $12\left(\dfrac{1}{2}m - \dfrac{5}{12}\right)$

70. $8\left(\dfrac{1}{8} + \dfrac{1}{2}m\right)$

71. $21\left(\dfrac{1}{3} + \dfrac{1}{7}x\right)$

72. $6\left(\dfrac{3}{2}y + \dfrac{1}{3}\right)$

73. $6\left(\dfrac{1}{2}x - \dfrac{1}{3}y\right)$

74. $12\left(\dfrac{1}{4}x + \dfrac{2}{3}y\right)$

75. $0.09(x + 2{,}000)$

76. $0.04(x + 7{,}000)$

77. $0.12(x + 500)$

78. $0.06(x + 800)$

79. $a\left(1 + \dfrac{1}{a}\right)$

80. $a\left(1 - \dfrac{1}{a}\right)$

81. $a\left(\dfrac{1}{a} - 1\right)$

82. $a\left(\dfrac{1}{a} + 1\right)$

C State the property or properties that justify the following.

83. $3 + 2 = 2 + 3$

84. $5 + 0 = 5$

85. $4\left(\dfrac{1}{4}\right) = 1$

86. $10(0.1) = 1$

87. $4 + x = x + 4$

88. $3(x - 10) = 3x - 30$

89. $2(y + 8) = 2y + 16$

90. $3 + (4 + 5) = (3 + 4) + 5$

91. $(3 + 1) + 2 = 1 + (3 + 2)$

92. $(5 + 2) + 9 = (2 + 5) + 9$

93. $(8 + 9) + 10 = (8 + 10) + 9$

94. $(7 + 6) + 5 = (5 + 6) + 7$

95. $3(x + 2) = 3(2 + x)$

96. $2(7y) = (7 \cdot 2)y$

97. If $\dfrac{4}{5} = a$, then $a = \dfrac{4}{5}$.

98. $a(5b) = 5(ab)$

99. $4(xy) = 4(yx)$

100. $3[2 + (-2)] = 3(0)$

101. $8[7 + (-7)] = 8(0)$

102. $7(1) = 7$

103. $\left(\dfrac{4}{3}\right)\left(\dfrac{3}{4}\right) = 1$

104. $\left(\dfrac{2}{3}\right)\left(\dfrac{3}{2}\right) = 1$

105. If $\dfrac{1}{8} = b$, then $b = \dfrac{1}{8}$.

106. $7 - 12 = -12 + 7$

Each of the following problems has a mistake in it. Correct the right-hand side.

107. $3(x + 2) = 3x + 2$

108. $5(4 + x) = 4 + 5x$

109. $9(a + b) = 9a + b$

110. $2(y + 1) = 2y + 1$

111. $3(0) = 3$

112. $5\left(\dfrac{1}{5}\right) = 5$

113. $3 + (-3) = 1$

114. $8(0) = 8$

115. $10(1) = 0$

116. $3 \cdot \dfrac{1}{3} = 0$

117. $\dfrac{2}{3}\left(\dfrac{3}{2}\right) = 0$

118. $6x - 6x = x$

Applying the Concepts

119. Getting Dressed While getting dressed for work, a man puts on his socks and puts on his shoes. Are the two statements "put on your socks" and "put on your shoes" commutative? That is, will changing the order of the events always produce the same result?

120. Skydiving A skydiver flying over the jump area is about to do two things: jump out of the plane and pull the rip cord. Are the two events "jump out of the plane" and "pull the rip cord" commutative?

121. Division Give an example that shows that division is not a commutative operation; that is, find two numbers for which changing the order of division gives two different answers.

122. Subtraction Simplify the expression $10 - (5 - 2)$ and the expression $(10 - 5) - 2$ to show that subtraction is not an associative operation.

123. Hours Worked Carlo works as a waiter. He works double shifts 4 days a week. The lunch shift is 2 hours and the dinner shift is 3 hours. Find the total number of hours he works per week using the numbers 2, 3, and 4. Do the calculation two different ways so that the results give further justification for the distributive property.

124. Jose earns $8 per hour regular pay and $12 per hour working overtime. He is paid overtime if he works more than 40 hours in a week. Find his pay if Jose works each of these weeks:

Week	Hours	Pay
1	36	
2	54	
3	60	

125. Quantitative Literacy and You What is quantitative literacy? Here is what the *Mathematical Association of America* has to say about quantitative literacy:

.... every college graduate should be able to apply simple mathematical methods to the solution of real-world problems. A quantitatively literate college graduate should be able to:

a. Interpret mathematical models such as formulas, graphs, tables, and schematics, and draw inferences from them.

b. Represent mathematical information symbolically, visually, numerically, and verbally.

c. Estimate and check answers to mathematical problems in order to determine reasonableness, identify alternatives, and select optimal results.

d. Recognize that mathematical and statistical methods have limits.

e. List two specific skills that you would like to acquire from this course that will improve your quantitative literacy. For example, you may want to better understand the financing options available for purchasing a new car.

Find the Mistake

Each sentence below contains a mistake. Circle the mistake and write the correct word(s) or numbers(s) on the line provided.

1. To simplify the problem $8\left(\frac{4}{7}x\right)$, we use the distributive property. _____

2. The distributive property states that multiplication distributes over division. _____

3. The additive inverse property states that $a(1) = a$ and $1(a) = a$. _____

4. The expression $35\left(\frac{2}{3}y + 18z\right)$ is equal to $35\left(\frac{2}{3}y\right) + 35(18z)$ because of the associative property of multiplication.

1.6 Simplifying Expressions

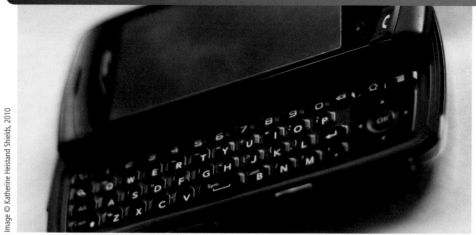

OBJECTIVES

A Simplify expressions by combining similar terms.

B Simplify expressions involving parentheses.

C Calculate the value of an expression for a given value of the variable.

KEY WORDS

expression

term

similar terms

parentheses

value of an expression

If a cellular phone company charges $50 per month plus $30 for a data package and $0.10 for each text, then the amount of your monthly bill is given by the expression $50 + 30 + 0.10t$. To find the amount you will pay in one month if you send 100 texts, you substitute 100 for t and simplify the resulting expression. For example,

$$\text{Monthly bill} = 50 + 30 + 0.10(100)$$
$$90 = 50 + 30 + 10 \quad \text{in dollars}$$

In this section, we will learn how to simplify algebraic expressions similar to this.

A Combining Similar Terms

An *expression* is a number and/or variable combined with math operations used to represent a number. Examples include

$$-7y, \quad x + 7, \quad 2t - 9, \quad 4st - \frac{1}{4}s$$

Notice in the expressions above, "terms" are being added or subtracted to form the expression. A *term* is a number or a number and one or more variables multiplied together. Some terms in the expressions above include x, $2t$, and $-\frac{1}{4}s$.

> **DEFINITION** similar terms
>
> Two terms with the same variable part (same variables raised to the same powers) are called *similar* (or *like*) *terms.*

The terms $3x$ and $4x$ are similar because their variable parts are identical. Likewise, the terms $18y$, and $-10y$ are similar terms as are $3x^2$ and $-4x^2$.

To simplify an algebraic expression, we simply reduce the number of terms in the expression. We accomplish this by applying the distributive property along with our knowledge of addition and subtraction of positive and negative real numbers. The following examples illustrate the procedure.

EXAMPLE 1 Simplify by combining similar terms.

a. $3x + 4x$ **b.** $7a - 10a$ **c.** $18y - 10y + 6y$

Solution We combine similar terms by applying the distributive property.

a. $3x + 4x = (3 + 4)x$ Distributive property
$ = 7x$ Add 3 and 4.

b. $7a - 10a = (7 - 10)a$ Distributive property
$ = -3a$ Add 7 and -10.

c. $18y - 10y + 6y = (18 - 10 + 6)y$ Distributive property
$ = 14y$ Add $18, -10,$ and 6.

When the expression we intend to simplify is more complicated, we use the commutative and associative properties first.

EXAMPLE 2 Simplify each expression.

a. $3x + 5 + 2x - 3$ **b.** $4a - 7b - 2a + 3b$ **c.** $5x + \dfrac{1}{2}y - x - \dfrac{3}{2}y$

Solution We combine similar terms by applying the distributive property.

a. $3x + 5 + 2x - 3 = 3x + 2x + 5 - 3$ Commutative property
$ = (3x + 2x) + (5 - 3)$ Associative property
$ = (3 + 2)x + (5 - 3)$ Distributive property
$ = 5x + 2$

b. $4a - 7b - 2a + 3b = (4a - 2a) + (-7b + 3b)$ Commutative and associative properties
$ = (4 - 2)a + b(-7 + 3)$ Distributive property
$ = 2a - 4b$

c. $5x + \dfrac{1}{2}y - x - \dfrac{3}{2}y = (5x - x) + \dfrac{1}{2}y - \dfrac{3}{2}y$ Commutative and associative properties
$\phantom{5x + \frac{1}{2}y - x - \frac{3}{2}y} = (5 - 1)x + \left(\dfrac{1}{2} - \dfrac{3}{2}\right)y$ Distributive property
$\phantom{5x + \frac{1}{2}y - x - \frac{3}{2}y} = 4x - y$

Notice that in each case the result has fewer terms than the original expression. Because there are fewer terms, the resulting expression is said to be simpler than the original expression.

B Simplifying Expressions Involving Parentheses

If an expression contains parentheses, it is often necessary to apply the distributive property to remove the parentheses before combining similar terms.

EXAMPLE 3 Simplify the expression $5(2x - 8) - 3$.

Solution We begin by distributing the 5 across $2x - 8$. We then combine similar terms.

$$5(2x - 8) - 3 = 10x - 40 - 3 \qquad \text{Distributive property}$$
$$ = 10x - 43 \qquad \text{Combine similar terms.}$$

EXAMPLE 4 Simplify $7 - 3(2y + 1)$.

Solution By the rule for order of operations, we must multiply before we add or subtract. For that reason, it would be incorrect to subtract 3 from 7 first. Instead, we multiply -3 and $2y + 1$ to remove the parentheses and then combine similar terms.

$$7 - 3(2y + 1) = 7 - 6y - 3 \quad \text{Distributive property}$$
$$= -6y + 4 \quad \text{Combine similar terms.}$$

4. Simplify $5 - 2(4y + 1)$.

EXAMPLE 5 Simplify $5(x - 2) - (3x + 4)$.

Solution We begin by applying the distributive property to remove the parentheses. The expression $-(3x + 4)$ can be thought of as $-1(3x + 4)$. Thinking of it in this way allows us to apply the distributive property.

$$-1(3x + 4) = -1(3x) + (-1)(4)$$
$$= -3x - 4$$

The complete solution looks like this:

$$5(x - 2) - (3x + 4) = 5x - 10 - 3x - 4 \quad \text{Distributive property}$$
$$= 2x - 14 \quad \text{Combine similar terms.}$$

5. Simplify $6(x - 3) - (7x + 2)$.

As you can see from the explanation in Example 5, we use the distributive property to simplify expressions in which parentheses are preceded by a negative sign. In general, we can write

$$-(a + b) = -1(a + b)$$
$$= -a + (-b)$$
$$= -a - b$$

The negative sign outside the parentheses ends up changing the sign of each term within the parentheses. In words, we say "the opposite of a sum is the sum of the opposites."

C The Value of an Expression

An algebraic expression like $3x + 2$ has a certain value depending on what number we assign to x. For instance, when x is 4, $3x + 2$ becomes $3(4) + 2$, or 14. When x is -8, $3x + 2$ becomes $3(-8) + 2$, or -22. The value of an expression is found by replacing or substituting the variable with a given number.

EXAMPLE 6 Find the value of the following expressions by replacing the variable with the given number.

6. Find the following:
 a. The value of $4x - 7$ when $x = 3$
 b. The value of $2a + 4$ when $a = -5$
 c. The value of $2x - 5 + 6x$ when $x = -5$
 d. The value of $7x - 3 - 4x$ when $x = 10$
 e. The value of $y^2 - 10y + 25$ when $y = -2$

Expression	The Variable	Value of the Expression
a. $3x - 1$	$x = 2$	$3(2) - 1 = 6 - 1 = 5$
b. $-7a + 4$	$a = -3$	$-7(-3) + 4 = 21 + 4 = 25$
c. $2x - 3 + 4x$	$x = -1$	$2(-1) - 3 + 4(-1) = -2 - 3 + (-4)$ $= -9$
d. $2x - 5 - 8x$	$x = 5$	$2(5) - 5 - 8(5) = 10 - 5 - 40$ $= -35$
e. $y^2 - 6y + 9$	$y = 4$	$4^2 - 6(4) + 9 = 16 - 24 + 9 = 1$

Answers
4. $-8y + 3$
5. $-x - 20$
6. a. 5 **b.** -6 **c.** -45 **d.** 27
 e. 49

Simplifying an expression should not change its value; that is, if an expression has a certain value when x is 5, then it will always have that value, no matter how much it has been simplified, as long as x is 5. If we were to simplify the expression in Example 6d first, it would look like

$$2x - 5 - 8x = -6x - 5$$

When x is 5, the simplified expression $-6x - 5$ is

$$-6(5) - 5 = -30 - 5 = -35$$

It has the same value as the original expression when x is 5.

We also can find the value of an expression that contains two variables if we know the values for both variables.

7. Find the value of $5a - 3b - 2$ when $a = 3$ and $b = 4$.

EXAMPLE 7 Find the value of the expression $2x - 3y + 4$ when $x = -5$ and $y = 6$.

Solution Substituting -5 for x and 6 for y, the expression becomes

$$2(-5) - 3(6) + 4 = -10 - 18 + 4$$
$$= -28 + 4$$
$$= -24$$

8. Find the value of $x^2 - 4xy - 4y^2$ when $x = 5$ and $y = 3$.

EXAMPLE 8 Find the value of the expression $x^2 - 2xy - y^2$ when $x = 3$ and $y = -4$.

Solution Replacing each x in the expression with the number 3 and each y in the expression with the number -4 gives us

$$3^2 - 2(3)(-4) - (-4)^2 = 9 - 2(3)(-4) - 16$$
$$= 9 - (-24) - 16$$
$$= 33 - 16$$
$$= 17$$

9. Find the value of $-7 - 5x + 9(x + y)$ when $x = 3$ and $y = -\frac{1}{3}$.

EXAMPLE 9 Find the value of the expression $-11 - 3x + 2(x + y)$ when $x = -3$ and $y = \frac{1}{2}$.

Solution Replacing x in the expression with -3, and replacing y with $\frac{1}{2}$ gives us.

$$-11 - 3(-3) + 2\left[(-3) + \frac{1}{2}\right] = -11 + 9 - 6 + 1$$
$$= -2 - 6 + 1$$
$$= -8 + 1$$
$$= -7$$

> ### GETTING READY FOR CLASS
>
> *After reading through the preceding section, respond in your own words and in complete sentences.*
>
> **A.** What is a term?
>
> **B.** What are similar terms?
>
> **C.** Explain how you would use the distributive property to combine similar terms.
>
> **D.** Explain how you would find the value of $5x + 3$ when x is 6.

Answers

7. -29

8. 1

9. 2

Vocabulary Review

Choose the correct words to fill in the blanks below.

parentheses similar expression value term variable

1. An _____ is a number and/or variable combined with math operations used to represent a number.
2. A _____ in an expression is a number or a number and one or more variables multiplied together.
3. Two monomials with the same variables raised to the same powers are called _____ terms.
4. The _____ of an expression is found by replacing or substituting the _____ with a given number.
5. To simplify an expression that contains _____, it may be necessary to apply the distributive property to remove the parentheses before combining similar terms.

Problems

A Simplify the following expressions.

1. $3x - 6x$

2. $7x - 5x$

3. $-2a + a$

4. $3a - a$

5. $7x + 3x + 2y$

6. $8x - 2y - y$

7. $3a - 2b + 5a$

8. $7a + 3b + 6c$

9. $4x - 3 + \frac{1}{2}x$

10. $5x + 6 - 3x$

11. $\frac{3}{5}a + 5 + \frac{2}{5}a$

12. $6a + 7b + 8 - 3b$

13. $2x - 3 + 3x - 2$

14. $6x + 5 - 2x + 3 - 4x$

15. $3a - 1 + a + 3$

16. $-a + 2 + 8a - 7$

17. $-4x + 8 - 5x - 10$

18. $-\frac{1}{3}a - 1 - \frac{1}{2}a + 5$

19. $0.25b - 11 + 0.50b$

20. $8 - 0.75x - 15 - 2.30x$

21. $0.08x + 0.09x$

22. $0.04x + 0.05x$

23. $0.04x - 0.2x$

24. $0.08x - 0.3x$

SCAN TO ACCESS

B Simplify the following expressions.

25. $5(2x - 1) + 4$

26. $2(4x - 3) + 2$

27. $7(3y + 2) - 8$

28. $6(4y + 2) - 7$

29. $-3(2x - 1) + 5$

30. $-4(3x - 2) - 6x$

31. $5a - 2(a + 1)$

32. $7 - 8(2a + 3a)$

33. $6 - 4(x - 5)$

34. $12 - 3\left(\dfrac{1}{3}x - 2\right)$

35. $-9 - (2 - y) + 1$

36. $-10 - 2\left(\dfrac{1}{2} - \dfrac{3}{2}y\right) + 3y$

37. $-6x + 2(2 - 3x) + x$

38. $-7 - (3 + x) - 1$

39. $(4x - 7) - (2x + 5)$

40. $(7x - 3) - (4x + 2)$

41. $8(2a + 4) - (6a - 1)$

42. $9(3a + 5) - (8a - 7)$

43. $0.10(x - 2) + (x - 3)$

44. $2(2x + 1) - (x + 4) + x$

45. $4(2y - 8) - 2(y + 7)$

46. $5(y - 3) - y(4 - 2)$

47. $-(2x + 1) - \dfrac{1}{2}(x + 5)$

48. $-3(3x - 2) - (2x + 3)$

49. $(x + y) + (x - y)$

50. $2(x - 2) + 3(5x)$

51. $x + (x + 3)(-3)$

52. $x - 2(x + 2)$

53. $(x + 3y) + 3(2x - y)$

54. $(2x - y) - 2(x + 3y)$

55. $-6\left(\frac{1}{2}x - \frac{1}{3}y\right) + 12\left(\frac{1}{4}x + \frac{2}{3}y\right)$

56. $6\left(\frac{1}{3}x + \frac{1}{2}y\right) - 4\left(x + \frac{3}{4}y\right)$

57. $0.08x + 0.09(x + 2,000)$

58. $0.06x + 0.04(x + 7,000)$

59. $0.12x + 0.10(4,000 - x)$

60. $0.08x + 0.07(10,000 - x)$

C Evaluate the following expressions when $x = 2$. (Find the value of the expressions if x is 2.)

61. $3x - 1$

62. $4x + 3$

63. $-2x - 5$

64. $-3x + 6$

65. $x^2 - 8x + 16$

66. $x^2 - 10x + 25$

67. $(x - 4)^2$

68. $(x - 5)^2$

Evaluate the following expressions when $x = -5$. Then simplify the expression, and check to see that it has the same value for $x = -5$.

69. $7x - 4 - x - 3$

70. $3x + 4 + 7x - 6$

71. $5(2x + 1) + 4$

72. $2\left(\frac{1}{2}x - 10\right) + 5$

Evaluate the following expressions when $x = -3$ and $y = 5$.

73. $x^2 - 2xy + y^2$

74. $x^2 + 2xy + y^2$

75. $(x - y)^2$

76. $(y - x)^2$

77. $x^2 + 6xy + 9y^2$

78. $x^2 - 10xy + 25y^2$

79. $(x + 3y)^2$

80. $(x - 5y)^2$

Find the value of $b^2 - 4ac$ for the given values of a, b, and c. (You will see these types of problems later in the book.)

81. $a = 1, b = -5, c = -6$ **82.** $a = 1, b = -6, c = 7$ **83.** $a = 2, b = 4, c = -3$ **84.** $a = 3, b = 4, c = -2$

Find the value of $12x - 3$ for each of the following values of x.

85. $\dfrac{1}{2}$ **86.** $\dfrac{1}{3}$ **87.** $\dfrac{1}{4}$ **88.** $\dfrac{1}{6}$

89. $\dfrac{3}{2}$ **90.** $\dfrac{2}{3}$ **91.** $\dfrac{3}{4}$ **92.** $\dfrac{5}{6}$

Applying the Concepts

93. Temperature and Altitude If the temperature on the ground is 70°F, then the temperature at A feet above the ground can be found from the expression $-0.0035A + 70$. Find the temperature at the following altitudes.

 a. 8,000 feet

 b. 12,000 feet

 c. 24,000 feet

 d. 29,035 feet (height of Mt. Everest)

94. Perimeter of a Rectangle The expression $2l + 2w$ gives the perimeter of a rectangle with length l and width w. Find the perimeter of the rectangles with the following lengths and widths.

 a. Length = 8 meters, width = 5 meters

8 m

 b. Length = 10 feet, width = 3 feet

10 ft

95. Cellular Phone Rates A cellular phone company has a plan that charges $50 per month plus $0.10 for each text that you send. The expression $50 + 0.10t$ gives the amount of money you will pay for sending t text messages a month. Find the monthly bill.

 a. 20 text messages in a month

 b. 50 text messages in a month

 c. 100 text messages in a month

96. Cost of Bottled Water A water bottling company charges $7.00 per month for their water dispenser and $1.10 for each gallon of water delivered. If you have g gallons of water delivered in a month, then the expression $7 + 1.1g$ gives the amount of your bill for that month. Find the monthly bill for each of the following deliveries.

 a. 10 gallons

 b. 20 gallons

 c. 30 gallons

97. The total pressure as you dive below the surface of the ocean increases by the formula $P = 14.7 + 0.4x$, where x is the depth (in feet) and P is the pressure (in lb/in²). Find the total pressure for each of the following diving depths.

a. 15 feet

b. 25 feet

c. 50 feet

98. A water company charges a flat monthly rate of $7.50 plus $0.0058 per cubic foot of water used. Find the total charges for each of the following monthly usages.

a. 4,000 ft³

b. 12,000 ft³

c. 20,000 ft³

99. Taxes You work H hours a day and you pay 21% of your income for federal income tax and another 8% for miscellaneous deductions. This is represented by the expression $0.21H + 0.08H$. Simplify this expression. If you work 8 hours a day under these conditions, how many hours do you work to pay for your taxes and miscellaneous deductions?

100. Apartment Furniture The chart shows the cost of some furniture items. Suppose you and some roommates are moving in to an apartment. Write an expression to answer the following questions, then simplify that expression.

Item	Cost
Sofa	$180
Sphere Chair	$40
Bookcase	$60
Computer Desk	$140
Storage Drawers	$11

a. You are buying sphere chairs and will get $10 off each one you buy.

b. You are buying computer desks and will get $20 off each one. You will also get a $50 rebate on the total purchase.

c. You are buying 2 sofas and are getting a discount of $x off each one. You will also be getting a $100 rebate off the total purchase.

Perimeter If the perimeter of a figure is defined as the sum of the outside lengths, find the perimeter of the following figures. Simplify your answer.

101.

102.

Find the Mistake

Each sentence below contains a mistake. Circle the mistake and write the correct word(s) or numbers(s) on the line provided.

1. Simplifying the expression $12z + 9z - 3z$ by combining similar terms gives us $21z^2 - 3z$. _____

2. The value of the expression $8y + \frac{1}{3}y$ when $y = 3$ is $14\frac{1}{3}$. _____

3. To simplify the expression $-3(5x - 4) - 8x$, first subtract $8x$ from $5x$ to combine similar terms.

4. The value of the expression $(2x - 4)^2$ when $x = \frac{1}{2}$ is the same as the value when $x = 4$. _____

Algebraic Fiction

In this chapter, we discussed variables and algebraic expressions using an intuitive approach. To reinforce that discussion, let's learn about a 20th century author named William Wallace Cook (1867-1933). As a prolific author, Cook contributed many reputable works to literary history. He once wrote fifty-four books in a single year, and was subsequently nicknamed "the man who deforested Canada." In 1928, he used his experience to catalogue possible plotlines in a single source entitled Plotto: The Master Book of All Plots. In Plotto, Cook listed every narrative he could think of, reaching a final count of 1,462. His simplest plots consisted of a character (A) who was connected to another character (B) by a clause (C). For example, A (female character) and B (male character) meet for the first time in a math class (C). For our purposes, we could translate Cook's format into an algebraic expression, such as one variable (A) is connected to another variable (B) by an operation (e.g., $+, -, \times, \div$, etc.), or $A + B$.

1. Let's extend Cook's approach for writing fiction and apply it to some algebraic expressions of our own. First, let's assign a character description to each of the following variables.

 Example:
 $A = Professor$ $B = Student$ $C = Scientist$ $D = Politician$
 $E = Musician$ $F = Writer$ $G = Waiter$

 A _____ E _____

 B _____ F _____

 C _____ G _____

 D _____

2. Insert the values (characters) you have assigned each variable to each of the following expressions.

 Example:
 $A + B + C = professor + student + scientist$
 $D + E - G = politician + musician - waiter$
 $A \cdot D + E \cdot F = professor \cdot politician + musician \cdot writer$
 $F \div C + B \cdot G = writer \div scientist + student \cdot waiter$

 $A + B + C$ _____

 $D + E - G$ _____

 $A \cdot D + E \cdot F$ _____

 $F \div C + B \cdot G$ _____

3. Replace the operation symbols with a corresponding relationship between the characters; in other words, insert an action or interaction between your characters (e.g., $A + B + C$ could translate to "A meets B who introduces him to C."). Some operation symbols may intuitively add personality traits to your characters (e.g., $A \cdot D$ could translate to "A is also D" or $A \div D$ could translate to "A used to be a D."). Don't forget to pay attention to the order of operations. Be creative and have fun.

 Example:
 $A + B + C = A$ professor introduces a student to a scientist.
 $D + E - G = A$ politician has lunch with a musician but cannot get the attention of the waiter.
 $A \cdot D + E \cdot F = A$ professor who teaches politics influences a musician to start writing song lyrics.
 $F \div C + B \cdot G = A$ writer who used to be a scientist finds inspiration upon meeting a student
 waiting tables to pay for tuition.

 $A + B + C$ _____

 $D + E - G$ _____

 $A \cdot D + E \cdot F$ _____

 $F \div C + B \cdot G$ _____

Chapter 1 Summary

Absolute Value [1.1]

The *absolute value* of a real number is its distance from zero on the real number line. Absolute value is never negative.

EXAMPLES

1. $|5| = 5$
 $|-5| = 5$

Opposites [1.1]

Any two real numbers the same distance from zero on the number line but in opposite directions from zero are called *opposites*. Opposites always add to zero.

2. The numbers 3 and -3 are opposites; their sum is 0.
 $$3 + (-3) = 0$$

Symbols [1.1]

$a = b$	a is equal to b.
$a \neq b$	a is not equal to b.
$a < b$	a is less than b.
$a \nless b$	a is not less than b.
$a > b$	a is greater than b.
$a \ngtr b$	a is not greater than b.
$a \geq b$	a is greater than or equal to b.
$a \leq b$	a is less than or equal to b.

Addition of Real Numbers [1.2]

To add two real numbers with

1. the same sign, simply add their absolute values and use the common sign.

2. different signs, subtract the smaller absolute value from the larger absolute value. The answer has the same sign as the number with the larger absolute value.

3. Add all combinations of positive and negative 10 and 13.
 $$10 + 13 = 23$$
 $$10 + (-13) = -3$$
 $$-10 + 13 = 3$$
 $$-10 + (-13) = -23$$

Subtraction of Real Numbers [1.2]

To subtract one number from another, simply add the opposite of the number you are subtracting; that is, if a and b represent real numbers, then

$$a - b = a + (-b)$$

4. Subtracting 2 is the same as adding -2.
 $$7 - 2 = 7 + (-2) = 5$$

Multiplication of Real Numbers [1.3]

To multiply two real numbers, simply multiply their absolute values. Like signs give a positive answer. Unlike signs give a negative answer.

5. $3(5) = 15$
 $3(-5) = -15$
 $-3(5) = -15$
 $-3(-5) = 15$

Division of Real Numbers [1.4]

Division by a number is the same as multiplication by its reciprocal. Like signs give a positive answer. Unlike signs give a negative answer.

6. $-\frac{6}{2} = -6\left(\frac{1}{2}\right) = -3$
 $\frac{-6}{-2} = -6\left(-\frac{1}{2}\right) = 3$

Properties of Real Numbers [1.5]

	For Addition	*For Multiplication*
Commutative:	$a + b = b + a$	$a \cdot b = b \cdot a$
Associative:	$a + (b + c) = (a + b) + c$	$a \cdot (b \cdot c) = (a \cdot b) \cdot c$
Distributive:	$a(b + c) = ab + ac$	
Identity:	$a + 0 = a$	$a \cdot 1 = a$
Inverse:	$a + (-a) = 0$	$a\left(\dfrac{1}{a}\right) = 1$
Symmetric:	$a = b$ then $b = a$	

7. The terms $2x$, $5x$, and $-7x$ are all similar because their variable parts are the same.

Similar Terms [1.6]

A *term* is a number or a number and one or more variables multiplied together. *Similar terms* are terms with the same variable part.

8. Simplify $3x + 4x$.
$$3x + 4x = (3 + 4)x$$
$$= 7x$$

Simplifying Expressions [1.6]

In this chapter, we simplified expressions that contained variables by using the distributive property to combine similar terms.

> **COMMON MISTAKE**
>
> 1. Interpreting absolute value as changing the sign of the number inside the absolute value symbols. $|-5| = +5, |+5| = -5$. (The first expression is correct; the second one is not.) To avoid this mistake, remember: Absolute value is a distance and distance is always measured in positive units.
>
> 2. Using the phrase "two negatives make a positive." This works only with multiplication and division. With addition, two negative numbers produce a negative answer. It is best not to use the phrase "two negatives make a positive" at all.

Translate into symbols. [1.1]

1. The difference of 25 and x is 9.

2. The product of 7 and a is 63.

3. The sum of 5 and n is greater than 10.

4. The quotient of 10 and y is less than or equal to 2.

Simplify. [1.1]

5. $|5 - 3| - |2 - 1|$

6. $20 - 5|4 - 3(2 - 1)| - 5 \cdot 3$

For each number, name the opposite, and absolute value. [1.1]

7. -4

8. $\dfrac{7}{3}$

Add. [1.2]

9. $6 + (-12)$

10. $|-3 + 7| + |4 + (-9)|$

Subtract. [1.2]

11. $-12 - 7$

12. $4 - (6 - 11) + 4$

Multiply. [1.3]

13. $6(-5)$

14. $4(-3)(-5)$

15. $12\left(-\dfrac{2}{3}\right)$

16. $\left(-\dfrac{3}{4}\right)^2$

Simplify. [1.1, 1.2, 1.3, 1.4]

17. $-4(3) - 5$

18. $4(2)^3 - 5(-3)^2$

19. $12 + 5(4 - 6)$

20. $7 - 4[-3(7 - 3) - (4)(-4)]$

21. $\dfrac{-7(2) - 3(4)}{6 - 8}$

22. $\dfrac{5(4 - 6) - 3(-3 + 5)}{5(3) - 11}$

Match each expression below with the letter of the property that justifies it. [1.5]

23. $3(2 - x) = 6 - 3x$

24. $(6 + a) + 4 = 6 + (a + 4)$

25. $5(3y) = (5 \cdot 3)y$

26. $y \cdot 4 = 4y$

 a. Commutative property of addition
 b. Commutative property of multiplication
 c. Associative property of addition
 d. Associative property of multiplication
 e. Distributive property

Simplify. [1.5, 1.6]

27. $6 + (9 - 2x)$

28. $4(-9x)$

29. $-3(4x - 5)$

30. $\dfrac{2}{3}(9x - 6)$

31. $b - 4(b - 3)$

32. $(4n - m) - 3(n + 4m)$

33. $-8\left(\dfrac{1}{2}x - \dfrac{1}{4}y\right) + 6\left(\dfrac{1}{3}x - \dfrac{1}{2}y\right)$

34. $0.02x + 0.06(x + 5000)$

Use the illustration below to answer the following questions.

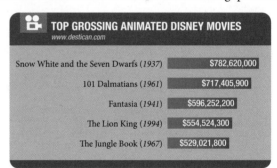

TOP GROSSING ANIMATED DISNEY MOVIES
www.destican.com

Movie	Gross
Snow White and the Seven Dwarfs (*1937*)	$782,620,000
101 Dalmatians (*1961*)	$717,405,900
Fantasia (*1941*)	$596,252,200
The Lion King (*1994*)	$554,524,300
The Jungle Book (*1967*)	$529,021,800

35. How much did *Fantasia* gross?

36. What did *101 Dalmatians* and *The Jungle Book* gross together?

37. How much more did *Snow White and the Seven Dwarfs* gross than *The Lion King*?

At the end of each chapter from here on, we will also provide a review that consists of problems similar to those found in the current and previous chapters. Use these reviews as a study tool for exams, and as a way to check your progress, attending to any difficulties you come across. Remember, that math concepts build on each other. It is important to master the skills that come before in order to find success as you proceed through the book.

Factor each number into a product of prime numbers.

1. 240

2. 234

3. 306

4. 495

Perform the indicated operations.

5. $\dfrac{9}{14} \div \dfrac{3}{7}$

6. $\dfrac{24}{28} \cdot \dfrac{7}{12}$

7. $\dfrac{7}{12} - \dfrac{3}{8}$

8. $\dfrac{9}{16} - \dfrac{5}{12}$

9. $0.743 + 6.502$

10. $3.027 - 1.86$

11. $(6.301)(3.21)$

12. $1.75 \div 0.7$

13. $15 - [6 + 3(5 - 2)]$

14. $\left(\dfrac{5}{2}\right)\left(\dfrac{6}{15}\right) - 0.94$

15. $6 + 5[12 \div (2 \cdot 3)]$

16. $\dfrac{42 \div 3(7)}{6 - 3}$

17. $6^2 - 3[15 - 3(7 - 4)]$

18. $3^2 + 2^3(7 - 2 \cdot 3)$

Simplify.

19. 3^2

20. 4^3

21. $5^2 - 2^3$

22. $6^2 - 3^3$

Perform the indicated operation.

23. $7 - 12$

24. $-6 - 3$

25. $5 + (-2)$

26. $-17 + 4$

27. $(-2)(5)(-6)$

28. $(3)(4)(-5)$

29. $\dfrac{(8)(-3)}{(-4)}$

30. $16 \div (-4)$

Simplify as much as possible.

31. $3(2x - 4)$

32. $-3(x + 7)$

33. $5(6x - 1)$

34. $\dfrac{3}{2}(4x - 6)$

35. $4(a + 7) - 18$

36. $10 - 4(a + 3)$

37. $3(x + 7) + 5x$

38. $4(2x + 5) - 9x$

39. $6x - 7y + 2y$

40. $2(4a - 5) - 3 - 4a$

41. $5x - 10 - 3x + 7$

42. $-(x - 3) - 2y - 6$

Perform the indicated operation.

43. $3(5 - 2) - 6(3)$

44. $7 - [3 - 3(4 - 7)]$

45. $9 \div (-3) + 2(5 - 2)$

46. $(4)(-5) - 3[4(3 - 6)]$

47. $24 \div 3(-4) + 3(6 - 4)$

48. $3(-4) \div 24(6)$

Translate into symbols. [1.1]

1. The difference of 15 and x is 12.

2. The product of 6 and a is 30.

3. The sum of 7 and a is greater than 15.

4. The quotient of 12 and n is less than or equal to 5.

Simplify according to the rule for order of operations. [1.1]

5. $|4 - 5| - |3 - 5|$

6. $15 - 3|2 - 4(3 - 1)| - 4 \cdot 2$

For each number, name the opposite and absolute value. [1.1]

7. -6

8. $\dfrac{5}{3}$

Add. [1.2]

9. $5 + (-8)$

10. $|-2 + 7| + |-3 + (-4)|$

Subtract. [1.2]

11. $-7 - 4$

12. $5 - (6 + 3) - 3$

Multiply. [1.3]

13. $7(-4)$

14. $3(-4)(-2)$

15. $9\left(-\dfrac{1}{3}\right)$

16. $\left(-\dfrac{3}{2}\right)^3$

Simplify. [1.1, 1.2, 1.3, 1.4]

17. $-2(3) - 7$

18. $2(3)^3 - 4(-2)^4$

19. $9 + 4(2 - 6)$

20. $5 - 3[-2(1 + 4) + 3(-3)]$

21. $\dfrac{-4(3) + 5(-2)}{-5 - 6}$

22. $\dfrac{4(3 - 5) - 2(-6 + 8)}{4(-2) + 10}$

Match each expression below with the letter of the property that justifies it. [1.5]

23. $4(2y) = (4 \cdot 2)y$

24. $5(x - 3) = 5x - 15$

25. $4 + x = x + 4$

26. $(a + 5) - 2 = a + (5 - 2)$

 a. Commutative property of addition
 b. Commutative property of multiplication
 c. Associative property of addition
 d. Associative property of multiplication
 e. Distributive property

Simplify. [1.5, 1.6]

27. $5 + (7 + 3x)$

28. $3(-5y)$

29. $-5(2x - 3)$

30. $-\dfrac{1}{3}(6x + 12)$

31. $n - 5(n - 2)$

32. $(3a - b) - 5(a + 3b)$

33. $-10\left(\dfrac{1}{5}x - \dfrac{1}{2}y\right) + 12\left(\dfrac{1}{4}x - \dfrac{1}{3}y\right)$

34. $0.03n + 0.08(n + 3000)$

Use the illustration below to answer the following questions.

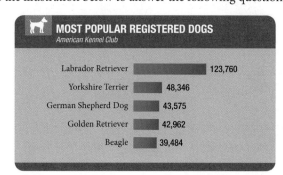

35. How many Labrador Retrievers are registered with the American Kennel Club?

36. What is the total number of Golden Retrievers and Beagles registered?

37. How many more Yorkshire Terriers are registered than German Shepherds?

Linear Equations and Inequalities in One Variable

2

Estádio Jornalista Mário Filho Stadium
Rio de Janiero, Brazil
© 2010 Google

The honor of hosting the 2014 FIFA World Cup for soccer has fallen to Brazil, making it just the fifth country to host two World Cup tournaments. Expected to be one of the main venues of the competition, Estádio Jornalista Mário Filho in Rio de Janiero is the sixth largest stadium in the world. The stadium was originally constructed for the 1950 World Cup, where it acquired the nickname "Maracana" after a nearby river. The Maracana hosted the 1950 tournament final where Uruguay upset the home team Brazil in a stunning 2-1 win. At that time, the crowd was reported to have been 200,000 soccer fans, including those who illegally stormed the gates and entered after the start of the match. That crowd is still the largest ever to watch a soccer match. Due to an accident in 1992 resulting in two deaths, renovations greatly reduced Maracana's capacity, which is currently listed at 103,045.

Suppose the 2014 FIFA World Cup lists their ticket prices in the same manner as the 2010 World Cup in South Africa. The following chart shows the 2010 prices by seating category in US dollars.

	Category 1	Category 2	Category 3	Wheelchair
Opening Match	$450.00	$300.00	$200.00	$70.00
Group Stage	$160.00	$120.00	$80.00	$20.00
Round of 16	$200.00	$150.00	$100.00	$50.00
Quarter Finals	$300.00	$200.00	$150.00	$75.00
Semi-Finals	$600.00	$400.00	$250.00	$100.00
3rd/4th Place	$300.00	$200.00	$150.00	$75.00
The Final	$900.00	$600.00	$400.00	$150.00

Suppose the Stadium Maracana sells out for the final game with 100,000 paid tickets. If the seating prices hold, and there are 50,000 Category 2 tickets sold, 15,000 Category 3 tickets sold, and 5,000 Wheelchair tickets sold, how many Category 1 tickets will be sold? An equation to answer this question looks like this:

$$100{,}000 = 50{,}000 + 15{,}000 + 5{,}000 + x$$

Solving the equation above requires the use of the addition property of equality, one of the topics we will cover in this chapter.

2.1 Addition Property of Equality

A Check the solution to an equation by substitution.

B Use the addition property of equality to solve an equation.

©iStockphoto.com/ivicaNS

KEY WORDS

solution set

equivalent equation

addition property of equality

Did you know you can roughly calculate the outside temperature by counting the number of times a cricket chirps? Researchers at the Library of Congress suggest that you focus on one cricket and count the number of chirps in 15 seconds. Then add 37 to that number and you'll get the approximate outside temperature in degrees Fahrenheit. We can also use this information to write the following linear equation:

$$t = n + 37$$

where t is the outside temperature in degrees Fahrenheit and n is the number of cricket chirps in 15 seconds. Suppose the temperature outside is 70 degrees Fahrenheit. After plugging this value into the equation, we can solve for n to find the number of chirps; that is, we must find all replacements for the variable that make the equation a true statement. To do so, we must first understand how to use the addition property of equality, which we will cover in this section.

A Solutions to Equations

> **DEFINITION** solution set
>
> The *solution set* for an equation is the set of all numbers that when used in place of the variable make the equation a true statement

Note The symbol { } is used in roster notation to represent a list of numbers in a set.

For example, the equation $x + 2 = 5$ has the solution set {3} because when x is 3 the equation becomes the true statement $3 + 2 = 5$, or $5 = 5$.

Practice Problems

1. Is 4 a solution to $2x + 3 = 7$?

EXAMPLE 1 Is 5 a solution to $2x - 3 = 7$?

Solution We substitute 5 for x in the equation, and then simplify to see if a true statement results. A true statement means we have a solution; a false statement indicates the number we are using is not a solution.

$$\begin{aligned}
\text{When} \rightarrow \qquad & x = 5 \\
\text{the equation} \rightarrow \qquad & 2x - 3 = 7 \\
\text{becomes} \rightarrow \qquad & 2(5) - 3 \overset{?}{=} 7 \\
& 10 - 3 \overset{?}{=} 7 \\
& 7 = 7 \qquad \textit{A true statement}
\end{aligned}$$

Note We use a question mark over the equal signs to show that we don't know yet whether the two sides of the equation are equal.

Because $x = 5$ turns the equation into the true statement $7 = 7$, we know 5 is a solution to the equation.

Answer

1. No

EXAMPLE 2 Is -2 a solution to $8 = 3x + 4$?

Solution Substituting -2 for x in the equation, we have

$$8 \overset{?}{=} 3(-2) + 4$$
$$8 \overset{?}{=} -6 + 4$$
$$8 = -2 \qquad \text{A false statement}$$
$$x \neq -2$$

Substituting -2 for x in the equation produces a false statement. Therefore, $x = -2$ is not a solution to the equation.

2. Is $\frac{4}{3}$ a solution to $8 = 3x + 4$?

Note The symbol \neq means "does not equal".

The important thing about an equation is its solution set. Therefore, we make the following definition to classify together all equations with the same solution set.

> **DEFINITION** equivalent equation
>
> Two or more equations with the same solution set are said to be *equivalent equations*.

Equivalent equations may look different but must have the same solution set.

EXAMPLE 3

a. $x + 2 = 5$ and $x = 3$ are equivalent equations because both have solution set $\{3\}$.

b. $a - 4 = 3$, $a - 2 = 5$, and $a = 7$ are equivalent equations because they all have solution set $\{7\}$.

c. $y + 3 = 4$, $y - 8 = -7$, and $y = 1$ are equivalent equations because they all have solution set $\{1\}$.

3. Answer the following.
 a. Are the equations $x + 3 = 9$ and $x = 6$ equivalent?
 b. Are the equations $a - 5 = 3$ and $a = 6$ equivalent?
 c. Are $y + 2 = 5$, $y - 1 = 2$, and $y = 3$ equivalent equations?

B Addition Property of Equality

If two numbers are equal and we increase (or decrease) both of them by the same amount, the resulting quantities are also equal. We can apply this concept to equations. Adding the same amount to both sides of an equation always produces an equivalent equation—one with the same solution set. This fact about equations is called the *addition property of equality* and can be stated more formally as follows:

> **PROPERTY** Addition Property of Equality
>
> For any three algebraic expressions A, B, and C,
>
> *In symbols*: if $A = B$
>
> then $A + C = B + C$
>
> *In words*: Adding the same quantity to both sides of an equation will not change the solution set.

Note We will use this property many times in the future. Be sure you understand it completely by the time you finish this section.

This property is just as simple as it seems. We can add any amount to both sides of an equation and always be sure we have not changed the solution set.

Consider the equation $x + 6 = 5$. We want to solve this equation for the value of x. To solve for x means we want to end up with x on one side of the equal sign and a number on the other side. Because we want x by itself, we will add -6 to both sides.

$$x + 6 + (-6) = 5 + (-6) \qquad \text{Addition property of equality}$$
$$x + 0 = -1$$
$$x = -1$$

All three equations say the same thing about x. They all say that x is -1. All three equations are equivalent. The last one is just easier to read.

Answers

2. Yes

3. a. Yes **b.** No **c.** Yes

Here are some further examples of how the addition property of equality can be used to solve equations.

4. Solve $x - 3 = 10$ for x.

EXAMPLE 4 Solve the equation $x - 5 = 12$ for x.

Solution Because we want x alone on the left side, we choose to add 5 to both sides.

$$x - 5 + 5 = 12 + 5 \qquad \textit{Addition property of equality}$$
$$x + 0 = 17$$
$$x = 17$$

To check our solution to Example 4, we substitute 17 for x in the original equation:

$$\text{When} \rightarrow \qquad x = 17$$
$$\text{the equation} \rightarrow \qquad x - 5 = 12$$
$$\text{becomes} \rightarrow \qquad 17 - 5 \overset{?}{=} 12$$
$$12 = 12 \qquad \textit{A true statement}$$

As you can see, our solution checks. The purpose for checking a solution to an equation is to catch any mistakes we may have made in the process of solving the equation.

5. Solve $a + \dfrac{2}{3} = -\dfrac{1}{6}$ for a.

EXAMPLE 5 Solve $a + \dfrac{3}{4} = -\dfrac{1}{2}$ for a.

Solution Because we want a by itself on the left side of the equal sign, we add the opposite of $\frac{3}{4}$ to each side of the equation.

$$a + \frac{3}{4} + \left(-\frac{3}{4}\right) = -\frac{1}{2} + \left(-\frac{3}{4}\right) \qquad \textit{Addition property of equality}$$
$$a + 0 = -\frac{1}{2} \cdot \frac{2}{2} + \left(-\frac{3}{4}\right) \qquad \textit{LCD on the right side is 4.}$$
$$a = -\frac{2}{4} + \left(-\frac{3}{4}\right) \qquad \textit{$\frac{2}{4}$ is equivalent to $\frac{1}{2}$.}$$
$$a = -\frac{5}{4} \qquad \textit{Add fractions.}$$

The solution is $a = -\frac{5}{4}$. To check our result, we replace a with $-\frac{5}{4}$ in the original equation. The left side then becomes $-\frac{5}{4} + \frac{3}{4}$, which reduces to $-\frac{1}{2}$, so our solution checks.

6. Solve $2.7 + x = 8.1$ for x.

EXAMPLE 6 Solve $7.3 + x = -2.4$ for x.

Solution Again, we want to isolate x, so we add the opposite of 7.3 to both sides.

$$7.3 + (-7.3) + x = -2.4 + (-7.3) \qquad \textit{Addition property of equality}$$
$$0 + x = -9.7$$
$$x = -9.7$$

Sometimes it is necessary to simplify each side of an equation before using the addition property of equality. The reason we simplify both sides first is that we want as few terms as possible on each side of the equation before we use the addition property of equality. The following examples illustrate this procedure.

7. Solve $-2x + 3 + 3x = 6 + 8$ for x.

EXAMPLE 7 Solve $-x + 2 + 2x = 7 + 5$ for x.

Solution We begin by combining similar terms on each side of the equation. Then we use the addition property to solve the simplified equation.

$$x + 2 = 12 \qquad \textit{Simplify both sides first.}$$
$$x + 2 + (-2) = 12 + (-2) \qquad \textit{Addition property of equality}$$
$$x + 0 = 10$$
$$x = 10$$

Answers

4. 13

5. $-\dfrac{5}{6}$

6. 5.4

7. 11

EXAMPLE 8 Solve $4(2a - 3) - 7a = 2 - 5$.

Solution We must begin by applying the distributive property to separate terms on the left side of the equation. Following that, we combine similar terms and then apply the addition property of equality.

$$4(2a - 3) - 7a = 2 - 5 \qquad \text{Original equation}$$
$$8a - 12 - 7a = 2 - 5 \qquad \text{Distributive property}$$
$$a - 12 = -3 \qquad \text{Simplify each side.}$$
$$a - 12 + 12 = -3 + 12 \qquad \text{Add 12 to each side.}$$
$$a = 9$$

To check our solution, we replace a with 9 in the original equation.

$$4(2 \cdot 9 - 3) - 7 \cdot 9 \stackrel{?}{=} 2 - 5$$
$$4(15) - 63 \stackrel{?}{=} -3$$
$$60 - 63 \stackrel{?}{=} -3$$
$$-3 = -3 \qquad \text{A true statement}$$

We can also add a term involving a variable to both sides of an equation.

EXAMPLE 9 Solve $3x - 5 = 2x + 7$.

Solution We can solve this equation in two steps. First, we add $-2x$ to both sides of the equation. When this has been done, x appears on the left side only. Second, we add 5 to both sides.

$$3x + (-2x) - 5 = 2x + (-2x) + 7 \qquad \text{Add } -2x \text{ to both sides.}$$
$$x - 5 = 7 \qquad \text{Simplify each side.}$$
$$x - 5 + 5 = 7 + 5 \qquad \text{Add 5 to both sides.}$$
$$x = 12 \qquad \text{Simplify each side.}$$

A Note on Subtraction

Although the addition property of equality is stated for addition only, we can subtract the same number from both sides of an equation as well. Because subtraction is defined as addition of the opposite, subtracting the same quantity from both sides of an equation does not change the solution.

$$x + 2 = 12 \qquad \text{Original equation}$$
$$x + 2 - 2 = 12 - 2 \qquad \text{Subtract 2 from each side.}$$
$$x = 10$$

GETTING READY FOR CLASS

After reading through the preceding section, respond in your own words and in complete sentences.

A. What is a solution set for an equation?

B. What are equivalent equations?

C. Explain in words the addition property of equality.

D. How do you check a solution to an equation?

8. Solve $5(3a - 4) - 14a = 25$.

Note Remember that terms cannot be combined until the parentheses are removed.

Note Again, we place a question mark over the equal sign because we don't know yet whether the expressions on the left and right side of the equal sign will be equal.

9. Solve $5x - 4 = 4x + 6$.

Note In my experience teaching algebra, I find that students make fewer mistakes if they think in terms of addition rather than subtraction. So, you are probably better off if you continue to use the addition property just the way we have used it in the examples in this section. But, if you are curious as to whether you can subtract the same number from both sides of an equation, the answer is yes.

Answers
8. 45
9. 10

Vocabulary Review

Choose the correct words to fill in the blanks below.

equivalent addition subtraction solution set

1. The set of numbers used in place of the variable in an equation are called the _____ for that equation.
2. Two or more equations with the same solution set are said to be _____ equations.
3. The _____ property of equality states that adding the same quantity to both sides of an equation will not change the solution set.
4. Since subtraction is defined as addition of the opposite, we can apply the addition property of equality to both sides of a _____ problem and the solution will not change.

Problems

A Check to see if the number to the right of each of the following equations is the solution to the equation.

1. $2x + 1 = 5; 2$

2. $4x + 3 = 7; 1$

3. $3x + 4 = 19; 5$

4. $3x + 8 = 14; 2$

5. $2x - 4 = 2; 4$

6. $5x - 6 = 9; 3$

7. $2x + 1 = 3x + 3; -2$

8. $4x + 5 = 2x - 1; -6$

9. $x - 4 = 2x + 1; -4$

10. $x - 8 = 3x + 2; -5$

11. $x - 2 = 3x + 1; \dfrac{-3}{2}$

12. $x + 4 = 4x - 1; \dfrac{5}{3}$

B Solve the following equations.

13. $x - 3 = 8$

14. $x - 2 = 7$

15. $x + 2 = 6$

16. $x + 5 = 4$

17. $a + \dfrac{1}{2} = -\dfrac{1}{4}$

18. $a + \dfrac{1}{3} = -\dfrac{5}{6}$

19. $x + 2.3 = -3.5$

20. $x + 7.9 = 23.4$

21. $y + 11 = -6$

22. $y - 3 = -1$

23. $x - \dfrac{5}{8} = -\dfrac{3}{4}$

24. $x - \dfrac{2}{5} = -\dfrac{1}{10}$

25. $m - 6 = -10$

26. $m - 10 = -6$

27. $6.9 + x = 3.3$

SCAN TO ACCESS

28. $7.5 + x = 2.2$

29. $5 = a + 4$

30. $12 = a - 3$

31. $-\dfrac{5}{9} = x - \dfrac{2}{5}$

32. $-\dfrac{7}{8} = x - \dfrac{4}{5}$

Simplify both sides of the following equations as much as possible, and then solve.

33. $4x + 2 - 3x = 4 + 1$

34. $5x + 2 - 4x = 7 - 3$

35. $8a - \dfrac{1}{2} - 7a = \dfrac{3}{4} + \dfrac{1}{8}$

36. $9a - \dfrac{4}{5} - 8a = \dfrac{3}{10} - \dfrac{1}{5}$

37. $3 - 4x + 5x = 18$

38. $10 - 3x + 4x = 20$

39. $-11x + 2 + 10x + 2x = 9$

40. $-10x + 5 - 4x + 15x = 0$

41. $-2.5 + 4.8 = 8x - 1.2 - 7x$

42. $-4.8 + 6.3 = 7x - 2.7 - 6x$

43. $2y - 10 + 3y - 4y = 18 - 6$

44. $15 - 21 = 8x + 3x - 10x$

The following equations contain parentheses. Apply the distributive property to remove the parentheses, then simplify each side before using the addition property of equality.

45. $2(x + 3) - x = 4$

46. $5(x + 1) - 4x = 2$

47. $-3(x - 4) + 4x = 3 - 7$

48. $-2(x - 5) + 3x = 4 - 9$

49. $5(2a + 1) - 9a = 8 - 6$

50. $4(2a - 1) - 7a = 9 - 5$

51. $-(x + 3) + 2x - 1 = 6$

52. $-(x - 7) + 2x - 8 = 4$

53. $4y - 3(y - 6) + 2 = 8$

54. $7y - 6(y - 1) + 3 = 9$

55. $-3(2m - 9) + 7(m - 4) = 12 - 9$

56. $-5(m - 3) + 2(3m + 1) = 15 - 8$

Solve the following equations by the method used in Example 9 in this section. Check each solution in the original equation.

57. $4x = 3x + 2$

58. $6x = 5x - 4$

59. $8a = 7a - 5$

60. $9a = 8a - 3$

61. $2x = 3x + 1$

62. $4x = 3x + 5$

63. $3y + 4 = 2y + 1$

64. $5y + 6 = 4y + 2$

65. $2m - 3 = m + 5$

66. $8m - 1 = 7m - 3$

67. $4x - 7 = 5x + 1$

68. $3x - 7 = 4x - 6$

69. $5x - \dfrac{2}{3} = 4x + \dfrac{4}{3}$

70. $3x - \dfrac{5}{4} = 2x + \dfrac{1}{4}$

71. $8a - 7.1 = 7a + 3.9$

72. $10a - 4.3 = 9a + 4.7$

Applying the Concepts

73. Movie Tickets A movie theater has a total of 300 seats. For a special Saturday night preview, they reserve 20 seats to give away to their VIP guests at no charge. If x represents the number of tickets they can sell for the preview, then x can be found by solving the equation $x + 20 = 300$.

a. Solve the equation for x.

b. If tickets for the preview are $7.50 each, what is the maximum amount of money they can make from ticket sales?

74. Movie Tickets A movie theater has a total of 250 seats. For a special Saturday night preview, they reserve 45 seats to give away to their VIP guests at no charge. If x represents the number of tickets they can sell for the preview, then x can be found by solving the equation $x + 45 = 250$.

a. Solve the equation for x.

b. If tickets for the preview are $8.00 each, what is the maximum amount of money they can make from ticket sales?

75. Geometry Two angles are complementary angles. If one of the angles is 23°, then solving the equation $x + 23° = 90°$ will give you the other angle. Solve the equation.

76. Geometry Two angles are supplementary angles. If one of the angles is 23°, then solving the equation $x + 23° = 180°$ will give you the other angle. Solve the equation.

77. Theater Tickets The El Portal Center for the Arts in North Hollywood, California, holds a maximum of 400 people. The two balconies hold 86 and 89 people each; the rest of the seats are at the stage level. Solving the equation $x + 86 + 89 = 400$ will give you the number of seats on the stage level.

a. Solve the equation for x.

b. If tickets on the stage level are $30 each, and tickets in either balcony are $25 each, what is the maximum amount of money the theater can bring in for a show?

78. Geometry The sum of the angles in the triangle on the swing set is 180°. Use this fact to write an equation containing x. Then solve the equation.

79. Using the equation $t = n + 40$ given in the introduction of this section, complete the following table.

n	t
32	
	90
44	
	136

80. Using the equation $T = 40 + O$ which relates overtime hours to total hours, complete the following table.

O	T
10	
	51
17	
	62

81. Geometry The three angles shown in the triangle at the front of the tent in the following figure add up to 180°. Use this fact to write an equation containing x, and then solve the equation to find the number of degrees in the angle at the top of the triangle.

82. Suppose Ava and Kendra are bridesmaids in a friend's wedding. Answer the following questions about the money they spend.

a. Suppose Kendra spent $210 for a bridesmaid dress, then bought new shoes to go with the dress. If the total bill was $289, does the equation $210 + x = 289$ describe this situation? If so, what does x represent?

b. Suppose Ava buys a bridesmaid dress on sale and then spends $120 on her makeup, nails, and hair. If the total bill is $219, and x represents a positive number, does the equation $219 + x = 120$ describe the situation?

Getting Ready for the Next Section

To understand all of the explanations and examples in the next section you must be able to work the problems below.

Simplify.

83. $\frac{3}{2}\left(\frac{2}{3}y\right)$

84. $\frac{5}{2}\left(-\frac{2}{5}y\right)$

85. $\frac{1}{5}(5x)$

86. $-\frac{1}{4}(-4a)$

87. $\frac{1}{5}(30)$

88. $-\frac{1}{4}(24)$

89. $\frac{3}{2}(4)$

90. $\frac{1}{26}(13)$

91. $12\left(-\frac{3}{4}\right)$

92. $12\left(\frac{1}{2}\right)$

93. $\frac{3}{2}\left(-\frac{5}{4}\right)$

94. $\frac{5}{3}\left(-\frac{6}{5}\right)$

95. $13 + (-5)$

96. $-13 + (-5)$

97. $-\dfrac{3}{4} + \left(-\dfrac{1}{2}\right)$

98. $-\dfrac{7}{10} + \left(-\dfrac{1}{2}\right)$

99. $7x + (-4x)$

100. $5x + (-2x)$

101. $3x + (-12x)$

102. $5x - 9x$

Find the Mistake

Each sentence below contains a mistake. Circle the mistake and write the correct word or expression on the line provided.

1. The number 2 is a solution to the equation $3x - 9 = 8x + 1$. _____

2. Saying that the number 12 is a solution to the equation $\dfrac{1}{6}y + 3 = 5$ is a false statement. _____

3. The addition property of equality states that $A(C) = B(C)$ if $A = B$. _____

4. To solve the equation $\dfrac{5}{6}b + 6 = 11$ for b, begin by adding -11 to both sides. _____

Navigation Skills: Prepare, Study, Achieve

Completing homework assignments in full is a key piece to succeeding in this class. To do this effectively, you must pay special attention to each set of instructions. When you do your homework, you usually work a number of similar problems at a time. But the problems may vary on a test. It is very important to make a habit of paying attention to the instructions to elicit correct answers on a test.

Secondly, to complete an assignment efficiently, you will need to memorize various definitions, properties, and formulas. Reading the definition in the book alone is not enough. There are many techniques for successful memorization. Spend some time doing the following:

- Reread the definition.
- Say the definition out loud.
- Explain the definition to another person.
- Write the definition down on a separate sheet of notes.
- Create a mnemonic device using key words from the definition.
- Analyze how the definition applies to your homework problems.

The above suggestions are ways to engage your senses when memorizing an abstract concept. This will help anchor it in your memory. For instance, it is easier to remember explaining to your friend a difficult math formula, than it is to simply recall it from a single read of the chapter.

Lastly, once you've completed an assignment, take any extra time you've allotted for studying to work more problems, and if you feel ready, read ahead and work problems you will encounter in the next section.

2.2 Multiplication Property of Equality

Image © Vandenwyngaert Stephane

OBJECTIVES

A Use the multiplication property of equality to solve an equation.

B Use the addition and multiplication properties of equality together to solve an equation.

KEY WORDS

multiplication property of equality

Henry Ford made famous the mass production assembly line in an effort to streamline his automobile manufacturing. He realized that an assembler was taking 8.5 hours to complete his entire task because the assembler had to retrieve parts on his own as needed. Ford asked that parts be delivered to an assembler, which reduced the time it took to complete the task to 2.5 minutes! Later, Ford installed the moving conveyor belt, reducing the completion time even further.

Imagine yourself working on an assembly line. If it takes you 2.5 minutes to assemble your item, how many items can you assemble in an hour's time? To answer this question, we must use the given information to set up the following equation:

$$60 = 2.5x$$

Notice we converted an hour into 60 minutes to match the units for your assembly time. In this section, we will introduce you to the multiplication property of equality. You will use this property to solve linear equations, such as the one for the assembly line.

A Multiplication Property of Equality

In the previous section, we found that adding the same number to both sides of an equation never changed the solution set. The same idea holds for multiplication by numbers other than zero. We can multiply both sides of an equation by the same nonzero number and always be sure we have not changed the solution set. (The reason we cannot multiply both sides by zero will become apparent later.) This fact about equations is called the *multiplication property of equality*, which can be stated formally as follows:

> **PROPERTY** Multiplication Property of Equality
>
> For any three algebraic expressions A, B, and C, where $C \neq 0$,
>
> *In symbols:* if $A = B$
>
> then $AC = BC$
>
> *In words:* Multiplying both sides of an equation by the same nonzero number will not change the solution set.

Note This property is also used many times throughout the book. Make sure you understand it completely.

Suppose we want to solve the equation $5x = 30$. We have $5x$ on the left side but would like to have just x. We choose to multiply both sides by $\frac{1}{5}$ which allows us to divide out the fives and isolate the variable. The solution follows.

$$5x = 30$$
$$\frac{1}{5}(5x) = \frac{1}{5}(30) \qquad \text{Multiplication property of equality}$$
$$\left(\frac{1}{5} \cdot 5\right)x = \frac{1}{5}(30) \qquad \text{Associative property of multiplication}$$
$$1x = 6 \qquad \text{Note: } 1x \text{ is the same as } x.$$
$$x = 6$$

We chose to multiply by $\frac{1}{5}$ because it is the reciprocal of 5 and $\left(\frac{1}{5}\right)(5) = 1$. We can see that multiplication by any number except zero will not change the solution set. If, however, we were to multiply both sides by zero, the result would always be $0 = 0$. Although the statement $0 = 0$ is true, we have lost our variable and cannot solve the equation. This is the only restriction of the multiplication property of equality. We are free to multiply both sides of an equation by any number except zero in an effort to isolate the variable.

Here are some more examples that use the multiplication property of equality.

EXAMPLE 1 Solve $-4a = 24$ for a.

Solution Because we want the variable a alone on the left side, we choose to multiply both sides by $-\frac{1}{4}$.

$$-\frac{1}{4}(-4a) = -\frac{1}{4}(24) \qquad \text{Multiplication property of equality}$$
$$\left[-\frac{1}{4}(-4)\right]a = -\frac{1}{4}(24) \qquad \text{Associative property}$$
$$a = -6$$

EXAMPLE 2 Solve $-\frac{t}{3} = 5$ for t.

Solution Because division by 3 is the same as multiplication by $\frac{1}{3}$, we can write $-\frac{t}{3}$ as $-\frac{1}{3}t$. To solve the equation, we multiply each side by the reciprocal of $-\frac{1}{3}$, which is -3.

$$-\frac{t}{3} = 5 \qquad \text{Original equation}$$
$$-\frac{1}{3}t = 5 \qquad \text{Dividing by 3 is equivalent to multiplying by } \frac{1}{3}.$$
$$-3\left(-\frac{1}{3}t\right) = -3(5) \qquad \text{Multiply each side by } -3.$$
$$t = -15$$

EXAMPLE 3 Solve $\frac{2}{3}y = \frac{4}{9}$.

Solution We can multiply both sides by $\frac{3}{2}$ and have $1y$ on the left side.

$$\frac{3}{2}\left(\frac{2}{3}y\right) = \frac{3}{2}\left(\frac{4}{9}\right) \qquad \text{Multiplication property of equality}$$
$$\left(\frac{3}{2}\cdot\frac{2}{3}\right)y = \frac{3}{2}\left(\frac{4}{9}\right) \qquad \text{Associative property}$$
$$y = \frac{2}{3} \qquad \text{Simplify } \frac{3}{2}\left(\frac{4}{9}\right) = \frac{12}{18} = \frac{2}{3}.$$

EXAMPLE 4 Solve $5 + 8 = 10x + 20x - 4x$.

Solution Our first step will be to simplify each side of the equation.

$$13 = 26x \qquad \text{Simplify by combining similar terms.}$$
$$\frac{1}{26}(13) = \frac{1}{26}(26x) \qquad \text{Multiplication property of equality}$$
$$\frac{13}{26} = x$$
$$\frac{1}{2} = x \qquad \text{Reduce to lowest terms.}$$
$$x = \frac{1}{2} \qquad \text{Symmetric property}$$

Practice Problems

1. Solve $3a = -27$ for a.

2. Solve $\frac{t}{4} = 6$ for t.

3. Solve $\frac{2}{5}y = \frac{3}{4}$.

Note Notice in Examples 1 through 3 that if the variable is being multiplied by a number like -4 or $\frac{2}{3}$, we always multiply by the number's reciprocal, $-\frac{1}{4}$ or $\frac{3}{2}$, to end up with just the variable on one side of the equation.

4. Solve $3 + 7 = 6x + 8x - 9x$.

Note Remember the symmetric property allows us to move our variable from the right to the left-hand side once we have solved an equation. For any real numbers, if $a = b$, then $b = a$.

Answers
1. -9
2. 24
3. $\frac{15}{8}$
4. 2

B Solving Equations Using Addition and Multiplication

In the next three examples, we will use both the addition property of equality and the multiplication property of equality.

EXAMPLE 5 Solve $6x + 5 = -13$ for x.

Solution We begin by adding -5 to both sides of the equation.

$$6x + 5 + (-5) = -13 + (-5) \qquad \text{Add } -5 \text{ to both sides.}$$
$$6x = -18 \qquad \text{Simplify.}$$
$$\frac{1}{6}(6x) = \frac{1}{6}(-18) \qquad \text{Multiply both sides by } \frac{1}{6}.$$
$$x = -3$$

EXAMPLE 6 Solve $5x = 2x + 12$ for x.

Solution We begin by adding $-2x$ to both sides of the equation.

$$5x + (-2x) = 2x + (-2x) + 12 \qquad \text{Add } -2x \text{ to both sides.}$$
$$3x = 12 \qquad \text{Simplify.}$$
$$\frac{1}{3}(3x) = \frac{1}{3}(12) \qquad \text{Multiply both sides by } \frac{1}{3}.$$
$$x = 4 \qquad \text{Simplify.}$$

EXAMPLE 7 Solve $3x - 4 = -2x + 6$ for x.

Solution We begin by adding $2x$ to both sides.

$$3x + 2x - 4 = -2x + 2x + 6 \qquad \text{Add } 2x \text{ to both sides.}$$
$$5x - 4 = 6 \qquad \text{Simplify.}$$

Now we add 4 to both sides.

$$5x - 4 + 4 = 6 + 4 \qquad \text{Add 4 to both sides.}$$
$$5x = 10 \qquad \text{Simplify.}$$
$$\frac{1}{5}(5x) = \frac{1}{5}(10) \qquad \text{Multiply by } \frac{1}{5}.$$
$$x = 2 \qquad \text{Simplify.}$$

The next example involves fractions. You will see that the properties we use to solve equations containing fractions are the same as the properties we used to solve the previous equations. Also, the LCD that we used previously to add fractions can be used with the multiplication property of equality to simplify equations containing fractions.

EXAMPLE 8 Solve $\frac{2}{3}x + \frac{1}{2} = -\frac{3}{4}$.

Solution We can solve this equation by applying our properties and working with the fractions, or we can begin by eliminating the fractions.

Method 1 Working with the fractions.

$$\frac{2}{3}x + \frac{1}{2} + \left(-\frac{1}{2}\right) = -\frac{3}{4} + \left(-\frac{1}{2}\right) \qquad \text{Add } -\frac{1}{2} \text{ to each side.}$$
$$\frac{2}{3}x = -\frac{5}{4} \qquad \text{Note that } -\frac{3}{4} + \left(-\frac{1}{2}\right) = -\frac{3}{4} + \left(-\frac{2}{4}\right).$$
$$\frac{3}{2}\left(\frac{2}{3}x\right) = \frac{3}{2}\left(-\frac{5}{4}\right) \qquad \text{Multiply each side by } \frac{3}{2}.$$
$$x = -\frac{15}{8}$$

5. Solve $4x + 3 = -13$ for x.

6. Solve $7x = 4x + 15$ for x.

> *Note* Notice that in Example 6 we used the addition property of equality first to combine all the terms containing x on the left side of the equation. Once this had been done, we used the multiplication property to isolate x on the left side.

7. Solve $2x - 5 = -3x + 10$ for x.

8. Solve $\frac{3}{5}x + \frac{1}{2} = -\frac{7}{10}$.

Answers
5. -4
6. 5
7. 3
8. -2

Note Our original equation
has denominators of 3, 2, and
4. The LCD for these three
denominators is 12, and it
has the property that all three
denominators will divide it
evenly. Therefore, if we multiply
both sides of our equation by 12,
each denominator will divide
into 12 and we will be left with
an equation that does not contain
any denominators other than 1.

Method 2 Eliminating the fractions in the beginning.

$$12\left(\frac{2}{3}x + \frac{1}{2}\right) = 12\left(-\frac{3}{4}\right) \quad \text{Multiply each side by the LCD 12.}$$

$$12\left(\frac{2}{3}x\right) + 12\left(\frac{1}{2}\right) = 12\left(-\frac{3}{4}\right) \quad \text{Distributive property on the left side}$$

$$8x + 6 = -9 \qquad \text{Multiply.}$$

$$8x = -15 \qquad \text{Add } -6 \text{ to each side.}$$

$$x = -\frac{15}{8} \qquad \text{Multiply each side by } \frac{1}{8}.$$

As the third line in Method 2 indicates, multiplying each side of the equation by the LCD eliminates all the fractions from the equation.

As you can see, both methods yield the same solution.

A Note on Division

Because *division* is defined as multiplication by the reciprocal, multiplying both sides of an equation by the same number is equivalent to dividing both sides of the equation by the reciprocal of that number; that is, multiplying each side of an equation by $\frac{1}{3}$ and dividing each side of the equation by 3 are equivalent operations. If we were to solve the equation $3x = 18$ using division instead of multiplication, the steps would look like this:

$$3x = 18 \qquad \text{Original equation}$$

$$\frac{3x}{3} = \frac{18}{3} \qquad \text{Divide each side by 3.}$$

$$x = 6$$

Using division instead of multiplication on a problem like this may save you some writing. However, with multiplication, it is easier to explain "why" we end up with just one x on the left side of the equation. (The "why" has to do with the associative property of multiplication.) Continue to use multiplication to solve equations like this one until you understand the process completely. Then, if you find it more convenient, you can use division instead of multiplication.

GETTING READY FOR CLASS

After reading through the preceding section, respond in your own words and in complete sentences.

A. Explain in words the multiplication property of equality.

B. If an equation contains fractions, how do you use the multiplication property of equality to clear the equation of fractions?

C. Why is it okay to divide both sides of an equation by the same nonzero number?

D. Give an example of using an LCD with the multiplication property of equality to simplify an equation.

Vocabulary Review

Choose the correct words to fill in the blanks below.

multiplication division nonzero reciprocal

1. Multiplying both sides of an equation by the same _____ number will not change the solution set.

2. $AC = BC$, where $C \neq 0$ is the _____ property of equality written in symbols.

3. _____ is defined as multiplication by the reciprocal.

4. Dividing both sides of the equation by the same number is equivalent to multiplying both sides of the equation by the _____ of that number.

Problems

A Solve the following equations using the multiplication property of equality. Be sure to show your work.

1. $5x = 10$

2. $6x = 12$

3. $7a = 28$

4. $4a = 36$

5. $-8x = 4$

6. $-6x = 2$

7. $8m = -16$

8. $5m = -25$

9. $-3x = -9$

10. $-9x = -36$

11. $-7y = -28$

12. $-15y = -30$

13. $2x = 0$

14. $7x = 0$

15. $-5x = 0$

16. $-3x = 0$

17. $\dfrac{x}{3} = 2$

18. $\dfrac{x}{4} = 3$

19. $-\dfrac{m}{5} = 10$

20. $-\dfrac{m}{7} = 1$

21. $-\dfrac{x}{2} = -\dfrac{3}{4}$

22. $-\dfrac{x}{3} = \dfrac{5}{6}$

23. $\dfrac{2}{3}a = 8$

24. $\dfrac{3}{4}a = 6$

25. $-\dfrac{3}{5}x = \dfrac{9}{5}$

26. $-\dfrac{2}{5}x = \dfrac{6}{15}$

27. $-\dfrac{5}{8}y = -20$

28. $-\dfrac{7}{2}y = -14$

SCAN TO ACCESS

Simplify both sides as much as possible, and then solve.

29. $-4x - 2x + 3x = 24$

30. $7x - 5x + 8x = 20$

31. $4x + 8x - 2x = 15 - 10$

32. $5x + 4x + 3x = 4 + 8$

33. $-3 - 5 = 3x + 5x - 10x$

34. $10 - 16 = 12x - 6x - 3x$

35. $18 - 13 = \dfrac{1}{2}a + \dfrac{3}{4}a - \dfrac{5}{8}a$

36. $20 - 14 = \dfrac{1}{3}a + \dfrac{5}{6}a - \dfrac{2}{3}a$

37. $\dfrac{1}{2}x - \dfrac{1}{3}x = 2 - 4$

38. $\dfrac{1}{3}x - \dfrac{1}{4}x = \dfrac{1}{2} + 2$

39. $\dfrac{1}{2}x + \dfrac{1}{3}x = \dfrac{1}{2} - 3$

40. $\dfrac{1}{4}y + \dfrac{1}{3}y = 2 - 16$

Solve the following equations by multiplying both sides by -1.

41. $-x = 4$

42. $-x = -3$

43. $-x = -4$

44. $-x = 3$

45. $15 = -a$

46. $-15 = -a$

47. $-y = \dfrac{1}{2}$

48. $-y = -\dfrac{3}{4}$

B Solve each of the following equations using the addition and multiplication properties where necessary.

49. $3x - 2 = 7$

50. $2x - 3 = 9$

51. $2a + 1 = 3$

52. $5a - 3 = 7$

53. $\dfrac{1}{8} + \dfrac{1}{2}x = \dfrac{1}{4}$

54. $\dfrac{1}{3} + \dfrac{1}{7}x = -\dfrac{8}{21}$

55. $6x = 2x - 12$

56. $8x = 3x - 10$

57. $2y = -4y + 18$

58. $3y = -2y - 15$

59. $-7x = -3x - 8$

60. $-5x = -2x - 12$

61. $8x + 4 = 2x - 5$

62. $5x + 6 = 3x - 6$

63. $x + \dfrac{1}{2} = \dfrac{1}{4}x - \dfrac{5}{8}$

64. $\dfrac{1}{3}x + \dfrac{2}{5} = \dfrac{1}{5}x - \dfrac{2}{5}$

65. $6m - 3 = m + 2$

66. $6m - 5 = m + 5$

67. $\dfrac{1}{2}m - \dfrac{1}{4} = \dfrac{1}{12}m + \dfrac{1}{6}$

68. $\dfrac{1}{2}m - \dfrac{5}{12} = \dfrac{1}{12}m + \dfrac{5}{12}$

69. $9y + 2 = 6y - 4$

70. $6y + 14 = 2y - 2$

71. $\dfrac{3}{2}y + \dfrac{1}{3} = y - \dfrac{2}{3}$

72. $\dfrac{3}{2}y + \dfrac{7}{2} = \dfrac{1}{2}y - \dfrac{1}{2}$

Applying the Concepts

73. Break-Even Point Movie theaters pay a certain price for the movies that you and I see. Suppose a theater pays $1,500 for each showing of a popular movie. If they charge $7.50 for each ticket they sell, then the equation

$$7.5x = 1,500$$

gives the number of tickets they must sell to equal the $1,500 cost of showing the movie. This number is called the break-even point. Solve the equation for x to find the break-even point.

74. Basketball Laura plays basketball for her community college. In one game she scored 13 points total, with a combination of free throws, field goals, and three-pointers. Each free throw is worth 1 point, each field goal is 2 points, and each three-pointer is worth 3 points. If she made 1 free throw and 3 field goals, then solving the equation

$$1 + 3(2) + 3x = 13$$

will give us the number of three-pointers she made. Solve the equation to find the number of three-point shots Laura made.

75. Taxes Suppose 21% of your monthly pay is withheld for federal income taxes and another 8% is withheld for Social Security, state income tax, and other miscellaneous items. If you are left with $987.50 a month in take-home pay, then the amount you earned before the deductions were removed from your check is given by the equation

$$G - 0.21G - 0.08G = 987.5$$

Solve this equation to find your gross income.

76. Rhind Papyrus The *Rhind Papyrus* is an ancient document that contains mathematical riddles. One problem asks the reader to find a quantity such that when it is added to one-fourth of itself the sum is 15. The equation that describes this situation is

$$x + \dfrac{1}{4}x = 15$$

Solve this equation.

Angles The sum of the measures of the interior angles in a triangle is 180°. Given the following angle measures, set up an equation and solve for x.

77.

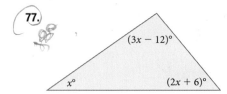

$(3x - 12)°$

$x°$ $(2x + 6)°$

78.

$x°$

$(x + 10)°$

Angles Solve for x.

79.

$(3x - 15)°$

$(2x)°$

80.

$(4x - 60)°$

$(x + 20)°$

$4x - 60 + x + 20 = 180$

$5x - 40 = 180$

$5x = 220$

$\frac{5x}{5} = \frac{220}{5}$

$x = 44$

81. Music Downloads Suppose you wish to purchase music through the internet for a personal playlist. It costs $10 to join and $1.50 per song. You have $40 in credit to use toward joining and making your playlist. How many songs can you purchase without going over budget?

82. Music Downloads Ellis and Celia wish to purchase music through the internet. Ellis finds a site that costs $20 to join and $1.25 per song thereafter. Celia finds one that costs $10 to join and $1.50 per song thereafter. How many songs must they purchase so that they have spent the same amount?

83. Recall from section 1.6 the formula $P = 14.7 + 0.4x$, which gives the pressure (in lb/in²) at a diving depth of x feet. Use this formula to find the diving depths which would result in the following pressures:

a. 22.7 lb/in²

b. 28.7 lb/in²

c. 44.7 lb/in²

84. Suppose the total bill from a water company is given by $B = 7.5 + 0.0058w$, where B is the total bill and w is the water usage (in cubic feet). Use this formula to find the water usage resulting in the following monthly water bills.

a. $56.80

b. $94.50

c. $167.00

85. The formula for Franco's pay is $P = 320 + 12(h - 40)$, where P is the pay (in $) and h is the hours worked. How many hours must Franco work to earn $500 in one week?

86. The formula for Esmeralda's weekly pay is $P = 400 + 15(h - 40)$, where P is the pay (in $) and h is the hours worked. How many hours must Esmeralda work to earn $520 in one week?

Getting Ready for the Next Section

To understand all of the explanations and examples in the next section you must be able to work the problems below.

Solve each equation.

87. $2x = 4$

88. $3x = 24$

89. $30 = 5x$

90. $0 = 5x$

91. $0.17x = 510$

92. $0.1x = 400$

93. $0.12x = 78$

94. $0.35x = 420$

Apply the distributive property and then simplify if possible.

95. $3(x - 5) + 4$

96. $5(x - 3) + 2$

97. $0.09(x + 2,000)$

98. $0.04(x + 7,000)$

99. $7 - 3(2y + 1)$

100. $4 - 2(3y + 1)$

101. $3(2x - 5) - (2x - 4)$

102. $4(3x - 2) - (6x - 5)$

Find the Mistake

Each sentence below contains a mistake. Circle the mistake and write the correct word or number on the line provided.

1. The multiplication property of equality states that multiplying one side of an equation by the same nonzero number will not change the solution set. _____

2. To solve $-\frac{2b}{9} = 3$ for b, we multiply both sides of the equation by the opposite of $-\frac{2}{9}$. _____

3. To begin solving the equation $\frac{5}{6}x + \frac{1}{2} = -\frac{1}{4}$, we can either add $\frac{5}{6}$ to each side of the equation or multiply both sides by the LCD 12. _____

4. To solve the equation $-\frac{8}{9}y = -3$, we begin by multiplying both sides of the equation by the LCD of 9 and 3. _____

A Solve a linear equation in one variable.

2.3 Solving Linear Equations in One Variable

©iStockphoto.com/mevans

In the spring of 2010, over 500 robots took over a California town for the seventh annual RoboGames, held at the San Mateo County Expo Center. Robots of all shapes and sizes competed in events, such as stair climbing, obstacle running, kung-fu, sumo wrestling, basketball, and even combat battles. Advance tickets to the games were $20, and entry fees for most events were $35 each. Suppose you purchased an advance ticket and entered a robot into x number of events, spending $125 total. We could write the following equation based on the above information, and then solve for x to determine how many events you entered.

$$35x + 20 = 125$$

In this section, we will use the material we have developed in the first two sections of this chapter to build a method for solving any linear equation such as the one above.

A Solving Linear Equations

The equations we have worked with in the first two sections of this chapter are called *linear equations in one variable*. Let's begin with a formal definition.

> **DEFINITION** linear equation in one variable
>
> A *linear equation in one variable* is any equation that can be put in the form $Ax + B = C$, where A, B, and C are real numbers and A is not zero.

Each of the equations we will solve in this section is a linear equation in one variable. The steps we use to solve a linear equation in one variable are listed here:

> **HOW TO** Strategy for Solving Linear Equations in One Variable
>
> **Step 1a:** Use the distributive property to separate terms, if necessary.
>
> **1b:** If fractions are present, consider multiplying both sides by the LCD to eliminate the fractions. If decimals are present, consider multiplying both sides by a power of 10 to clear the equation of decimals.
>
> **1c:** Combine similar terms on each side of the equation.
>
> **Step 2:** Use the addition property of equality to get all variable terms on one side of the equation and all constant terms on the other side. A variable term is a term that contains the variable (for example, 5x). A constant term is a term that does not contain the variable (the number 3, for example).
>
> **Step 3:** Use the multiplication property of equality to get x (that is, 1x) by itself on one side of the equation.
>
> **Step 4:** Check your solution in the original equation to be sure that you have not made a mistake in the solution process.

Note You may have some previous experience solving equations. Even so, you should solve the equations in this section using the method developed here. Your work should look like the examples in the text. If you have learned shortcuts or a different method of solving equations somewhere else, you can always go back to them later. What is important now is that you are able to solve equations by the methods shown here.

As you work through the examples in this section, it is not always necessary to use all four steps when solving equations. The number of steps used depends on the equation. In Example 1, there are no fractions or decimals in the original equation, so step 1b will not be used. Likewise, after applying the distributive property to the left side of the equation in Example 1, there are no similar terms to combine on either side of the equation, making step 1c also unnecessary.

EXAMPLE 1 Solve $2(x + 3) = 10$.

Solution To begin, we apply the distributive property to the left side of the equation to separate terms.

Step 1a: $\qquad\qquad 2x + 6 = 10$ $\qquad\qquad$ Distributive property

Step 2: $\qquad 2x + 6 + (-6) = 10 + (-6)$ \qquad Addition property of equality

$\qquad\qquad\qquad\quad 2x = 4$

Step 3: $\qquad\qquad \dfrac{1}{2}(2x) = \dfrac{1}{2}(4)$ $\qquad\qquad$ Multiply each side by $\dfrac{1}{2}$.

$\qquad\qquad\qquad\quad x = 2$ $\qquad\qquad$ The solution is 2.

The solution to our equation is 2. We check our work (to be sure we have not made either a mistake in applying the properties or an arithmetic mistake) by substituting 2 into our original equation and simplifying each side of the result separately.

Step 4: \qquad When \rightarrow $\qquad\qquad\qquad x = 2$

\qquad the equation \rightarrow $\qquad 2(x + 3) = 10$

\qquad becomes \rightarrow $\qquad\quad 2(2 + 3) \overset{?}{=} 10$

$\qquad\qquad\qquad\qquad\qquad 2(5) \overset{?}{=} 10$

$\qquad\qquad\qquad\qquad\qquad 10 = 10$ \quad A true statement

Our solution checks.

The general method of solving linear equations is actually very simple. We can add any number to both sides of the equation and multiply both sides by any nonzero number. The equation may change in form, but the solution set will not.

The examples that follow show a variety of equations and their solutions. When you have finished this section and worked the problems in the problem set, the steps in the solution process should be a description of how you operate when solving equations.

EXAMPLE 2 Solve $3(x - 5) + 4 = 13$ for x.

Solution Our first step will be to apply the distributive property to the left side of the equation.

Step 1a: $\qquad 3x - 15 + 4 = 13$ $\qquad\qquad$ Distributive property

Step 1c: $\qquad\quad 3x - 11 = 13$ $\qquad\qquad$ Simplify the left side.

Step 2: $\qquad 3x - 11 + 11 = 13 + 11$ \qquad Add 11 to both sides.

$\qquad\qquad\qquad\quad 3x = 24$

Step 3: $\qquad\qquad \dfrac{1}{3}(3x) = \dfrac{1}{3}(24)$ $\qquad\qquad$ Multiply both sides by $\dfrac{1}{3}$.

$\qquad\qquad\qquad\quad x = 8$ $\qquad\qquad$ The solution is 8.

Step 4: \qquad When \rightarrow $\qquad\qquad\qquad x = 8$

\qquad the equation \rightarrow $\qquad 3(x - 5) + 4 = 13$

\qquad becomes \rightarrow $\qquad 3(8 - 5) + 4 \overset{?}{=} 13$

$\qquad\qquad\qquad\qquad\qquad 3(3) + 4 \overset{?}{=} 13$

$\qquad\qquad\qquad\qquad\qquad 9 + 4 \overset{?}{=} 13$

$\qquad\qquad\qquad\qquad\qquad 13 = 13$ \qquad A true statement

Practice Problems

1. Solve $3(x + 4) = 6$.

2. Solve $2(x - 3) + 3 = 9$ for x.

Answers

1. -2

2. 6

3. Solve.

$$7(x - 3) + 5 = 4(3x - 2) - 8$$

EXAMPLE 3 Solve $5(x - 3) + 2 = 5(2x - 8) - 3$.

Solution In this case, we first apply the distributive property to each side of the equation.

Step 1a:	$5x - 15 + 2 = 10x - 40 - 3$	*Distributive property*
Step 1c:	$5x - 13 = 10x - 43$	*Simplify each side.*
Step 2:	$5x + (-5x) - 13 = 10x + (-5x) - 43$	*Add $-5x$ to both sides.*
	$-13 = 5x - 43$	
	$-13 + 43 = 5x - 43 + 43$	*Add 43 to both sides.*
	$30 = 5x$	
Step 3:	$\dfrac{1}{5}(30) = \dfrac{1}{5}(5x)$	*Multiply both sides by $\frac{1}{5}$.*
	$6 = x$	*The solution is 6.*

Step 4: Replacing x with 6 in the original equation, we have

$$5(6 - 3) + 2 \stackrel{?}{=} 5(2 \cdot 6 - 8) - 3$$

$$5(3) + 2 \stackrel{?}{=} 5(12 - 8) - 3$$

$$5(3) + 2 \stackrel{?}{=} 5(4) - 3$$

$$15 + 2 \stackrel{?}{=} 20 - 3$$

$$17 = 17 \qquad \textit{A true statement}$$

Note It makes no difference on which side of the equal sign x ends up. Most people prefer to have x on the left side because we read from left to right, and it seems to sound better to say x is 6 rather than 6 is x. Both expressions, however, have exactly the same meaning.

4. Solve the equation.

$$0.06x + 0.04(x + 7{,}000) = 680$$

EXAMPLE 4 Solve the equation $0.08x + 0.09(x + 2{,}000) = 690$.

Solution We can solve the equation in its original form by working with the decimals, or we can eliminate the decimals first by using the multiplication property of equality and solving the resulting equation. Both methods follow.

Method 1
Working with the decimals.

	$0.08x + 0.09(x + 2{,}000) = 690$	*Original equation*
Step 1a:	$0.08x + 0.09x + 0.09(2{,}000) = 690$	*Distributive property*
Step 1c:	$0.17x + 180 = 690$	*Simplify the left side.*
Step 2:	$0.17x + 180 + (-180) = 690 + (-180)$	*Add -180 to each side.*
	$0.17x = 510$	
Step 3:	$\dfrac{0.17x}{0.17} = \dfrac{510}{0.17}$	*Divide each side by 0.17.*
	$x = 3{,}000$	

Note that we divided each side of the equation by 0.17 to obtain the solution. This is still an application of the multiplication property of equality because dividing by 0.17 is equivalent to multiplying by $\frac{1}{0.17}$.

Method 2
Eliminating the decimals in the beginning.

	$0.08x + 0.09(x + 2{,}000) = 690$	*Original equation*
Step 1a:	$0.08x + 0.09x + 180 = 690$	*Distributive property*
Step 1b:	$100(0.08x + 0.09x + 180) = 100(690)$	*Multiply both sides by 100.*
	$8x + 9x + 18{,}000 = 69{,}000$	

Answers

3. 0

4. 4,000

Step 1c:	$17x + 18{,}000 = 69{,}000$	Simplify the left side.
Step 2:	$17x = 51{,}000$	Add $-18{,}000$ to each side.
Step 3:	$\dfrac{17x}{17} = \dfrac{51{,}000}{17}$	Divide each side by 17.
	$x = 3{,}000$	

Substituting 3,000 for x in the original equation, we have

Step 4:	$0.08(3{,}000) + 0.09(3{,}000 + 2{,}000) \overset{?}{=} 690$
	$0.08(3{,}000) + 0.09(5{,}000) \overset{?}{=} 690$
	$240 + 450 \overset{?}{=} 690$
	$690 = 690$ A true statement

EXAMPLE 5 Solve $7 - 3(2y + 1) = 16$.

Solution We begin by multiplying -3 times the sum of $2y$ and 1.

Step 1a:	$7 - 6y - 3 = 16$	Distributive property
Step 1c:	$-6y + 4 = 16$	Simplify the left side.
Step 2:	$-6y + 4 + (-4) = 16 + (-4)$	Add -4 to both sides.
	$-6y = 12$	
Step 3:	$-\dfrac{1}{6}(-6y) = -\dfrac{1}{6}(12)$	Multiply both sides by $-\dfrac{1}{6}$.
	$y = -2$	

There are two things to notice about the example that follows. First, the distributive property is used to remove parentheses that are preceded by a negative sign. Second, the addition property and the multiplication property are not shown in as much detail as in the previous examples.

EXAMPLE 6 Solve $3(2x - 5) - (2x - 4) = 6 - (4x + 5)$.

Solution When we apply the distributive property to remove the grouping symbols and separate terms, we have to be careful with the signs. Remember, we can think of $-(2x - 4)$ as $-1(2x - 4)$, so that

$$-(2x - 4) = -1(2x - 4) = -2x + 4$$

It is not uncommon for students to make a mistake with this type of simplification and write the result as $-2x - 4$, which is incorrect. Here is the complete solution to our equation:

$3(2x - 5) - (2x - 4) = 6 - (4x + 5)$	Original equation
$6x - 15 - 2x + 4 = 6 - 4x - 5$	Distributive property
$4x - 11 = -4x + 1$	Simplify each side.
$8x - 11 = 1$	Add $4x$ to each side.
$8x = 12$	Add 11 to each side.
$x = \dfrac{12}{8}$	Multiply each side by $\dfrac{1}{8}$.
$x = \dfrac{3}{2}$	Reduce to lowest terms.

The solution, $\frac{3}{2}$, checks when replacing x in the original equation.

5. Solve $4 - 2(3y + 1) = -16$.

6. Solve.
$4(3x - 2) - (6x - 5) = 6 - (3x + 1)$

Answers

5. 3

6. $\dfrac{8}{9}$

7. Solve $2(t + 2) - 2 = 2(t + 3) - 4$.

EXAMPLE 7 Solve $-2(3t - 2) = 2(4 - 3t) - 4$.

Solution We apply the distributive property to simplify both sides of the equation.

$$-2(3t - 2) = 2(4 - 3t) - 4 \qquad \textit{Original equation}$$

$$-6t + 4 = 8 - 6t - 4 \qquad \textit{Distributive property}$$

$$-6t + 4 = -6t + 4 \qquad \textit{Simplify each side.}$$

$$-6t = -6t \qquad \textit{Add } -4 \textit{ to each side.}$$

$$0 = 0 \qquad \textit{Add 6t to each side.}$$

We have eliminated the variable and are left with a true statement. This means that any real number that we use in place of the variable will be a solution to the equation. The solution is all real numbers.

8. Solve $-2(3y - 2) = 3(3 - 2y)$.

EXAMPLE 8 Solve $4(y - 5) + 3 = 2(2y - 3)$.

Solution Again, we apply the distributive property.

$$4(y - 5) + 3 = 2(2y - 3) \qquad \textit{Original equation}$$

$$4y - 20 + 3 = 4y - 6 \qquad \textit{Distributive property}$$

$$4y - 17 = 4y - 6 \qquad \textit{Simplify each side.}$$

$$-17 = -6 \qquad \textit{Add } -4y \textit{ to each side.}$$

We have eliminated the variable and are left with a false statement. This means that no real numbers used in place of the variable will be a solution to the equation. There is no solution.

GETTING READY FOR CLASS

After reading through the preceding section, respond in your own words and in complete sentences.

A. What is the first step in solving a linear equation containing parentheses?

B. When solving a linear equation, why should you get all variable terms on one side and all constant terms on the other before using the multiplication property of equality?

C. What is the last step in solving a linear equation?

D. If an equation contains decimals, what can you do to eliminate the decimals?

Answers

7. All real numbers
8. No solution

Vocabulary Review

The following is a list of steps for solving a linear equation in one variable. Choose the correct words to fill in the blanks below.

solution	addition	multiplication	distributive
LCD	decimals	similar	constant

Step 1a: Use the _____ property to separate terms.

 1b: To clear any fractions, multiply both sides of the equation by the _____. To clear _____, multiply both sides by a power of 10.

 1c: Combine _____ terms.

Step 2: Use the _____ property of equality to get all variable terms on one side of the equation and all _____ terms on the other side.

Step 3: Use the _____ property of equality to get x by itself on one side of the equation.

Step 4: Check your _____ in the original equation.

Problems

A Solve each of the following equations using the four steps shown in this section.

1. $2(x + 3) = 12$

2. $3(x - 2) = 6$

3. $6(x - 1) = -18$

4. $4(x + 5) = 16$

5. $2(4a + 1) = -6$

6. $3(2a - 4) = 12$

7. $14 = 2(5x - 3)$

8. $-25 = 5(3x + 4)$

9. $-2(3y + 5) = 14$

10. $-3(2y - 4) = -6$

11. $-5(2a + 4) = 0$

12. $-3(3a - 6) = 0$

13. $1 = \frac{1}{2}(4x + 2)$

14. $1 = \frac{1}{3}(6x + 3)$

15. $3(t - 4) + 5 = -4$

16. $5(t - 1) + 6 = -9$

SCAN TO ACCESS

Solve each equation.

17. $4(2x + 1) - 7 = 1$

18. $6(3y + 2) - 8 = -2$

19. $\frac{1}{2}(x - 3) = \frac{1}{4}(x + 1)$

20. $\frac{1}{3}(x - 4) = \frac{1}{2}(x - 6)$

21. $-0.7(2x - 7) = 0.3(11 - 4x)$

22. $-0.3(2x - 5) = 0.7(3 - x)$

23. $-2(3y + 3) = 3(1 - 2y) - 9$

24. $-5(4y - 3) = 2(2 - 10y) + 11$

25. $\frac{3}{4}(8x - 4) + 3 = \frac{2}{5}(5x + 10) - 1$

26. $\frac{5}{6}(6x + 12) + 1 = \frac{2}{3}(9x - 3) + 5$

27. $0.06x + 0.08(100 - x) = 6.5$

28. $0.05x + 0.07(100 - x) = 6.2$

29. $6 - 5(2a - 3) = 1$

30. $-8 - 2(3 - a) = 0$

31. $0.2x - 0.5 = 0.5 - 0.2(2x - 13)$

32. $0.4x - 0.1 = 0.7 - 0.3(6 - 2x)$

33. $2(t - 3) + 3(t - 2) = 28$

34. $-3(t - 5) - 2(2t + 1) = -8$

35. $5(x - 2) - (3x + 4) = 3(6x - 8) + 10$

36. $3(x - 1) - (4x - 5) = 2(5x - 1) - 7$

37. $2(5x - 3) - (2x - 4) = 5 - (6x + 1)$

38. $3(4x - 2) - (5x - 8) = 8 - (2x + 3)$

39. $-(3x + 1) + 7 = 4 - (3x + 2)$

40. $-(6x + 2) = 8 - 2(3x + 1)$

41. $x + (2x - 1) = 2$

42. $x + (5x + 2) = 20$

43. $x - (3x + 5) = -3$

44. $x - (4x - 1) = 7$

45. $15 = 3(x - 1)$

46. $12 = 4(x - 5)$

47. $4x - (-4x + 1) = 5$

48. $-2x - (4x - 8) = -1$

49. $5x - 8(2x - 5) = 7$

50. $3x + 4(8x - 15) = 10$

51. $2(4x - 1) = 4(2x + 3)$

52. $2(3y - 3) = 3(2y + 4)$

53. $0.2x + 0.5(12 - x) = 3.6$

54. $0.3x + 0.6(25 - x) = 12$

55. $0.5x + 0.2(18 - x) = 5.4$

56. $0.1x + 0.5(40 - x) = 32$

57. $x + (x + 3)(-3) = x - 3$

58. $x - 2(x + 2) = x - 2$

59. $5(x + 2) + 3(x - 1) = -9$

60. $4(x + 1) + 3(x - 3) = 2$

61. $3(x - 3) + 2(2x) = 5$

62. $2(x - 2) + 3(5x) = 30$

63. $-5(y + 3) = -5y - 15$

64. $3(4y - 1) - 5 = 4(3y - 2)$

65. $3x + 2(x - 2) = 6$

66. $5x - (x - 5) = 25$

67. $50(x - 5) = 30(x + 5)$

68. $34(x - 2) = 26(x + 2)$

69. $0.08x + 0.09(x + 2,000) = 860$

70. $0.11x + 0.12(x + 4,000) = 940$

71. $0.10x + 0.12(x + 500) = 214$

72. $0.08x + 0.06(x + 800) = 104$

73. $5x + 10(x + 8) = 245$

74. $5x + 10(x + 7) = 175$

75. $5x + 10(x + 3) + 25(x + 5) = 435$

76. $5(x + 3) + 10x + 25(x + 7) = 390$

The next two problems are intended to give you practice reading, and paying attention to, the instructions that accompany the problems you are working. Working these problems is an excellent way to get ready for a test or a quiz.

77. Work each problem according to the instructions.

 a. Solve $4x - 5 = 0$.

 b. Solve $4x - 5 = 25$.

 c. Add $(4x - 5) + (2x + 25)$.

 d. Solve $4x - 5 = 2x + 25$.

 e. Multiply $4(x - 5)$.

 f. Solve $4(x - 5) = 2x + 25$.

78. Work each problem according to the instructions.

 a. Solve $3x + 6 = 0$.

 b. Solve $3x + 6 = 4$.

 c. Add $(3x + 6) + (7x + 4)$.

 d. Solve $3x + 6 = 7x + 4$.

 e. Multiply $3(x + 6)$.

 f. Solve $3(x + 6) = 7x + 4$.

79. Solve the equation $35x + 20 = 125$, given at the introduction of this section, to find the number of robot events that were entered.

80. A traveling circus with thrill rides charges $6 for admission plus $1.50 for each ride you take. If you pay $16.50 total, how many rides did you take?

81. A stock trading company charges a commission of $15 per trade plus $0.025 per dollar value of stock traded. If your commission is $120, what was the value of the stock you traded?

82. Brad earns a base monthly salary of $2,400 plus $0.018 commission on his total sales. If his monthly pay is $3,246, what were his sales that month?

Getting Ready for the Next Section

To understand all of the explanations and examples in the next section you must be able to work the problems below.

Solve each equation.

83. $40 = 2x + 12$

84. $80 = 2x + 12$

85. $12 + 2y = 6$

86. $3x + 18 = 6$

87. $24x = 6$

88. $45 = 0.75x$

89. $70 = x \cdot 210$

90. $15 = x \cdot 80$

Apply the distributive property.

91. $\dfrac{1}{2}(-3x + 6)$

92. $\dfrac{2}{3}(-6x + 9)$

93. $-\dfrac{1}{4}(-5x + 20)$

94. $-\dfrac{1}{3}(-4x + 12)$

95. $12\left(\dfrac{1}{3}x - \dfrac{1}{4}\right)$

96. $-8\left(-\dfrac{1}{4}x + \dfrac{1}{8}\right)$

97. $-24\left(-\dfrac{1}{3}x - \dfrac{1}{8}\right)$

98. $-15\left(\dfrac{3}{5}x - \dfrac{2}{3}\right)$

Find the Mistake

Each sentence below contains a mistake. Circle the mistake and write the correct word(s) or equation on the line provided.

1. A linear equation in one variable is any equation that can be put in the form of $y = Ax + B$. _____

2. $12 + 6y - 4 = 3y$ is not a linear equation in one variable because it cannot be simplified to the form $Ax + B = C$.

3. To solve the linear equation $3(2x + 7) = 2(x - 1)$, begin by dividing both sides of the equation by $(x - 1)$.

4. To solve a linear equation, make sure to keep similar terms on opposite sides of the equation.

2.4 Formulas

OBJECTIVES

A Find the value of a variable in a formula given replacements for the other variables.

B Solve a formula for one of its variables.

C Continue to find the complement and supplement of an angle.

D Solve basic percent problems.

KEY WORDS

formula

solving for a variable

Recall our discuss from Section 1 about finding the outside temperature by listening to a cricket chirp. What if we needed to find the number of chirps based on a Celsius temperature? The formula to find the outside temperature in Celsius is

$$t = \left(\frac{n}{3}\right) + 4$$

where t is the outside temperature in degrees Celsius and n is the number of cricket chirps in 25 seconds. In this section, we will learn about formulas and how to solve a formula for one of its variables. For instance, we will learn how the Celsius temperature formula could be solved for n and become

$$n = 3(t - 4)$$

Then we could provide a numerical replacement for t and easily calculate a value for n.

A Solving Formulas

In this section, we continue solving equations by working with formulas. Here is the definition of a formula.

> **DEFINITION** formula
>
> In mathematics, a **formula** is an equation that contains more than one variable.

The equation $P = 2l + 2w$, which tells us how to find the perimeter of a rectangle, is an example of a formula.

To begin our work with formulas, we will consider some examples in which we are given numerical replacements for all but one of the variables.

Practice Problems

1. Suppose the livestock pen in Example 1 has a perimeter of 80 feet. If the width is still 6 feet, what is the new length?

EXAMPLE 1 The perimeter P of a rectangular livestock pen is 40 feet. If the width w is 6 feet, find the length.

Solution First, we substitute 40 for P and 6 for w in the formula $P = 2l + 2w$. Then we solve for l.

$$
\begin{aligned}
\text{When} \rightarrow \qquad & P = 40 \text{ and } w = 6 \\
\text{the formula} \rightarrow \qquad & P = 2l + 2w \\
\text{becomes} \rightarrow \qquad & 40 = 2l + 2(6) \\
& 40 = 2l + 12 \qquad && \text{Multiply 2 and 6.} \\
& 28 = 2l \qquad && \text{Add} -12 \text{ to each side.} \\
& 14 = l \qquad && \text{Multiply each side by } \tfrac{1}{2}.
\end{aligned}
$$

To summarize our results, if a rectangular pen has a perimeter of 40 feet and a width of 6 feet, then the length must be 14 feet.

2. Find x when y is 9 in the formula $3x + 2y = 6$.

EXAMPLE 2 Find y when $x = 4$ in the formula $3x + 2y = 6$.

Solution We substitute 4 for x in the formula and then solve for y.

$$
\begin{aligned}
\text{When} \rightarrow \qquad & x = 4 \\
\text{the formula} \rightarrow \qquad & 3x + 2y = 6 \\
\text{becomes} \rightarrow \qquad & 3(4) + 2y = 6 \\
& 12 + 2y = 6 \qquad && \text{Multiply 3 and 4.} \\
& 2y = -6 \qquad && \text{Add} -12 \text{ to each side.} \\
& y = -3 \qquad && \text{Multiply each side by } \tfrac{1}{2}.
\end{aligned}
$$

B Solving for a Variable

In the next examples we will solve a formula for one of its variables without being given numerical replacements for the other variables.

Consider the formula for the area of a triangle:

$A = \tfrac{1}{2}bh$

where A = area, b = length of the base, and h = height of the triangle.

Suppose we want to solve this formula for h. What we must do is isolate the variable h on one side of the equal sign. We begin by multiplying both sides by 2, because it is the reciprocal of $\tfrac{1}{2}$.

$$2 \cdot A = 2 \cdot \frac{1}{2}bh$$

$$2A = bh$$

Then we divide both sides by b.

$$\frac{2A}{b} = \frac{bh}{b}$$

$$h = \frac{2A}{b}$$

Answers

1. 34 feet
2. -4

The original formula $A = \tfrac{1}{2}bh$ and the final formula $h = \frac{2A}{b}$ both give the same relationship among A, b, and h. The first one has been solved for A and the second one has been solved for h.

> **RULE** Solving for a Variable
>
> To solve a formula for one of its *variables*, we must isolate that variable on either side of the equal sign. All other variables and constants will appear on the other side.

EXAMPLE 3 Solve $3x + 2y = 6$ for y.

Solution To solve for y, we must isolate y on the left side of the equation. To begin, we use the addition property of equality to add $-3x$ to each side.

$$3x + 2y = 6 \qquad \text{Original formula}$$

$$3x + (-3x) + 2y = (-3x) + 6 \qquad \text{Add } -3x \text{ to each side.}$$

$$2y = -3x + 6 \qquad \text{Simplify the left side.}$$

$$\frac{1}{2}(2y) = \frac{1}{2}(-3x + 6) \qquad \text{Multiply each side by } \tfrac{1}{2}.$$

$$y = -\frac{3}{2}x + 3$$

EXAMPLE 4 Solve $h = vt - 16t^2$ for v.

Solution Let's begin by interchanging the left and right sides of the equation. That way, the variable we are solving for, v, will be on the left side.

$$vt - 16t^2 = h \qquad \text{Symmetric property}$$

$$vt - 16t^2 + 16t^2 = h + 16t^2 \qquad \text{Add } 16t^2 \text{ to each side.}$$

$$vt = h + 16t^2$$

$$\frac{vt}{t} = \frac{h + 16t^2}{t} \qquad \text{Divide each side by } t.$$

$$v = \frac{h + 16t^2}{t}$$

We know we are finished because we have isolated the variable we are solving for on the left side of the equation and it does not appear on the other side.

EXAMPLE 5 Solve $\dfrac{y - 1}{x} = \dfrac{3}{2}$ for y.

Solution Although we will do more extensive work with formulas of this form later in the book, we need to know how to solve this particular formula for y in order to understand some things in the next chapter. We begin by multiplying each side of the formula by x. Doing so will simplify the left side of the equation, and make the rest of the solution process simple.

$$\frac{y - 1}{x} = \frac{3}{2} \qquad \text{Original formula}$$

$$x \cdot \frac{y - 1}{x} = \frac{3}{2} \cdot x \qquad \text{Multiply each side by } x.$$

$$y - 1 = \frac{3}{2}x \qquad \text{Simplify each side.}$$

$$y = \frac{3}{2}x + 1 \qquad \text{Add 1 to each side.}$$

This is our solution. If we look back to the first step, we can justify our result on the left side of the equation this way: Dividing by x is equivalent to multiplying by its reciprocal $\frac{1}{x}$. Here is what it looks like when written out completely:

$$x \cdot \frac{y - 1}{x} = x \cdot \frac{1}{x}(y - 1) = 1(y - 1) = (y - 1)$$

3. Solve $5x - 4y = 20$ for y.

4. Solve $P = 2w + 2l$ for l.

5. Solve $\dfrac{y - 2}{x} = \dfrac{4}{3}$ for y.

Answers

3. $y = \frac{5}{4}x - 5$

4. $l = \frac{P - 2w}{2}$

5. $y = \frac{4}{3}x + 2$

C Complementary and Supplementary Angles

> **FACTS FROM GEOMETRY** More on Complementary and Supplementary Angles
>
> In Chapter 1 we defined complementary angles as angles that add to 90°; that is, if x and y are complementary angles, then
>
> $$x + y = 90°$$
>
> If we solve this formula for y, we obtain a formula equivalent to our original formula:
>
> $$y = 90° - x$$
>
> Because y is the complement of x, we can generalize by saying that the complement of angle x is the angle $90° - x$. By a similar reasoning process, we can say that the supplement of angle x is the angle $180° - x$. To summarize, if x is an angle, then
>
> The complement of x is $90° - x$, and
>
> The supplement of x is $180° - x$
>
> If you go on to take a trigonometry class, you will see this formula again.

6. Find the complement and the supplement of 35°.

EXAMPLE 6 Find the complement and the supplement of 25°.

Solution We can use the formulas above with $90° - x$ and $180° - x$.

The complement of 25° is $90° - 25° = 65°$.

The supplement of 25° is $180° - 25° = 155°$.

D Basic Percent Problems

The last examples in this section show how basic percent problems can be translated directly into equations. To understand these examples, you must recall that *percent* means "per hundred" that is, 75% is the same as $\frac{75}{100}$, 0.75, and, in reduced fraction form, $\frac{3}{4}$. Likewise, the decimal 0.25 is equivalent to 25%. To change a decimal to a percent, we move the decimal point two places to the right and write the % symbol. To change from a percent to a decimal, we drop the % symbol and move the decimal point two places to the left. The table that follows gives some of the most commonly used fractions and decimals and their equivalent percents.

Fraction	Decimal	Percent
$\frac{1}{2}$	0.5	50%
$\frac{1}{4}$	0.25	25%
$\frac{3}{4}$	0.75	75%
$\frac{1}{3}$	$0.33\overline{3}$	$33\frac{1}{3}$%
$\frac{2}{3}$	0.667	$66\frac{2}{3}$%
$\frac{1}{5}$	0.2	20%
$\frac{2}{5}$	0.4	40%

Answer

6. 55°; 145°

EXAMPLE 7 What number is 25% of 60?

Solution To solve a problem like this, we let $x =$ the number in question (that is, the number we are looking for). Then, we translate the sentence directly into an equation by using an equal sign for the word "is" and multiplication for the word "of." Here is how it is done:

What number is 25% of 60?

$$x = 0.25 \cdot 60$$
$$= 15$$

Notice that we must write 25% as a decimal in order to do the arithmetic in the problem. The number 15 is 25% of 60.

7. What number is 25% of 74?

EXAMPLE 8 What percent of 24 is 6?

Solution Translating this sentence into an equation, as we did in Example 7, we have

What percent of 24 is 6?

$$x \cdot 24 = 6$$
$$24x = 6$$

Next, we multiply each side by $\dfrac{1}{24}$. (This is the same as dividing each side by 24.)

$$\frac{1}{24}(24x) = \frac{1}{24}(6)$$
$$x = \frac{6}{24}$$
$$= \frac{1}{4}$$
$$= 0.25, \text{ or } 25\%$$

25% of 24 is 6, or in other words, the number 6 is 25% of 24.

8. What percent of 84 is 21?

EXAMPLE 9 45 is 75% of what number?

Solution Again, we translate the sentence directly.

45 is 75% of what number?

$$45 = 0.75 \cdot x$$

Next, we multiply each side by $\frac{1}{0.75}$ (which is the same as dividing each side by 0.75).

$$\frac{1}{0.75}(45) = \frac{1}{0.75}(0.75x)$$
$$\frac{45}{0.75} = x$$
$$60 = x$$

The number 45 is 75% of 60.

9. 35 is 40% of what number?

Answers
7. 18.5
8. 25%
9. 87.5

10. In another pizza ingredient, pepperoni, 75% of the calories are from fat. If one serving of pepperoni contains 132 calories, how many of those calories are from fat?

EXAMPLE 10 The American Dietetic Association (ADA) recommends eating foods in which the calories from fat are less than 30% of the total calories. The nutrition labels from two ingredients you may use for a pizza are shown in Figure 1. For each ingredient, what percent of the total calories come from fat?

PIZZA DOUGH

Nutrition Facts

Serving Size 1/6 of package (65g)
Servings Per Container: 6

Amount Per Serving

Calories 160	Calories from fat 18

	% Daily Value*
Total Fat 2g	3%
Saturated Fat 0.5g	3%
Poly unsaturated Fat 0g	
Monounsaturated Fat 0g	
Cholesterol 0mg	0%
Sodium 470mg	20%
Total Carbohydrate 31g	10%
Dietary Fiber 1g	4%
Sugars 4g	
Protein 5g	

Vitamin A 0%	•	Vitamin C 0%
Calcium 0%	•	Iron 10%

*Percent Daily Values are based on a 2,000 calorie diet

MOZZARELLA CHEESE

Nutrition Facts

Serving Size 1 oz (28.3g)
Servings Per Container: 12

Amount Per Serving

Calories 72	Calories from fat 41

	% Daily Value*
Total Fat 4.5g	7%
Saturated Fat 2.9g	14%
Cholesterol 18mg	6%
Sodium 175mg	7%
Total Carbohydrate 0.8g	0%
Fiber 0g	0%
Sugars 0.3g	
Protein 6.9g	

Vitamin A 3%	•	Vitamin C 0%
Calcium 22%	•	Iron 0%

*Percent Daily Values (DV) are based on a 2,000 calorie diet

FIGURE 1

Solution The information needed to solve this problem is located towards the top of each label. Each serving of the pizza dough contains 160 calories, of which 18 calories come from fat. To find the percent of total calories that come from fat, we must answer this question:

18 is what percent of 160?

For the mozzarella cheese, one serving contains 72 calories, of which 41 calories come from fat. To find the percent of total calories that come from fat, we must answer this question:

41 is what percent of 72?

Translating each equation into symbols, we have

18 is what percent of 160

$$18 = x \cdot 160$$

$$x = \frac{18}{160}$$

$x = 0.11$ to the nearest hundredth

$$x = 11\%$$

41 is what percent of 72

$$41 = x \cdot 72$$

$$x = \frac{41}{72}$$

$x = 0.57$ to the nearest hundredth

$$x = 57\%$$

Comparing the two ingredients, 11% of the calories in the pizza dough are fat calories, whereas 57% of the calories in the mozzarella cheese are fat calories. According to the ADA, pizza dough is the healthier ingredient.

GETTING READY FOR CLASS

After reading through the preceding section, respond in your own words and in complete sentences.

A. What is a formula?

B. How do you solve a formula for one of its variables?

C. What are complementary angles?

D. Translate the following question into an equation: what number is 40% of 20?

Answer

10. 99 calories

Vocabulary Review

Choose the correct words to fill in the blanks below.

isolate　　　　　　variable　　　　　　complementary　　　　　　formula　　　　　　supplement

1. A formula is an equation with more than one _____.

2. $P = 2l + 2w$ is the _____ for finding the perimeter of a rectangle.

3. To solve for a variable in a formula, _____ that variable on either side of the equals sign.

4. If x and y are _____ angles, then $x + y = 90°$.

5. The _____ of angle x is the angle $180° - x$.

Problems

A Use the formula $P = 2l + 2w$ to find the length l of a rectangular lot if

1. The width w is 50 feet and the perimeter P is 300 feet.　　　　**2.** The width w is 75 feet and the perimeter P is 300 feet.

Use the formula $2x + 3y = 6$ to find y when

3. $x = 3$　　　　　　**4.** $x = -2$　　　　　　**5.** $x = 0$　　　　　　**6.** $x = -3$

Use the formula $2x - 5y = 20$ to find x when

7. $y = 2$　　　　　　**8.** $y = -4$　　　　　　**9.** $y = 0$　　　　　　**10.** $y = -6$

Use the equation $y = (x + 1)^2 - 3$ to find the value of y when

11. $x = -2$　　　　　　**12.** $x = -1$　　　　　　**13.** $x = 1$　　　　　　**14.** $x = 2$

15. Use the formula $y = \dfrac{20}{x}$ to find y when

　　a. $x = 10$

　　b. $x = 5$

16. Use the formula $y = 2x^2$ to find y when

　　a. $x = 5$

　　b. $x = -6$

17. Use the formula $y = Kx$ to find K when

　　a. $y = 15$ and $x = 3$

　　b. $y = 72$ and $x = 4$

18. Use the formula $y = Kx^2$ to find K when

　　a. $y = 32$ and $x = 4$

　　b. $y = 45$ and $x = 3$

SCAN TO ACCESS

B Solve each of the following for the indicated variable.

19. $A = lw$ for l

20. $d = rt$ for r

21. $V = lwh$ for h

22. $PV = nRT$ for P

23. $P = a + b + c$ for a

24. $P = a + b + c$ for b

25. $x - 3y = -1$ for x

26. $x + 3y = 2$ for x

27. $-3x + y = 6$ for y

28. $2x + y = -17$ for y

29. $2x + 3y = 6$ for y

30. $4x + 5y = 20$ for y

31. $y - 3 = -2(x + 4)$ for y

32. $y + 5 = 2(x + 2)$ for y

33. $y - 3 = -\frac{2}{3}(x + 3)$ for y

34. $y - 1 = -\frac{1}{2}(x + 4)$ for y

35. $P = 2l + 2w$ for w

36. $P = 2l + 2w$ for l

37. $h = vt + 16t^2$ for v

38. $h = vt - 16t^2$ for v

39. $A = \pi r^2 + 2\pi rh$ for h

40. $A = 2\pi r^2 + 2\pi rh$ for h

41. $I = prt$ for t

42. $PV = nrT$ for T

43. Solve for y.

 a. $\dfrac{y - 1}{x} = \dfrac{3}{5}$

 b. $\dfrac{y - 2}{x} = \dfrac{1}{2}$

 c. $\dfrac{y - 3}{x} = 4$

44. Solve for y.

 a. $\dfrac{y + 1}{x} = -\dfrac{3}{5}$

 b. $\dfrac{y + 2}{x} = -\dfrac{1}{2}$

 c. $\dfrac{y + 3}{x} = -4$

45. Solve for y.

 a. $\dfrac{y + 2}{x} = -\dfrac{1}{2}$

 b. $\dfrac{y - 2}{x} = -3$

 c. $\dfrac{y + 1}{x} = -\dfrac{1}{3}$

46. Solve for y.

 a. $\dfrac{y - 1}{x} = -\dfrac{3}{4}$

 b. $\dfrac{y + 5}{x} = -7$

 c. $\dfrac{y + 2}{x} = -3$

Solve each formula for y.

47. $\dfrac{x}{7} - \dfrac{y}{3} = 1$

48. $\dfrac{x}{5} - \dfrac{y}{9} = 1$

49. $-\dfrac{1}{4}x + \dfrac{1}{8}y = 1$

50. $-\dfrac{1}{9}x + \dfrac{1}{3}y = 1$

C Find the complement and the supplement of each angle.

51. $30°$

52. $60°$

53. $45°$

54. $15°$

D Translate each of the following into an equation, and then solve that equation.

55. What number is 25% of 40?

56. What number is 75% of 40?

57. What number is 12% of 2,000?

58. What number is 9% of 3,000?

59. What percent of 28 is 7?

60. What percent of 28 is 21?

61. What percent of 40 is 14?

62. What percent of 20 is 14?

63. 32 is 50% of what number?

64. 16 is 50% of what number?

65. 240 is 12% of what number?

66. 360 is 12% of what number?

Applying the Concepts

More About Temperatures As we mentioned in Chapter 1, in the U.S. system, temperature is measured on the Fahrenheit scale. In the metric system, temperature is measured on the Celsius scale. On the Celsius scale, water boils at 100 degrees and freezes at 0 degrees. To denote a temperature of 100 degrees on the Celsius scale, we write 100°C, which is read "100 degrees Celsius."

Table 1 is intended to give you an intuitive idea of the relationship between the two temperature scales. Table 2 gives the formulas, in both symbols and words, that are used to convert between the two scales.

Table 1

Situation	Temperature	
	Fahrenheit	Celsius
Water freezes	32°F	0°C
Room temperature	68°F	20°C
Normal body temperature	98.6°F	37°C
Water boils	212°F	100°C

Table 2

To convert from	Formula in symbols	Formula in words
Fahrenheit to Celsius	$C = \dfrac{5}{9}(F - 32)$	Subtract 32, then multiply by $\dfrac{5}{9}$.
Celsius to Fahrenheit	$F = \dfrac{9}{5}C + 32$	Multiply by $\dfrac{9}{5}$, then add 32.

67. Let $F = 212$ in the formula $C = \frac{5}{9}(F - 32)$, and solve for C. Does the value of C agree with the information in Table 1?

68. Let $C = 100$ in the formula $F = \frac{9}{5}C + 32$, and solve for F. Does the value of F agree with the information in Table 1?

69. Let $F = 68$ in the formula $C = \frac{5}{9}(F - 32)$, and solve for C. Does the value of C agree with the information in Table 1?

70. Let $C = 37$ in the formula $F = \frac{9}{5}C + 32$, and solve for F. Does the value of F agree with the information in Table 1?

71. Solve the formula $F = \frac{9}{5}C + 32$ for C.

72. Solve the formula $C = \frac{5}{9}(F - 32)$ for F.

Nutrition Labels The nutrition label in Figure 2 is from a caffe latte. The label in Figure 3 is from a nonfat caffe latte. Use the information on these labels for problems 73–76. Round your answers to the nearest tenth of a percent.

CAFFE LATTE

NONFAT CAFFE LATTE

Nutrition Facts

Serving Size 16 fl oz (453.0g)
Servings Per Container: 1

Amount Per Serving

Calories 272	Calories from fat 126

	% Daily Value*
Total Fat 14.0g	22%
Saturated Fat 9.0g	45%
Cholesterol 60mg	20%
Sodium 208mg	9%
Total Carbohydrate 22.0g	7%
Fiber 0g	0%
Sugars 19.0g	
Protein 14.0g	

Vitamin A 12%	•	Vitamin C 0%
Calcium 48%	•	Iron 0%

*Percent Daily Values (DV) are based on a 2,000 calorie diet

Nutrition Facts

Serving Size 16 fl oz (453.0g)
Servings Per Container: 1

Amount Per Serving

Calories 168	Calories from fat 0

	% Daily Value*
Total Fat 0.3g	0%
Saturated Fat 0.0g	0%
Cholesterol 8mg	3%
Sodium 232mg	10%
Total Carbohydrate 24.0g	8%
Fiber 0g	0%
Sugars 21.0g	
Protein 16.0g	

Vitamin A 16%	•	Vitamin C 0%
Calcium 48%	•	Iron 0%

*Percent Daily Values (DV) are based on a 2,000 calorie diet

FIGURE 2

FIGURE 3

73. What percent of the calories in the caffe latte are calories from fat?

74. What percent of the calories in the nonfat caffe latte are calories from fat?

75. One nonfat caffe latte weighs 453 grams, of which 24 grams are carbohydrates. What percent are carbohydrates?

76. One caffe latte weighs 453 grams. What percent is sugar?

Circumference The circumference of a circle is given by the formula $C = 2\pi r$. Find r using the following values. (The values given for π are approximations.)

77. The circumference C is 44 meters and π is $\frac{22}{7}$.

78. The circumference C is 176 meters and π is $\frac{22}{7}$.

79. The circumference is 9.42 inches and π is 3.14.

80. The circumference is 12.56 inches and π is 3.14.

Volume The volume of a cylinder is given by the formula $V = \pi r^2 h$. Find the height h if

81. The volume V is 42 cubic feet, the radius is $\frac{7}{22}$ feet, and π is $\frac{22}{7}$.

82. The volume V is 84 cubic inches, the radius is $\frac{7}{11}$ inches, and π is $\frac{22}{7}$.

83. The volume is 6.28 cubic centimeters, the radius is 3 centimeters, and π is 3.14.

84. The volume is 12.56 cubic centimeters, the radius is 2 centimeters, and π is 3.14.

Baseball Major league baseball has various player awards at the end of the year. One of them is the Rolaids Relief Man of the Year. To compute the Relief Man standings, points are gained or taken away based on the number of wins, losses, saves, and blown saves a relief pitcher has at the end of the year. The pitcher with the most Rolaids Points is the Rolaids Relief Man of the Year. The formula $P = 3s + 2w + t - 2l - 2b$ gives the number of Rolaids points a pitcher earns, where s = saves, w = wins, t = tough saves, l = losses, and b = blown saves. Use this formula to complete the following tables.

85. National League 2010

Pitcher, Team	W	L	Saves	Tough Saves	Blown Saves	Rolaids Points
Heath Bell, San Diego	6	1	47	3	5	
Brian Wilson, San Francisco	3	3	48	5	7	
Carlos Marmol, Chicago	2	3	38	5	5	
Billy Wagner, Atlanta	7	2	37	7	0	
Francisco Cordero, Cincinati	6	5	40	8	0	

86. American League 2010

Pitcher, Team	W	L	Saves	Tough Saves	Blown Saves	Rolaids Points
Rafael Soriano, Tampa Bay	3	2	45	0	3	
Joakim Soria, Kansas City	1	2	43	0	3	
Neftali Feliz, Texas	4	3	40	1	3	
Jonathan Papelbon, Boston	5	7	37	1	8	
Kevin Gregg, Toronto	2	6	37	1	6	

Getting Ready for the Next Section

To understand all of the explanations and examples in the next section you must be able to work the problems below.

Write an equivalent expression in English. Include the words *sum* and *difference* when possible.

87. $4 + 1 = 5$

88. $7 + 3 = 10$

89. $6 - 2 = 4$

90. $8 - 1 = 7$

91. $x - 15 = -12$

92. $2x + 3 = 7$

93. $x + 3 = 4(x - 3)$

94. $2(2x - 5) = 2x - 34$

For each of the following expressions, write an equivalent equation.

95. Twice the sum of 6 and 3 is 18.

96. Four added to the product of 5 and -1 is -1.

97. The sum of twice 5 and 3 is 13.

98. Twice the difference of 8 and 2 is 12.

99. The sum of a number and five is thirteen.

100. The difference of ten and a number is negative eight.

101. Five times the sum of a number and seven is thirty.

102. Five times the difference of twice a number and six is negative twenty.

Find the Mistake

Each sentence below contains a mistake. Circle the mistake and write the correct word(s) or equation on the line provided.

1. A formula contains only one variable. _____

2. To solve for a variable in a formula, make sure the variable appears on both sides of the equation.

3. To solve the formula $2ab = a + 8$ for b, multiply each side of the equation by $2a$. _____

4. The question "What percent of 108 is 9?" can be translated into the equation $\frac{x}{9} = 108$. _____

Landmark Review: Checking Your Progress

Solve each of the following equations.

1. $x + 3 = 15$

2. $x - 5 = 20$

3. $5.6 + y = 10.8$

4. $-\frac{1}{5} + b = \frac{2}{3}$

5. $4x = 16$

6. $\frac{1}{5}y = 2$

7. $-5x + 3 + 3x + 5 = 0$

8. $3y + 4 - 6y = -10 + 5$

9. $4(x + 3) = 24$

10. $-4 - 3(x + 2) - 3 = -25$

Solve each of the following for the indicated variable.

11. $A = lw$ for l

12. $P = a + b + c$ for c

13. $x - 4 + 3y = 14$ for x

14. $\frac{y - 3}{x} = -\frac{4}{3}$ for y

2.5 Applications: Number, Age, Perimeter, Coin Problems

OBJECTIVES

Apply the Blueprint for Problem Solving to solve:

A Number Problems

B Age Problems

C Perimeter Problems

D Coin Problems

KEY WORDS

Blueprint for Problem Solving

An artist in Vermont built a sculpture of a dinosaur called Vermontasaurus that is 25 feet tall. The artist used found junk and scrap wood to construct his controversial work of art. The controversy lies in the structure's adherence to state building codes. The state's fire safety board has demanded that Vermontasaurus visitors not walk under the structure because of its questionable structural integrity. The area has been roped off, pending approval of a building permit.

Suppose you were tasked with constructing a permanent rectangular fence around the legs of the Vermontasaurus. You know that the length of the fence is 4 feet more than three times the width, and the perimeter is 168 feet. How would you find the length and width? In this section, we will use our knowledge of linear equations to solve similar application problems.

Problem Solving

To begin this section, we list the steps used in solving application problems. We call this strategy the *Blueprint for Problem Solving*. It is an outline that will overlay the solution process we use on all application problems.

BLUEPRINT FOR PROBLEM SOLVING

Step 1: *Read* the problem, and then mentally *list* the items that are known and the items that are unknown. Make sure to include units of measurement where appropriate.

Step 2: *Assign a variable* to one of the unknown items. (In most cases this will amount to letting $x =$ the item that is asked for in the problem.) Then *translate* the other *information* in the problem to expressions involving the variable. Draw a diagram or create a table, where appropriate, to help convey this data.

Step 3: *Reread* the problem, and then *write an equation*, using the items and variables listed in steps 1 and 2, that describes the situation.

Step 4: *Solve the equation* found in step 3.

Step 5: *Write* your *answer* using a complete sentence.

Step 6: *Reread* the problem, and *check* your solution with the original words in the problem.

There are a number of substeps within each of the steps in our blueprint. For instance, with steps 1 and 2 it is always a good idea to draw a diagram or picture if it helps visualize the relationship between the items in the problem. In other cases a table helps organize the information. As you gain more experience using the blueprint to solve application problems, you will find additional techniques that expand the blueprint.

To help with problems of the type shown next in Examples 1 and 2, here are some common English words and phrases and their mathematical translations.

English	Algebra
The sum of a and b	$a + b$
The difference of a and b	$a - b$
The product of a and b	$a \cdot b$
The quotient of a and b	$\frac{a}{b}$
of	\cdot (multiply)
is	$=$ (equals)
A number	x
4 more than x	$x + 4$
4 times x	$4x$
4 less than x	$x - 4$

A Number Problems

EXAMPLE 1 The sum of twice a number and three is seven. Find the number.

Solution Using the Blueprint for Problem Solving as an outline, we solve the problem as follows:

Step 1: *Read* the problem, and then mentally *list* the items that are known and the items that are unknown.

> *Known items:* The numbers 3 and 7

> *Unknown items:* The number in question

Step 2: *Assign a variable* to one of the unknown items. Then *translate the other information* in the problem to expressions involving the variable.

> Let $x =$ the number asked for in the problem, then "The sum of twice a number and three" translates to $2x + 3$.

Step 3: *Reread* the problem, and then *write an equation,* using the items and variables listed in steps 1 and 2, that describes the situation.
> With all word problems, the word *is* translates to $=$.

> The sum of twice x and 3 is 7.
$$2x + 3 \qquad = 7$$

Step 4: *Solve the equation* found in step 3.

$$2x + 3 = 7$$
$$2x + 3 + (-3) = 7 + (-3)$$
$$2x = 4$$
$$\frac{1}{2}(2x) = \frac{1}{2}(4)$$
$$x = 2$$

Practice Problems

1. The sum of twice a number and five is eleven. Find the number.

Note Although many problems in this section may seem contrived, they provide excellent practice for our problem solving strategy. This practice will prove useful when we encounter more difficult application problems later in the book.

Answer

1. 3

Step 5: *Write* your *answer* using a complete sentence.

The number is 2.

Step 6: *Reread* the problem, and *check* your solution with the original words in the problem.

The sum of twice 2 and 3 is 7; a true statement.

You may find some examples and problems in this section that you can solve without using algebra or our blueprint. It is very important that you solve these problems using the methods we are showing here. The purpose behind these problems is to give you experience using the blueprint as a guide to solving problems written in words. Your answers are much less important than the work that you show to obtain your answer. You will be able to condense the steps in the blueprint later in the course. For now, though, you need to show your work in the same detail that we are showing in the examples in this section.

EXAMPLE 2 One number is three more than twice another; their sum is eighteen. Find the numbers.

Solution

Step 1: *Read and list.*

Known items: Two numbers that add to 18. One is 3 more than twice the other.
Unknown items: The numbers in question.

Step 2: *Assign a variable, and translate information.*

Let x = the first number. The other is $2x + 3$.

Step 3: *Reread and write an equation.*

$$\underbrace{\text{Their sum}} \;\; \overset{\downarrow}{\text{is}} \; \overset{\downarrow}{18}.$$

$$x + (2x + 3) = 18$$

Step 4: *Solve the equation.*

$$x + (2x + 3) = 18$$
$$3x + 3 = 18$$
$$3x + 3 + (-3) = 18 + (-3)$$
$$3x = 15$$
$$x = 5$$

Step 5: *Write the answer.*

The first number is 5. The other is $2 \cdot 5 + 3 = 13$.

Step 6: *Reread and check.*

The sum of 5 and 13 is 18, and 13 is 3 more than twice 5.

B Age Problems

Remember as you read through the solutions to the examples in this section, step 1 is done mentally. Read the problem, and then mentally list the items that you know and the items that you don't know. The purpose of step 1 is to give you direction as you begin to work application problems. Finding the solution to an application problem is a process; it doesn't happen all at once. The first step is to read the problem with a purpose in mind. That purpose is to mentally note the items that are known and the items that are unknown.

2. If three times the difference of a number and two were decreased by six, the result would be three. Find the number.

Note Although these applications contain more than one variable quantity, we are choosing to write our equation in terms of a single variable to make the process of finding solutions easier.

Answer

2. 5

3. Pete is five years older than Mary. Five years ago Pete was twice as old as Mary. Find their ages now.

EXAMPLE 3 Bill is 6 years older than Tom. Three years ago Bill's age was four times Tom's age. Find the age of each boy now.

Solution Applying the Blueprint for Problem Solving, we have

Step 1: *Read and list.*

Known items: Bill is 6 years older than Tom. Three years ago Bill's age was four times Tom's age.

Unknown items: Bill's age and Tom's age

Step 2: *Assign a variable, and translate information.*

Let T = Tom's age now. That makes Bill $T + 6$ years old now. A table like the one shown here can help organize the information in an age problem. Notice how we placed the T in the box that corresponds to Tom's age now.

	Three Years Ago	Now
Bill		$T + 6$
Tom		T

If Tom is T years old now, 3 years ago he was $T - 3$ years old. If Bill is $T + 6$ years old now, 3 years ago he was $T + 6 - 3 = T + 3$ years old. We use this information to fill in the remaining spaces in the table.

	Three Years Ago	Now
Bill	$T + 3$	$T + 6$
Tom	$T - 3$	T

Step 3: *Reread and write an equation.*

Reading the problem again, we see that 3 years ago Bill's age was four times Tom's age. Writing this as an equation, we have

Bill's age 3 years ago = $4 \cdot$ (Tom's age 3 years ago):

$$T + 3 = 4(T - 3)$$

Step 4: *Solve the equation.*

$$T + 3 = 4(T - 3)$$
$$T + 3 = 4T - 12$$
$$T + (-x) + 3 = 4T + (-T) - 12$$
$$3 = 3T - 12$$
$$3 + 12 = 3T - 12 + 12$$
$$15 = 3T$$
$$T = 5$$

Step 5: *Write the answer.*

Tom is 5 years old. Bill is 11 years old.

Step 6: *Reread and check.*

If Tom is 5 and Bill is 11, then Bill is 6 years older than Tom. Three years ago Tom was 2 and Bill was 8. At that time, Bill's age was four times Tom's age. As you can see, the answers check with the original problem.

Answers

3. Mary is 10; Pete is 15.

C Perimeter Problems

To understand Example 4 completely, you need to recall that the perimeter of a rectangle is the sum of the lengths of the sides. The formula for the perimeter is $P = 2l + 2w$.

EXAMPLE 4 The length of a rectangle is 5 inches more than twice the width. The perimeter is 34 inches. Find the length and width.

Solution When working problems that involve geometric figures, a sketch of the figure helps organize and visualize the problem.

Step 1: *Read and list.*

> *Known items:* The figure is a rectangle. The length is 5 inches more than twice the width. The perimeter is 34 inches.
> *Unknown items:* The length and the width

Step 2: *Assign a variable, and translate information.*

> Because the length is given in terms of the width (the length is 5 more than twice the width), we let x = the width of the rectangle. The length is 5 more than twice the width, so it must be $2x + 5$. The diagram below is a visual description of the relationships we have listed so far.

$$x$$

$$2x + 5$$

Step 3: *Reread and write an equation.*

> The equation that describes the situation is

> Twice the length + twice the width is the perimeter

$$2(2x + 5) \quad + \quad 2x \quad = \quad 34$$

Step 4: *Solve the equation.*

> $2(2x + 5) + 2x = 34$ 	*Original equation*
>
> $4x + 10 + 2x = 34$ 	*Distributive property*
>
> $6x + 10 = 34$ 	*Add $4x$ and $2x$.*
>
> $6x = 24$ 	*Add -10 to each side.*
>
> $x = 4$ 	*Divide each side by 6.*

Step 5: *Write the answer.*

> The width x is 4 inches. The length is $2x + 5 = 2(4) + 5 = 13$ inches.

Step 6: *Reread and check.*

> If the length is 13 and the width is 4, then the perimeter must be $2(13) + 2(4) = 26 + 8 = 34$, which checks with the original problem.

D Coin Problems

EXAMPLE 5 Jennifer has $2.45 in dimes and nickels. If she has 8 more dimes than nickels, how many of each coin does she have?

Solution

Step 1: *Read and list.*

> *Known items:* The type of coins, the total value of the coins, and that there are 8 more dimes than nickels.
> *Unknown items:* The number of nickels and the number of dimes

Step 2: *Assign a variable, and translate information.*

If we let N = the number of nickels, then $N + 8$ = the number of dimes. Because the value of each nickel is 5 cents, the amount of money in nickels is $5N$. Similarly, because each dime is worth 10 cents, the amount of money in dimes is $10(N + 8)$. Here is a table that summarizes the information we have so far:

	Nickels	Dimes
Number	N	$N + 8$
Value (in cents)	$5N$	$10(N + 8)$

Step 3: *Reread and write an equation.*

Because the total value of all the coins is 245 cents, the equation that describes this situation is

Amount of money in nickels		Amount of money in dimes		Total amount of money
$5N$	$+$	$10(N + 8)$	$=$	245

Step 4: *Solve the equation.*

To solve the equation, we apply the distributive property first.

$$5N + 10N + 80 = 245 \qquad \text{Distributive property}$$
$$15N + 80 = 245 \qquad \text{Add } 5N \text{ and } 10N.$$
$$15N = 165 \qquad \text{Add } -80 \text{ to each side.}$$
$$N = 11 \qquad \text{Divide each side by 15.}$$

Step 5: *Write the answer.*

The number of nickels is $N = 11$.
The number of dimes is $N + 8 = 11 + 8 = 19$.

Step 6: *Reread and check.*

To check our results

11 nickels are worth $5(11) = 55$ cents
19 dimes are worth $10(19) = 190$ cents

The total value is 245 cents = $2.45

When you begin working the problems in the problem set that follows, there are a few things to remember. The first is that you may have to read the problems a number of times before you begin to see how to solve them. The second thing to remember is that word problems are not always solved correctly the first time you try them. Sometimes it takes a few attempts and some wrong answers before you can set up and solve these problems correctly. Don't give up.

GETTING READY FOR CLASS

After reading through the preceding section, respond in your own words and in complete sentences.

A. Why is the first step in the Blueprint for Problem Solving done mentally?

B. What is the last thing you do when solving an application problem?

C. Why is it important to still use the blueprint and show your work even if you can solve a problem without using algebra?

D. Write an application problem whose solution depends on solving the equation $4x + 2 = 10$.

Vocabulary Review

The following is a list of steps for the Blueprint for Problem Solving. Choose the correct words to fill in the blanks below.

equation variable known sentence unknown problem

Step 1: Read the problem and mentally list the _____ and unknown items.

Step 2: Assign a variable to one of the _____ items, and then translate the rest of the information into expressions involving the variable.

Step 3: Reread the problem and write an equation using the _____ from step 2.

Step 4: Solve the _____.

Step 5: Write your answer using a complete _____.

Step 6: Reread the _____ and check your solution.

Problems

Solve the following word problems using the steps given in the Blueprint for Problem Solving.

A Number Problems

1. The sum of a number and five is thirteen. Find the number.

2. The difference of ten and a number is negative eight. Find the number.

3. The sum of twice a number and four is fourteen. Find the number.

4. The difference of four times a number and eight is sixteen. Find the number.

5. Five times the sum of a number and seven is thirty. Find the number.

6. Five times the difference of twice a number and six is negative twenty. Find the number.

7. One number is two more than another. Their sum is eight. Find both numbers.

8. One number is three less than another. Their sum is fifteen. Find the numbers.

9. One number is four less than three times another. If their sum is increased by five, the result is twenty-five. Find the numbers.

10. One number is five more than twice another. If their sum is decreased by ten, the result is twenty-two. Find the numbers.

SCAN TO ACCESS

B Age Problems

11. Shelly is 3 years older than Michele. Four years ago the sum of their ages was 67. Find the age of each person now.

	Four Years Ago	Now
Shelly	$x - 1$	$x + 3$
Michele	$x - 4$	x

12. Cary is 9 years older than Dan. In 7 years the sum of their ages will be 93. Find the age of each man now. (Begin by filling in the table.)

	Now	In Seven Years
Cary	$x + 9$	
Dan	x	$x + 7$

13. Cody is twice as old as Evan. Three years ago the sum of their ages was 27. Find the age of each boy now.

	Three Years Ago	Now
Cody		
Evan	$x - 3$	x

14. Justin is 2 years older than Ethan. In 9 years the sum of their ages will be 30. Find the age of each boy now.

	Now	In Nine Years
Justin		
Ethan	x	

15. Fred is 4 years older than Barney. Five years ago the sum of their ages was 48. How old are they now?

	Five Years Ago	Now
Fred		
Barney		x

16. Tim is 5 years older than JoAnn. Six years from now the sum of their ages will be 79. How old are they now?

	Now	Six Years From Now
Tim		
JoAnn	x	

17. Jack is twice as old as Lacy. In 3 years the sum of their ages will be 54. How old are they now?

18. John is 4 times as old as Martha. Five years ago the sum of their ages was 50. How old are they now?

19. Pat is 20 years older than his son Patrick. In 2 years Pat will be twice as old as Patrick. How old are they now?

20. Diane is 23 years older than her daughter Amy. In 6 years Diane will be twice as old as Amy. How old are they now?

c Perimeter Problems

21. The perimeter of a square is 36 inches. Find the length of one side.

22. The perimeter of a square is 44 centimeters. Find the length of one side.

23. The perimeter of a square is 60 feet. Find the length of one side.

24. The perimeter of a square is 84 meters. Find the length of one side.

25. One side of a triangle is three times the shortest side. The third side is 7 feet more than the shortest side. The perimeter is 62 feet. Find all three sides.

26. One side of a triangle is half the longest side. The third side is 10 meters less than the longest side. The perimeter is 45 meters. Find all three sides.

27. One side of a triangle is half the longest side. The third side is 12 feet less than the longest side. The perimeter is 53 feet. Find all three sides.

28. One side of a triangle is 6 meters more than twice the shortest side. The third side is 9 meters more than the shortest side. The perimeter is 75 meters. Find all three sides.

29. The length of a rectangle is 5 inches more than the width. The perimeter is 34 inches. Find the length and width.

$x + 5$

30. The width of a rectangle is 3 feet less than the length. The perimeter is 10 feet. Find the length and width.

31. The length of a rectangle is 7 meters more than twice the width. The perimeter is 68 meters. Find the length and width.

32. The length of a rectangle is 4 inches more than three times the width. The perimeter is 72 inches. Find the length and width.

33. The length of a rectangle is 6 feet more than three times the width. The perimeter is 36 feet. Find the length and width.

34. The length of a rectangle is 3 centimeters less than twice the width. The perimeter is 54 centimeters . Find the length and width.

D Coin Problems

35. Marissa has $4.40 in quarters and dimes. If she has 5 more quarters than dimes, how many of each coin does she have?

	Dimes	Quarters
Number	x	$x + 5$
Value (in cents)	$10(x)$	$25(x + 5)$

36. Kendra has $2.75 in dimes and nickels. If she has twice as many dimes as nickels, how many of each coin does she have? (Completing the table may help you get started.)

	Nickels	Dimes
Number	x	$2x$
Value (in cents)	$5(x)$	

37. Tanner has $4.35 in nickels and quarters. If he has 15 more nickels than quarters, how many of each coin does he have?

	Nickels	Quarters
Number	$x + 15$	x
Value (in cents)		

38. Connor has $9.00 in dimes and quarters. If he has twice as many quarters as dimes, how many of each coin does he have?

	Dimes	Quarters
Number	x	$2x$
Value (in cents)		

39. Sue has $2.10 in dimes and nickels. If she has 9 more dimes than nickels, how many of each coin does she have?

	Dimes	Nickels
Number		x
Value (in cents)		

40. Mike has $1.55 in dimes and nickels. If he has 7 more nickels than dimes, how many of each coin does he have?

	Dimes	Nickels
Number	x	
Value (in cents)		

41. Katie has a collection of nickels, dimes, and quarters with a total value of $4.35. There are 3 more dimes than nickels and 5 more quarters than nickels. How many of each coin is in her collection? (*Hint:* Let N = the number of nickels.)

	Nickels	Dimes	Quarters
Number	N		
Value			

42. Mary Jo has $3.90 worth of nickels, dimes, and quarters. The number of nickels is 3 more than the number of dimes. The number of quarters is 7 more than the number of dimes. How many of each coin does she have? (*Hint:* Let D = the number of dimes.)

	Nickels	Dimes	Quarters
Number			
Value			

43. Corey has a collection of nickels, dimes, and quarters with a total value of $2.55. There are 6 more dimes than nickels and twice as many quarters as nickels. How many of each coin is in her collection?

	Nickels	Dimes	Quarters
Number	x		
Value			

44. Kelly has a collection of nickels, dimes, and quarters with a total value of $7.40. There are four more nickels than dimes and twice as many quarters as nickels. How many of each coin is in her collection?

	Nickels	Dimes	Quarters
Number			
Value			

Getting Ready for the Next Section

To understand all of the explanations and examples in the next section you must be able to work the problems below. Simplify the following expressions.

45. $x + 2x + 2x$

46. $x + 2x + 3x$

47. $x + 0.075x$

48. $x + 0.065x$

49. $0.09(x + 2,000)$

50. $0.06(x + 1,500)$

Find the Mistake

Each application problem below is followed by a corresponding equation. Write true or false if the equation accurately represents the problem. If false, write the correct equation on the line provided.

1. The sum of 7 and the quotient of m and n is 9 more than n.

Equation: $\dfrac{7}{(m+n)} = n + 9$ _____

2. Danny is 30 years older than his daughter Emma. Four years ago, Danny's age was six times Emma's age. Find Danny and Emma's age.

Equation: $x + 26 = 6(x - 4)$ _____

3. The width of a rectangle is 6 centimeters more than three times the length. The perimeter is 42 centimeters. Find the length and width.

Equation: $3x + 6 + x = 42$ _____

4. Amelie has $3.10 in nickels and dimes. If the number of dimes she has is 4 less than twice the number of nickels, how many of each coin does she have?

Equation: $2x - 4 + x = 310$ _____

2.6 Applications: Integer, Interest, Triangle, Miscellaneous Problems

©iStockphoto.com/polygrafix

In Las Vegas, the Stratosphere's hotel and casino has opened another attraction for thrill-seekers. SkyJump is a death-defying controlled free-fall from the 108th floor of the hotel to the ground. Jump Package 1 includes the jump cost plus a DVD of the jump for $114.99. A jump without the DVD costs $99.99. Suppose in one weekend, SkyJump had 15 jumpers, and receipts for that weekend's jumps totalled $1574.85. In this section, we will practice more application problems using linear equations, such as the one needed to determine how many jumpers purchased Jump Package 1 and how many paid just for the jump.

A Consecutive Integer Problems

Our first example involves consecutive integers. Here is the definition:

> **DEFINITION** consecutive integers
>
> When we ask for *consecutive integers*, we mean integers that are next to each other on the number line, like 5 and 6, or 13 and 14, or -4 and -3.

In the dictionary, consecutive is defined as following one another in uninterrupted order. If we ask for consecutive odd integers, then we mean odd integers that follow one another on the number line. For example, 3 and 5, 11 and 13, and -9 and -7 are consecutive odd integers. As you can see, to get from one odd integer to the next consecutive odd integer we add 2.

If we are asked to find two consecutive integers and we let x equal the first integer, the next one must be $x + 1$, because consecutive integers always differ by 1. Likewise, if we are asked to find two consecutive odd or even integers, and we let x equal the first integer, then the next one will be $x + 2$ because consecutive even or odd integers always differ by 2. Here is a table that summarizes this information.

Note When assigning a variable or an expression to two consecutive integers, we could choose the approach shown here, or we could assign x to the larger integer and $x - 1$ to the smaller integer. Either case would yield the same answers. But for the sake of simplicity, we will continue to assign x to the first and smaller of the two consecutive integers.

In Words	Using Algebra	Example
Two consecutive integers	$x, x + 1$	The sum of two consecutive integers is 15. $x + (x + 1) = 15$ or $7 + 8 = 15$
Three consecutive integers	$x, x + 1, x + 2$	The sum of three consecutive integers is 24. $x + (x + 1) + (x + 2) = 24$ or $7 + 8 + 9 = 24$
Two consecutive odd integers	$x, x + 2$	The sum of two consecutive odd integers is 16. $x + (x + 2) = 16$ or $7 + 9 = 16$
Two consecutive even integers	$x, x + 2$	The sum of two consecutive even integers is 18. $x + (x + 2) = 18$ or $8 + 10 = 18$

EXAMPLE 1 The sum of two consecutive odd integers is 28. Find the two integers.

Solution

Step 1: *Read and list.*

Known items: Two consecutive odd integers. Their sum is equal to 28.
Unknown items: The numbers in question.

Step 2: *Assign a variable, and translate information.*

If we let $x =$ the first of the two consecutive odd integers, then $x + 2$ is the next consecutive one.

Step 3: *Reread and write an equation.*

Their sum is 28.

$$x + (x + 2) = 28$$

Step 4: *Solve the equation.*

$2x + 2 = 28$	*Simplify the left side.*
$2x = 26$	*Add -2 to each side.*
$x = 13$	*Multiply each side by $\frac{1}{2}$.*

Step 5: *Write the answer.*

The first of the two integers is 13. The second of the two integers will be two more than the first, which is 15.

Step 6: *Reread and check.*

Suppose the first integer is 13. The next consecutive odd integer is 15. The sum of 15 and 13 is 28.

B Interest Problems

EXAMPLE 2 Suppose you invest a certain amount of money in an account that earns 8% in annual interest. At the same time, you invest $2,000 more than that in an account that pays 9% in annual interest. If the total interest from both accounts at the end of the year is $690, how much is invested in each account? (Hint: The formula for simple interest is $I = Prt$, where I is interest, P is principal amount invested, r is interest rate, and t is time in years.)

Solution

Step 1: *Read and list.*

Known items: The interest rates, the total interest earned, and how much more is invested at 9%
Unknown items: The amounts invested in each account.

Step 2: *Assign a variable, and translate information.*

Let $P =$ the amount of money invested at 8%. From this, $P + 2,000 =$ the amount of money invested at 9%. The interest earned on P dollars invested at 8% is $0.08P$. The interest earned on $P + 2,000$ dollars invested at 9% is $0.09(P + 2,000)$.

Here is a table that summarizes this information:

	Dollars invested at 8%	Dollars invested at 9%
Number of	P	$P + 2,000$
Interest on	$0.08P$	$0.09(P + 2,000)$

Practice Problems

1. The sum of two consecutive integers is 27. What are the two integers?

Note Now that you have worked through a number of application problems using our blueprint, you probably have noticed that step 3, in which we write an equation that describes the situation, is the key step. Anyone with experience solving application problems will tell you that there will be times when your first attempt at step 3 results in the wrong equation. Remember, mistakes are part of the process of learning to do things correctly.

2. On July 1, 1992, Alfredo opened a savings account that earned 4% in annual interest. At the same time, he put $3,000 more than he had in his savings into a mutual fund that paid 5% annual interest. One year later, the total interest from both accounts was $510. How much money did Alfredo put in his savings account?

Answers

1. 13 and 14
2. $4,000

Step 3: *Reread and write an equation.*

Because the total amount of interest earned from both accounts is $690, the equation that describes the situation is

Interest earned at 8%	+	Interest earned at 9%	=	Total interest earned
$0.08P$	+	$0.09(P + 2{,}000)$	=	690

Step 4: *Solve the equation.*

$$0.08P + 0.09(P + 2{,}000) = 690$$

$$0.08P + 0.09P + 180 = 690 \qquad \text{Distributive property}$$

$$0.17P + 180 = 690 \qquad \text{Add } 0.08P \text{ and } 0.09P.$$

$$0.17P = 510 \qquad \text{Add } -180 \text{ to each side.}$$

$$P = 3{,}000 \qquad \text{Divide each side by } 0.17.$$

Step 5: *Write the answer.*

The amount of money invested at 8% is $3,000, whereas the amount of money invested at 9% is $P + 2,000 = 3,000 + 2,000 = \$5,000$.

Step 6: *Reread and check.*

The interest at 8% is 8% of 3,000 = 0.08(3,000) = $240
The interest at 9% is 9% of 5,000 = 0.09(5,000) = $450

The total interest = $690

C Triangle Angle Problems

FACTS FROM GEOMETRY Labeling Triangles and the Sum of the Angles

One way to label the important parts of a triangle is to label the vertices with capital letters and the sides with small letters, as shown in Figure 1.

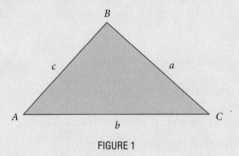

FIGURE 1

In Figure 1, notice that side a is opposite vertex A, side b is opposite vertex B, and side c is opposite vertex C. Also, because each vertex is the vertex of one of the angles of the triangle, we refer to the three interior angles as A, B, and C.

In any triangle, the sum of the interior angles is 180°. For the triangle shown in Figure 1, the relationship is written

$$A + B + C = 180°$$

EXAMPLE 3 The angles in a triangle are such that one angle is twice the smallest angle, whereas the third angle is three times as large as the smallest angle. Find the measure of all three angles.

Solution

Step 1: *Read and list.*

 Known items: The sum of all three angles is 180°, one angle is twice the smallest angle, the largest angle is three times the smallest angle.

 Unknown items: The measure of each angle

Step 2: *Assign a variable, and translate information.*

 Let A be the smallest angle, then $2A$ will be the measure of another angle and $3A$ will be the measure of the largest angle.

Step 3: *Reread and write an equation.*

 When working with geometric objects, drawing a generic diagram sometimes will help us visualize what it is that we are asked to find. In Figure 2, we draw a triangle with angles A, B, and C.

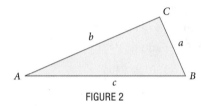

FIGURE 2

 We can write the value of angle B in terms of A as $B = 2A$, and the value of angle C as $C = 3A$. We know that the sum of angles A, B, and C will be 180°, so our equation becomes

$$A + 2A + 3A = 180°$$

Step 4: *Solve the equation.*

$$A + 2A + 3A = 180°$$
$$6A = 180°$$
$$A = 30°$$

Step 5: *Write the answer.*

 The smallest angle A measures 30°
 Angle B measures $2A$, or $2(30°) = 60°$
 Angle C measures $3A$, or $3(30°) = 90°$

Step 6: *Reread and check.*

 The angles must add to 180°.
$$A + B + C = 180°$$
$$30° + 60° + 90° \stackrel{?}{=} 180°$$
$$180° = 180° \qquad \textit{Our answers check.}$$

D Miscellaneous Problems

EXAMPLE 4 Tickets to a school play cost $6 for adults and $4 for children. A total of 200 tickets were sold, and $940 was collected in receipts. How many adult tickets and how many child tickets were sold?

Solution

Step 1: *Read and list.*

 Known items: 200 tickets were sold, $940 was collected

 Unknown items: Number of adult tickets, number of child tickets

3. The angles in a triangle are such that one angle is three times the smallest angle, whereas the largest angle is five times the smallest angle. Find the measure of all three angles.

4. Tickets to a community theater cost $10 for adults and $6 for children. A total of $680 was collected for one evening performance. If 20 more adults attended than children, how many children and how many adults attended the theater?

Answers

3. 20°, 60°, and 100°
4. 50 adults, 30 children

Step 2: *Assign a variable, and translate information.*

Let x be the number of adult tickets sold, then $200 - x$ will be the number of child tickets sold.

Step 3: *Reread and write an equation.*

$$\text{Price of adult} \times \text{\# Adult} + \text{Price of child} \times \text{\# Child} = \text{Total collected}$$
$$6 \cdot (x) \qquad + \qquad 4 \cdot (200 - x) \qquad = \qquad 940$$

Step 4: *Solve the equation.*

$$6(x) + 4(200 - x) = 940$$
$$6x + 800 - 4x = 940$$
$$2x + 800 = 940$$
$$2x = 140$$
$$x = 70$$

Step 5: *Write the answer.*

There were 70 adult tickets sold, and $200 - 70 = 130$ child tickets sold.

Step 6: *Reread and check.*

The total receipts must add to $940.

$$6 \cdot (70) + 4 \cdot (130) \stackrel{?}{=} 940$$
$$\underline{420 + 520 \stackrel{?}{=} 940}$$
$$940 = 940$$

GETTING READY FOR CLASS

After reading through the preceding section, respond in your own words and in complete sentences.

A. Write an equation for the sum of two consecutive integers being 13.

B. Write an application problem whose solution depends on solving the equation $x + 0.075x = 500$.

C. How do you label triangles?

D. What rule is always true about the three angles in a triangle?

Vocabulary Review

Choose the correct words to fill in the blanks below.

even sum consecutive integers

1. _____ integers are integers that are next to each other on the number line.

2. Two consecutive _____ integers would be given by x and $x + 2$.

3. Two consecutive _____ would be given by x and $x + 1$.

4. In any triangle, the _____ of the interior angles is 180°, and written by $A + B + C = 180°$.

A Consecutive Integer Problems

1. The sum of two consecutive integers is 11. Find the numbers.

2. The sum of two consecutive integers is 15. Find the numbers.

3. The sum of two consecutive integers is -9. Find the numbers.

4. The sum of two consecutive integers is -21. Find the numbers.

5. The sum of two consecutive odd integers is 28. Find the numbers.

6. The sum of two consecutive odd integers is 44. Find the numbers.

7. The sum of two consecutive even integers is 106. Find the numbers.

8. The sum of two consecutive even integers is 66. Find the numbers.

9. The sum of two consecutive even integers is -30. Find the numbers.

10. The sum of two consecutive odd integers is -76. Find the numbers.

SCAN TO ACCESS

11. The sum of three consecutive odd integers is 57. Find the numbers.

12. The sum of three consecutive odd integers is −51. Find the numbers.

13. The sum of three consecutive even integers is 132. Find the numbers.

14. The sum of three consecutive even integers is −108. Find the numbers.

B Interest Problems

15. Suppose you invest money in two accounts. One of the accounts pays 8% annual interest, whereas the other pays 9% annual interest. If you have $2,000 more invested at 9% than you have invested at 8%, how much do you have invested in each account if the total amount of interest you earn in a year is $860? (Begin by completing the following table.)

	Dollars invested at 8%	Dollars invested at 9%
Number of	x	
Interest on		

16. Suppose you invest a certain amount of money in an account that pays 11% interest annually, and $4,000 more than that in an account that pays 12% annually. How much money do you have in each account if the total interest for a year is $940?

	Dollars invested at 11%	Dollars invested at 12%
Number of	x	
Interest on		

17. Tyler has two savings accounts that his grandparents opened for him. The two accounts pay 10% and 12% in annual interest; there is $500 more in the account that pays 12% than there is in the other account. If the total interest for a year is $214, how much money does he have in each account?

18. Travis has a savings account that his parents opened for him. It pays 6% annual interest. His uncle also opened an account for him, but it pays 8% annual interest. If there is $800 more in the account that pays 6%, and the total interest from both accounts is $104, how much money is in each of the accounts?

19. A stockbroker has money in three accounts. The interest rates on the three accounts are 8%, 9%, and 10%. If she has twice as much money invested at 9% as she has invested at 8%, three times as much at 10% as she has at 8%, and the total interest for the year is $280, how much is invested at each rate? (*Hint:* Let x = the amount invested at 8%.)

20. An accountant has money in three accounts that pay 9%, 10%, and 11% in annual interest. He has twice as much invested at 9% as he does at 10% and three times as much invested at 11% as he does at 10%. If the total interest from the three accounts is $610 for the year, how much is invested at each rate? (*Hint:* Let x = the amount invested at 10%.)

c Triangle Angle Problems

21. Two angles in a triangle are equal and their sum is equal to the third angle in the triangle. What are the measures of each of the three interior angles?

22. One angle in a triangle measures twice the smallest angle, whereas the largest angle is six times the smallest angle. Find the measures of all three angles.

23. The smallest angle in a triangle is $\frac{1}{5}$ as large as the largest angle. The third angle is twice the smallest angle. Find the three angles.

24. One angle in a triangle is half the largest angle but three times the smallest. Find all three angles.

25. A right triangle has one 37° angle. Find the other two angles.

26. In a right triangle, one of the acute angles is twice as large as the other acute angle. Find the measure of the two acute angles.

27. One angle of a triangle measures 20° more than the smallest, while a third angle is twice the smallest. Find the measure of each angle.

28. One angle of a triangle measures 50° more than the smallest, while a third angle is three times the smallest. Find the measure of each angle.

D Miscellaneous Problems

29. Ticket Prices Miguel is selling tickets to a barbecue. Adult tickets cost $6.00 and children's tickets cost $4.00. He sells six more children's tickets than adult tickets. The total amount of money he collects is $184. How many adult tickets and how many children's tickets did he sell?

	Adult	Child
Number	x	$x + 6$
Income	$6(x)$	$4(x + 6)$

30. Working Two Jobs Maggie has a job working in an office for $10 an hour and another job driving a tractor for $12 an hour. One week she works in the office twice as long as she drives the tractor. Her total income for that week is $416. How many hours did she spend at each job?

Job	Office	Tractor
Hours Worked	$2x$	x
Wages Earned	$10(2x)$	$12x$

31. Phone Bill The cost of a long-distance phone call is $0.41 for the first minute and $0.32 for each additional minute. If the total charge for a long-distance call is $5.21, how many minutes was the call?

32. Phone Bill Danny, who is 1 year old, is playing with the telephone when he accidentally presses one of the buttons his mother has programmed to dial her friend Sue's number. Sue answers the phone and realizes Danny is on the other end. She talks to Danny, trying to get him to hang up. The cost for a call is $0.23 for the first minute and $0.14 for every minute after that. If the total charge for the call is $3.73, how long did it take Sue to convince Danny to hang up the phone?

33. Hourly Wages JoAnn works in the publicity office at the state university. She is paid $12 an hour for the first 35 hours she works each week and $18 an hour for every hour after that. If she makes $492 one week, how many hours did she work?

34. Hourly Wages Diane has a part-time job that pays her $8.50 an hour. During one week she works 26 hours and is paid $230.10. She realizes when she sees her check that she has been given a raise. How much per hour is that raise?

35. Office Numbers Professors Wong and Gil have offices in the mathematics building at Miami Dade College. Their office numbers are consecutive odd integers with a sum of 14,660. What are the office numbers of these two professors?

36. Cell Phone Numbers Diana and Tom buy two cell phones. The phone numbers assigned to each are consecutive integers with a sum of 11,109,295. If the smaller number is Diana's, what are their phone numbers?

37. Age Marissa and Kendra are 2 years apart. Their ages are two consecutive even integers. Kendra is the younger of the two. If Marissa's age is added to twice Kendra's age, the result is 26. How old is each girl?

38. Age Justin's and Ethan's ages form two consecutive odd integers. What is the difference of their ages?

39. Arrival Time Jeff and Carla Cole are driving separately from San Luis Obispo, California, to the north shore of Lake Tahoe, a distance of 425 miles. Jeff leaves San Luis Obispo at 11:00 a.m. and averages 55 miles per hour on the drive, Carla leaves later, at 1:00 p.m. but averages 65 miles per hour. Which person arrives in Lake Tahoe first?

40. Piano Lessons Tyler's parents pay him $0.50 to do the laundry and $1.25 to mow the lawn. In one month, he does the laundry 6 more times than he mows the lawn. If his parents pay him $13.50 that month, how many times did he mow the lawn?

At one time, the Texas Junior College Teachers Association annual conference was held in Austin. At that time a taxi ride in Austin was $1.25 for the first $\frac{1}{5}$ of a mile and $0.25 for each additional $\frac{1}{5}$ of a mile. Use this information for Problems 41 and 42.

41. Cost of a Taxi Ride If the distance from one of the convention hotels to the airport is 7.5 miles, how much will it cost to take a taxi from that hotel to the airport?

42. Cost of a Taxi Ride Suppose the distance from one of the hotels to one of the western dance clubs in Austin is 12.4 miles. If the fare meter in the taxi gives the charge for that trip as $16.50, is the meter working correctly?

43. Geometry The length and width of a rectangle are consecutive even integers. The perimeter is 44 meters. Find the length and width.

44. Geometry The length and width of a rectangle are consecutive odd integers. The perimeter is 128 meters. Find the length and width.

45. Geometry The angles of a triangle are three consecutive integers. Find the measure of each angle.

46. Geometry The angles of a triangle are three consecutive even integers. Find the measure of each angle.

47. Boards A 78-centimeter-long board is cut into 3 pieces. The longest piece is twice as long as the shortest piece. The other piece is 6 cm longer than the shortest piece. How long is each piece?

48. Rope A 52-meter-long rope is cut into three segments. The shortest piece is 6 meters shorter than the next piece, which is 13 meters shorter than the longest piece. How long is each piece?

Ike and Nancy give western dance lessons at the Elk's Lodge on Sunday nights. The lessons cost $3.00 for members of the lodge and $5.00 for nonmembers. Half of the money collected for the lesson is paid to Ike and Nancy. The Elk's Lodge keeps the other half. One Sunday night Ike counts 36 people in the dance lesson. Use this information to work Problems 47 through 50.

49. Dance Lessons What is the least amount of money Ike and Nancy will make?

50. Dance Lessons What is the largest amount of money Ike and Nancy will make?

51. Dance Lessons At the end of the evening, the Elk's Lodge gives Ike and Nancy a check for $80 to cover half of the receipts. Can this amount be correct?

52. Dance Lessons Besides the number of people in the dance lesson, what additional information does Ike need to know to always be sure he is being paid the correct amount?

53. Solve the problem given at the introduction to this section to find how many jumpers purchased package 1.

54. Solve the same problem as the previous problem, but this time assume Sky Jump had 30 jumpers with total receipts of $3,269.70.

Getting Ready for the Next Section

To understand all the explanations and examples in the next section you must be able to work the problems below.

Solve the following equations.

55. a. $x - 3 = 6$

b. $x + 3 = 6$

c. $-x - 3 = 6$

d. $-x + 3 = 6$

56. a. $x - 7 = 16$

b. $x + 7 = 16$

c. $-x - 7 = 16$

d. $-x + 7 = 16$

57. a. $\dfrac{x}{4} = -2$

b. $-\dfrac{x}{4} = -2$

c. $\dfrac{x}{4} = 2$

d. $-\dfrac{x}{4} = 2$

58. a. $3a = 15$

b. $3a = -15$

c. $-3a = 15$

d. $-3a = -15$

59. $2.5x - 3.48 = 4.9x + 2.07$

60. $2(1 - 3x) + 4 = 4x - 14$

61. $3(x - 4) = -2$

62. $-2(x - 3) = 8$

63. Solve $2x - 3y = 6$ for y.

64. Solve $3x + 2y = 7$ for y.

65. Solve $3x - 2y = 5$ for x.

66. Solve $2x - 5y = -10$ for x.

Find the Mistake

Each application problem below is followed by a corresponding equation. Write true or false if the equation accurately represents the problem. If false, write the correct equation on the line provided. (Hint: Some equations may appear in a simplified form.)

1. The sum of three consecutive integers divided by 2 is 36.

 Equation: $\dfrac{(3x + 3)}{2} = 36$ _____

2. You have money in two savings accounts that pay 7% and 8% in annual interest. If you have four times as much money in the account that pays 7%, and the total interest of both accounts is $396, how much money is in each account?

 Equation: $0.07x + 0.08(4x) = 396$ _____

3. One angle of a triangle measures 4 degrees more than the smallest angle. The largest angle is 16 degrees less than twice the smallest angle. Find the measure of each angle.

 Equation: $4x - 12 = 180$ _____

4. Tickets for a dinner fundraiser cost $15 per plate with dessert and $12 per plate without dessert. 150 tickets were sold for a total of $2070. How many of each ticket was sold?

 Equation: $15x + 12(150 - x) = 2070$ _____

OBJECTIVES

A Use the addition property for inequalities to solve an inequality and graph the solution set.

B Use the multiplication property for inequalities to solve inequalities.

C Use both the addition and multiplication properties to solve and graph inequalities.

D Solve and graph compound inequalities.

E Translate and solve application problems involving inequalities.

KEY WORDS

linear inequality

addition property for inequalities

multiplication property for inequalities

compound inequality

continued inequality

double inequality

Image © sxc.hu/ckforjc, 2004

Suppose we know that the state of New Mexico has no more than 10 organic stores. We can write the following inequality to represent this information:

$$x \leq 10$$

If we know for certain of three stores in New Mexico, we can add to the original inequality and say $x + 3 \leq 10$. Up to now we have solved equalities. For example, when we say $x + 3 = 10$, we know x must be 7 since the statement is only true when $x = 7$. But if we say that $x + 3 \leq 10$, then there are many numbers that make this statement true. If x is 3, $3 + 3$ is less than 10. If x is 2.5, $2.5 + 3$ is less than 10. If x is any real number less than or equal to 7, then the statement $x + 3 \leq 10$ is true.

We also know that any number greater than 7 mark the statement false. If x is 8, then $x + 3$ is not less than 10.

When we solve an inequality, we look for all possible solutions—and there will usually be many of them. Any number that makes an inequality true is a *solution* to that inequality. All the numbers that make the inequality true are called the *solution set*.

We can express a solution set symbolically. If the solution set to an inequality is all the real numbers less than 7, we say the solution is $x < 7$. If the solution set to this inequality is all numbers less than or equal to 7, we say the solution is $x \leq 7$.

We can also express the solution set graphically using the number line. When a solution set is all numbers less than a certain number, we make that number with an open circle.

If we want to include the number (7, in this case) in the solution set, we mark that number with a closed circle.

Linear inequalities are solved by a method similar to the one used in solving linear equations. The only real differences between the methods are in the multiplication property for inequalities and in graphing the solution set.

A Addition Property for Inequalities

An inequality differs from an equation only with respect to the comparison symbol between the two quantities being compared. In place of the equal sign, we use < (less than), ≤ (less than or equal to), > (greater than), or ≥ (greater than or equal to). The addition property for inequalities is almost identical to the addition property for equality.

> **PROPERTY** Addition Property for Inequalities
>
> For any three algebraic expressions A, B, and C,
>
> if $\qquad A < B$
>
> then $\qquad A + C < B + C$
>
> *In words*: Adding the same quantity to both sides of an inequality will not change the solution set.

It makes no difference which inequality symbol we use to state the property. Adding the same amount to both sides always produces an inequality equivalent to the original inequality. Also, because subtraction can be thought of as addition of the opposite, this property holds for subtraction as well as addition.

EXAMPLE 1 Solve $x + 5 < 7$ and graph the solution.

Solution To isolate x, we add -5 to both sides of the inequality

$$x + 5 < 7$$

$$x + 5 + (-5) < 7 + (-5) \qquad \text{Addition property for inequalities}$$

$$x < 2$$

We can go one step further here and graph the solution set. The solution set is all real numbers less than 2. To graph this set, we simply draw a straight line and label the center 0 (zero) for reference. Then we label the 2 on the right side of zero and extend an arrow beginning at 2 and pointing to the left. We use an open circle at 2 because it is not included in the solution set. Here is the graph.

EXAMPLE 2 Solve $x - 6 \leq -3$ and graph the solution.

Solution Adding 6 to each side will isolate x on the left side

$$x - 6 \leq -3$$

$$x - 6 + 6 \leq -3 + 6 \qquad \text{Add 6 to both sides.}$$

$$x \leq 3$$

The graph of the solution set is

Notice that the dot at the 3 in Example 2 is darkened because 3 is included in the solution set. In this book we will use open circles on the graphs of solution sets with $<$ or $>$ and closed (darkened) circles on the graphs of solution sets with \leq or \geq.

B Multiplication Property for Inequalities

To see the idea behind the multiplication property for inequalities, we will consider three true inequality statements and explore what happens when we multiply both sides by a positive number and then what happens when we multiply by a negative number.

Consider the following three true statements:

$$3 < 5 \qquad -3 < 5 \qquad -5 < -3$$

Now multiply both sides by the positive number 4.

$$4(3) < 4(5) \qquad 4(-3) < 4(5) \qquad 4(-5) < 4(-3)$$

$$12 < 20 \qquad\quad -12 < 20 \qquad\quad -20 < -12$$

In each case, the inequality symbol in the result points in the same direction it did in the original inequality. We say the "sense" of the inequality doesn't change when we multiply both sides by a positive quantity.

Notice what happens when we go through the same process but multiply both sides by -4 instead of 4:

$$3 < 5 \qquad\qquad -3 < 5 \qquad\qquad -5 < -3$$

$$-4(3) > -4(5) \quad -4(-3) > -4(5) \quad -4(-5) > -4(-3)$$

$$-12 > -20 \qquad\qquad 12 > -20 \qquad\qquad 20 > 12$$

In each case, we have to change the direction in which the inequality symbol points to keep each statement true. Multiplying both sides of an inequality by a negative quantity always reverses the sense of the inequality. Our results are summarized in the multiplication property for inequalities.

> **PROPERTY** **Multiplication Property for Inequalities**
>
> For any three algebraic expressions A, B, and C,
>
> $$\text{if} \qquad\qquad A < B$$
> $$\text{then} \qquad\quad AC < BC \qquad \text{when } C \text{ is positive}$$
> $$\text{and} \qquad\quad AC > BC \qquad \text{when } C \text{ is negative}$$
>
> *In words:* Multiplying both sides of an inequality by a positive number does not change the solution set. When multiplying both sides of an inequality by a negative number, it is necessary to reverse the inequality symbol to produce an equivalent inequality.

We can multiply both sides of an inequality by any nonzero number we choose. If that number happens to be negative, we must also reverse the sense of the inequality.

3. Solve $4a < 12$ and graph the solution.

EXAMPLE 3 Solve $3a < 15$ and graph the solution.

Solution We begin by multiplying each side by $\frac{1}{3}$. Because $\frac{1}{3}$ is a positive number, we do not reverse the direction of the inequality symbol.

$$3a < 15$$

$$\frac{1}{3}(3a) < \frac{1}{3}(15) \qquad \text{Multiply each side by } \frac{1}{3}.$$

$$a < 5$$

4. Solve $-2x \le 10$ and graph the solution.

EXAMPLE 4 Solve $-3a \le 18$ and graph the solution.

Solution We begin by multiplying both sides by $-\frac{1}{3}$. Because $-\frac{1}{3}$ is a negative number, we must reverse the direction of the inequality symbol at the same time that we multiply by $-\frac{1}{3}$.

$$-3a \le 18$$

$$-\frac{1}{3}(-3a) \ge -\frac{1}{3}(18) \qquad \text{Multiply both sides by } -\frac{1}{3} \text{ and reverse the direction of the inequality symbol.}$$

$$a \ge -6$$

Answers

3. $a < 3$

4. $x \ge -5$

EXAMPLE 5 Solve $-\frac{x}{4} > 2$ and graph the solution.

Solution To isolate x, we multiply each side by -4. Because -4 is a negative number, we also must reverse the direction of the inequality symbol.

$$-\frac{x}{4} > 2$$

$$-4\left(-\frac{x}{4}\right) < -4(2) \qquad \text{\textit{Multiply each side by} -4\textit{, and reverse the direction of the inequality symbol.}}$$

$$x < -8$$

5. Solve $-\frac{x}{3} > 2$ and graph the solution.

C Solving Linear Inequalities in One Variable

To solve more complicated inequalities, we use the following steps:

> **HOW TO** Solving Linear Inequalities in One Variable
>
> **Step 1a:** Use the distributive property to separate terms, if necessary.
>
> **1b:** If fractions are present, consider multiplying both sides by the LCD to eliminate the fractions. If decimals are present, consider multiplying both sides by a power of 10 to clear the inequality of decimals.
>
> **1c:** Combine similar terms on each side of the inequality.
>
> **Step 2:** Use the addition property for inequalities to get all variable terms on one side of the inequality and all constant terms on the other side.
>
> **Step 3:** Use the multiplication property for inequalities to get x by itself on one side of the inequality.
>
> **Step 4:** Graph the solution set.

EXAMPLE 6 Solve and graph: $2.5x - 3.48 < -4.9x + 2.07$.

Solution We have two methods we can use to solve this inequality. We can simply apply our properties to the inequality the way it is currently written and work with the decimal numbers, or we can eliminate the decimals to begin with and solve the resulting inequality.

Method 1 Working with the decimals.

$$2.5x - 3.48 < -4.9x + 2.07 \qquad \text{\textit{Original inequality}}$$

$$2.5x + 4.9x - 3.48 < -4.9x + 4.9x + 2.07 \qquad \text{\textit{Add} 4.9x \textit{to each side.}}$$

$$7.4x - 3.48 < 2.07$$

$$7.4x - 3.48 + 3.48 < 2.07 + 3.48 \qquad \text{\textit{Add} 3.48 \textit{to each side.}}$$

$$7.4x < 5.55$$

$$\frac{7.4x}{7.4} < \frac{5.55}{7.4} \qquad \text{\textit{Divide each side by} 7.4.}$$

$$x < 0.75$$

6. Solve and graph:
$4.5x + 2.31 > 6.3x - 4.89$.

Answers

5. $x < -6$

6. $x < 4$

Method 2 Eliminating the decimals in the beginning.

Because the greatest number of places to the right of the decimal point in any of the numbers is 2, we can multiply each side of the inequality by 100 and we will be left with an equivalent inequality that contains only whole numbers.

$$2.5x - 3.48 < -4.9x + 2.07$$ *Original inequality*

$$100(2.5x - 3.48) < 100(-4.9x + 2.07)$$ *Multiply each side by 100.*

$$100(2.5x) - 100(3.48) < 100(-4.9x) + 100(2.07)$$ *Distributive property*

$$250x - 348 < -490x + 207$$ *Multiply.*

$$740x - 348 < 207$$ *Add 490x to each side.*

$$740x < 555$$ *Add 348 to each side.*

$$\frac{740x}{740} < \frac{555}{740}$$ *Divide each side by 740.*

$$x < 0.75$$

The solution by either method is $x < 0.75$. Here is the graph:

EXAMPLE 7 Solve and graph: $3(x - 4) \geq -2$.

Solution $3x - 12 \geq -2$ *Distributive property*

$$3x - 12 + 12 \geq -2 + 12$$ *Add 12 to both sides.*

$$3x \geq 10$$

$$\frac{1}{3}(3x) \geq \frac{1}{3}(10)$$ *Multiply both sides by $\frac{1}{3}$.*

$$x \geq \frac{10}{3}$$

EXAMPLE 8 Solve and graph: $2(1 - 3x) + 4 < 4x - 14$.

Solution $2 - 6x + 4 < 4x - 14$ *Distributive property*

$$-6x + 6 < 4x - 14$$ *Simplify.*

$$-6x + 6 + (-6) < 4x - 14 + (-6)$$ *Add −6 to both sides.*

$$-6x < 4x - 20$$

$$-6x + (-4x) < 4x + (-4x) - 20$$ *Add −4x to both sides.*

$$-10x < -20$$

$$\left(-\frac{1}{10}\right)(-10x) > \left(-\frac{1}{10}\right)(-20)$$ *Multiply by $-\frac{1}{10}$, reverse the direction of the inequality.*

$$x > 2$$

7. Solve and graph
$2(x - 5) \geq -6$.

8. Solve and graph
$3(1 - 2x) + 5 < 3x - 1$.

7. $x \geq 2$

8. $x > 1$

EXAMPLE 9 Solve $2x - 3y < 6$ for y.

Solution We can solve this formula for y by first adding $-2x$ to each side and then multiplying each side by $-\frac{1}{3}$. When we multiply by $-\frac{1}{3}$ we must reverse the direction of the inequality symbol. Because this is a formula, we will not graph the solution.

$$2x - 3y < 6 \qquad\qquad \text{Original formula}$$
$$2x + (-2x) - 3y < (-2x) + 6 \qquad\qquad \text{Add } -2x \text{ to each side.}$$
$$-3y < -2x + 6$$
$$-\frac{1}{3}(-3y) > -\frac{1}{3}(-2x + 6) \qquad\qquad \text{Multiply each side by } -\frac{1}{3}.$$
$$y > \frac{2}{3}x - 2$$

9. Solve $2x - 3y < 6$ for x.

D Solving Compound Inequalities

Two inequalities considered together and connected by the words *and* or *or* are called *compound inequalities*.

EXAMPLE 10 Solve the compound inequality.

$$3t + 7 \le -4 \qquad \text{or} \qquad 3t + 7 \ge 4$$

Solution We solve each half of the compound inequality separately, then we graph the solution set.

$$3t + 7 \le -4 \qquad \text{or} \qquad 3t + 7 \ge 4$$
$$3t \le -11 \qquad \text{or} \qquad 3t \ge -3 \qquad \text{Add } -7.$$
$$t \le -\frac{11}{3} \qquad \text{or} \qquad t \ge -1 \qquad \text{Multiply by } \frac{1}{3}.$$

The solution set can be graphed in the following way:

$$-\frac{11}{3} \qquad\qquad -1$$

10. Solve and graph the solution for the compound inequality
$3t - 6 \le -3$ or $3t - 6 \ge 3$.

Sometimes compound inequalities that use the word *and* as the connecting word can be written in a shorter form. For example, the compound inequality $-3 \le x$ and $x \le 4$ can be written $-3 \le x \le 4$. Inequalities of the form $-3 \le x \le 4$ are called *continued inequalities* or *double inequalities*. The graph of $-3 \le x \le 4$ is

$$-3 \qquad\qquad 4$$

EXAMPLE 11 Solve and graph $-3 \le 2x - 5 \le 3$.

Solution We can extend our properties for addition and multiplication to cover this situation. If we add a number to the middle expression, we must add the same number to the outside expressions. We do the same for multiplication, remembering to reverse the direction of the inequality symbols if we multiply by a negative number.

$$-3 \le 2x - 5 \le 3$$
$$2 \le 2x \le 8 \qquad \text{Add 5 to all three members.}$$
$$1 \le x \le 4 \qquad \text{Multiply through by } \frac{1}{2}.$$

Here is how we graph this solution set:

$$1 \qquad\qquad 4$$

11. Solve and graph $-7 \le 2x + 1 \le 7$.

E Applications

When working application problems that involve inequalities, the phrases "at least" and "at most" translate as follows:

In Words	In Symbols
x is at least 30	$x \geq 30$
x is at most 20	$x \leq 20$

Our next example is similar to an example done earlier in this chapter. This time it involves an inequality instead of an equation.

We can modify our Blueprint for Problem Solving to solve application problems whose solutions depend on writing and then solving inequalities.

12. The sum of two consecutive integers is at least 27. What are the possibilities for the first of the two integers?

EXAMPLE 12 The sum of two consecutive odd integers is at most 28. What are the possibilities for the first of the two integers?

Solution When we use the phrase "their sum is at most 28," we mean that their sum is less than or equal to 28.

Step 1: *Read and list.*
 Known items: Two consecutive odd integers. Their sum is less than or equal to 28.
 Unknown items: The numbers in question.

Step 2: *Assign a variable, and translate information.*
 If we let $x =$ the first of the two consecutive odd integers, then $x + 2$ is the next consecutive one.

Step 3: *Reread and write an inequality.*
 Their sum is at most 28.

$$x + (x + 2) \leq 28$$

Step 4: *Solve the inequality.*

$2x + 2 \leq 28$	Simplify the left side.
$2x \leq 26$	Add -2 to each side.
$x \leq 13$	Multiply each side by $\frac{1}{2}$.

Step 5: *Write the answer.*
 The first of the two integers must be an odd integer that is less than or equal to 13. The second of the two integers will be two more than whatever the first one is.

Step 6: *Reread and check.*
 Suppose the first integer is 13. The next consecutive odd integer is 15. The sum of 15 and 13 is 28. If the first odd integer is less than 13, the sum of it and the next consecutive odd integer will be less than 28.

13. Re-work Example 13 if she gets $700 or more in interest.

EXAMPLE 13 Monica wants to invest money into an account that pays $6\frac{1}{4}\%$ annual interest. She wants to be able to count on at least $500 in interest at the end of the year. How much money should she invest to get $500 or more at the end of the year?

Solution The phrase "or more" tells us this is an inequality problem. Once we find out the amount needed to get $500 interest, we can solve the problem and answer the question.

Step 1: *Read and list.*
 Known items: The interest rate if $6\frac{1}{4}\%$. The minimum acceptable interest amount is $500.
 Unknown items: The investment amount.

Answers
12. The first integer is at least 13;
 $x \geq 13$
13. $x \geq 11{,}200$

Step 2: *Assign a variable, and translate information.*
We will let x = the amount to invest.
In words: $6\frac{1}{4}$% of x must be at least \$500.

Step 3: *Reread and write an inequality.*
Our sentence from step 2 can be translated directly into the inequality.

$$0.0625x \geq 500$$

Step 4: *Solve the inequality.*

$0.0625x \geq 500$	*Original inequality*
$x \geq 8{,}000$	*Divide both sides by 0.0625.*

Step 5: *Write the answer.*
We now know that if Monica invests \$8,000, her interest will be \$500. Any more than \$8,000 will give her more than \$500 so the answer is that Monica must invest at least \$8,000.

Step 6: *Reread and check.*
Put in any numbers greater than or equal to 8,000 in the inequality and check to make sure the result is at least 500.

$$x = 9{,}000$$

$$0.0625(9{,}000) = 562.5, \text{ which is } > \$500.$$

GETTING READY FOR CLASS

After reading through the preceding section, respond in your own words and in complete sentences.

A. State the addition property for inequalities.

B. When do you darken the circle on the graph of a solution set?

C. How is the multiplication property for inequalities different from the multiplication property of equality?

D. When do you reverse the direction of an inequality symbol?

Vocabulary Review

Choose the correct words to fill in the blanks below.

graph compound positive inequalities negative compound

1. The addition property for _____ states that $A + C < B + C$ if $A < B$.

2. The multiplication property for inequalities states that multiplying both sides of an inequality by a _____ number does not change the solution set.

3. When multiplying both sides of an inequality by a _____ number, it is necessary to reverse the inequality symbol.

4. The last step to solving a linear inequality in one variable is to _____ the solution set.

5. A _____ inequality is more than one inequality considered together using the words *and* or *or*.

Problems

A Solve the following inequalities using the addition property of inequalities. Graph each solution set.

1. $x - 5 < 7$

2. $x + 3 < -5$

3. $a - 4 \leq 8$

4. $a + 3 \leq 10$

5. $x - 4.3 > 8.7$

6. $x - 2.6 > 10.4$

7. $y + 6 \geq 10$

8. $y + 3 \geq 12$

9. $2 < x - 7$

10. $3 < x + 8$

11. $-4 > x + 2$

12. $-5 \geq x - 2$

B Solve the following inequalities using the multiplication property of inequalities. If you multiply both sides by a negative number, be sure to reverse the direction of the inequality symbol. Graph the solution set.

13. $3x < 6$

14. $2x < 14$

15. $5a \le 25$

16. $4a \le 16$

17. $\dfrac{x}{3} > 5$

18. $\dfrac{x}{7} > 1$

19. $-2x > 6$

20. $-3x \ge 9$

21. $-3x \ge -18$

22. $-8x \ge -24$

23. $-\dfrac{x}{5} \le 10$

24. $-\dfrac{x}{9} \ge -1$

25. $-\dfrac{2}{3}y > 4$

26. $-\dfrac{3}{4}y > 6$

27. $-\dfrac{x}{4} > 1$

28. $-\dfrac{1}{3}x \le -2$

29. $-\dfrac{5}{6}x < 10$

30. $-\dfrac{2}{3}x < -6$

C Solve the following inequalities. Graph the solution set in each case.

31. $2x - 3 < 9$

32. $3x - 4 < 17$

33. $-\dfrac{1}{5}y - \dfrac{1}{3} \le \dfrac{2}{3}$

34. $-\dfrac{1}{6}y - \dfrac{1}{2} \le \dfrac{2}{3}$

35. $-7.2x + 1.8 > -19.8$

36. $-7.8x - 1.3 > 22.1$

37. $\dfrac{2}{3}x - 5 \le 7$

38. $\dfrac{3}{4}x - 8 \le 1$

39. $-\dfrac{2}{5}a - 3 > 5$

40. $-\dfrac{4}{5}a - 2 > 10$

41. $5 - \dfrac{3}{5}y > -10$

42. $4 - \dfrac{5}{6}y > -11$

43. $0.3(a + 1) \le 1.2$

44. $0.4(a - 2) \le 0.4$

45. $2(5 - 2x) \le -20$

46. $7(8 - 2x) > 28$

47. $3x - 5 > 8x$

48. $8x - 4 > 6x$

49. $\dfrac{1}{3}y - \dfrac{1}{2} \le \dfrac{5}{6}y + \dfrac{1}{2}$

50. $\dfrac{7}{6}y + \dfrac{4}{3} \le \dfrac{11}{6}y - \dfrac{7}{6}$

51. $-2.8x + 8.4 < -14x - 2.8$

52. $-7.2x - 2.4 < -2.4x + 12$

53. $3(m - 2) - 4 \ge 7m + 14$

54. $2(3m - 1) + 5 \ge 8m - 7$

55. $3 - 4(x - 2) \le -5x + 6$

56. $8 - 6(x - 3) \le -4x + 12$

57. $4 - 3(x - 2) \ge -x + 4$

58. $5 - 2(2x - 1) \ge -2x + 4$

59. $-3x + 9 < 4 - 6(2x - 5)$

60. $-2x + 4 < 5 - 3(2x - 6)$

Solve each of the following formulas for y.

61. $3x + 2y < 6$

62. $-3x + 2y < 6$

63. $2x - 5y > 10$

64. $-2x - 5y > 5$

65. $-3x + 7y \le 21$

66. $-7x + 3y \le 21$

67. $2x - 4y \ge -4$

68. $4x - 2y \ge -8$

The next two problems are intended to give you practice reading, and paying attention to, the instructions that accompany the problems you are working.

69. Work each problem according to the instructions given.

 a. Evaluate $-5x + 3$ when $x = 0$.

 b. Solve: $-5x + 3 = -7$.

 c. Is 0 a solution to $-5x + 3 < -7$?

 d. Solve: $-5x + 3 < -7$.

70. Work each problem according to the instructions given.

 a. Evaluate $-2x - 5$ when $x = 0$.

 b. Solve: $-2x - 5 = 1$.

 c. Is 0 a solution to $-2x - 5 > 1$?

 d. Solve: $-2x - 5 > 1$.

For each graph below, write an inequality whose solution is the graph.

71.

 0 3

 $(-\infty, 3)$

72.

 0 3

 $[3, \infty)$

73.

 0 3

 $(-\infty, 3]$

74.

 0 3

 $(3, \infty)$

D Solve and graph the solution sets for the following compound inequalities.

75. $x + 5 \leq -2$ or $x + 5 \geq 2$

$x - 5 - 2 - 5$ $x \geq 2 - 5$

$x \leq -7$ $x \geq -3$

$(-\infty, 7] \cup [-3, \infty)$

76. $3x + 2 < -3$ or $3x + 2 > 3$

77. $5y + 1 \leq -4$ or $5y + 1 \geq 4$

78. $7y - 5 \leq -2$ or $7y - 5 \geq 2$

79. $2x + 5 < 3x - 1$ or $x - 4 > 2x + 6$

80. $3x - 1 > 2x + 4$ or $5x - 2 < 3x + 4$

81. $3x + 1 < -8$ or $-2x + 1 \leq -3$

82. $2x - 5 \leq -1$ or $-3x - 6 < -15$

 2 3

 $(-\infty, 2] \cup (3, \infty)$

$2x \leq +5 - 1$ $-2x - 6 \leq -15$

$2x \leq 4$ $-3x \leq -9$

$x \leq 2$ $x \geq 3$

Solve the following continued inequalities. Use a line graph to write each solution set.

83. $-2 \leq m - 5 \leq 2$

84. $-3 \leq m + 1 \leq 3$

85. $-40 < 20a + 20 < 20$

86. $-60 < 50a - 40 < 60$

87. $0.5 \le 0.3a - 0.7 \le 1.1$

88. $0.1 \le 0.4a + 0.1 \le 0.3$

89. $3 < \frac{1}{2}x + 5 < 6$

90. $5 < \frac{1}{4}x + 1 < 9$

91. $4 < 6 + \frac{2}{3}x < 8$

92. $3 < 7 + \frac{4}{5}x < 15$

93. $-2 < -\frac{1}{2}x + 1 \le 1$

94. $-3 \le -\frac{1}{3}x - 1 < 2$

95. $-\frac{1}{2} \le \frac{3x + 1}{2} \le \frac{1}{2}$

96. $-\frac{5}{6} \le \frac{2x + 5}{3} \le \frac{5}{6}$

97. $-1.5 \le \frac{2x - 3}{4} \le 3.5$

98. $-1.25 \le \frac{2x + 3}{4} \le .75$

99. $-\frac{3}{4} \le \frac{4x - 3}{2} \le 1.5$

100. $-\frac{5}{6} \le \frac{5x + 2}{3} \le \frac{5}{6}$

Applying the Concepts

101. Consecutive Integers The sum of two consecutive integers is at least 583. What are the possibilities for the first of the two integers?

102. Consecutive Integers The sum of two consecutive integers is at most 583. What are the possibilities for the first of the two integers?

103. Number Problems The sum of twice a number and six is less than ten. Find all solutions.

104. Number Problems Twice the difference of a number and three is greater than or equal to the number increased by five. Find all solutions.

105. Number Problems The product of a number and four is greater than the number minus eight. Find the solution set.

106. Number Problems The quotient of a number and five is less than the sum of seven and two. Find the solution set.

107. Geometry Problems The length of a rectangle is 3 times the width. If the perimeter is to be at least 48 meters, what are the possible values for the width? (If the perimeter is at least 48 meters, then it is greater than or equal to 48 meters.)

108. Geometry Problems The length of a rectangle is 3 more than twice the width. If the perimeter is to be at least 51 meters, what are the possible values for the width? (If the perimeter is at least 51 meters, then it is greater than or equal to 51 meters.)

109. Geometry Problems The numerical values of the three sides of a triangle are given by three consecutive even integers. If the perimeter is greater than 24 inches, what are the possibilities for the shortest side?

110. Geometry Problems The numerical values of the three sides of a triangle are given by three consecutive odd integers. If the perimeter is greater than 27 inches, what are the possibilities for the shortest side?

111. Car Heaters If you have ever gotten in a cold car early in the morning you know that the heater does not work until the engine warms up. This is because the heater relies on the heat coming off the engine. Write an equation using an inequality sign to express when the heater will work if the heater works only after the engine is 100°F.

112. Exercise When Kate exercises, she either swims or runs. She wants to spend a minimum of 8 hours a week exercising, and she wants to swim 3 times the amount she runs. What is the minimum amount of time she must spend doing each exercise?

113. Profit and Loss Movie theaters pay a certain price for the movies that you and I see. Suppose a theater pays $1,500 for each showing of a popular movie. If they charge $7.50 for each ticket they sell, then they will lose money if ticket sales are less than $1,500. However, they will make a profit if ticket sales are greater than $1,500. What is the range of tickets they can sell and still lose money? What is the range of tickets they can sell and make a profit?

114. Stock Sales Suppose you purchase x shares of a stock at $12 per share. After 6 months you decide to sell all your shares at $20 per share. Your broker charges you $15 for the trade. If your profit is at least $3,985, how many shares did you purchase in the first place?

Maintaining Your Skills

The problems that follow review some of the more important skills you have learned in previous sections and chapters. You can consider the time you spend working these problems as time spent studying for exams.

Answer the following percent problems.

115. What number is 25% of 32?

116. What number is 15% of 75?

117. What number is 20% of 120?

118. What number is 125% of 300?

119. What percent of 36 is 9?

120. What percent of 16 is 9?

121. What percent of 50 is 5?

122. What percent of 140 is 35?

123. 16 is 20% of what number?

124. 6 is 3% of what number?

125. 8 is 2% of what number?

126. 70 is 175% of what number?

Simplify each expression.

127. $-|-5|$

128. $\left(-\dfrac{2}{3}\right)^3$

129. $-3 - 4(-2)$

130. $2^4 + 3^3 \div 9 - 4^2$

131. $5\,|3 - 8| - 6\,|2 - 5|$

132. $7 - 3(2 - 6)$

133. $5 - 2[-3(5 - 7) - 8]$

134. $\dfrac{5 + 3(7 - 2)}{2(-3) - 4}$

135. Find the difference of -3 and -9.

136. If you add -4 to the product of -3 and 5, what number results?

137. Apply the distributive property to $\dfrac{1}{2}(4x - 6)$.

138. Use the associative property to simplify $-6\left(\dfrac{1}{3}x\right)$.

Find the Mistake

Each sentence below contains a mistake. Circle the mistake and write the correct word(s) or number on the line provided.

1. If the solution set to an inequality is all the real numbers greater than 9, the solution is $x \leq 9$. _____

2. To include the number 14 in the solution to the inequality $x \geq 14$, we mark the number on the graph with an open circle.

3. The last step in solving an inequality is to use the multiplication property for inequalities. _____

4. Upon solving the inequality $-5y < 4x + 9$ for y, the inequality symbol remains the same. _____

A Resources for research (e.g., internet, library, or interviews)

B Pen or pencil

Show Me the Money

Starting a business involves making important financial decisions. These decisions require a lot of math. For this project, create a business model for a new business. The business can be one that you operate out of your home, or one that requires renting a commercial space from which you offer services or sell your inventory. Research the financial information you would need to start this business. What mathematical concepts would you use to interpret this information? Create linear equations that will give you calculated quantities for cost, revenue, and profit. How would linear inequalities give you knowledge about your business?

Chapter 2 Summary

Solution Set [2.1]

1. The solution set for the equation $x + 2 = 5$ is $\{3\}$ because when x is 3 the equation is $3 + 2 = 5$, or $5 = 5$.

The *solution set* for an equation (or inequality) is all the numbers that, when used in place of the variable, make the equation (or inequality) a true statement.

Equivalent Equations [2.1]

2. The equation $a - 4 = 3$ and $a - 2 = 5$ are equivalent because both have solution set $\{7\}$.

Two equations are called *equivalent* if they have the same solution set.

Addition Property of Equality [2.1]

3. Solve $x - 5 = 12$.
$$x - 5\,(+\,5) = 12\,(+\,5)$$
$$x + 0 = 17$$
$$x = 17$$

When the same quantity is added to both sides of an equation, the solution set for the equation is unchanged. Adding the same amount to both sides of an equation produces an equivalent equation.

Multiplication Property of Equality [2.2]

4. Solve $3x = 18$.
$$\frac{1}{3}(3x) = \frac{1}{3}(18)$$
$$x = 6$$

If both sides of an equation are multiplied by the same nonzero number, the solution set is unchanged. Multiplying both sides of an equation by a nonzero quantity produces an equivalent equation.

Linear Equations [2.3]

A linear equation in one variable is any equation that can be put in the form $Ax + B = C$, where A, B and C are real numbers and a is not zero.

Strategy for Solving Linear Equations in One Variable [2.3]

5. Solve $2(x + 3) = 10$.
$$2x + 6 = 10$$
$$2x + 6 + (-6) = 10 + (-6)$$
$$2x = 4$$
$$\frac{1}{2}(2x) = \frac{1}{2}(4)$$
$$x = 2$$

Step 1a: Use the distributive property to separate terms, if necessary.

1b: If fractions are present, consider multiplying both sides by the LCD to eliminate the fractions. If decimals are present, consider multiplying both sides by a power of 10 to clear the equation of decimals.

1c: Combine similar terms on each side of the equation.

Step 2: Use the addition property of equality to get all variable terms on one side of the equation and all constant terms on the other side. A variable term is a term that contains the variable (for example, $5x$). A constant term is a term that does not contain the variable (the number 3, for example).

Step 3: Use the multiplication property of equality to get x (that is, $1x$) by itself on one side of the equation.

Step 4: Check your solution in the original equation to be sure that you have not made a mistake in the solution process.

Formulas [2.4]

A formula is an equation with more than one variable. To solve a formula for one of its variables, we use the addition and multiplication properties of equality to move everything except the variable in question to one side of the equal sign so the variable in question is alone on the other side.

6. Solving $P = 2l + 2w$ for l, we have

$$P - 2w = 2l$$

$$\frac{P - 2w}{2} = l$$

Blueprint for Problem Solving [2.5, 2.6]

Step 1: *Read* the problem, and then mentally *list* the items that are known and the items that are unknown.

Step 2: *Assign a variable* to one of the unknown items. (In most cases this will amount to letting $x =$ the item that is asked for in the problem.) Then *translate* the other *information* in the problem to expressions involving the variable.

Step 3: *Reread* the problem, and then *write an equation,* using the items and variables listed in steps 1 and 2, that describes the situation.

Step 4: *Solve the equation* found in step 3.

Step 5: *Write* your *answer* using a complete sentence.

Step 6: *Reread* the problem, and *check* your solution with the original words in the problem.

Addition Property for Inequalities [2.7]

Adding the same quantity to both sides of an inequality produces an equivalent inequality, one with the same solution set.

7. Solve $x + 5 < 7$.

$$x + 5 + (-5) < 7 + (-5)$$

$$x < 2$$

Multiplication Property for Inequalities [2.7]

Multiplying both sides of an inequality by a positive number never changes the solution set. If both sides are multiplied by a negative number, the sign of the inequality must be reversed to produce an equivalent inequality.

8. Solve $-3a \le 18$.

$$-\frac{1}{3}(-3a) \ge -\frac{1}{3}(18)$$

$$a \ge -6$$

Strategy for Solving Linear Inequalities in One Variable [2.7]

Step 1a: Use the distributive property to separate terms, if necessary.

1b: If fractions are present, consider multiplying both sides by the LCD to eliminate the fractions. If decimals are present, consider multiplying both sides by a power of 10 to clear the inequality of decimals.

1c: Combine similar terms on each side of the inequality.

Step 2: Use the addition property for inequalities to get all variable terms on one side of the inequality and all constant terms on the other side.

Step 3: Use the multiplication property for inequalities to get x by itself on one side of the inequality.

Step 4: Graph the solution set.

9. Solve $3(x - 4) \ge -2$.

$$3x - 12 \ge -2$$

$$3x - 12\,(+\,12) \ge -2\,(+\,12)$$

$$3x \ge 10$$

$$\frac{1}{3}(3x) \ge \frac{1}{3}(10)$$

$$x \ge \frac{10}{3}$$

Is 2 a solution to each of the following equations? [2.1]

1. $4x + 7 = 8$ **2.** $5y - 3 = 7$

Is 3 a solution to each of the following equations? [2.1]

3. $2n + 4 = 10$ **4.** $4a - 3 = 5$

Solve the following equations. [2.1]

5. $x - 5 = 10$

6. $y + 4 = 15$

7. $n - 5.4 = 10.6$

Solve the following equations. [2.2, 2.3]

8. $4x - 3 = -1$ **9.** $5y - 24 = y$

10. $\frac{1}{4}x - \frac{1}{16} = \frac{1}{8} + \frac{1}{2}x$ **11.** $-4(2x - 7) + 6 = -6$

12. $4x - 9 = 7$

13. $0.04 - 0.03(200 - x) = 5.2$

14. $3(t - 7) + 4(t + 5) = 5t + 5$

15. $5x - 3(4x - 5) = -9x + 11$

16. $3(x + 5) - 5x + 15(x + 7) = 211$

17. $2(n + 4) + 8n - 5(n + 3) = 43$

18. If $6x - 8y = 24$, find y when $x = 3$. [2.4]

19. If $6x - 8y = 24$, find x when $y = -3$. [2.4]

20. Solve $3x - 2y = 6$ for y. [2.4]

21. Solve $9 = \dfrac{k(T_1 - T_2)}{x}$ for T_1. [2.4]

Solve each word problem. [2.5, 2.6]

22. Age Problem Brian is 4 years older than Michelle. Five years ago the sum of their ages was 48. How old are they now?

23. Geometry The length of a rectangle is 1 inch more than twice the width. The perimeter is 44 inches. What are the length and width?

24. Coin Problem A man has $1.15 in quarters and dimes in his pocket. If he has one more dime than he has quarters, how many of each coin does he have?

25. Investing A woman has money in two accounts. One account pays 7% annual interest, whereas the other pays 15% annual interest. If she has $1,500 more invested at 15% than she does at 7% and her total interest for a year is $1,545, how much does she have in each account?

Solve each inequality, and graph the solution. [2.7]

26. $\frac{2}{3}x + 5 \le -3$ **27.** $-5y > 30$

28. $0.5 - 0.4x < 2.1$ **29.** $2 + 3(n + 3) \ge 17$

30. $3x - 4 < 2x$ **31.** $3x - 6 \le 15$

Write an inequality whose solution is the given graph. [2.7]

32.

33.

34.

The illustration below shows the number of iPad only apps in the most popular categories. If there were 3,500 apps, what percentage of apps were in the categories below? Round to the nearest tenth of a percent.

TOP iPAD APPLICATION CATEGORIES
techcrunch.com, April 2010

Games	833
Entertainment	260
Education	205
Utilities	165
Productivity	152
Books	133

*Number of Applications Compatible with iPad Only

35. Games **36.** Entertainment

Simplify.

1. $6 - 13 + 2$

2. $-5 + 19 + (-3)$

3. $9 - 3 \cdot 4$

4. $24 \div 6 \cdot (-3)$

5. $7 - 2(7 + 3)$

6. $-3(5 - 7) + 2(12 - 9)$

7. $\left(-\dfrac{3}{2}\right)^3$

8. $\left[\dfrac{1}{3}(-15)\right]^2$

9. $\dfrac{40}{(-120)}$

10. $\dfrac{216}{270}$

11. $-\dfrac{9}{14} \div \dfrac{3}{7}$

12. $\dfrac{3}{7} \cdot \dfrac{2}{5}$

13. $\dfrac{-2(9)}{(-24)}$

14. $\dfrac{3(-2) + 4(5)}{-6 - 2}$

15. $\dfrac{4}{15} + \dfrac{3}{5}$

16. $\dfrac{3}{24} - \dfrac{11}{32}$

17. $\dfrac{3}{7}(14x)$

18. $-4(8x)$

19. $\dfrac{2}{3}(9x - 12)$

20. $4(3x - 2) - 3(5x - 3)$

Solve each equation.

21. $9x = 8x - 3$

22. $x - 9 = -5$

23. $\dfrac{8}{5}x = -40$

24. $\dfrac{x}{6} = \dfrac{3}{2}$

25. $4x - 9 = 23$

26. $5x - 12 = 2x + 27$

27. $2(2x - 3) - 3(5x - 1) = -25$

28. $6 - 4(3t - 5) = 2$

29. $3(4a - 3) - 2(a + 2) = 15$

30. $\dfrac{1}{3}(6x - 12) = \dfrac{1}{2}(10x - 6)$

31. Solve $P = 2l + 2w$ for w.

32. Solve $8x + 9y = 24$ for y.

Solve each inequality, and graph the solution.

33. $4(x - 2) \geq 12$

34. $-3x - 15 > 3$

35. $3x - 2 \geq 19$

36. $-5 < 2x - 3$

Translate into symbols.

37. The difference of y and 7 is 17.

38. The sum of x and 3 is -5.

39. Evaluate $y^2 + 3y - 9$ when $y = -4$.

40. Evaluate $2a^2 + 3ab - 4b^2$ when $a = -3$, $b = 2$.

41. What is 40% of 70?

42. What percent of 25 is 9?

43. Number Problem The sum of a number and 11 is 32. Find the number.

44. Number Problem Twice a number increased by 9 is 3. Find the number.

45. Geometry A right triangle has one 37° angle. Find the other two angles.

46. Geometry Two angles are complementary. If one is 49°, find the other one.

47. Hourly Pay Zach tutors in the math lab. He gets paid $9 per hour for the first 12 hours and $12 per hour for each hour after that. He made $180 one week. How many hours did he work?

48. Cost of a Letter The cost of mailing a letter was 44¢ for the first ounce and 17¢ for each additional ounce. If the cost of mailing a letter was $1.12, how many ounces did the letter weigh?

The illustration shows the number of viewers for Fox's most popular shows.

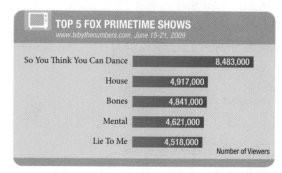

49. Find the total number of viewers for *So You Think You Can Dance* and *House*.

50. How many more viewers for *Bones* are there than for *Lie To Me*?

Is 4 a solution to each of the following equations? [2.1]

1. $5x + 4 = 8$ **2.** $3y - 5 = 7$

Is 7 a solution to each of the following equations? [2.1]

3. $4n + 3 = 31$ **4.** $6a - 6 = 5$

Solve the following equations. [2.1]

5. $x - 7 = 10$

6. $y + 2 = 7$

7. $n + 4.2 = 9.4$

Solve the following equations. [2.2, 2.3]

8. $3x - 2 = 7$ **9.** $4y + 15 = y$

10. $\dfrac{1}{4}x - \dfrac{1}{12} = \dfrac{1}{3}x + \dfrac{1}{6}$ **11.** $-3(3 - 2x) - 7 = 8$

12. $3x - 9 = -6$

13. $0.05 + 0.07(100 - x) = 3.2$

14. $4(t - 3) + 2(t + 4) = 2t - 16$

15. $4x - 2(3x - 1) = 2x - 8$

16. $5(x + 2) - 3x + 3(x + 4) = 32$

17. $3(a + 7) + 6a - 3(a + 2) = 45$

18. If $3x - 4y = 16$, find y when $x = 4$. [2.4]

19. If $3x - 4y = 16$, find x when $y = 2$. [2.4]

20. Solve $2x + 6y = 12$ for y. [2.4]

21. Solve $x^2 = v^2 + 2ad$ for a. [2.4]

Solve each word problem. [2.5, 2.6]

22. Age Problem Paul is twice as old as Becca. Five years ago the sum of their ages was 44. How old are they now?

23. Geometry The length of a rectangle is 5 centimeters less than 3 times the width. The perimeter is 150 centimeters. What are the length and width?

24. Coin Problem A man has a collection of dimes and nickels with a total value of $1.70. If he has 8 more dimes than he has nickels, how many of each coin does he have?

25. Investing A woman has money in two accounts. One account pays 6% annual interest, whereas the other pays 12% annual interest. If she has $500 more invested at 12% than she does at 6% and her total interest for a year is $186, how much does she have in each account?

Solve each inequality, and graph the solution. [2.7]

26. $\dfrac{1}{2}x - 2 > 3$ **27.** $-6y \le 24$

28. $0.3 - 0.2x < 1.1$ **29.** $3 - 2(n - 1) \ge 9$

30. $5x - 3 < 2x$ **31.** $-3 \le 2x - 7$

Write an inequality whose solution is the given graph. [2.7]

32.
$-3 \qquad 0$

33.
$-1 \qquad 0$

34.
$-2 \qquad 0$

The snapshot shows the number of hours per day the average commute is for several cities throughout the United States. Use the information to determine the percentage of each day (24 hours) are spent commuting in a car for the following cities. Round to the nearest tenth of a percent.

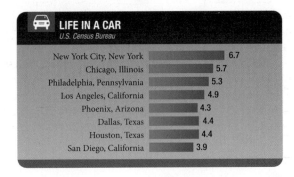

LIFE IN A CAR
U.S. Census Bureau

New York City, New York	6.7
Chicago, Illinois	5.7
Philadelphia, Pennsylvania	5.3
Los Angeles, California	4.9
Phoenix, Arizona	4.3
Dallas, Texas	4.4
Houston, Texas	4.4
San Diego, California	3.9

35. New York City **36.** Los Angeles

Graphing Linear Equations and Inequalities in Two Variables

3

Ligurian Sea, Italy
Image © 2010 European Space Imaging

©iStockphoto.com/danbreckwoldt

Italy's coast of the Ligurian Sea is known for its rugged cliffs and historic settlements. If you are lucky enough to visit, you will find five fishing villages connected by a walking trail, known as Sentiero Azzurro ("Light Blue Trail"). Constructed over the centuries as a connection between these villages, the trail has become a must-see for visitors. The walk varies in difficulty and takes approximately 5 hours. Some sections are easy strolls, others are very challenging hikes. One leg of the trail begins with an easy walk and finishes with a climb of 368 stairs. It winds through olive orchards and vineyards and can be rough in places, but offers spectacular views of the bay and the unique coastline. The difference in height for the Sentiero Azzurro from bottom to top is more than 600 meters and the length is more than 10 km.

From the above information, we know that the steepness of the trail varies along its length. However, we can use the knowledge of the overall change in height from the bottom of the trail to the top to potentially determine an average steepness, or slope. In mathematics, slope is the difference between the rise of a path compared to the distance over which that rise takes place. Calculation of slope is one of the topics we will explore in this chapter.

A Create a scatter diagram or line graph from a table of data.

B Graph ordered pairs on a rectangular coordinate system.

KEY WORDS

scatter diagram

line graph

x-coordinate

y-coordinate

axis

quadrant

3.1 The Rectangular Coordinate System

Image © Katherine Heistand Shields, 2010

In New Jersey, a fast food restaurant called Cluck-U Chicken markets a wide range of hot sauces for its famous buffalo wings. Their 911 hot sauce is so spicy, the restaurant requires all tasters to sign a waiver upon order. According to Cluck-U's corporate office, the Scoville rating (a scale to measure the spicy heat of the chili pepper) of the 911 sauce is 450,000 units. Look at Table 1 below, which compares other peppers with the 911 hot sauce. Figure 1 is a bar graph, which is a visual representation of the same information. The ratings in the table and graph are approximate because each entry on the Scoville scale typically appears as a range.

Table 1

Spicy Peppers

Pepper or Pepper-Infused Item	Scoville Units
Green pepper	0
Carolina cayenne pepper	100,000
Bird's eye pepper	200,000
Habanero pepper	300,000
Cluck-U's 911 sauce	450,000

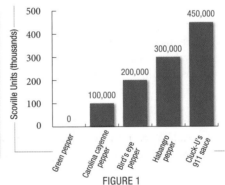

FIGURE 1

A Scatter Diagrams and Line Graphs

The information in the table in the introduction can be visualized with a *scatter diagram*, also called a *scatter plot* and *line graph* as well. Figure 2 is a scatter diagram of the information in Table 1. We use dots instead of the bars shown in Figure 1 to show the Scoville rating of each pepper item. Figure 3 is called a *line graph*. It is constructed by taking the dots in Figure 2 and connecting each one to the next with a straight line. Notice that we have labeled the axes in these two figures a little differently than we did with the bar chart by making the axes intersect at the number 0.

FIGURE 2

FIGURE 3

EXAMPLE 1 The tables below give the first five odd numbers and the first five counting numbers squared as paired data. In each case construct a scatter diagram.

Odd Numbers	
Position	Value
1	1
2	3
3	5
4	7
5	9

Squares	
Position	Value
1	1
2	4
3	9
4	16
5	25

Solution The two scatter diagrams are based on the data from these tables shown here. Notice how the dots in Figure 4 seem to line up in a straight line, whereas the dots in Figure 5 give the impression of a curve. We say the points in Figure 4 suggest a linear relationship between the two sets of data, whereas the points in Figure 5 suggest a nonlinear relationship.

FIGURE 4

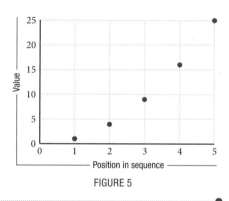

FIGURE 5

B Graphing Ordered Pairs

As you know, each dot in Figures 4 and 5 corresponds to a pair of numbers, one of which is associated with the horizontal axis and the other with the vertical axis. Paired data play a very important role in the equations we will solve in the next section. To prepare ourselves for those equations, we need to expand the concept of paired data to include negative numbers. At the same time, we want to standardize the position of the axes in the diagrams that we use to visualize paired data.

Our ability to graph paired data is due to the invention of the rectangular coordinate system also caled the *Cartesian coordinate system* after the French philosopher René Descartes (1595–1650). Descartes is the person usually credited with the invention of the rectangular coordinate system. As a philosopher, Descartes is responsible for the statement "I think, therefore I am." Until Descartes invented his coordinate system in 1637, algebra and geometry were treated as separate subjects. The rectangular coordinate system allows us to connect algebra and geometry by associating geometric shapes with algebraic equations.

Here is the formal definition for a graph's coordinates:

DEFINITION *x-coordinate, y-coordinate*

A pair of numbers enclosed in parentheses and separated by a comma, such as $(-2, 1)$, is called an ordered pair of numbers. The first number in the pair is called the *x*-coordinate of the ordered pair; the second number is called the *y*-coordinate. For the ordered pair $(-2, 1)$, the *x*-coordinate is -2 and the *y*-coordinate is 1.

Ordered pairs of numbers are important in the study of mathematics because they give us a way to visualize solutions to equations. To see the visual component of ordered pairs, we need the diagram shown in Figure 6. It is called the *rectangular coordinate system.*

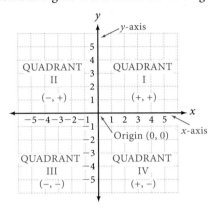

FIGURE 6

The rectangular coordinate system is built from two number lines oriented perpendicular to each other. The horizontal number line is exactly the same as our real number line and is called the *x-axis*. The vertical number line is also the same as our real number line with the positive direction up and the negative direction down. It is called the *y-axis*. The point where the two axes intersect is called the *origin*. As you can see from Figure 6, the axes divide the plane into four *quadrants*, which are numbered I through IV in a counterclockwise direction.

To graph the ordered pair (a, b), we start at the origin and move a units forward or back (forward if a is positive and back if a is negative). Then we move b units up or down (up if b is positive, down if b is negative). The point where we end up is the graph of the ordered pair (a, b). To graph the ordered pair $(5, 2)$, we start at the origin and move 5 units to the right. Then, from that position, we move 2 units up.

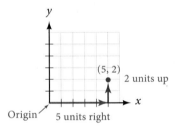

Note It is very important that you graph ordered pairs quickly and accurately. Remember, the first coordinate goes with the horizontal axis and the second coordinate goes with the vertical axis.

2. Graph the ordered pairs $(2, 3)$, $(2, -3)$, $(-2, 3)$, and $(-2, -3)$.

EXAMPLE 2 Graph the ordered pairs $(3, 4)$, $(3, -4)$, $(-3, 4)$, and $(-3, -4)$.

Solution

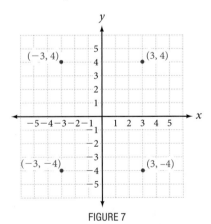

FIGURE 7

We can see in Figure 7 that when we graph ordered pairs, the x-coordinate corresponds to movement parallel to the x-axis (horizontal) and the y-coordinate corresponds to movement parallel to the y-axis (vertical).

EXAMPLE 3 Graph the ordered pairs $(-1, 3)$, $(2, 5)$, $(0, 0)$, $(0, -3)$, and $(4, 0)$.

Solution

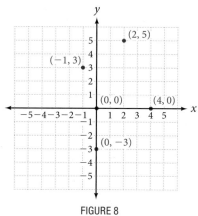

FIGURE 8

3. Graph the ordered pairs $(-2, 1)$, $(3, 5)$, $(0, 2)$, $(-5, 0)$, and $(-3, -3)$.

Note If we do not label the axes of a coordinate system, we assume that each square is one unit long and one unit wide.

Review the three points in Example 3 that contain a zero as either an *x*- or a *y*-coordinate. It is important to note that ordered pairs that sit on an axis or at the origin are not associated with any quadrant.

GETTING READY FOR CLASS

After reading through the preceding section, respond in your own words and in complete sentences.

A. What is an ordered pair of numbers?

B. Explain in words how you would graph the ordered pair $(3, 4)$.

C. How does a scatter diagram differ from a line graph?

D. Where is the origin on a rectangular coordinate system and why is it important?

Vocabulary Review

Choose the correct words to fill in the blanks below.

x-axis	line graph	ordered pair	scatter diagram/plot
y-axis	origin	quadrants	

1. A _____ uses dots on a graph to identify data.

2. Connecting the dots of a scatter diagram gives a_____ .

3. A pair of numbers enclosed in parentheses and separated by a comma is called an _____ .

4. A rectangular coordinate system has a horizontal number line called the _____ and a vertical number line called the _____ .

5. The x-axis and y-axis divide the rectangular coordinate system into four _____ .

6. The point where the x-and y-axis intersect is called the _____ .

Problems

A 1. Non-Camera Phone Sales The table here shows what are the estimated sales of non-camera phones for the years 2006–2010. Use the information from the table and chart to construct a line graph.

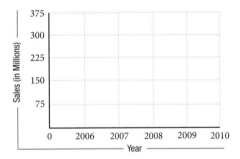

Year	Sales (in millions)
2006	300
2007	250
2008	175
2009	150
2010	125

2. Camera Phone Sales The table here shows the estimated sales of camera phones from 2006 to 2010. Use the information from the table to construct a line graph.

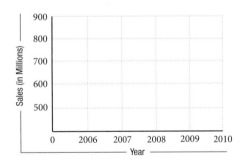

Year	Sales (in millions)
2006	500
2007	650
2008	750
2009	875
2010	900

SCAN TO ACCESS

B Graph the following ordered pairs.

3. (3, 2) **4.** (3, −2) **5.** (−3, 2) **6.** (−3, −2) **7.** (5, 1) **8.** (5, −1)

X Y \ X Y

9. (1, 5) **10.** (1, −5) **11.** (−1, 5) **12.** (−1, −5) **13.** $\left(2, \dfrac{1}{2}\right)$ **14.** $\left(3, \dfrac{3}{2}\right)$

X Y X Y

15. $\left(-4, -\dfrac{5}{2}\right)$ **16.** $\left(-5, -\dfrac{3}{2}\right)$ **17.** (3, 0) **18.** (−2, 0) **19.** (0, 5) **20.** (0, 0)

X Y

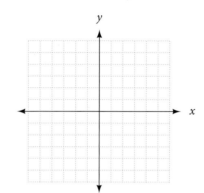

Odd-Numbered Problems Even-Numbered Problems

Give the coordinates of each numbered point in the figure and list which quadrant each point is in.

21. −4, 4 **22.**

23. **24.**

25. −3, 0 **26.**

27. **28.**

29. −5, −5 **30.**

Graph the points (4, 3) and (−4, −1), and draw a straight line that passes through both of them. Then answer the following questions.

31. Does the graph of (2, 2) lie on the line?

32. Does the graph of (−2, 0) lie on the line?

33. Does the graph of (0, −2) lie on the line?

34. Does the graph of (−6, 2) lie on the line?

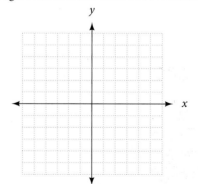

Graph the points $(-2, 4)$ and $(2, -4)$, and draw a straight line that passes through both of them. Then answer the following questions.

35. Does the graph of $(0, 0)$ lie on the line?

36. Does the graph of $(-1, 2)$ lie on the line?

37. Does the graph of $(2, -1)$ lie on the line?

38. Does the graph of $(1, -2)$ lie on the line?

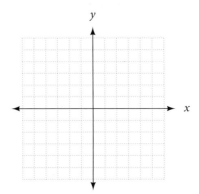

Draw a straight line that passes through the points $(3, 4)$ and $(3, -4)$. Then answer the following questions.

39. Is the graph of $(3, 0)$ on this line?

40. Is the graph of $(0, 3)$ on this line?

41. Is there any point on this line with an x-coordinate other than 3?

42. If you extended the line, would it pass through a point with a y-coordinate of 10?

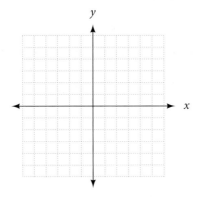

Draw a straight line that passes through the points $(3, 4)$ and $(-3, 4)$. Then answer the following questions.

43. Is the graph of $(4, 0)$ on this line?

44. Is the graph of $(0, 4)$ on this line?

45. Is there any point on this line with a y-coordinate other than 4?

46. If you extended the line, would it pass through a point with an x-coordinate of 10?

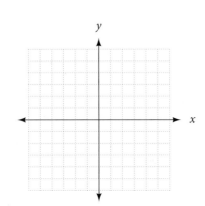

Applying the Concepts

47. Hourly Wages Jane takes a job at the local Marcy's department store. Her job pays $8.00 per hour. The graph shows how much Jane earns for working from 0 to 40 hours in a week.

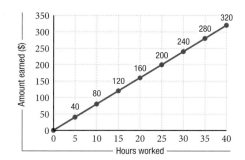

a. List three ordered pairs that lie on the line graph.

b. How much will she earn for working 40 hours?

c. If her check for one week is $240, how many hours did she work?

d. She works 35 hours one week, but her paycheck before deductions are subtracted out is for $260. Is this correct? Explain.

48. Hourly Wages Judy takes a job at Gigi's boutique. Her job pays $6.00 per hour plus $50 per week in commission. The graph shows how much Judy earns for working from 0 to 40 hours in a week.

a. List three ordered pairs that lie on the line graph.

b. How much will she earn for working 40 hours?

c. If her check for one week is $230, how many hours did she work?

d. She works 35 hours one week, but her paycheck before deductions are subtracted out is for $260. Is this correct? Explain.

49. Baseball Attendance The graph gives the attendance at Major League Baseball games over a ten-year period. If x represents the year in question and y represents attendance in millions, write five ordered pairs that describe the information in the table.

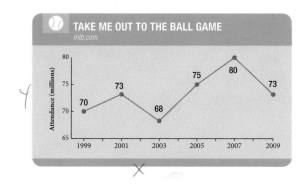

50. FIFA World Cup Goals The graph shows the number of goals scored by the US Men's National Team in FIFA World Cup competition. Write five ordered pairs that lie on the graph.

51. Right triangle ABC has legs of length 5. Point C is the ordered pair $(6, 2)$. Find the coordinates of A and B.

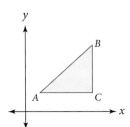

52. Right triangle ABC has legs of length 7. Point C is the ordered pair $(-8, -3)$. Find the coordinates of A and B.

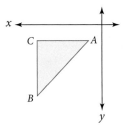

53. Rectangle $ABCD$ has a length of 5 and a width of 3. Point D is the ordered pair $(7, 2)$. Find points A, B, and C.

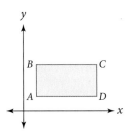

54. Rectangle $ABCD$ has a length of 5 and a width of 3. Point D is the ordered pair $(-1, 1)$. Find points A, B, and C.

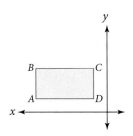

Getting Ready for the Next Section

55. Let $2x + 3y = 6$.
 a. Find x if $y = 4$.

 b. Find x if $y = -2$.

 c. Find y if $x = 3$.

 d. Find y if $x = 9$.

56. Let $2x - 5y = 20$.
 a. Find x if $y = 0$.

 b. Find x if $y = -6$.

 c. Find y if $x = 0$.

 d. Find y if $x = 5$.

57. Let $y = 2x - 1$.
 a. Find x if $y = 7$.

 b. Find x if $y = 3$.

 c. Find y if $x = 0$.

 d. Find y if $x = 5$.

58. Let $y = 3x - 2$.
 a. Find x if $y = 4$.

 b. Find x if $y = 3$.

 c. Find y if $x = 2$.

 d. Find y if $x = -3$.

59. Find y when x is 4 in the formula $3x + 2y = 6$.

60. Find y when x is 0 in the formula $3x + 2y = 6$.

61. Find y when x is 0 in $y = -\dfrac{1}{3}x + 2$.

62. Find y when x is 3 in $y = -\dfrac{1}{3}x + 2$.

63. Find y when x is 2 in $y = \dfrac{3}{2}x - 3$.

64. Find y when x is 4 in $y = \dfrac{3}{2}x - 3$.

65. Solve $5x + y = 4$ for y.

66. Solve $-3x + y = 5$ for y.

67. Solve $3x - 2y = 6$ for y.

68. Solve $2x - 3y = 6$ for y.

Find the Mistake

Each sentence below contains a mistake. Circle the mistake and write the correct word or number on the line provided.

1. A scatter diagram uses bars to show data in the form of order pairs. _____

2. In the ordered pair $(-3, 6)$, the x–coordinate is 6. _____

3. To graph the ordered pair $(7, 3)$, start at the origin, move 7 units to the left, and 3 units up. _____

4. The ordered pair $(-2, 5)$ appears in the first quadrant. _____

Navigation Skills: Prepare, Study, Achieve

Preparation is another key component for success in this course. What does it mean to be prepared for this class? Before you come to class, make sure to

- Complete the homework from the previous section.
- Read the upcoming section.
- Answer the Getting Ready for Class questions.
- Rework example problems.
- Work practice problems.
- Make a point to attend every class session and arrive on time.
- Prepare a list of questions to ask your instructor or fellow students that will help you work through difficult problems.
- Commit to being an engaged, attentive, and active listener while in class.
- Take notes as the instructor speaks and don't hesitate to ask questions if you need clarification.

Research note-taking techniques online if you find that you aren't quite sure how to take notes or your current method of taking notes is not helpful.

You shouldn't expect to master a new topic the first time you read about it or learn about it in class. Mastering mathematics takes a lot of practice, so make the commitment to come to class prepared and practice, practice, practice.

OBJECTIVES

A Find solutions to linear equations in two variables.

B Determine whether an ordered pair is a solution to a linear equation in two variables.

C Graph a linear equation in two variables.

D Graph horizontal lines, vertical lines, and lines through the origin.

KEY WORDS

ordered pair

graph

linear equation in two variables

horizontal

vertical

origin

A new robot named EMILY (Emergency Integrated Lifesaving Lanyard) patrols dangerous ocean waters off the coast of Malibu, California. Currently, a lifeguard controls EMILY by remote. However, developers will soon equip the robot with sonar and an autonomous system able to detect swimmers in distress without the help of a lifeguard on shore. EMILY's electric-powered impeller drives the robot through even the roughest surf at a speed 6 times faster than that of a human. A swimmer in distress can use EMILY as a flotation device until further help arrives, or EMILY can tow the swimmer back to shore.

Using the information about EMILY's speed, we can write the following equation, which uses two variables:

$$y = 6x$$

where x is a human's swimming speed and y is EMILY's speed. In this section, we will begin to investigate equations in two variables, such as the one above. As you will see, equations in two variables have pairs of numbers for solutions. Because we know how to use paired data to construct tables, histograms, and other charts, we can take our work with paired data further by using equations in two variables to construct tables of paired data.

A Solving Linear Equations

Let's begin this section by reviewing the relationship between equations in one variable and their solutions.

If we solve the equation $3x - 2 = 10$, the solution is $x = 4$. If we graph this solution, we simply draw the real number line and place a dot at the point whose coordinate is 4. The relationship between linear equations in one variable, their solutions, and the graphs of those solutions look like this:

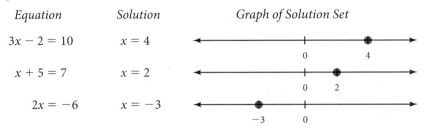

Equation	Solution	Graph of Solution Set
$3x - 2 = 10$	$x = 4$	
$x + 5 = 7$	$x = 2$	
$2x = -6$	$x = -3$	

When the equation has one variable, the solution is a single number whose graph is a point on a line.

Now, consider the equation $2x + y = 3$. The first thing we notice is that there are two variables instead of one. Therefore, a solution to the equation $2x + y = 3$ will be not a single number but a pair of numbers, one for x and one for y, that makes the equation

a true statement. One pair of numbers that works is $x = 2$, $y = -1$ because when we substitute them for x and y in the equation, we get a true statement.

$$2(2) + (-1) \overset{?}{=} 3$$

$$4 - 1 = 3$$

$$3 = 3 \qquad \text{A true statement}$$

The pair of numbers $x = 2$, $y = -1$ is written as $(2, -1)$. As you know from Section 3.1, $(2, -1)$ is called an *ordered pair* because it is a pair of numbers written in a specific order. The first number is always associated with the variable x, and the second number is always associated with the variable y. We call the first number in the ordered pair the *x-coordinate* and the second number the *y-coordinate* of the ordered pair.

Let's look back to the equation $2x + y = 3$. The ordered pair $(2, -1)$ is not the only solution. Another solution is $(0, 3)$ because when we substitute 0 for x and 3 for y we get

$$2(0) + 3 \overset{?}{=} 3$$

$$0 + 3 = 3$$

$$3 = 3 \qquad \text{A true statement}$$

Still another solution is the ordered pair $(5, -7)$ because

$$2(5) + (-7) \overset{?}{=} 3$$

$$10 - 7 = 3$$

$$3 = 3 \qquad \text{A true statement}$$

As a matter of fact, for any number we want to use for x, there is another number we can use for y that will make the equation a true statement. There is an infinite number of ordered pairs that satisfy (are solutions to) the equation $2x + y = 3$; we have listed just a few of them.

EXAMPLE 1 Given the equation $2x + 3y = 6$, complete the following ordered pairs so they will be solutions to the equation: $(0, \)$, $(\ , 1)$, $(3, \)$.

Solution To complete the ordered pair $(0, \)$, we substitute 0 for x in the equation and then solve for y.

$$2(0) + 3y = 6$$

$$3y = 6$$

$$y = 2$$

The ordered pair is $(0, 2)$.

To complete the ordered pair $(\ , 1)$, we substitute 1 for y in the equation and solve for x.

$$2x + 3(1) = 6$$

$$2x + 3 = 6$$

$$2x = 3$$

$$x = \frac{3}{2}$$

The ordered pair is $\left(\frac{3}{2}, 1\right)$.

To complete the ordered pair $(3, \)$, we substitute 3 for x in the equation and solve for y.

$$2(3) + 3y = 6$$

$$6 + 3y = 6$$

$$3y = 0$$

$$y = 0$$

The ordered pair is $(3, 0)$.

Note If this discussion seems a little long and confusing, you may want to look over some of the examples first and then come back and read this. Remember, it isn't always easy to read material in mathematics. What is important is that you understand what you are doing when you work problems. The reading is intended to assist you in understanding what you are doing. It is important to read everything in the book, but you don't always have to read it in the order it is written.

Practice Problems

1. For the equation $2x + 5y = 10$, complete the ordered pairs $(0, \)$, $(\ , 1)$, and $(5, \)$.

Answers

1. $(0, 2)$, $\left(\frac{5}{2}, 1\right)$, $(5, 0)$

Notice in each case that once we have used a number in place of one of the variables, the equation becomes a linear equation in one variable. We then use the method explained in Chapter 2 to solve for that variable.

EXAMPLE 2 Complete the following table for the equation $2x - 5y = 20$.

x	y
0	−4
15	2
10	0
−5	−6

Solution Filling in the table is equivalent to completing the following ordered pairs: $(0, \), (\ , 2), (\ , 0), (-5, \)$. So we proceed as in Example 1.

When $x = 0$, we have

$$2(0) - 5y = 20$$
$$0 - 5y = 20$$
$$-5y = 20$$
$$y = -4$$

When $y = 2$, we have

$$2x - 5(2) = 20$$
$$2x - 10 = 20$$
$$2x = 30$$
$$x = 15$$

When $y = 0$, we have

$$2x - 5(0) = 20$$
$$2x - 0 = 20$$
$$2x = 20$$
$$x = 10$$

When $x = -5$, we have

$$2(-5) - 5y = 20$$
$$-10 - 5y = 20$$
$$-5y = 30$$
$$y = -6$$

The completed table looks like this:

x	y
0	−4
15	2
10	0
−5	−6

The table above is equivalent to the ordered pairs $(0, -4)$, $(15, 2)$, $(10, 0)$, and $(-5, -6)$.

2. Complete the table for $3x - 2y = 12$.

x	y
0	
	3
	0
−3	

3. Complete the table for $y = 3x - 2$.

x	y
0	−2
2	4
3	7
	3

EXAMPLE 3 Complete the following table for the equation $y = 2x - 1$.

x	y
0	−1
5	9
4	7
2	3

Solution

When $x = 0$, we have

$$y = 2(0) - 1$$
$$y = 0 - 1$$
$$y = -1$$

When $x = 5$, we have

$$y = 2(5) - 1$$
$$y = 10 - 1$$
$$y = 9$$

When $y = 7$, we have When $y = 3$, we have

$$7 = 2x - 1$$ $$3 = 2x - 1$$
$$8 = 2x$$ $$4 = 2x$$
$$4 = x$$ $$2 = x$$

The completed table is

x	y
0	−1
5	9
4	7
2	3

which means the ordered pairs $(0, -1)$, $(5, 9)$, $(4, 7)$, and $(2, 3)$ are among the solutions to the equation $y = 2x - 1$.

B Determining if an Ordered Pair is a Solution

EXAMPLE 4 Which of the ordered pairs $(2, 3)$, $(1, 5)$, and $(-2, -4)$ are solutions to the equation $y = 3x + 2$?

Solution If an ordered pair is a solution to the equation, then it must satisfy the equation; that is, when the coordinates are used in place of the variables in the equation, the equation becomes a true statement.

Try $(2, 3)$ in $y = 3x + 2$:

$$3 \stackrel{?}{=} 3(2) + 2$$
$$3 = 6 + 2$$
$$3 = 8 \qquad \text{A false statement}$$

Try $(1, 5)$ in $y = 3x + 2$:

$$5 \stackrel{?}{=} 3(1) + 2$$
$$5 = 3 + 2$$
$$5 = 5 \qquad \text{A true statement}$$

Try $(-2, -4)$ in $y = 3x + 2$:

$$-4 \stackrel{?}{=} 3(-2) + 2$$
$$-4 = -6 + 2$$
$$-4 = -4 \qquad \text{A true statement}$$

The ordered pairs $(1, 5)$ and $(-2, -4)$ are solutions to the equation $y = 3x + 2$, and $(2, 3)$ is not.

C Graphing Linear Equations

At the end of the previous section we used a line graph to obtain a visual picture of *some* of the solutions to the equations $y = 1.5x + 15$. In this section we will use the rectangular coordinate system introduced in Section 3.1 to obtain a visual picture of *all* solutions to a linear equation in two variables. The process we use to obtain a visual picture of all solutions to an equation is called *graphing*. The picture itself is called the *graph* of the equation.

4. Which of the following ordered pairs is a solution to $y = 4x + 1$? $(0, 1), (3, 11), (2, 9)$

Answers

4. $(0, 1)$ and $(2, 9)$

5. Graph the equation $x + y = 3$.

EXAMPLE 5 Graph the solution set for $x + y = 5$.

Solution We know from the previous section that an infinite number of ordered pairs are solutions to the equation $x + y = 5$. We can't possibly list them all. What we can do is list a few of them and see if there is any pattern to their graphs.

Some ordered pairs that are solutions to $x + y = 5$ are $(0, 5)$, $(2, 3)$, $(3, 2)$, $(5, 0)$. The graph of each is shown in Figure 1.

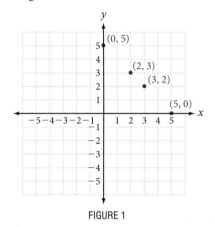

FIGURE 1

Now, by passing a straight line through these points we can graph the solution set for the equation $x + y = 5$. Linear equations in two variables always have graphs that are straight lines. The graph of the solution set for $x + y = 5$ is shown in Figure 2.

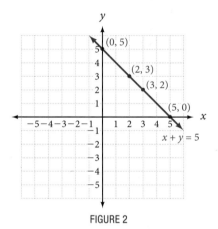

FIGURE 2

Every ordered pair that satisfies $x + y = 5$ has its graph on the line, and any point on the line has coordinates that satisfy the equation. So there is a one-to-one correspondence between points on the line and solutions to the equation. To summarize, the set of all points of a graph that satisfy an equation is the equation's solution set.

Here is the precise definition for a linear equation in two variables.

> **DEFINITION** linear equation in two variables
>
> Any equation that can be put in the form $Ax + By = C$, where A, B, and C are real numbers and A and B are not both 0, is called a ***linear equation in two variables***. The graph of any equation of this form is a straight line (that is why these equations are called "linear"). The form $Ax + By = C$ is called ***standard form***.

To graph a linear equation in two variables, as we did in Example 5, we simply graph its solution set; that is, we draw a line through all the points whose coordinates satisfy the equation. Here are the steps to follow:

> **HOW TO** To Graph a Linear Equation in Two Variables
>
> **Step 1:** Find any three ordered pairs that satisfy the equation. This can be done by using a convenient number for one variable and solving for the other variable.
>
> **Step 2:** Graph the three ordered pairs found in step 1. Actually, we need only two points to graph a straight line. The third point serves as a check. If all three points do not line up, there is a mistake in our work.
>
> **Step 3:** Draw a straight line through the three points graphed in step 2.

EXAMPLE 6 Graph the equation $y = 3x - 1$.

Solution Because $y = 3x - 1$ can be put in the form $ax + by = c$, it is a linear equation in two variables. Hence, the graph of its solution set is a straight line. We can find some specific solutions by substituting numbers for x and then solving for the corresponding values of y. We are free to choose any numbers for x, so let's use 0, 2, and -1.

		In table form

Let $x = 0$: $y = 3(0) - 1$
$y = 0 - 1$
$y = -1$

The ordered pair $(0, -1)$ is one solution.

Let $x = 2$: $y = 3(2) - 1$
$y = 6 - 1$
$y = 5$

The ordered pair $(2, 5)$ is a second solution.

Let $x = -1$: $y = 3(-1) - 1$
$y = -3 - 1$
$y = -4$

The ordered pair $(-1, -4)$ is a third solution.

x	y
0	-1
2	5
-1	-4

Next, we graph the ordered pairs $(0, -1)$, $(2, 5)$, $(-1, -4)$ and draw a straight line through them. The line we have drawn in Figure 3 is the graph of $y = 3x - 1$.

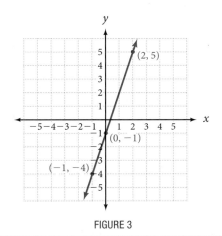

FIGURE 3

Example 6 illustrates the connection between algebra and geometry that we mentioned previously. Descartes' rectangular coordinate system allows us to associate the equation $y = 3x - 1$ (an algebraic concept) with a specific straight line (a geometric concept). The study of the relationship between equations in algebra and their associated geometric figures is called *analytic geometry*.

Note The meaning of the convenient numbers referred to in step 1 will become clear as you read the next two examples.

6. Graph the equation $y = 2x + 3$.

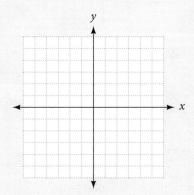

Note It may seem that we have simply picked the numbers 0, 2, and -1 out of the air and used them for x. In fact we have done just that. Could we have used numbers other than these? The answer is yes, we can substitute any number for x; for a linear equation in two variables, there will always be a value of y to go with it.

7. Graph $y = \frac{3}{2}x - 3$.

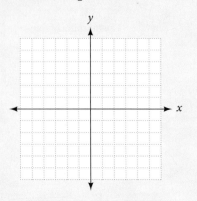

Note In Example 7 the values of x we used, $-3, 0$, and 3, are referred to as convenient values of x because they are easier to work with than some other numbers. For instance, if we let $x = 2$ in our original equation, we would have to add $-\frac{2}{3}$ and 2 to find the corresponding value of y. Not only would the arithmetic be more difficult but also the ordered pair we obtained would have a fraction for its y-coordinate, making it more difficult to graph accurately.

EXAMPLE 7 Graph the equation $y = -\frac{1}{3}x + 2$.

Solution We need to find three ordered pairs that satisfy the equation. To do so, we can let x equal any numbers we choose and find corresponding values of y. But, every value of x we substitute into the equation is going to be multiplied by $-\frac{1}{3}$. For our convenience, let's use numbers for x that are divisible by 3, like $-3, 0$, and 3. That way, when we multiply them by $-\frac{1}{3}$, the result will be an integer.

$$\text{Let } x = -3: \quad y = -\frac{1}{3}(-3) + 2$$
$$y = 1 + 2$$
$$y = 3$$

The ordered pair $(-3, 3)$ is one solution.

$$\text{Let } x = 0: \quad y = -\frac{1}{3}(0) + 2$$
$$y = 0 + 2$$
$$y = 2$$

The ordered pair $(0, 2)$ is a second solution.

$$\text{Let } x = 3: \quad y = -\frac{1}{3}(3) + 2$$
$$y = -1 + 2$$
$$y = 1$$

In table form

x	y
-3	3
0	2
3	1

The ordered pair $(3, 1)$ is a third solution.

Graphing the ordered pairs $(-3, 3)$, $(0, 2)$, and $(3, 1)$ and drawing a straight line through their graphs, we have the graph of the equation $y = -\frac{1}{3}x + 2$, as shown in Figure 4.

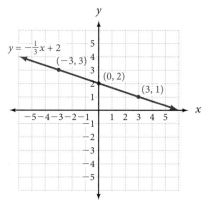

FIGURE 4

8. Graph the solution set for $2x - 4y = 8$.

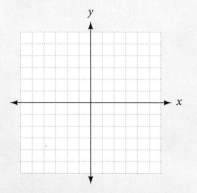

EXAMPLE 8 Graph the solution set for $3x - 2y = 6$.

Solution It will be easier to find convenient values of x to use in the equation if we first solve the equation for y. To do so, we add $-3x$ to each side, and then we multiply each side by $-\frac{1}{2}$.

$$3x - 2y = 6 \qquad \text{\textit{Original equation}}$$
$$-2y = -3x + 6 \qquad \text{\textit{Add } } -3x \text{ \textit{to each side.}}$$
$$-\frac{1}{2}(-2y) = -\frac{1}{2}(-3x + 6) \qquad \text{\textit{Multiply each side by } } -\frac{1}{2}.$$
$$y = \frac{3}{2}x - 3 \qquad \text{\textit{Simplify each side.}}$$

Now, because each value of x will be multiplied by $\frac{3}{2}$, it will be to our advantage to choose values of x that are divisible by 2. That way, we will obtain values of y that do not contain fractions. This time, let's use 0, 2, and 4 for x.

When $x = 0$: $\quad y = \frac{3}{2}(0) - 3$

$$y = 0 - 3$$
$$y = -3$$

The ordered pair $(0, -3)$ is one solution.

When $x = 2$: $\quad y = \frac{3}{2}(2) - 3$

$$y = 3 - 3$$
$$y = 0$$

The ordered pair $(2, 0)$ is a second solution.

When $x = 4$: $\quad y = \frac{3}{2}(4) - 3$

$$y = 6 - 3$$
$$y = 3$$

The ordered pair $(4, 3)$ is a third solution.

Graphing the ordered pairs $(0, -3)$, $(2, 0)$, and $(4, 3)$ and drawing a line through them, we have the graph shown in Figure 5.

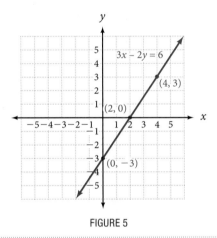

FIGURE 5

Note After reading through Example 8, many students ask why we didn't use -2 for x when we were finding ordered pairs that were solutions to the original equation. The answer is, we could have. If we were to let $x = -2$, the corresponding value of y would have been -6. As you can see by looking at the graph in Figure 5, the ordered pair $(-2, -6)$ is on the graph.

D Horizontal and Vertical Lines; Lines Through The Origin

EXAMPLE 9 Graph each of the following lines.

a. $y = \frac{1}{2}x$ **b.** $x = 3$ **c.** $y = -2$

Solution

a. The line $y = \frac{1}{2}x$ passes through the origin because $(0, 0)$ satisfies the equation. To sketch the graph we need at least one more point on the line. When x is 2, we obtain the point $(2, 1)$, and when x is -4, we obtain the point $(-4, -2)$. The graph of $y = \frac{1}{2}x$ is shown in Figure 6A.

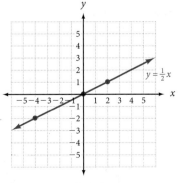

FIGURE 6A

9. Graph each of the following lines.

a. $y = \frac{2}{3}x$

b. $x = 3$

c. $y = -2$

Note The equations in parts
b and c are linear equations in
two variables where the second
variable is not explicit; in other
words, $x = 3$ is the same as

$$0y + x = 3$$

and $y = -2$ is the same as

$$0x + y = -2$$

b. The line $x = 3$ is the set of all points whose
x-coordinate is 3. The variable y does not
appear in the equation, so the y-coordinate
can be any number. Note that we can write
our equation as a linear equation in two
variables by writing it as $x + 0y = 3$. Because
the product of 0 and y will always be 0, y can
be any number. The graph of $x = 3$ is the
vertical line shown in Figure 6B.

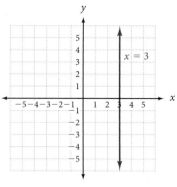

FIGURE 6B

c. The line $y = -2$ is the set of all points whose
y-coordinate is -2. The variable x does not
appear in the equation, so the x-coordinate
can be any number. Again, we can write our
equation as a linear equation in two variables
by writing it as $0x + y = -2$. Because the
product of 0 and x will always be 0, x can
be any number. The graph of $y = -2$ is the
horizontal line shown in Figure 6C.

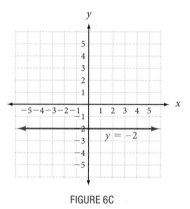

FIGURE 6C

FACTS FROM GEOMETRY Special Equations and Their Graphs

For the equations below, m, a, and b are real numbers.

Through the Origin Vertical Line Horizontal Line

FIGURE 7A *Any equation of
the form $y = mx$ has a graph
that passes through the origin.*

FIGURE 7B *Any equation of
the form $x = a$ has a vertical
line for its graph.*

FIGURE 7C *Any equation of
the form $y = b$ has a horizontal
line for its graph.*

GETTING READY FOR CLASS

*After reading through the preceding section, respond in your own words and in
complete sentences.*

A. How can you tell if an ordered pair is a solution to an equation?

B. How would you find a solution to $y = 3x - 5$?

C. How many solutions are there to an equation that contains two variables?

D. What type of equation has a line that passes through the origin?

Vocabulary Review

Choose the correct words to fill in the blanks below.

solution	graphing	graph	first	vertical
linear	ordered pairs	straight line	second	horizontal

1. The _____ number of an ordered pair is called the x-coordinate.

2. The _____ number of an ordered pair is called the y-coordinate.

3. An ordered pair is a _____ to an equation if it satisfies the equation.

4. The process of obtaining a visual picture of all solutions to an equation is called _____.

5. A _____ equation in two variables is any equation that can be put into the form $ax + bx = c$.

6. The first step to graph a linear equation in two variables is to find any three _____ that satisfy the equation.

7. The second step to graph a linear equation in two variables is to _____ the three ordered pairs that satisfy the equation.

8. The third step to graph a linear equation in two variables is to draw a _____ through the three ordered pairs that satisfy the equation.

9. Any equation of the form $y = b$ has a _____ line for it's graph, whereas any equation of the form $x = a$ has a _____ line for it's graph.

Problems

A For each equation, complete the given ordered pairs.

1. $2x + y = 6$ (0,), (, 0), (, −6)

2. $3x − y = 5$ (0,), (1,), (, 5)

3. $3x + 4y = 12$ (0,), (, 0), (−4,)

4. $5x − 5y = 20$ (0,), (, −2), (1,)

5. $y = 4x − 3$ (1,), (, 0), (5,)
$6 = 4x-3$ x,y x,y x,y
$3=4x$

6. $y = 3x − 5$ (, 13), (0,), (−2,)

7. $y = 7x − 1$ (2,), (, 6), (0,)

8. $y = 8x + 2$ (3,), (, 0), (, −6)

9. $x = −5$ (, 4), (, −3), (, 0)
$0=-5$ x,y x,y x,y

10. $y = 2$ (5,), (−8,), $\left(\frac{1}{2}, \right)$

For each of the following equations, complete the given table.

11. $y = 3x$

x	y
1	3
-3	*9*
4	12
6	18

12. $y = -2x$

x	y
-4	
0	
	10
	12

13. $y = 4x$

x	y
0	
	-2
-3	
	12

14. $y = -5x$

x	y
3	
	0
-2	
	-20

15. $x + y = 5$

x	y
2	
3	
	0
	-4

16. $x - y = 8$

x	y
0	
4	
	-3
	-2

17. $2x - y = 4$

x	y
	0
	2
1	
-3	

18. $3x - y = 9$

x	y
	0
	-9
5	
-4	

19. $y = 6x - 1$

x	y
0	
	-7
-3	
	8

20. $y = 5x + 7$

x	y
0	
-2	
-4	
	-8

21. $y = -2x + 3$

x	y
0	
-2	
2	
	-7

22. $y = -3x + 1$

x	y
0	
	7
	-5
-3	

B For the following equations, tell which of the given ordered pairs are solutions.

23. $2x - 5y = 10$ $(2, 3), (0, -2), \left(\frac{5}{2}, 1\right)$

24. $3x + 7y = 21$ $(0, 3), (7, 0), (1, 2)$

25. $y = 7x - 2$ $(1, 5), (0, -2), (-2, -16)$

26. $y = 8x - 3$ $(0, 3), (5, 16), (1, 5)$

27. $y = 6x$ $(1, 6), (-2, 12), (0, 0)$

28. $y = -4x$ $(0, 0), (2, 4), (-3, 12)$

29. $x + y = 0$ $(1, 1), (2, -2), (3, 3)$

30. $x - y = 1$ $(0, 1), (0, -1), (1, 2)$

31. $x = 3$ $(3, 0), (3, -3), (5, 3)$

32. $y = -4$ $(3, -4), (-4, 4), (0, -4)$

C D For the following equations, complete the given ordered pairs, and use the results to graph the solution set for the equation.

33. $x + y = 4$ $(0, \), (2, \), (\ , 0)$

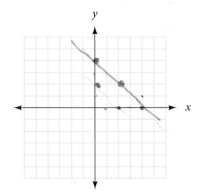

34. $x - y = 3$ $(0, \), (2, \), (\ , 0)$

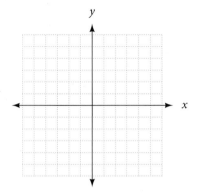

35. $x + y = 3$ $(0, 3), (2, 1), (4, -1)$

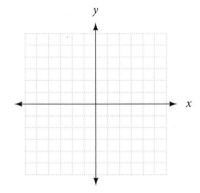

36. $x - y = 4$ $(1, \), (-1, \), (\ , 0)$

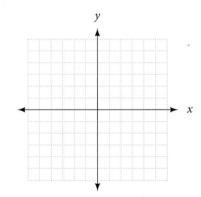

37. $y = 2x$ $(0, \), (-2, \), (2, \)$

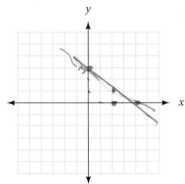

38. $y = \dfrac{1}{2}x$ $(0, \), (-2, \), (2, \)$

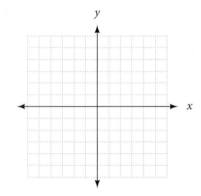

39. $y = \dfrac{1}{3}x$ $(-3, \), (0, \), (3, \)$

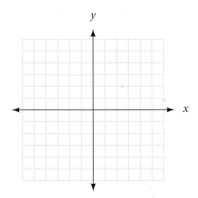

40. $y = 3x$ $(-2, \), (0, \), (2, \)$

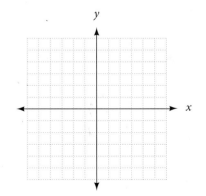

41. $y = 2x + 1$ $(0, \quad), (-1, \quad), (1, \quad)$

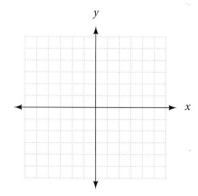

42. $y = -2x + 1$ $(0, \quad), (-1, \quad), (1, \quad)$

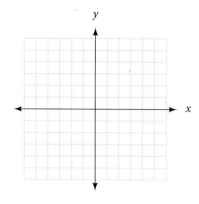

43. $y = 4$ $(0, \quad), (-1, \quad), (2, \quad)$

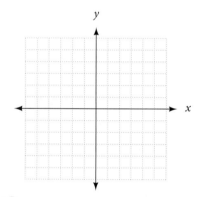

44. $x = 3$ $(\quad, -2), (\quad, 0), (\quad, 5)$

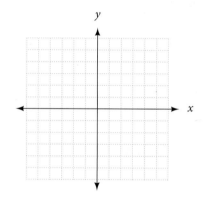

45. $y = \frac{1}{2}x + 3$ $(-2, \quad), (0, \quad), (2, \quad)$

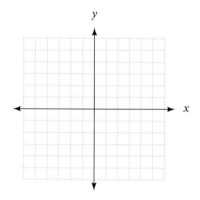

46. $y = \frac{1}{2}x - 3$ $(-2, \quad), (0, \quad), (2, \quad)$

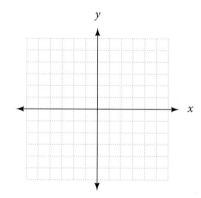

47. $y = -\frac{2}{3}x + 1$ $(-3, \quad), (0, \quad), (3, \quad)$

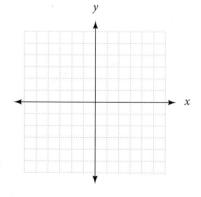

48. $y = -\frac{2}{3}x - 1$ $(-3, \quad), (0, \quad), (3, \quad)$

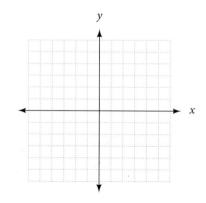

C Solve each equation for y. Then, complete the given ordered pairs, and use them to draw the graph.

49. $2x + y = 3$ $\quad (-1, \), (0, \), (1, \)$

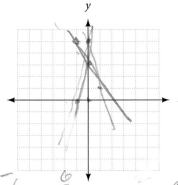

50. $3x + y = 2$ $\quad (-1, \), (0, \), (1, \)$

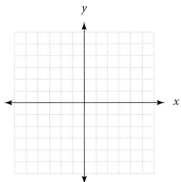

51. $3x + 2y = 6$ $\quad (0, \), (2, \), (4, \)$

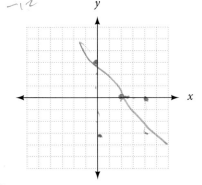

52. $2x + 3y = 6$ $\quad (0, \), (3, \), (6, \)$

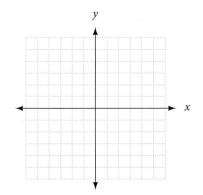

53. $-x + 2y = 6$ $\quad (-2, \), (0, \), (2, \)$

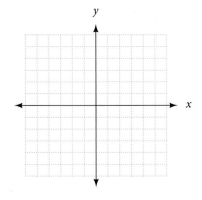

54. $-x + 3y = 6$ $\quad (-3, \), (0, \), (3, \)$

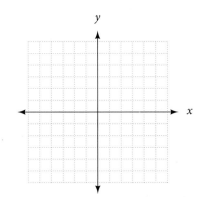

Find three solutions to each of the following equations, and then graph the solution set.

55. $y = -\dfrac{1}{2}x$

56. $y = -2x$

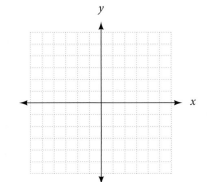

57. $y = 3x - 1$

58. $y = -3x - 1$

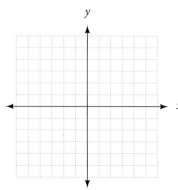

59. $-2x + y = 1$

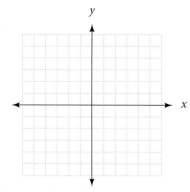

60. $-3x + y = 1$

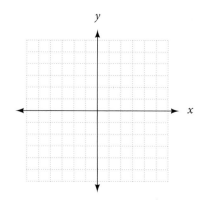

61. $3x + 4y = 8$

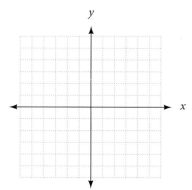

62. $3x - 4y = 8$

63. $x = -2$

64. $y = 3$

65. $y = 2$

66. $x = -3$

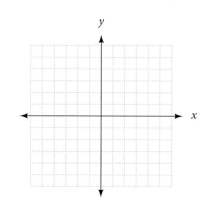

Graph each equation.

67. $y = \dfrac{3}{4}x + 1$

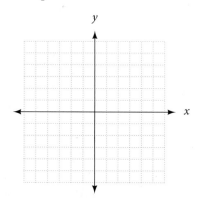

68. $y = \dfrac{2}{3}x + 1$

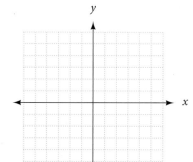

69. $y = \dfrac{1}{3}x + \dfrac{2}{3}$

70. $y = \frac{1}{2}x + \frac{1}{2}$

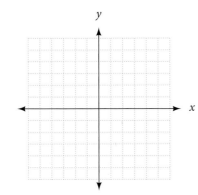

71. $y = \frac{2}{3}x + \frac{2}{3}$

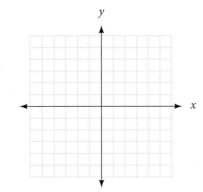

72. $y = -\frac{3}{4}x + \frac{3}{2}$

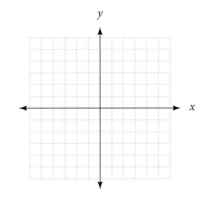

D For each equation in each table below, indicate whether the graph is horizontal (H), or vertical (V), or whether it passes through the origin (O).

73.

Equation	H, V, and/or O
$x = 3$	
$y = 3$	
$y = 3x$	
$y = 0$	

74.

Equation	H, V, and/or O
$x = \frac{1}{2}$	
$y = \frac{1}{2}$	
$y = \frac{1}{2}x$	
$x = 0$	

75.

Equation	H, V, and/or O
$x = -\frac{3}{5}$	
$y = -\frac{3}{5}$	
$y = -\frac{3}{5}x$	
$x = 0$	

76.

Equation	H, V, and/or O
$x = -4$	
$y = -4$	
$y = -4x$	
$y = 0$	

77. Use the graph to complete the table.

x	y
	-3
-2	
0	
	0
6	

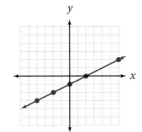

78. Use the graph at the right to complete the table. (*Hint:* Some parts have two answers.)

x	y
-3	6
	4
0	3
	1
6	

Applying the Concepts

79. Perimeter If the perimeter of a rectangle is 30 inches, then the relationship between the length l and the width w is given by the equation

$$2l + 2w = 30$$

What is the length when the width is 3 inches?

80. Perimeter The relationship between the perimeter P of a square and the length of its side s is given by the formula

$$P = 4s$$

If each side of a square is 5 inches, what is the perimeter? If the perimeter of a square is 28 inches, how long is a side?

81. Janai earns $12 per hour working as a math tutor. We can express the amount she earns each week, y, for working x hours with the equation $y = 12x$. Indicate with a yes or no, which of the following could be one of Janai's paychecks before deductions are taken out. If you answer no, explain your answer.

a. $60 for working five hours

b. $100 for working nine hours

c. $80 for working seven hours

d. $168 for working 14 hours

82. Erin earns $15 per hour working as a graphic designer. We can express the amount she earns each week, y, for working x hours with the equation $y = 15x$. Indicate with a yes or no which of the following could be one of Erin's paychecks before deductions are taken out. If you answer no, explain your answer.

a. $75 for working five hours

b. $125 for working nine hours

c. $90 for working six hours

d. $500 for working 35 hours

83. The equation $V = -45,000t + 600,000$, can be used to find the value, V, of a small crane at the end of t years.

a. What is the value of the crane at the end of five years?

b. When is the crane worth $330,000?

c. Is it true that the crane with be worth $150,000 after nine years?

d. How much did the crane cost?

84. The equation $V = -400t + 2,500$, can be used to find the value, V, of a notebook computer at the end of t years.

a. What is the value of the notebook computer at the end of four years?

b. When is the notebook computer worth $1,700?

c. Is it true that the notebook computer with be worth $100 after five years?

d. How much did the notebook computer cost?

Getting Ready for the Next Section

85. Let $3x + 2y = 6$.

 a. Find x when $y = 0$.

 b. Find y when $x = 0$.

86. Let $2x - 5y = 10$.

 a. Find x when $y = 0$.

 b. Find y when $x = 0$.

87. Let $-x + 2y = 4$.

 a. Find x when $y = 0$.

 b. Find y when $x = 0$.

88. Let $3x - y = 6$.

 a. Find x when $y = 0$.

 b. Find y when $x = 0$.

89. Let $y = -\dfrac{1}{3}x + 2$.

 a. Find x when $y = 0$.

 b. Find y when $x = 0$.

90. Let $y = \dfrac{3}{2}x - 3$.

 a. Find x when $y = 0$.

 b. Find y when $x = 0$.

Find the Mistake

Each sentence below contains a mistake. Circle the mistake and write the correct word(s) or number on the line provided.

1. A solution to a linear equation in two variables will be a single number that makes the equation a true statement.

2. A solution to the equation $3y = 6x + 9$ is $(2, 6)$. _____

3. To graph a linear equation in two variables, you must find only one solution. _____

4. The line $y = \dfrac{1}{2}x$ crosses the x–axis but not the y–axis. _____

A Find the intercepts of a line
 from the equation of the line.

B Use intercepts to graph
 a line.

3.3 Graphing Linear Equations Using Intercepts

©iStockphoto.com/dlanier

Some talented water skiers ski on their bare feet while being towed behind a motor boat. During training, the barefoot skiers may use foot skis, which are small thin skis just a few inches longer and a little wider than the skier's foot. With foot skis, the skier can learn maneuvers faster and with fewer wipeouts than without. Normal water skiing speeds reach 30-45 miles per hour. An appropriate speed for barefoot water skiing needs to be faster and depends on the skier's body weight using the following linear equation:

$$y = \frac{x}{10} + 20$$

where x is the skier's weight in pounds. In this section, we will practice finding x- and y-intercepts for equations similar to the barefoot skier's equation.

A Intercepts

In this section we continue our work with graphing lines by finding the points where a line crosses the axes of our coordinate system. To do so, we use the fact that any point on the x-axis has a y-coordinate of 0 and any point on the y-axis has an x-coordinate of 0. We begin with the following definition:

> **DEFINITION** *x*-intercept, *y*-intercept
>
> The *x-intercept* of a straight line is the x-coordinate of the point where the graph intersects the x-axis and the y-coordinate is zero. The *y-intercept* is defined similarly. It is the y-coordinate of the point where the graph intersects the y-axis and the x-coordinate is zero.

B Using Intercepts to Graph Lines

If the x-intercept is a, then the point $(a, 0)$ lies on the graph. (This is true because any point on the x-axis has a y-coordinate of 0.)

If the y-intercept is b, then the point $(0, b)$ lies on the graph. (This is true because any point on the y-axis has an x-coordinate of 0.)

Graphically, the relationship is shown in Figure 1.

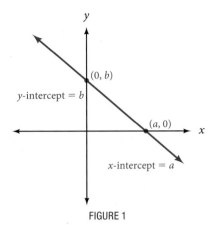

FIGURE 1

EXAMPLE 1 Find the x- and y-intercepts for $3x - 2y = 6$, and then use them to draw the graph.

Solution To find where the graph crosses the x-axis, we let $y = 0$. (The y-coordinate of any point on the x-axis is 0.)

x-intercept:

$$
\begin{aligned}
\text{When} \rightarrow && y &= 0 \\
\text{the equation} \rightarrow && 3x - 2y &= 6 \\
\text{becomes} \rightarrow && 3x - 2(0) &= 6 \\
&& 3x - 0 &= 6 \\
&& x &= 2 \qquad \text{Multiply each side by } \tfrac{1}{3}.
\end{aligned}
$$

The graph crosses the x-axis at $(2, 0)$, which means the x-intercept is 2.

y-intercept:

$$
\begin{aligned}
\text{When} \rightarrow && x &= 0 \\
\text{the equation} \rightarrow && 3x - 2y &= 6 \\
\text{becomes} \rightarrow && 3(0) - 2y &= 6 \\
&& 0 - 2y &= 6 \\
&& -2y &= 6 \\
&& y &= -3 \qquad \text{Multiply each side by } -\tfrac{1}{2}.
\end{aligned}
$$

The graph crosses the y-axis at $(0, -3)$, which means the y-intercept is -3.

Plotting the x- and y-intercepts and then drawing a line through them, we have the graph of $3x - 2y = 6$, as shown in Figure 2.

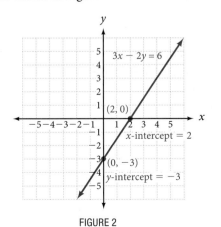

FIGURE 2

Practice Problems

1. Find the x- and y-intercepts for $2x - 5y = 10$, and use them to draw the graph.

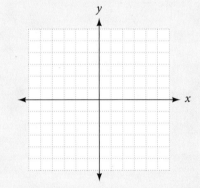

2. Graph $3x - y = 6$ by first finding the intercepts.

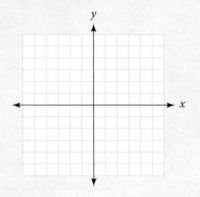

EXAMPLE 2 Graph $-x + 2y = 4$ by finding the intercepts and using them to draw the graph.

Solution Again, we find the x-intercept by letting $y = 0$ in the equation and solving for x. Similarly, we find the y-intercept by letting $x = 0$ and solving for y.

x-intercept:

$$\text{When} \rightarrow \qquad\qquad y = 0$$
$$\text{the equation} \rightarrow \quad -x + 2y = 4$$
$$\text{becomes} \rightarrow \quad\; -x + 2(0) = 4$$
$$-x + 0 = 4$$
$$-x = 4$$
$$x = -4 \qquad \textit{Multiply each side by} -1.$$

The x-intercept is -4, indicating that the point $(-4, 0)$, is on the graph of $-x + 2y = 4$.

y-intercept:

$$\text{When} \rightarrow \qquad\qquad x = 0$$
$$\text{the equation} \rightarrow \quad -x + 2y = 4$$
$$\text{becomes} \rightarrow \quad\; -0 + 2y = 4$$
$$2y = 4$$
$$y = 2 \qquad \textit{Multiply each side by } \tfrac{1}{2}.$$

The y-intercept is 2, indicating that the point $(0, 2)$ is on the graph of $-x + 2y = 4$.

Plotting the intercepts and drawing a line through them, we have the graph of $-x + 2y = 4$, as shown in Figure 3.

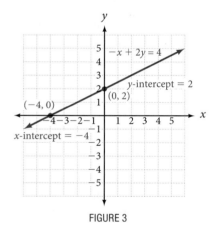

FIGURE 3

Graphing a line by finding the intercepts, as we have done in Examples 1 and 2, is an easy method of graphing if the equation has the form $ax + by = c$ and both the numbers a and b divide the number c evenly.

In our next example, we use the intercepts to graph a line in which y is given in terms of x.

EXAMPLE 3 Use the intercepts for $y = -\frac{1}{3}x + 2$ to draw its graph.

Solution We graphed this line previously in Example 7 of Section 3.2 by substituting three different values of x into the equation and solving for y. This time we will graph the line by finding the intercepts.

x-intercept:

$$\text{When} \rightarrow \qquad y = 0$$

$$\text{the equation} \rightarrow \qquad y = -\frac{1}{3}x + 2$$

$$\text{becomes} \rightarrow \qquad 0 = -\frac{1}{3}x + 2$$

$$-2 = -\frac{1}{3}x \qquad \text{Add } -2 \text{ to each side.}$$

$$6 = x \qquad \text{Multiply each side by } -3.$$

The x-intercept is 6, which means the graph passes through the point (6, 0).

y-intercept:

$$\text{When} \rightarrow \qquad x = 0$$

$$\text{the equation} \rightarrow \qquad y = -\frac{1}{3}x + 2$$

$$\text{becomes} \rightarrow \qquad y = -\frac{1}{3}(0) + 2$$

$$y = 2$$

The y-intercept is 2, which means the graph passes through the point (0, 2). The graph of $y = -\frac{1}{3}x + 2$ is shown in Figure 4. Compare this graph, and the method used to obtain it, with Example 7 in Section 3.2.

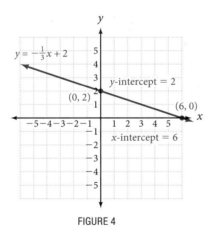

FIGURE 4

3. Use the intercepts to graph $y = \frac{3}{2}x - 3$.

After reading through the preceding section, respond in your own words and in complete sentences.

A. What is the x-intercept for a graph?

B. What is the y-intercept for a graph?

C. How do you find the y-intercept for a line from the equation $3x - y = 6$?

D. How do you graph a line for $y = -\frac{1}{2}x + 3$ using its intercepts?

Vocabulary Review

Choose the correct words to fill in the blanks below.

x-axis	y-axis	x-intercept	y-intercept

1. The _____ of a straight line is the x-coordinate of the point where the graph crosses the x-axis.

2. The _____ of a straight line is the y-coordinate of the point where the graph crosses the y-axis.

3. Any point on the _____ has a y-coordinate of 0.

4. Any point on the _____ has a x-coordinate of 0.

Problems

A Complete each table.

1.

Equation	x-intercept	y-intercept
$3x + 4y = 12$		
$3x + 4y = 4$		
$3x + 4y = 3$		
$3x + 4y = 2$		

2.

Equation	x-intercept	y-intercept
$-2x + 3y = 6$		
$-2x + 3y = 3$		
$-2x + 3y = 2$		
$-2x + 3y = 1$		

3.

Equation	x-intercept	y-intercept
$x - 3y = 2$		
$y = \frac{1}{3}x - \frac{2}{3}$		
$x - 3y = 0$		
$y = \frac{1}{3}x$		

4.

Equation	x-intercept	y-intercept
$x - 2y = 1$		
$y = \frac{1}{2}x - \frac{1}{2}$		
$x - 2y = 0$		
$y = \frac{1}{2}x$		

SCAN TO ACCESS

B Find the *x*- and *y*-intercepts for the following equations. Then use the intercepts to graph each equation.

5. $2x + y = 4$

6. $2x + y = 2$

7. $-x + y = 3$

8. $-x + y = 4$

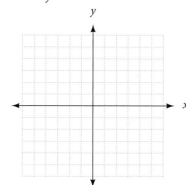

9. $-x + 2y = 2$

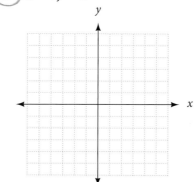

10. $-x + 2y = 4$

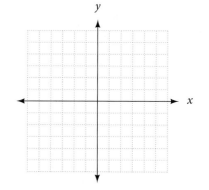

11. $5x + 2y = 10$

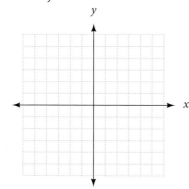

12. $2x + 5y = 10$

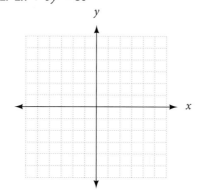

13. $4x - 2y = 8$

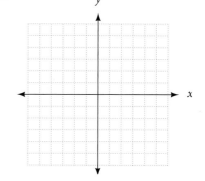

14. $2x - 4y = 8$

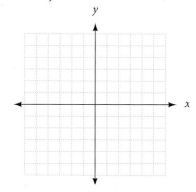

15. $-4x + 5y = 20$

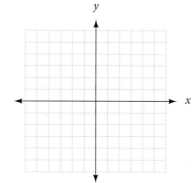

16. $-5x + 4y = 20$

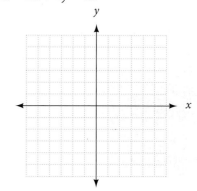

17. $y = 2x - 6$

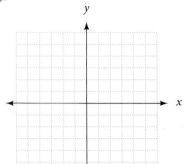

18. $y = 2x + 6$

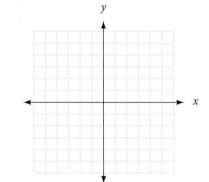

19. $y = 2x + 2$

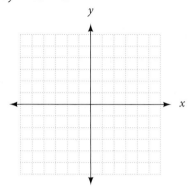

20. $y = -2x + 2$

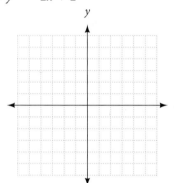

21. $y = 2x - 1$

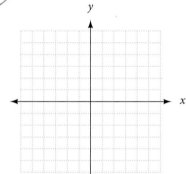

22. $y = -2x - 1$

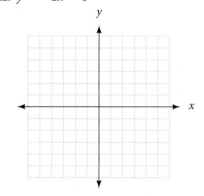

23. $y = \dfrac{1}{2}x + 3$

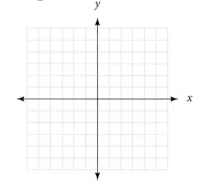

24. $y = \dfrac{1}{2}x - 3$

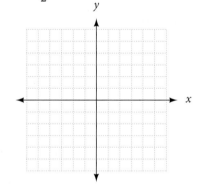

25. $y = -\dfrac{1}{3}x - 2$

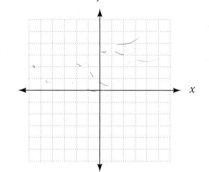

26. $y = -\dfrac{1}{3}x + 2$

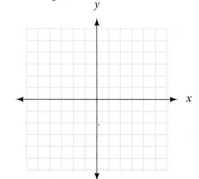

27. $y = -\dfrac{2}{3}x + 4$

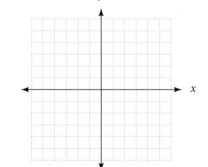

28. $y = -\dfrac{1}{2}x + 2$

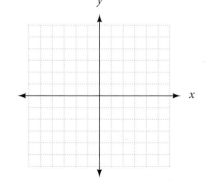

For each of the following lines the *x*-intercept and the *y*-intercept are both 0, which means the graph of each will go through the origin, (0, 0). Graph each line by finding a point on each, other than the origin, and then drawing a line through that point and the origin.

29. $y = -2x$

30. $y = \frac{1}{2}x$

31. $y = -\frac{1}{3}x$

32. $y = -3x$

33. $y = \frac{2}{3}x$

34. $y = \frac{3}{2}x$

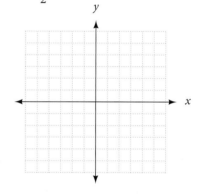

The next two problems are intended to give you practice reading, and paying attention to, the instructions that accompany the problems you are working. Working these problems is an excellent way to get ready for a test or a quiz.

35. Work each problem according to the instructions.

 a. Solve: $2x - 3 = -3$.

 b. Find the *x*-intercept: $2x - 3y = -3$.

 c. Find *y* when *x* is 0: $2x - 3y = -3$.

 d. Graph: $2x - 3y = -3$.

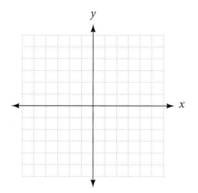

 e. Solve for *y*: $2x - 3y = -3$.

36. Work each problem according to the instructions.

 a. Solve: $3x - 4 = -4$.

 b. Find the *y*-intercept: $3x - 4y = -4$.

 c. Find *x* when *y* is 0: $3x - 4y = -4$.

 d. Graph: $3x - 4y = -4$.

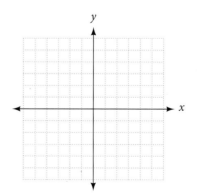

 e. Solve for *y*: $3x - 4y = -4$.

From the graphs below, find the *x*- and *y*-intercepts for each line.

37.

(3,5)

38.

39.

40.
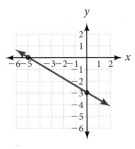

41. Graph the line that passes through the point $(-4, 4)$ and has an *x*-intercept of -2. What is the *y*-intercept of this line?

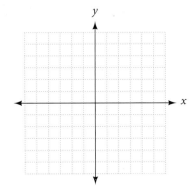

42. Graph the line that passes through the point $(-3, 4)$ and has a *y*-intercept of 1. What is the *x*-intercept of this line?

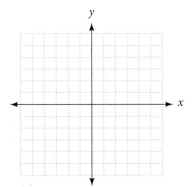

43. A line passes through the point $(1, 4)$ and has a *y*-intercept of 3. Graph the line and name its *x*-intercept.

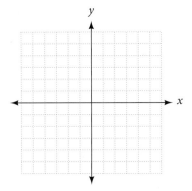

44. A line passes through the point $(3, 4)$ and has an *x*-intercept of 1. Graph the line and name its *y*-intercept.

45. Graph the line that passes through the points $(-2, 5)$ and $(5, -2)$. What are the *x*- and *y*-intercepts for this line?

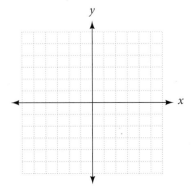

46. Graph the line that passes through the points $(5, 3)$ and $(-3, -5)$. What are the *x*- and *y*-intercepts for this line?

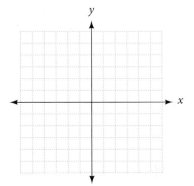

47. The vertical line $x = 3$ has only one intercept. Graph $x = 3$, and name its intercept. [Remember, ordered pairs (x, y) that are solutions to the equation $x = 3$ are ordered pairs with an x-coordinate of 3 and any y-coordinate.]

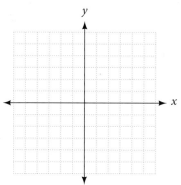

48. Graph the vertical line $x = -2$. Then name its intercept.

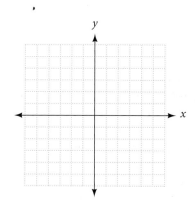

49. The horizontal line $y = 4$ has only one intercept. Graph $y = 4$, and name its intercept. [Ordered pairs (x, y) that are solutions to the equation $y = 4$ are ordered pairs with a y-coordinate of 4 and any x-coordinate.]

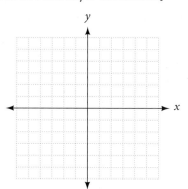

50. Graph the horizontal line $y = -3$. Then name its intercept.

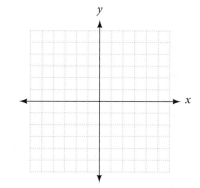

Applying the Concepts

51. Seating Capacity A theater has a capacity of 500 people. Because a different price is charged for students than for general admission, the owners would like to know what combinations of students and general tickets can be sold when the show is sold out. The relationship between student and general tickets can be written as

$$x + y = 500$$

Graph this equation where the x-axis is student tickets and the y-axis is general tickets. Find the intercepts first, and limit your graph to the first quadrant.

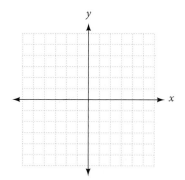

52. Ladder Angles A 13-foot ladder is placed against a barn. As the top of the ladder moves up and down the wall, the angles made by the ladder and the wall, α, and the ladder and the ground, β, can be written as

$$\alpha + \beta = 90.$$

Graph this equation on a coordinate system where α is the y-axis and β is the x-axis. Find the intercepts first, and limit your graph to the first quadrant only.

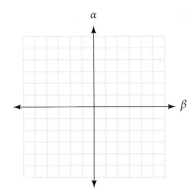

53. Working Two Jobs Maggie has a job working in an office for $10 an hour and another job driving a tractor pays $12 an hour. Maggie works at both jobs and earns $480 in one week.

 a. Write an equation that gives the number of hours x, she worked for $10 per hour and the number of hours, y, she worked for $12 per hour.

 b. Find the x- and y-intercepts for this equation.

 c. Graph this equation from the intercepts, using only the first quadrant.

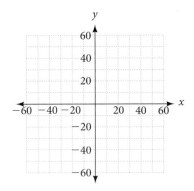

 d. From the graph, find how many hours she worked at $12, if she worked 36 hours at $10 per hour?

 e. From the graph, find how many hours she worked at $10, if she worked 25 hours at $12 per hour?

54. Ticket prices Devin is selling tickets to a BBQ. Adult tickets cost $6.00 and children's tickets cost $4.00. The total amount of money he collects is $240.

 a. Write an equation that gives the number of adult tickets x, he sold for $6 and the children's tickets, y, he sold for $4

 b. Find the x- and y-intercepts for this equation.

 c. Graph this equation from the intercepts, using only the first quadrant.

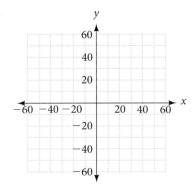

 d. From the graph, find how many children's tickets were sold if 30 adult tickets sold?

 e. From the graph, find how many adult tickets were sold if 30 children's tickets sold?

55. Brian drives a certain amount of time at 50 mph and another amount at 60 mph. Together he drives 450 miles.

 a. Write an equation involving x (time driving at 50 mph) and y (time driving at 60 mph).

 b. Find the x- and y- intercepts for the equation in part a.

 c. Graph this equation in the first quadrant only.

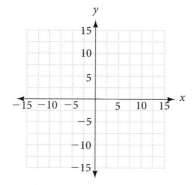

 d. From the graph, find how many hours Brian drove at 60 mph if he drove a total of 8 hours?

 e. If Brian drove at 50 mph for 4 hours, how long did he drive at 60 mph?

56. Juan tutors in the math lab for $8 per hour and privately for $12 per hour. He earns $240 in one week.

 a. Write and equation involving x (number of math lab hours) and y (number of private hours).

 b. Find the x- and y- intercepts for this equation.

 c. Graph this equation in the first quadrant only.

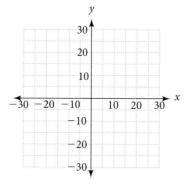

 d. From the graph, find how many hours did Juan tutor in the math lab if he tutored privately for 10 hours?

 e. If Juan is limited to tutoring 19 hours in the math lab, how many hours must he tutor privately?

Getting Ready for the Next Section

57. Evaluate

 a. $\dfrac{5-2}{3-1}$

 b. $\dfrac{2-5}{1-3}$

58. Evaluate

 a. $\dfrac{-4-1}{5-(-2)}$

 b. $\dfrac{1+4}{-2-5}$

59. Evaluate the following expressions when $x = 3$, and $y = 5$.

 a. $\dfrac{y-2}{x-1}$

 b. $\dfrac{2-y}{1-x}$

60. Evaluate the following expressions when $x = 4$, and $y = -1$.

 a. $\dfrac{-4-y}{5-x}$

 b. $\dfrac{y+4}{x-5}$

Find the Mistake

Each sentence below contains a mistake. Circle the mistake and write the correct word or number on the line provided.

1. The y–intercept is where the graph of a line crosses the x–axis at the point $(0, y)$. _____

2. The x–intercept for the linear equation $9x = \frac{1}{3}y + 5$ is $(-15, 0)$. _____

3. The y–intercept for the linear equation $2x + 7y = 3$ is $\left(0, \frac{3}{2}\right)$. _____

4. The intercepts for the linear equation $15 = 3y - 5x$ are $(5, 0)$ and $(0, -3)$. _____

3.4 Graphing Linear Equations Using Slope

Heli-skiing is a sport where a helicopter transports a skier to the top of the mountain he will ski down. The skier leaps out of the helicopter's cabin and onto an untracked slope, often inaccessible from the ground.

Let's suppose for every 30 feet of horizontal distance the heli-skier covers, his elevation drops 10 feet. In this section, we will learn how to use these numbers to assign a numerical value to the slope of the hill.

In defining the slope of a straight line, we are looking for a number to associate with a straight line that does two things. First of all, we want the slope of a line to measure the "steepness" of the line; that is, in comparing two lines, the slope of the steeper line should have the larger numerical value. Second, we want a line that *rises* going from left to right to have a *positive* slope. We want a line that *falls* going from left to right to have a *negative* slope. (A horizontal line that neither rises nor falls going from left to right must, therefore, have 0 slope; whereas, a vertical line has an undefined slope.) These are illustrated in Figure 1.

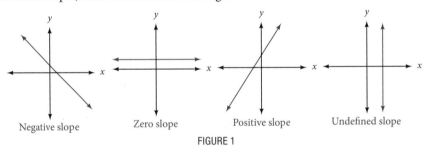

Negative slope Zero slope Positive slope Undefined slope

FIGURE 1

Suppose we know the coordinates of two points on a line. Because we are trying to develop a general formula for the slope of a line, we will use general points—let's call the two points $P_1(x_1, y_1)$ and $P_2(x_2, y_2)$. They represent the coordinates of any two different points on our line. We define the *slope* of our line to be the ratio of the vertical change of *y*-coordinates to the horizontal change of *x*-coordinates as we move from point (x_1, y_1) to point (x_2, y_2) on the line. (See Figure 2 on next page.)

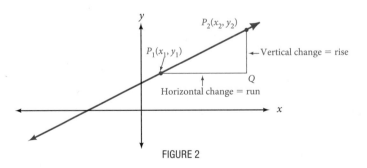

FIGURE 2

We call the vertical change the *rise* in the graph and the horizontal change the *run* in the graph. The slope, then, is

$$\text{Slope} = \frac{\text{vertical change}}{\text{horizontal change}} = \frac{\text{rise}}{\text{run}}$$

We would like to have a numerical value to associate with the rise in the graph and a numerical value to associate with the run in the graph. A quick study of Figure 2 shows that the coordinates of point Q must be (x_2, y_1), because Q is directly below point P_2 and right across from point P_1. We can draw our diagram again in the manner shown in Figure 3. It is apparent from this graph that the rise can be expressed as $(y_2 - y_1)$ and the run as $(x_2 - x_1)$. We usually denote the slope of a line by the letter m. The complete definition of slope follows along with a diagram (Figure 3) that illustrates the definition.

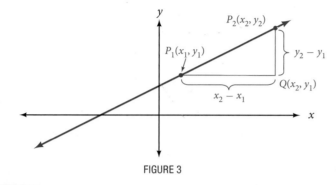

FIGURE 3

DEFINITION slope

If points (x_1, y_1) and (x_2, y_2) are any two different points and $x_1 \neq x_2$, then the *slope* of the line on which they lie is, denoted by m,

$$\text{Slope} = m = \frac{\text{rise}}{\text{run}} = \frac{y_2 - y_1}{x_2 - x_1}$$

This definition of the *slope* of a line does just what we want it to do. If the line rises going from left to right, the slope will be positive. If the line falls from left to right, the slope will be negative. Also, the steeper the line, the larger numerical value the slope will have.

A Finding the Slope Given Two Points on a Line

EXAMPLE 1 Find the slope of the line between the points $(1, 2)$ and $(3, 5)$.

Solution We can let

$$(x_1, y_1) = (1, 2)$$

and

$$(x_2, y_2) = (3, 5)$$

then

$$m = \frac{y_2 - y_1}{x_2 - x_1} = \frac{5 - 2}{3 - 1} = \frac{3}{2}$$

The slope is $\frac{3}{2}$. For every vertical change of 3 units, there will be a corresponding horizontal change of 2 units. (See Figure 4.)

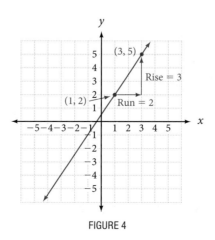

FIGURE 4

2. Find the slope of the line through (3, 4) and (1, −2).

EXAMPLE 2 Find the slope of the line through (−2, 1) and (5, −4).

Solution It makes no difference which ordered pair we call (x_1, y_1) and which we call (x_2, y_2).

$$\text{Slope} = m = \frac{y_2 - y_1}{x_2 - x_1} = \frac{-4 - 1}{5 - (-2)} = -\frac{5}{7}$$

The slope is $-\frac{5}{7}$. Every vertical change of −5 units (down 5 units) is accompanied by a horizontal change of 7 units (to the right 7 units). (See Figure 5.)

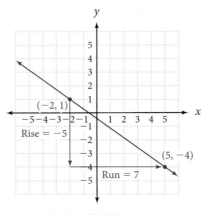

FIGURE 5

B Graphing a Line from Its Slope

3. Graph the line with slope $\frac{1}{4}$ and y-intercept −3.

EXAMPLE 3 Graph the line with slope $\frac{3}{2}$ and y-intercept 1.

Solution Because the y-intercept is 1, we know that one point on the line is (0, 1). So, we begin by plotting the point (0, 1), as shown in Figure 6.

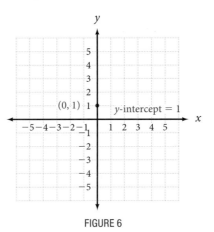

FIGURE 6

There are many lines that pass through the point shown in Figure 6, but only one of those lines has a slope of $\frac{3}{2}$. The slope, $\frac{3}{2}$, can be thought of as the rise in the graph divided by the run in the graph. Therefore, if we start at the point (0, 1) and move 3 units up (that's a rise of 3) and then 2 units to the right (a run of 2), we will be at another point on the graph. Figure 7 shows that the point we reach by doing so is the point (2, 4).

$$\text{Slope} = m = \frac{\text{rise}}{\text{run}} = \frac{3}{2}$$

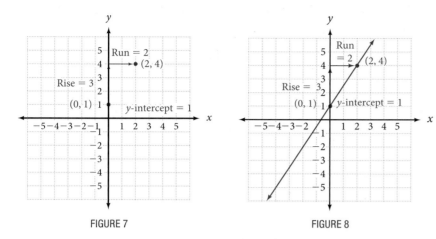

FIGURE 7 FIGURE 8

To graph the line with slope $\frac{3}{2}$ and y-intercept 1, we simply draw a line through the two points in Figure 7 to obtain the graph shown in Figure 8.

EXAMPLE 4 Find the slope of the line containing $(3, -1)$ and $(3, 4)$.

Solution Using the definition for slope, we have

$$m = \frac{y_2 - y_1}{x_2 - x_1} = \frac{4 - (-1)}{3 - 3} = \frac{5}{0}$$

The expression $\frac{5}{0}$ is undefined; that is, there is no real number to associate with it. In this case, we say the line *has an undefined slope.*

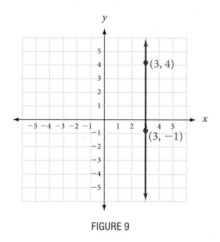

FIGURE 9

The graph of our line is shown in Figure 9. All vertical lines have an undefined slope. (And all horizontal lines, as we mentioned earlier, have a slope of zero.)

As a final note, the following summary reminds us that all horizontal lines have equations of the form $y = b$ and slopes of 0. Because they cross the y-axis at b, the y-intercept is b; there is no x-intercept. Vertical lines have an undefined slope and equations of the form $x = a$. Each will have an x-intercept at a and no y-intercept. Finally, equations of the form $y = mx$ have graphs that pass through the origin. The slope is always m and both the x-intercept and the y-intercept are 0.

4. Find the slope of the line through $(2, -3)$ and $(-1, -3)$.

Note Exceptions to this discussion are the lines $y = 0$ and $x = 0$. For $y = 0$, the x-intercepts are infinite; whereas $x = 0$ has infinite y-intercepts.

Answer

4. 0

FACTS FROM GEOMETRY Special Equations and Their Graphs, Slopes, and Intercepts

For the equations below, m, a, and b are real numbers.

Through the Origin	Vertical Line	Horizontal Line

 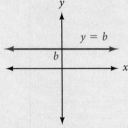

FIGURE 10A	FIGURE 10B	FIGURE 10C

Equation: $y = mx$

Slope $= m$

x-intercept $= 0$

y-intercept $= 0$

Equation: $x = a$

Undefined

x-intercept $= a$

No y-intercept

Equation: $y = b$

Slope $= 0$

No x-intercept

y-intercept $= b$

GETTING READY FOR CLASS

After reading through the preceding section, respond in your own words and in complete sentences.

A. Using x- and y-coordinates, how do you find the slope of a line?

B. Would you rather climb a hill with a slope of 1 or a slope of 3? Explain why.

C. Describe how you would graph a line from its slope and y-intercept.

D. Describe how to obtain the slope of a line if you know the coordinates of two points on the line.

Vocabulary Review

Choose the correct words to fill in the blanks below.

vertical undefined ratio zero horizontal slope

1. The slope of a line is the _____ of the vertical change of y-coordinates to the horizontal change of x-coordinates as we move from point (x_1, y_1) to point (x_2, y_2) on the line.

2. The _____ change between one point on a straight line to another point is called the rise.

3. The _____ change between one point on a straight line to another point is called the run.

4. The _____ of a line is given by the equation $m = \frac{y_2 - y_1}{x_2 - x_1}$.

5. A horizontal line has a(n) _____ slope.

6. A vertical line has a(n) _____ slope.

Problems

A Find the slope of the line through the following pairs of points. Then plot each pair of points, draw a line through them, and indicate the rise and run in the graph in the same manner shown in Examples 1 and 2.

1. $(2, 1), (4, 4)$

2. $(3, 1), (5, 4)$

3. $(1, 4), (5, 2)$

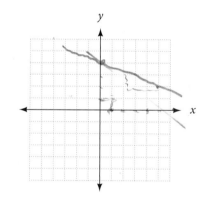

4. $(1, 3), (5, 2)$

5. $(1, -3), (4, 2)$

6. $(2, -3), (5, 2)$

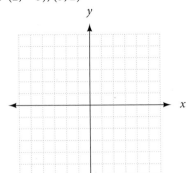

7. $(-3, -2), (1, 3)$

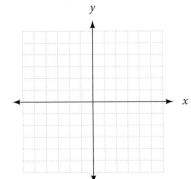

8. $(-3, -1), (1, 4)$

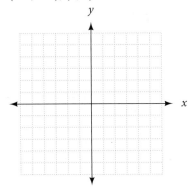

9. $(-3, 2), (3, -2)$

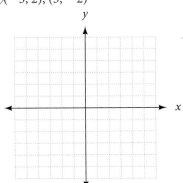

10. $(-3, 3), (3, -1)$

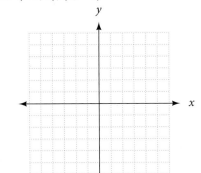

11. $(2, -5), (3, -2)$

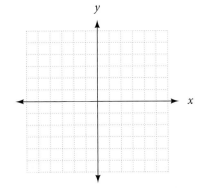

12. $(2, -4), (3, -1)$

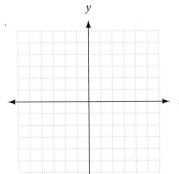

13. $(2, 4), (2, -3)$

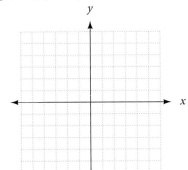

14. $(2, 3), (-1, 3)$

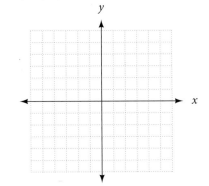

B In each of the following problems, graph the line with the given slope and y-intercept b.

15. $m = \dfrac{2}{3}, b = 1$

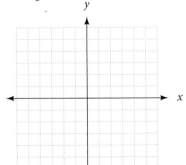

16. $m = \dfrac{3}{4}, b = -2$

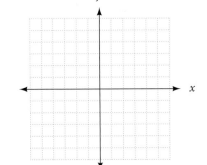

17. $m = \dfrac{3}{2}, b = -3$

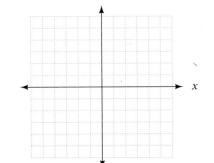

18. $m = \frac{4}{3}, b = 2$

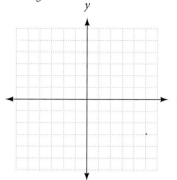

19. $m = -\frac{4}{3}, b = 5$

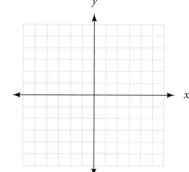

20. $m = -\frac{3}{5}, b = 4$

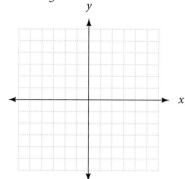

21. $m = 2, b = 1$

$y = mx + b$ $y = 2x + 1$

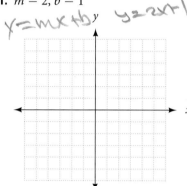

22. $m = -2, b = 4$

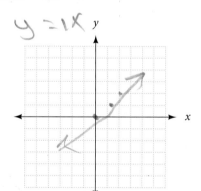

23. $m = 3, b = -1$

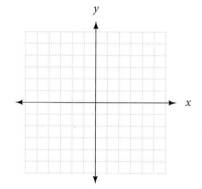

24. $m = 3, b = -2$

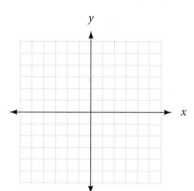

25. $m = 1, b = 0$

$y = 1x$

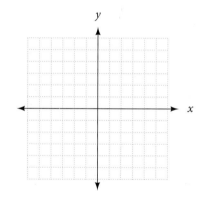

26. $m = -1, b = 0$

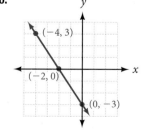

Find the slope and *y*-intercept for each line.

27.

28.

29.

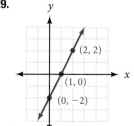

30.

31. Graph the line that has an x-intercept of 3 and a y-intercept of -2. What is the slope of this line?

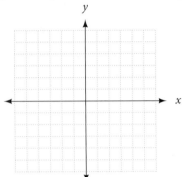

32. Graph the line that has an x-intercept of 2 and a y-intercept of -3. What is the slope of this line?

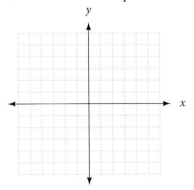

33. Graph the line with x-intercept 4 and y-intercept 2. What is the slope of this line?

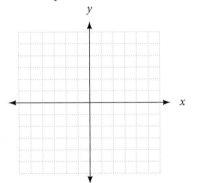

34. Graph the line with x-intercept -4 and y-intercept -2. What is the slope of this line?

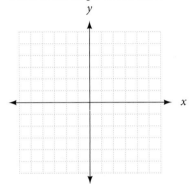

35. Graph the line $y = 2x - 3$, then name the slope and y-intercept by looking at the graph.

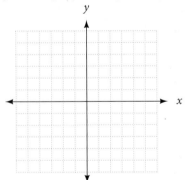

36. Graph the line $y = -2x + 3$, then name the slope and y-intercept by looking at the graph.

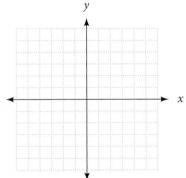

37. Graph the line $y = \frac{1}{2}x + 1$, then name the slope and y-intercept by looking at the graph.

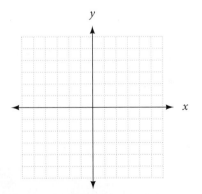

38. Graph the line $y = -\frac{1}{2}x - 2$, then name the slope and y-intercept by looking at the graph.

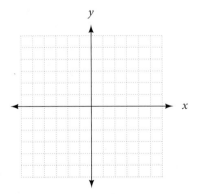

39. Find y if the line through $(4, 2)$ and $(6, y)$ has a slope of 2.

40. Find y if the line through $(1, y)$ and $(7, 3)$ has a slope of 6.

41. Find y if the line through $(3, 2)$ and $(3, y)$ has an undefined slope.

when $x_2 = x_1$

$$m = \frac{x_2 - y_1}{x_2 - x_1} = \frac{y - 2}{3 - 3} = \frac{y - 2}{0}$$

42. Find y if the line through $(2, 3)$ and $(5, y)$ has a slope of 0.

For each equation in each table, give the slope of the graph.

43.

Equation	Slope
$x = 3$	
$y = 3$	
$y = 3x$	

44.

Equation	Slope
$y = \frac{3}{2}$	
$x = \frac{3}{2}$	
$y = \frac{3}{2}x$	

45.

Equation	Slope
$y = -\frac{2}{3}$ ⟵ 0 b=k	
$x = -\frac{2}{3}$ x = k ↑ undifind	
$y = -\frac{2}{3}x = -\frac{2}{3}$	

46.

Equation	Slope
$x = -2$	
$y = -2$	
$y = -2x$	

Applying the Concepts

47. **Garbage Production** The table and completed line graph gives the annual production of garbage in the United States for some specific years. Find the slope of each of the four line segments, A, B, C, and D.

Year	Garbage (millions of tons)
1960	88
1970	121
1980	152
1990	205
2000	224
2007	254

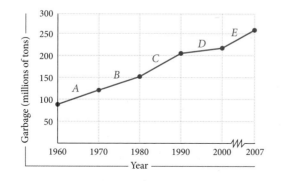

48. **Grass Height** The table and completed line graph gives the growth of a certain grass species over time. Find the slopes of the line segments labeled A, B, and C.

Day	Plant Height
0	0
2	1
4	3
6	6
8	13
10	23

49. Non-Camera Phone Sales The table and line graph here each show the estimated non-camera phone sales each year through 2010. Find the slope of each of the three line segments, *A*, *B*, and *C*.

Year	Sales (in millions)
2006	300
2007	250
2008	175
2009	150
2010	125

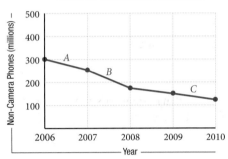

Estimated Non-Camera Phone Sales

50. Camera Phone Sales The table and line graph here each show the estimated sales of camera phones from 2006 to 2010. Find the slopes of line segments *A*, *B*, and *C*.

Year	Sales (in millions)
2006	500
2007	650
2008	750
2009	875
2010	900

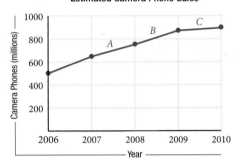

Estimated Camera Phone Sales

51. The federal budget deficit has become a very common political topic. The following data lists the total national outstanding debt over some selected years. Use this data to answer the questions that follow.

Year	Debt (trillions of $)
1980	0.9
1990	3.2
2000	5.7
2010	13.6

a. Using the table, sketch a line graph of *A*, *B*, and *C*. Add a graph.

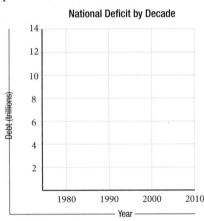

National Deficit by Decade

b. Calculate the slopes of the line segments.

c. Based on these slopes, is the rate of change of the debt increasing or decreasing?

Nike Women's Marathon The following chart shows the number of finishers at the Nike Women's Marathon in San Francisco, CA from 2005 to 2010.

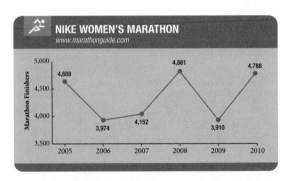

52. Find the slope of the line from 2005 to 2010. Explain in words what the slope represents.

53. Find the slope of the line from 2009 to 2010. Explain in words what the slope represents.

54. Heating a Block of Ice A block of ice with an initial temperature of $-20°C$ is heated at a steady rate. The graph shows how the temperature changes as the ice melts to become water and the water boils to become steam and water.

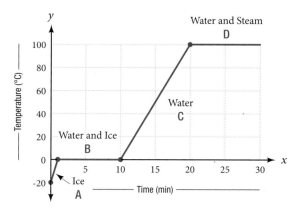

a. How long does it take all the ice to melt?

b. From the time the heat is applied to the block of ice, how long is it before the water boils?

c. Find the slope of the line segment labeled A. What units would you attach to this number?

d. Find the slope of the line segment labeled C. Be sure to attach units to your answer.

e. Is the temperature changing faster during the 1st minute or the 16th minute?

55. Slope of a Highway A sign at the top of the Cuesta Grade, outside of San Luis Obispo, reads "7% downgrade next 3 miles." The following diagram is a model of the Cuesta Grade that illustrates the information on that sign.

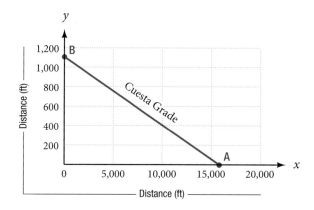

a. At point B, the graph crosses the y-axis at 1,106 feet. How far is it from the origin to point A?

b. What is the slope of the Cuesta Grade?

56. Facebook Users The graph below shows the number of Facebook users over a six year period. Using the graph, find the slope of the line connecting the first (2004, 1) and last (2010, 400) points on the line. Explain in words what the slope represents.

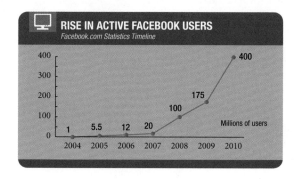

57. Mac Users The graph shows the average increase in the percentage of computer users who use Macs since 2003. Find the slope of the line connecting the first (2003, 2.2) and last (2010, 6.7) points on the line. Explain in words what the slope represents.

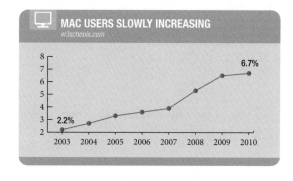

Getting Ready for the Next Section

Solve each equation for y.

58. $-2x + y = 4$ **59.** $-4x + y = -2$ **60.** $2x + y = 3$ **61.** $3x + 2y = 6$

62. $x + 2y = 4$ **63.** $x - 3y = 6$ **64.** $2x - 3y = 12$ **65.** $3x + 4y = 12$

66. $4x - 5y = 20$ **67.** $-2x - 5y = 10$ **68.** $-y - 3 = -2(x + 4)$ **69.** $-y + 5 = 2(x + 2)$

70. $-y - 3 = -\dfrac{2}{3}(x + 3)$ **71.** $-y - 1 = -\dfrac{1}{2}(x + 4)$ **72.** $-\dfrac{y - 1}{x} = \dfrac{3}{2}$ **73.** $-\dfrac{y + 1}{x} = \dfrac{3}{2}$

Find the Mistake

Each sentence below contains a mistake. Circle the mistake and write the correct word or number on the line provided.

1. A horizontal line has a positive slope. _____

2. A line that rises from the left to the right has a negative slope. _____

3. The slope of the line that passes through the points $(-1, 6)$ and $(4, -9)$ is $-\dfrac{1}{3}$. _____

4. Use the point $(3, -2)$ to graph a line with a y–intercept -6 and a slope $\dfrac{5}{2}$. _____

Landmark Review: Checking Your Progress

1. What is the name of the point with the coordinates $(0, 0)$?

2. Name the x-coordinate of any point that is on the line passing through the points $(4, 3)$ and $(4, 8)$.

3. Is the point $(5, 3)$ on the line that passes through $(2, 4)$ and $(4, 4)$?

For each equation, complete the ordered pairs.

4. $4x - y = 3$ $(0,\), (\ , 0), (3,\)$

5. $2y - 3x = 15$ $(\ , 3), (1,\), (6,\)$

Find the x- and y-intercepts for the following equations.

6. $x + y = 1$

7. $x - 2y = 6$

Find the slope of the line through the following pairs of points.

8. $(1, 3), (5, 4)$ **9.** $(4, 2), (3, 4)$ **10.** $(3, 4), (3, 2)$

11. Would you expect a line with a positive slope to rise or fall moving from left to right?

12. If a line has a slope of 0 and doesn't pass through the origin, what do you know about the x-intercept?

OBJECTIVES

A Find the equation of a line given the slope and y-intercept of the line.

B Find the slope and y-intercept of a line given the equation of the line.

C Find the equation of a line given a point on the line and the slope of the line.

D Find the equation of a line given two points on the line.

E Find the equations of parallel and perpendicular lines.

KEY WORDS

slope-intercept form

point-slope form

parallel

perpendicular

Every winter, the Angel Fire Resort in New Mexico hosts the World Shovel Racing Championships. At the top of a snow-covered slope, contestants sit straddling the handle of a stock grain scoop shovel. Then they speed down the 1,000-foot course using the handle to steer. Helmets are required because wipeouts are common. Let's say the fastest racer finished the course in 17 seconds and the runner up finished in 18 seconds. That gives us the two points (17, 1000) and (18, 1000) that could help us draw the graph for the race. In this section, we will learn ways to find the equation of a line using slope, intercepts, and given points.

To this point in the chapter, most of the problems we have worked have used the equation of a line to find different types of information about the line. For instance, given the equation of a line, we can find points on the line, the graph of the line, the intercepts, and the slope of the line. In this section, we reverse things somewhat and use information about a line, such as its slope and y-intercept, to find the line's equation.

There are three main types of problems to solve in this section.

1. Find the equation of a line from the slope and the y-intercept.

2. Find the equation of a line given one point on the line and the slope of the line.

3. Find the equation of a line given two points on the line.

Examples 1 and 2 illustrate the first type of problem. Example 5 solves the second type of problem. We solve the third type of problem in Example 6.

A The Slope-Intercept Form of an Equation of a Straight Line

EXAMPLE 1 Find the equation of the line with slope $\frac{3}{2}$ and y-intercept 1.

Solution We graphed the line with slope $\frac{3}{2}$ and y-intercept 1 in Example 3 of the previous section. Figure 1 shows that graph.

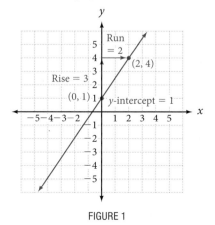

FIGURE 1

Practice Problems

1. Find the equation of the line with slope $\frac{4}{3}$ and y-intercept 2.

Answer

1. $y = \frac{4}{3}x + 2$

What we want to do now is find the equation of the line shown in Figure 1. To do so, we take any other point (x, y) on the line and apply our slope formula to that point and the point $(0, 1)$. We set that result equal to $\frac{3}{2}$, because $\frac{3}{2}$ is the given slope of our line.

$$m = \frac{\text{rise}}{\text{run}} = \frac{y - 1}{x - 0} = \frac{3}{2} \qquad \text{Slope} = \frac{\text{vertical change}}{\text{horizontal change}}.$$

$$\frac{y - 1}{x} = \frac{3}{2} \qquad\qquad x - 0 = x.$$

$$y - 1 = \frac{3}{2}x \qquad\qquad \text{Multiply each side by } x.$$

$$y = \frac{3}{2}x + 1 \qquad\qquad \text{Add 1 to each side.}$$

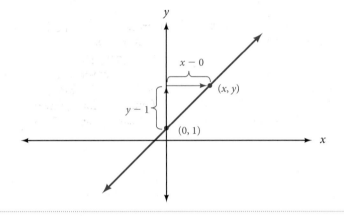

What is interesting and useful about the equation we have just found is that the number in front of x is the slope of the line and the constant term is the y-intercept. It is no coincidence that it turned out this way. Whenever an equation has the form $y = mx + b$, the graph is always a straight line with slope m and y-intercept b. To see that this is true in general, suppose we want the equation of a line with slope m and y-intercept b. Because the y-intercept is b, then the point $(0, b)$ is on the line. If (x, y) is any other point on the line, then we apply our slope formula to get

$$\frac{y - b}{x - 0} = m \qquad\qquad \text{Slope} = \frac{\text{vertical change}}{\text{horizontal change}}.$$

$$\frac{y - b}{x} = m \qquad\qquad x - 0 = x.$$

$$y - b = mx \qquad\qquad \text{Multiply each side by } x.$$

$$y = mx + b \qquad\qquad \text{Add } b \text{ to each side.}$$

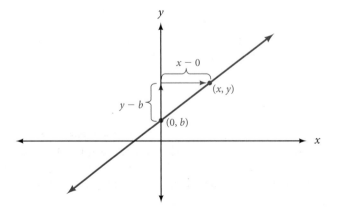

Here is a summary of what we have just found:

> **Slope-Intercept Form of the Equation of a Line**
>
> The equation of the non-vertical line with slope m and y-intercept b is always given by
>
> $$y = mx + b$$

EXAMPLE 2 Find the equation of the line with slope $-\frac{4}{3}$ and y-intercept 5. Then, graph the line.

Solution Substituting $m = -\frac{4}{3}$ and $b = 5$ into the equation $y = mx + b$, we have

$$y = -\frac{4}{3}x + 5$$

Finding the equation from the slope and y-intercept is just that easy. If the slope is m and the y-intercept is b, then the equation is always $y = mx + b$.

Because the y-intercept is 5, the graph goes through the point $(0, 5)$. To find a second point on the graph, we start at $(0, 5)$ and move 4 units down (that's a rise of -4) and 3 units to the right (a run of 3). The point we reach is $(3, 1)$. Drawing a line that passes through $(0, 5)$ and $(3, 1)$, we have the graph of our equation. (Note that we could also let the rise $= 4$ and the run $= -3$ and obtain the same graph.) The graph is shown in Figure 2.

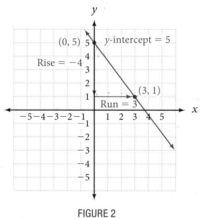

FIGURE 2

B Finding the Slope and y-Intercept

EXAMPLE 3 Find the slope and y-intercept for $-2x + y = -4$. Then, use them to draw the graph.

Solution To identify the slope and y-intercept from the equation, the equation must be in the form $y = mx + b$ (slope-intercept form). To write our equation in this form, we must solve the equation for y. To do so, we simply add $2x$ to each side of the equation.

$$-2x + y = -4 \qquad \textit{Original equation}$$

$$y = 2x - 4 \qquad \textit{Add 2x to each side.}$$

The equation is now in slope-intercept form, so the slope must be 2 and the y-intercept must be -4. The graph, therefore, crosses the y-axis at $(0, -4)$. Because the slope is 2, we can let the rise $= 2$ and the run $= 1$ because $y = 2x - 4$ can also be written as $y = \frac{2}{1}x - 4$. We can then find a second point on the graph, which is $(1, -2)$. The graph is shown in Figure 3.

2. Find the equation of the line with slope $\frac{2}{3}$ and y-intercept 1. Then graph the line.

3. Find the slope and y-intercept for $3x - 2y = -6$. Then use them to draw the graph.

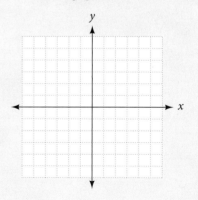

Answers

2. $y = \frac{2}{3}x + 1$

3. Slope is $\frac{3}{2}$; y-intercept is 3.

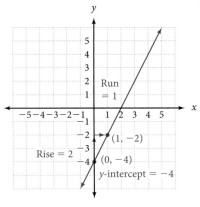

FIGURE 3

4. Find the slope and y-intercept for $4x - 2y = 12$.

EXAMPLE 4 Find the slope and y-intercept for $3x - 2y = 6$.

Solution To find the slope and y-intercept from the equation, we must write the equation in the form $y = mx + b$. This means we must solve the equation $3x - 2y = 6$ for y.

$$3x - 2y = 6 \qquad \text{\textit{Original equation}}$$

$$-2y = -3x + 6 \qquad \text{\textit{Add} } -3x \text{ \textit{to each side.}}$$

$$-\frac{1}{2}(-2y) = -\frac{1}{2}(-3x + 6) \qquad \text{\textit{Multiply each side by} } -\frac{1}{2}.$$

$$y = \frac{3}{2}x - 3 \qquad \text{\textit{Simplify each side.}}$$

Now that the equation is written in slope-intercept form, we can identify the slope as $\frac{3}{2}$ and the y-intercept as -3. We use the same process as in Example 3 to graph the line. The graph is shown in Figure 4.

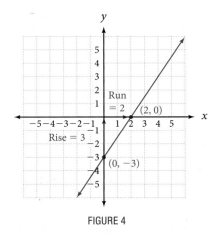

FIGURE 4

C The Point-Slope Form of an Equation of a Straight Line

A second useful form of the equation of a non-vertical straight line is the point-slope form.

Let line l contain the point (x_1, y_1) and have slope m. If (x, y) is any other point on l, then by the definition of slope we have

$$\frac{y - y_1}{x - x_1} = m$$

Multiplying both sides by $(x - x_1)$ gives us

$$(x - x_1) \cdot \frac{y - y_1}{x - x_1} = m(x - x_1)$$

$$y - y_1 = m(x - x_1)$$

Answers

4. Slope is 2; y-intercept is -6.

This last equation is known as the *point-slope form* of the equation of a straight line.

Point-Slope Form of the Equation of a Line

The point-slope form of the equation of the non-vertical line through (x_1, y_1) with slope m is given by

$$y - y_1 = m(x - x_1)$$

This form is used to find the equation of a line, either given one point on the line and the slope, or given two points on the line.

EXAMPLE 5 Using the point-slope form, find the equation of the line with slope -2 that contains the point $(-4, 3)$. Write the answer in slope-intercept form.

Solution Using $\quad (x_1, y_1) = (-4, 3)$ and $m = -2$

in $\qquad y - y_1 = m(x - x_1)$ \qquad Point-slope form

gives us $\quad y - 3 = -2(x + 4)$ \qquad Note: $x - (-4) = x + 4.$

$\qquad\qquad\quad y - 3 = -2x - 8$ \qquad Multiply out right side.

$\qquad\qquad\qquad\quad y = -2x - 5$ \qquad Add 3 to each side.

Figure 5 is the graph of the line that contains $(-4, 3)$ and has a slope of -2. Notice that the y-intercept on the graph matches that of the equation we found, which we wrote in slope-intercept form.

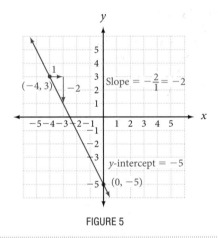

FIGURE 5

D Finding an Equation Given Two Points

EXAMPLE 6 Find the equation of the line that passes through the points $(-3, 3)$ and $(3, -1)$.

Solution We begin by finding the slope of the line:

$$m = \frac{3 - (-1)}{-3 - 3} = \frac{4}{-6} = -\frac{2}{3}$$

Using $(x_1, y_1) = (3, -1)$ and $m = -\frac{2}{3}$ in $y - y_1 = m(x - x_1)$ yields

$$y + 1 = -\frac{2}{3}(x - 3)$$

$$y + 1 = -\frac{2}{3}x + 2 \qquad \text{Multiply out right side.}$$

$$y = -\frac{2}{3}x + 1 \qquad \text{Add} -1 \text{ to each side.}$$

5. Find the equation of the line with slope 3 that contains the point $(-1, 2)$.

6. Find the equation of the line that passes through $(2, 5)$, and $(6, -3)$.

Answers

5. $y = 3x + 5$

6. $y = -2x + 9$

Figure 6 shows the graph of the line that passes through the points $(-3, 3)$ and $(3, -1)$. As you can see, the slope and y-intercept are $-\frac{2}{3}$ and 1, respectively.

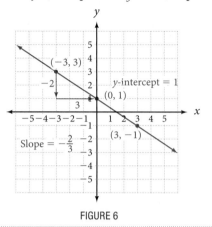

FIGURE 6

Methods of Graphing Lines

Here is a list of the graphing methods we have discovered in the last few sections.

1. Substitute convenient values of x into the equation, and find the corresponding values of y. We used this method first for equations like $y = 2x - 3$. To use this method for equations that looked like $2x - 3y = 6$, we first solved them for y.

2. Find the x- and y-intercepts. This method works best for equations of the form $3x + 2y = 6$ where the numbers in front of x and y divide the constant term evenly.

3. Find the slope and y-intercept. This method works best when the equation has the form $y = mx + b$ and b is an integer.

E Parallel and Perpendicular Lines

Now we can use what we have learned about linear equations to find the equations for parallel and perpendicular lines. Parallel lines are two lines that are nonintersecting and rise or fall at the same rate. In other words, two lines are parallel if and only if they have the same slope.

Here is a formal definition for parallel lines:

> **DEFINITION** parallel
>
> Two straight lines are ***parallel*** if they have the same slope. In other words, for $y = mx + b$, all equations of the form $y = mx + c$ are parallel if $b \neq c$.

7. Find the slope of a line parallel to the line that passes through $(-1, 3)$ and has a y-intercept of 1.

EXAMPLE 7 Find the slope of a line parallel to the line that passes through $(-3, 1)$ and has a y-intercept of 3.

Solution Since the y-intercept of the line is 3, we can determine that another point on the line is $(0, 3)$. We begin by finding the slope of the line.

$$m = \frac{1 - 3}{-3 - 0} = \frac{-2}{-3} = \frac{2}{3}$$

Since we know the y-intercept is $(0, 3)$ and using the slope $m = \frac{2}{3}$, we can write an equation for our line in slope-intercept form.

$$y = \frac{2}{3}x + 3$$

Two lines are parallel if they have the same slope, therefore, any line with the slope $m = \frac{2}{3}$ will be parallel to our original line.

Answer

7. -2

Although it is not as obvious, it is also true that two nonvertical lines are perpendicular if and only if the product of their slopes is −1. This is the same as saying their slopes are negative reciprocals. The following are graphs of perpendicular lines. Notice the slopes of the corresponding equations; the slope of one is the negative reciprocal of the slope of the other.

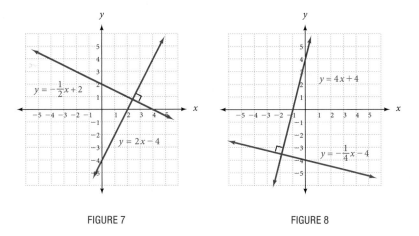

FIGURE 7 FIGURE 8

The following definition can be assumed for perpendicular lines:

> **DEFINITION** perpendicular
>
> Two non-vertical lines are *perpendicular* if the product of their slopes is −1. That is, $y = mx + b$ is perpendicular to $y = \left(-\frac{1}{m}\right)x + c$, if m is any real number except 0.

Note Also remember that a vertical line and a horizontal line are perpendicular.

EXAMPLE 8 Find the slope of a line perpendicular to the line $y = 3x - 2$.

Solution Since the equation is already written in slope-intercept form, we know that the slope of the first line is 3. Any line with a slope that, when multiplied by 3, results in a product of −1 is perpendicular to this line. For this to be true, the slope must be the negative reciprocal of 3, which is $-\frac{1}{3}$.

Furthermore, any line of the form $y = -\frac{1}{3}x + a$, where a is any real number, will provide a graph for our solution.

●

8. Find the slope of a line perpendicular to the line $y = \frac{1}{3}x - 2$.

> **GETTING READY FOR CLASS**
>
> *After reading through the preceding section, respond in your own words and in complete sentences.*
>
> **A.** What are m and b in the equation $y = mx + b$? What form is this equation?
>
> **B.** How would you find the slope and y-intercept for the line $4x - 3y = 9$?
>
> **C.** What is the point-slope form of the equation of a line?
>
> **D.** Find the equation of the line that passes through the points $(8, 1)$ and $(4, -2)$.

Answer

8. −3

EXERCISE SET 3.5

Vocabulary Review

Choose the correct words to fill in the blanks below.

substitute	slope	slope-intercept	y-intercept
parallel	perpendicular	point-slope	

1. The equation of a line with a slope m and y-intercept b given in _____ form is $y = mx + b$.
2. The equation of a line through (x_1, y_1) with slope m given in _____ form is $y - y_1 = m(x - x_1)$.
3. The first method used to graph lines is to _____ convenient values of x into the equation, and find the corresponding values of y.
4. The second method used to graph lines is to find the x-intercept and then _____.
5. The third method used to graph lines is to find the _____ and the y-intercept.
6. Two straight lines are _____ if they have the same slope.
7. Two straight non-vertical lines are _____ if the product of their slopes is -1.

Problems

A In each of the following problems, give the equation of the line with the given slope and y-intercept.

1. $m = \frac{2}{3}, b = 1$
2. $m = \frac{3}{4}, b = -2$
3. $m = \frac{3}{2}, b = -1$
4. $m = \frac{4}{3}, b = 2$

5. $m = -\frac{2}{3}, b = 3$
6. $m = -\frac{3}{5}, b = 4$
7. $m = 2, b = -4$
8. $m = -2, b = 4$

B Find the slope and y-intercept for each of the following equations by writing them in the form $y = mx + b$. Then, graph each equation.

9. $-2x + y = 4$

10. $-2x + y = 2$

11. $3x + y = 3$

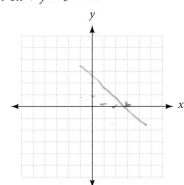

12. $3x + y = 6$

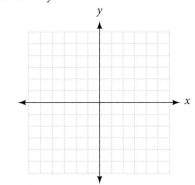

13. $3x + 2y = 6$

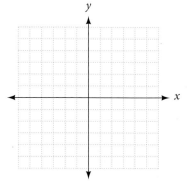

14. $2x + 3y = 6$

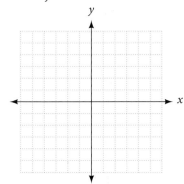

15. $4x - 5y = 20$

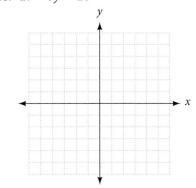

16. $2x - 5y = 10$

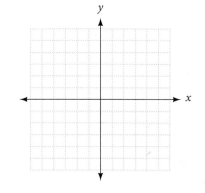

17. $-2x - 5y = 10$

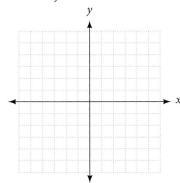

18. $-4x + 5y = 20$

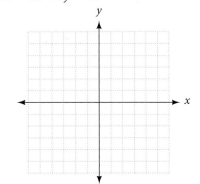

19. $-3x + 2y = -6$

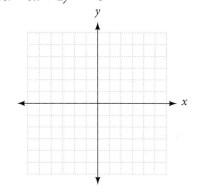

20. $2x - 3y = -6$

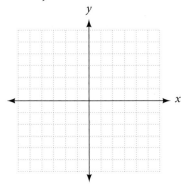

C For each of the following problems, the slope and one point on a line are given. In each case use the point-slope form to find the equation of that line. Write your answers in slope-intercept form.

21. $(-2, -5)$, $m = 2$ **22.** $(-1, -5)$, $m = 2$ **23.** $(-4, 1)$, $m = -\dfrac{1}{2}$ **24.** $(-2, 1)$, $m = -\dfrac{1}{2}$

25. $(2, -3)$, $m = \dfrac{3}{2}$ **26.** $(3, -4)$, $m = \dfrac{4}{3}$ **27.** $(-1, 4)$, $m = -3$ **28.** $(-2, 5)$, $m = -3$

29. $(3, -4)$, $m = 0$ **30.** $(-5, 2)$, $m = 0$ **31.** $(-4, -3)$, m is undefined **32.** $(-5, 1)$, m is undefined

$y = mx + b$

D Find the equation of the line that passes through each pair of points. Write your answers in slope-intercept form, if possible.

33. $(-2, -4)$, $(1, -1)$ **34.** $(2, 4)$, $(-3, -1)$ **35.** $(-1, -5)$, $(2, 1)$ **36.** $(-1, 6)$, $(1, 2)$

$$y - y_1 = m(x - x_1)$$
$$y - 1 = 2(x - 2)$$
$$y - 1 = 2x - 4 \quad = y = 2x - 3$$

37. $(-3, -2)$, $(3, 6)$ **38.** $(-3, 6)$, $(3, -2)$ **39.** $(-3, -1)$, $(3, -5)$ **40.** $(-3, -5)$, $(3, 1)$

$$\dfrac{-1 - (-5)}{-3 - 3} = \dfrac{4}{-6} \quad \dfrac{-2}{3} m$$

41. $(-4, 2)$, $(3, 2)$ **42.** $(-1, 7)$, $(1, 7)$ **43.** $(-5, 1)$, $(-5, 6)$ **44.** $(2, -3)$, $(2, 1)$

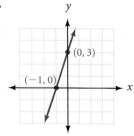

$$\dfrac{2 - 2}{4 - 3} = \dfrac{0}{-7} = 0$$
$$y = 2$$

$x = -5$
$$\dfrac{6 - 1}{-5 - 5} = \dfrac{5}{0} \text{ undefined}$$

$y = mx + b$

Find the slope and y-intercept for each line. Then write the equation of each line in slope-intercept form.

45.

46.

47.

48.

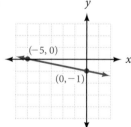

$y = mx + y$
Slope y inte

$$y - y_1 = m(x - x_1)$$
$$y + 5 = \dfrac{-2}{3}(x - 3) \quad = y = -\dfrac{2}{3}x + 2$$

The next two problems are intended to give you practice reading, and paying attention to, the instructions that accompany the problems you are working. Working these problems is an excellent way to get ready for a test or a quiz.

49. Work each problem according to the instructions given.

 a. Solve: $-2x + 1 = 6$.

 b. Write in slope-intercept form: $-2x + y = 6$.

 c. Find the y-intercept: $-2x + y = 6$.

 d. Find the slope: $-2x + y = 6$.

 e. Graph: $-2x + y = 6$.

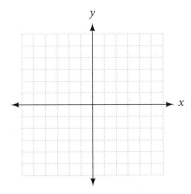

50. Work each problem according to the instructions given.

 a. Solve: $x + 3 = -6$.

 b. Write in slope-intercept form: $x + 3y = -6$.

 c. Find the y-intercept: $x + 3y = -6$.

 d. Find the slope: $x + 3y = -6$.

 e. Graph: $x + 3y = -6$.

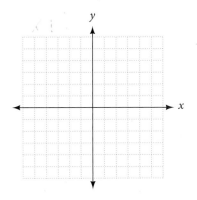

51. Find the equation of the line with x-intercept 3 and y-intercept 2.

52. Find the equation of the line with x-intercept 2 and y-intercept 3.

53. Find the equation of the line with x-intercept -2 and y-intercept -5.

$$m = \frac{-5}{2}$$
$$y = \frac{-5}{2}x - 5$$

54. Find the equation of the line with x-intercept -3 and y-intercept -5.

55. The equation of the vertical line that passes through the points $(3, -2)$ and $(3, 4)$ is either $x = 3$ or $y = 3$. Which one is it?

56. The equation of the horizontal line that passes through the points $(2, 3)$ and $(-1, 3)$ is either $x = 3$ or $y = 3$. Which one is it?

E 57. Parallel Lines Provide an equation for a line parallel to the line given by $y = 4x - 2$.

58. Parallel Lines Find the slope of a line parallel to the line that crosses the point $(3, 1)$ and has a y-intercept of 2.

59. Parallel Lines Find the slope of a line parallel to the line that passes through (2, 3) and (4, 1).

60. Perpendicular Lines Find the equation of a line perpendicular to the line given by $y = \frac{1}{3}x - 2$.

61. Perpendicular Lines If a line contains the x-intercept 4 and the y-intercept –2, find the slope of any perpendicular line.

62. Perpendicular Lines One line passes through the points (3, 4) and (–3, 1). Find the slope of a second line that is perpendicular to the first.

Applying the Concepts

63. Value of a Copy Machine Cassandra buys a new color copier for her small business. It will cost $21,000 and will decrease in value each year. The graph below shows the value of the copier after the first 5 years of ownership.

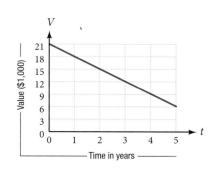

a. How much is the copier worth after 5 years?

b. After how many years is the copier worth $12,000?

c. Find the slope of this line.

d. By how many dollars per year is the copier decreasing in value?

e. Find the equation of this line where V is the value after t years.

64. Value of a Forklift Elliot buys a new forklift for his business. It will cost $140,000 and will decrease in value each year. The graph below shows the value of the forklift after the first 6 years of ownership.

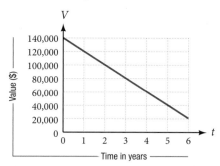

a. How much is the forklift worth after 6 years?

b. After how many years is the forklift worth $80,000?

c. Find the slope of this line.

d. By how many dollars per year is the forklift decreasing in value?

e. Find the equation of this line where V is the value after t years.

65. Salesperson's Income Kevin starts a new job in sales next month. He will earn $1,000 per month plus a certain amount for each shirt he sells. The graph below shows the amount Kevin will earn per month based on how many shirts he sells.

a. How much will he earn for selling 1,000 shirts?

b. How many shirts must he sell to earn $2,000 for a month?

c. Find the slope of this line.

d. How much money does Kevin earn for each shirt he sells?

e. Find the equation of this line where y is the amount he earns for selling x number of shirts.

66. Movie Prices The chart shows the price of movie tickets over a nine-year period. Let x represent the number of years past 2000 and y represent the prices shown on the graph. Use the first and last points on the graph to find the equation of the line that connects them. Then use the equation to estimate the price in 2011.

Find the Mistake

Each sentence below contains a mistake. Circle the mistake and write the correct word(s) on the line provided.

1. The equation $y = -\frac{4}{7}x + 2$ is in point-slope form. _____

2. For the equation $-8x + 2y = -4$, the slope is -2 and the y-intercept is 4. _____

3. The line represented by the equation $(y - 2) = 6(x - 4)$ has a y-intercept of 6. _____

4. The lines represented by the equations $y = \frac{7}{2}x + \frac{1}{3}$ and $y = -\frac{2}{7}x - 5$ are parallel. _____

3.6 Graphing Linear Inequalities in Two Variables

©iStockphoto.com/tarantas

At the U.S. National Handcar Races in Truckee, California, five-person teams face off on a railroad track. One team member gives the 1000-pound handcar a push while the other four members, standing on its flatbed, use the hand pump to propel the handcar down a 1000-foot stretch of track. The team with the fastest time wins the race. The races are held each year in honor of California's great railroad history, which aside from a few local lines began in 1869 with the completion of the First Continental Railroad. The handcar was once used for inspection and maintenance of the tracks, before an engine-powered version replaced it.

Let's assume that in 2009 the fastest handcar race time was 34 seconds, which was 7 seconds slower than the fastest time in 2010. Using this information, we can create an inequality that includes all the race times for both years. In this section, we will work with graphs of linear inequalities, such as this one.

A Graphing Linear Inequalities in Two Variables

A linear inequality in two variables is any expression that can be put in the form

$$Ax + By < C$$

where A, B, and C are real numbers (A and B not both 0). The inequality symbol can be any of the following four: $<, \leq, >, \geq$.

Some examples of linear inequalities are

$$2x + 3y < 6 \qquad y \geq 2x + 1 \qquad x - y \leq 0$$

Although not all of these inequalities have the form $Ax + By < C$, each one can be put in that form with the appropriate symbol.

In an earlier chapter, we graphed the solution set for a linear inequality in one variable on a number line. The solution set for a linear inequality in two variables is graphed as a section of the coordinate plane. The boundary for the section is found by replacing the inequality symbol with an equals sign and graphing the resulting equation using the methods summarized in the previous section. The boundary is included in the solution set (and represented with a solid line) if the inequality symbol used originally is \leq or \geq. The boundary is not included (and is represented with a broken line) if the original symbol is $<$ or $>$.

EXAMPLE 1 Graph the linear inequality $x + y \leq 4$.

Solution The boundary for the graph is the graph of $x + y = 4$. The boundary (a solid line) is included in the solution set because the inequality symbol is \leq.

The graph of the boundary is shown in Figure 1.

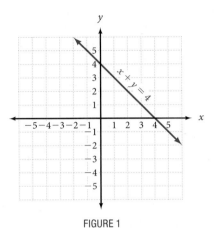

FIGURE 1

The boundary separates the coordinate plane into two sections, or regions: the region above the boundary and the region below the boundary. The solution set for $x + y \leq 4$ is one of these two regions along with the boundary. To find the correct region, we simply choose any convenient point that is *not* on the boundary. We then substitute the coordinates of the point into the original inequality $x + y \leq 4$. If the point we choose satisfies the inequality, then it is a member of the solution set, and we can assume that all points on the same side of the boundary as the chosen point are also in the solution set. If the coordinates of our point do not satisfy the original inequality, then the solution set lies on the other side of the boundary.

In this example, a convenient point not on the boundary is the origin. Substituting $(0, 0)$ into $x + y \leq 4$ gives us

$$0 + 0 \overset{?}{\leq} 4$$

$$0 \leq 4 \qquad \text{A true statement}$$

Because the origin is a solution to the inequality $x + y \leq 4$, and the origin is below the boundary, all other points below the boundary are also solutions.

The graph of $x + y \leq 4$ is shown in Figure 2.

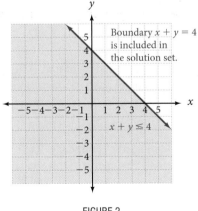

FIGURE 2

The region above the boundary is described by the inequality $x + y > 4$.

Practice Problems

1. Graph the linear inequality $x + y \geq 3$.

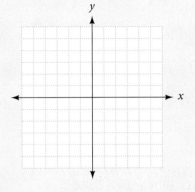

Here is a list of steps to follow when graphing the solution set for linear inequalities in two variables:

> **HOW TO** Graph the Solution Set for Linear Inequalities in Two Variables
>
> **Step 1:** Replace the inequality symbol with an equals sign. The resulting equation represents the boundary for the solution set.
>
> **Step 2:** Graph the boundary found in step 1 using a *solid line* if the boundary is included in the solution set (that is, if the original inequality symbol was either ≤ or ≥). Use a *broken line* to graph the boundary if it is *not* included in the solution set; that is, if the original inequality was either < or >.
>
> **Step 3:** Choose any convenient point not on the boundary and substitute the coordinates into the *original* inequality. If the resulting statement is *true*, the graph lies on the *same* side of the boundary as the chosen point. If the resulting statement is *false*, the solution set lies on the *opposite* side of the boundary.

EXAMPLE 2 Graph the linear inequality $y < 2x - 3$.

Solution The boundary is the graph of $y = 2x - 3$. The boundary is not included because the original inequality symbol is $<$. We therefore use a broken line to represent the boundary, as shown in Figure 3.

A convenient test point is again the origin. Using (0, 0) in $y < 2x - 3$, we have

$$0 \overset{?}{<} 2(0) - 3$$

$$0 < -3 \qquad \text{A false statement}$$

FIGURE 3

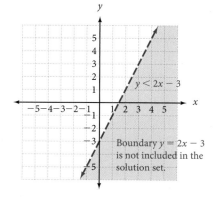

FIGURE 4

Because our test point gives us a false statement and it lies above the boundary, the solution set must lie on the other side of the boundary, as shown in Figure 4.

EXAMPLE 3 Graph the linear inequality $2x + 3y \le 6$.

Solution We begin by graphing the boundary $2x + 3y = 6$. The boundary is included in the solution because the inequality symbol is \le.

If we use (0, 0) as our test point, we see that it yields a true statement when its coordinates are substituted into $2x + 3y \le 6$. The graph, therefore, lies below the boundary, as shown in Figure 5.

2. Graph the linear inequality for
$y < \frac{1}{2}x + 3$.

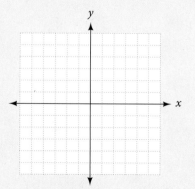

3. Graph the linear inequality
$3x - 2y \le 6$.

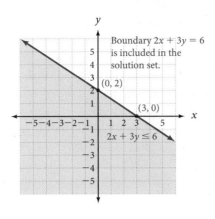

FIGURE 5

The ordered pair (0, 0) is a solution to $2x + 3y \leq 6$; all points on the same side of the boundary as (0, 0) also must be solutions to the inequality $2x + 3y \leq 6$.

EXAMPLE 4 Graph the linear inequality $x \leq 5$.

Solution The boundary is $x = 5$, which is a vertical line. All points to the left have x-coordinates less than 5, and all points to the right have x-coordinates greater than 5, as shown in Figure 6.

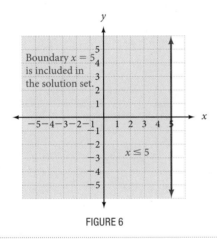

FIGURE 6

4. Graph the linear inequality $y > -2$.

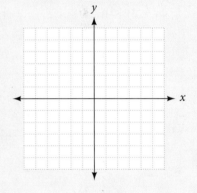

GETTING READY FOR CLASS

After reading through the preceding section, respond in your own words and in complete sentences.

A. When graphing a linear inequality in two variables, how do you find the equation of the boundary line?

B. What is the significance of a broken line in the graph of an inequality?

C. When graphing a linear inequality in two variables, how do you know which side of the boundary line to shade?

D. Why is the coordinate (0, 0) a convenient test point?

Vocabulary Review

Choose the correct words to fill in the blanks below.

| boundary | solution set | broken | linear inequality | solid |

1. A _____ in two variables is any expression that can be put in the form $Ax + Bx < C$, where the inequality symbol can be $<$, $>$, \geq, or \leq.

2. The _____ for a linear inequality is a section of the coordinate plane.

3. The _____ for the solution set of a linear inequality can be found by replacing the inequality symbol with an equals sign and graphing the resulting equation.

4. Use a _____ line for the boundary of a linear inequality if the original inequality symbol was \geq or \leq.

5. Use a _____ line for the boundary of a linear inequality if the original inequality symbol was $>$ or $<$.

Problems

A Graph the following linear inequalities.

1. $2x - 3y < 6$

2. $3x + 2y \geq 6$

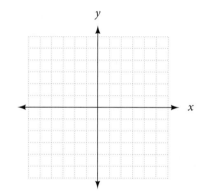

3. $x - 2y \leq 4$

4. $2x + y > 4$

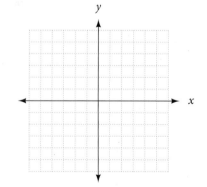

5. $x - y \leq 2$

6. $x - y \leq 1$

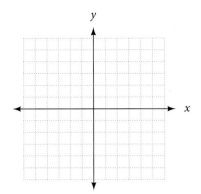

7. $3x - 4y \geq 12$

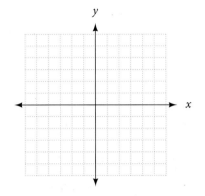

8. $4x + 3y < 12$

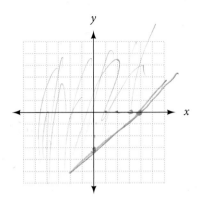

9. $5x - y \leq 5$

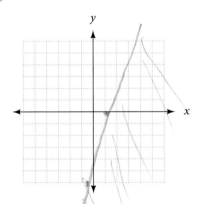

10. $4x + y > 4$

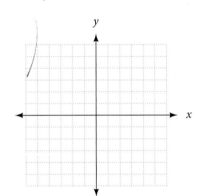

11. $2x + 6y \leq 12$

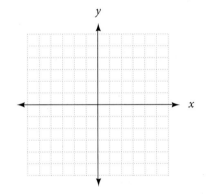

12. $x - 5y > 5$

13. $x \geq 1$

14. $x < 5$

15. $x \geq -3$

16. $y \leq -4$

17. $y < 2$

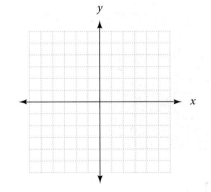

18. $3x - y > 1$

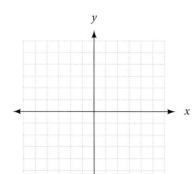

19. $2x + y > 3$

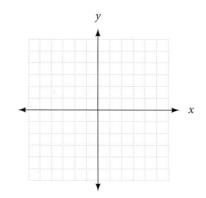

20. $5x + 2y < 2$

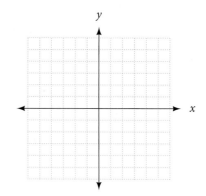

21. $y \leq 3x - 1$

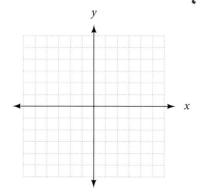

22. $y \geq 3x + 2$

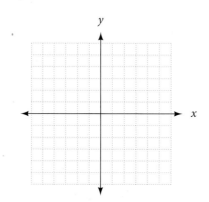

23. $y \leq -\frac{1}{2}x + 2$

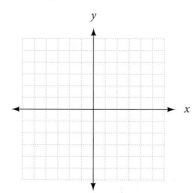

24. $y < \frac{1}{3}x + 3$

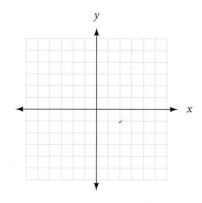

The next two problems are intended to give you practice reading, and paying attention to, the instructions that accompany the problems you are working.

25. Work each problem according to the instructions.

 a. Solve: $4 + 3y < 12$.

 b. Solve: $4 - 3y < 12$.

 c. Solve for y: $4x + 3y = 12$.

 d. Graph: $y < -\dfrac{4}{3}x + 4$.

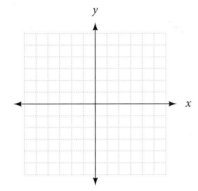

26. Work each problem according to the instructions.

 a. Solve: $3x + 2 \geq 6$.

 b. Solve: $-3x + 2 \geq 6$.

 c. Solve for y: $3x + 2y = 6$.

 d. Graph: $y \geq -\dfrac{3}{2}x + 3$.

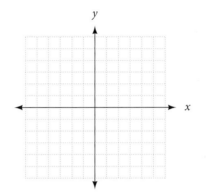

27. Find the equation of the line in part a, then use this information to find the inequalities for the graphs on parts b and c.

a.

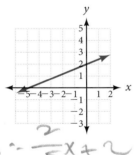

$$y = \frac{2}{5}x + 2$$

b.

c.

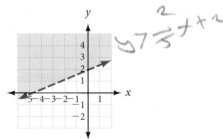

$$y \geq \frac{2}{5}x + 2$$

28. Find the equation of the line in part a, then use this information to find the inequalities for the graphs on parts b and c.

a.

b.

c.

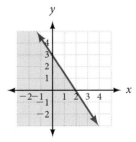

Review Problems

29. Simplify the expression $7 - 3(2x - 4) - 8$.

30. Find the value of $x^2 - 2xy + y^2$ when $x = 3$ and $y = -4$.

Solve each equation.

31. $-\dfrac{3}{2}x = 12$

32. $2x - 4 = 5x + 2$

33. $8 - 2(x + 7) = 2$

34. $3(2x - 5) - (2x - 4) = 6 - (4x + 5)$

35. Solve the formula $P = 2l + 2w$ for w.

36. Solve the formula $C = 2\pi r$ for r.

Solve each inequality, and graph the solution on a number line.

37. $-4x < 20$

38. $3 - 2x > 5$

39. $3 - 4(x - 2) \geq -5x + 6$

40. $4 - 6(x + 1) < -2x + 3$

41. Solve the inequality $3x - 2y \leq 12$ for y.

42. What number is 12% of 2,000?

43. **Geometry** The length of a rectangle is 5 inches more than 3 times the width. If the perimeter is 26 inches, find the length and width.

44. The length of a rectangle is 1 inch less than twice the width. If the perimeter is 28 inches, find the length and width.

Find the Mistake

Each sentence below contains a mistake. Circle the mistake and write the correct word(s) on the line provided.

1. The solution set for the linear inequality $y < 6x - \dfrac{3}{5}$ contains the points through which the boundary line passes.

2. The boundary for the linear inequality $2 - 8y \geq x + 6$ is represented by a broken line. _____

3. If a test point substituted into an original inequality results in a false statement, the solution set for the inequality lies on the same side as the chosen point. _____

4. The solution set for the linear inequality $x < -8$, includes all points to the right of the line. _____

A Place to Call Home

Supplies Needed

A Piece of graph paper

B A pen or pencil

C A ruler

D Access to the internet

Building a town from scratch is an intricate project. Research zoning codes for your hometown. Based on your research, what qualities would contribute to a successful town layout? Now imagine you have been commissioned to build and live in a new town. You will need to include the following businesses and community services:

Ambulance station Bank

Fire station Grocery store

Police station Post office

Hospital Community park

Courthouse Your house

On a piece of graph paper, draw an x- and y-axis with the origin in the center of the paper. Each square is equivalent to one standard city block. Plot each of the above services on your graph. Log the coordinates of each service. Explain your reasoning behind the placement of each service.

Now find the distance between the locations of the following pairings, as well as the equation of the line that passes through each:

1. Police Station and Courthouse

2. Ambulance and Hospital

3. Grocery Store and Community Park

4. Bank and Post Office

5. Fire Station and Your House

EXAMPLES

1. The equation $3x + 2y = 6$ is an example of a linear equation in two variables.

Linear Equation in Two Variables [3.2]

A linear equation in two variables is any equation that can be put in the form $Ax + By = C$. The graph of every linear equation is a straight line.

Strategy for Graphing Linear Equations in Two Variables By Plotting Points [3.2]

2. The graph of $y = -\frac{2}{3}x - 1$ is shown below.

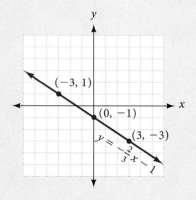

Step 1: Find any three ordered pairs that satisfy the equation. This can be done by using a convenient number for one variable and solving for the other variable.

Step 2: Graph the three ordered pairs found in step 1. Actually, we need only two points to graph a straight line. The third point serves as a check. If all three points do not line up, there is a mistake in our work.

Step 3: Draw a straight line through the three points graphed in step 2.

Intercepts [3.3]

3. To find the x-intercept for $3x + 2y = 6$, we let $y = 0$ and get

$$3x = 6$$
$$x = 2$$

In this case the x-intercept is 2, and the graph crosses the x-axis at $(2, 0)$.

The x-intercept of an equation is the x-coordinate of the point where the graph intersects the x-axis. The y-intercept is the y-coordinate of the point where the graph intersects the y-axis. We find the y-intercept by substituting $x = 0$ into the equation and solving for y. The x-intercept is found by letting $y = 0$ and solving for x.

Slope of a Line [3.4]

4. The slope of the line through $(3, -5)$ and $(-2, 1)$ is

$$m = \frac{-5 - 1}{3 - (-2)} = \frac{-6}{5} = -\frac{6}{5}$$

The *slope* of the line containing the points (x_1, y_1) and (x_2, y_2) is given by

$$\text{Slope} = m = \frac{y_2 - y_1}{x_2 - x_1} = \frac{\text{rise}}{\text{run}}$$

Slope-Intercept Form of a Straight Line [3.5]

5. The equation of the line with a slope of 2 and a y-intercept of 5 is

$$y = 2x + 5$$

The equation of the non-vertical line with a slope of m and a y-intercept of b is

$$y = mx + b$$

Point-Slope Form of a Straight Line [3.5]

If a non-vertical line has a slope of m and contains the point (x_1, y_1), the equation can be written as

$$y - y_1 = m(x - x_1)$$

6. The equation of the line through $(1, 2)$ with a slope of 3 is
$$y - 2 = 3(x - 1)$$
$$y - 2 = 3x - 3$$
$$y = 3x - 1$$

To Graph a Linear Inequality in Two Variables [3.6]

Step 1: Replace the inequality symbol with an equals sign. The resulting equation represents the boundary for the solution set.

Step 2: Graph the boundary found in step 1, using a *solid line* if the original inequality symbol was either \leq, or \geq. Use a *broken line* otherwise.

Step 3: Choose any convenient point not on the boundary and substitute the coordinates into the *original* inequality. If the resulting statement is *true*, the graph lies on the *same* side of the boundary as the chosen point. If the resulting statement is *false*, the solution set lies on the *opposite* side of the boundary.

7. Graph $x - y \geq 3$.

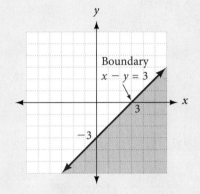

Graph the ordered pairs. [3.1]

1. $(2, -1)$

2. $\left(\frac{1}{2}, 0\right)$

3. $(-3, 5)$

4. $(1.2, -2)$

5. Fill in the following ordered pairs for the equation $2x - 5y = 10$. [3.2]

$$\left(3, \quad\right) \left(-2, \quad\right) \left(\quad, \frac{2}{3}\right) \left(\frac{1}{2}, \quad\right)$$

6. Which of the following ordered pairs are solutions to $y = 0.8x - 8$? [3.2]

$(10, 0) \quad (7, -2.2) \quad (8, -1.6) \quad (3, -5.8)$

Graph each line. [3.3]

7. $3x - y = 2$

8. $-y = 3$

Find the x- and y-intercepts. [3.3]

9. $2x - 12y = 12$

10. $0.3x - 2 = y$

11.

12.

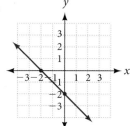

Find the slope of the line through each pair of points. [3.4]

13. $(2, 4), (-1, 3)$

14. $(0, 2), (-3, -4)$

Find the slope of each line. [3.4]

15.

16.

17.

18.

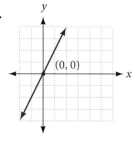

19. Find the equation of the line through $(3, -1)$ with a slope of 2. [3.5]

20. Find the equation of the line with a slope of -3 and y-intercept 6. [3.5]

21. Find the equation of the line passing through the points $(2, 2)$ and $(-3, 1)$. [3.5]

22. A straight line has an x-intercept 3 and contains the point $(-2, 6)$. Find its equation. [3.5]

23. Write the equation, in slope-intercept form, of the line whose graph is shown below. [3.5]

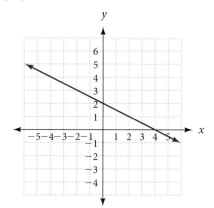

Give the slope and y-intercept of each equation. [3.5]

24. $4x - y = 3$

25. $4x - 5y = 10$

Find the equation of the line that contains the given point and has the given slope. [3.5]

26. $(3, 5), m = 3$

27. $(-2, 5), m = \frac{3}{2}$

Graph each linear inequality in two variables. [3.6]

28. $y < 4x + 2$

29. $3x - 8y \geq 12$

Simplify.

1. $9 - 5 \cdot 2$

2. $3 + 2(5 - 7)$

3. $\dfrac{2}{5}(15) - \dfrac{1}{3}(12)$

4. $2 \cdot 3 + 18 \div 9 - 2^4$

5. $(3 - 7)(-2 - 6)$

6. $2[3 + (-7)] - 4(8)$

7. $\dfrac{19 + (-41)}{8 - 8}$

8. $\dfrac{(6 - 3)^2}{6^2 - 3^2}$

9. $\dfrac{90}{126}$

10. $\dfrac{0}{-14}$

11. $4(9x)$

12. $6a + 9 - 3a - 15$

Solve each equation.

13. $7x - 3 - 6x + 9 = 4$

14. $6x + 9 = -15$

15. $-3x + 8 = 2x + 3$

16. $3(3 - t) - 2 = -5$

17. $2 = -\dfrac{1}{3}(2x - 3)$

18. $0.06 + 0.03(300 - x) = 12$

Solve each inequality, and graph the solution.

19. $-7x < 42$

20. $7 - 3x \le -14$

Graph on a rectangular coordinate system.

21. $y = 3x - 2$

22. $2x + 3y = 6$

23. $y = 3x$

24. $y = -\dfrac{1}{4}x$

25. $3x - 4y > 6$

26. $y < 5$

27. Graph the points $(-4, -1)$ and $(1, 4)$, and draw a line that passes through both of them. Does $(2, -5)$ lie on the line? Does $(5, 8)$?

28. Graph the line through $(3, -12)$ with x-intercept 1. What is the y-intercept?

29. Find the x- and y-intercepts for $3x + 2y = 12$, and then graph the line.

30. Graph the line with slope $\dfrac{3}{5}$ and y-intercept -3.

31. Find the slope of the line through $(1, 2)$ and $(5, -2)$.

32. Find the equation of the line with slope -2 and y-intercept 5.

33. Find the slope of the line $6x - 2y = 18$.

34. Find the equation of the line with slope $\dfrac{2}{3}$ that passes through $(-3, -2)$. Write your answer in slope-intercept form.

35. Find the equation of the line through $(4, -2)$ and $(2, -5)$. Write your answer in slope-intercept form.

36. Find the equation of the line with x-intercept 5, y-intercept -3. Write your answer in slope-intercept form.

37. Complete the ordered pairs (, 1) and (0,) for $4x - 5y = 15$.

38. Which of the ordered pairs $(2, 3)$, $(-2, -5)$, and , $(6, 1)$ are solutions to $2x + 4y = 16$?

39. Write the expression, and then simplify: the difference of 24 and twice 6.

40. Twelve added to the product of 8 and -4 is what number?

Give the opposite, reciprocal, and absolute value of the given number.

41. $-\dfrac{3}{5}$

42. 4

43. Evaluate $x^2 - 3x - 9$ when $x - 3$.

44. Use the formula $6x + 5y = 11$ to find y when $x = 1$.

45. Geometry The length of a rectangle is 6 inches more than the width. The perimeter is 40 inches. Find the length and width.

46. Coin Problem Brandon has $3.05 in dimes and quarters. If he has 6 more dimes than quarters, how many of each coin does he have?

Use the line graph to answer the following questions.

DISNEYLAND ATTENDANCE OVER THE YEARS
TEA/ERA Attendance Report

47. Find the slope of the line segment connecting the data point for 1955 and 1970, and explain its significance.

48. Find the slope of the line segment connecting the data point for 1970 and 2010, and explain its significance.

Graph the ordered pairs. [3.1]

1. $(2, -1)$ **2.** $(-4, 3)$

3. $(-3, -2)$ **4.** $(0, -4)$

5. Fill in the following ordered pairs for the equation
$3x - 2y = 6$. [3.2]

$$(0, \quad) \; (\quad, 0) \; (4, \quad) \; (\quad, -6)$$

6. Which of the following ordered pairs are solutions to
$y = -3x + 7$? [3.2]

$$(0, 7) \; (2, -1) \; (4, -5) \; (-5, -3)$$

Graph each line. [3.3]

7. $y = -\dfrac{1}{2}x + 4$ **8.** $x = -3$

Find the x- and y-intercepts. [3.3]

9. $8x - 4y = 16$ **10.** $y = \dfrac{3}{2}x + 6$

11.

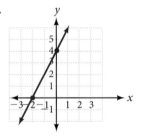

Find the slope of the line through each pair of points. [3.4]

12. $(3, 2), (-5, 6)$ **13.** $(0, 9), (7, 1)$

Find the slope of each line. [3.4]

14. **15.**

16.

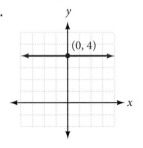

17. Find the equation of the line through $(4, 1)$ with a slope
of $-\frac{1}{2}$. [3.5]

18. Find the equation of the line with a slope of 3 and
y-intercept -5. [3.5]

19. Find the equation of the line passing through the points
$(3, -4)$ and $(-6, 2)$. [3.5]

20. A straight line has an x-intercept -2 and passes through
the point $(-4, 6)$. Find its equation. [3.5]

21. Write the equation, in slope-intercept form, of the line
whose graph is shown below. [3.5]

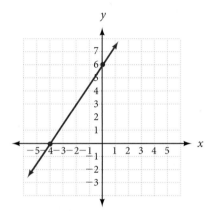

Find the equation for each line. [3.5]

22. The line through $(-6, 1)$ that has slope $m = -3$

23. The line through $(6, 8)$ and $(-3, 2)$

24. The line which contains the point $(-4, 1)$ and is
perpendicular to the line $y = -\frac{2}{3}x + 4$

25. The line which contains the point $(0, -6)$ and is parallel
to the line $4x - 2y = 3$

Graph each linear inequality in two variables. [3.6]

26. $y > x - 6$ **27.** $6x - 9y \le 18$

Exponents and Polynomials

4

Lake Baikal, Siberia
Image © TerraMetrics
© 2010 Cnes/ Spot Image

Google

Lake Baikal is the largest natural lake in the world. Located in the southeastern part of Siberia, it is also the deepest freshwater lake on Earth and contains the largest amount of fresh surface water. Experts estimate that nearly 20% of the world's fresh water lies within its banks. With its beginnings dating back 31 million years, it is also considered one of the oldest lakes on the planet. Despite its age, Lake Baikal is still expanding at a rate of 2 centimeters per year. Experts believe that in several million years this expansion will result in a new great ocean covering Asia.

©iStockphoto.com/himn

Lake Baikal has 26 islands of varying sizes. Suppose one of these islands was the shape of a rectangle; one side was x^2 meters long, and the other side was x^3 meters long. How would you find the surface area of this island? The solution to this problem requires multiplication of exponents, which is one of the topics we will discuss in this chapter.

4.1 Multiplication with Exponents and Scientific Notation

OBJECTIVES

A Use the definition of integer exponents to evaluate expressions containing exponents.

B Use the product property of exponents to simplify expressions.

C Use the power property of exponents to simplify expressions.

D Use the distributive property of exponents to simplify expressions.

E Write numbers greater than 10 in scientific notation and standard form.

KEY WORDS

product property of exponents

power property of exponents

distributive property of exponents

volume

rectangular solid

scientific notation

Image © Katherine Heistand Shields, 2010

During a hike up Bishop's Peak in San Luis Obispo, CA, a thirty-year-old hiker's heart beat 145 times per minute. If he hiked up the mountain and back down again in 4.5 hours, his heart beat approximately 39,150 times. In this section, we will learn how to write this large number in scientific notation. For instance, 39,150 can be written as 3.915×10^4. But first, we must learn to evaluate exponents and work with properties of exponents to further understand their meaning in this notation.

A Evaluating Exponents

Recall that an *exponent* is a number written just above and to the right of another number, which is called the *base*. In the expression 5^2, for example, the exponent is 2 and the base is 5. the expression 5^2 is read "5 to the second power" or "5 squared." The meaning of the expression is

$$5^2 = 5 \cdot 5 = 25$$

In the expression 5^3, the exponent is 3 and the base is 5. The expression 5^3 is read "5 to the third power" or "5 cubed." The meaning of the expression is

$$5^3 = 5 \cdot 5 \cdot 5 = 125$$

Here are some further examples:

EXAMPLE 1 Write each expression as a single number.

a. 4^3 **b.** -3^4 **c.** $(-2)^5$ **d.** $\left(-\dfrac{3}{4}\right)^2$

Solution

a. $4^3 = 4 \cdot 4 \cdot 4 = 16 \cdot 4 = 64$ Exponent 3, base 4

b. $-3^4 = -3 \cdot 3 \cdot 3 \cdot 3 = -81$ Exponent 4, base 3

c. $(-2)^5 = (-2)(-2)(-2)(-2)(-2) = -32$ Exponent 5, base -2

d. $\left(-\dfrac{3}{4}\right)^2 = \left(-\dfrac{3}{4}\right)\left(-\dfrac{3}{4}\right) = \dfrac{9}{16}$ Exponent 2, base $-\dfrac{3}{4}$

Question: In what way are $(-5)^2$ and -5^2 different?
Answer: In the first case, the base is -5. In the second case, the base is 5. The answer to the first is 25. The answer to the second is -25. Can you tell why? Would there be a difference in the answers if the exponent in each case were changed to 3? To clarify whether the base of an exponential expression contains a negative, it is helpful to enclose the base in parenthesis as we have done with $(-5)^2$.

We can simplify our work with exponents by developing some properties of exponents. We want to list the things we know are true about exponents and then use these properties to simplify expressions that contain exponents.

B Product Property of Exponents

The first property of exponents applies to products with the same base. We can use the definition of exponents, as indicating repeated multiplication, to simplify expressions like $7^4 \cdot 7^2$.

$$7^4 \cdot 7^2 = (7 \cdot 7 \cdot 7 \cdot 7)(7 \cdot 7)$$
$$= (7 \cdot 7 \cdot 7 \cdot 7 \cdot 7 \cdot 7)$$
$$= 7^6 \qquad \text{Note: } 4 + 2 = 6.$$

As you can see, multiplication with the same base resulted in addition of exponents. We can summarize this result with the following property.

> **PROPERTY** Product Property of Exponents
>
> If a is any real number and r and s are integers, then
>
> $$a^r \cdot a^s = a^{r+s}$$
>
> *In words:* To multiply two expressions with the same base, add exponents and use the common base.

Here is an example using the product property.

EXAMPLE 2 Use the product property to simplify the following expressions. Leave your answers in terms of exponents.
a. $5^3 \cdot 5^6$ **b.** $x^7 \cdot x^8$ **c.** $3^4 \cdot 3^8 \cdot 3^5$

Solution

a. $5^3 \cdot 5^6 = 5^{3+6} = 5^9$

b. $x^7 \cdot x^8 = x^{7+8} = x^{15}$

c. $3^4 \cdot 3^8 \cdot 3^5 = 3^{4+8+5} = 3^{17}$

C Power Property of Exponents

Another common type of expression involving exponents is one in which an expression containing an exponent is raised to another power. The expression $(5^3)^2$ is an example.

$$(5^3)^2 = (5^3)(5^3)$$
$$= 5^{3+3}$$
$$= 5^6 \qquad \text{Note: } 3 \cdot 2 = 6.$$

This result offers justification for the power property of exponents.

> **PROPERTY** Power Property of Exponents
>
> If a is any real number and r and s are integers, then
>
> $$(a^r)^s = a^{r \cdot s}$$
>
> *In words:* A power raised to another power is the base raised to the product of the powers.

2. Simplify, leaving your answers in terms of exponents.
a. $5^4 \cdot 5^5$
b. $x^3 \cdot x^7$
c. $4^5 \cdot 4^2 \cdot 4^6$

Note In Example 2, notice that each base in the original problem is the same base that appears in the answer and that it is written only once in the answer. A very common mistake that people make when they first begin to use the product property is to write a 2 in front of the base in the answer. For example, people making this mistake would get $2x^{15}$ or $(2x)^{15}$ as the result in Example 2b. To avoid this mistake, you must be sure you understand the meaning of the product property exactly as it is written.

3. Simplify.

 a. $(3^4)^5$

 b. $(x^7)^2$

EXAMPLE 3 Simplify the following expressions:

a. $(4^5)^6$ **b.** $(x^3)^5$

Solution

a. $(4^5)^6 = 4^{5 \cdot 6} = 4^{30}$

b. $(x^3)^5 = x^{3 \cdot 5} = x^{15}$

D Distributive Property of Exponents

The third property of exponents applies to expressions in which the product of two or more numbers or variables is raised to a power. Let's look at how the expression $(2x)^3$ can be simplified:

$$(2x)^3 = (2x)(2x)(2x)$$

$$= (2 \cdot 2 \cdot 2)(x \cdot x \cdot x)$$

$$= 2^3 \cdot x^3 \qquad \text{Note: The exponent 3 distributes over the product } 2x.$$

$$= 8x^3$$

We can generalize this result into a third property of exponents.

> **PROPERTY** **Distributive Property of Exponents**
>
> If a and b are any two real numbers and r is an integer, then
>
> $$(ab)^r = a^r b^r$$
>
> *In words:* The power of a product is the product of the powers.

> *Note* A common mistake that students make when using this property is to not distribute the exponent to both parts of the base as in
>
> $$(2x)^3 \neq 2x^3$$
>
> Be careful to follow the property correctly, as we have shown here.

Here are some examples using the distributive property of exponents to simplify expressions.

4. Simplify.

 a. $\left(-\dfrac{1}{3}x^3y^2\right)^2$

 b. $(x^5)^2(x^4)^3$

 c. $(3y)^2(2y^3)$

 d. $(3x^3y^2)^2(2xy^4)^3$

EXAMPLE 4 Simplify the following expressions:

a. $\left(-\dfrac{1}{4}x^2y^3\right)^2$ **b.** $(x^4)^3(x^2)^5$ **c.** $(2y)^3(3y^2)$ **d.** $(2x^2y^5)^3(3x^4y)^2$

Solution

a. $\left(-\dfrac{1}{4}x^2y^3\right)^2 = \left(-\dfrac{1}{4}\right)^2(x^2)^2(y^3)^2$ Distributive property of exponents

$$= \dfrac{1}{16}x^4y^6 \qquad\qquad\qquad \text{Power property of exponents}$$

b. $(x^4)^3(x^2)^5 = x^{12} \cdot x^{10}$ Power property of exponents

$$= x^{22} \qquad\qquad\qquad \text{Product property of exponents}$$

c. $(2y)^3(3y^2) = 2^3y^3(3y^2)$ Distributive property of exponents

$$= 8 \cdot 3(y^3 \cdot y^2) \qquad \text{Commutative and associative properties}$$

$$= 24y^5 \qquad\qquad\quad \text{Product property of exponents}$$

d. $(2x^2y^5)^3(3x^4y)^2 = 2^3(x^2)^3(y^5)^3 \cdot 3^2(x^4)^2y^2$ Distributive property of exponents

$$= 8x^6y^{15} \cdot 9x^8y^2 \qquad\qquad\qquad \text{Power property of exponents}$$

$$= (8 \cdot 9)(x^6x^8)(y^{15}y^2) \qquad\quad \text{Commutative and associative properties}$$

$$= 72x^{14}y^{17} \qquad\qquad\qquad\qquad \text{Product property of exponents}$$

Answers

3. a. 3^{20} **b.** x^{14}

4. a. $\dfrac{1}{9}x^6y^4$ **b.** x^{22} **c.** $18y^5$

 d. $72x^9y^{16}$

FACTS FROM GEOMETRY Volume of a Rectangular Solid

It is easy to see why the phrase "five squared" is associated with the expression 5^2. Simply find the area of the square shown in Figure 1 with a side of 5.

FIGURE 1 FIGURE 2

To see why the phrase "five cubed" is associated with the expression 5^3, we have to find the *volume* of a cube for which all three dimensions are 5 units long. The volume of a cube is a measure of the space occupied by the cube. To calculate the volume of the cube shown in Figure 2, we multiply the three dimensions together to get $5 \cdot 5 \cdot 5 = 5^3$.

The cube shown in Figure 2 is a special case of a general category of three dimensional geometric figures called *rectangular solids*. Rectangular solids have rectangles for sides, and all connecting sides meet at right angles. The three dimensions are length, width, and height. To find the volume of a rectangular solid, we find the product of the three dimensions, as shown in Figure 2.

> *Note* If we include units with the dimensions of the diagrams, then the units for the area will be square units and the units for volume will be cubic units. More specifically,
>
> If a square has a side 5 inches long, then its area will be
> $$A = (5 \text{ inches})^2 = 25 \text{ inches}^2$$
> where the unit inches2 stands for square inches.
>
> If a cube has a single side 5 inches long, then its volume will be
> $$V = (5 \text{ inches})^3 = 125 \text{ inches}^3$$
> where the unit inches3 stands for cubic inches.
>
> If a rectangular solid has a length of 5 inches, a width of 4 inches, and a height of 3 inches, then its volume is
> $$V = (5 \text{ inches})(4 \text{ inches})(3 \text{ inches})$$
> $$= 60 \text{ inches}^3$$

E Scientific Notation for Large Numbers

Many branches of science require working with very large numbers. In astronomy, for example, distances commonly are given in light-years. A light-year is the distance light travels in a year. It is approximately

$$5{,}880{,}000{,}000{,}000 \text{ miles}$$

This number is difficult to use in calculations because of the number of zeros it contains. Scientific notation provides a way of writing very large numbers in a more manageable form.

DEFINITION scientific notation

A number is in *scientific notation* when it is written as the product of a number between 1 and 10 and an integer power of 10. A number written in scientific notation has the form

$$n \times 10^r$$

where $1 \le n < 10$ and $r =$ an integer.

EXAMPLE 5 Write 376,000 in scientific notation.

Solution We must rewrite 376,000 as the product of a number between 1 and 10 and a power of 10. To do so, we move the decimal point 5 places to the left so that it appears between the 3 and the 7. Then we multiply this number by 10^5. The number that results has the same value as our original number and is written in scientific notation.

$$376{,}000 = 3.76 \times 10^5$$

Moved 5 places.

Decimal point was originally here.

Keeps track of the 5 places we moved the decimal point.

5. Write 4,810,000 in scientific notation.

Answer

5. 4.81×10^6

6. Write 5,473,000 in scientific notation.

EXAMPLE 6 Write 4,758,000 in scientific notation.

Solution As with the previous example, we must move the decimal to the left but in this case, we move it 6 places. It now appears between the 4 and the 7. We round to the nearest thousandth and multiply our result by 10^6.

$$4{,}758{,}000 = 4.758 \times 10^6$$

Moved 6 places.

Decimal point was originally here.

Keeps track of the 6 places we moved the decimal point.

Our answer has the same value as our original number and is written in scientific notation.

7. Write 3.05×10^5 in expanded form.

EXAMPLE 7 Write 4.52×10^3 in expanded form.

Solution Since 10^3 is 1,000, we can think of this as simply a multiplication problem; that is,

$$4.52 \times 10^3 = 4.52 \times 1{,}000 = 4{,}520$$

On the other hand, we can think of the exponent 3 as indicating the number of places we need to move the decimal point to write our number in expanded form. Since our exponent is positive 3, we move the decimal point three places to the right.

$$4.52 \times 10^3 = 4{,}520$$

In the next section, we will see how a negative power of 10 affects our answers in scientific notation.

GETTING READY FOR CLASS

After reading through the preceding section, respond in your own words and in complete sentences.

A. Explain the difference between -5^2 and $(-5)^2$.

B. How do you multiply two expressions containing exponents when they each have the same base?

C. Explain in words how you would use the product property and the distributive property of exponents to simplify $(2x^3)^2$.

D. How would you write 5,190,000 in scientific notation?

Answers

6. 5.473×10^6

7. 305,000

Vocabulary Review

Choose the correct words to fill in the blanks below.

 power distributive scientific notation

 base product exponent

1. In the expression 7^3, the 7 is called the _____ and the 3 is called the _____.

2. The _____ property for exponents states that you should add exponents and use the common base to multiply two expressions with the same base.

3. The _____ property for exponents states that a power raised to another power is the base raised to the product of the powers.

4. The _____ property for exponents states that the power of a product is the product of the powers.

5. _____ is a way of writing very large numbers that is more manageable and takes the form $n \times 10^r$.

Problems

A Name the base and exponent in each of the following expressions. Then use the definition of exponents as repeated multiplication to simplify.

1. 4^2

2. 6^2

3. $(0.3)^2$

4. $(0.03)^2$

5. 4^3

6. 10^3

7. $(-5)^2$

8. -5^2

9. -2^3

10. $(-2)^3$

11. 3^4

12. $(-3)^4$

13. $\left(\dfrac{2}{3}\right)^2$

14. $\left(\dfrac{2}{3}\right)^3$

15. $\left(\dfrac{1}{2}\right)^4$

16. $\left(\dfrac{4}{5}\right)^2$

17. $\left(-\dfrac{3}{4}\right)^2$

18. $\left(-\dfrac{2}{3}\right)^4$

19. $\left(\dfrac{1}{6}\right)^2$

20. $-\left(\dfrac{1}{9}\right)^2$

21. a. Complete the following table.

Number x	1	2	3	4	5	6	7
Square x^2							

b. Using the results of part a, fill in the blank in the following statement: For numbers larger than 1, the square of the number is _____ than the number.

22. a. Complete the following table.

Number x	$\dfrac{1}{2}$	$\dfrac{1}{3}$	$\dfrac{1}{4}$	$\dfrac{1}{5}$	$\dfrac{1}{6}$	$\dfrac{1}{7}$	$\dfrac{1}{8}$
Square x^2							

b. Using the results of part a, fill in the blank in the following statement: For numbers between 0 and 1, the square of the number is _____ than the number.

B Use the product property of exponents to simplify the following expressions.

23. $x^4 \cdot x^5$

24. $x^7 \cdot x^3$

25. $y^{10} \cdot y^{20}$

26. $y^{30} \cdot y^{30}$

27. $2^5 \cdot 2^4 \cdot 2^3$

28. $4^2 \cdot 4^3 \cdot 4^4$

29. $x^4 \cdot x^6 \cdot x^8 \cdot x^{10}$

30. $x^{20} \cdot x^{18} \cdot x^{16} \cdot x^{14}$

C Use the power property of exponents to write each of the following problems with a single exponent. (Assume all variables are positive numbers.)

31. $(x^2)^5$

32. $(x^5)^2$

33. $(5^4)^3$

34. $(5^3)^4$

35. $(y^3)^3$

36. $(y^2)^2$

37. $(2^5)^{10}$

38. $(10^5)^2$

39. $(a^3)^x$

40. $(a^5)^x$

41. $(b^x)^y$

42. $(b^r)^s$

D Use the distributive property of exponents to simplify each of the following expressions.

43. $(4x)^2$

44. $(2x)^4$

45. $(2y)^5$

46. $(5y)^2$

47. $(-3x)^4$

48. $(-3x)^3$

49. $(0.5ab)^2$

50. $(0.4ab)^2$

51. $(4xyz)^3$

52. $(5xyz)^3$

53. $(-4xy)^3$

54. $(-3xy)^4$

Simplify the following expressions by using the properties of exponents.

55. $(2x^4)^3$

56. $(3x^5)^2$

57. $(4a^3)^2$

58. $(5a^2)^2$

59. $(x^2)^3(x^4)^2$

60. $(x^5)^2(x^3)^5$

61. $(a^3)^1(a^2)^4$

62. $(a^4)^1(a^1)^3$

63. $(2x)^3(2x)^4$

64. $(3x)^2(3x)^3$

65. $(3x^2)^3(2x)^4$

66. $(3x)^3(2x^3)^2$

67. $(4x^2y^3)^2$

68. $(9x^3y^5)^2$

69. $\left(\dfrac{2}{3}a^4b^5\right)^3$

70. $\left(\dfrac{3}{4}ab^7\right)^3$

71. Complete the following table, and then construct a line graph of the information in the table

Number x	-3	-2	-1	0	1	2	3
Square x^2							

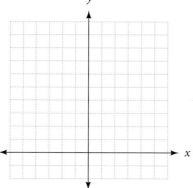

72. Complete the table, and then construct a line graph of the information in the table.

Number x	-3	-2	-1	0	1	2	3
Cube x^3							

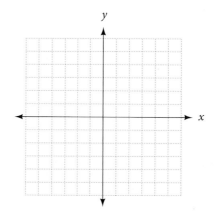

73. Complete the table. When you are finished, notice how the points in this table could be used to refine the line graph you created in Problem 71.

Number x	-2.5	-1.5	-0.5	0	0.5	1.5	2.5
Square x^2							

74. Complete the following table. When you are finished, notice that this table contains exactly the same entries as the table from Problem 73. This table uses fractions, whereas the table from Problem 73 uses decimals.

Number x	$-\dfrac{5}{2}$	$-\dfrac{3}{2}$	$-\dfrac{1}{2}$	0	$\dfrac{1}{2}$	$\dfrac{3}{2}$	$\dfrac{5}{2}$
Square x^2							

E Write each number in scientific notation.

75. 43,200

76. 432,000

77. 570 $= 5.7 \times 10^2$

78. 5,700

79. 238,000

80. 2,380,000

81. 10,000,000,000

82. 20

Write each number in standard form.

83. 2.49×10^3

84. 2.49×10^4

85. 3.52×10^2

86. 3.52×10^5

87. 2.8×10^4

88. 2.8×10^3

89. -3.1×10^5

90. -4.2×10^6

Applying the Concepts

91. The five largest lakes in the world (by area) are listed below in the table. Complete the table by writing each area in scientific notation. Round your answer to two decimal places.

Name	Location	Area (sq. mi.)	Scientific Notation
Caspian Sea	5 countries	152,239	
Superior	US, Canada	31,820	
Victoria	Tanzania, Uganda	26,828	
Huron	US, Canada	23,010	
Michigan	US	22,400	

The table below lists the five outer planets (from Earth), and their average distance to Earth in miles. Write each distance in scientific notation. Round your answer to two decimal places.

Planet	Average Distance	Scientific Notation
Mars	48.60 million	
Jupiter	390.60 million	
Saturn	777.00 million	
Uranus	1.69 billion	
Neptune	2.70 billion	

92. Volume of a Cube Find the volume of a cube if each side is 3 inches long.

93. Volume of a Cube Find the volume of a cube if each side is 3 feet long.

94. Volume of a Cube A bottle of perfume is packaged in a box that is in the shape of a cube. Find the volume of the box if each side is 2.5 inches long. Round to the nearest tenth.

95. Volume of a Cube A television set is packaged in a box that is in the shape of a cube. Find the volume of the box if each side is 18 inches long.

96. Volume of a Box A DVD box set of your favorite TV series is in the shape of a rectangular solid. Find the volume of the container if the height is 8 inches, the width is 4.5 inches, and the depth is 1 inch.

97. Volume of a Box Your textbook is in the shape of a rectangular solid. Find the volume in cubic inches.

98. Volume of a Box If a box has a volume of 42 cubic feet, is it possible for you to fit inside the box? Explain your answer.

99. Volume of a Box A box has a volume of 45 cubic inches. Will a can of soup fit inside the box? Explain your answer.

100. Age in seconds If you are 21 years old, you have been alive for more than 650,000,000 seconds. Write the number of seconds in scientific notation.

101. Distance Around the Earth The distance around the Earth at the equator is more than 130,000,000 feet. Write the number of feet in scientific notation.

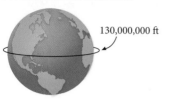

130,000,000 ft

102. Lifetime Earnings If you earn at least $12 an hour and work full-time for 30 years, you will make at least 7.4×10^5 dollars. Write your total earnings in expanded form.

103. Heart Beats per Year If your pulse is 72, then in one year your heart will beat at least 3.78×10^7 times. Write the number of heart beats in expanded form.

104. Investing If you put $1,000 into a savings account every year from the time you are 25 years old until you are 55 years old, you will have more than 1.8×10^5 dollars in the account when you reach 55 years of age (assuming 10% annual interest). Write 1.8×10^5 in expanded form.

105. Investing If you put $20 into a savings account every month from the time you are 20 years old until you are 30 years old, you will have more than 3.27×10^3 dollars in the account when you reach 30 years of age (assuming 6% annual interest compounded monthly). Write 3.27×10^3 in expanded form.

106. The value of an investment of $8,000 growing at 7% per year is given by the formula $V = 8000(1.07)^t$, where t represents the number of years the money is invested. Find the value of the investment after 20 years.

107. The value on an investment of $15,000 growing at 5% per year is given by the formula $V = 15000(1.05)^t$, where t represents the number of years the money is invested. Find the value of this investment after 25 years.

Displacement The displacement, in cubic inches, of a car engine is given by the formula

$$d = \pi \cdot s \cdot c \cdot \left(\frac{1}{2} \cdot b \right)^2$$

where s is the stroke, b is the bore, and c is the number of cylinders. (Note: The bore is the diameter of a cylinder in a piston engine and the stroke is the distance the cylinder travels.)

Calculate the engine displacement for each of the following cars. Use 3.14 to approximate π and round to the nearest cubic inch.

108. Ferrari Modena 8 cylinders, 3.35 inches of bore, 3.11 inches of stroke

109. Audi A8 8 cylinders, 3.32 inches of bore, 3.66 inches of stroke

110. Mitsubishi Eclipse 6 cylinders, 3.59 inches of bore, 2.99 inches of stroke

111. Porsche 911 GT3 6 cylinders, 3.94 inches of bore, 3.01 inches of stroke

Getting Ready for the Next Section

Subtract.

112. $4 - 7$

113. $-4 - 7$

114. $4 - (-7)$

115. $-4 - (-7)$

116. $15 - 20$

117. $15 - (-20)$

118. $-15 - (-20)$

119. $-15 - 20$

Simplify each expression.

120. $2(3) - 4$

121. $5(3) - 10$

122. $4(3) - 3(2)$

123. $-8 - 2(3)$

124. $2(5 - 3)$

125. $2(3) - 4 - 3(-4)$

126. $5 + 4(-2) - 2(-3)$

127. $2(3) + 4(5) - 5(2)$

Find the Mistake

Each sentence below contains a mistake. Circle the mistake and write the correct word(s) on the line provided.

1. Simplifying the expression $\left(-\frac{12}{32}\right)^3$ will give a positive answer. _____

2. You would use the distributive property for exponents to simplify the expression $x^8 \cdot x^4 \cdot x^{10}$.

3. To simplify the expression $(x^3 \cdot y)^4 (y \cdot z^2)^8$, we must first use the power property of exponents.

4. 5.83×10^5 in an example of a number written in expanded form. _____

Navigation Skills: Prepare, Study, Achieve

Expect to encounter problems you find difficult when taking this course. Also expect to make mistakes. Mistakes highlight possible difficulties you are having and help you learn how to overcome them. We suggest making a list of problems you find difficult. As the course progresses, add new problems to the list, rework the problems on your list, and use the list to study for exams. Be aware of the mistakes you make and what you need to do to ensure you will not make that same mistake twice.

Image © Katherine Heistand Shields, 2010

Imagine you are a sculpture artist specializing in glass blowing. Your most demanded pieces are multicolored glass balls used as Christmas ornaments or sun-catchers hung in windows. Each glass ball fits into a cube-shaped box with a side of x inches long. Suppose you need to figure out how many small cube-shaped boxes will fit into a larger shipping box, also shaped like a cube but with a side of $5x$ inches long. First, you would need to find the volume of each box.

Volume for the smaller box: $x \cdot x \cdot x = x^3$

Volume for the larger box: $5x \cdot 5x \cdot 5x = (5x)^3$

To find how many smaller boxes fit in the larger box, we use division.

$$\frac{(5x)^3}{x^3}$$

In this section, we will learn a property of exponents under division that will help us simplify this problem.

Previously, we found that multiplication with the same base results in addition of exponents; that is, $a^r \cdot a^s = a^{r+s}$. Since division is the inverse operation of multiplication, we can expect division with the same base to result in subtraction of exponents.

A Negative Exponents

To develop the properties for exponents under division, we again apply the definition of exponents.

$$\frac{x^5}{x^3} = \frac{x \cdot x \cdot x \cdot x \cdot x}{x \cdot x \cdot x}$$

$$= \frac{x \cdot x \cdot x}{x \cdot x \cdot x} (x \cdot x)$$

$$= 1(x \cdot x)$$

$$= x^2 \qquad \text{Note: } 5 - 3 = 2.$$

$$\frac{2^4}{2^7} = \frac{2 \cdot 2 \cdot 2 \cdot 2}{2 \cdot 2 \cdot 2 \cdot 2 \cdot 2 \cdot 2 \cdot 2}$$

$$= \frac{2 \cdot 2 \cdot 2 \cdot 2}{2 \cdot 2 \cdot 2 \cdot 2} \cdot \frac{1}{2 \cdot 2 \cdot 2}$$

$$= \frac{1}{2 \cdot 2 \cdot 2}$$

$$= \frac{1}{2^3} \qquad \text{Note: } 7 - 4 = 3.$$

In both cases, division with the same base resulted in subtraction of the smaller exponent from the larger. The problem is deciding whether the answer is a fraction. The problem is resolved easily by the following property.

PROPERTY Negative Exponent Property

If r is a positive integer, then $a^{-r} = \dfrac{1}{a^r} = \left(\dfrac{1}{a}\right)^r$ where $(a \neq 0)$.

The following examples illustrate how we use this property to simplify expressions that contain negative exponents.

EXAMPLE 1 Write each expression with a positive exponent and then simplify.

a. 2^{-3} **b.** 5^{-2} **c.** $3x^{-6}$

Solution

a. $2^{-3} = \dfrac{1}{2^3} = \dfrac{1}{8}$ *Note: Negative exponents do not indicate negative numbers. they indicate reciprocals.*

b. $5^{-2} = \dfrac{1}{5^2} = \dfrac{1}{25}$

c. $3x^{-6} = 3 \cdot \dfrac{1}{x^6} = \dfrac{3}{x^6}$ *Note: Be careful not to apply the negative exponents to the 3.*

B Quotient Property of Exponents

Now let us look back to our original problem and try to work it again with the help of a negative exponent. We know that $\frac{2^4}{2^7} = \frac{1}{2^3}$. Let us decide now that with division of the same base, we will always subtract the exponent in the denominator from the exponent in the numerator and see if this conflicts with what we know is true.

$$\frac{2^4}{2^7} = 2^{4-7} \qquad \textit{Subtract the bottom exponent from the top exponent.}$$

$$= 2^{-3} \qquad \textit{Subtract.}$$

$$= \frac{1}{2^3} \qquad \textit{Definition of negative exponents}$$

Subtracting the exponent in the denominator from the exponent in the numerator and then using the negative exponent property gives us the same result we obtained previously. We can now continue the list of properties of exponents we started in Section 4.1.

PROPERTY Quotient Property of Exponents

If a is any real number and r and s are integers, then

$$\frac{a^r}{a^s} = a^{r-s} \qquad (a \neq 0)$$

In words: To divide with the same base, subtract the exponent in the denominator from the exponent in the numerator and raise the base to the exponent that results.

The following examples show how we use the quotient property of exponents and the negative exponent property to simplify expressions involving division.

EXAMPLE 2 Simplify the following expressions.

a. $\dfrac{x^9}{x^6}$ **b.** $\dfrac{x^4}{x^{10}}$ **c.** $\dfrac{2^{15}}{2^{20}}$

Solution

a. $\dfrac{x^9}{x^6} = x^{9-6} = x^3$

b. $\dfrac{x^4}{x^{10}} = x^{4-10} = x^{-6} = \dfrac{1}{x^6}$

c. $\dfrac{2^{15}}{2^{20}} = 2^{15-20} = 2^{-5} = \dfrac{1}{2^5} = \dfrac{1}{32}$

Practice Problems

1. Write each expression with a positive exponent and then simplify.

 a. 3^{-2}

 b. 4^{-3}

 c. $5x^{-4}$

2. Simplify each expression.

 a. $\dfrac{x^{10}}{x^4}$

 b. $\dfrac{x^5}{x^7}$

 c. $\dfrac{2^{21}}{2^{25}}$

Answers

1. **a.** $\dfrac{1}{9}$ **b.** $\dfrac{1}{64}$ **c.** $\dfrac{5}{x^4}$

2. **a.** x^6 **b.** $\dfrac{1}{x^2}$ **c.** $\dfrac{1}{16}$

C Expanded Distributive Property of Exponents

Our final property of exponents is similar to the distributive property of exponents from Section 4.1, but it involves division instead of multiplication. After we have stated the property, we will give a proof of it. The proof shows why this property is true.

> **PROPERTY** Expanded Distributive Property of Exponents
>
> If a and b are any two real numbers ($b \neq 0$) and r is an integer, then
>
> $$\left(\frac{a}{b}\right)^r = \frac{a^r}{b^r}$$
>
> *In words:* A quotient raised to a power is the quotient of the powers.

Note This property may also lead to division by zero (undefined) if a is zero and r is less than zero.

Proof

$$\left(\frac{a}{b}\right)^r = \left(a \cdot \frac{1}{b}\right)^r \qquad \text{Definition of division}$$

$$= a^r \cdot \left(\frac{1}{b}\right)^r \qquad \text{Distributive property of exponents}$$

$$= a^r \cdot b^{-r} \qquad \text{Negative exponent property}$$

$$= a^r \cdot \frac{1}{b^r} \qquad \text{Negative exponent property}$$

$$= \frac{a^r}{b^r} \qquad \text{Multiply.}$$

EXAMPLE 3 Simplify the following expressions:

a. $\left(\frac{x}{2}\right)^3$ **b.** $\left(\frac{5}{y}\right)^2$ **c.** $\left(\frac{2}{3}\right)^4$

Solution

a. $\left(\frac{x}{2}\right)^3 = \frac{x^3}{2^3} = \frac{x^3}{8}$

b. $\left(\frac{5}{y}\right)^2 = \frac{5^2}{y^2} = \frac{25}{y^2}$

c. $\left(\frac{2}{3}\right)^4 = \frac{2^4}{3^4} = \frac{16}{81}$

3. Simplify each expression.

a. $\left(\frac{x}{5}\right)^2$

b. $\left(\frac{2}{a}\right)^3$

c. $\left(\frac{3}{4}\right)^3$

D Zero and One as Exponents

Recall the definition of 0 and 1 as exponents from Chapter 0. We will review the concept here and justify why we use them as defined.

To obtain an expression for x^1, we will solve a problem two different ways:

$$\left.\begin{array}{l} \dfrac{x^3}{x^2} = \dfrac{x \cdot x \cdot x}{x \cdot x} = x \\[2em] \dfrac{x^3}{x^2} = x^{3-2} = x^1 \end{array}\right\} \qquad \text{Hence } x^1 = x.$$

Stated generally, this rule says that $a^1 = a$. This seems reasonable and we will use it since it is consistent with our property of division using the same base.

We use the same procedure to obtain an expression for x^0

$$\left.\begin{array}{l} \dfrac{5^2}{5^2} = \dfrac{25}{25} = 1 \\[2em] \dfrac{5^2}{5^2} = 5^{2-2} = 5^0 \end{array}\right\} \qquad \text{Hence } 5^0 = 1.$$

Answers

3. **a.** $\frac{x^2}{25}$ **b.** $\frac{8}{a^3}$ **c.** $\frac{27}{64}$

It seems, therefore, that the best definition of x^0 is 1 for all bases x except $x = 0$. In the case of $x = 0$, we have 0^0, which we will not define. This definition will probably seem awkward at first. Most people would like to define x^0 as 0 when they first encounter it. Remember, the zero in this expression is an exponent, so x^0 does not mean to multiply by zero. Thus, we can make the general statement that $a^0 = 1$ for all real numbers except $a = 0$.

Here are some examples involving the exponents 0 and 1.

EXAMPLE 4 Simplify the following expressions.

a. 8^0 **b.** 8^1 **c.** $4^0 + 4^1$ **d.** $(2x^2y)^0$

Solution

a. $8^0 = 1$

b. $8^1 = 8$

c. $4^0 + 4^1 = 1 + 4 = 5$

d. $(2x^2y)^0 = 1$

E Combinations of Properties

Here is a summary of the definitions and properties of exponents we have developed so far. For each definition or property in the list, a and b are real numbers, and r and s are integers.

Definitions of Exponents	Properties of Exponents	
$a^1 = a$	**1.** $a^r \cdot a^s = a^{r+s}$	Product property of exponents
$a^0 = 1 \quad a \neq 0$	**2.** $(a^r)^s = a^{rs}$	Power property of exponents
	3. $(ab)^r = a^r b^r$	Distributive property of exponents
	4. $a^{-r} = \dfrac{1}{a^r} = \left(\dfrac{1}{a}\right)^r \quad a \neq 0$	Negative exponent property
	5. $\dfrac{a^r}{a^s} = a^{r-s} \quad a \neq 0$	Quotient property of exponents
	6. $\left(\dfrac{a}{b}\right)^r = \dfrac{a^r}{b^r} \quad b \neq 0$	Expanded distributive property of exponents

Here are some additional examples. These examples use a combination of the preceding properties and definitions.

EXAMPLE 5 Simplify each expression. Write all answers with positive exponents.

a. $\dfrac{(5x^3)^2}{x^4} = \dfrac{25x^6}{x^4}$ *Power and distributive properties of exponents*

$= 25x^2$ *Quotient property of exponents*

b. $\dfrac{x^{-8}}{(x^2)^3} = \dfrac{x^{-8}}{x^6}$ *Power property of exponents*

$= x^{-8-6}$ *Quotient property of exponents*

$= x^{-14}$ *Subtract.*

$= \dfrac{1}{x^{14}}$ *Negative exponent property*

4. Simplify each expression.

a. 10^0

b. 10^1

c. $6^1 + 6^0$

d. $(3x^5y^2)^0$

5. Simplify, and write your answers with positive exponents only.

a. $\dfrac{x^{-6}}{(x^3)^4}$

b. $\dfrac{(2x^3)^2}{x^4}$

c. $\left(\dfrac{y^8}{y^3}\right)^2$

d. $(2x^4)^{-2}$

e. $x^{-6} \cdot x^2$

f. $\dfrac{a^5(a^{-2})^4}{(a^{-3})^2}$

Answers

4. a. 1 **b.** 10 **c.** 7 **d.** 1

5. a. $\frac{1}{x^{18}}$ b. $4x^2$ **c.** y^{10} **d.** $\frac{1}{4x^8}$

e. $\frac{1}{x^4}$ **f.** a^3

c. $\left(\dfrac{y^5}{y^3}\right)^2 = \dfrac{(y^5)^2}{(y^3)^2}$ Expanded distributive property of exponents

$= \dfrac{y^{10}}{y^6}$ Power property of exponents

$= y^4$ Quotient property of exponents

Notice in Example 5c that we could have simplified inside the parentheses first and then raised the result to the second power:

$$\left(\dfrac{y^5}{y^3}\right)^2 = (y^{5-3})^2 = (y^2)^2 = y^4$$

d. $(3x^5)^{-2} = \dfrac{1}{(3x^5)^2}$ Negative exponent property

$= \dfrac{1}{9x^{10}}$ Power and distributive properties of exponents

e. $x^{-8} \cdot x^5 = x^{-8+5}$ Product property of exponents

$= x^{-3}$ Add.

$= \dfrac{1}{x^3}$ Negative exponent property

f. $\dfrac{(a^3)^2 a^{-4}}{(a^{-4})^3} = \dfrac{a^6 a^{-4}}{a^{-12}}$ Power property of exponents

$= \dfrac{a^2}{a^{-12}}$ Product property of exponents

$= a^{14}$ Quotient property of exponents

In the next two examples we use division to compare the area and volume of geometric figures.

EXAMPLE 6 Suppose you have two squares, one of which is larger than the other. If the length of a side of the larger square is 3 times as long as the length of a side of the smaller square, how many of the smaller squares will it take to cover up the larger square?

Solution If we let x represent the length of a side of the smaller square, then the length of a side of the larger square is $3x$. The area of each square, along with a diagram of the situation, is given in Figure 1.

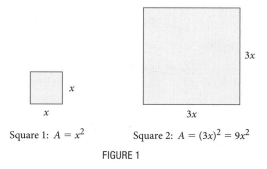

Square 1: $A = x^2$ Square 2: $A = (3x)^2 = 9x^2$

FIGURE 1

To find out how many smaller squares it will take to cover up the larger square, we divide the area of the larger square by the area of the smaller square.

$$\frac{\text{Area of square 2}}{\text{Area of square 1}} = \frac{9x^2}{x^2} = 9$$

It will take 9 of the smaller squares to cover the larger square.

6. Suppose the side of the larger square in Example 6 is 4 times as long as the side of the smaller square. How many smaller squares will it take to cover the larger square?

Answer

6. It will take 16 smaller squares to cover the larger square.

7. Suppose the side of the larger box in Example 7 is 4 times as long as the side of the smaller box. How many of the smaller boxes can be fit into the larger box?

EXAMPLE 7 Suppose you have two boxes, each of which is a cube. If the length of a side in the second box is 3 times as long as the length of a side of the first box, how many of the smaller boxes will fit inside the larger box?

Solution If we let x represent the length of a side of the smaller box, then the length of a side of the larger box is $3x$. The volume of each box, along with a diagram of the situation, is given in Figure 2.

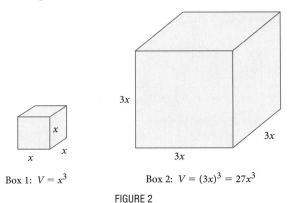

Box 1: $V = x^3$ Box 2: $V = (3x)^3 = 27x^3$

FIGURE 2

To find out how many smaller boxes will fit inside the larger box, we divide the volume of the larger box by the volume of the smaller box.

$$\frac{\text{Volume of box 2}}{\text{Volume of box 1}} = \frac{27x^3}{x^3} = 27$$

We can fit 27 of the smaller boxes inside the larger box.

F Scientific Notation for Small Numbers

Now that we have completed our list of definitions and properties of exponents, we can expand the work we did previously with scientific notation.

Recall that a number is in scientific notation when it is written in the form

$$n \times 10^r$$

where $1 \leq n < 10$ and r is an integer.

Since negative exponents give us reciprocals, we can use negative exponents to write very small numbers in scientific notation. For example, the number 0.00057, when written in scientific notation, is equivalent to 5.7×10^{-4}. Here's why:

$$5.7 \times 10^{-4} = 5.7 \times \frac{1}{10^4} = 5.7 \times \frac{1}{10{,}000} = \frac{5.7}{10{,}000} = 0.00057$$

The table below lists some other numbers in both scientific notation and expanded form.

Number written the long way		Same number written in scientific notation
376,000	=	3.76×10^5
49,500	=	4.95×10^4
3,200	=	3.2×10^3
591	=	5.91×10^2
46	=	4.6×10^1
8	=	8×10^0
0.47	=	4.7×10^{-1}
0.093	=	9.3×10^{-2}
0.00688	=	6.88×10^{-3}
0.0002	=	2×10^{-4}
0.000098	=	9.8×10^{-5}

Answer

7. The larger box will hold 64 smaller boxes.

Notice that in each case, when the number is written in scientific notation, the decimal point in the first number is placed so that the number is between 1 and 10. The exponent on 10 in the second number keeps track of the number of places we moved the decimal point in the original number to get a number between 1 and 10.

$$376,000 = 3.76 \times 10^5$$

Moved 5 places.

Decimal point was originally here.

Keeps track of the 5 places we moved the decimal point.

$$0.00688 = 6.88 \times 10^{-3}$$

Moved 3 places.

Keeps track of the 3 places we moved the decimal point.

EXAMPLE 8 Write 0.000584 in scientific notation.

Solution We move the decimal point 4 place to the right so that 0.000584 will be a product of a number between 1 and 10 and a power of 10. The decimal point appears between the 5 and the 8, and we multiply the result by 10^{-4}.

$$0.000584 = 5.84 \times 10^{-4}$$

Decimal point was originally here.

Moved 4 places.

Keeps track of the 4 places we moved the decimal point.

8. Write 0.00273 in scientific notation.

GETTING READY FOR CLASS

After reading through the preceding section, respond in your own words and in complete sentences.

A. How do you divide two expressions containing exponents when they each have the same base?

B. Explain the difference between 3^2 and 3^{-2}.

C. Explain what happens when you use 0 as an exponent.

D. What does a negative exponent mean in scientific notation?

Answer

8. 2.73×10^{-3}

Vocabulary Review

Choose the correct words to fill in the blanks below.

exponents quotient expanded negative

1. The _____ exponent property states that $a^{-r} = \frac{1}{a^r} = \left(\frac{1}{a}\right)^r$ if r is a positive integer and a does not equal zero.

2. The _____ property for exponents allows you to divide with the same base by subtracting the exponent in the denominator from the exponent in the numerator and raise the base to the exponent that results.

3. The _____ distributive property for exponents states that $\left(\frac{1}{a}\right)^r = \frac{a}{b^r}$ if a and b are any two real numbers, b does not equal zero, and r is an integer.

4. Negative _____ used in scientific notation represent very small numbers.

Problems

A Write each of the following with positive exponents, and then simplify, when possible.

1. 3^{-2}
2. 3^{-3}
3. 6^{-2}
4. 2^{-6}

5. 8^{-2}
6. 3^{-4}
7. 5^{-3}
8. 9^{-2}

9. $2x^{-3}$
10. $5x^{-1}$
11. $(2x)^{-3}$
12. $(5x)^{-1}$

13. $(5y)^{-2}$
14. $5y^{-2}$
15. 10^{-2}
16. 10^{-3}

17. Complete the following table.

Number x	Square x^2	Power of 2 2^x
-3		
-2		
-1		
0		
1		
2		
3		

18. Complete the following table.

Number x	Cube x^3	Power of 3 3^x
-3		
-2		
-1		
0		
1		
2		
3		

B Use the quotient property of exponents to simplify each of the following expressions. Write all answers that contain exponents with positive exponents only.

19. $\dfrac{5^1}{5^3}$

20. $\dfrac{7^6}{7^8}$

21. $\dfrac{x^{10}}{x^4}$

22. $\dfrac{x^4}{x^{10}}$

23. $\dfrac{4^3}{4^0}$

24. $\dfrac{4^0}{4^3}$

25. $\dfrac{(2x)^7}{(2x)^4}$

26. $\dfrac{(2x)^4}{(2x)^7}$

27. $\dfrac{6^{11}}{6}$ $=\dfrac{6^{10}}{1}=6^{10}$

28. $\dfrac{8^7}{8}$

29. $\dfrac{6}{6^{11}}$

30. $\dfrac{8}{8^7}$

31. $\dfrac{2^{-5}}{2^3}$

32. $\dfrac{2^{-5}}{2^{-3}}$

33. $\dfrac{2^5}{2^{-3}}$

34. $\dfrac{2^{-3}}{2^{-5}}$

35. $\dfrac{(3x)^{-5}}{(3x)^{-8}}$

36. $\dfrac{(2x)^{-10}}{(2x)^{-15}}$

37. $\dfrac{(-4x)^{10}}{(-4x)^8}$ $=(-4x)^2$

38. $\dfrac{(-3x)^4}{(-3x)^7}$

$\dfrac{(3x)^8}{(3x)^5}=\dfrac{(3x)^3}{1}=(3x)^3$

$(-4x)^2$

C Use the expanded distributive property of exponents to simplify each of the following expressions. Write all answers with positive exponents only.

39. $\left(\dfrac{x}{3}\right)^3$

40. $\left(\dfrac{x}{2}\right)^4$

41. $\left(\dfrac{4}{y}\right)^2$

42. $\left(\dfrac{-6}{y}\right)^2$

43. $\left(-\dfrac{5}{x}\right)^3$

44. $\left(\dfrac{-4}{y}\right)^2$

45. $\left(\dfrac{-8}{y}\right)^2$

46. $\left(\dfrac{-3}{a}\right)^4$

D Simplify the following expressions.

47. $(-27x)^0$

48. $(-9y^2)^0$

49. $(-2y)^1$

50. $(-8y^2)^1$

51. $(-3ab^2)^1$

52. $(-4x^2y)^1$

53. $(-2xy^2)^0$

54. $\left(\dfrac{x^3}{y^4}\right)^0$

E Simplify the following expressions. Any answers that contain exponents should contain positive exponents only.

55. $(3xy)^4$

56. $(4xy)^3$

57. 10^0

58. 10^1

59. $(2a^2b)^1$

60. $(2a^2b)^0$

61. $(7y^3)^{-2}$

62. $(5y^4)^{-2}$

63. $x^{-3}x^{-5}$

64. $x^{-6} \cdot x^8$

65. $y^7 \cdot y^{-10}$

66. $y^{-4} \cdot y^{-6}$

67. $\dfrac{(x^2)^3}{x^4}$

68. $\dfrac{(x^5)^3}{x^{10}}$

69. $\dfrac{(a^4)^3}{(a^3)^2}$

70. $\dfrac{(a^5)^3}{(a^5)^2}$

71. $\dfrac{y^7}{(y^2)^8}$

72. $\dfrac{y^2}{(y^3)^4}$

73. $\left(\dfrac{y^7}{y^2}\right)^8$

74. $\left(\dfrac{y^2}{y^3}\right)^4$

75. $\dfrac{(x^{-2})^3}{x^{-5}}$

76. $\dfrac{(x^2)^{-3}}{x^{-5}}$

77. $\left(\dfrac{x^{-2}}{x^{-5}}\right)^3$

78. $\left(\dfrac{x^2}{x^{-5}}\right)^{-3}$

79. $\left(\dfrac{2a^{-3}b^2}{4a^{-5}b^6}\right)^{-2}$

80. $\left(\dfrac{3a^{-4}b^3}{2a^{-3}b^4}\right)^{-3}$

81. $\dfrac{(a^3)^2(a^4)^5}{(a^5)^2}$

82. $\dfrac{(a^4)^8(a^2)^5}{(a^3)^4}$

83. $\dfrac{(a^{-2})^3(a^4)^2}{(a^{-3})^{-2}}$

84. $\dfrac{(a^{-5})^{-3}(a^7)^{-1}}{(a^{-3})^5}$

85. $\dfrac{(3x^{-2}y^{-4})(4x^3y^{-2})^{-2}}{(3x^3y^{-2})^{-2}}$

86. $\dfrac{(4x^{-7}y^{-3})(2x^4y^{-3})^{-1}}{(2x^2y^{-2})^{-2}}$

87. Complete the following table, and then construct a line graph of the information in the table.

Number x	Power of x 2^x
−3	
−2	
−1	
0	
1	
2	
3	

88. Complete the following table, and then construct a line graph of the information in the table.

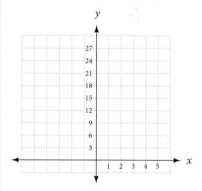

Number x	Power of x 3^x
−3	
−2	
−1	
0	
1	
2	
3	

F Write each of the following numbers in scientific notation.

89. 0.0048

90. 0.000048

91. 25

92. 35

93. 0.000009

94. 0.0009

95. −0.000087

96. −0.00000514

Write each of the following numbers in standard form.

97. 4.23×10^{-3}

98. 4.23×10^{3}

99. 8×10^{-5}

100. 8×10^{5}

101. 4.2×10^{0}

102. 4.2×10^{1}

103. -3.6×10^{-3}

104. -2.05×10^{-4}

105. Complete the following table.

Standard Form	Scientific Notation $n \times 10^r$
0.000357	3.57×10^{-4}
0.00357	
0.0357	
0.357	
3.57	
35.7	
357	
3,570	
35,700	

106. Complete the following table.

Standard Form	Scientific Notation $n \times 10^r$
0.000123	1.23×10^{-4}
	1.23×10^{-3}
	1.23×10^{-2}
	1.23×10^{-1}
	1.23×10^{0}
	1.23×10^{1}
	1.23×10^{2}
	1.23×10^{3}
	1.23×10^{4}

Applying the Concepts

107. Some home computers can do a calculation in 2×10^{-3} seconds. Write this number in standard form.

108. Some of the cells in the human body have a radius of 3×10^{-5} inches. Write this number in standard form.

109. Margin of Victory The chart below shows some close finishes from the 2008 Beijing Olympic games. Write each number in scientific notation.

110. Some cameras used in scientific research can take one picture every 0.000000167 second. Write this number in scientific notation.

111. The number 25×10^3 is not in scientific notation because 25 is larger than 10. Write 25×10^3 in scientific notation.

112. The number 0.25×10^3 is not in scientific notation because 0.25 is less than 1. Write 0.25×10^3 in scientific notation.

113. The number 23.5×10^4 is not in scientific notation because 23.5 is not between 1 and 10. Rewrite 23.5×10^4 in scientific notation.

114. The number 375×10^3 is not in scientific notation because 375 is not between 1 and 10. Rewrite 375×10^3 in scientific notation.

115. The number 0.82×10^{-3} is not in scientific notation because 0.82 is not between 1 and 10. Rewrite 0.82×10^{-3} in scientific notation.

116. The number 0.93×10^{-2} is not in scientific notation because 0.93 is not between 1 and 10. Rewrite 0.93×10^{-2} in scientific notation.

Comparing Areas Suppose you have two squares, one of which is larger than the other. Suppose further that the side of the larger square is twice as long as the side of the smaller square.

117. If the length of the side of the smaller square is 10 inches, give the area of each square. Then find the number of smaller squares it will take to cover the larger square.

118. How many smaller squares will it take to cover the larger square if the length of the side of the smaller square is 1 foot?

119. If the length of the side of the smaller square is x, find the area of each square. Then find the number of smaller squares it will take to cover the larger square.

120. Suppose the length of the side of the larger square is 1 foot. How many smaller squares will it take to cover the larger square?

Comparing Volumes Suppose you have two boxes, each of which is a cube. Suppose further that the length of a side of the second box is twice as long as the length of a side of the first box.

121. If the length of a side of the first box is 6 inches, give the volume of each box. Then find the number of smaller boxes that will fit inside the larger box.

122. How many smaller boxes can be placed inside the larger box if the length of a side of the second box is 1 foot?

123. If the length of a side of the first box is x, find the volume of each box. Then find the number of smaller boxes that will fit inside the larger box.

124. Suppose the length of a side of the larger box is 12 inches. How many smaller boxes will fit inside the larger box?

125. The radius of Earth is 6,378 km, while the radius of Jupiter is 71,472 km. How many Earths would fit inside of Jupiter? Answer this by comparing volume. The volume can be calculated using the formula $\frac{4}{3}\pi r^3$.

126. The radius of Jupiter is 71,472 km, while the radius of the sun is 695,000 km. How many Jupiters would fit inside the sun? Answer this by comparing their volumes. The volume can be calculated using the formula $\frac{4}{3}\pi r^3$.

Getting Ready for the Next Section

Simplify.

127. $3(4.5)$

128. $\dfrac{1}{2} \cdot \dfrac{5}{7}$

129. $\dfrac{4}{5}(10)$

130. $\dfrac{9.6}{3}$

131. $6.8(3.9)$

132. $9 - 20$

133. $-3 + 15$

134. $2x \cdot x \cdot \dfrac{1}{2}x$

135. $x^5 \cdot x^3$

136. $y^2 \cdot y$

137. $\dfrac{x^3}{(x^2)}$

138. $\dfrac{x^2}{x}$

139. $\dfrac{y^3}{y^5}$

140. $\dfrac{x^2}{x^5}$

141. $\dfrac{x^4}{(x^2)^2}$

142. $\dfrac{(y^2)^3}{y^6}$

Write in expanded form.

143. 3.4×10^2

144. 6.0×10^{-4}

145. 3.5×10^{-5}

146. -2.1×10^4

Find the Mistake

Each sentence below contains a mistake. Circle the mistake and write the correct expression on the line provided.

1. The negative exponent property says that $x^{-y} = \dfrac{y}{x}$. _____

2. When using the quotient property of exponents to simplify the expression $\dfrac{y^{12}}{y^8}$, we get y^{20}. _____

3. When using the expanded distributive property of exponents to simplify the expression $\left(\dfrac{x}{3}\right)^4$, we get $\dfrac{x^4}{3}$. _____

4. The number 0.00238 written in scientific notation is 2.38×10^{-2}. _____

4.3 Operations with Monomials

©iStockphoto.com/KonovalikovAndrey

OBJECTIVES

A Use rules of exponents to multiply monomials.

B Use rules of exponents to divide monomials.

C Multiply and divide numbers written in scientific notation.

D Add and subtract monomials.

KEY WORDS

monomial

numerical coefficient

terms

similar terms

A preschool teacher needs to cover the wood floors in two of her home's bedrooms with carpet to provide a softer play surface for her students. She hires a carpet installer who creates a formula to know how much carpet two bedrooms in the house needs. The first room is in the shape of a square with a wall the length of x. The second room is the shape of a rectangle, with one wall the same length as the first room but a second wall twice as long. Using our knowledge of area, we can write two monomials that each represent the area of the rooms.

Square room area: $x \cdot x = x^2$

Rectangular room area: $x \cdot 2x = 2x^2$

Later in this section, we can add these monomials to determine how much carpet the installer will need. But first, we will discuss multiplication and division of monomials.

> **DEFINITION** monomial
>
> A *monomial* is a one-term expression that is either a constant (number) or the product of a constant and one or more variables raised to whole number exponents.

The following are examples of monomials:

$$-3 \qquad 15x \qquad -23x^2y \qquad 49x^4y^2z^4 \qquad \frac{3}{4}a^2b^3$$

The numerical part of each monomial is called the *numerical coefficient*, or just *coefficient*. Monomials are also called *terms*.

A Multiplying Monomials

There are two basic steps involved in the multiplication of monomials. First, we rewrite the products using the commutative and associative properties. Then, we simplify by multiplying coefficients and adding exponents of like bases.

EXAMPLE 1 Multiply each expression.

a. $(-3x^2)(4x^3)$ **b.** $\left(\frac{4}{5}x^5 \cdot y^2\right)(10x^3 \cdot y)$

Solution

a. $(-3x^2)(4x^3) = (-3 \cdot 4)(x^2 \cdot x^3)$ *Commutative and associative properties*

$\qquad\qquad\qquad = -12x^5$ *Multiply coefficients, add exponents.*

b. $\left(\frac{4}{5}x^5 \cdot y^2\right)(10x^3 \cdot y) = \left(\frac{4}{5} \cdot 10\right)(x^5 \cdot x^3)(y^2 \cdot y)$ *Commutative and associative properties*

$\qquad\qquad\qquad\qquad = 8x^8y^3$ *Multiply coefficients, add exponents.*

Practice Problems

1. Multiply.

 a. $(-2x^3)(5x^2)$

 b. $(4x^6y^3)(2xy^2)$

Answers

1. a. $-10x^5$ **b.** $8x^7y^5$

You can see that in each case the work was the same—multiply coefficients and add exponents of the same base.

B Dividing Monomials

We can expect division of monomials to proceed in a similar way. Since our properties are consistent, division of monomials will result in division of coefficients and subtraction of exponents of like bases.

2. Divide.

 a. $\dfrac{27x^4y^3}{9xy^2}$

 b. $\dfrac{25x^4}{5x}$

EXAMPLE 2 Divide each expression.

a. $\dfrac{15x^3}{3x^2}$ **b.** $\dfrac{39x^2y^3}{3xy^5}$

Solution

a. $\dfrac{15x^3}{3x^2} = \dfrac{15}{3} \cdot \dfrac{x^3}{x^2}$ *Write as separate fractions.*

$= 5x$ *Divide coefficients, subtract exponents.*

b. $\dfrac{39x^2y^3}{3xy^5} = \dfrac{39}{3} \cdot \dfrac{x^2}{x} \cdot \dfrac{y^3}{y^5}$ *Write as separate fractions.*

$= 13x \cdot \dfrac{1}{y^2}$ *Divide coefficients, subtract exponents.*

$= \dfrac{13x}{y^2}$ *Write answer as a single fraction.*

In Example 2b, the expression $\frac{y^3}{y^5}$ simplifies to $\frac{1}{y^2}$ because of the quotient property of exponents and the negative exponent property. If we were to show all the work in this simplification process, it would look like this:

$$\dfrac{y^3}{y^5} = y^{3-5} \quad \textit{Quotient property of exponents}$$

$$= y^{-2} \quad \textit{Subtract.}$$

$$= \dfrac{1}{y^2} \quad \textit{Negative exponent property}$$

The point of this explanation is this: Even though we may not show all the steps when simplifying an expression involving exponents, the result we obtain still can be justified using the properties of exponents. We have not introduced any new properties in Example 2; we have just not shown the details of each simplification.

3. Divide $\dfrac{13a^6b^2}{39a^4b^7}$.

EXAMPLE 3 Divide $25a^5b^3$ by $50a^2b^7$.

Solution

$$\dfrac{25a^5b^3}{50a^2b^7} = \dfrac{25}{50} \cdot \dfrac{a^5}{a^2} \cdot \dfrac{b^3}{b^7} \quad \textit{Write as separate fractions.}$$

$$= \dfrac{1}{2} \cdot a^3 \cdot \dfrac{1}{b^4} \quad \textit{Divide coefficients, subtract exponents.}$$

$$= \dfrac{a^3}{2b^4} \quad \textit{Write answer as a single fraction.}$$

Notice in Example 3 that dividing 25 by 50 results in $\frac{1}{2}$. This is the same result we would obtain if we reduced the fraction $\frac{25}{50}$ to lowest terms, and there is no harm in thinking of it that way. Also, notice that the expression $\frac{b^3}{b^7}$ simplifies to $\frac{1}{b^4}$ by the quotient property of exponents and the negative exponent property, even though we have not shown the steps involved in doing so.

Answers

2. a. $3x^3y$ **b.** $5x^3$

3. $\dfrac{a^2}{3b^5}$

C Multiplication and Division of Numbers Written in Scientific Notation

We multiply and divide numbers written in scientific notation using the same steps we used to multiply and divide monomials.

EXAMPLE 4 Multiply $(4 \times 10^7)(2 \times 10^{-4})$.

Solution Since multiplication is commutative and associative, we can rearrange the order of these numbers and group them as follows:

$$(4 \times 10^7)(2 \times 10^{-4}) = (4 \times 2)(10^7 \times 10^{-4})$$
$$= 8 \times 10^3$$

Notice that we add exponents, $7 + (-4) = 3$, when we multiply with the same base.

EXAMPLE 5 Divide $\dfrac{9.6 \times 10^{12}}{3 \times 10^4}$.

Solution We group the numbers between 1 and 10 separately from the powers of 10 and proceed as we did in Example 4:

$$\frac{9.6 \times 10^{12}}{3 \times 10^4} = \frac{9.6}{3} \times \frac{10^{12}}{10^4}$$
$$= 3.2 \times 10^8$$

Notice that the procedure we used in both of these examples is very similar to multiplication and division of monomials, for which we multiplied or divided coefficients and added or subtracted exponents.

D Addition and Subtraction of Monomials

Addition and subtraction of monomials will be almost identical since subtraction is defined as addition of the opposite. With multiplication and division of monomials, the key was rearranging the numbers and variables using the commutative and associative properties. With addition, the key is application of the distributive property. We sometimes use the phrase *combine monomials* to describe addition and subtraction of monomials.

> **DEFINITION** similar terms
>
> Two monomials with the same variable part (same variables raised to the same powers) are called *similar* (or *like*) *terms.*

You can add only similar terms. This is because the distributive property (which is the key to addition of monomials) cannot be applied to terms that are not similar.

EXAMPLE 6 Combine the following monomials.

a. $-3x^2 + 15x^2$ **b.** $9x^2y - 20x^2y$ **c.** $5x^2 + 8y^2$

solution

a. $-3x^2 + 15x^2 = (-3 + 15)x^2$ Distributive property
$$= 12x^2$$ Add coefficients.

b. $9x^2y - 20x^2y = (9 - 20)x^2y$ Distributive property
$$= -11x^2y$$ Add coefficients.

c. $5x^2 + 8y^2$ In this case, we cannot apply the distributive property, so we cannot add the monomials.

4. Multiply $(3 \times 10^6)(3 \times 10^{-8})$.

5. Divide $\dfrac{4.8 \times 10^{20}}{2 \times 10^{12}}$.

6. Combine if possible.

 a. $-4x^2 + 9x^2$

 b. $12x^2y - 15x^2y$

 c. $6x^2 + 9x^3$

Answers
4. 9×10^{-2}
5. 2.4×10^8
6. a. $5x^2$ **b.** $-3x^2y$
 c. Cannot be combined

The next examples show how we simplify expressions containing monomials when more than one operation is involved.

EXAMPLE 7 Apply the distributive property to simplify.

a. $x^2\left(1 - \dfrac{6}{x}\right)$ **b.** $ab\left(\dfrac{1}{b} - \dfrac{1}{a}\right)$

7. Apply the distributive property.

a. $a^2\left(2 - \dfrac{4}{a}\right)$

b. $xy\left(\dfrac{3}{x} + \dfrac{2}{y}\right)$

Solution

a. $x^2\left(1 - \dfrac{6}{x}\right) = x^2 \cdot 1 - x^2 \cdot \dfrac{6}{x} = x^2 - \dfrac{6x^2}{x} = x^2 - 6x$

b. $ab\left(\dfrac{1}{b} - \dfrac{1}{a}\right) = ab \cdot \dfrac{1}{b} - ab \cdot \dfrac{1}{a} = \dfrac{ab}{b} - \dfrac{ab}{a} = a - b$

8. Simplify $\dfrac{(10x^3y^2)(6xy^4)}{12x^2y^3}$.

EXAMPLE 8 Simplify $\dfrac{(6x^4y)(3x^7y^5)}{9x^5y^2}$.

Solution We begin by multiplying the two monomials in the numerator.

$$\dfrac{(6x^4y)(3x^7y^5)}{9x^5y^2} = \dfrac{18x^{11}y^6}{9x^5y^2} \qquad \text{Simplify numerator.}$$

$$= 2x^6y^4 \qquad \text{Divide.}$$

9. Simplify $\dfrac{(1.2 \times 10^6)(6.3 \times 10^{-5})}{6 \times 10^{-10}}$.

EXAMPLE 9 Simplify $\dfrac{(6.8 \times 10^5)(3.9 \times 10^{-7})}{7.8 \times 10^{-4}}$.

Solution We group the numbers between 1 and 10 separately from the powers of 10.

$$\dfrac{(6.8)(3.9)}{7.8} \times \dfrac{(10^5)(10^{-7})}{10^{-4}} = 3.4 \times 10^{5+(-7)-(-4)}$$

$$= 3.4 \times 10^2$$

10. Simplify $\dfrac{24x^7}{3x^2} + \dfrac{14x^9}{7x^4}$.

EXAMPLE 10 Simplify $\dfrac{14x^5}{2x^2} + \dfrac{15x^8}{3x^5}$.

Solution Simplifying each expression separately and then combining similar terms gives

$$\dfrac{14x^5}{2x^2} + \dfrac{15x^8}{3x^5} = 7x^3 + 5x^3 \qquad \text{Divide.}$$

$$= 12x^3 \qquad \text{Add.}$$

GETTING READY FOR CLASS

After reading through the preceding section, respond in your own words and in complete sentences.

A. What is a monomial?

B. Describe how you would multiply $3x^2$ and $5x^2$.

C. Describe how you would add $3x^2$ and $5x^2$.

D. Describe how you would multiply two numbers written in scientific notation.

Answers

7. a. $2a^2 - 4a$ **b.** $3y + 2x$

8. $5x^2y^3$

9. 1.26×10^{11}

10. $10x^5$

Vocabulary Review

Choose the correct words to fill in the blanks below.

terms coefficient subtraction variable constant

1. A monomial is a one-term expression that is either a _____ or the product of a constant and one or more variables raised to whole number exponents.

2. Monomials are also called _____.

3. The numerical _____ of $7xy^2$ is 7.

4. Two monomials with the same _____ part are called similar terms.

5. Division of monomials will result in division of coefficients and _____ of exponents of like bases.

Problems

A Multiply each expression and simplify.

1. $(3x^4)(4x^3)$

2. $(6x^5)(-2x^2)$

3. $(-2y^4)(8y^7)$

4. $(5y^{10})(2y^5)$

5. $(8x)(4x)$

6. $(7x)(5x)$

7. $(10a^3)(10a)(2a^2)$

8. $(5a^4)(10a)(10a^4)$

9. $(6ab^2)(-4a^2b)$

10. $(-5a^3b)(4ab^4)$

11. $(4x^2y)(3x^3y^3)(2xy^4)$

12. $(5x^6)(-10xy^4)(-2x^2y^6)$

B Divide each expression and simplify. Write all answers with positive exponents only.

13. $\dfrac{15x^3}{5x^2}$

14. $\dfrac{25x^5}{5x^4}$

15. $\dfrac{18y^9}{3y^{12}}$

16. $\dfrac{24y^4}{8y^7}$

17. $\dfrac{32a^3}{64a^4}$

18. $\dfrac{25a^5}{75a^6}$

19. $\dfrac{21a^2b^3}{-7ab^5}$

20. $\dfrac{32a^5b^6}{8ab^5}$

21. $\dfrac{3x^3y^2z}{27xy^2z^3}$ $= \dfrac{x^2}{9z^2}$

22. $\dfrac{5x^5y^4z}{30x^3yz^2}$

23. $\dfrac{-4x^2y^3z}{16x^5yz^2}$

24. $\dfrac{-16x^3yz^2}{24xy^7z}$

25. Fill in the table.

a	b	ab	$\dfrac{a}{b}$	$\dfrac{b}{a}$
10	$5x$			
$20x^3$	$6x^2$			
$25x^5$	$5x^4$			
$3x^{-2}$	$3x^2$			
$-2y^4$	$8y^7$			

26. Fill in the table.

a	b	ab	$\dfrac{a}{b}$	$\dfrac{b}{a}$
$10y$	$2y^2$			
$10y^2$	$2y$			
$5y^3$	15			
5	$15y^3$			
$4y^{-3}$	$4y^3$			

SCAN TO ACCESS

Simplify. Write all answers with positive exponents only.

27. $\dfrac{(3x^2)(8x^5)}{6x^4}$

28. $\dfrac{(7x^3)(6x^8)}{14x^5}$

29. $\dfrac{(9a^2b)(2a^3b^4)}{18a^5b^7}$

30. $\dfrac{(21a^5b)(2a^8b^4)}{14ab}$

31. $\dfrac{(4x^3y^2)(9x^4y^{10})}{(3x^5y)(2x^6y)}$

32. $\dfrac{(5x^4y^4)(10x^3y^3)}{(25xy^5)(2xy^7)}$

33. $\dfrac{(-2xy^2)(-4x^3y)}{(-6xy^3)(xy)}$

34. $\dfrac{(-12a^3b^2)(3ab^3)}{(-6a^2b^2)(ab)}$

Find each product. Write all answers in scientific notation.

35. $(3 \times 10^3)(2 \times 10^5)$

36. $(4 \times 10^8)(1 \times 10^6)$

37. $(3.5 \times 10^4)(5 \times 10^{-6})$

38. $(7.1 \times 10^5)(2 \times 10^{-8})$

39. $(5.5 \times 10^{-3})(2.2 \times 10^{-4})$

40. $(3.4 \times 10^{-2})(4.5 \times 10^{-6})$

Find each quotient. Write all answers in scientific notation.

41. $\dfrac{8.4 \times 10^5}{2 \times 10^2}$

42. $\dfrac{9.6 \times 10^{20}}{3 \times 10^6}$

43. $\dfrac{6 \times 10^8}{2 \times 10^{-2}}$

44. $\dfrac{8 \times 10^{12}}{4 \times 10^{-3}}$

45. $\dfrac{2.5 \times 10^{-6}}{5 \times 10^{-4}}$

46. $\dfrac{4.5 \times 10^{-8}}{9 \times 10^{-4}}$

47. $\dfrac{-4.0 \times 10^{-6}}{8.0 \times 10^{-3}}$

48. $\dfrac{-1.5 \times 10^{-3}}{-6.0 \times 10^2}$

Simplify each expression, and write all answers in scientific notation.

49. $\dfrac{(6 \times 10^8)(3 \times 10^5)}{9 \times 10^7}$

50. $\dfrac{(8 \times 10^4)(5 \times 10^{10})}{2 \times 10^7}$

51. $\dfrac{(5 \times 10^3)(4 \times 10^{-5})}{2 \times 10^{-2}}$

52. $\dfrac{(7 \times 10^6)(4 \times 10^{-4})}{1.4 \times 10^{-3}}$

53. $\dfrac{(2.8 \times 10^{-7})(3.6 \times 10^4)}{2.4 \times 10^3}$

54. $\dfrac{(5.4 \times 10^2)(3.5 \times 10^{-9})}{4.5 \times 10^6}$

Combine by adding or subtracting similar terms as indicated.

55. $3x^2 + 5x^2$

56. $4x^3 + 8x^3$

57. $8x^5 - 19x^5$

58. $75x^6 - 50x^6$

59. $2a + a - 3a$

60. $5a + a - 6a$

61. $10x^3 - 8x^3 + 2x^3$

62. $7x^5 + 8x^5 - 12x^5$

63. $20ab^2 - 19ab^2 + 30ab^2$

64. $18a^3b^2 - 20a^3b^2 + 10a^3b^2$

65. $-3x^2y - 7x^2y - x^2y$

66. $-2a^3b + a^3b - 12a^3b$

67. Fill in the table.

a	b	ab	a + b
$5x$	$3x$		
$4x^2$	$2x^2$		
$3x^3$	$6x^3$		
$2x^4$	$-3x^4$		
x^5	$7x^5$		

68. Fill in the table.

a	b	ab	a − b
$2y$	$3y$		
$-2y$	$3y$		
$4y^2$	$5y^2$		
y^3	$-3y^3$		
$5y^4$	$7y^4$		

Apply the distributive property to simplify each expression.

69. $xy\left(x + \dfrac{1}{y}\right)$

70. $xy\left(y + \dfrac{1}{x}\right)$

71. $xy\left(\dfrac{1}{y} + \dfrac{1}{x}\right)$

72. $xy\left(\dfrac{1}{x} - \dfrac{1}{y}\right)$

73. $x^2\left(1 - \dfrac{4}{x^2}\right)$

74. $x^2\left(1 - \dfrac{9}{x^2}\right)$

75. $x^2\left(1 - \dfrac{1}{x} - \dfrac{6}{x^2}\right)$

76. $x^2\left(1 - \dfrac{5}{x} + \dfrac{6}{x^2}\right)$

77. $x^2\left(1 - \dfrac{5}{x}\right)$

78. $x^2\left(1 - \dfrac{3}{x}\right)$

79. $x^2\left(1 - \dfrac{8}{x}\right)$

80. $x^2\left(1 - \dfrac{6}{x}\right)$

Simplify.

81. $\dfrac{18x^4}{3x} + \dfrac{21x^7}{7x^4}$

82. $\dfrac{24x^{10}}{6x^4} + \dfrac{32x^7}{8x}$

83. $\dfrac{45a^6}{9a^4} - \dfrac{50a^8}{2a^6}$

84. $\dfrac{16a^9}{4a} - \dfrac{28a^{12}}{4a^4}$

85. $\dfrac{6x^7y^4}{3x^2y^2} + \dfrac{8x^5y^8}{2y^6}$

86. $\dfrac{40x^{10}y^{10}}{8x^2y^5} + \dfrac{10x^8y^8}{5y^3}$

Getting Ready for the Next Section

Simplify.

87. $3 - 8$

88. $-5 + 7$

89. $-1 + 7$

90. $1 - 8$

91. $3(5)^2 + 1$

92. $3(-2)^2 - 5(-2) + 4$

93. $2x^2 + 4x^2$

94. $3x^2 - x^2$

95. $-5x + 7x$

96. $x - 2x$

97. $-(2x + 9)$

98. $-(4x^2 - 2x - 6)$

99. Find the value of $2x + 3$ when $x - 4$.

100. Find the value of $(3x)^2$ when $x = 3$.

101. Find the value of $(-2x)^2$ when $x = -3$.

102. Find the value of $(-5y)^3$ when $y = -1$.

Find the Mistake

Each sentence below contains a mistake. Circle the mistake and write the correct word(s) or expression on the line provided.

1. The coefficient of the monomial $12x^4y^2z$ is 4. _____

2. Multiplying the monomials $-15x^3y$ and $2x^2$ gives us $30x^6y$. _____

3. When dividing $28x^6y^2$ by $4x^2y$, write the problem as the separate fractions $\frac{28x^6}{4x^2} \cdot \frac{y^2}{y}$. _____

4. Monomials with similar terms have the same coefficients but different variable parts.

Landmark Review: Checking Your Progress

Simplify the following expressions. Assume all variables are positive numbers and any answers containing exponents should contain positive exponents only.

1. $x^5 \cdot x^3$

2. $y^{15} \cdot y^{12}$

3. $(x^3)^2$

4. $(a^x)^y$

5. $(3a)^3$

6. $(2y)^2(4y)^3$

7. $\frac{5^4}{5^2}$

8. $\frac{(3x)^5}{(3x)^2}$

9. $\frac{2^{-4}}{2^{-2}}$

10. $\frac{(a^2)^3}{(a^4)^2}$

11. $(4xy)^1$

12. $(14x^ny^m)^0$

13. $\frac{9 \times 10^5}{3 \times 10^3}$

14. $\frac{(3 \times 10^5)(6 \times 10^4)}{9 \times 10^6}$

15. $(6x^2y)(3x^3y^3)$

16. $\frac{3a^3b^2c}{6abc^4}$

4.4 Addition and Subtraction of Polynomials

OBJECTIVES

A Add and subtract polynomials.

B Find the value of a polynomial for a given value of the variable.

Suppose we left the television and the ceiling fan on when we left the house for the day. The amount of energy wasted by the ceiling fan was 5 less watts per hour than the television that wasted x watts per hour. The next day, we left on the television, the ceiling fan, and the stereo, which wasted three times the amount of the television. How much total energy did we waste over the two days?

To help us solve this problem, we will arrange all our given information into a table.

KEY WORDS

polynomial

binomial

trinomial

degree

| | Energy Use (Watts/Hour) | | |
	Television	Ceiling fan	Stereo
Day 1	x	$x - 5$	
Day 2	x	$x - 5$	$3x$

Next we need to learn how to add these two polynomials together to get our answer. Once you have read through this section, return to this problem and try to solve it.

In this section we will extend what we learned in Section 4.3 to expressions called polynomials. We begin this section with the definition of a polynomial.

> **DEFINITION** polynomial
>
> A *polynomial* is a finite sum of monomials.

Here are some examples of polynomials:

$$3x^2 + 2x + 1 \qquad 15x^2y + 21xy^2 - y^2 \qquad 3a - 2b + 4c - 5d$$

Polynomials can be further classified by the number of terms they contain. A polynomial with two terms is called a binomial. If it has three terms, it is a trinomial. As stated before, a monomial has only one term.

> **DEFINITION** degree
>
> The degree of a monomial is the sum of the exponents of the variables; whereas the *degree* of a polynomial in one variable is the highest power to which the variable is raised.

Various degrees of polynomials:

$$3x^5 + 2x^3 + 1 \qquad \text{A trinomial of degree 5}$$
$$2x + 1 \qquad \text{A binomial of degree 1}$$
$$3x^2 + 2x + 1 \qquad \text{A trinomial of degree 2}$$
$$3x^5 \qquad \text{A monomial of degree 5}$$
$$-9 \qquad \text{A monomial of degree 0}$$

Practice Problems

1. Add
 $(2x^2 + 3x + 4) + (5x^2 - 6x - 3)$.

A Adding and Subtracting Polynomials

There are no new rules for adding one or more polynomials. We rely only on our previous knowledge. Here are some examples.

EXAMPLE 1 Add $(2x^2 - 5x + 3) + (4x^2 + 7x - 8)$.

Solution We use the commutative and associative properties to group similar terms together and then apply the distributive property to add

$$(2x^2 - 5x + 3) + (4x^2 + 7x - 8)$$
$$= (2x^2 + 4x^2) + (-5x + 7x) + (3 - 8) \qquad \text{Commutative and associative properties}$$
$$= (2 + 4)x^2 + (-5 + 7)x + (3 - 8) \qquad \text{Distributive property}$$
$$= 6x^2 + 2x - 5 \qquad \text{Add.}$$

The results here indicate that to add two polynomials, we add coefficients of similar terms.

2. Add $x^2 + 7x + 4x + 28$.

EXAMPLE 2 Add $x^2 + 3x + 2x + 6$.

Solution The only similar terms here are the two middle terms. We combine them as usual to get

$$x^2 + 3x + 2x + 6 = x^2 + 5x + 6$$

The opposite of a polynomial is the opposite of each of its terms. When you subtract one polynomial from another you subtract each of its terms.

3. Subtract
 $(5x^2 + x + 2) - (x^2 + 3x + 7)$.

EXAMPLE 3 Subtract $(3x^2 + x + 4) - (x^2 + 2x + 3)$.

Solution To subtract $x^2 + 2x + 3$, we change the sign of each of its terms and add. If you are having trouble remembering why we do this, remember that we can think of $-(x^2 + 2x + 3)$ as $-1(x^2 + 2x + 3)$. If we distribute the -1 across $x^2 + 2x + 3$, we get $-x^2 - 2x - 3$.

$$(3x^2 + x + 4) - (x^2 + 2x + 3)$$
$$= 3x^2 + x + 4 - x^2 - 2x - 3 \qquad \text{Take the opposite of each term in the second polynomial.}$$
$$= (3x^2 - x^2) + (x - 2x) + (4 - 3)$$
$$= 2x^2 - x + 1$$

4. Subtract $-4x^2 - 3x + 8$ from
 $x^2 + x + 1$.

EXAMPLE 4 Subtract $-4x^2 + 5x - 7$ from $x^2 - x - 1$.

Solution The polynomial $x^2 - x - 1$ comes first, then the subtraction sign, and finally the polynomial $-4x^2 + 5x - 7$ in parentheses.

$$(x^2 - x - 1) - (-4x^2 + 5x - 7)$$
$$= x^2 - x - 1 + 4x^2 - 5x + 7 \qquad \text{Take the opposite of each term in the second polynomial.}$$
$$= (x^2 + 4x^2) + (-x - 5x) + (-1 + 7)$$
$$= 5x^2 - 6x + 6$$

Answers

1. $7x^2 - 3x + 1$
2. $x^2 + 11x + 28$
3. $4x^2 - 2x - 5$
4. $5x^2 + 4x - 7$

B Finding the Value of a Polynomial

The last topic we want to consider in this section is finding the value of a polynomial for a given value of the variable.

To find the value of the polynomial $3x^2 + 1$ when x is 5, we replace x with 5 and simplify the result.

$$\text{When} \rightarrow \qquad x = 5$$
$$\text{the polynomial} \rightarrow \quad 3x^2 + 1$$
$$\text{becomes} \rightarrow \qquad 3(5)^2 + 1 = 3(25) + 1$$
$$= 75 + 1$$
$$= 76$$

EXAMPLE 5 Find the value of $3x^2 - 5x + 4$ when $x = -2$.

Solution

$$\text{When} \rightarrow \qquad x = -2$$
$$\text{the polynomial} \rightarrow \quad 3x^2 - 5x + 4$$
$$\text{becomes} \rightarrow \qquad 3(-2)^2 - 5(-2) + 4 = 3(4) + 10 + 4$$
$$= 12 + 10 + 4$$
$$= 26$$

EXAMPLE 6 Find the value of $2x^2 + 3x + 5$ when $x = 3$.

Solution

$$\text{When} \rightarrow \qquad x = 3$$
$$\text{the polynomial} \rightarrow \quad 2x^2 + 3x + 5$$
$$\text{becomes} \rightarrow \qquad 2(3)^2 + 3(3) + 5 = 2(9) + 9 + 5$$
$$= 18 + 9 + 5$$
$$= 32$$

5. Find the value of $2x^2 - x + 3$ when $x = -3$.

6. Find the value of $4x^2 + 2x - 3$ when $x = 5$.

GETTING READY FOR CLASS

After reading through the preceding section, respond in your own words and in complete sentences.

A. What is the difference between a binomial and a trinomial?

B. What is the degree of a polynomial?

C. Describe how you would subtract one polynomial from another.

D. How you would find the value of $3x^2 - 5x + 4$ when x is -2?

Answers

5. 24
6. 107

Vocabulary Review

Choose the correct words to fill in the blanks below.

 degree similar polynomial opposite

1. A _____ is a finite sum of monomials.

2. The _____ of a polynomial in one variable is the highest power to which the variable is raised.

3. To add and subtract polynomials, first group _____ terms.

4. The _____ of a polynomial is the opposite of each of its terms.

Problems

Identify each of the following polynomials as a trinomial, binomial, or monomial, and give the degree in each case.

1. $2x^3 - 3x^2 + 1$

2. $4x^2 - 4x + 1$

3. $5 + 8a - 9a^3$

4. $6 + 12x^3 + x^4$

5. $2x - 1$

6. $4 + 7x$

7. $45x^2 - 1$

8. $3a^3 + 8$

9. $7a^2$

10. $90x$

11. -4

12. 56

A Perform the following additions and subtractions.

13. $(2x^2 + 3x + 4) + (3x^2 + 2x + 5)$

14. $(x^2 + 5x + 6) + (x^2 + 3x + 4)$

15. $(3a^2 - 4a + 1) + (2a^2 - 5a + 6)$

16. $(5a^2 - 2a + 7) + (4a^2 - 3a + 2)$

17. $x^2 + 4x + 2x + 8$

18. $x^2 + 5x - 3x - 15$

19. $6x^2 - 3x - 10x + 5$

20. $10x^2 + 30x - 2x - 6$

SCAN TO ACCESS

21. $x^2 - 3x + 3x - 9$

22. $x^2 - 5x + 5x - 25$

23. $3y^2 - 5y - 6y + 10$

24. $y^2 - 18y + 2y - 12$

25. $(6x^3 - 4x^2 + 2x) + (9x^2 - 6x + 3)$

26. $(5x^3 + 2x^2 + 3x) + (2x^2 + 5x + 1)$

27. $\left(\frac{2}{3}x^2 - \frac{1}{5}x - \frac{3}{4}\right) + \left(\frac{4}{3}x^2 - \frac{4}{5}x + \frac{7}{4}\right)$

$2x^2 - x + 1$

28. $\left(\frac{3}{8}x^3 - \frac{5}{7}x^2 - \frac{2}{5}\right) + \left(\frac{5}{8}x^3 - \frac{2}{7}x^2 + \frac{7}{5}\right)$

29. $(a^2 - a - 1) - (-a^2 + a + 1)$

30. $(5a^2 - a - 6) - (-3a^2 - 2a + 4)$

31. $\left(\frac{5}{9}x^3 + \frac{1}{3}x^2 - 2x + 1\right) - \left(\frac{2}{3}x^3 + x^2 + \frac{1}{2}x - \frac{3}{4}\right)$

$-\frac{1}{9}x^3 - \frac{2}{3}x^2 - \frac{5}{2}x + \frac{7}{4}$

32. $\left(4x^3 - \frac{2}{5}x^2 + \frac{3}{8}x - 1\right) - \left(\frac{9}{2}x^3 + \frac{1}{4}x^2 - x + \frac{5}{6}\right)$

33. $(4y^2 - 3y + 2) + (5y^2 + 12y - 4) - (13y^2 - 6y + 20)$

$+ \quad -$

$-4y^2 + 15y - 22$

34. $(2y^2 - 7y - 8) - (6y^2 + 6y - 8) + (4y^2 - 2y + 3)$

35. Subtract $10x^2 + 23x - 50$ from $11x^2 - 10x + 13$.

-3

36. Subtract $2x^2 - 3x + 5$ from $4x^2 - 5x + 10$.

37. Subtract $3y^2 + 7y - 15$ from $11y^2 + 11y + 11$.

$-3y^2 - 7y + 15$

$8y^2 + 4y + 26$

38. Subtract $15y^2 - 8y - 2$ from $3y^2 - 3y + 2$.

39. Add $50x^2 - 100x - 150$ to $25x^2 - 50x + 75$.

40. Add $7x^2 - 8x + 10$ to $-8x^2 + 2x - 12$.

41. Subtract $2x + 1$ from the sum of $3x - 2$ and $11x + 5$.

42. Subtract $3x - 5$ from the sum of $5x + 2$ and $9x - 1$.

B Find the value of the following polynomials for the given variables.

43. Find the value of the polynomial $x^2 - 2x + 1$ when x is 3. **44.** Find the value of the polynomial $(x - 1)^2$ when x is 3.

Applying the Concepts

45. Packaging A crystal ball with a diameter of 6 inches is being packaged for shipment. If the crystal ball is placed inside a circular cylinder with radius 3 inches and height 6 inches, how much volume will need to be filled with padding? Leave your answer in terms of π. (The volume of a sphere with radius r is $\frac{4}{3}\pi r^3$, and the volume of a right circular cylinder with radius r and height h is $\pi r^2 h$.)

3 in.

6 in.

6 in.

46. Packaging Suppose the circular cylinder of Problem 45 has a radius of 4 inches and a height of 7 inches. How much volume will need to be filled with padding? Leave your answer in terms of π.

Getting Ready for the Next Section

Simplify each expression.

47. $(-5)(-1)$

48. $3(-4)$

49. $(-1)(6)$

50. $(-7) \cdot 8$

51. $(5x)(-4x)$

52. $(3x)(2x)$

53. $3x(-7)$

54. $3x(-1)$

55. $5x + (-3x)$

56. $-3x - 10x$

57. $7x - 15x$

58. $-5x - x$

Multiply.

59. $3(2x - 6)$

60. $5(2xy - 7)$

61. $-3(xy - 4)$

62. $-4(a^2 - 9)$

Find the Mistake

Each sentence below contains a mistake. Circle the mistake and write the correct expression on the line provided.

1. The degree of the polynomial $x^2 + 6 + 4x - x^7$ is 2. _____

2. Adding $x^3 + 4x^2 - 8 + x^2$ gives us $6x^5 - 8$. _____

3. The first step when subtracting $(x - 5)^2$ from $(x + 5)^2$ is to use the distributive property on the operation symbols so the problem looks like $(x + 5)^2 + (x + 5)^2$. _____

4. The value of the polynomial $7a^2 + 5a - 10$ when a is 4 is 794. _____

4.5 Multiplication of Polynomials

OBJECTIVES

A Multiply a monomial with a polynomial.

B Multiply two binomials using the FOIL method.

C Multiply two polynomials using the Column method.

KEY WORDS

FOIL method

column method

revenue

Swamp soccer is an exhilarating team sport played just like a standard game of soccer, except the field is a muddy swamp or bog. By the end of the game, most if not all players are soaked and covered head to toe in mud. Suppose you are a coach of your local swamp soccer team. You have coached the team for x years and plan to spend two more years coaching before retiring. If the amount of money you spend on uniforms each year can be given by the binomial $30x + 10$, how much will you have spent by the end of your coaching career?

We can set up this problem as the multiplication of two binomials. Therefore, if $x + 2$ equals the number of total coaching years, then we have

$$(x + 2)(30x + 10)$$

Before we explore the multiplication of two binomials further, let's look at an example of multiplying monomials with polynomials.

A Multiplying Monomials with Polynomials

We begin our discussion of multiplication of polynomials by finding the product of a monomial and a trinomial.

EXAMPLE 1 Multiply $3x^2(2x^2 + 4x + 5)$.

Solution Applying the distributive property gives us

$$3x^2(2x^2 + 4x + 5) = 3x^2(2x^2) + 3x^2(4x) + 3x^2(5) \qquad \text{Distributive property}$$
$$= 6x^4 + 12x^3 + 15x^2 \qquad \text{Multiply.}$$

B Multiplying Binomials

The distributive property is the key to multiplication of polynomials. We can use it to find the product of any two polynomials. There are some shortcuts we can use in certain situations, however. Let's look at an example that involves the product of two binomials.

EXAMPLE 2 Multiply $(3x - 5)(2x - 1)$.

Solution $(3x - 5)(2x - 1) = 3x(2x - 1) - 5(2x - 1)$
$$= 3x(2x) + 3x(-1) + (-5)(2x) + (-5)(-1)$$
$$= 6x^2 - 3x - 10x + 5$$
$$= 6x^2 - 13x + 5$$

Practice Problems

1. Multiply $2x^3(4x^2 - 5x + 3)$.

2. Multiply $(2x - 4)(3x + 2)$.

Answers

1. $8x^5 - 10x^4 + 6x^3$

2. $6x^2 - 8x - 8$

If we look closely at the second and third lines of work in this example, we can see that the terms in the answer come from all possible products of terms in the first binomial with terms in the second binomial. This result is generalized as follows.

> **RULE** Multiplication of Two Polynomials
>
> To multiply any two polynomials, multiply each term in the first with each term in the second.

There are two ways we can put this rule to work, the first of which is called the FOIL method.

FOIL Method

If we look at the original problem in Example 2 and then at the answer, we see that the first term in the answer came from multiplying the first terms in each binomial.

$$3x \cdot 2x = 6x^2 \qquad \text{First}$$

The middle term in the answer came from adding the products of the two outside terms with the two inside terms in each binomial.

$$3x(-1) = -3x \qquad \text{Outside}$$
$$\underline{-5(2x) = -10x} \qquad \text{Inside}$$
$$= -13x$$

The last term in the answer came from multiplying the two last terms.

$$-5(-1) = 5 \qquad \text{Last}$$

To summarize the FOIL method, we will multiply another two binomials.

3. Multiply $(3x + 5)(2x - 1)$.

EXAMPLE 3 Multiply $(2x + 3)(5x - 4)$.

Solution $(2x + 3)(5x - 4) = \underline{2x(5x)} + \underline{2x(-4)} + \underline{3(5x)} + \underline{3(-4)}$

$$\qquad\qquad \uparrow \qquad \uparrow \qquad \uparrow \qquad \uparrow$$
$$\qquad\quad \text{First} \quad \text{Outside} \quad \text{Inside} \quad \text{Last}$$

$$= 10x^2 - 8x + 15x - 12$$
$$= 10x^2 + 7x - 12$$

With practice, adding the products of the outside and inside terms $(-8x + 15x = 7x)$ can be done mentally.

C Column Method

The FOIL method can be applied only when multiplying two binomials. To find products of polynomials with more than two terms, we use what is called the Column method.

The column method of multiplying two polynomials is very similar to long multiplication with whole numbers. It is just another way of finding all possible products of terms in one polynomial with terms in another polynomial.

4. Multiply $(3x - 2)(2x^2 - 3x + 4)$.

EXAMPLE 4 Multiply $(2x + 3)(3x^2 - 2x + 1)$.

Solution
$$3x^2 - 2x + 1$$
$$\underline{2x + 3}$$
$$6x^3 - 4x^2 + 2x \qquad \leftarrow 2x(3x^2 - 2x + 1)$$
$$\underline{\qquad\quad 9x^2 - 6x + 3} \qquad \leftarrow 3(3x^2 - 2x + 1)$$
$$6x^3 + 5x^2 - 4x + 3 \qquad \text{Add similar terms.}$$

It will be to your advantage to become very fast and accurate at multiplying polynomials. You should be comfortable using either method. The following examples illustrate the three types of multiplication.

Answers

3. $6x^2 + 7x - 5$

4. $6x^3 - 13x^2 + 18x - 8$

EXAMPLE 5 Multiply.

a. $4a^2(2a^2 - 3a + 5)$

$$\begin{array}{r} 2a^2 - 3a + 5 \\ 4a^2 \\ \hline 8a^4 - 12a^3 + 20a \end{array}$$

b. $(x - 2)(y + 3) = x(y) + x(3) + (-2)(y) + (-2)(3)$

$\qquad\qquad\qquad$ F \qquad O \qquad I \qquad L

$\qquad\quad = xy + 3x - 2y - 6$

c. $(x + y)(a - b) = x(a) + x(-b) + y(a) + y(-b)$

$\qquad\qquad\qquad$ F \qquad O \qquad I \qquad L

$\qquad\quad = xa - xb + ya - yb$

d. $(5x - 1)(2x + 6) = 5x(2x) + 5x(6) + (-1)(2x) + (-1)(6)$

$\qquad\qquad\qquad\qquad$ F \qquad O \qquad I \qquad L

$\qquad\qquad\quad = 10x^2 + 30x + (-2x) + (-6)$

$\qquad\qquad\quad = 10x^2 + 28x - 6$

Applications

EXAMPLE 6 The length of a rectangle is 3 more than twice the width. Write an expression for the area of the rectangle.

Solution We begin by drawing a rectangle and labeling the width with x. Since the length is 3 more than twice the width, we label the length with $2x + 3$.

$$\begin{array}{c} 2x + 3 \\ x\ \boxed{} \end{array}$$

Since the area A of a rectangle is the product of the length and width, we write our formula for the area of this rectangle as

$\qquad A = x(2x + 3)$

$\qquad A = 2x^2 + 3x \qquad$ Multiply.

Suppose that a store sells x items at p dollars per item. The total amount of money obtained by selling the items is called the *revenue*. It can be found by multiplying the number of items sold, x, by the price per item, p. For example, if 100 items are sold for $6 each, the revenue is $100(6) = \$600$. Similarly, if 500 items are sold for $8 each, the total revenue is $500(8) = \$4,000$. If we denote the revenue with the letter R, then the formula that relates R, x, and p is

$\qquad\qquad$ Revenue = (number of items sold)(price of each item)

In symbols: $R = xp$.

5. Multiply.

a. $5a^2(3a^2 - 2a + 4)$

b. $(x + 3)(y - 5)$

c. $(x + 2y)(a + b)$

d. $(4x + 1)(3x + 5)$

6. The length of a rectangle is 5 more than three times the width. Write an expression for the area.

Answers

5. a. $15a^4 - 10a^3 + 20a^2$

\quad **b.** $xy - 5x + 3y - 15$

\quad **c.** $ax + bx + 2ay + 2by$

\quad **d.** $12x^2 + 23x + 5$

6. $3x^2 + 5x$

7. Suppose the store in Example 7 can sell x flash drives for p dollars according to the equation $x = 700 - 110p$. Find the revenue.

EXAMPLE 7 A store selling flash drives for home computers knows from past experience that it can sell x flash drives each day at a price of p dollars each, according to the equation $x = 800 - 100p$. Write a formula for the daily revenue that involves only the variables R and p.

Solution From our previous discussion we know that the revenue R is given by the formula

$$R = xp$$

But, since $x = 800 - 100p$, we can substitute $800 - 100p$ for x in the revenue equation to obtain

$$R = (800 - 100p)p$$
$$R = 800p - 100p^2$$

This last formula gives the revenue R in terms of the price p.

GETTING READY FOR CLASS

After reading through the preceding section, respond in your own words and in complete sentences.

A. Describe how you would use the Column method to multiply two polynomials.

B. Describe how the distributive property is used to multiply a monomial and a polynomial.

C. Describe how you would use the FOIL method to multiply two binomials.

D. Show how the product of two binomials can be a trinomial.

Vocabulary Review

Choose the correct words to fill in the blanks below.

term distributive products column binomials

1. To multiply a monomial with a polynomial, apply the _____ property.
2. To multiply any two polynomials, multiply each _____ in the first with each term in the second.
3. The FOIL method allows us to multiply any two _____ by finding the sum of the products of the first, outside, inside, and last two terms.
4. Use the column method to find the _____ of polynomials with more than two terms.
5. The _____ method of multiplying two polynomials is similar to long multiplication with whole numbers.

Problems

A Multiply the following by applying the distributive property.

1. $2x(3x + 1)$

2. $4x(2x - 3)$

3. $2x^2(3x^2 - 2x + 1)$

4. $5x(4x^3 - 5x^2 + x)$

5. $2ab(a^2 - ab + 1)$

6. $3a^2b(a^3 + a^2b^2 + b^3)$

7. $y^2(3y^2 + 9y + 12)$

8. $5y(2y^2 - 3y + 5)$

9. $4x^2y(2x^3y + 3x^2y^2 + 8y^3)$

10. $6xy^3(2x^2 + 5xy + 12y^2)$

B C Multiply the following binomials. You should do about half the problems using the FOIL method and the other half using the column method. Remember, you want to be comfortable using both methods.

11. $(x + 3)(x + 4)$

12. $(x + 2)(x + 5)$

13. $(x + 6)(x + 1)$

14. $(x + 1)(x + 4)$

15. $\left(x + \dfrac{1}{2}\right)\left(x + \dfrac{3}{2}\right)$

16. $\left(x + \dfrac{3}{5}\right)\left(x + \dfrac{2}{5}\right)$

17. $(a + 5)(a - 3)$

18. $(a - 8)(a + 2)$

19. $(x - a)(y + b)$

20. $\left(y + \dfrac{5}{6}\right)\left(y - \dfrac{5}{6}\right)$　　**21.** $\left(y - \dfrac{4}{7}\right)\left(y + \dfrac{4}{7}\right)$　　**22.** $(2x - 3)(x - 4)$　　**23.** $(3x - 5)(x - 2)$

24. $(a + 2)(2a - 1)$　　**25.** $(a - 6)(3a + 2)$　　**26.** $(2x - 5)(3x - 2)$　　**27.** $(3x + 6)(2x - 1)$

28. $(2x + 3)(a + 4)$　　**29.** $(2x - 3)(a - 4)$　　**30.** $(5x - 4)(5x + 4)$　　**31.** $(6x + 5)(6x - 5)$

32. $\left(2x - \dfrac{1}{2}\right)\left(x + \dfrac{3}{2}\right)$　　**33.** $\left(4x - \dfrac{3}{2}\right)\left(x + \dfrac{1}{2}\right)$　　**34.** $(1 - 2a)(3 - 4a)$　　**35.** $(1 - 3a)(3 + 2a)$

For each of the following problems, fill in the area of each small rectangle and square, and then add the results together to find the indicated product. Also, show that you get the same answer by multiplying the binomials and simplifying the product.

36. $(x + 2)(x + 3)$

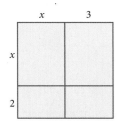

37. $(x + 4)(x + 5)$

38. $(x + 1)(2x + 2)$

39. $(2x + 1)(2x + 2)$

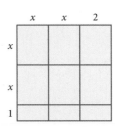

Multiply the following polynomials, and simplify your answer.

40. $(a - 3)(a^2 - 3a + 2)$

41. $(a + 5)(a^2 + 2a + 3)$

42. $(x + 2)(x^2 - 2x + 4)$

43. $(x + 3)(x^2 - 3x + 9)$

44. $(2x + 1)(x^2 + 8x + 9)$

45. $(3x - 2)(x^2 - 7x + 8)$

46. $(5x^2 + 2x + 1)(x^2 - 3x + 5)$

47. $(2x^2 + x + 1)(x^2 - 4x + 3)$

48. $(x^2 + 3)(2x^2 - 5)$

49. $(4x^3 - 8)(5x^3 + 4)$

50. $(3a^4 + 2)(2a^2 + 5)$

51. $(7a^4 - 8)(4a^3 - 6)$

52. $(x + 3)(x + 4)(x + 5)$

53. $(x - 3)(x - 4)(x - 5)$

Simplify the following expressions.

54. $(x - 3)(x - 2) + 2$

55. $(2x - 5)(3x + 2) - 4$

56. $(2x - 3)(4x + 3) + 4$

57. $(3x + 8)(5x - 7) + 52$

58. $(x + 4)(x - 5) + (-5)(2)$

59. $(x + 3)(x - 4) + (-4)(2)$

60. $2(x - 3) + x(x + 2)$

61. $5(x + 3) + 1(x + 4)$

62. $3x(x + 1) - 2x(x - 5)$

63. $4x(x - 2) - 3x(x - 4)$

64. $x(x + 2) - 3$

65. $2x(x - 4) + 6$

66. $a(a - 3) + 6$

67. $a(a - 4) + 8$

Applying the Concepts

68. Area The length of a rectangle is 5 units more than twice the width. Write an expression for the area of the rectangle.

69. Area The length of a rectangle is 2 more than three times the width. Write an expression for the area of the rectangle.

70. Area The width and length of a rectangle are given by two consecutive integers. Write an expression for the area of the rectangle.

71. Area The width and length of a rectangle are given by two consecutive even integers. Write an expression for the area of the rectangle.

72. A rectangular swimming pool has a length which is three times as long as its width. A five foot pathway is built to surround the entire pool.

 a. Write and simplify an expression for the total area of the pool plus the pathway.

 b. Write and simplify an expression for the area of the pathway alone.

73. A circular garden has a radius of r feet. A 2 foot wide rock border is placed around the entire garden.

 a. Write and simplify an expression in terms of π for the total area of the garden plus the rock border.

 b. Write and simplify an expression for the area of the rock border alone.

74. A rectangular piece of cardboard has a length that is 1 inch more than twice the width w. Squares of sides 2 inches are cut from the corners, and the resulting flaps are folded up to create the sides of a box.

a. Write and simplify an expression for the volume of the box.

b. Write and simplify an expression for the surface area of the box. Assume the box is open (with no top).

75. Revenue A store selling hair ribbons knows that the number of ribbons it can sell each week, x, is related to the price per ribbon, p, by the equation $x = 1,200 - 100p$. Write an expression for the weekly revenue that involves only the variables R and p. (*Remember:* The equation for revenue is $R = xp$.)

76. Revenue A store selling headphones knows from past experience that the number of headphones it can sell each week, x, is related to the price per set, p, by the equation $x = 1,300 - 100p$. Write an expression for the weekly revenue that involves only the variables R and p.

Getting Ready for the Next Section

Simplify each expression.

77. $13 \cdot 13$

78. $3x \cdot 3x$

79. $2(x)(-5)$

80. $2(2x)(-3)$

81. $6x + (-6x)$

82. $3x + (-3x)$

83. $(2x)(-3) + (2x)(3)$

84. $(2x)(-5y) + (2x)(5y)$

Multiply the polynomials and simplify your answer.

85. $-4(3x - 4)$

86. $-2x(2x + 7)$

87. $(x - 1)(x + 2)$

88. $(x + 5)(x - 6)$

89. $(x + 3)(x + 3)$

90. $(3x - 2)(3x - 2)$

91. $(5x + 3)(5x - 3)$

92. $(4x - 2)(4x + 2)$

Find the Mistake

Each sentence below contains a mistake. Circle the mistake and write the correct word(s) or expression on the line provided.

1. To find the product of two binomials, multiply the first terms and then multiply the second terms.

2. When using the FOIL method to multiply the binomials $(9x - 2)$ and $(5x + 3)$, the product of the two outside terms is -6.

3. For the problem $(x^2 - 4)(3x^2 + 6)$, adding the products of the two outside terms with the two inside terms gives us $4x^2 + 2$.

4. For the problem $4x^2(3x^2 + 4x + 9)$, the degree of the answer is 2.

OBJECTIVES

A Find the square of a binomial.

B Multiply expressions of the form $(a + b)(a - b)$.

KEY WORDS

square of a binomial

difference of two squares

Suppose an extreme whitewater kayaker decides to spend a weekend making runs down the Class III Gorge run of the South Fork American River in California. The kayaker considers a successful run one where he paddles the length of the river without flipping over. If the number of successful runs for the kayaker during the weekend equals $(x + 2)^2$, how would we square this binomial to help find a solution? Let's explore this concept further.

A Squaring Binomials

In this section, we will combine the results of the last section with our definition of exponents to find some special products.

EXAMPLE 1 Find the square of $(3x - 2)$.

Solution To square $(3x - 2)$, we multiply it by itself.

$$(3x - 2)^2 = (3x - 2)(3x - 2) \qquad \text{Definition of exponents}$$
$$= 9x^2 - 6x - 6x + 4 \qquad \text{FOIL method}$$
$$= 9x^2 - 12x + 4 \qquad \text{Combine similar terms.}$$

Notice that the first and last terms in the answer are the square of the first and last terms in the original problem and that the middle term is twice the product of the two terms in the original binomial.

EXAMPLE 2 Expand and multiply.

a. $(a + b)^2$ **b.** $(a - b)^2$

Solution

a. $(a + b)^2 = (a + b)(a + b)$
$$= a^2 + 2ab + b^2$$

b. $(a - b)^2 = (a - b)(a - b)$
$$= a^2 - 2ab + b^2$$

Practice Problems

1. Find the square of $(4x - 5)$.

2. Multiply.

 a. $(x + y)^2$

 b. $(x - y)^2$

Answers

1. $16x^2 - 40x + 25$

2. a. $x^2 + 2xy + y^2$

 b. $x^2 - 2xy + y^2$

Binomial squares having the form of Example 2 occur very frequently in algebra. It will be to your advantage to memorize the following rule for squaring a binomial.

Note A very common mistake when squaring binomials is to write

$$(a + b)^2 = a^2 + b^2$$

which just isn't true. The mistake becomes obvious when we substitute 2 for a and 3 for b:

$$(2 + 3)^2 \neq 2^2 + 3^2$$
$$25 \neq 13$$

Exponents do not distribute over addition or subtraction.

> **RULE** Square of a Binomial
>
> The square of a binomial is the sum of the square of the first term, the square of the last term, and twice the product of the two original terms. In symbols this rule is written as follows:
>
$(a + b)^2$	$=$	a^2	$+$	$2ab$	$+$	b^2
> | $(a + b)(a + b)$ | $=$ | Square of first term | | Twice product of the two terms | | Square of last term |

This rule also applies to squares of differences, such as $(a - b)^2$, because the difference can be thought of as the addition of a negative.

EXAMPLE 3 Multiply using the preceding rule.

	First term squared	Twice their product	Last term squared	Answer
a. $(x - 5)^2 =$	x^2	$+ \ 2(x)(-5) \ +$	25	$= \ x^2 - 10x + 25$
b. $(x + 2)^2 =$	x^2	$+ \ 2(x)(2) \ +$	4	$= \ x^2 + 4x + 4$
c. $(2x - 3)^2 =$	$4x^2$	$+ \ 2(2x)(-3) \ +$	9	$= \ 4x^2 - 12x + 9$
d. $(5x - 4)^2 =$	$25x^2$	$+ \ 2(5x)(-4) \ +$	16	$= \ 25x^2 - 40x + 16$

3. Multiply using the method shown in Example 3.
 a. $(x - 3)^2$
 b. $(x + 5)^2$
 c. $(3x - 2)^2$
 d. $(4x + 5)^2$

B More Special Products

Another special product that occurs frequently is $(a + b)(a - b)$. The only difference in the two binomials is the sign between the two terms. The interesting thing about this type of product is that the middle term is always zero. Here are some examples.

EXAMPLE 4 Multiply using the FOIL method.

a. $(2x - 3)(2x + 3) = 4x^2 + 6x - 6x - 9$ FOIL method

$$= 4x^2 - 9$$

b. $(x - 5)(x + 5) = x^2 + 5x - 5x - 25$ FOIL method

$$= x^2 - 25$$

c. $(3x - 1)(3x + 1) = 9x^2 + 3x - 3x - 1$ FOIL method

$$= 9x^2 - 1$$

4. Multiply using the FOIL method.
 a. $(3x - 5)(3x + 5)$
 b. $(x - 4)(x + 4)$
 c. $(4x - 1)(4x + 1)$

Notice that in each case the middle term is zero and therefore doesn't appear in the answer. The answers all turn out to be the difference of two squares. Here is a rule to help you memorize the result:

> **RULE** Difference of Two Squares
>
> When multiplying two binomials that differ only in the sign between their terms, subtract the square of the last term from the square of the first term.
>
> $$(a - b)(a + b) = a^2 - b^2$$

Answers
3. **a.** $x^2 - 6x + 9$
 b. $x^2 + 10x + 25$
 c. $9x^2 - 12x + 4$
 d. $16x^2 + 40x + 25$
4. **a.** $9x^2 - 25$ **b.** $x^2 - 16$
 c. $16x^2 - 1$

Here are some problems that result in the difference of two squares:

EXAMPLE 5 Multiply using the preceding rule.

a. $(x + 3)(x - 3) = x^2 - 9$

b. $(a + 2)(a - 2) = a^2 - 4$

c. $(9a + 1)(9a - 1) = 81a^2 - 1$

d. $(2x - 5y)(2x + 5y) = 4x^2 - 25y^2$

e. $(3a - 7b)(3a + 7b) = 9a^2 - 49b^2$

Although all the problems in this section can be worked correctly using the methods in the previous section, they can be done much faster if the two rules are *memorized.* Here is a summary of the two rules:

$$\left. \begin{array}{l} (a + b)^2 = (a + b)(a + b) = a^2 + 2ab + b^2 \\ (a - b)^2 = (a - b)(a - b) = a^2 - 2ab + b^2 \end{array} \right\} \text{Binomial Squares}$$

$$(a - b)(a + b) = a^2 - b^2 \qquad \text{Difference of two squares}$$

EXAMPLE 6 Write an expression in symbols for the sum of the squares of three consecutive even integers. Then, simplify that expression.

Solution If we let $x =$ the first of the even integers, then $x + 2$ is the next consecutive even integer, and $x + 4$ is the one after that. An expression for the sum of their squares is

$$x^2 + (x + 2)^2 + (x + 4)^2 \qquad \text{Sum of squares}$$

$$= x^2 + (x^2 + 4x + 4) + (x^2 + 8x + 16) \qquad \text{Expand squares.}$$

$$= 3x^2 + 12x + 20 \qquad \text{Add similar terms.}$$

GETTING READY FOR CLASS

After reading through the preceding section, respond in your own words and in complete sentences.

A. Describe how you would square the binomial $(a + b)$.

B. Explain why $(x + 3)^2$ cannot be $x^2 + 9$.

C. Why is the middle term in the product of $(a + b)$ and $(a - b)$ equal to zero?

D. When multiplied out, how will $(x + 3)^2$ and $(x - 3)^2$ differ?

5. Multiply using the formula
$(a + b)(a - b) = a^2 - b^2$.
a. $(x + 2)(x - 2)$
b. $(a + 7)(a - 7)$
c. $(6a + 1)(6a - 1)$
d. $(8x - 2y)(8x + 2y)$
e. $(4a - 3b)(4a + 3b)$

6. Expand and simplify.
$x^2 + (x - 1)^2 + (x + 1)^2$

Answers

5. a. $x^2 - 4$ **b.** $a^2 - 49$ **c.** $36a^2 - 1$
d. $64x^2 - 4y^2$ **e.** $16a^2 - 9b^2$
6. $3x^2 + 2$

Vocabulary Review

Choose the correct words to fill in the blanks below.

| subtract | sum | square of a binomial | difference of two squares |

1. The square of a binomial is the _____ of the square of the first term, the square of the last term, and twice the product of the two original terms.
2. The _____ is given by $(x + y)^2 = x^2 + 2xy + y^2$.
3. When multiplying two binomials that differ only in the sign between their terms, _____ the square of the last term from the square of the first term.
4. The _____ is given by $(a - b)(a + b) = a^2 - b^2$.

Problems

A Perform the indicated operations for each binomial.

1. $(x - 2)^2$

2. $(x + 2)^2$

3. $(a + 3)^2$

4. $(a - 3)^2$

5. $(x - 5)^2$

6. $(x - 4)^2$

7. $\left(a - \dfrac{1}{2}\right)^2$

8. $\left(a + \dfrac{1}{2}\right)^2$

9. $(x + 10)^2$

10. $(x - 10)^2$

11. $(a + 0.8)^2$

12. $(a - 0.4)^2$

13. $(2x - 1)^2$

14. $(3x + 2)^2$

15. $(4a + 5)^2$

16. $(4a - 5)^2$

17. $(3x - 2)^2$

18. $(2x - 3)^2$

19. $(3a + 5b)^2$

20. $(5a - 3b)^2$

21. $(4x - 5y)^2$

22. $(5x + 4y)^2$

23. $(7m + 2n)^2$

24. $(2m - 7n)^2$

25. $(6x - 10y)^2$

26. $(10x + 6y)^2$

27. $(x^2 + 5)^2$

28. $(x^2 + 3)^2$

29. $(a^2 + 1)^2$

30. $(a^2 - 2)^2$

31. $(x^3 - 4)^2$

32. $(y^3 + 2)^2$

SCAN TO ACCESS

Comparing Expressions The following problems are meant to show that $(x + y)^2 \neq x^2 + y^2$ and $(x - y)^2 \neq x^2 - y^2$. Fill in each table.

33.

x	$(x + 3)^2$	$x^2 + 9$	$x^2 + 6x + 9$
1			
2			
3			
4			

34.

x	$(x - 5)^2$	$x^2 - 25$	$x^2 - 10x + 25$
1			
2			
3			
4			

35.

a	1	3	3	4
b	1	5	4	5
$(a + b)^2$				
$a^2 + b^2$				
$a^2 + ab + b^2$				
$a^2 + 2ab + b^2$				

36.

a	2	5	2	4
b	1	2	5	3
$(a - b)^2$				
$a^2 - b^2$				
$a^2 - 2ab + b^2$				

B Multiply the following binomials.

37. $(a + 5)(a - 5)$

38. $(a - 6)(a + 6)$

39. $(y - 1)(y + 1)$

40. $(y - 2)(y + 2)$

41. $(9 + x)(9 - x)$

42. $(10 - x)(10 + x)$

43. $(2x + 5)(2x - 5)$

44. $(3x + 5)(3x - 5)$

45. $\left(4x + \dfrac{1}{3}\right)\left(4x - \dfrac{1}{3}\right)$

46. $\left(6x + \dfrac{1}{4}\right)\left(6x - \dfrac{1}{4}\right)$

47. $(2a + 7)(2a - 7)$

48. $(3a + 10)(3a - 10)$

49. $(6 - 7x)(6 + 7x)$

50. $(7 - 6x)(7 + 6x)$

51. $(x^2 + 3)(x^2 - 3)$

52. $(x^2 + 2)(x^2 - 2)$

53. $(a^2 + 4)(a^2 - 4)$

54. $(a^2 + 9)(a^2 - 9)$

55. $(5y^4 - 8)(5y^4 + 8)$

56. $(7y^5 + 6)(7y^5 - 6)$

Simplify the following expressions.

57. $(x + 3)(x - 3) + (x - 5)(x + 5)$

58. $(x - 7)(x + 7) + (x - 4)(x + 4)$

59. $(2x + 3)^2 - (4x - 1)^2$

60. $(3x - 5)^2 - (2x + 3)^2$

61. $(a + 1)^2 - (a + 2)^2 + (a + 3)^2$

62. $(a - 1)^2 + (a - 2)^2 - (a - 3)^2$

63. $(2x + 3)^3$

64. $(3x - 2)^3$

65. $(x + 5)^2 - (x - 5)^2$

66. $(2x + 1)^2 - (2x - 1)^2$

67. $(x - y)^2 - (x + y)^2$

68. $(a + b)^2 - (a - b)^2$

69. $(x - 1)^2 + x^2 + (x + 1)^2$

70. $x^2 + (x + 1)^2 + (x + 2)^2$

71. $(2a - 1)^3$

72. $(4a - 3)^3$

Applying the Concepts

73. Shortcut The formula for the difference of two squares can be used as a shortcut to multiplying certain whole numbers if they have the correct form. Use the difference of two squares formula to multiply 49(51) by first writing 49 as $(50 - 1)$ and 51 as $(50 + 1)$.

74. Shortcut Use the difference of two squares formula to multiply 101(99) by first writing 101 as $(100 + 1)$ and 99 as $(100 - 1)$.

75. Comparing Expressions Evaluate the expression $(x + 3)^2$ and the expression $x^2 + 6x + 9$ for $x = 2$.

76. Comparing Expressions Evaluate the expression $x^2 - 25$ and the expression $(x - 5)(x + 5)$ for $x = 6$.

77. Number Problem Write an expression for the sum of the squares of two consecutive integers. Then, simplify that expression.

78. Number Problem Write an expression for the sum of the squares of two consecutive odd integers. Then, simplify that expression.

79. Number Problem Write an expression for the sum of the squares of three consecutive integers. Then, simplify that expression.

80. Number Problem Write an expression for the sum of the squares of three consecutive odd integers. Then, simplify that expression.

81. Area We can use the concept of area to further justify our rule for squaring a binomial. The length of each side of the square shown in the figure is $a + b$. (The longer line segment has length a and the shorter line segment has length b.) The area of the whole square is $(a + b)^2$. However, the whole area is the sum of the areas of the two smaller squares and the two smaller rectangles that make it up. Write the area of the two smaller squares and the two smaller rectangles and then add them together to verify the formula $(a + b)^2 = a^2 + 2ab + b^2$.

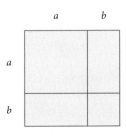

82. Area The length of each side of the large square shown in the figure is $x + 5$. Therefore, its area is $(x + 5)^2$. Find the area of the two smaller squares and the two smaller rectangles that make up the large square, then add them together to verify the formula $(x + 5)^2 = x^2 + 10x + 25$.

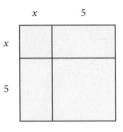

83. You have an x foot square available to plant a lawn. You want a 2 foot border around the lawn. Write and simplify an expression for the actual amount of grass to plant.

84. You have a lawn which is 20 feet by 20 feet square. You want to plant x feet of flowers around the lawn. Write an expression for the total area of lawn plus flowers, then simplify.

Getting Ready for the Next Section

Simplify.

85. $\dfrac{10x^3}{5x}$

86. $\dfrac{15x^2}{5x}$

87. $\dfrac{3x^2}{3}$

88. $\dfrac{4x^2}{2}$

89. $\dfrac{9x^2}{3x}$

90. $\dfrac{3x^4}{3x^2}$

91. $\dfrac{24x^3y^2}{8x^2y}$

92. $\dfrac{4x^2y^3}{8x^2y}$

Find the Mistake

Each sentence below contains a mistake. Circle the mistake and write the correct word(s) or expression on the line provided.

1. To find the square of the binomial $(5a^2 - 9a)$, we multiply $5a^2$ by $-9a$. _____

2. The second term in a binomial square is the square of the last term. _____

3. Simplifying $(7x - 3)^2$ gives us $7x^2 - 21x + 9$. _____

4. The problem $(4a - 6)(4a + 6) = 16a^2 - 36$ is an example of a binomial square. _____

4.7 Dividing a Polynomial by a Monomial

Image © sxc.hu, canoncan, 2006

The number of tubes an artist used to paint his masterpieces in one month may be given by $3x^2 + 2x$. If $3x$ represents the number of paintings he completed in that month, how would we find the number of tubes he used per painting? We would set this up as a division problem. To solve it, we need to understand how to divide a polynomial by a monomial.

A Dividing a Polynomial by a Monomial

To divide a polynomial by a monomial, we will use the definition of division and apply the distributive property. Follow the steps in this example closely.

EXAMPLE 1 Divide $10x^3 - 15x^2$ by $5x$.

Solution

$$\frac{10x^3 - 15x^2}{5x} = (10x^3 - 15x^2)\frac{1}{5x} \qquad \text{Division by 5x is the same as multiplication by } \tfrac{1}{5x}.$$

$$= 10x^3\left(\frac{1}{5x}\right) - 15x^2\left(\frac{1}{5x}\right) \quad \text{Distribute } \tfrac{1}{5x} \text{ to both terms.}$$

$$= \frac{10x^3}{5x} - \frac{15x^2}{5x} \qquad \text{Multiplication by } \tfrac{1}{5x} \text{ is the same as division by 5x.}$$

$$= 2x^2 - 3x \qquad \text{Divide monomials.}$$

If we were to leave out the first steps in Example 1, the problem would look like this:

$$\frac{10x^3 - 15x^2}{5x} = \frac{10x^3}{5x} - \frac{15x^2}{5x}$$

$$= 2x^2 - 3x$$

The problem is much shorter and clearer this way. You may leave out the first two steps from Example 1 when working problems in this section. They are part of Example 1 only to help show you why the following rule is true.

> **RULE** Division of a Polynomial by a Monomial
>
> To divide a polynomial by a monomial, simply divide each term in the polynomial by the monomial.

On the next page are some further examples using our rule for division of a polynomial by a monomial.

Practice Problems

1. Divide $8x^3 - 12x^2$ by $4x$.

Answer

1. $2x^2 - 3x$

2. Divide $\dfrac{5x^2 - 10}{5}$.

EXAMPLE 2 Divide $\dfrac{3x^2 - 6}{3}$.

Solution We begin by writing the 3 in the denominator under each term in the numerator. Then we simplify the result.

$$\frac{3x^2 - 6}{3} = \frac{3x^2}{3} - \frac{6}{3} \qquad \textit{Divide each term in the numerator by 3.}$$

$$= x^2 - 2 \qquad \textit{Simplify.}$$

3. Divide $\dfrac{8x^2 - 2}{2}$.

EXAMPLE 3 Divide $\dfrac{4x^2 - 2}{2}$.

Solution Dividing each term in the numerator by 2, we have

$$\frac{4x^2 - 2}{2} = \frac{4x^2}{2} - \frac{2}{2} \qquad \textit{Divide each term in the numerator by 2.}$$

$$= 2x^2 - 1 \qquad \textit{Simplify.}$$

4. Find the quotient of $15x^4 - 10x^3$ and $5x$.

EXAMPLE 4 Find the quotient of $27x^3 - 9x^2$ and $3x$.

Solution We again are asked to divide the first polynomial by the second one.

$$\frac{27x^3 - 9x^2}{3x} = \frac{27x^3}{3x} - \frac{9x^2}{3x} \qquad \textit{Divide each term by 3x.}$$

$$= 9x^2 - 3x \qquad \textit{Simplify.}$$

5. Divide $\dfrac{21x^3y^2 + 14x^2y^2 - 7x^2y^3}{7x^2y}$.

EXAMPLE 5 Divide $\dfrac{24x^3y^2 + 16x^2y^2 - 4x^2y^3}{8x^2y}$.

Solution Writing $8x^2y$ under each term in the numerator and then simplifying, we have

$$\frac{24x^3y^2 + 16x^2y^2 - 4x^2y^3}{8x^2y} = \frac{24x^3y^2}{8x^2y} + \frac{16x^2y^2}{8x^2y} - \frac{4x^2y^3}{8x^2y}$$

$$= 3xy + 2y - \frac{y^2}{2}$$

From the examples in this section, it is clear that to divide a polynomial by a monomial, we must divide each term in the polynomial by the monomial. Often, students taking algebra for the first time will make the following mistake:

$$\frac{x + 2}{2} = x + 1 \qquad \textit{Mistake}$$

The mistake here is in not dividing both terms in the numerator by 2. The correct way to divide $x + 2$ by 2 looks like this:

$$\frac{x + 2}{2} = \frac{x}{2} + \frac{2}{2} = \frac{x}{2} + 1 \qquad \textit{Correct}$$

GETTING READY FOR CLASS

After reading through the preceding section, respond in your own words and in complete sentences.

A. What property of real numbers is the key to dividing a polynomial by a monomial?

B. Describe how you would divide the polynomial $x^2 + 10x$ by $5x$.

C. Describe how you would divide a polynomial by a negative monomial.

D. In your own words, explain the mistake in the problem $\dfrac{x + 2}{2} = x + 1$.

Answers

2. $x^2 - 2$
3. $4x^2 - 1$
4. $3x^3 - 2x^2$
5. $3xy + 2y - y^2$

Vocabulary Review

Choose the correct words to fill in the blanks below.

distributive term polynomial division

1. Use the definition of _____ and apply the _____ property to divide a polynomial by a monomial.

2. To divide a _____ by a monomial, divide each _____ in the polynomial by the monomial.

Problems

A Divide the following polynomials by $5x$.

1. $5x^2 - 10x$

2. $10x^3 - 15x$

3. $15x - 10x^3$

4. $50x^3 - 20x^2$

5. $25x^2y - 10xy$

6. $15xy^2 + 20x^2y$

7. $35x^5 - 30x^4 + 25x^3$

8. $40x^4 - 30x^3 + 20x^2$

9. $50x^5 - 25x^3 + 5x$

10. $75x^6 + 50x^3 - 25x$

11. $-20x^3 + 15x^2 - 5x$

12. $100x^5 - 80x^4 - 40x^3$

Divide the following by $-2a$.

13. $8a^2 - 4a$

14. $a^3 - 6a^2$

15. $16a^5 + 24a^4$

16. $30a^6 + 20a^3$

17. $8ab + 10a^2$

18. $6a^2b - 10ab^2$

19. $12a^3b - 6a^2b^2 + 14ab^3$

20. $4ab^3 - 16a^2b^2 - 22a^3b$

21. $a^2 + 2ab + b^2$

22. $a^2b - 2ab^2 + b^3$

23. $-3a^4 - 8a^3 + 12a^2$

24. $-5a^5 - 4a^4 - 16a^2$

Find the following quotients.

25. $\dfrac{6x + 8y}{2}$

26. $\dfrac{9x - 3y}{3}$

27. $\dfrac{7y - 21}{-7}$

28. $\dfrac{14y - 12}{2}$

29. $\dfrac{10xy - 8x}{2x}$

30. $\dfrac{12xy^2 - 18x}{-6x}$

31. $\dfrac{x^2y - x^3y^2}{x}$

32. $\dfrac{x^2y - x^3y^2}{x^2}$

33. $\dfrac{x^2y - x^3y^2}{-x^2y}$

34. $\dfrac{ab + a^2b^2}{ab}$

35. $\dfrac{a^2b^2 - ab^2}{-ab^2}$

36. $\dfrac{a^2b^2c - ab^2c^2}{abc}$

37. $\dfrac{x^3 - 3x^2y + xy^2}{x}$

38. $\dfrac{x^2 - 3xy^2 + xy^3}{x}$

39. $\dfrac{10a^2 - 15a^2b + 25a^2b^2}{5a^2}$

40. $\dfrac{11a^2b^2 - 33ab}{-11ab}$

41. $\dfrac{26x^2y^2 - 13xy}{-13xy}$

42. $\dfrac{6x^2y^2 - 3xy}{6xy}$

43. $\dfrac{4x^2y^2 - 2xy}{4xy}$

44. $\dfrac{6x^2a + 12x^2b - 6x^2c}{36x^2}$

45. $\dfrac{5a^2x - 10ax^2 + 15a^2x^2}{20a^2x^2}$

46. $\dfrac{12ax - 9bx + 18cx}{6x^2}$

47. $\dfrac{16x^5 + 8x^2 + 12x}{12x^3}$

48. $\dfrac{27x^2 - 9x^3 - 18x^4}{-18x^3}$

Divide. Assume all variables represent positive numbers.

49. $\dfrac{9a^{5m} - 27a^{3m}}{3a^{2m}}$

50. $\dfrac{26a^{3m} - 39a^{5m}}{13a^{3m}}$

51. $\dfrac{10x^{5m} - 25x^{3m} + 35x^m}{5x^m}$

52. $\dfrac{18x^{2m} + 24x^{4m} - 30x^{6m}}{6x^{2m}}$

Simplify each numerator, and then divide.

53. $\dfrac{2x^3(3x + 2) - 3x^2(2x - 4)}{2x^2}$

54. $\dfrac{5x^2(6x - 3) + 6x^3(3x - 1)}{3x}$

55. $\dfrac{(x + 2)^2 - (x - 2)^2}{2x}$

56. $\dfrac{(x - 3)^2 - (x + 3)^2}{3x}$

57. $\dfrac{(x + 5)^2 + (x + 5)(x - 5)}{2x}$

58. $\dfrac{(x - 4)^2 + (x + 4)(x - 4)}{2x}$

59. Comparing Expressions Evaluate the expression $\dfrac{10x + 15}{5}$ and the expression $2x + 3$ when $x = 2$.

60. Comparing Expressions Evaluate the expression $\dfrac{6x^2 + 4x}{2x}$ and the expression $3x + 2$ when $x = 5$.

61. Comparing Expressions Show that the expression $\dfrac{3x + 8}{2}$ is not the same as the expression $3x + 4$ by replacing x with 10 in both expressions and simplifying the results.

62. Comparing Expressions Show that the expression $\dfrac{x + 10}{x}$ is not equal to 10 by replacing x with 5 and simplifying.

Getting Ready for the Next Section

Divide.

63. $27\overline{)3,962}$

64. $13\overline{)18,780}$

65. $\dfrac{2x^2 + 5x}{x}$

66. $\dfrac{7x^5 + 9x^3 + 3x^7}{x^3}$

Multiply each expression.

67. $x(x - 3)$

68. $(x - 3)(-2)$

69. $2x^2(x - 5)$

70. $10x(x - 5)$

71. $x^2(2x - 3)$

72. $2x(4x - 1)$

73. $-3(x - 7)$

74. $-5(x + 2)$

Subtract each expression.

75. $(x^2 - 5x) - (x^2 - 3x)$

76. $(2x^3 + 0x^2) - (2x^3 - 10x^2)$

77. $(-2x + 8) - (-2x + 6)$

78. $(4x - 14) - (4x - 10)$

79. $(-4x + 3) - (-4x - 2)$

80. $(-8x - 12) - (-8x + 3)$

81. $(-4x^2 - 2x) - (-4x^2 + 3x)$

82. $(-x^2 + 8x) - (-x^2 + 12x)$

Find the Mistake

Each sentence below contains a mistake. Circle the mistake and write the correct word(s) or expression on the line provided.

1. To divide a polynomial by a monomial, divide only the first term of the polynomial by the monomial. _____

2. Dividing $15x^3 - 6x$ by $-3x$ gives us $15x^3 - 2$. _____

3. The quotient of $12x^2y + 4x^4y^2 - x^3y^2$ and x^2y gives us $12xy + 4x^2y - 1$. _____

4. If you divide $-9x^2 - 12x$ by $-3x$, the answer is $-3x - 4$. _____

Image © Katherine Heistand Shields, 2010

4.8 Dividing a Polynomial by a Polynomial

OBJECTIVES

A Divide a polynomial by a polynomial.

Suppose you earn an annual salary of $42,500. You are considering buying a luxurious house and need to know your monthly gross income to further calculate if you can afford the mortgage payments. To determine your monthly income, we need to divide 42,500 by 12. How would you set up a long division problem to find this answer? This section will cover how to do so and then transition to *long division of polynomials.*

KEY WORDS

long division of polynomials

remainder

A Dividing a Polynomial by a Polynomial

Since long division for polynomials is very similar to long division with whole numbers, we will begin by reviewing a division problem with whole numbers. You may realize when looking at Example 1 that you don't have a very good idea why you proceed as you do with long division. What you do know is that the process always works. We are going to approach the explanations in this section in much the same manner; that is, we won't always be sure why the steps we will use are important, only that they always produce the correct result.

EXAMPLE 1 Divide $27\overline{)3,962}$.

Solution

$$
\begin{array}{r}
1 \\
27\overline{)3962} \\
27 \\
\hline
12
\end{array}
$$
Estimate 27 into 39.
Multiply 1 × 27 = 27.
Subtract 39 − 27 = 12.

$$
\begin{array}{r}
1 \\
27\overline{)3962} \\
27 \downarrow \\
\hline
126
\end{array}
$$
Bring down the 6.

These are the four basic steps in long division. Estimate, multiply, subtract, and bring down the next term. To finish the problem, we simply perform the same four steps again.

$$
\begin{array}{r}
14 \\
27\overline{)3962} \\
27 \downarrow \\
\hline
126 \\
108 \downarrow \\
\hline
182
\end{array}
$$
4 is the estimate.
Multiply to get 108.
Subtract to get 18, then bring down the 2.

Practice Problems

1. Divide $35\overline{)4281}$.

Answer

1. $122\frac{11}{35}$

One more time.

$$\begin{array}{r} 146 \\ 27\overline{)3962} \\ \underline{27} \\ 126 \\ \underline{108} \\ 182 \\ \underline{162} \\ 20 \end{array}$$ *6 is the estimate.*

Multiply to get 162.

Subtract to get 20.

Since there is nothing left to bring down, we have our answer.

$$\frac{3962}{27} = 146 + \frac{20}{27} \quad \text{or} \quad 146\frac{20}{27}$$

To check out our answer, we evaluate the product of 146 and 27, then add the remainder.

$$(146)(27) + 20 = 3{,}942 + 20 = 3{,}962$$

Here is how it works with polynomials:

EXAMPLE 2 Divide $\dfrac{x^2 - 5x + 8}{x - 3}$

Solution

$$\begin{array}{r} x \\ x - 3 \overline{)x^2 - 5x + 8} \\ x^2 - 3x \\ \hline -2x \\ -2x + 8 \end{array}$$

Estimate $x^2 \div x = x$.

Multiply $x(x - 3) = x^2 - 3x$.

Subtract $(x^2 - 5x) - (x^2 - 3x) = -2x$.

Bring down the 8.

Notice that to subtract one polynomial from another, we add its opposite. That is why we change the signs on $x^2 - 3x$ and add what we get to $x^2 - 5x$. (To subtract the second polynomial, simply change the signs and add.)

We perform the same four steps again.

$$\begin{array}{r} x - 2 \\ x - 3 \overline{)x^2 - 5x + 8} \\ x^2 - 3x \\ \hline -2x + 8 \\ 2x - 6 \\ \hline 2 \end{array}$$

−2 is the estimate ($-2x \div x = -2$).

Multiply $-2(x - 3) = -2x + 6$.

Subtract $(-2x + 8) - (-2x + 6) = 2$.

Since there is nothing left to bring down, we have our answer.

$$\frac{x^2 - 5x + 8}{x - 3} = x - 2 + \frac{2}{x - 3}$$

To check our answer, we multiply $(x - 3)(x - 2)$ to get $x^2 - 5x + 6$. Then, adding on the remainder 2 we have $x^2 - 5x + 8$.

EXAMPLE 3 Divide $\dfrac{6x^2 - 11x - 14}{2x - 5}$.

Solution

$$\begin{array}{r} 3x + 2 \\ 2x - 5 \overline{)6x^2 - 11x - 14} \\ 6x^2 - 15x \\ \hline 4x - 14 \\ 4x - 10 \\ \hline -4 \end{array}$$

$$\frac{6x^2 - 11x - 14}{2x - 5} = 3x + 2 + \frac{-4}{2x - 5}$$

To check our answer, we multiply $(2x - 5)(3x + 2)$ to get $6x^2 - 11x - 10$. Then, adding on the remainder −4, we have $6x^2 - 11x - 14$.

2. Divide $\dfrac{x^2 - 5x + 8}{x - 2}$.

3. Divide $\dfrac{8x^2 - 6x - 5}{2x - 3}$.

Answers

2. $x - 3 + \dfrac{2}{x - 2}$

3. $4x + 3 + \dfrac{4}{2x - 3}$

One last step is sometimes necessary. The two polynomials in a division problem must both be in descending powers of the variable and cannot skip any powers from the highest power down to the constant term. It is also beneficial to keep similar powers vertically aligned as we divide.

EXAMPLE 4 Divide $\dfrac{2x^3 - 3x + 2}{x - 5}$.

Solution The problem will be much less confusing if we write $2x^3 - 3x + 2$ as $2x^3 + 0x^2 - 3x + 2$. Adding $0x^2$ does not change our original problem.

$$
\begin{array}{r}
2x^2 \\
x - 5 \overline{)\, 2x^3 + 0x^2 - 3x + 2} \\
\underline{\mp\, 2x^3 \mp 10x^2} \\
+ 10x^2 - 3x
\end{array}
$$

Estimate $2x^3 \div x = 2x^2$.

Multiply $2x^2(x - 5) = 2x^3 - 10x^2$.

Subtract
$(2x^3 + 0x^2) - (2x^3 - 10x^2) = 10x^2$.
Bring down the next term.

Adding the term $0x^2$ gives us a column in which to write $10x^2$. (Remember, you can add and subtract only similar terms.)

Here is the completed problem:

$$
\begin{array}{r}
2x^2 + 10x + 47 \\
x - 5 \overline{)\, 2x^3 + 0x^2 - 3x + 2} \\
\underline{\mp\, 2x^3 \mp 10x^2} \\
+ 10x^2 - 3x \\
\underline{\mp\, 10x^2 \mp 50x} \\
+ 47x + 2 \\
\underline{\mp\, 47x \mp 235} \\
237
\end{array}
$$

Our answer is $\dfrac{2x^3 - 3x + 2}{x - 5} = 2x^2 + 10x + 47 + \dfrac{237}{x - 5}$

You can check with multiplication that this quotient is correct.

As you can see, long division with polynomials is a mechanical process. Once you have done it correctly a couple of times, it becomes very easy to produce the correct answer.

GETTING READY FOR CLASS

After reading through the preceding section, respond in your own words and in complete sentences.

A. What are the four steps used in long division with whole numbers?

B. How is division of two polynomials similar to long division with whole numbers?

C. How would you check your answer when dividing polynomials?

D. How do you use 0 when dividing the polynomial $2x^3 - 3x + 2$ by $x - 5$?

4. Divide $\dfrac{3x^3 - 2x + 1}{x - 3}$.

Answer

4. $3x^2 + 9x + 25 + \dfrac{76}{x - 3}$

EXERCISE SET 4.8

Vocabulary Review

Choose the correct words to fill in the blanks below.

opposite estimate powers variable long division

1. Dividing a polynomial by a polynomial is very similar to _____ with whole numbers.
2. The four basic steps in long division are _____, multiply, subtract, and bring down the next term.
3. To subtract one polynomial from another, add its _____.
4. The two polynomials in a division problem must both be in descending powers of the _____.
5. The two polynomials in a division problem must also not skip any _____ from the highest power down to the constant term.

Problems

A Divide the following polynomials.

1. $\dfrac{x^2 - 5x + 6}{x - 3}$

2. $\dfrac{x^2 - 5x + 6}{x - 2}$

3. $\dfrac{a^2 + 9a + 20}{a + 5}$

4. $\dfrac{a^2 + 9a + 20}{a + 4}$

5. $\dfrac{x^2 - 6x + 9}{x - 3}$

6. $\dfrac{x^2 + 10x + 25}{x + 5}$

7. $\dfrac{2x^2 + 5x - 3}{2x - 1}$

8. $\dfrac{4x^2 + 4x - 3}{2x - 1}$

9. $\dfrac{2a^2 - 9a - 5}{2a + 1}$

10. $\dfrac{4a^2 - 8a - 5}{2a + 1}$

11. $\dfrac{x^2 + 5x + 8}{x + 3}$

12. $\dfrac{x^2 + 5x + 4}{x + 3}$

13. $\dfrac{a^2 + 3a + 2}{a + 5}$

14. $\dfrac{a^2 + 4a + 3}{a + 5}$

15. $\dfrac{x^2 + 2x + 1}{x - 2}$

16. $\dfrac{x^2 + 6x + 9}{x - 3}$

17. $\dfrac{x^2 + 5x - 6}{x + 1}$

18. $\dfrac{x^2 - x - 6}{x + 1}$

19. $\dfrac{a^2 + 3a + 1}{a + 2}$

SCAN TO ACCESS

20. $\dfrac{a^2 - a + 3}{a + 1}$

21. $\dfrac{2x^2 - 2x + 5}{2x + 4}$

22. $\dfrac{15x^2 + 19x - 4}{3x + 8}$

23. $\dfrac{6a^2 + 5a + 1}{2a + 3}$

24. $\dfrac{4a^2 + 4a + 3}{2a + 1}$

25. $\dfrac{6a^3 - 13a^2 - 4a + 15}{3a - 5}$

26. $\dfrac{2a^3 - a^2 + 3a + 2}{2a + 1}$

27. $\dfrac{8x^3 + 2x^2 - 4x - 5}{2x - 1}$

28. $\dfrac{9x^3 - 3x^2 + x - 2}{3x + 1}$

29. $\dfrac{2x^3 - 9x^2 + 11x - 6}{2x^2 - 3x + 2}$

30. $\dfrac{6x^3 + 7x^2 - x + 3}{3x^2 - x + 1}$

31. $\dfrac{x^4 - 3x^3 + 2x^2 - 3x - 9}{x^2 - 2x - 3}$

32. $\dfrac{x^3 + 3x^2 - 4x - 12}{x^2 + x - 6}$

33. $\dfrac{x^4 - x^3 - 4x^2 - x + 1}{x^2 - 3x + 1}$

34. $\dfrac{x^4 - x^3 - x + 1}{x^2 - 2x + 1}$

Fill in the missing terms in the numerator, and then use long division to find the quotients (see Example 4).

35. $\dfrac{x^3 + 4x + 5}{x + 1}$

36. $\dfrac{x^3 + 4x^2 - 8}{x + 2}$

37. $\dfrac{x^3 - 1}{x - 1}$

38. $\dfrac{x^3 + 1}{x + 1}$

39. $\dfrac{x^3 - 8}{x - 2}$

40. $\dfrac{x^3 + 27}{x + 3}$

41. $\dfrac{x^3 + 64}{x + 4}$

42. $\dfrac{x^3 - 27}{x - 3}$

Applying the Concepts

Long Division The chart shows the number of goals scored in the World Cup by the highest scoring countries of all time. If there have been 18 World Cup tournaments since the inaugural games in 1930, how many goals per World Cup have the following countries scored? Round to the nearest tenth.

43. Argentina

44. Spain

45. Brazil

46. Italy

FIFA WORLD CUP - MOST POINTS OF ALL-TIME
www.fifa.com

Brazil	212
Germany	187
Italy	153
Argentina	121
England	94
France	86
Spain	81

Maintaining Your Skills

Compute the following.

47. $6(3 + 4) + 5$

48. $[(1 + 2)(2 + 3)] + (4 \div 2)$

49. $1^2 + 2^2 + 3^2$

50. $(1 + 2 + 3)^2$

51. $5(6 + 3 \cdot 2) + 4 + 3 \cdot 2$

52. $(1 + 2)^3 + [(2 \cdot 3) + (4 \cdot 5)]$

53. $(1^3 + 2^3) + [(2 \cdot 3) + (4 \cdot 5)]$

54. $[2(3 + 4 + 5)] \div 3$

55. $(2 \cdot 3 + 4 + 5) \div 3$

56. $10^4 + 10^3 + 10^2 + 10^1$

57. $6 \cdot 10^3 + 5 \cdot 10^2 + 4 \cdot 10^1$

58. $5 \cdot 10^3 + 2 \cdot 10^2 + 8 \cdot 10^1$

59. $1 \cdot 10^3 + 7 \cdot 10^2 + 6 \cdot 10^1 + 0$

60. $4(2 - 1) + 5(3 - 2)$

61. $4 \cdot 2 - 1 + 5 \cdot 3 - 2$

62. $2^3 + 3^2 \cdot 4 - 5$

63. $(2^3 + 3^2) \cdot 4 - 5$

64. $2(2^2 + 3^2) + 3(3^2)$

Find the Mistake

Each sentence below contains a mistake. Circle the mistake and write the correct word(s) or expression on the line provided.

1. For a division of polynomials problem, you have found the answer when you are left with only a whole number to bring down. _____

2. To begin the problem $\frac{x^2 - 8x + 4}{x - 4}$, we first estimate $8x \div x = 8$. _____

3. The quotient of $x^3 - 9x + 2$ and $x - 1$ has no remainder. _____

4. When dividing $9x^3 - 7x - 4$ by $3x - 2$, we first write the problem as $3x - 2 \overline{)9x^3 - 7x - 4}$. _____

Universal Mass

Research the history of mass and common units used to measure it. For example, when the French invented the metric system toward the end of the 18th century, what item did they use as an equivalent for 1 kilogram? Explain how our knowledge of mass has changed over time. Then find the mass in kilograms of the following items. Write the values in expanded form and in scientific notation.

	Mass	
	Expanded Form	Scientific Notation
Mercury		
Venus		
Earth		
Mars		
Jupiter		
Saturn		
Uranus		
Neptune		
Sun		
Earth's Moon		

Use the above masses to answer the following questions.

1. A tonne is a unit of measurement for mass in the metric system that equals 1000 kg. If oxygen takes up 23.1% of the earth's mass, how much is this quantity in tonnes?

2. Find the difference in mass between the following pairings. Write your answer in scientific notation.

 a. Moon vs. Earth _____

 b. Mars vs. Jupiter and Saturn _____

 c. Uranus and Neptune vs. Mercury and Venus _____

 d. Sun vs. all seven planets combined _____

Exponents: Definition and Properties [4.1, 4.2]

EXAMPLES

1. a. $2^3 = 2 \cdot 2 \cdot 2 = 8$

b. $x^5 \cdot x^3 = x^{5+3} = x^8$

c. $\frac{x^5}{x^3} = x^{5-3} = x^2$

d. $(3x)^2 = 3^2 \cdot x^2 = 9x^2$

e. $\left(\frac{2}{3}\right)^3 = \frac{2^3}{3^3} = \frac{8}{27}$

f. $(x^5)^3 = x^{5 \cdot 3} = x^{15}$

g. $3^{-2} = \frac{1}{3^2} = \frac{1}{9}$

Integer exponents indicate repeated multiplications.

$$a^r \cdot a^s = a^{r+s} \qquad \text{Product property for exponents}$$

$$\frac{a^r}{a^s} = a^{r-s} \qquad \text{Quotient property for exponents}$$

$$(ab)^r = a^r \cdot b^r \qquad \text{Distributive property for exponents}$$

$$\left(\frac{a}{b}\right)^r = \frac{a^r}{b^r} \qquad \text{Expanded distributive property for exponents}$$

$$(a^r)^s = a^{r \cdot s} \qquad \text{Power property for exponents}$$

$$a^{-r} = \frac{1}{a^r} \qquad \text{Negative exponents property}$$

Multiplication of Monomials [4.3]

2. $(5x^2)(3x^4) = 15x^6$

To multiply two monomials, multiply coefficients and add exponents.

Division of Monomials [4.3]

3. $\frac{12x^9}{4x^5} = 3x^4$

To divide two monomials, divide coefficients and subtract exponents.

Scientific Notation [4.1, 4.2, 4.3]

4. $768,000 = 7.68 \times 10^5$
$0.00039 = 3.9 \times 10^{-4}$

A number is in scientific notation when it is written as the product of a number between 1 and 10 and an integer power of 10.

Addition of Polynomials [4.4]

5. $(3x^2 - 2x + 1) + (2x^2 + 7x - 3)$
$= 5x^2 + 5x - 2$

To add two polynomials, add coefficients of similar terms.

Subtraction of Polynomials [4.4]

6. $(3x + 5) - (4x - 3)$
$= 3x + 5 - 4x + 3$
$= -x + 8$

To subtract one polynomial from another, add the opposite of the second to the first.

Multiplication of Polynomials [4.5]

To multiply a polynomial by a monomial, we apply the distributive property. To multiply two binomials we use the FOIL method. In other situations we use the Column method. Each method achieves the same result: To multiply any two polynomials, we multiply each term in the first polynomial by each term in the second polynomial.

7. a. $2a^2(5a^2 + 3a - 2)$
$= 10a^4 + 6a^3 - 4a^2$

b. $(x + 2)(3x - 1)$
$= 3x^2 - x + 6x - 2$
$= 3x^2 + 5x - 2$

c.
$$\begin{array}{r} 2x^2 - 3x + 4 \\ 3x - 2 \\ \hline 6x^3 - 9x^2 + 12x \\ -4x^2 + 6x - 8 \\ \hline 6x^3 - 13x^2 + 18x - 8 \end{array}$$

Special Products [4.6]

$(a + b)^2 = a^2 + 2ab + b^2$

$(a - b)^2 = a^2 - 2ab + b^2$ } Binomial squares

$(a + b)(a - b) = a^2 - b^2$ Difference of two squares

8. $(x + 3)^2 = x^2 + 6x + 9$
$(x - 3)^2 = x^2 - 6x + 9$
$(x + 3)(x - 3) = x^2 - 9$

Dividing a Polynomial by a Monomial [4.7]

To divide a polynomial by a monomial, divide each term in the polynomial by the monomial.

9. $\frac{12x^3 - 18x^2}{6x} = 2x^2 - 3x$

Long Division with Polynomials [4.8]

Division with polynomials is similar to long division with whole numbers. The steps in the process are estimate, multiply, subtract, and bring down the next term. The divisors in all the long-division problems in this chapter were binomials.

10. $x^2 - 5x - 8 \div x - 3$

$= x - 2 + \frac{2}{x - 3}$

$$\begin{array}{r} x - 2 \\ x - 3 \overline{)\, x^2 - 5x + 8} \\ \mp x^2 \pm 3x \downarrow \\ \hline -2x + 8 \\ \pm 2x \mp 6 \\ \hline 2 \end{array}$$

Simplify each of the following expressions. [4.1]

1. $-\left(-\dfrac{1}{2}\right)^2$

2. $\left(\dfrac{2}{3}\right)^3$

3. $(3x^2)^3(2x^3)^2$

4. Write 3.47×10^4 in expanded form. [4.1]

Simplify each expression. Write all answers with positive exponents only. [4.2]

5. $(-3)^{-3}$

6. $(17a^2b^7)^0$

7. $\dfrac{a^{-3}}{a^{-1}}$

8. $\dfrac{(x^{-2})^3(x^{-3})^{-5}}{(x^{-4})^{-2}}$

9. Write 0.03478 in scientific notation. [4.2]

Simplify. Write all answers with positive exponents only. [4.3]

10. $\dfrac{35x^3y^7z^{-2}}{14x^2y^7z^{-3}}$

11. $\dfrac{(6a^2b)^2(9a^3b^3)}{18a^4b^7}$

12. $\dfrac{6x^2}{3x^{-1}} + \dfrac{18x^5}{6x^2}$

13. $\dfrac{(2.7 \times 10^3)(3.2 \times 10^{-2})}{7.2 \times 10^5}$

Add or subtract as indicated. [4.4]

14. $8x^2 - 4x + 3x - 2$

15. $(5x^2 - 3x + 4) - (3x^2 + 4x - 7)$

16. Subtract $2x + 8$ from $3x - 6$.

17. Find the value of $2y^2 - 4y + 8$ when y is -2.

Multiply. [4.5]

18. $2a^2(4a^2 + 5a + 7)$

19. $(x - 2)(x + 3)$

20. $(3x - 3)(4x - 4)$

21. $(x + 3)(x^2 + 3x + 9)$

Multiply. [4.6]

22. $(x + 1)^3$

23. $(3a - 2)^2$

24. $(2x - b)(2x + b)$

25. $(b^2 - 2)(b^2 + 2)$

26. Divide $-10x^3 + 15x^2 + 15x$ by $5x$. [4.7]

Divide. [4.8]

27. $\dfrac{8x^2 - 4x - 24}{2x + 3}$

28. $\dfrac{3x^2 - 11x - 4}{x - 4}$

29. Volume Find the volume of a cube if the length of a side is 1.5 miles. [4.1]

30. Volume Find the volume of a rectangular solid if the length is one third the width, and the height is six times the width. [4.3]

31. The illustration below shows prices for some of the most expensive cars in the world.

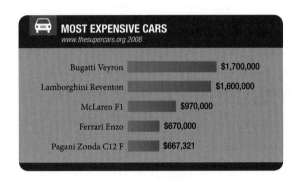

Write the prices of the cars in scientific notation. Round your answer to the nearest tenth.

32. The illustration below shows some of the close finishes from the 2010 Vancouver Olympic games.

Write the margin of victory for each race in scientific notation.

Simplify.

1. $\left(-\dfrac{1}{2}\right)^4$

2. $-|8|$

3. $2 \cdot 8 + 10$

4. $6 \cdot 8 + 3 \cdot 8$

5. $-10 + (-15)$

6. $-3 - (-15)$

7. $7 - 7\left(\dfrac{1}{-7}\right)$

8. $(-3)(-2)(-5)$

9. $6\left(-\dfrac{1}{12}x + \dfrac{1}{6}y\right)$

10. $3(2a - 4) + 6(-a + 3)$

11. $\dfrac{y^8 \cdot y^2}{y^3}$

12. $(3y)^3$

13. $(2a^2b^3)^3(3ab^4)^2$

14. $\dfrac{(8xy^3)^2(12x^2y^3)^2}{(16x^2y^3)^3(9x^2y)}$

Simplify and write your answer in scientific notation.

15. $(3.5 \times 10^4)(7 \times 10^{-5})$

16. $\dfrac{8.1 \times 10^{-7}}{2.7 \times 10^{-5}}$

Perform the indicated operations.

17. $3ab - 2a^2b + 7ab$

18. $(-4x^2 - 5x + 3) + (3x^2 - 2x + 1) - (x^2 + 3x + 7)$

19. $(5x + 1)^2$

20. $(x - 2)(x^2 + 2x + 4)$

Solve each equation.

21. $3 - 2y + 8y = -9$

22. $6a - 5 = 3a$

23. $18 = 4(-x + 2.5)$

24. $3(2x + 5) + 3 = 2x - 10$

Solve the inequality, and graph the solution set.

25. $-\dfrac{a}{3} < -4$

26. $7(2t + 8) - 4 \le -(3t - 1)$

27. $-3 > -x + 3 > -4$

28. $-4 \le 3x + 2 \le 14$

Divide.

29. $\dfrac{6x^2 + 5x - 6}{3x - 2}$

30. $\dfrac{x^5 - x^4 - 6x^3}{x^2 - 3x}$

Graph on a rectangular coordinate system.

31. $2x + 2y = 8$

32. $x - y = 5$

33. Find the value of $3x - 5$ when $x = -2$.

34. Find x when y is 4 in the equation $y = 3x - 8$.

35. Solve $A = \dfrac{1}{2}bh$ for h.

36. Solve $\dfrac{2y - 1}{x} = -2$ for y.

For the set $\{-3, -\dfrac{\pi}{2} -\sqrt{2}, 0.999, \dfrac{3}{2}, 2\}$ list all of the

37. Whole numbers

38. Irrational numbers

Give the property that justifies the statement.

39. $5 + (-5) = 0$

40. $a + 2 = 2 + a$

41. Find the x- and y-intercepts for the equation $y = 3x - 2$.

42. Find the slope of the line $3x - 6y = 7$.

Use the illustration to answer the following questions.

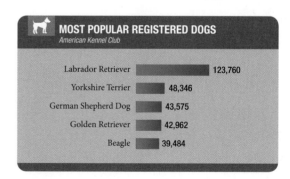

43. How many more Labrador Retrievers are registered than German Shepherds?

44. How many fewer Beagles are registered than Yorkshire Terriers?

Simplify each of the following expressions. [4.1]

1. $(-3)^5$

2. $\left(\dfrac{2}{5}\right)^2$

3. $(2x^2)^3(3x^2)^2$

4. Write 2.47×10^5 in expanded form. [4.1]

Simplify each expression. Write all answers with positive exponents only. [4.2]

5. -3^{-2}

6. $(7a^2b^2)^1$

7. $\dfrac{a^{-4}}{a^{-1}}$

8. $\dfrac{(x^{-2})^3(x^{-4})^{-1}}{x}$

9. Write 0.0278×10^3 in scientific notation. [4.2]

Simplify. Write all answers with positive exponents only. [4.3]

10. $\dfrac{35x^3y^3z}{7xyz^3}$

11. $\dfrac{(3a^2b)(12a^2b)}{4ab^2}$

12. $\dfrac{24x^3}{3x^2} - \dfrac{33x^{-1}}{11x^{-2}}$

13. $\dfrac{(3.2 \times 10^{-3})(1.8 \times 10^2)}{3.6 \times 10^{-5}}$

Add or subtract as indicated. [4.4]

14. $8x^2 - 4x + 3x - 12$

15. $(3x^2 - x + 3) - (2x^2 - 8x + 3)$

16. Subtract $5x + 8$ from $-(x + 2)$.

17. Find the value of $2y^2 - y - 4$ when y is 3.

Multiply. [4.5]

18. $2a^2(3a^2 + 5a - 4)$

19. $\left(x - \dfrac{3}{2}\right)\left(x + \dfrac{1}{2}\right)$

20. $(4x + 3)(2x - 2)$

21. $(x + 3)(x - 3)$

Multiply. [4.6]

22. $(x - 3)^2(x + 3)^2$

23. $(2a - b)^2$

24. $(\sqrt{3}x - \sqrt{8}y)(\sqrt{3}x + \sqrt{8}y)$

25. $(a^2 + \sqrt{3})(a^2 - \sqrt{3})$

Divide. [4.8]

26. $\dfrac{3x^2 + 7x + 2}{x + 2}$

27. $\dfrac{6x^2 - 2x - 1}{3x - 1}$

28. **Volume** Find the volume of a cube if the length of a side is 0.5 feet. [4.1]

29. **Volume** Find the volume of a rectangular solid if the length is two thirds the width, and the height is one and a half times the width. [4.3]

The illustration below shows the top grossing Disney animated movies of all time.

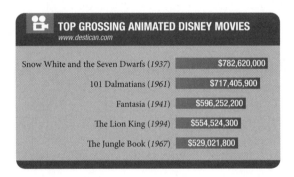

Write the total revenue for the given movies in scientific notation, rounded to the nearest hundredth.

30. Snow White and the Seven Dwarfs

31. The Lion King

The illustration below shows the number of stars received by the top rated comic strips of all time at an online rating site.

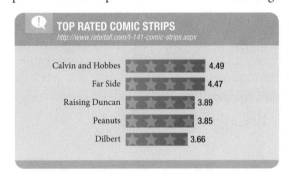

Write the difference in the number of stars received by the following strips in scientific notation.

32. Calvin and Hobbes compared to Far Side

33. Raising Duncan compared to Peanuts

Factoring Polynomials

Millau Viaduct Bridge
Millau-Creissels, France
Image © 2010 IGN-France

Google

U pon its grand opening in 2004, the Millau Viaduct in France was the world's tallest bridge. It has a cable-stayed design, where cables hanging from columns attach to and support the road. The bridge stands at an impressive height of 1,125 feet at one column's summit. Slightly taller than the Eiffel Tower and only a few feet short of the Empire State Building, the viaduct is an architectural marvel. This huge structure crosses the valley of the River Tran and has greatly decongested traffic in the region since it opened.

Suppose you could stand at the top of one of the bridge's columns and toss a rock into the air with an upward velocity of 30 feet per second. As it falls to the River Tran below, the equation that gives the rock's height h at any time t is

$$h = 1125 + 30t - 16t^2$$

Solving the equation to find the height of the rock requires factoring the trinomial in the equation. We will work similar problems in this chapter.

©iStockphoto.com/JoeGough

Note As you will see as we progress through the book, factoring is a tool that is used in solving a number of problems. Before seeing how it is used, however, we first must learn how to do it. So, in this section and the sections that follow, we will be developing our factoring skills.

In flight archery, a person uses a bow to shoot an arrow into the air with a goal of achieving the greatest distance, as opposed to aiming for a specific target. The curved path that the arrow takes from the bow to the ground creates the shape of a parabola, which we will discuss further in Chapter 9. For now, let's say the arrow's path can be represented by the polynomial $-16x^2 + 62x + 8$. To begin factoring this polynomial, we need to find the greatest common factor, which is the primary focus of this section.

Recall the following diagram to illustrate the relationship between multiplication and factoring.

Multiplication

Factors $\rightarrow \ 3 \cdot 5 = 15 \leftarrow$ Product

Factoring

A similar relationship holds for multiplication of polynomials. Reading the following diagram from left to right, we say the product of the binomials $x + 2$ and $x + 3$ is the trinomial $x^2 + 5x + 6$. However, if we read in the other direction, we can say that $x^2 + 5x + 6$ factors into the product of $x + 2$ and $x + 3$.

Multiplication

Factors $\rightarrow (x + 2)(x + 3) = x^2 + 5x + 6 \leftarrow$ Product

Factoring

In this chapter, we develop a systematic method of factoring polynomials.

A Factoring The Greatest Common Factor

In this section, we will apply the distributive property to polynomials to factor from them what is called the greatest common factor.

> **DEFINITION** greatest common factor
>
> The *greatest common factor* for a polynomial is the largest monomial that divides (is a factor of) each term of the polynomial.

We use the term *largest monomial* to mean the monomial with the greatest coefficient and highest power of the variable that is a common factor of each term in the polynomial.

EXAMPLE 1 Find the greatest common factor for the polynomial $3x^5 + 12x^2$.

Solution The terms of the polynomial are $3x^5$ and $12x^2$. The largest number that divides the coefficients is 3, and the highest power of x that is a factor of x^5 and x^2 is x^2. Therefore, the greatest common factor for $3x^5 + 12x^2$ is $3x^2$; that is, $3x^2$ is the largest monomial that divides each term of $3x^5 + 12x^2$.

EXAMPLE 2 Find the greatest common factor for $8a^3b^2 + 16a^2b^3 + 20a^3b^3$.

Solution The largest number that divides each of the coefficients is 4. The highest power of the variable that is a factor of a^3b^2, a^2b^3, and a^3b^3 is a^2b^2. The greatest common factor for $8a^3b^2 + 16a^2b^3 + 20a^3b^3$ is $4a^2b^2$. It is the largest monomial that is a factor of each term.

Once we have recognized the greatest common factor of a polynomial, we can apply the distributive property and factor it out of each term. We rewrite the polynomial as the product of its greatest common factor with the polynomial that remains after the greatest common factor has been factored from each term in the original polynomial.

EXAMPLE 3 Factor the greatest common factor from $3x - 15$.

Solution The greatest common factor for the terms $3x$ and 15 is 3. We can rewrite both $3x$ and 15 so that the greatest common factor 3 is showing in each term. It is important to realize that $3x$ means $3 \cdot x$. The 3 and the x are not "stuck" together.

$$3x - 15 = 3 \cdot x - 3 \cdot 5$$

Now, applying the distributive property, we have

$$3 \cdot x - 3 \cdot 5 = 3(x - 5)$$

To check a factoring problem like this, we can multiply 3 and $x - 5$ to get $3x - 15$, which is what we started with. Factoring is simply a procedure by which we change sums and differences into products. In this case, we changed the difference $3x - 15$ into the product $3(x - 5)$. Note, however, that we have not changed the meaning or value of the expression. The expression we end up with is equivalent to the expression we started with.

EXAMPLE 4 Factor the greatest common factor from $-5x^3 - 15x^2$.

Solution The greatest common factor is $-5x^2$. We rewrite the polynomial as

$$-5x^3 - 15x^2 = -5x^2 \cdot x - (-5x^2) \cdot 3$$

Then we apply the distributive property to get

$$-5x^2 \cdot x - (-5x^2) \cdot 3 = -5x^2(x + 3)$$

To check our work, we simply multiply $-5x^2$ and $(x + 3)$ to get $-5x^3 - 15x^2$, which is our original polynomial.

EXAMPLE 5 Factor the greatest common factor from $16x^5 - 20x^4 + 8x^3$.

Solution The greatest common factor is $4x^3$. We rewrite the polynomial so we can see the greatest common factor $4x^3$ in each term; then we apply the distributive property to factor it out.

$$16x^5 - 20x^4 + 8x^3 = 4x^3 \cdot 4x^2 - 4x^3 \cdot 5x + 4x^3 \cdot 2$$

$$= 4x^3(4x^2 - 5x + 2)$$

Practice Problems

1. Find the greatest common factor for $10x^6 + 15x^4$.

2. Find the greatest common factor for $27a^4b^5 + 18a^3b^3 + 9a^3b^2$.

3. Factor the greatest common factor from $5x + 15$.

4. Factor the greatest common factor from $-25x^4 - 35x^3$.

5. Factor the greatest common factor from $20x^8 - 12x^7 + 16x^6$.

Answers

1. $5x^4$
2. $9a^3b^2$
3. $5(x + 3)$
4. $-5x^3(5x + 7)$
5. $4x^6(5x^2 - 3x + 4)$

6. Factor the greatest common factor from $8xy^3 - 16x^2y^2 + 8x^3y$.

EXAMPLE 6 Factor the greatest common factor from $6x^3y - 18x^2y^2 + 12xy^3$.

Solution The greatest common factor is $6xy$. We rewrite the polynomial in terms of $6xy$ and then apply the distributive property as follows:

$$6x^3y - 18x^2y^2 + 12xy^3 = 6xy \cdot x^2 - 6xy \cdot 3xy + 6xy \cdot 2y^2$$
$$= 6xy(x^2 - 3xy + 2y^2)$$

7. Factor the greatest common factor from $2ab^2 - 6a^2b^2 + 8a^3b^2$.

EXAMPLE 7 Factor the greatest common factor from $3a^2b - 6a^3b^2 + 9a^3b^3$.

Solution The greatest common factor is $3a^2b$.

$$3a^2b - 6a^3b^2 + 9a^3b^3 = 3a^2b(1) - 3a^2b(2ab) + 3a^2b(3ab^2)$$
$$= 3a^2b(1 - 2ab + 3ab^2)$$

B Factoring by Grouping

To develop our next method of factoring, called *factoring by grouping*, we start by examining the polynomial $xc + yc$. The greatest common factor for the two terms is c. Factoring c from each term we have

$$xc + yc = c(x + y)$$

But suppose that c itself was a more complicated expression, such as $a + b$, so that the expression we were trying to factor was $x(a + b) + y(a + b)$, instead of $xc + yc$. The greatest common factor for $x(a + b) + y(a + b)$ is $(a + b)$. Factoring this common factor from each term looks like this:

$$x(a + b) + y(a + b) = (a + b)(x + y)$$

To see how all of this applies to factoring polynomials, consider the polynomial

$$xy + 3x + 2y + 6$$

There is no greatest common factor other than the number 1. However, if we group the terms together two at a time, we can factor an x from the first two terms and a 2 from the last two terms.

$$xy + 3x + 2y + 6 = x(y + 3) + 2(y + 3)$$

The expression on the right can be thought of as having two terms: $x(y + 3)$ and $2(y + 3)$. Each of these expressions contains the common factor $y + 3$, which can be factored out using the distributive property in reverse.

$$x(y + 3) + 2(y + 3) = (y + 3)(x + 2)$$

This last expression is in factored form. The process we used to obtain it is called factoring by grouping. Here are some additional examples:

8. Factor $ax - bx + ay - by$.

EXAMPLE 8 Factor $ax + bx + ay + by$.

Solution We begin by factoring the greatest common factor x from the first two terms and the greatest common factor y from the last two terms.

$$ax + bx + ay + by = x(a + b) + y(a + b)$$
$$= (a + b)(x + y)$$

To convince yourself that this is factored correctly, multiply the two factors $(a + b)$ and $(x + y)$.

Answers

6. $8xy(y^2 - 2xy + x^2)$
7. $2ab^2(1 - 3a + 4a^2)$
8. $(a - b)(x + y)$

EXAMPLE 9 Factor $3ax - 2a + 15x - 10$ by grouping.

Solution First, we factor a from the first two terms and 5 from the last two terms. Then, we factor $3x - 2$ from the remaining two expressions.

$$3ax - 2a + 15x - 10 = a(3x - 2) + 5(3x - 2)$$
$$= (3x - 2)(a + 5)$$

Again, multiplying $(3x - 2)$ and $(a + 5)$ will convince you that these are the correct factors.

EXAMPLE 10 Factor $2x^2 + 5ax - 2xy - 5ay$.

Solution From the first two terms we factor x. From the second two terms we must factor $-y$ so that the binomial that remains after we do so matches the binomial produced by the first two terms.

$$2x^2 + 5ax - 2xy - 5ay = x(2x + 5a) - y(2x + 5a)$$
$$= (2x + 5a)(x - y)$$

Another way to accomplish the same result is to use the commutative property to interchange the middle two terms, and then factor by grouping.

$$2x^2 + 5ax - 2xy - 5ay = 2x^2 - 2xy + 5ax - 5ay \qquad \textit{Commutative property}$$
$$= 2x(x - y) + 5a(x - y)$$
$$= (x - y)(2x + 5a)$$

This is the same result we obtained previously.

GETTING READY FOR CLASS

After reading through the preceding section, respond in your own words and in complete sentences.

A. What is the greatest common factor for a polynomial?

B. How would you use the distributive property when factoring a polynomial?

C. When would you try to factor by grouping?

D. After factoring a polynomial, how can you check your result?

9. Factor $5ax - 3ay + 10x - 6y$ by grouping.

10. Factor $3x^2 + 7bx - 3xy - 7by$.

Answers

9. $(5x - 3y)(a + 2)$

10. $(3x + 7b)(x - y)$

Vocabulary Review

Choose the correct words to fill in the blanks below.

 grouping coefficient monomial factors

1. The greatest common factor for a polynomial is the largest _____ that divides each term of the polynomial.

2. The largest monomial is the monomial with the greatest _____ and highest power of the variable.

3. Factoring by _____ is a method of factoring a polynomial that involves grouping terms two at a time to find common _____.

Problems

A Factor the following polynomials by using the greatest common factor.

1. $15x + 25$

2. $14x + 21$

3. $6a + 9$

4. $8a + 10$

5. $4x - 8y$

6. $9x - 12y$

7. $3x^2 - 6x + 9$

8. $2x^2 + 6x + 12$

9. $3a^2 - 3a + 60$

10. $2a^2 - 18a + 32$

11. $24y^2 - 52y + 36$

12. $18y^2 + 48y + 40$

13. $9x^2 - 8x^3$

14. $7x^3 - 4x^2$

15. $13a^2 - 26a^3$

16. $5a^2 - 10a^3$

17. $21x^2y - 28xy^2$

18. $30xy^2 - 25x^2y$

19. $22a^2b^2 - 11ab^2$

20. $15x^3 - 25x^2 + 30x$

21. $7x^3 + 21x^2 + 28x$

22. $16x^4 - 20x^2 - 16x$

23. $121y^4 + 11x^4$

24. $25a^4 + 5b^4$

SCAN TO ACCESS

25. $100x^4 - 50x^3 + 25x^2$

26. $36x^5 + 72x^3 - 81x^2$

27. $8a^2 + 16b^2 + 32c^2$

28. $9a^2 - 18b^2 - 27c^2$

29. $4a^2b - 16ab^2 + 32a^2b^2$

30. $5ab^2 + 10a^2b^2 + 15a^2b$

31. $6a^3b^2 + 6a^3b - 6a^2b^2$

32. $6ab^2 + 3a^2b^2 - 3b^3$

33. $121a^3b^2 - 22a^2b^3 + 33a^3b^3$

34. $20a^4b^3 - 18a^3b^4 + 22a^4b^4$

35. $12x^2y^3 - 72x^5y^3 - 36x^4y^4$

36. $49xy - 21x^2y^2 + 35x^3y^3$

B Factor the following polynomials by grouping.

37. $xy + 5x + 3y + 15$

$x(y + 5) \quad 3(x + 5)$

38. $xy + 2x + 4y + 8$

39. $xy + 6x + 2y + 12$

40. $xy + 2y + 6x + 12$

41. $ab + 7a - 3b - 21$

$a(b + 7) \quad 3(b + 7)$

42. $ab + 3b - 7a - 21$

43. $ax - bx + ay - by$

44. $ax - ay + bx - by$

45. $2ax + 6x - 5a - 15$

46. $3ax + 21x - a - 7$

47. $3xb - 4b - 6x + 8$

48. $3xb - 4b - 15x + 20$

49. $x^2 + ax + 2x + 2a$

50. $x^2 + ax + 3x + 3a$

51. $x^2 - ax - bx + ab$

52. $x^2 + ax - bx - ab$

Factor the following polynomials by grouping. You can group the terms together two at a time or three at a time. Either way will produce the same result.

53. $ax + ay + bx + by + cx + cy$

$a(x + y) \quad b(x + y) \quad c(x + y)$

$a + b + c (x + y)$

54. $ax + bx + cx + ay + by + cy$

Factor the following polynomials by grouping the terms together two at a time.

55. $6x^2 + 9x + 4x + 6$

56. $6x^2 - 9x - 4x + 6$

57. $20x^2 - 2x + 50x - 5$

$2x(10x - 1) \quad 5(10x - 1)$

58. $20x^2 + 25x + 4x + 5$

59. $20x^2 + 4x + 25x + 5$

60. $20x^2 + 4x - 25x - 5$

61. $x^3 + 2x^2 + 3x + 6$

62. $x^3 - 5x^2 + 4x - 20$

63. $6x^3 - 4x^2 + 15x - 10$

64. $8x^3 - 12x^2 + 14x - 21$

65. $3x^3 + 2x^2 - 15x - 10$

66. $4x^3 - 3x^2 + 8x - 6$

67. The greatest common factor of the binomial $3x + 6$ is 3, and the greatest common factor of the binomial $2x + 4$ is 2. What is the greatest common factor of their product $(3x + 6)(2x + 4)$ when it has been multiplied out?

68. The greatest common factors of the binomials $4x + 2$ and $5x + 10$ are 2 and 5, respectively. What is the greatest common factor of their product $(4x + 2)(5x + 10)$ when it has been multiplied out?

69. The following factorization is incorrect. Find the mistake, and correct the right-hand side.

$$12x^2 + 6x + 3 = 3(4x^2 + 2x)$$

70. Find the mistake in the following factorization, and then rewrite the right-hand side correctly.

$$10x^2 + 2x + 6 = 2(5x^2 + 3)$$

Applying the Concepts

71. Investing If you invest $1,000 in an account with an annual interest rate of r compounded annually, the amount of money you have in the account after one year is

$$A = 1,000 + 1,000r$$

Write this formula again with the right side in factored form. Then, find the amount of money in this account at the end of one year if the interest rate is 12%.

72. Investing If you invest P dollars in an account with an annual interest rate of 8% compounded annually, then the amount of money in that account after one year is given by the formula

$$A = P + 0.08P$$

Rewrite this formula with the right side in factored form, and then find the amount of money in the account at the end of one year if $500 was the initial investment.

73. Biological Growth If 1,000,000 bacteria are placed in a petri dish and the bacteria have a growth rate of r (a percent expressed as a decimal) per hour, then 1 hour later the amount of bacteria will be given by

$$A = 1,000,000 + 1,000,000r$$

a. Factor the right side of the equation.

b. If $r = 30\%$, find the number of bacteria present after one hour.

74. Biological Growth If there are B E. coli bacteria present initially in a petri dish and their growth rate is r (a percent expressed as a decimal) per hour, then after one hour the amount of bacteria present is given by

$$A = B + Br$$

a. Factor the right side of this equation.

b. The following bar graph shows the number of E. coli bacteria present initially and the number of bacteria present hours later. Use the bar chart to find B and A in the preceding equation.

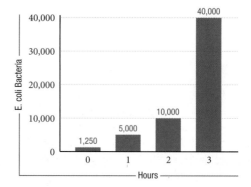

Getting Ready for the Next Section

Multiply.

75. $(x - 7)(x + 2)$

76. $(x - 7)(x - 2)$

77. $(x - 3)(x + 2)$

78. $(x + 3)(x - 2)$

79. $(x + 3)(x^2 - 3x + 9)$

80. $(x - 2)(x^2 + 2x + 4)$

81. $(2x + 1)(x^2 + 4x - 3)$

82. $(3x + 2)(x^2 - 2x - 4)$

83. $3x^4(6x^3 - 4x^2 + 2x)$

84. $2x^4(5x^3 + 4x^2 - 3x)$

85. $\left(x + \dfrac{1}{3}\right)\left(x + \dfrac{2}{3}\right)$

86. $\left(x + \dfrac{1}{4}\right)\left(x + \dfrac{3}{4}\right)$

87. $(6x + 4y)(2x - 3y)$

88. $(8a - 3b)(4a - 5b)$

89. $(9a + 1)(9a - 1)$

90. $(7b + 1)(7b + 1)$

91. $(x - 9)(x - 9)$

92. $(x - 8)(x - 8)$

93. $(x + 2)(x^2 - 2x + 4)$

94. $(x - 3)(x^2 + 3x + 9)$

Find the Mistake

Each sentence below contains a mistake. Circle the mistake and write the correct word(s) or expression(s) on the line provided.

1. The greatest common factor for a polynomial is the monomial with the smallest coefficient and highest power of the variable that divides each term of the polynomial. _____

2. The greatest common factor for the polynomial $12x^4 + 3x^3 - 6x$ is $12x^4$. _____

3. Using the process of factoring by grouping, we can factor the polynomial $8xy + 2x - 12y - 3$ to get $y(8x + 2x - 12 - 3)$.

4. To factor the polynomial $9xy - 27x - 2y + 6$ by grouping, factor the number 3 from the first, second, and third term. _____

Navigation Skills: Prepare, Study, Achieve

Think about your current study routine. Has it been successful? There are many things that you must consider when creating a routine. One important aspect is the environment in which you choose to study. Think about the location you typically study. Are you able to focus there without distraction? Consider what things may distract you (e.g., cell phone, television, noise, friends who socialize rather than study) and find a place to study where these things are absent. Other important aspects of a productive study routine are time of day you choose to study, and sights and sounds around you during your study time. A study environment that is not distracting will help you focus and foster further success in this course.

A Factor a trinomial whose leading coefficient is the number 1.

B Factor a polynomial by first factoring out the greatest common factor and then factoring the polynomial that remains.

KEY WORDS

leading coefficients

5.2 Factoring Trinomials of the Form $x^2 + bx + c$

Image © sxc.hu, bugdog, 2006

The Peregrine falcon is the fastest-moving animal on Earth, with a diving speed of 200 miles per hour. The falcon has incredible eyesight, 6 times more powerful than a human, and can spot its prey scurrying on the ground from hundreds of feet in the air. The path of the falcon's dive to catch her prey takes the form of the trinomial $x^2 + bx + c$, where the values of the variables depend on the dive's height, length, and speed. In this section, we will factor this type of trinomial in which the coefficient of the squared term is 1.

The more familiar we are with multiplication of binomials the easier factoring trinomials will be. Therefore, recall multiplication of binomials from Chapter 4:

$$(x + 3)(x + 4) = x^2 + 7x + 12$$

$$(x - 5)(x + 2) = x^2 - 3x - 10$$

Using the FOIL method we learned in the last chapter, the first term in the answer is the product of the first terms in each binomial. The last term in the answer is the product of the last terms in each binomial. The middle term in the answer comes from adding the product of the outside terms to the product of the inside terms.

Let's have a and b represent real numbers and look at the product of $(x + a)$ and $(x + b)$.

$$(x + a)(x + b) = x^2 + ax + bx + ab$$

$$= x^2 + (a + b)x + ab$$

The coefficient of the middle term is the sum of a and b. The last term is the product of a and b. Writing this as a factoring problem, we have

$$x^2 + \underset{\underset{\text{Sum}}{\downarrow}}{(a + b)x} + \underset{\underset{\text{Product}}{\downarrow}}{ab} = (x + a)(x + b)$$

A Factoring with a Leading Coefficient of 1

To factor a trinomial in which the coefficient of x^2 is 1, we need only find the numbers a and b whose sum is the coefficient of the middle term and whose product is the constant term (last term).

EXAMPLE 1 Factor $x^2 + 8x + 12$.

Solution The coefficient of x^2 is 1. We need two numbers whose sum is 8 and whose product is 12. The numbers are 6 and 2.

$$x^2 + 8x + 12 = (x + 6)(x + 2)$$

We can easily check our work by multiplying $(x + 6)$ and $(x + 2)$.

Check: $(x + 6)(x + 2) = x^2 + 6x + 2x + 12$
$$= x^2 + 8x + 12$$

EXAMPLE 2 Factor $x^2 - 2x - 15$.

Solution The coefficient of x^2 is again 1. We need to find a pair of numbers whose sum is -2 and whose product is -15. It can be helpful to create a table of products and sums to find our correct pair of numbers. Here are all the possibilities for products that are -15.

Products	Sums
$-1(15) = -15$	$-1 + 15 = 14$
$1(-15) = -15$	$1 + (-15) = -14$
$-5(3) = -15$	$-5 + 3 = -2$
$5(-3) = -15$	$5 + (-3) = 2$

The third line gives us what we want. The factors of $x^2 - 2x - 15$ are $(x - 5)$ and $(x + 3)$.

$$x^2 - 2x - 15 = (x - 5)(x + 3)$$

B Factoring Trinomials Using the Greatest Common Factor

In Example 3, we begin by factoring out the greatest common factor. The first step in factoring any trinomial is to look for the greatest common factor. If the trinomial in question has a greatest common factor other than 1, we factor it out first and then try to factor the trinomial that remains.

EXAMPLE 3 Factor $2x^2 + 10x - 28$.

Solution We begin by factoring out the greatest common factor, which is 2.

$$2x^2 + 10x - 28 = 2(x^2 + 5x - 14)$$

Now, we factor the remaining trinomial by finding a pair of numbers whose sum is 5 and whose product is -14. Here are the possibilities:

Products	Sums
$-1(14) = -14$	$-1 + 14 = 13$
$1(-14) = -14$	$1 + (-14) = -13$
$-7(2) = -14$	$-7 + 2 = -5$
$7(-2) = -14$	$7 + (-2) = 5$

From the last line, we see that the factors of $x^2 + 5x - 14$ are $(x + 7)$ and $(x - 2)$. Here is the complete problem:

$$2x^2 + 10x - 28 = 2(x^2 + 5x - 14)$$
$$= 2(x + 7)(x - 2)$$

Practice Problems

1. Factor $x^2 + 5x + 6$.

2. Factor $x^2 + 2x - 15$.

Note Again, we can check our results by multiplying our factors to see if their product is the original polynomial.

3. Factor $2x^2 - 2x - 24$.

Answers
1. $(x + 2)(x + 3)$
2. $(x + 5)(x - 3)$
3. $2(x - 4)(x + 3)$

4. Factor $-3x^3 - 18x^2 - 15x$.

5. Factor $x^2 + 7xy + 12y^2$.

EXAMPLE 4 Factor $-3x^3 + 3x^2 + 18x$.

Solution We begin by factoring out the greatest common factor, which is $-3x$. Note that when the leading term is negative, we use the negative sign in our greatest common factor. Then we factor the remaining trinomial. Without showing the table of products and sums as we did in Examples 2 and 3, here is the complete problem.

$$-3x^3 + 3x^2 + 18x = -3x(x^2 - x - 6)$$
$$= -3x(x - 3)(x + 2)$$

This approach may also be used to factor trinomials with two variables, as we do in the next example.

EXAMPLE 5 Factor $x^2 + 8xy + 12y^2$.

Solution This time we need two expressions whose product is $12y^2$ and whose sum is $8y$. The two expressions are $6y$ and $2y$ (see Example 1 in this section).

$$x^2 + 8xy + 12y^2 = (x + 6y)(x + 2y)$$

You should convince yourself that these factors are correct by finding their product.

GETTING READY FOR CLASS

After reading through the preceding section, respond in your own words and in complete sentences.

A. When the leading coefficient of a trinomial is 1, what is the relationship between the other two coefficients and the factors of the trinomial?

B. In words, explain how you would factor a trinomial with a leading coefficient other than 1.

C. How can you check to see that you have factored a trinomial correctly?

D. Describe how you would find the factors of $x^2 + 8x + 12$.

Answers

4. $-3x(x + 1)(x + 5)$
5. $(x + 4y)(x + 3y)$

Vocabulary Review

Choose the correct words to fill in the blanks in the paragraph below.

 greatest common factor coefficient leading

 trinomial sum product

To factor a trinomial in which the _____ of x^2 is 1, find the numbers a and b whose _____ is the coefficient of the middle term and whose _____ is the constant term. If a trinomial has a _____ coefficient other than 1, factor out the _____ first (if possible) and then factor the _____ that remains.

Problems

A Factor the following trinomials.

1. $x^2 + 7x + 12$ **2.** $x^2 + 7x + 10$ **3.** $x^2 + 3x + 2$ **4.** $x^2 + 7x + 6$

5. $a^2 + 10a + 21$ **6.** $a^2 - 7a + 12$ **7.** $x^2 - 7x + 10$ **8.** $x^2 - 3x + 2$

9. $y^2 - 10y + 21$ **10.** $y^2 - 7y + 6$ **11.** $x^2 - x - 12$ **12.** $x^2 - 4x - 5$

13. $y^2 + y - 12$ **14.** $y^2 + 3y - 18$ **15.** $x^2 + 5x - 14$ **16.** $x^2 - 5x - 24$

17. $r^2 - 8r - 9$ **18.** $r^2 - r - 2$ **19.** $x^2 - x - 30$ **20.** $x^2 + 8x + 12$

21. $a^2 + 15a + 56$ **22.** $a^2 - 9a + 20$ **23.** $y^2 - y - 42$ **24.** $y^2 + y - 42$

25. $x^2 + 13x + 42$ **26.** $x^2 - 13x + 42$ **27.** $a^2 - 2a - 63$ **28.** $a^2 - a - 132$

B Factor the following polynomials completely. First factor out the greatest common factor, and then factor the remaining trinomial.

29. $2x^2 + 6x + 4$

30. $3x^2 - 6x - 9$

31. $3a^2 - 3a - 60$

32. $2a^2 - 18a + 28$

33. $100x^2 - 500x + 600$

34. $100x^2 - 900x + 2,000$

35. $100p^2 - 1,300p + 4,000$

36. $100p^2 - 1,200p + 3,200$

37. $x^4 - x^3 - 12x^2$

38. $x^4 - 11x^3 + 24x^2$

39. $2r^3 + 4r^2 - 30r$

40. $5r^3 + 45r^2 + 100r$

41. $2y^4 - 6y^3 - 8y^2$

42. $3r^3 - 3r^2 - 6r$

43. $x^5 + 4x^4 + 4x^3$

44. $x^5 + 13x^4 + 42x^3$

45. $3y^4 - 12y^3 - 15y^2$

$3y^2(y^2 - 4y - 5)$

46. $5y^4 - 10y^3 + 5y^2$

47. $4x^4 - 52x^3 + 144x^2$

48. $3x^3 - 3x^2 - 18x$

49. $-2x^3 + 10x^2 - 12x$

50. $-5x^3 + 15x^2 + 20x$

51. $-7x^4 - 21x^3 - 14x^2$

52. $-x^5 + 6x^4 - 8x^3$

Factor the following trinomials.

53. $x^2 + 5xy + 6y^2$

54. $x^2 - 5xy + 6y^2$

55. $x^2 - 9xy + 20y^2$

56. $x^2 + 9xy + 20y^2$

57. $a^2 + 2ab - 8b^2$

58. $a^2 - 2ab - 8b^2$

59. $a^2 - 10ab + 25b^2$

60. $a^2 + 6ab + 9b^2$

61. $a^2 + 10ab + 25b^2$

62. $a^2 - 6ab + 9b^2$

63. $x^2 + 2xa - 48a^2$

64. $x^2 - 3xa - 10a^2$

65. $x^2 - 5xb - 36b^2$

66. $x^2 - 13xb + 36b^2$

67. $2x^3 - 4x^2y - 6xy^2$

68. $3x^3 - 15x^2y + 18xy^2$

69. $x^3y - 7x^2y^2 + 12xy^3$

70. $a^3b - a^2b^2 - 20ab^3$

71. $-2a^3 - 4a^2b + 16ab^2$

72. $-3a^2b - 15ab^2 - 18b^3$

73. $x^4 - 5x^2 + 6$

74. $x^6 - 2x^3 - 15$

75. $x^2 - 80x - 2,000$

76. $x^2 - 190x - 2,000$

77. $x^2 - x + \dfrac{1}{4}$

78. $x^2 - \dfrac{2}{3}x + \dfrac{1}{9}$

79. $x^2 + 0.6x + 0.08$

80. $x^2 + 0.8x + 0.15$

81. If one of the factors of $x^2 + 24x + 128$ is $x + 8$, what is the other factor?

82. If one factor of $x^2 + 260x + 2,500$ is $x + 10$, what is the other factor?

83. What polynomial, when factored, gives $(4x + 3)(x - 1)$?

84. What polynomial factors to $(4x - 3)(x + 1)$?

85. If $x + 3$ is a factor of $x^2 + bx - 15$, what is the value of b?

86. If $x - 4$ is a factor of $x^2 + bx - 20$, what is the value of b?

87. If $x + 5$ is a factor of $x^2 + 2x + c$, what is the value of c?

88. If $x - 4$ is a factor of $x^2 - 12x + c$, what is the value of c?

Maintaining Your Skills

89. iPod Memory The chart shows the memory capacity of each generation of new iPod. Use it to answer the following questions.

a. How much more memory was there in the 2003 generation as compared to the 2001?

b. How much more memory was there in the 2007 generation as compared to the 2003?

c. If the increase from 2007 to 2009 continued, how much capacity was there in the 2011 generation iPod?

90. Mac Users The chart shows the percentage increase in Mac users over a seven-year period.

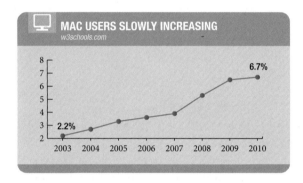

Let x represent the number of years past 2003 and y represent the percentages shown on the graph. Use the first and last points on the graph to find the equation of the line that connects them. Then use the equation to estimate the percentage increase in 2011.

Getting Ready for the Next Section

Multiply using the FOIL method.

91. $(6a + 1)(a + 2)$

92. $(6a - 1)(a - 2)$

93. $(3a + 2)(2a + 1)$

94. $(3a - 2)(2a - 1)$

95. $(6a + 2)(a + 1)$

96. $(3a + 1)(2a + 2)$

97. $2a(a + 4)(2a - 3)$

98. $3a(2a + 1)(3a - 1)$

Find the Mistake

Each sentence below contains a mistake. Circle the mistake and write the correct expression on the line provided.

1. The leading coefficient of the polynomial $x^2 - 2x - 35$ is 2. _____

2. Factoring the polynomial in the above sentence gives us $(x + 7)(x - 5)$. _____

3. For the polynomial $8x^3 - 30x^2 - 108x$, the greatest common factor is $8x^3$. _____

4. Factoring the polynomial in the above sentences gives us $(x - 6)(4x + 9)$. _____

5.3 Factoring Trinomials of the Form $ax^2 + bx + c$

OBJECTIVES

A Use trial and error to factor a trinomial whose leading coefficient is other than 1.

B Factor a polynomial by first factoring out the greatest common factor and then factoring the polynomial that remains.

KEY WORDS

trial and error

The sport of fierljeppen originates in the Netherlands and requires a person to jump over a canal using a long pole. The pole is typically between 8 and 13 meters long and has a round plate at the bottom. This plate prevents the pole from sinking into the bottom of the body of water being jumped. The person sprints to the pole that is sticking out of a canal at an angle. Then he grabs the pole and tries to climb to the top to control its swing over the water. If well-trained, the person lands in a sand pit on the opposite side, instead of in the water.

Suppose the path of the jumper, once she grabs onto the pole, is given by the trinomial $-x^2 + 40x - 370$. In this section, we will consider similar trinomials whose greatest common factor is 1 and whose leading coefficient (the coefficient of the squared term) is a number other than 1.

A Factoring $ax^2 + bx + c$ by Trial and Error

Suppose we want to factor the trinomial $2x^2 - 5x - 3$. We know the factors (if they exist) will be a pair of binomials. The product of their first terms is $2x^2$ and the product of their last term is -3. Let us list all the possible factors along with the trinomial that would result if we were to multiply them together. Remember, the middle term comes from the product of the inside terms plus the product of the outside terms.

Binomial Factors	First Term	Middle Term	Last Term
$(2x - 3)(x + 1)$	$2x^2$	$-x$	-3
$(2x + 3)(x - 1)$	$2x^2$	$+x$	-3
$(2x - 1)(x + 3)$	$2x^2$	$+5x$	-3
$(2x + 1)(x - 3)$	$2x^2$	$-5x$	-3

We can see from the last line that the factors of $2x^2 - 5x - 3$ are $(2x + 1)$ and $(x - 3)$. There is no straightforward way, as there was in the previous section, to find the factors, other than by trial and error or by simply listing all the possibilities. We look for possible factors that, when multiplied, will give the correct first and last terms, and then we see if we can adjust them to give the correct middle term.

1. Factor $2a^2 + 7a + 6$.

EXAMPLE 1 Factor $6a^2 + 7a + 2$.

Solution We list all the possible pairs of factors that, when multiplied together, give a trinomial whose first term is $6a^2$ and whose last term is 2.

Binomial Factors	First Term	Middle Term	Last Term
$(6a + 1)(a + 2)$	$6a^2$	$+13a$	$+2$
$(6a - 1)(a - 2)$	$6a^2$	$-13a$	$+2$
$(3a + 2)(2a + 1)$	$6a^2$	$+7a$	$+2$
$(3a - 2)(2a - 1)$	$6a^2$	$-7a$	$+2$

Note Remember, we can always check our results by multiplying the factors we have and comparing that product with our original polynomial.

The factors of $6a^2 + 7a + 2$ are $(3a + 2)$ and $(2a + 1)$ since these give the correct middle term.

Check: $(3a + 2)(2a + 1) = 6a^2 + 7a + 2$

Notice that in the preceding list we did not include the factors $(6a + 2)$ and $(a + 1)$. We do not need to try these since the first factor has a 2 common to each term and so could be factored again, giving $2(3a + 1)(a + 1)$. Since our original trinomial, $6a^2 + 7a + 2$, did *not* have a greatest common factor of 2, neither of its factors will.

2. Factor $4x^2 - x - 5$.

EXAMPLE 2 Factor $4x^2 - x - 3$.

Solution We list all the possible factors that, when multiplied, give a trinomial whose first term is $4x^2$ and whose last term is -3.

Binomial Factors	First Term	Middle Term	Last Term
$(4x + 1)(x - 3)$	$4x^2$	$-11x$	-3
$(4x - 1)(x + 3)$	$4x^2$	$+11x$	-3
$(4x + 3)(x - 1)$	$4x^2$	$-x$	-3
$(4x - 3)(x + 1)$	$4x^2$	$+x$	-3
$(2x + 1)(2x - 3)$	$4x^2$	$-4x$	-3
$(2x - 1)(2x + 3)$	$4x^2$	$+4x$	-3

The third line shows that the factors are $(4x + 3)$ and $(x - 1)$.

Check: $(4x + 3)(x - 1) = 4x^2 - x - 3$

You will find that the more practice you have at factoring this type of trinomial, the faster you will get the correct factors. You will pick up some shortcuts along the way, or you may come across a system of eliminating some factors as possibilities. Whatever works best for you is the method you should use. Factoring is a very important tool, and you must be good at it.

B Factoring Trinomials Using the Greatest Common Factor

Once again, the first step in any factoring problem is to factor out the greatest common factor if it is other than 1.

EXAMPLE 3 Factor $12y^3 + 10y^2 - 12y$.

Solution We begin by factoring out the greatest common factor, $2y$.

$$12y^3 + 10y^2 - 12y = 2y(6y^2 + 5y - 6)$$

We now list all possible factors of a trinomial with the first term $6y^2$ and last term -6, along with the associated middle terms.

Possible Factors	Middle Term When Multiplied
$(3y + 2)(2y - 3)$	$-5y$
$(3y - 2)(2y + 3)$	$+5y$
$(6y + 1)(y - 6)$	$-35y$
$(6y - 1)(y + 6)$	$+35y$

The second line gives the correct factors. The complete problem is

$$12y^3 + 10y^2 - 12y = 2y(6y^2 + 5y - 6)$$
$$= 2y(3y - 2)(2y + 3)$$

EXAMPLE 4 Factor $30x^2y - 5xy^2 - 10y^3$.

Solution The greatest common factor is $5y$.

$$30x^2y - 5xy^2 - 10y^3 = 5y(6x^2 - xy - 2y^2)$$
$$= 5y(2x + y)(3x - 2y)$$

EXAMPLE 5 A ball is tossed into the air with an upward velocity of 16 feet per second from the top of a building 32 feet high. The equation that gives the height of the ball above the ground at any time t is

$$h = 32 + 16t - 16t^2$$

Factor the right side of this equation and then find h when t is 2.

Solution We begin by factoring out the greatest common factor, 16. Then, we factor the trinomial that remains.

$$h = 32 + 16t - 16t^2$$
$$h = 16(2 + t - t^2)$$
$$h = 16(2 - t)(1 + t)$$

Letting $t = 2$ in the equation, we have

$$h = 16(0)(3) = 0$$

When t is 2, h is 0.

GETTING READY FOR CLASS

After reading through the preceding section, respond in your own words and in complete sentences.

A. Why would you use the trial and error method of factoring a trinomial?

B. Describe the criteria you would use to set up a table of possible factors of a trinomial.

C. What does it mean if you factor a trinomial and one of your factors has a greatest common factor of 3?

D. Describe how you would look for possible factors of $6a^2 + 7a + 2$.

3. Factor $30y^2 + 5y - 5$.

4. Factor $10x^2y^2 + 5xy^3 - 30y^4$.

5. Assume the ball in Example 5 is tossed from a building that is 48 feet tall and has an upward velocity of 32 feet per second. The equation that gives the height of the ball above the ground at any time t is $h = 48 + 32t - 16t^2$. Factor the right side of this equation and then find h when t is 3.

Answers
3. $5(3y - 1)(2y + 1)$
4. $5y^2(2x - 3y)(x + 2y)$
5. $h = 16(3 - t)(1 + t)$; When t is 3, h is 0.

Vocabulary Review

Choose the correct words to fill in the blanks in the paragraph below.

trial and error factors products polynomial multiply leading coefficient

To factor a polynomial whose _____ is something other than 1, list all the possibilities
in a process called _____. This process involves looking for possible _____
whose products will give the correct first and last terms, and whose sum of the _____ will
give the middle term. To check your results, _____ the factors and compare the product to the
original _____.

Problems

A Factor the following trinomials.

1. $2x^2 + 7x + 3$

2. $2x^2 + 5x + 3$

3. $2a^2 - a - 3$

4. $2a^2 + a - 3$

5. $3x^2 + 2x - 5$

6. $3x^2 - 2x - 5$

7. $3y^2 - 14y - 5$

8. $3y^2 + 14y - 5$

9. $6x^2 + 13x + 6$

10. $6x^2 - 13x + 6$

11. $4x^2 - 12xy + 9y^2$

12. $4x^2 + 12xy + 9y^2$

13. $4y^2 - 11y - 3$

14. $4y^2 + y - 3$

15. $20x^2 - 41x + 20$

16. $20x^2 + 9x - 20$

17. $20a^2 + 48ab - 5b^2$

18. $20a^2 + 29ab + 5b^2$

19. $20x^2 - 21x - 5$

SCAN TO ACCESS

20. $20x^2 - 48x - 5$

21. $12m^2 + 16m - 3$

22. $12m^2 + 20m + 3$

23. $20x^2 + 37x + 15$

24. $20x^2 + 13x - 15$

25. $12a^2 - 25ab + 12b^2$

26. $12a^2 + 7ab - 12b^2$

27. $3x^2 - xy - 14y^2$

28. $3x^2 + 19xy - 14y^2$

29. $14x^2 + 29x - 15$

30. $14x^2 + 11x - 15$

31. $6x^2 - 43x + 55$

32. $6x^2 - 7x - 55$

33. $15t^2 - 67t + 38$

34. $15t^2 - 79t - 34$

35. $28t^2 - 15t - 27$

36. $36t^2 + 33t - 20$

B Factor each of the following polynomials completely. Use the greatest common factor for your first step, then factor the remaining trinomial.

37. $4x^2 + 2x - 6$

38. $6x^2 - 51x + 63$

39. $24a^2 - 50a + 24$

40. $18a^2 + 48a + 32$

41. $10x^3 - 23x^2 + 12x$

42. $10x^4 + 7x^3 - 12x^2$

43. $6x^4 - 11x^3 - 10x^2$

44. $6x^3 + 19x^2 + 10x$

45. $10a^3 - 6a^2 - 4a$

46. $6a^3 + 15a^2 + 9a$

47. $15x^3 - 102x^2 - 21x$

48. $2x^4 - 24x^3 + 64x^2$

49. $35y^3 - 60y^2 - 20y$

50. $14y^4 - 32y^3 + 8y^2$

51. $15a^4 - 2a^3 - a^2$

52. $10a^5 - 17a^4 + 3a^3$

53. $24x^2y - 6xy - 45y$

54. $8x^2y^2 + 26xy^2 + 15y^2$

55. $12x^2y - 34xy^2 + 14y^3$

56. $12x^2y - 46xy^2 + 14y^3$

57. Evaluate the expression $2x^2 + 7x + 3$ and the expression $(2x + 1)(x + 3)$ for $x = 2$.

58. Evaluate the expression $2a^2 - a - 3$ and the expression $(2a - 3)(a + 1)$ for $a = 5$.

59. What polynomial factors to $(2x + 3)(2x - 3)$?

60. What polynomial factors to $(5x + 4)(5x - 4)$?

61. What polynomial factors to $(x + 3)(x - 3)(x^2 + 9)$?

62. What polynomial factors to $(x + 2)(x - 2)(x^2 + 4)$?

63. If $x + 3$ is a factor of $2x^2 + bx - 15$, what is the value of b?

64. If $2x - 5$ is a factor of $6x^2 + bx + 10$, what is the value of b?

65. If $3x + 4$ is a factor of $6x^2 - 7x + c$, what is the value of c?

66. If $5x + 3$ is a factor of $20x^2 - 13x + c$, what is the value of c?

Applying the Concepts

67. Archery Margaret shoots an arrow into the air. The equation for the height (in feet) of the tip of the arrow is

$$h = 8 + 62t - 16t^2$$

Factor the right side of this equation. Then fill in the table for various heights of the arrow, using the factored form of the equation.

Time t (seconds)	0	1	2	3	4
Height h (feet)					

68. Coin Toss At the beginning of every football game, the referee flips a coin to see who will kick off. The equation that gives the height (in feet) of the coin tossed in the air is

$$h = 6 + 29t - 16t^2$$

a. Factor this equation.

b. Use the factored form of the equation to find the height of the quarter after 0 seconds, 1 second, and 2 seconds.

69. Constructing a Box Yesterday I was experimenting with how to cut and fold a certain piece of cardboard to make a box with different volumes. Unfortunately, today I have lost both the cardboard and most of my notes. I remember that I made the box by cutting equal squares from the corners then folding up the side flaps.

 I don't remember how big the cardboard was, and I can only find the last page of notes, which says that if x is the length of a side of a small square (in inches), then the volume is $V = 99x - 40x^2 + 4x^3$.

 a. Factor the right side of this expression completely.

 b. What were the dimensions of the original piece of cardboard?

70. Constructing a Box Repeat Problem 69 if the remaining formula is $V = 15x - 16x^2 + 4x^3$.

Getting Ready for the Next Section

Multiply each of the following.

71. $(x + 3)(x - 3)$

72. $(x - 4)(x + 4)$

73. $(x + 5)(x - 5)$

74. $(x - 6)(x + 6)$

75. $(x + 7)(x - 7)$

76. $(x - 8)(x + 8)$

77. $(x + 9)(x - 9)$

78. $(x - 10)(x + 10)$

79. $(2x - 3y)(2x + 3y)$

80. $(5x - 6y)(5x + 6y)$

81. $(x^2 + 4)(x + 2)(x - 2)$

82. $(x^2 + 9)(x + 3)(x - 3)$

83. $(x + 3)^2$

84. $(x - 4)^2$

85. $(x + 5)^2$

86. $(x - 6)^2$

87. $(x + 7)^2$

88. $(x - 8)^2$

89. $(x + 9)^2$

90. $(x - 10)^2$

91. $(2x + 3)^2$

92. $(3x - y)^2$

93. $(4x - 2y)^2$

94. $(5x - 6y)^2$

Find the Mistake

Each sentence below contains a mistake. Circle the mistake and write the correct word(s) or expression(s) on the line provided.

1. $(2x + 2)$ is a factor of the polynomial $6x^2 + 7x + 12$. _____

2. The first step in factoring $8a^3 - 4a^2 - 6a$ is to list all possible factors of a trinomial with the first term $8a^3$ and the last term $-6a$, along with the associated middle terms. _____

3. All of the factors for $8a^3 - 4a^2 - 6a$ are $(4a - 1)(a + 3)$. _____

4. The factors of the polynomial $40x^3z - 16x^2z - 12xz$ are $(5xz + 3)(2xz - 2)$ _____

Landmark Review: Checking Your Progress

1. What is the first thing you should look for when you begin to factor a polynomial?

Factor the following polynomials.

2. $4a^3 + 8a^2 - 10a$

3. $40x^2y + 24xy^2 + 16x^2y^2$

4. $x^2 - 8x - x + 8$

5. $xy + 6x - 2y - 12$

6. $x^3 - 7x^2 - 4x + 28$

7. $x^3 - 8x^2 + 6x - 48$

8. $x^2 + 10x + 21$

9. $x^2 - 3x - 28$

10. $x^2 - 7x + 12$

11. $x^2 + 2x - 24$

12. $6x^2 - 5xy - 6y^2$

13. $3a^2 + 11a + 10$

14. $6y^2 - 7y + 2$

15. $20x^2 - 49xy + 30y^2$

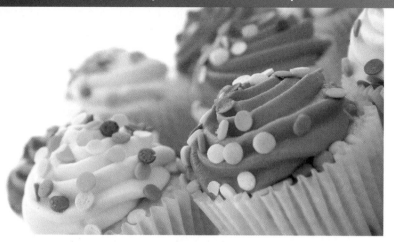

The cupcake craze is in full force with bakeries specializing in the small treats popping up all over the country. Suppose a bakery can bake x hundred cupcakes for a total cost given by the trinomial $100x^2 + 600x + 900$. Upon factoring out the greatest common factor of 100, we find the remaining expression contains a special trinomial called a perfect square trinomial, which we will discuss in this section.

$$100x^2 + 600x + 900 = 100(x^2 + 6x + 9)$$

But first, let's examine how three special products we listed in the previous chapter can also be considered as three special factors.

Recall the following three special products:

$$(a + b)^2 = (a + b)(a + b) = a^2 + 2ab + b^2$$
$$(a - b)^2 = (a - b)(a - b) = a^2 - 2ab + b^2$$
$$(a + b)(a - b) = a^2 - b^2$$

Since factoring is the reverse of multiplication, we can also consider the three special products as three special factorings:

$$a^2 + 2ab + b^2 = (a + b)^2$$
$$a^2 - 2ab + b^2 = (a - b)^2$$
$$a^2 - b^2 = (a + b)(a - b)$$

Any trinomial of the form $a^2 + 2ab + b^2$ or $a^2 - 2ab + b^2$ can be factored by the methods of Section 4.6. The last line is the factoring to obtain the *difference of two squares*. The difference of two squares always factors in this way. Again, these are patterns you must be able to recognize on sight.

A The Difference of Two Squares

EXAMPLE 1 Factor $16x^2 - 25$.

Solution We can see that the first term is a perfect square, and the last term is also. This fact becomes even more obvious if we rewrite the problem as

$$16x^2 - 25 = (4x)^2 - (5)^2$$

The first term is the square of the quantity $4x$, and the last term is the square of 5. The completed problem looks like this:

$$16x^2 - 25 = (4x)^2 - (5)^2$$
$$= (4x + 5)(4x - 5)$$

To check our results, we multiply

$$(4x + 5)(4x - 5) = 16x^2 + 20x - 20x - 25$$
$$= 16x^2 - 25$$

Practice Problems

1. Factor $9x^2 - 25$.

Answer

1. $(3x + 5)(3x - 5)$

2. Factor $49a^2 - 1$.

3. Factor $a^4 - b^4$.

Note If you think the sum of two squares $x^2 + y^2$ factors, you should try it. Write down the factors you think it has, and then multiply them using the FOIL method. You won't get $x^2 + y^2$. This type of polynomial that cannot be factored is considered *prime*.

4. Factor $4x^2 + 12x + 9$.

Note As we have indicated before, perfect square trinomials like the ones in Examples 4 and 5 can be factored by the methods developed in previous sections. Recognizing that they factor to binomial squares simply saves time in factoring.

5. Factor $6x^2 + 24x + 24$.

EXAMPLE 2 Factor $36a^2 - 1$.

Solution We rewrite the two terms to show they are perfect squares and then factor. Remember, 1 is its own square, $1^2 = 1$.

$$36a^2 - 1 = (6a)^2 - (1)^2$$
$$= (6a + 1)(6a - 1)$$

To check our results, we multiply

$$(6a + 1)(6a - 1) = 36a^2 + 6a - 6a - 1$$
$$= 36a^2 - 1$$

EXAMPLE 3 Factor $x^4 - y^4$.

Solution x^4 is the perfect square $(x^2)^2$, and y^4 is $(y^2)^2$.

$$x^4 - y^4 = (x^2)^2 - (y^2)^2$$
$$= (x^2 - y^2)(x^2 + y^2)$$

Notice the factor $(x^2 - y^2)$ is itself the difference of two squares and therefore can be factored again. The factor $(x^2 + y^2)$ is the *sum* of two squares and cannot be factored again. The complete problem is this:

$$x^4 - y^4 = (x^2)^2 - (y^2)^2$$
$$= (x^2 - y^2)(x^2 + y^2)$$
$$= (x + y)(x - y)(x^2 + y^2)$$

B Perfect Square Trinomials

EXAMPLE 4 Factor $25x^2 - 60x + 36$.

Solution Although this trinomial can be factored by the method we used in Section 5.3, we notice that the first and last terms are the perfect squares $(5x)^2$ and $(6)^2$. Before going through the method for factoring trinomials by listing all possible factors, we can check to see if $25x^2 - 60x + 36$ factors to $(5x - 6)^2$. We need only multiply to check.

$$(5x - 6)^2 = (5x - 6)(5x - 6)$$
$$= 25x^2 - 30x - 30x + 36$$
$$= 25x^2 - 60x + 36$$

The trinomial $25x^2 - 60x + 36$ factors to $(5x - 6)(5x - 6) = (5x - 6)^2$ and therefore, it is called a *perfect square trinomial*.

EXAMPLE 5 Factor $5x^2 + 30x + 45$.

Solution We begin by factoring out the greatest common factor, which is 5. Then we notice that the trinomial that remains is a perfect square trinomial.

$$5x^2 + 30x + 45 = 5(x^2 + 6x + 9)$$
$$= 5(x + 3)^2$$

Answers

2. $(7a + 1)(7a - 1)$
3. $(a^2 + b^2)(a + b)(a - b)$
4. $(2x + 3)^2$
5. $6(x + 2)^2$

EXAMPLE 6 Factor $(x - 3)^2 - 25$.

Solution This example has the form $a^2 - b^2$, where a is $x - 3$ and b is 5. We factor it according to the formula for the difference of two squares.

$$(x - 3)^2 - 25 = (x - 3)^2 - 5^2 \qquad \text{Write 25 as } 5^2.$$

$$= [(x - 3) - 5][(x - 3) + 5] \qquad \text{Factor.}$$

$$= (x - 8)(x + 2) \qquad \text{Simplify.}$$

Notice in this example we could have expanded $(x - 3)^2$, subtracted 25, and then factored to obtain the same result.

$$(x - 3)^2 - 25 = x^2 - 6x + 9 - 25 \qquad \text{Expand } (x \cdot 3)^2.$$

$$= x^2 - 6x - 16 \qquad \text{Simplify.}$$

$$= (x - 8)(x + 2) \qquad \text{Factor.}$$

6. Factor $(x - 4)^2 - 36$.

EXAMPLE 7 Factor each of the following.

a. $16x^2 + 56x + 49$ **b.** $16x^2 - 56x + 49$ **c.** $16x^2 - 49$ **d.** $16x^2 + 49$

Solution We use the formulas for perfect square trinomials and the difference of two squares whenever we can.

a. $16x^2 + 56x + 49 = (4x + 7)(4x + 7)$
$$= (4x + 7)^2$$

b. $16x^2 - 56x + 49 = (4x - 7)(4x - 7)$
$$= (4x - 7)^2$$

c. $16x^2 - 49 = (4x)^2 - (7)^2$
$$= (4x - 7)(4x + 7)$$

d. $16x^2 + 49$ represents the sum of two squares and cannot be factored. We say that it is prime.

7. Factor each of the following:
 a. $9x^2 + 30x + 25$
 b. $9x^2 - 30x + 25$
 c. $9x^2 - 25$
 d. $9x^2 + 25$

GETTING READY FOR CLASS

After reading through the preceding section, respond in your own words and in complete sentences.

A. Describe how you factor the difference of two squares.

B. What is a perfect square trinomial?

C. How do you know when you've factored completely?

D. Describe how you would factor $25x^2 - 60x + 36$.

Answers

6. $(x - 10)(x + 2)$
7. a. $(3x + 5)^2$ **b.** $(3x - 5)^2$
 c. $(3x + 5)(3x - 5)$ **d.** Prime

Vocabulary Review

Choose the correct words to fill in the blanks below.

sum square perfect squares difference of two squares trinomial

1. If the first and last terms of a binomial are _____ and the operation symbol that separates the two terms implies subtraction, then the binomial is a special factoring called a _____ .

2. The _____ of two squares $(x^2 + y^2)$ cannot be factored.

3. A perfect square _____ has first and last terms that are perfect squares.

4. If you _____ a binomial, your result will be a perfect square trinomial.

Problems

A Factor the following binomials.

1. $x^2 - 9$

2. $x^2 - 25$

3. $a^2 - 36$

4. $a^2 - 64$

5. $x^2 - 49$

6. $x^2 - 121$

7. $4a^2 - 16$

8. $4a^2 + 16$

9. $9x^2 + 25$

10. $16x^2 - 36$

11. $25x^2 - 169$

12. $x^2 - y^2$

13. $9a^2 - 16b^2$

14. $49a^2 - 25b^2$

15. $9 - m^2$

16. $16 - m^2$

17. $25 - 4x^2$

18. $36 - 49y^2$

19. $2x^2 - 18$

20. $3x^2 - 27$

SCAN TO ACCESS

21. $32a^2 - 128$

22. $3a^3 - 48a$

23. $8x^2y - 18y$

24. $50a^2b - 72b$

25. $a^4 - b^4$

26. $a^4 - 16$

27. $16m^4 - 81$

28. $81 - m^4$

29. $3x^3y - 75xy^3$

30. $2xy^3 - 8x^3y$

31. $-3x^3 + 48x$

32. $-5y^3 + 45y$

B Factor the following trinomials.

33. $x^2 - 2x + 1$

34. $x^2 - 6x + 9$

35. $x^2 + 2x + 1$

36. $x^2 + 6x + 9$

37. $a^2 - 10a + 25$

38. $a^2 + 10a + 25$

39. $y^2 + 4y + 4$

40. $y^2 - 8y + 16$

41. $x^2 - 4x + 4$

42. $x^2 + 8x + 16$

43. $m^2 - 12m + 36$

44. $m^2 + 12m + 36$

45. $4a^2 + 12a + 9$

46. $9a^2 - 12a + 4$

47. $49x^2 - 14x + 1$

48. $64x^2 - 16x + 1$

49. $9y^2 - 30y + 25$

50. $25y^2 + 30y + 9$

51. $x^2 + 10xy + 25y^2$

52. $25x^2 + 10xy + y^2$

53. $9a^2 + 6ab + b^2$

54. $9a^2 - 6ab + b^2$

55. $4x^2 - 12xy + 9y^2$

56. $9x^2 + 24xy + 16y^2$

Factor the following by first factoring out the greatest common factor.

57. $3a^2 + 18a + 27$

58. $4a^2 - 16a + 16$

59. $2x^2 + 20xy + 50y^2$

60. $3x^2 + 30xy + 75y^2$

61. $5x^3 + 30x^2y + 45xy^2$

62. $12x^2y - 36xy^2 + 27y^3$

63. $-4x^3 + 24x^2 - 36x$

64. $-3x^2y + 12xy^2 - 2y^3$

Factor by grouping the first three terms together.

65. $x^2 + 6x + 9 - y^2$

66. $x^2 + 10x + 25 - y^2$

67. $x^2 + 2xy + y^2 - 9$

68. $a^2 + 2ab + b^2 - 25$

69. Find a value for b so that the polynomial $x^2 + bx + 49$ factors to $(x + 7)^2$.

70. Find a value of b so that the polynomial $x^2 + bx + 81$ factors to $(x + 9)^2$.

71. Find the value of b for which the polynomial $x^2 + bx + 25$ factors to $(x - 5)^2$.

72. Find the value of b for which the polynomial $x^2 + bx + 64$ factors to $(x - 8)^2$.

73. Find the value of c for which the polynomial $x^2 + 10x + c$ factors to $(x + 5)^2$.

74. Find the value of a for which the polynomial $ax^2 + 12x + 9$ factors to $(2x + 3)^2$.

75. Find the value of c for which the polynomial $x^2 - 8x + c$ factors to $(x - 4)^2$.

76. Find the value of c for which the polynomial $x^2 - 12x + c$ factors to $(x - 6)^2$.

77. Find the value of a for which the polynomial $ax^2 - 12x + 4$ factors to $(3x - 2)^2$.

78. Find the value of a for which the polynomial $ax^2 - 8x + 1$ factors to $(4x - 1)^2$.

Applying the Concepts

79. Area

 a. What is the area of the following figure?

 b. Factor the answer from part a.

 c. Find a way to cut the figure into two pieces and put them back together to show that the factorization in part b is correct.

80. Area

 a. What is the area of the following figure?

 b. Factor the expression from part a.

 c. Cut and rearrange the figure to show that the factorization is correct.

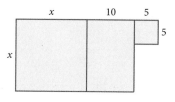

Find the area for the shaded regions; then write your result in factored form.

81.

82.

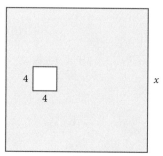

Getting Ready for the Next Section

Multiply each of the following:

83. a. 1^3

 b. 2^3

 c. 3^3

 d. 4^3

 e. 5^3

84. a. $(-1)^3$

 b. $(-2)^3$

 c. $(-3)^3$

 d. $(-4)^3$

 e. $(-5)^3$

85. a. $x(x^2 - x + 1)$

 b. $1(x^2 - x + 1)$

 c. $(x + 1)(x^2 - x + 1)$

86. a. $x(x^2 + x + 1)$

b. $-1(x^2 + x + 1)$

c. $(x - 1)(x^2 + x + 1)$

87. a. $x(x^2 - 2x + 4)$

b. $2(x^2 - 2x + 4)$

c. $(x + 2)(x^2 - 2x + 4)$

88. a. $x(x^2 + 2x + 4)$

b. $-2(x^2 + 2x + 4)$

c. $(x - 2)(x^2 + 2x + 4)$

89. a. $x(x^2 - 3x + 9)$

b. $3(x^2 - 3x + 9)$

c. $(x + 3)(x^2 - 3x + 9)$

90. a. $x(x^2 + 3x + 9)$

b. $-3(x^2 + 3x + 9)$

c. $(x - 3)(x^2 + 3x + 9)$

91. a. $x(x^2 - 4x + 16)$

b. $4(x^2 - 4x + 16)$

c. $(x + 4)(x^2 - 4x + 6)$

92. a. $x(x^2 + 4x + 16)$

b. $-4(x^2 + 4x + 16)$

c. $(x - 4)(x^2 + 4x + 16)$

93. a. $x(x^2 - 5x + 25)$

b. $5(x^2 - 5x + 25)$

c. $(x + 5)(x^2 - 5x + 25)$

94. a. $x(x^2 + 5x + 25)$

b. $-5(x^2 + 5x + 25)$

c. $(x - 5)(x^2 + 5x + 25)$

Find the Mistake

Each sentence below contains a mistake. Circle the mistake and write the correct word(s) or expression on the line provided.

1. The difference of two squares says that $a^2 - b^2 = (a - b)^2$. _____

2. Factoring $81y^6 - 196$ gives us $(9y^3 - 14)(9y^3 - 14)$. _____

3. The polynomial $18x^2 - 84x + 98$ is an example of a difference of two squares. _____

4. The polynomial in the above sentence factors to $3x - 7$. _____

A company that sells mountain biking apparel has just received an order for a new helmet and gloves. The helmet is packaged in a box with a volume of a^3 and the gloves are packaged in a box with a volume of b^3. In order to figure out the minimum volume needed for a shipping box to contain both items, we need to set up the following expression:

$$a^3 + b^3$$

This expression can also be called the sum of two cubes, which we were learn how to factor in this section.

Previously, we factored a variety of polynomials. Among the polynomials we factored were polynomials that were the difference of two squares. The formula we used to factor the difference of two squares looks like this:

$$a^2 - b^2 = (a + b)(a - b)$$

If we ran across a binomial that had the form of the difference of two squares, we factored it by applying this formula. For example, to factor $x^2 - 25$, we simply notice that it can be written in the form $x^2 - 5^2$, which looks like the difference of two squares. According to the formula above, this binomial factors into $(x + 5)(x - 5)$.

A Factoring the Sum or Difference of Two Cubes

In this section, we want to use two new formulas that will allow us to factor the sum and difference of two cubes. For example, we want to factor the binomial $x^3 - 8$, which is the difference of two cubes. (To see that it is the difference of two cubes, notice that it can be written $x^3 - 2^3$.) We also want to factor $y^3 + 27$, which is the sum of two cubes. (To see this, notice that $y^3 + 27$ can be written as $y^3 + 3^3$.)

The formulas that allow us to factor the sum of two cubes and the difference of two cubes are not as simple as the formula for factoring the difference of two squares. Here is what they look like:

$$a^3 + b^3 = (a + b)(a^2 - ab + b^2) \qquad \text{Sum of two cubes}$$
$$a^3 - b^3 = (a - b)(a^2 + ab + b^2) \qquad \text{Difference of two cubes}$$

Let's begin our work with these two formulas by showing that they are true. To do so, we multiply out the right side of each formula.

EXAMPLE 1 Verify the formula for a sum of two cubes and the formula for a difference of two cubes.

Solution We verify the formulas by multiplying the right sides and comparing the results with the left sides.

$$
\begin{array}{r}
a^2 - ab + b^2 \\
a + b \\
\hline
a^3 - a^2b + ab^2 \\
a^2b - ab^2 + b^3 \\
\hline
a^3 \qquad\qquad + b^3
\end{array}
\qquad\qquad
\begin{array}{r}
a^2 + ab + b^2 \\
a - b \\
\hline
a^3 + a^2b + ab^2 \\
- a^2b - ab^2 - b^3 \\
\hline
a^3 \qquad\qquad - b^3
\end{array}
$$

The first formula is correct. The second formula is correct.

Here are some examples that use the formulas for factoring the sum and difference of two cubes.

EXAMPLE 2 Factor $x^3 - 8$.

Solution Since the two terms are prefect cubes, we write them as such and apply the formula.

$$
\begin{aligned}
x^3 - 8 &= x^3 - 2^3 \\
&= (x - 2)(x^2 + 2x + 2^2) \\
&= (x - 2)(x^2 + 2x + 4)
\end{aligned}
$$

EXAMPLE 3 Factor $y^3 + 27$.

Solution Proceeding as we did in Example 2, we first write 27 as 3^3. Then, we apply the formula for factoring the sum of two cubes, which is $a^3 + b^3 = (a + b)(a^2 - ab + b^2)$.

$$
\begin{aligned}
y^3 + 27 &= y^3 + 3^3 \\
&= (y + 3)(y^2 - 3y + 3^2) \\
&= (y + 3)(y^2 - 3y + 9)
\end{aligned}
$$

EXAMPLE 4 Factor $a^3 - \dfrac{1}{8}$.

Solution The first term is the cube of a, the second term is the cube of $\frac{1}{2}$.

$$
\begin{aligned}
a^3 - \frac{1}{8} &= a^3 - \left(\frac{1}{2}\right)^3 \\
&= \left(a - \frac{1}{2}\right)\left(a^2 + \frac{1}{2}a + \frac{1}{4}\right)
\end{aligned}
$$

EXAMPLE 5 Factor $x^6 - y^6$.

Solution To begin, we have a choice of how we want to write the two terms. We can write the expression as the difference of two squares, $(x^3)^2 - (y^3)^2$, or as the difference of two cubes, $(x^2)^3 - (y^2)^3$. It is easier to use the difference of two squares if we have a choice.

$$
\begin{aligned}
x^6 - y^6 &= (x^3)^2 - (y^3)^2 \\
&= (x^3 - y^3)(x^3 + y^3) \\
&= (x - y)(x^2 + xy + y^2)(x + y)(x^2 - xy + y^2)
\end{aligned}
$$

Practice Problems

2. Factor $x^3 - 27$.

3. Factor $y^3 + 64$.

4. Factor $x^3 + \dfrac{1}{125}$.

5. Factor $a^6 + b^6$.

Answers

2. $(x - 3)(x^2 + 3x + 9)$

3. $(y + 4)(y^2 - 4y + 16)$

4. $\left(x + \dfrac{1}{5}\right)\left(x^2 - \dfrac{x}{5} + \dfrac{1}{25}\right)$

5. $(a^2 + b^2)(a^4 - a^2b^2 + b^4)$

Vocabulary Review

Choose the correct words to fill in the blanks below.

 sum difference cubes factored

1. The sum of two _____ is given by $a^3 + b^3 = (a + b)(a^2 - ab + b^2)$.

2. The _____ of two cubes is given by $a^3 - b^3 = (a - b)(a^2 + ab + b^2)$.

3. The sum of two cubes can be _____, whereas the sum of two squares cannot.

4. The expression $27 + x^6$ is an example of a _____ of two cubes $3^3 + (x^2)^3$.

Problems

A Factor each of the following polynomials.

1. $x^3 - y^3$

2. $x^3 + y^3$

3. $a^3 + 8$

4. $a^3 - 8$

5. $27 + x^3$

6. $27 - x^3$

7. $y^3 - 1$

8. $y^3 + 1$

9. $y^3 - 64$

10. $y^3 + 64$

11. $125h^3 - t^3$

12. $t^3 + 125h^3$

13. $x^3 - 216$

14. $216 + x^3$

15. $2y^3 - 54$

16. $81 + 3y^3$

17. $2a^3 - 128b^3$

18. $128a^3 + 2b^3$

19. $2x^3 + 432y^3$

20. $432x^3 - 2y^3$

21. $10a^3 - 640b^3$

22. $640a^3 + 10b^3$

23. $10r^3 - 1{,}250$

24. $10r^3 + 1{,}250$

25. $64 + 27a^3$

26. $27 - 64a^3$

27. $8x^3 - 27y^3$

28. $27x^3 - 8y^3$

29. $t^3 + \dfrac{1}{27}$

30. $t^3 - \dfrac{1}{27}$

31. $27x^3 - \dfrac{1}{27}$

32. $8x^3 + \dfrac{1}{8}$

33. $64a^3 + 125b^3$

34. $125a^3 - 27b^3$

35. $\dfrac{1}{8}x^3 - \dfrac{1}{27}y^3$

36. $\dfrac{1}{27}x^3 + \dfrac{1}{8}y^3$

37. $a^6 - b^6$

38. $x^6 - 64y^6$

39. $64x^6 - y^6$

40. $x^6 - (3y)^6$

41. $x^6 - (5y)^6$

42. $(4x)^6 - (7y)^6$

43. $x^{27} - 27$

44. $x^{15} + 8$

45. Find a value for c so that $x + c$ is a factor of $x^3 + 8$.

46. Find a value for c so that $x + c$ is a factor of $x^3 - 27$.

47. Find a value for c so that $x + 5$ is a factor of $x^3 + c$.

48. Find a value for c so that $x - 4$ is a factor of $x^3 + c$.

Getting Ready for the Next Section

Multiply each of the following.

49. $2x^3(x + 2)(x - 2)$

50. $3x^2(x + 3)(x - 3)$

51. $3x^2(x - 3)^2$

52. $2x^2(x + 5)^2$

53. $y(y^2 + 25)$

54. $y^3(y^2 + 36)$

55. $(5a - 2)(3a + 1)$

56. $(3a - 4)(2a - 1)$

57. $4x^2(x - 5)(x + 2)$

58. $6x(x - 4)(x + 2)$

59. $2ab^3(b^2 - 4b + 1)$

60. $2a^3b(a^2 + 3a + 1)$

Find the Mistake

Each sentence below contains a mistake. Circle the mistake and write the correct word or expression(s) on the line provided.

1. The difference of two squares says that $a^3 - b^3 = (a - b)(a^2 + ab + b^2)$. _____

2. The expression $a^3 + 8b^3$ is an example of a difference of two cubes. _____

3. The factors of the sum of two cubes in the above sentence are $(a + 8)$ and $(a^2 - ab + 64)$.

4. Factoring $x^{12} - 343$ gives us $(x^4 - 7)(x^8 - 49)$. _____

5.6 Factoring: A General Review

©iStockphoto.com/xof711

OBJECTIVES

A Factor a variety of polynomials.

The Norwegian fjords are a popular location for wingsuit flying. Many of the thrill-seekers opt to leap, flip, or be thrown off the edge of a sheer cliff to begin their flight. A wingsuit flyer wears a special jumpsuit that adds surface area to his body to create lift during a flight. The suit contains extra fabric that stretches from each arm to the flyer's body, as well as between the flyer's legs. At a designated altitude, the flyer may deploy a parachute from a backpack to ensure a safe landing.

Suppose the wingsuit flyer's flight path from the time he jumps to the time he opens the parachute can be given by the following expression:

$$-9x^3 - 12x^2 + 4x + 40$$

In this section, we will review the different methods of factoring that we presented in the previous sections of the chapter. This section is important because it will give you an opportunity to factor a variety of polynomials such as the one above. Prior to this section, the polynomials you worked with were grouped together according to the method used to factor them; that is, in Section 5.4 all the polynomials you factored were either the difference of two squares or perfect square trinomials. What usually happens in a situation like this is that you become proficient at factoring the kind of polynomial you are working with at the time but have trouble when given a variety of polynomials to factor.

A Factoring Polynomials of Any Type

We begin this section with a checklist that can be used in factoring polynomials of any type. When you have finished this section and the problem set that follows, you want to be proficient enough at factoring that the checklist is second nature to you.

> **HOW TO** Factor a Polynomial
>
> **Step 1:** If the polynomial has a greatest common factor other than 1, then factor out the greatest common factor.
>
> **Step 2:** If the polynomial has two terms (a binomial), then see if it is the difference of two squares or the sum or difference of two cubes, then factor accordingly. Remember, if it is the sum of two squares, it will not factor.
>
> **Step 3:** If the polynomial has three terms (a trinomial), then either it is a perfect square trinomial, which will factor into the square of a binomial, or it is not a perfect square trinomial, in which case you can try to use the trial and error method developed in Section 5.3.
>
> **Step 4:** If the polynomial has more than three terms, try to factor it by grouping.
>
> **Step 5:** As a final check, see if any of the factors you have written can be factored further. If you have overlooked a common factor, you can catch it here. Remember, some polynomials cannot be factored, in which case they are considered prime.

5.6 Factoring: A General Review

419

Practice Problems

1. Factor $3x^4 - 27x^2$.

Here are some examples illustrating how we use the checklist.

EXAMPLE 1 Factor $2x^5 - 8x^3$.

Solution First, we check to see if the greatest common factor is other than 1. Since the greatest common factor is $2x^3$, we begin by factoring it out. Once we have done so, we notice that the binomial that remains is the difference of two squares.

$$2x^5 - 8x^3 = 2x^3(x^2 - 4) \qquad \text{Factor out the greatest common factor, } 2x^3.$$

$$= 2x^3(x + 2)(x - 2) \qquad \text{Factor the difference of two squares.}$$

Note that the greatest common factor $2x^3$ that we factored from each term in the first step of Example 1 remains as part of the answer to the problem, because it is one of the factors of the original binomial. Remember, the expression we end up with when factoring must be equal to the expression we start with. We can't just drop a factor and expect the resulting expression to equal the original expression.

2. Factor $2x^5 + 20x^4 + 50x^3$.

EXAMPLE 2 Factor $3x^4 - 18x^3 + 27x^2$.

Solution Step 1 is to factor out the greatest common factor, $3x^2$. After we have done so, we notice that the trinomial that remains is a perfect square trinomial, which will factor as the square of a binomial.

$$3x^4 - 18x^3 + 27x^2 = 3x^2(x^2 - 6x + 9) \quad \text{Factor out } 3x^2.$$

$$= 3x^2(x - 3)^2 \qquad x^2 - 6x + 9 \text{ is the square of } x - 3.$$

3. Factor $y^5 + 36y^3$.

EXAMPLE 3 Factor $y^3 + 25y$.

Solution We begin by factoring out the y that is common to both terms. The binomial that remains after we have done so is the sum of two squares, which does not factor, so after the first step we are finished.

$$y^3 + 25y = y(y^2 + 25) \qquad \text{Factor out the greatest common factor, } y; \text{ then,}$$
$$\text{notice that } y^2 + 25 \text{ cannot factored further.}$$

4. Factor $15a^2 - a - 2$.

EXAMPLE 4 Factor $6a^2 - 11a + 4$.

Solution Here we have a trinomial that does not have a greatest common factor other than 1. Since it is not a perfect square trinomial, we factor it by trial and error; that is, we look for binomial factors of the product whose first terms is $6a^2$ and whose last terms is 4. Then we look for the combination of these types of binomials whose product gives us a middle term of $-11a$. Without showing all the different possibilities, here is the answer:

$$6a^2 - 11a + 4 = (3a - 4)(2a - 1)$$

5. Factor $4x^4 - 12x^3 - 40x^2$.

EXAMPLE 5 Factor $6x^3 - 12x^2 - 48x$.

Solution This trinomial has a greatest common factor of $6x$. The trinomial that remains after the $6x$ has been factored from each term must be factored by trial and error.

$$6x^3 - 12x^2 - 48x = 6x(x^2 - 2x - 8)$$
$$= 6x(x - 4)(x + 2)$$

6. Factor $3ab + 9a + 2b + 6$.

EXAMPLE 6 Factor $xy + 8x + 3y + 24$.

Solution Since our polynomial has four terms, we try factoring by grouping.

$$xy + 8x + 3y + 24 = x(y + 8) + 3(y + 8)$$
$$= (y + 8)(x + 3)$$

Answers

1. $3x^2(x + 3)(x - 3)$
2. $2x^3(x + 5)^2$
3. $y^3(y^2 + 36)$
4. $(5a - 2)(3a + 1)$
5. $4x^2(x - 5)(x + 2)$
6. $(b + 3)(3a + 2)$

Vocabulary Review

The following is a list of steps for factoring. Choose the correct words to fill in the blanks below.

factors cubes greatest common factor trial and error

grouping perfect square trinomial squares

Step 1: If the polynomial has a _____ other than 1, then factor out the greatest common factor.

Step 2: If the polynomial has two terms, then see if it is the difference of two _____ or the sum or difference of two _____, then factor accordingly.

Step 3: If the polynomial has three terms, then either it is a _____, which will factor into the square of a binomial, or it is not a perfect square trinomial, which will factor using _____.

Step 4: If the polynomial has more than three terms, try to factor it by _____.

Step 5: As a final check, see if any _____ can be factored further.

Problems

A Factor each of the following polynomials completely; that is, once you are finished factoring, none of the factors you obtain should be factorable. Also, note that the even-numbered problems are not necessarily similar to the odd-numbered problems that precede them in this exercise set.

1. $x^2 - 81$

2. $x^2 - 18x + 81$

3. $x^2 + 2x - 15$

4. $15x^2 + 11x - 6$

5. $x^2 + 6x + 9$

6. $12x^2 - 11x + 2$

7. $y^2 - 10y + 25$

8. $21y^2 - 25y - 4$

9. $2a^3b + 6a^2b + 2ab$

10. $6a^2 - ab - 15b^2$

11. $x^2 + x + 1$

12. $2x^2 - 4x + 2$

13. $12a^2 - 75$

14. $18a^2 - 50$

15. $9x^2 - 12xy + 4y^2$

16. $x^3 - x^2$

17. $4x^3 + 16xy^2$

18. $16x^2 + 49y^2$

19. $2y^3 + 20y^2 + 50y$

20. $3y^2 - 9y - 30$

21. $a^6 + 4a^4b^2$

SCAN TO ACCESS

22. $5a^2 - 45b^2$

23. $xy + 3x + 4y + 12$

24. $xy + 7x + 6y + 42$

25. $x^4 - 16$

26. $x^4 - 81$

27. $xy - 5x + 2y - 10$

28. $xy - 7x + 3y - 21$

29. $5a^2 + 10ab + 5b^2$

30. $3a^3b^2 + 15a^2b^2 + 3ab^2$

31. $x^2 + 49$

32. $16 - x^4$

33. $3x^2 + 15xy + 18y^2$

34. $3x^2 + 27xy + 54y^2$

35. $2x^2 + 15x - 38$

36. $2x^2 + 7x - 85$

37. $100x^2 - 300x + 200$

38. $100x^2 - 400x + 300$

39. $x^2 - 64$

40. $9x^2 - 4$

41. $x^2 + 3x + ax + 3a$

42. $x^2 + 4x + bx + 4b$

43. $49a^7 - 9a^5$

44. $a^4 - 1$

45. $49x^2 + 9y^2$

46. $12x^4 - 62x^3 + 70x^2$

47. $25a^3 + 20a^2 + 3a$

48. $36a^4 - 100a^2$

49. $xa - xb + ay - by$

50. $xy - bx + ay - ab$

51. $48a^4b - 3a^2b$

52. $18a^4b^2 - 12a^3b^3 + 8a^2b^4$

53. $20x^4 - 45x^2$

54. $16x^3 + 16x^2 + 3x$

55. $3x^2 + 35xy - 82y^2$

56. $3x^2 + 37xy - 86y^2$

57. $16x^5 - 44x^4 + 30x^3$

58. $16x^2 + 16x - 1$

59. $2x^2 + 2ax + 3x + 3a$

60. $2x^2 + 2ax + 5x + 5a$

61. $y^4 - 1$

62. $25y^7 - 16y^5$

63. $12x^4y^2 + 36x^3y^3 + 27x^2y^4$

64. $16x^3y^2 - 4xy^2$

65. $-4x^8 - 32x^5$

66. $-3y^4 + 81y$

The following problems require you to factor out a common binomial factor first.
 For example,

the expression \rightarrow	$(x + 2)^3 - 3(x + 2)^2$
can be factored as \rightarrow	$(x + 2)^2 [(x + 2) - 3]$
which simplifies to \rightarrow	$(x + 2)^2(x - 1)$.

Use this technique in problems 67–74.

67. $(x + 4)^3 - 5(x + 4)^2$

68. $(x - 5)^3 + 2(x - 5)^2$

69. $(x + 2)^4 + 5(x + 2)^2$

70. $(x - 3)^4 + 4(x - 3)^2$

71. $(2x + 1)^3 - (2x + 1)$

72. $(3x - 2)^3 - (3x - 2)$

73. $(2x - 3)^3 - 2(2x - 3)$

74. $(3x - 1)^3 - (3x - 1)$

Maintaining Your Skills

75. Facebook Users The chart shows the rise in Facebook users over a six-year period. Give five ordered pairs shown on the chart.

76. Fox Shows The chart lists the top primetime shows on the Fox network. Write a mathematical statement using one of the symbols $<$ or $>$ to compare the following:

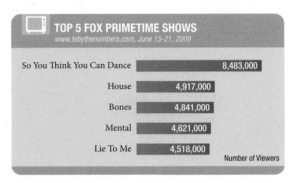

a. *House* to *Lie to Me*

b. *Mental* to *So You Think You Can Dance*

c. *Bones* to *Lie to Me*

Getting Ready for the Next Section

Solve each equation.

77. $3x - 6 = 9$

78. $5x - 1 = 14$

79. $2x + 3 = 0$

80. $4x - 5 = 0$

81. $4x + 3 = 0$

82. $3x - 1 = 0$

Find the Mistake

Each sentence below contains a mistake. Circle the mistake and write the correct word(s) or expression on the line provided.

1. If a binomial is the sum of two squares, it will factor into two consecutive integers. _____

2. To factor $14y^4 - 126y^2$, we first factor out the greatest common factor y^2. _____

3. Once you factored out the greatest common factor from the polynomial in the previous sentence, you are left with a perfect square trinomial. _____

4. To factor the polynomial $9x^2 + 3xy - 4y + 16x^2$, group the first and last terms together and the middle two terms together. _____

5.7 Solving Quadratic Equations by Factoring

OBJECTIVES

A Solve a quadratic equation by writing it in standard form and then factoring.

In this section, we will use the methods of factoring developed in previous sections, along with a special property of 0, to solve quadratic equations.

Picture a football player throwing a ball straight up in the air. The distance the ball is from the ground can be represented by the quadratic equation $d = -16t^2 + 80t$ where t is the time the ball is in the air. If given a value for d, we can put this equation into standard form, which will enable us to solve for t. Here is a formal definition for a quadratic equation and standard form.

KEY WORDS

quadratic equation

standard form

quadratic term

linear term

constant term

zero-factor property

A Solving a Quadratic Equation by Factoring

> **DEFINITION** quadratic equation
>
> Any equation that can be put in the form $ax^2 + bx + c = 0$, where a, b, and c are real numbers ($a \neq 0$), is called a **quadratic equation.** The equation $ax^2 + bx + c = 0$ is called **standard form** for a quadratic equation.
>
> $$\underset{a(\text{variable})^2}{x^2 \text{ term}} + \underset{b(\text{variable})}{x \text{ term}} + \underset{c(\text{absence of the variable})}{\text{Constant term}} = 0$$

The x^2 term is called the *quadratic term* and the x term is called the *linear term*.

The number 0 has a special property. If we multiply two numbers and the product is 0, then one or both of the original two numbers must be 0. In symbols, this property looks like this:

> **PROPERTY** Zero-Factor Property
>
> Let a and b represent real numbers. If $a \cdot b = 0$, then $a = 0$ or $b = 0$.

Suppose we want to solve the quadratic equation $x^2 + 5x + 6 = 0$. We can factor the left side into $(x + 2)(x + 3)$. Then we have

$$x^2 + 5x + 6 = 0$$
$$(x + 2)(x + 3) = 0$$

Now, $(x + 2)$ and $(x + 3)$ both represent real numbers when we replace x with a number. Their product is 0; therefore, either $(x + 3)$ is 0 or $(x + 2)$ is 0. Either way we have a solution to our equation. We use the zero-factor property stated to finish the problem.

$$x^2 + 5x + 6 = 0$$
$$(x + 2)(x + 3) = 0$$
$$x + 2 = 0 \quad \text{or} \quad x + 3 = 0$$
$$x = -2 \qquad\qquad x = -3$$

Note A quadratic equation in standard form may also use a variable other than x, or be missing the linear or constant term, such as the equation about the football in our opener.

Our solution set is $\{-2, -3\}$. Our equation has two solutions. To check our solutions we have to check each one separately to see that they both produce a true statement when used in place of the variable.

When → $x = -3$

the equation → $x^2 + 5x + 6 = 0$

becomes → $(-3)^2 + 5(-3) + 6 \stackrel{?}{=} 0$

$$9 + (-15) + 6 = 0$$

$$0 = 0$$

When → $x = -2$

the equation → $x^2 + 5x + 6 = 0$

becomes → $(-2)^2 + 5(-2) + 6 \stackrel{?}{=} 0$

$$4 + (-10) + 6 = 0$$

$$0 = 0$$

We have solved a quadratic equation by replacing it with two linear equations in one variable. Based on this process, we have developed the following strategy to solve a quadratic equation by factoring.

HOW TO **Strategy for Solving a Quadratic Equation by Factoring**

Step 1: Put the equation in standard form; that is, 0 on one side and decreasing powers of the variable on the other.

Step 2: Factor completely.

Step 3: Use the zero-factor property to set each variable factor from Step 2 equal to 0.

Step 4: Solve each equation produced in Step 3.

Step 5: Check each solution, if necessary.

Practice Problems

1. Solve the equation $2x^2 + 5x = 3$.

EXAMPLE 1 Solve the equation $2x^2 - 5x = 12$.

Solution

Step 1: Begin by adding -12 to both sides, so the equation is in standard form.

$$2x^2 - 5x = 12$$

$$2x^2 - 5x - 12 = 0$$

Step 2: Factor the left side completely.

$$(2x + 3)(x - 4) = 0$$

Step 3: Set each factor equal to 0.

$$2x + 3 = 0 \qquad \text{or} \qquad x - 4 = 0$$

Step 4: Solve each of the equations from step 3.

$$2x + 3 = 0 \qquad\qquad x - 4 = 0$$

$$2x = -3 \qquad\qquad\quad x = 4$$

$$x = -\frac{3}{2}$$

Answers

1. $\frac{1}{2}$, -3

Step 5: Substitute each solution into $2x^2 - 5x = 12$ to check.

$$\text{Check: } -\frac{3}{2} \qquad\qquad \text{Check: } 4$$

$$2\left(-\frac{3}{2}\right)^2 - 5\left(-\frac{3}{2}\right) \overset{?}{=} 12 \qquad 2(4)^2 - 5(4) \overset{?}{=} 12$$

$$2\left(\frac{9}{4}\right) + 5\left(\frac{3}{2}\right) = 12 \qquad\qquad 2(16) - 20 = 12$$

$$\frac{9}{2} + \frac{15}{2} = 12 \qquad\qquad 32 - 20 = 12$$

$$\frac{24}{2} = 12 \qquad\qquad\qquad 12 = 12$$

$$12 = 12$$

EXAMPLE 2 Solve the equation $16a^2 - 25 = 0$.

Solution The equation is already in standard form.

$$16a^2 - 25 = 0$$

$$(4a - 5)(4a + 5) = 0 \qquad\qquad \text{\textit{Factor the left side.}}$$

$$4a - 5 = 0 \quad \text{or} \quad 4a + 5 = 0 \qquad \text{\textit{Set each factor equal to 0.}}$$

$$4a = 5 \qquad\qquad 4a = -5 \qquad \text{\textit{Solve the resulting equations.}}$$

$$a = \frac{5}{4} \qquad\qquad a = -\frac{5}{4}$$

2. Solve the equation $49a^2 - 16 = 0$.

EXAMPLE 3 Solve the equation $4x^2 = 8x$.

Solution We begin by adding $-8x$ to each side of the equation to put it in standard form. Then we factor the left side of the equation by factoring out the greatest common factor.

$$4x^2 = 8x$$

$$4x^2 - 8x = 0 \qquad\qquad \text{\textit{Add } -8x \text{ to each side.}}$$

$$4x(x - 2) = 0 \qquad\qquad \text{\textit{Factor the left side.}}$$

$$4x = 0 \quad \text{or} \quad x - 2 = 0 \qquad \text{\textit{Set each factor equal to 0.}}$$

$$x = 0 \qquad\qquad x = 2 \qquad \text{\textit{Solve the resulting equations.}}$$

The solutions are 0 and 2.

3. Solve the equation $5x^2 = -2x$.

EXAMPLE 4 Solve the equation $x(2x + 3) = 44$.

Solution We must multiply out the left side first and then put the equation in standard form.

$$x(2x + 3) = 44$$

$$2x^2 + 3x = 44 \qquad\qquad \text{\textit{Multiply out the left side.}}$$

$$2x^2 + 3x - 44 = 0 \qquad\qquad \text{\textit{Add } -44 \text{ to each side.}}$$

$$(2x + 11)(x - 4) = 0 \qquad\qquad \text{\textit{Factor the left side.}}$$

$$2x + 11 = 0 \quad \text{or} \quad x - 4 = 0 \qquad \text{\textit{Set each factor equal to 0.}}$$

$$2x = -11 \qquad\qquad x = 4 \qquad \text{\textit{Solve the resulting equations.}}$$

$$x = -\frac{11}{2}$$

The two solutions are $-\frac{11}{2}$ and 4.

4. Solve the equation $x(13 - x) = 40$.

Answers

2. $-\frac{4}{7}, \frac{4}{7}$

3. $0, -\frac{2}{5}$

4. $5, 8$

5. Solve the equation
$(x + 2)^2 = (x + 1)^2 + x^2$.

EXAMPLE 5 Solve the equation $5^2 = x^2 + (x + 1)^2$.

Solution Before we can put this equation in standard form we must square the binomial. Remember, to square a binomial, we use the formula $(a + b)^2 = a^2 + 2ab + b^2$.

$$5^2 = x^2 + (x + 1)^2$$

$$25 = x^2 + x^2 + 2x + 1 \qquad \text{Expand } 5^2 \text{ and } (x + 1)^2.$$

$$25 = 2x^2 + 2x + 1 \qquad \text{Simplify the right side.}$$

$$0 = 2x^2 + 2x - 24 \qquad \text{Add } -25 \text{ to each side.}$$

$$0 = 2(x^2 + x - 12) \qquad \text{Begin factoring.}$$

$$0 = 2(x + 4)(x - 3) \qquad \text{Factor completely.}$$

$$x + 4 = 0 \quad \text{or} \quad x - 3 = 0 \qquad \text{Set each variable factor equal to 0.}$$

$$x = -4 \qquad\qquad x = 3$$

Note, in the second to the last line, that we do not set 2 equal to 0. That is because 2 can never be 0. It is always 2. We only use the zero-factor property to set variable factors equal to 0 because they are the only factors that can possibly be 0.

 Also notice that it makes no difference which side of the equation is 0 when we write the equation in standard form.

Although the equation in the next example is not a quadratic equation, it can be solved by the method shown in the first five examples.

6. Solve the equation
$4x^3 = x^2 + 5x$.

EXAMPLE 6 Solve the equation $24x^3 = -10x^2 + 6x$.

Solution First, we write the equation in standard form.

$$24x^3 + 10x^2 - 6x = 0 \qquad \text{Standard form}$$

$$2x(12x^2 + 5x - 3) = 0 \qquad \text{Factor out } 2x.$$

$$2x(3x - 1)(4x + 3) = 0 \qquad \text{Factor remaining trinomial.}$$

$$2x = 0 \quad \text{or} \quad 3x - 1 = 0 \quad \text{or} \quad 4x + 3 = 0 \qquad \text{Set factors equal to 0 to find solutions.}$$

$$x = 0 \qquad\qquad x = \frac{1}{3} \qquad\qquad x = -\frac{3}{4}$$

GETTING READY FOR CLASS

After reading through the preceding section, respond in your own words and in complete sentences.

A. When is an equation in standard form?

B. What is the first step in solving an equation by factoring?

C. Describe the zero-factor property in your own words.

D. Describe how you would solve $2x^2 - 5x = 12$ for x.

Answers

5. $-1, 3$

6. $0, \frac{5}{4}, -1$

Vocabulary Review

Choose the correct words to fill in the blanks below.

constant standard form zero-factor property quadratic equation

1. A _____ is any equation that can be put in the form $ax^2 + bx + c = 0$, where a, b, and c are real numbers and a does not equal zero.

2. The equation $ax^2 + bx + c = 0$ is called _____ for a quadratic equation.

3. The variable c is considered the _____ term of a quadratic equation.

4. The _____ states that if $a \cdot b = 0$, then $a = 0$ or $b = 0$.

The following is a list of steps for solving a quadratic equation by factoring. Write the correct step number in the blanks.

_____ Use the zero-factor property to set each variable factor equal to zero.

_____ Solve the equation.

_____ Factor completely.

_____ Check each solution.

_____ Put the equation in standard form.

Problems

A The following equations are already in factored form. Use the zero-factor property to set the factors equal to 0 and solve.

1. $(x + 2)(x - 1) = 0$

2. $(x + 3)(x + 2) = 0$

3. $(a - 4)(a - 5) = 0$

4. $(a + 6)(a - 1) = 0$

5. $x(x + 1)(x - 3) = 0$

6. $x(2x + 1)(x - 5) = 0$

7. $(3x + 2)(2x + 3) = 0$

8. $(4x - 5)(x - 6) = 0$

9. $m(3m + 4)(3m - 4) = 0$

10. $m(2m - 5)(3m - 1) = 0$

11. $2y(3y + 1)(5y + 3) = 0$

12. $3y(2y - 3)(3y - 4) = 0$

SCAN TO ACCESS

Solve the following equations.

13. $x^2 + 3x + 2 = 0$

14. $x^2 - x - 6 = 0$

15. $x^2 - 9x + 20 = 0$

16. $x^2 + 2x - 3 = 0$

17. $a^2 - 2a - 24 = 0$

18. $a^2 - 11a + 30 = 0$

19. $100x^2 - 500x + 600 = 0$

20. $100x^2 - 300x + 200 = 0$

21. $x^2 = -6x - 9$

22. $x^2 = 10x - 25$

23. $a^2 - 16 = 0$

24. $a^2 - 36 = 0$

25. $2x^2 + 5x - 12 = 0$

26. $3x^2 + 14x - 5 = 0$

27. $9x^2 + 12x + 4 = 0$

28. $12x^2 - 24x + 9 = 0$

29. $a^2 + 25 = 10a$

30. $a^2 + 16 = 8a$

31. $2x^2 = 3x + 20$

32. $6x^2 = x + 2$

33. $3m^2 = 20 - 7m$

34. $2m^2 = -18 + 15m$

35. $4x^2 - 49 = 0$

36. $16x^2 - 25 = 0$

37. $x^2 + 6x = 0$

38. $x^2 - 8x = 0$

39. $x^2 - 3x = 0$

40. $x^2 + 5x = 0$

41. $2x^2 = 8x$

42. $2x^2 = 10x$

43. $3x^2 = 15x$

44. $5x^2 = 15x$

45. $1,400 = 400 + 700x - 100x^2$

46. $2,700 = 700 + 900x - 100x^2$

47. $6x^2 = -5x + 4$

48. $9x^2 = 12x - 4$

49. $x(2x - 3) = 20$

50. $x(3x - 5) = 12$

51. $t(t + 2) = 80$

52. $t(t + 2) = 99$

53. $4,000 = (1,300 - 100p)p$

54. $3,200 = (1,200 - 100p)p$

55. $x(14 - x) = 48$

56. $x(12 - x) = 32$

57. $(x + 5)^2 = 2x + 9$

58. $(x + 7)^2 = 2x + 13$

59. $(y - 6)^2 = y - 4$

60. $(y + 4)^2 = y + 6$

61. $10^2 = (x + 2)^2 + x^2$

62. $15^2 = (x + 3)^2 + x^2$

63. $2x^3 + 11x^2 + 12x = 0$

64. $3x^3 + 17x^2 + 10x = 0$

65. $4y^3 - 2y^2 - 30y = 0$

66. $9y^3 + 6y^2 - 24y = 0$

67. $8x^3 + 16x^2 = 10x$

68. $24x^3 - 22x^2 = -4x$

69. $20a^3 = -18a^2 + 18a$

70. $12a^3 = -2a^2 + 10a$

71. $16t^2 - 32t + 12 = 0$

72. $16t^2 - 64t + 48 = 0$

Simplify each side as much as possible, then solve the equation.

73. $(a - 5)(a + 4) = -2a$

74. $(a + 2)(a - 3) = -2a$

75. $3x(x + 1) - 2x(x - 5) = -42$

76. $4x(x - 2) - 3x(x - 4) = -3$

77. $2x(x + 3) = x(x + 2) - 3$

78. $3x(x - 3) = 2x(x - 4) + 6$

79. $a(a - 3) + 6 = 2a$

80. $a(a - 4) + 8 = 2a$

81. $15(x + 20) + 15x = 2x(x + 20)$

82. $15(x + 8) + 15x = 2x(x + 8)$

83. $15 = a(a + 2)$

84. $6 = a(a - 5)$

Use factoring by grouping to solve the following equations.

85. $x^3 + 3x^2 - 4x - 12 = 0$

86. $x^3 + 5x^2 - 9x - 45 = 0$

87. $x^3 + x^2 - 16x - 16 = 0$

88. $4x^3 + 12x^2 - 9x - 27 = 0$

Getting Ready for the Next Section

Write each sentence as an algebraic equation.

89. The product of two consecutive integers is 72.

90. The product of two consecutive even integers is 80.

91. The product of two consecutive odd integers is 99.

92. The product of two consecutive odd integers is 63.

93. The product of two consecutive even integers is 10 less than 5 times their sum.

94. The product of two consecutive odd integers is 1 less than 4 times their sum.

The following word problems are taken from the book *Academic Algebra*, written by William J. Milne and published by the American Book Company in 1901. Solve each problem.

95. Cost of a Bicycle and a Suit A bicycle and a suit cost $90. How much did each cost, if the bicycle cost 5 times as much as the suit?

96. Cost of a Cow and a Calf A man bought a cow and a calf for $36, paying 8 times as much for the cow as for the calf. What was the cost of each?

97. Cost of a House and a Lot A house and a lot cost $3,000. If the house cost 4 times as much as the lot, what was the cost of each?

98. Daily Wages A plumber and two helpers together earned $7.50 per day. How much did each earn per day, if the plumber earned 4 times as much as each helper?

Find the Mistake

Each sentence below contains a mistake. Circle the mistake and write the correct word(s) or equation(s) on the line provided.

1. Standard form for a linear equation is $a^2 + bx + c = 0$. _____

2. The zero-factor property says that if $(a + 8)(b - 3) = 0$, then $a + 8 = b - 3$. _____

3. To solve the equation $x^2 = -3x + 108$, we must first factor both sides. _____

4. The factors for the equation $3x^2 + 77 = 40x$ are $x = 7$ and $x = -11$. _____

5.8 Applications of Quadratic Equations

Photo courtesy of Megan Silva

OBJECTIVES

A Solve number problems by factoring.

B Solve geometry problems by factoring.

C Solve business problems by factoring.

D Use the Pythagorean theorem to solve problems.

KEY WORDS

Pythagorean theorem

hypotenuse

Imagine you went backpacking through Evolution Valley in Kings Canyon National Park, California. You hiked directly east along one leg of a trail for 2 miles, then turned and hiked north for 3 more miles. If you were to draw a straight line from where you began to where you are now, what would be the distance? In order to solve this problem, we need to learn about the Pythagorean theorem which we will introduce in this section.

The Pythagorean theorem is one of the oldest theorems in the history of mathematics. The person whose name we associate with the theorem, Pythagoras (of Samos), was a Greek philosopher and mathematician who lived from about 560 B.C. to 480 B.C. According to the British philosopher Bertrand Russell, Pythagoras was "intellectually one of the most important men that ever lived."

Also in this section, the solutions to the examples show only the essential steps from our Blueprint for Problem Solving. Recall that step 1 is done mentally; we read the problem and mentally list the items that are known and the items that are unknown. This is an essential part of problem solving. However, now that you have had experience with application problems, you should be doing step 1 automatically.

A Number Problems

EXAMPLE 1 The product of two consecutive odd integers is 63. Find the integers.

Solution Let $x =$ the first odd integer; then $x + 2 =$ the second odd integer. An equation that describes the situation is

$$x(x + 2) = 63 \qquad \textit{Their product is 63.}$$

Then solve the equation.

$$x(x + 2) = 63$$
$$x^2 + 2x = 63$$
$$x^2 + 2x - 63 = 0$$
$$(x - 7)(x + 9) = 0$$
$$x - 7 = 0 \qquad \text{or} \qquad x + 9 = 0$$
$$x = 7 \qquad\qquad x = -9$$

If the first odd integer is 7, the next odd integer is $7 + 2 = 9$. If the first odd integer is -9, the next consecutive odd integer is $-9 + 2 = -7$. We have two pairs of consecutive odd integers that are solutions. They are 7 and 9, or -9 and -7.

We check to see that their products are 63.

$$7(9) = 63$$
$$-7(-9) = 63$$

Practice Problems

1. The product of two consecutive even integers is 48. Find the integers.

Answers

1. -8 and -6, or 6 and 8

Suppose we know that the sum of two numbers is 50. We want to find a way to represent each number using only one variable. If we let x represent one of the two numbers, how can we represent the other? Let's suppose for a moment that x turns out to be 30. Then the other number will be 20, because their sum is 50; that is, if two numbers add up to 50 and one of them is 30, then the other must be $50 - 30 = 20$. Generalizing this to any number x, we see that if two numbers have a sum of 50 and one of the numbers is x, then the other must be $50 - x$. The table that follows shows some additional examples.

If two numbers have a sum of	and one of them is	then the other must be
50	x	$50 - x$
100	x	$100 - x$
10	y	$10 - y$
12	n	$12 - n$

Now, let's look at an example that uses this idea.

EXAMPLE 2 The sum of two numbers is 13. Their product is 40. Find the numbers.

Solution If we let x represent one of the numbers, then $13 - x$ must be the other number because their sum is 13. Since their product is 40, we can write

$$x(13 - x) = 40 \qquad \text{The product of the two numbers is 40.}$$

$$13x - x^2 = 40 \qquad \text{Multiply the left side.}$$

$$x^2 - 13x = -40 \qquad \text{Multiply both sides by } -1 \text{ and reverse the order of the terms on the left side.}$$

$$x^2 - 13x + 40 = 0 \qquad \text{Add 40 to each side.}$$

$$(x - 8)(x - 5) = 0 \qquad \text{Factor the left side.}$$

$$x - 8 = 0 \quad \text{or} \quad x - 5 = 0$$

$$x = 8 \qquad\qquad x = 5$$

The two solutions are 8 and 5. If x is 8, then the other number is $13 - x = 13 - 8 = 5$. Likewise, if x is 5, the other number is $13 - x = 13 - 5 = 8$. Therefore, the two numbers we are looking for are 8 and 5. Their sum is 13 and their product is 40.

B Geometry Problems

Many word problems dealing with area can best be described algebraically by quadratic equations.

EXAMPLE 3 The length of a rectangle is 3 more than twice the width. The area is 44 square inches. Find the dimensions (find the length and width).

Solution As shown in Figure 1, let $x =$ the width of the rectangle. Then $2x + 3 =$ the length of the rectangle because the length is three more than twice the width.

FIGURE 1

Since the area is 44 square inches, an equation that describes the situation is

$$x(2x + 3) = 44 \qquad \text{Length} \cdot \text{width} = \text{area}$$

2. The sum of two numbers is 14. Their product is 48. Find the two numbers.

3. The length of a rectangle is 2 more than twice the width. The area is 60 square inches. Find the dimensions.

Answers
2. 6 and 8
3. Width is 5 inches; length is 12 inches.

We now solve the equation.

$$x(2x + 3) = 44$$
$$2x^2 + 3x = 44$$
$$2x^2 + 3x - 44 = 0$$
$$(2x + 11)(x - 4) = 0$$
$$2x + 11 = 0 \quad \text{or} \quad x - 4 = 0$$
$$x = -\frac{11}{2} \qquad\qquad x = 4$$

The solution $x = -\frac{11}{2}$ cannot be used since length and width are always given in positive units. The width is 4. The length is 3 more than twice the width or $2(4) + 3 = 11$.

Width = 4 inches

Length = 11 inches

The solutions check in the original problem since $4(11) = 44$.

EXAMPLE 4 The numerical value of the area of a square is twice its perimeter. What is the length of its side?

Solution As shown in Figure 2, let $x =$ the length of its side. Then $x^2 =$ the area of the square and $4x =$ the perimeter of the square.

FIGURE 2

An equation that describes the situation is

$$x^2 = 2(4x) \qquad \textit{The area is 2 times the perimeter.}$$
$$x^2 = 8x$$
$$x^2 - 8x = 0$$
$$x(x - 8) = 0$$
$$x = 0 \quad \text{or} \quad x = 8$$

Since $x = 0$ does not make sense in our original problem, we use $x = 8$. If the side has length 8, then the perimeter is $4(8) = 32$ and the area is $8^2 = 64$. Since 64 is twice 32, our solution is correct.

4. The height of a triangle is 3 inches more than the base. The area is 20 square inches. Find the base.

C Business Problems

5. Using the information in Example 5, find the number of items that should be made if the total cost is to be $700.

EXAMPLE 5 A company can manufacture x hundred items for a total cost of $C = 300 + 500x - 100x^2$. How many items were manufactured if the total cost is $900?

Solution We are looking for x when C is 900. We begin by substituting 900 for C in the cost equation. Then we solve for x.

$$\text{When} \rightarrow \qquad C = 900$$

$$\text{the equation} \rightarrow \qquad C = 300 + 500x - 100x^2$$

$$\text{becomes} \rightarrow \qquad 900 = 300 + 500x - 100x^2$$

We can write this equation in standard form by adding -300, $-500x$, and $100x^2$ to each side. The result looks like this:

$$900 = 300 + 500x - 100x^2$$

$100x^2 - 500x + 600 = 0$	Standard form
$100(x^2 - 5x + 6) = 0$	Begin factoring.
$100(x - 2)(x - 3) = 0$	Factor completely.
$x - 2 = 0 \quad \text{or} \quad x - 3 = 0$	Set variable factors equal to 0.
$x = 2 \qquad\qquad x = 3$	

Our solutions are 2 and 3, which means that the company can manufacture 2 hundred items or 3 hundred items for a total cost of $900.

6. Use the information in Example 6 to find the selling price that should be charged if the weekly revenue is to be $4,200.

EXAMPLE 6 A manufacturer of small portable radios knows that the number of radios she can sell each week is related to the price of the radios by the equation $x = 1,300 - 100p$ (x is the number of radios and p is the price per radio). What price should she charge for the radios to have a weekly revenue of $4,000?

Solution First, we must find the revenue equation. The equation for total revenue is $R = xp$, where x is the number of units sold and p is the price per unit. Since we want R in terms of p, we substitute $1,300 - 100p$ for x in the equation $R = xp$.

$$\text{If} \qquad R = xp$$

$$\text{and} \qquad x = 1,300 - 100p$$

$$\text{then} \qquad R = (1,300 - 100p)p$$

We want to find p when R is 4,000. Substituting 4,000 for R in the equation gives us

$$4,000 = (1,300 - 100p)p$$

If we multiply out the right side, we have

$$4,000 = 1,300p - 100p^2$$

To write this equation in standard form, we add $100p^2$ and $-1,300p$ to each side.

$100p^2 - 1,300p + 4,000 = 0$	Add $100p^2$ and $-1,300p$.
$100(p^2 - 13p + 40) = 0$	Begin factoring.
$100(p - 5)(p - 8) = 0$	Factor completely.
$p - 5 = 0 \quad \text{or} \quad p - 8 = 0$	Set variable factors equal to 0.
$p = 5 \qquad\qquad p = 8$	

If she sells the radios for $5 each or for $8 each, she will have a weekly revenue of $4,000.

D The Pythagorean Theorem

Next, we will work some problems involving the Pythagorean theorem, which uses variables to represent the three sides of a triangle. When calculating the lengths of these sides, the theorem often translates into a quadratic equation we can factor and solve.

> **FACTS FROM GEOMETRY** Pythagorean Theorem
>
> In any right triangle (Figure 3), the square of the longer side (called the hypotenuse) is equal to the sum of the squares of the other two sides (called legs).
>
>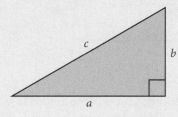
>
> $c^2 = a^2 + b^2$
>
> FIGURE 3

EXAMPLE 7 The three sides of a right triangle are three consecutive integers. Find the lengths of the three sides.

Solution Let $x =$ the first integer (shortest side)

then $\quad x + 1 =$ the next consecutive integer

and $\quad x + 2 =$ the last consecutive integer (longest side)

A diagram of the triangle is shown in Figure 4.

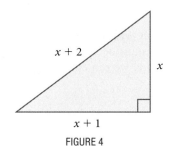

FIGURE 4

The Pythagorean theorem tells us that the square of the longest side $(x + 2)^2$ is equal to the sum of the squares of the two shorter sides, $(x + 1)^2 + x^2$. Here is the equation:

$$(x + 2)^2 = (x + 1)^2 + x^2$$

$x^2 + 4x + 4 = x^2 + 2x + 1 + x^2$ Expand squares.

$x^2 - 2x - 3 = 0$ Standard form

$(x - 3)(x + 1) = 0$ Factor.

$x - 3 = 0 \quad$ or $\quad x + 1 = 0$ Set factors equal to 0.

$x = 3 \qquad\qquad x = -1$

Since a triangle cannot have a side with a negative number for its length, we must not use -1 for a solution to our original problem; therefore, the shortest side is 3. The other two sides are the next two consecutive integers, 4 and 5.

7. The longer leg of a right triangle is 2 more than twice the shorter leg. The hypotenuse is 3 more than twice the shorter leg. Find the length of each side. (*Hint*: Let $x =$ the length of the shorter leg. Then the longer leg is $2x + 2$. Now write an expression for the hypotenuse, and you will be on your way.)

8. The hypotenuse of a right triangle is 10 inches, and the lengths of the two sides are given by two consecutive even integers. Find the two sides.

EXAMPLE 8 The hypotenuse of a right triangle is 5 inches, and the lengths of the two legs (the other two sides) are given by two consecutive integers. Find the lengths of the two legs.

Solution If we let x = the length of the shortest side, then the other side must be $x + 1$. A diagram of the triangle is shown in Figure 5.

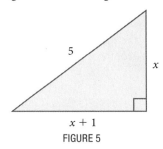

FIGURE 5

The Pythagorean theorem tells us that the square of the longest side, 5^2, is equal to the sum of the squares of the two shorter sides, $x^2 + (x + 1)^2$. Here is the equation:

$5^2 = x^2 + (x + 1)^2$	Pythagorean theorem
$25 = x^2 + x^2 + 2x + 1$	Expand 5^2 and $(x + 1)^2$.
$25 = 2x^2 + 2x + 1$	Simplify the right side.
$0 = 2x^2 + 2x - 24$	Add -25 to each side.
$0 = 2(x^2 + x - 12)$	Begin factoring.
$0 = 2(x + 4)(x - 3)$	Factor completely.
$x + 4 = 0$ or $x - 3 = 0$	Set variable factors equal to 0.
$x = -4$ $x = 3$	

Since a triangle cannot have a side with a negative number for its length, we cannot use -4; therefore, the shortest side must be 3 inches. The next side is $x + 1 = 3 + 1 = 4$ inches. Since the hypotenuse is 5, we can check our solutions with the Pythagorean theorem as shown in Figure 6.

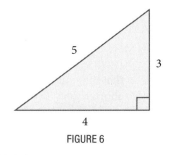

FIGURE 6

GETTING READY FOR CLASS

After reading through the preceding section, respond in your own words and in complete sentences.

A. What are consecutive integers?

B. How would you use factoring to solve a geometry problem?

C. Explain the Pythagorean theorem in words.

D. Write an application problem for which the solution depends on solving the equation $x(x + 1) = 12$.

Answers

8. The lengths of the sides are 6 inches and 8 inches.

Vocabulary Review

Choose the correct words to fill in the blanks below.

hypotenuse standard form Pythagorean theorem consecutive

1. The quadratic equation $20 = x^2 + x$ represents two _____ integers with a product of 20.

2. To write a quadratic equation in _____, have zero on one side of the equation and decreasing powers of the variable on the other.

3. In any right triangle, the square of the _____ is equal to the sum of the squares of the legs.

4. The _____ states that $c^2 = a^2 + b^2$, where c is the hypotenuse and a and b are the legs of a right triangle.

Problems

Solve the following word problems. Be sure to show the equation used.

A Number Problems

1. The product of two consecutive even integers is 80. Find the two integers.

2. The product of two consecutive integers is 72. Find the two integers.

3. The product of two consecutive odd integers is 99. Find the two integers.

4. The product of two consecutive integers is 132. Find the two integers.

5. The product of two consecutive even integers is 10 less than 5 times their sum. Find the two integers.

6. The product of two consecutive odd integers is 1 less than 4 times their sum. Find the two integers.

7. The sum of two numbers is 14. Their product is 48. Find the numbers.

8. The sum of two numbers is 12. Their product is 32. Find the numbers.

SCAN TO ACCESS

9. One number is 2 more than 5 times another. Their product is 24. Find the numbers.

10. One number is 1 more than twice another. Their product is 55. Find the numbers.

11. One number is 4 times another. Their product is 4 times their sum. Find the numbers.

12. One number is 2 more than twice another. Their product is 2 more than twice their sum. Find the numbers.

B Geometry Problems

13. The length of a rectangle is 1 more than the width. The area is 12 square inches. Find the dimensions.

14. The length of a rectangle is 3 more than twice the width. The area is 44 square inches. Find the dimensions.

15. The height of a triangle is twice the base. The area is 9 square inches. Find the base.

16. The height of a triangle is 2 more than twice the base. The area is 20 square feet. Find the base.

17. The hypotenuse of a right triangle is 10 inches. The lengths of the two legs are given by two consecutive even integers. Find the lengths of the two legs.

18. The hypotenuse of a right triangle is 15 inches. One of the legs is 3 inches more than the other. Find the lengths of the two legs.

19. The shorter leg of a right triangle is 5 meters. The hypotenuse is 1 meter longer than the longer leg. Find the length of the longer leg.

20. The shorter leg of a right triangle is 12 yards. If the hypotenuse is 20 yards, how long is the other leg?

C Business Problems

21. A company can manufacture x hundred items for a total cost of $C = 400 + 700x - 100x^2$. How many items are manufactured if the total cost is $1,400?

22. If the total cost C of manufacturing x hundred items is given by the equation $C = 700 + 900x - 100x^2$, how many items are manufactured when C is $2,700?

23. The total cost C of manufacturing x hundred DVDs is given by the equation $C = 600 + 1,000x - 100x^2$. How many DVDs are manufactured if the total cost is $2,200?

24. The total cost C of manufacturing x hundred pen and pencil sets is given by the equation $C = 500 + 800x - 100x^2$. How many pen and pencil sets are manufactured when C is $1,700?

25. A company that manufactures hair ribbons knows that the number of ribbons it can sell each week, x, is related to the price p per ribbon by the equation $x = 1,200 - 100p$. At what price should the company sell the ribbons if it wants the weekly revenue to be $3,200? (*Remember:* The equation for revenue is $R = xp$.)

26. A company manufactures flash drives for home computers. It knows from experience that the number of drives it can sell each day, x, is related to the price p per drive by the equation $x = 800 - 100p$. At what price should the company sell the flash drives if it wants the daily revenue to be $1,200?

27. The relationship between the number of calculators a company sells per week, x, and the price p of each calculator is given by the equation $x = 1,700 - 100p$. At what price should the calculators be sold if the weekly revenue is to be $7,000?

28. The relationship between the number of pencil sharpeners a company can sell each week, x, and the price p of each sharpener is given by the equation $x = 1,800 - 100p$. At what price should the sharpeners be sold if the weekly revenue is to be $7,200?

D Solve the following problems using the Pythagorean theorem.

29. **Pythagorean Theorem** A 13-foot ladder is placed so that it reaches to a point on the wall that is 2 feet higher than twice the distance from the base of the wall to the base of the ladder.

 a. How far from the wall is the base of the ladder?

 b. How high does the ladder reach?

30. **Pythagorean Theorem** A 25-foot ladder is placed so that its height on the wall is 3 feet more than triple the distance from the base to the wall.

 a. How far from the wall is the base of the ladder?

 b. How high does the ladder reach?

31. Constructing a Box I have a piece of cardboard that is twice as long as it is wide. If I cut a 2-inch by 2-inch square from each corner and fold up the resulting flaps, I get a box with a volume of 32 cubic inches. What are the dimensions of the cardboard?

32. Constructing a Box A cardboard rectangle has a length which is 3 inches more than its width. If 1-inch by 1-inch squares are cut out, and the resulting flaps are folded up, the volume would be 70 cubic inches. What are the dimensions of the cardboard?

33. Projectile Motion A gun fires a bullet almost straight up from the edge of a 100-foot cliff. If the bullet leaves the gun with a speed of 396 feet per second, its height at time t is given by $h(t) = -16t^2 + 396t + 100$, measured from the ground below the cliff.

a. When will the bullet land on the ground below the cliff? (*Hint:* What is its height when it lands? Remember that we are measuring from the ground below, not from the cliff.)

b. Make a table showing the bullet's height every five seconds, from the time it is fired ($t = 0$) to the time it lands. (*Note:* It is faster to substitute into the factored form.)

t	h

34. Height of a Projectile If a rocket is fired vertically into the air with a speed of 240 feet per second, its height at time t seconds is given by $h(t) = -16t^2 + 240t$. At what time(s) will the rocket be the following number of feet above the ground?

a. 704 feet

b. 896 feet

c. Why do parts a and b each have two answers?

d. How long will the rocket be in the air? (*Hint:* How high is it when it hits the ground?)

e. When the equation for part d is solved, one of the answers is $t = 0$ second. What does this represent?

Maintaining Your Skills

Simplify each expression. (Write all answers with positive exponents only.)

35. $(5x^3)^2(2x^6)^3$

36. 2^{-3}

37. $\dfrac{x^4}{x^{-3}}$

38. $\dfrac{(20x^2y^3)(5x^4y)}{(2xy^5)(10x^2y^3)}$

39. $(2 \times 10^{-4})(4 \times 10^5)$

40. $\dfrac{9 \times 10^{-3}}{3 \times 10^{-2}}$

41. $20ab^2 - 16ab^2 + 6ab^2$

42. Subtract $6x^2 - 5x - 7$ from $9x^2 + 3x - 2$.

Multiply.

43. $2x^2(3x^2 + 3x - 1)$

44. $(2x + 3)(5x - 2)$

45. $(3y - 5)^2$

46. $(a - 4)(a^2 + 4a + 16)$

47. $(2a^2 + 7)(2a^2 - 7)$

48. Divide $15x^{10} - 10x^8 + 25x^6$ by $5x^6$.

Find the Mistake

Each sentence below contains a mistake. Circle the mistake and write the correct word(s) or equation on the line provided.

1. The sum of two consecutive even integers is 34. Their product is 288. To find the two integers, you must solve for x in the equation $x^2 - 34x = 288$. _____

2. The width of a rectangle is $2x$ and its length is three times the width plus 5 more inches. The area of the rectangle is 781 square inches. The longest side of the rectangle measures 11 inches. _____

3. The Pythagorean theorem states that the square of the longest leg is equal to the square of the hypotenuse and the shortest leg. _____

4. The hypotenuse of a triangle is 13 feet. The longest leg of the triangle is twice the shortest leg plus 2 more feet. To find the lengths of the two legs, we must find the square root of the hypotenuse.

On the Surface

Below we have provided polygons with unknown measurements. We would like you to find the surface area of each polygon using a given set of smaller polygons. To begin this project, cut out and label the following rectangles from a piece of engineering graph paper (5 squares per inch):

A Piece of engineering graph paper

B A pen or pencil

C Scissors

1. 2x2 square and label each side with the variable a

2. 3x3 square and label each side with the variable b

3. 2x4 rectangle and label the short side a and the long side $2a$

4. 3x6 rectangle and label the short side b and the long side $2b$

5. 8x8 square and label each side with the expression $4a$

6. 1x1 square and label each side with a length of 3

Use two or more of your cut-outs to calculate the total surface area for each of the following. Answers should be in terms of the given side lengths using variables a and b.

Total Surface Area:

Polygon 1:

Polygon 2:

Polygon 3:

Polygon 4:

Polygon 5:

Polygon 6:

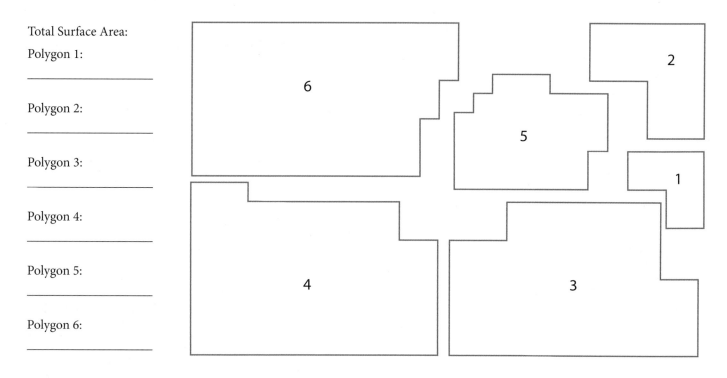

We have shaded parts of the following polygons. Use your cutouts to determine the surface area of the shaded portions only.

Total Surface Area:

Polygon 7:

Polygon 8:

Polygon 9:

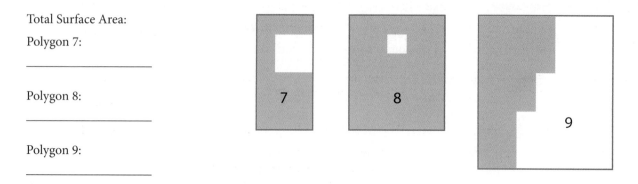

Suppose we give the values of $a = 5$ and $b = 12$. Use the polynomial expressions you found above to find the total surface areas of polygons 1-6 and the shaded surface areas of polygons 7-9.

Chapter 5 Summary

Greatest Common Factor [5.1]

The largest monomial that divides each term of a polynomial is called the greatest common factor for that polynomial. We begin all factoring by factoring out the greatest common factor.

EXAMPLES

1. $8x^4 - 10x^3 + 6x^2$
$= 2x^2 \cdot 4x^2 - 2x^2 \cdot 5x + 2x^2 \cdot 3$
$= 2x^2(4x^2 - 5x + 3)$

Factoring Trinomials [5.2, 5.3]

One method of factoring a trinomial is to list all pairs of binomials whose product of the first terms gives the first term of the trinomial, and whose product of the last terms gives the last term of the trinomial. Then we choose the pair that gives the correct middle term for the original trinomial.

2. $x^2 + 5x + 6 = (x + 2)(x + 3)$
$x^2 - 5x + 6 = (x - 2)(x - 3)$
$6x^2 - x - 2 = (2x + 1)(3x - 2)$
$6x^2 + 7x + 2 = (2x + 1)(3x + 2)$

Special Factoring [5.4]

$$a^2 + 2ab + b^2 = (a + b)^2$$
$$a^2 - 2ab + b^2 = (a - b)^2$$ *Perfect square trinomials*

$$a^2 - b^2 = (a + b)(a - b)$$ *Difference of two squares*

Remember, $a^2 + b^2$ is a prime sum because it does not factor.

3. $x^2 + 10x + 25 = (x + 5)^2$
$x^2 - 10x + 25 = (x - 5)^2$
$x^2 - 25 = (x + 5)(x - 5)$

Sum and Difference of Two Cubes [5.5]

$$a^3 - b^3 = (a - b)(a^2 + ab + b^2)$$ *Difference of two cubes*

$$a^3 + b^3 = (a + b)(a^2 - ab + b^2)$$ *Sum of two cubes*

4. $x^3 - 27 = (x - 3)(x^2 + 3x + 9)$
$x^3 + 27 = (x + 3)(x^2 - 3x + 9)$

Strategy for Factoring a Polynomial [5.6]

Step 1: If the polynomial has a greatest common factor other than 1, then factor out the greatest common factor.

Step 2: If the polynomial has two terms (a binomial), then see if it is the difference of two squares or the sum or difference of two cubes, and then factor accordingly. Remember, if it is the sum of two squares, it will not factor.

Step 3: If the polynomial has three terms (a trinomial), then it is either a perfect square trinomial that will factor into the square of a binomial, or it is not a perfect square trinomial, in which case you can try to use the trial and error method developed in Section 5.3.

Step 4: If the polynomial has more than three terms, then try to factor it by grouping.

Step 5: As a final check, see if any of the factors you have written can be factored further. If you have overlooked a common factor, you can catch it here. Remember, some polynomials cannot be factored, in which case they are considered prime.

5. a. $2x^5 - 8x^3 = 2x^3(x^2 - 4)$
$= 2x^3(x + 2)(x - 2)$

b. $3x^4 - 18x^3 + 27x^2$
$= 3x^2(x^2 - 6x + 9)$
$= 3x^2(x - 3)^2$

c. $6x^3 - 12x^2 - 48x$
$= 6x(x^2 - 2x - 8)$
$= 6x(x - 4)(x + 2)$

d. $x^2 + ax + bx + ab$
$= x(x + a) + b(x + a)$
$= (x + a)(x + b)$

Strategy for Solving a Quadratic Equation [5.7]

6. Solve $x^2 - 6x = -8$.
$$x^2 - 6x + 8 = 0$$
$$(x - 4)(x - 2) = 0$$
$$x - 4 = 0 \quad \text{or} \quad x - 2 = 0$$
$$x = 4 \qquad\qquad x = 2$$
Both solutions check.

Step 1: Write the equation in standard form $ax^2 + bx + c = 0$.

Step 2: Factor completely.

Step 3: Set each variable factor equal to 0.

Step 4: Solve the equations found in step 3.

Step 5: Check solutions, if necessary.

The Pythagorean Theorem [5.8]

7. The hypotenuse of a right triangle is 5 inches, and the lengths of the two legs (the other two sides) are given by two consecutive integers. Find the lengths of the two legs.

In any right triangle, the square of the longest side (called the hypotenuse) is equal to the sum of the squares of the other two sides (called legs).

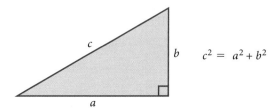

If we let $x =$ the length of the shortest side, then the other side must be $x + 1$. The Pythagorean theorem tells us that

$$5^2 = x^2 + (x + 1)^2$$
$$25 = x^2 + x^2 + 2x + 1$$
$$25 = 2x^2 + 2x + 1$$
$$0 = 2x^2 + 2x - 24$$
$$0 = 2(x^2 + x - 12)$$
$$0 = 2(x + 4)(x - 3)$$
$$x + 4 = 0 \quad \text{or} \quad x - 3 = 0$$
$$x = -4 \qquad\qquad x = 3$$

Since a triangle cannot have a side with a negative number for its length, we cannot use -4. One leg is $x = 3$ and the other leg is $x + 1 = 3 + 1 = 4$.

COMMON MISTAKE

It is a mistake to apply the zero-factor property to numbers other than zero. For example, consider the equation $(x - 3)(x + 4) = 18$. A fairly common mistake is to attempt to solve it with the following steps:

$$(x - 3)(x + 4) = 18$$
$$x - 3 = 18 \quad \text{or} \quad x + 4 = 18 \leftarrow \text{Mistake}$$
$$x = 21 \qquad\qquad x = 14$$

These are obviously not solutions, as a quick check will verify.

Check: $x = 21$ Check: $x = 14$

$(21 - 3)(21 + 4) \overset{?}{=} 18$ $(14 - 3)(14 + 4) \overset{?}{=} 18$

$18 \cdot 25 = 18$ $11 \cdot 18 = 18$

$450 \neq 18 \quad \xleftarrow{\text{False statements}} \quad 198 \neq 18$

The mistake is in setting each factor equal to 18. It is not necessarily true that when the product of two numbers is 18, either one of them is itself 18. The correct solution looks like this:

$$(x - 3)(x + 4) = 18$$
$$x^2 + x - 12 = 18$$
$$x^2 + x - 30 = 0$$
$$(x + 6)(x - 5) = 0$$
$$x + 6 = 0 \qquad \text{or} \qquad x - 5 = 0$$
$$x = -6 \qquad\qquad\qquad x = 5$$

To avoid this mistake, remember that before you factor a quadratic equation, you must write it in standard form. It is in standard form only when 0 is on one side and decreasing powers of the variable are on the other.

Factor out the greatest common factor. [5.1]

1. $4x - 24$

2. $14a^2b - 28ab + 21ab^2$

Factor by grouping. [5.1]

3. $2x^2 + 8ax - bx - 4ab$

4. $6xy - 3y - 8x + 4$

Factor the following completely. [5.2–5.6]

5. $x^2 + 3x - 28$

6. $x^2 - 5x + 6$

7. $x^2 - 49$

8. $x^4 - 81$

9. $x^2 + 9$

10. $12x^2 - 27y^2$

11. $x^3 - 5x^2 + 4x - 20$

12. $2x^2 - bx + 8ax - 4ab$

13. $6x^2 + 15x - 36$

14. $3x^2 - x - 10$

15. $16x^2 + 8x - 3$

16. $12x^3 - 26x^2 + 12x$

17. $x^3 - 216$

18. $81x^3 - 24$

Solve the following equations. [5.7]

19. $x^2 + 3x - 18 = 0$

20. $x^2 + 7x + 10 = 0$

21. $x^2 - 36 = 0$

22. $x^2 = 35 - 2x$

23. $x^2 + 4x = 45$

24. $y^3 = 16y$

25. $3x^2 = -2x + 8$

26. $4x^3 + x^2 - 3x = 0$

Solve the following word problems. Be sure to show the equation used. [5.8]

27. Number Problem Two numbers have a sum of 20. Their product is 91. Find the numbers.

28. Consecutive Integers The product of two positive consecutive even integers is 8 more than 4 times their sum. Find the integers.

29. Geometry The length of a rectangle is 3 feet more than the width. The area is 54 square feet. Find the dimensions.

30. Geometry One leg of a right triangle is 1 foot less than twice the other. The hypotenuse is 17 feet. Find the lengths of the two legs.

31. Production Cost A company can manufacture x hundred items for a total cost, C, given the equation $C = 1,400 + 500x - 100x^2$. How many items can be manufactured if the total cost is to be $800?

32. Price and Revenue A manufacturer knows that the number of items he can sell each week, x, is related to the price p of each item by the equation $x = 900 - 100p$. What price should he charge for each item to have a weekly revenue of $1,800? (*Remember: $R = xp$.*)

33. Pythagorean Theorem A 17-foot ladder is placed so that it leans against a barn at a point that is one foot less than twice the distance from the wall of the barn to the base of the ladder.

17 ft

a. How far from the wall is the ladder?

b. How high does the ladder reach?

34. Projectile Motion A ball is thrown into the air with an upward velocity of 4 feet per second from a building that is 20 feet high. The equation that gives the height of the ball above the ground at time t is $h = 20 + 4t - 16t^2$

a. At what time(s) will the ball be 18 feet above the ground?

b. When will the ball hit the ground?

Simplify.

1. $12 + (-3) + (-5)$

2. $26 - (-11)$

3. $\dfrac{48}{-6}$

4. $\dfrac{7(-3)}{3}$

5. $(-2)^5$

6. 3^{-3}

7. $\dfrac{-2(5-6) - 3(5-2)}{-6 - 5 + 4}$

8. $\dfrac{3^2 - 6^2}{(3-6)^2}$

9. $-3a + 6 + 5a - 4$

10. $5 - 3(5a - 4) - 21$

11. $(x^3)^7$

12. $(3xy)^0$

13. $(2x - 4)(3x - 5)$

14. $(x^2 - 3)(x^2 + 3)$

Solve each equation.

15. $3x = -21$

16. $\dfrac{x}{4} = 3$

17. $\dfrac{x}{5} = -3$

18. $\dfrac{1}{3}(9x - 2) + \dfrac{1}{2} = \dfrac{35}{6}$

19. $2x(x - 3)(3x + 4) = 0$

20. $9t^2 - 81 = 0$

Solve each inequality.

21. $15 < -3x$

22. $2 - 4x < -14$

Graph on a rectangular coordinate system.

23. $y = -\dfrac{2}{3}x$

24. $y = -3$

25. $3x - 9y \geq 18$

26. $x > -3$

27. Which of the ordered pairs $(3, 0)$, $(6, -2)$, and $\left(1, \frac{4}{3}\right)$ are solutions to the equation $6x - 9y = 18$?

28. Find the x- and y-intercepts for the equation $8x - 16y = 32$.

29. Find the slope of the line that passes through the points $(9, -6)$ and $(14, 4)$.

30. Find the equation for the line with slope $\frac{1}{4}$ that passes through $(8, -4)$.

31. Find the equation for the line with slope $-\frac{3}{7}$ and y-intercept $\frac{1}{3}$.

32. Find the slope and y-intercept for the equation $2x + 5y = -15$.

Factor completely.

33. $x^2 - 11x + 28$

34. $15x^2 - 14xy - 8y^2$

35. $36 - x^2$

36. $36x^2 - 12x + 1$

37. $4x^3 - 2x^2y - 12xy^2$

38. $16x^2y + 4xy - 72y$

39. $15xy - 12x - 5y + 4$

40. $x^3 - 27$

Give the opposite, reciprocal, and absolute value of the given number.

41. $\dfrac{3}{2}$

42. -5

Divide using long division.

43. $\dfrac{5x^2 - 13x - 9}{x - 4}$

44. $\dfrac{x^3 + 64}{x + 4}$

45. Cutting a Board into Two Pieces A 36-inch board is to be cut into two pieces. One piece is to be 8 inches longer than the other. How long is each piece?

46. Hamburgers and Fries Paula bought burgers and fries for some friends. The burgers cost $2.15 each, and the fries are $0.65 each. She bought a total of 8 items, for a total cost of $9.70. How many of each did she buy?

Use the pie chart to answer the following questions.

There were 3,500 iPad apps in April. How many iPad apps would fall into the following categories?

47. Free

48. Paid

Factor out the greatest common factor. [5.1]

1. $6x + 18$

2. $12a^2b - 24ab + 8ab^2$

Factor by grouping. [5.1]

3. $x^2 + 3ax - 2bx - 6ab$

4. $15y - 5xy - 12 + 4x$

Factor the following completely. [5.2–5.6]

5. $x^2 + x - 12$

6. $x^2 - 4x - 21$

7. $x^2 - 25$

8. $x^4 - 16$

9. $x^2 + 36$

10. $18x^2 - 32y^2$

11. $x^3 + 4x^2 - 3x - 12$

12. $x^2 + bx - 3x - 3b$

13. $4x^2 - 6x - 10$

14. $4n^2 + 13n - 12$

15. $12c^2 + c - 6$

16. $12x^3 + 12x^2 - 9x$

17. $x^3 + 125y^3$

18. $54b^3 - 128$

Solve the following equations. [5.7]

19. $x^2 - 2x - 15 = 0$

20. $x^2 - 7x + 12 = 0$

21. $x^2 - 25 = 0$

22. $x^2 = 5x + 14$

23. $x^2 + x = 30$

24. $y^3 = 9y$

25. $2x^2 = -5x + 12$

26. $15x^3 - 65x^2 - 150x = 0$

Solve the following word problems. Be sure to show the equation used. [5.8]

27. Number Problem Two numbers have a sum of 18. Their product is 72. Find the numbers.

28. Consecutive Integers The product of two consecutive even integers is 14 more than their sum. Find the integers.

29. Geometry The length of a rectangle is 1 foot more than 3 times the width. The area is 52 square feet. Find the dimensions.

30. Geometry One leg of a right triangle is 2 feet more than the other. The hypotenuse is 10 feet. Find the lengths of the two legs.

31. Production Cost A company can manufacture x hundred items for a total cost, C, given the equation $C = 100 + 500x - 100x^2$. How many items can be manufactured if the total cost is to be $700?

32. Price and Revenue A manufacturer knows that the number of items he can sell each week, x, is related to the price p of each item by the equation $x = 800 - 100p$. What price should he charge for each item to have a weekly revenue of $1,500? (*Remember: R = xp.*)

33. Pythagorean Theorem A 13-foot ladder is placed so that it leans against a wall at a point that is two feet more than twice the distance from the wall to the base of the ladder.

13 ft

a. How far from the wall is the ladder?

b. How high does the ladder reach?

34. Projectile Motion A ball is thrown into the air with an upward velocity of 26 feet per second from a building that is 12 feet high. The equation that gives the height of the ball above the ground at time t is $h = 12 + 26t - 16t^2$

h

a. At what time(s) will the ball be 22 feet above the ground?

b. When will the ball hit the ground?

Rational Expressions and Equations

Independence National Historical Park
Philadelphia, Pennsylvania
Gray Buildings © 2008 Sanborn

I f you travel to Philadelphia, Pennsylvania, you should not miss seeing Independence National Historical Park, home of Independence Hall. Upon its completion in 1753, the location served as the Pennsylvania State House for the Province of Pennsylvania. The Second Continental Congress used it as their primary meeting place from 1775 to 1783, and the United States Declaration of Independence and United States Constitution were both signed there. Its significance is shown by the fact that it is pictured on the back of the US $100 bill.

Interest in the building has remained high over the years, sparking a series of replicas made as collector's items. One version is a cast iron coin bank measuring 7 inches tall. Two of these banks recently sold on eBay, one for $850 and the other for $760. If the real Independence Hall is 135 feet tall, what is the ratio of the height of the replica to the original? Answering this question requires the use of ratios, one of the topics of this chapter.

A Find the restrictions on the variable in a rational expression.

B Reduce a rational expression to lowest terms.

C Finding ratios and rates.

KEY WORDS

rational expression

ratio

rate equation

©iStockphoto.com/fasloff

Recall our previous discussion regarding wingsuit flying. Carbon fiber is an extremely strong material in relation to its weight. Because of this quality, wingsuit flyers have utilized carbon fiber in a cutting edge piece of equipment called a wingpack. The wing of the wingpack is made of rigid carbon fiber. Strapped to his back, the wingpack allows the flyer to travel longer horizontal distances than with the wingsuit alone. A rating scale called the glide ratio is used to categorize the equipment's aptitude for increasing the flyer's horizontal distance. The glide ratio compares the amount of lift L from a wingpack divided by the amount of drag D it causes while moving through the air. Therefore, this ratio can be written as $\frac{L}{D}$, and is an example of a rational expression. The wingpack has a glide ratio value of 6, whereas the wingsuit is only 2.5.

Recall from Chapter 0 that we define rational numbers as any number that could be put in the form $\frac{a}{b}$, where a and b are integers ($b \neq 0$).

$$\text{Rational numbers} = \left\{ \frac{a}{b} \mid a \text{ and } b \text{ are integers}, b \neq 0 \right\}$$

Rational expressions have a similar definition. Here is a formal definition of a rational expression:

DEFINITION rational expression

A **rational expression** is any expression that can be put in the form $\frac{P}{Q}$, where P and Q are polynomials and $Q \neq 0$.

$$\text{Rational expressions} = \left\{ \frac{P}{Q} \mid P \text{ and } Q \text{ are polynomials}, Q \neq 0 \right\}$$

Each of the following is an example of a rational expression.

$$\frac{2x + 3}{x} \qquad \frac{x^2 - 6x + 9}{x^2 - 4} \qquad \frac{5}{x^2 + 6} \qquad \frac{2x^2 + 3x + 4}{2}$$

For the rational expression

$$\frac{x^2 - 6x + 9}{x^2 - 4} \qquad \begin{array}{l} \leftarrow \text{ Numerator} \\ \leftarrow \text{ Denominator} \end{array}$$

the polynomial on top, $x^2 - 6x + 9$, is called the numerator, and the polynomial on the bottom, $x^2 - 4$, is called the denominator. The same is true of the other rational expressions.

A Restrictions on Variables

We must be careful that we do not use a value of the variable that will give us a denominator of zero. Remember, division by zero is not defined.

EXAMPLE 1 State the restrictions on the variable.

$$\frac{x + 2}{x - 3}$$

Solution The variable x can be any real number except $x = 3$ since, when $x = 3$, the denominator is $3 - 3 = 0$. We state this restriction by writing $x \neq 3$.

EXAMPLE 2 State the restrictions on the variable.

$$\frac{5}{x^2 - x - 6}$$

Solution If we factor the denominator, we have $x^2 - x - 6 = (x - 3)(x + 2)$. By applying our knowledge of the zero-factor property, if either of the factors is zero, then the whole denominator is zero. Our restrictions are $x \neq 3$ and $x \neq -2$ since either one makes $x^2 - x - 6 = 0$.

We will not always list each restriction on a rational expression, but we should be aware of them and keep in mind that no rational expression can have a denominator of zero.

The two fundamental properties of rational expressions are listed next. We will use these two properties many times in this chapter.

> **PROPERTY** Multiplication Property of Rational Expressions
>
> *Multiplying* the numerator and denominator of a rational expression by the same nonzero quantity will not change the value of the rational expression.

> **PROPERTY** Division Property of Rational Expressions
>
> *Dividing* the numerator and denominator of a rational expression by the same nonzero quantity will not change the value of the rational expression.

B Reducing Rational Expressions

We can use the division property to reduce rational expressions to lowest terms. This is called simplifying a rational expression. Since this process is almost identical to the process of reducing fractions to lowest terms, let's recall how the fraction $\frac{6}{15}$ is reduced to lowest terms.

$$\frac{6}{15} = \frac{2 \cdot 3}{5 \cdot 3} \qquad \text{Factor numerator and denominator.}$$

$$= \frac{2 \cdot \cancel{3}}{5 \cdot \cancel{3}} \qquad \text{Divide out the common factor, 3.}$$

$$= \frac{2}{5} \qquad \text{Reduce to lowest terms.}$$

The same procedure applies to simplifying rational expressions. The process is summarized in the following rule:

> **RULE**
>
> To simplify a rational expression, first factor the numerator and denominator completely and then divide both the numerator and denominator by any factors they have in common.

Practice Problems

1. State the restrictions on the variable.

$$\frac{x + 5}{x - 4}$$

2. State the restrictions on the variable.

$$\frac{6}{x^2 + x - 6}$$

Answers

1. $x \neq 4$
2. $x \neq -3, x \neq 2$

3. Simplify the expression
$$\frac{x^2 - 4}{x^2 - 2x - 8}.$$

EXAMPLE 3 Simplify the expression $\dfrac{x^2 - 9}{x^2 + 5x + 6}$.

Solution We begin by factoring.

$$\frac{x^2 - 9}{x^2 + 5x + 6} = \frac{(x - 3)(x + 3)}{(x + 2)(x + 3)}$$

Notice that both polynomials contain the factor $(x + 3)$. If we divide the numerator by $(x + 3)$, we are left with $(x - 3)$. If we divide the denominator by $(x + 3)$, we are left with $(x + 2)$. The complete problem looks like this:

$$\frac{x^2 - 9}{x^2 + 5x + 6} = \frac{(x - 3)(x + 3)}{(x + 2)(x + 3)} \quad \text{Factor the numerator and denominator completely.}$$

> *Note* It is convenient to draw a line through the factors as we divide them out. It is especially helpful when the problems become longer.

$$= \frac{(x - 3)(x + 3)}{(x + 2)(x + 3)} \quad \text{Divide out the common factor, } x + 3.$$

$$= \frac{x - 3}{x + 2}$$

4. Simplify the expression
$$\frac{5a + 15}{10a^2 - 90}.$$

EXAMPLE 4 Simplify the expression $\dfrac{10a + 20}{5a^2 - 20}$.

Solution We begin by factoring out the greatest common factor from the numerator and denominator.

$$\frac{10a + 20}{5a^2 - 20} = \frac{10(a + 2)}{5(a^2 - 4)} \quad \text{Factor the greatest common factor from the numerator and denominator.}$$

$$= \frac{10(a + 2)}{5(a + 2)(a - 2)} \quad \text{Factor the denominator as the difference of two squares.}$$

$$= \frac{2}{a - 2} \quad \text{Divide out the common factors 5 and } a + 2.$$

5. Simplify the expression
$$\frac{x^3 - x^2 - 2x}{x^3 + 4x^2 + 3x}.$$

EXAMPLE 5 Simplify the expression $\dfrac{2x^3 + 2x^2 - 24x}{x^3 + 2x^2 - 8x}$.

Solution We begin by factoring the numerator and denominator completely. Then we divide out all factors common to the numerator and denominator. Here is what it looks like:

$$\frac{2x^3 + 2x^2 - 24x}{x^3 + 2x^2 - 8x} = \frac{2x(x^2 + x - 12)}{x(x^2 + 2x - 8)} \quad \text{Factor out the greatest common factor first.}$$

$$= \frac{2x(x - 3)(x + 4)}{x(x - 2)(x + 4)} \quad \text{Factor the remaining trinomials.}$$

$$= \frac{2(x - 3)}{x - 2} \quad \text{Divide out the factors common to the numerator and denominator.}$$

6. Simplify the expression
$$\frac{x + 3}{x^2 - 9}.$$

EXAMPLE 6 Simplify the expression $\dfrac{-x + 5}{-x^2 + 25}$.

Solution First, we factor out the greatest common factor of -1 from the numerator and the denominator. Then we can factor the denominator, recognizing it as a difference of squares.

$$\frac{-x + 5}{-x^2 + 25} = \frac{-1(x - 5)}{-1(x - 5)(x + 5)} \quad \text{Factor numerator and denominator completely.}$$

$$= \frac{1}{x + 5} \quad \text{Divide out the common factor, } x - 5.$$

Answers

3. $\dfrac{x - 2}{x - 4}$

4. $\dfrac{1}{2(a - 3)}$

5. $\dfrac{x - 2}{x + 3}$

6. $\dfrac{1}{x - 3}$

C Ratios

For the rest of this section we will concern ourselves with ratios, a topic closely related to reducing the fractions and rational expressions to lowest terms. Let's start with a definition.

> **DEFINITION** ratio
>
> If a and b are any two numbers, $b \neq 0$, then the *ratio* of a and b is $\dfrac{a}{b}$.

As you can see, ratios are similar to fractions or rational numbers. They are a way of comparing quantities that have the same units. Since we also can think of $\frac{a}{b}$ as the quotient of a and b, ratios are also quotients. The following table gives some ratios in words and as fractions.

Ratio	As a Fraction	In Lowest Terms	
25 to 75	$\frac{25}{75}$	$\frac{1}{3}$	
8 to 2	$\frac{8}{2}$	$\frac{4}{1}$	← With ratios it is common to leave the 1 in the denominator
20 to 16	$\frac{20}{16}$	$\frac{5}{4}$	

EXAMPLE 7 A solution of hydrochloric acid (HCl) and water contains 49 milliliters of water and 21 milliliters of HCl. Find the ratio of HCl to water and of HCl to the total volume of the solution.

Solution The ratio of HCl to water is 21 to 49, or

$$\frac{21}{49} = \frac{3}{7}$$

The amount of total solution volume is $49 + 21 = 70$ milliliters. Therefore, the ratio of HCl to total solution is 21 to 70, or

$$\frac{21}{70} = \frac{3}{10}$$

7. A solution of alcohol and water contains 27 milliliters of alcohol and 54 milliliters of water. Find the ratio of alcohol to water and the ratio of alcohol to total volume.

D Rate Equation

Many of the problems in this chapter will use what is called the *rate equation*. You use this equation on an intuitive level when you are estimating how long it will take you to drive long distances. For example, if you drive at 50 miles per hour for 2 hours, you will travel 100 miles. Here is the rate equation:

$$\text{Distance} = \text{rate} \cdot \text{time}$$

$$d = r \cdot t$$

The rate equation has two equivalent forms, the most common of which is obtained by solving for r. Here it is:

$$r = \frac{d}{t}$$

The rate r in the rate equation is the ratio of distance to time and also is referred to as *average speed*. The units for rate are miles per hour, feet per second, kilometers per hour, and so on.

Answers

7. $\frac{1}{2}$ and $\frac{1}{3}$

8. The Bear Paw chair lift at the Northstar ski resort in Lake Tahoe is 772 feet long. If a ride on this chair lift takes 2.2 minutes, what is the average speed of the lift in feet per minute? (Round to the nearest whole number.)

EXAMPLE 8 The Forest chair lift at the Northstar ski resort in Lake Tahoe is 5,603 feet long. If a ride on this chair lift takes 11 minutes, what is the average speed of the lift in feet per minute?

Solution To find the speed of the lift, we find the ratio of distance covered to time. (Our answer is rounded to the nearest whole number.)

$$\text{Rate} = \frac{\text{distance}}{\text{time}}$$

$$= \frac{5,603 \text{ feet}}{11 \text{ minutes}}$$

$$= \frac{5,603}{11} \text{ feet/minute}$$

$$= 509 \text{ feet/minute}$$

Note how we separate the numerical part of the problem from the units. In the next section, we will convert this rate to miles per hour.

GETTING READY FOR CLASS

After reading through the preceding section, respond in your own words and in complete sentences.

A. Explain what it means to have a restriction on a rational expression.

B. What are the properties we use to manipulate rational expressions?

C. How do you reduce a rational expression to lowest terms?

D. What is a ratio?

Answer

8. 351 feet per minute

Vocabulary Review

Choose the correct words to fill in the blanks below.

numerator undefined nonzero denominator ratio

distance rate equation simplify rational expression

1. A _____ is any expression that can be put in the form $\frac{P}{Q}$, where P and Q are polynomials and Q does not equal zero.

2. In the rational expression $\frac{3x + 4}{x^2 - 4}$, the _____ is $3x + 4$ and the _____ is $x^2 - 4$.

3. A value for the variable in a rational expression must not make the denominator equal zero since division by zero is _____.

4. Multiplying and dividing the numerator and denominator of a rational expression by the same _____ quantity will not change the value of the rational expression.

5. To _____ a rational expression, factor the numerator and denominator completely and then divide both by any factors they have in common.

6. A _____ is another name for a fraction or rational number.

7. The _____ compares three variable quantities: distance, rate, and time.

8. The rate r in the rate equation is the ratio of _____ to time.

Problems

A Simplify the following rational expressions, if possible. Also, specify any restrictions on the variable in problems 1 through 10.

1. $\dfrac{5}{5x - 10}$

2. $\dfrac{-4}{2x - 8}$

3. $\dfrac{a - 3}{a^2 - 9}$

4. $\dfrac{a + 4}{a^2 - 16}$

5. $\dfrac{x + 5}{x^2 - 25}$

6. $\dfrac{x - 2}{x^2 - 4}$

7. $\dfrac{2x^2 - 8}{4}$

8. $\dfrac{5x - 10}{x - 2}$

9. $\dfrac{2x - 10}{3x - 6}$

10. $\dfrac{4x - 8}{x - 2}$

11. $\dfrac{10a + 20}{5a + 10}$

12. $\dfrac{11a + 33}{6a + 18}$

13. $\dfrac{5x^2 - 5}{4x + 4}$

14. $\dfrac{7x^2 - 28}{2x + 4}$

SCAN TO ACCESS

15. $\dfrac{x - 3}{x^2 - 6x + 9}$

16. $\dfrac{x^2 - 10x + 25}{x - 5}$

17. $\dfrac{3x + 15}{3x^2 + 24x + 45}$

18. $\dfrac{5x + 15}{5x^2 + 40x + 75}$

19. $\dfrac{a^2 - 3a}{a^3 - 8a^2 + 15a}$

20. $\dfrac{a^2 - 3a}{a^3 + 2a^2 - 15a}$

21. $\dfrac{3x - 2}{9x^2 - 4}$

22. $\dfrac{2x - 3}{4x^2 - 9}$

23. $\dfrac{x^2 + 8x + 15}{x^2 + 5x + 6}$

24. $\dfrac{x^2 - 8x + 15}{x^2 - x - 6}$

25. $\dfrac{2m^3 - 2m^2 - 12m}{m^2 - 5m + 6}$

26. $\dfrac{2m^3 + 4m^2 - 6m}{m^2 - m - 12}$

27. $\dfrac{x^3 + 3x^2 - 4x}{x^3 - 16x}$

28. $\dfrac{3a^2 - 8a + 4}{9a^3 - 4a}$

29. $\dfrac{4x^3 - 10x^2 + 6x}{2x^3 + x^2 - 3x}$

30. $\dfrac{3a^3 - 8a^2 + 5a}{4a^3 - 5a^2 + 1a}$

31. $\dfrac{4x^2 - 12x + 9}{4x^2 - 9}$

32. $\dfrac{5x^2 + 18x - 8}{5x^2 + 13x - 6}$

33. $\dfrac{x + 3}{x^4 - 81}$

34. $\dfrac{x^2 + 9}{x^4 - 81}$

35. $\dfrac{3x^2 + x - 10}{x^4 - 16}$

36. $\dfrac{5x^2 - 16x + 12}{x^4 - 16}$

37. $\dfrac{42x^3 - 20x^2 - 48x}{6x^2 - 5x - 4}$

38. $\dfrac{36x^3 + 132x^2 - 135x}{6x^2 + 25x - 9}$

39. $\dfrac{x^3 - y^3}{x^2 - y^2}$

40. $\dfrac{x^3 + y^3}{x^2 - y^2}$

41. $\dfrac{x^3 + 8}{x^2 - 4}$

42. $\dfrac{x^3 - 125}{x^2 - 25}$

43. $\dfrac{x^3 + 8}{x^2 + x - 2}$

44. $\dfrac{x^2 - 2x - 3}{x^3 - 27}$

45. $\dfrac{3a^2 + 10ax - 25x^2}{25x^2 - 9a^2}$

46. $\dfrac{6x^2 + xy - 12y^2}{16y^2 - 9x^2}$

47. $\dfrac{6a^2 - 19ab + 15b^2}{12b^2 + ab - 6a^2}$

48. $\dfrac{9x^2 - 4y^2}{2y^2 + 3xy - 9x^2}$

B To simplify each of the following rational expressions, you will have to use factoring by grouping. Be sure to factor each numerator and denominator completely before dividing out any common factors. (Remember, factoring by grouping requires two steps.)

49. $\dfrac{xy + 3x + 2y + 6}{xy + 3x + 5y + 15}$

50. $\dfrac{xy + 7x + 4y + 28}{xy + 3x + 4y + 12}$

51. $\dfrac{x^2 - 3x + ax - 3a}{x^2 - 3x + bx - 3b}$

52. $\dfrac{x^2 - 6x + ax - 6a}{x^2 - 7x + ax - 7a}$

The next two problems are intended to give you practice reading, and paying attention to, the instructions that accompany the problems you're working. Working these problems is an excellent way to get ready for a test or quiz.

53. Work each problem according to the instructions.

 a. Add: $(x^2 - 4x) + (4x - 16)$.

 b. Subtract: $(x^2 - 4x) - (4x - 16)$.

 c. Multiply: $(x^2 - 4x)(4x - 16)$.

 d. Simplify: $\dfrac{x^2 - 4x}{4x - 16}$.

54. Work each problem according to the instructions.

 a. Add: $(9x^2 - 3x) + (6x - 2)$.

 b. Subtract: $(9x^2 - 3x) - (6x - 2)$.

 c. Multiply: $(9x^2 - 3x)(6x - 2)$.

 d. Simplify: $\dfrac{9x^2 - 3x}{6x - 2}$.

C Write each ratio as a fraction in lowest terms.

55. 8 to 6

56. 6 to 8

57. 200 to 250

58. 250 to 200

59. 32 to 4

60. 4 to 32

61. 25 to 125

62. 125 to 25

Applying the Concepts

63. Cost and Average Cost If a bank charges $2.00 per month and $0.15 per check for one of its checking accounts, then the total monthly cost to write x checks is $C = 2.00 + 0.15x$, and the average cost of each of the x checks written is $A = \dfrac{2.00 + 0.15x}{x}$. Compare these two formulas by filling in the following table. Round to the nearest cent.

Checks Written x	Total Cost $2.00 + 0.15x$	Cost per Check $\dfrac{2.00 + 0.15x}{x}$
0		
5		
10		
15		
20		

64. Cost and Average Cost Suppose an inkjet printer costs $150. A cartridge of black ink for the printer costs $10 and can print 2,200 pages. The total cost for using x cartridges is $C = 150 + 10x$ dollars. The average cost per printed page is given by $A = \dfrac{150 + 10x}{2,200x}$. Compare the total cost and average cost per page by completing the following table. Round all answers to the nearest tenth of a cent.

Cartridges Used x	Total Cost $150 + 10x$	Cost per Page $\dfrac{150 + 10x}{2200x}$
0		
5		
10		
15		
20		

65. Speed of a Car A car travels 122 miles in 3 hours. Find the average speed of the car in miles per hour. Round to the nearest tenth.

66. Speed of a Bullet A bullet fired from a gun travels a distance of 4,500 feet in 3 seconds. Find the average speed of the bullet in feet per second.

67. Soda Consumption The following chart shows soda consumption in gallons per person per year in different countries. Calculate the consumption in each country per person per month. Round to the nearest hundredth.

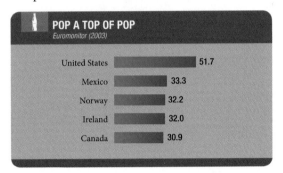

POP A TOP OF POP
Euromonitor (2003)

United States	51.7
Mexico	33.3
Norway	32.2
Ireland	32.0
Canada	30.9

68. Ferris Wheel A person riding a Ferris Wheel travels once around the wheel, a distance of 188 feet, in 30 seconds. What is the average speed of the rider in feet per second? Round to the nearest tenth.

69. Jogging A jogger covers a distance of 3 miles in 24 minutes. Find the average speed of the jogger in miles per minute.

70. Fuel Consumption An economy car travels 168 miles on 3.5 gallons of gas. Give the average fuel consumption of the car in miles per gallon.

71. Stock Market One method of comparing stock market is the price to earnings ratio, or P/E.

P/E = Current Stock Price/Earnings per Share

Most stocks have a P/E between 25 and 40. A stock with a P/E of less than 25 may be undervalued, while a stock with a P/E greater than 40 may be overvalued. Fill in the P/E for each stock listed in the table below. Based on your results, are any of the stocks undervalued?

Stock	Price	Earnings per Share	P/E
Yahoo	37.80	1.07	
Google	381.24	4.51	
Disney	24.96	1.34	
Nike	85.46	4.88	
Ebay	40.96	0.73	

72. Planet Diameters The chart shows the ratio of the diameters of planets in our solar system to the diameter of Earth. If the diameter of Earth is approximately 7,900 miles, use the information to answer the following questions.

DIAMETERS OUT OF THIS WORLD
www.nssclc.gsfc.nasa.gov

*Ratio of diameters planet to Earth

Mercury	0.4
Venus	0.9
Earth	1.0
Mars	0.5
Jupiter	11.2
Saturn	9.5
Uranus	4.0
Neptune	3.9

a. What is the diameter of Jupiter?

b. What is the diameter of Neptune?

c. What is the diameter of Mercury?

Getting Ready for the Next Section

Perform the indicated operations.

73. $\dfrac{3}{4} \cdot \dfrac{10}{21}$

74. $\dfrac{2}{9} \cdot \dfrac{15}{22}$

75. $\dfrac{4}{5} \div \dfrac{8}{9}$

76. $\dfrac{3}{5} \div \dfrac{15}{7}$

Simplify each rational expression.

77. $\dfrac{a(a+5)(a-5)(a+4)}{a^2+5a}$

78. $\dfrac{a(a+2)(a-4)(a+5)}{a^2-4a}$

Factor each polynomial completely.

79. x^2-9

80. x^2-25

81. $3x-9$

82. $2x-4$

83. x^2-x-20

84. $x^2+7x+12$

85. a^2+5a

86. a^2-4a

Multiply. Give the answers as decimals rounded to the nearest tenth.

87. $\dfrac{5603}{11}\cdot\dfrac{1}{2580}\cdot\dfrac{60}{1}$

88. $\dfrac{772}{2.2}\cdot\dfrac{1}{2580}\cdot\dfrac{60}{1}$

Maintaining Your Skills

Perform the following operations, and simplify.

89. $\dfrac{27x^5}{9x^2}-\dfrac{27x^5}{9x^2}$

90. $\dfrac{36x^9}{4x}-\dfrac{45x^3}{5x^{-5}}$

91. $\dfrac{72a^3b^7}{9ab^5}+\dfrac{64a^5b^3}{8a^3b}$

92. $\dfrac{80a^5b^3}{8a^3b}+\dfrac{33a^6b^{12}}{11a^3b^2}$

Divide each polynomial by the given monomial.

93. $\dfrac{38x^7+42x^5-84x^3}{2x^3}$

94. $\dfrac{49x^6-63x^4-35x^2}{7x^2}$

95. $\dfrac{28a^5b^5+36ab^4-44a^4b}{4ab}$

96. $\dfrac{30a^3b-12a^2b^2+6ab^3}{6ab}$

Find the Mistake

Each sentence below contains a mistake. Circle the mistake and write the correct word(s) or expression(s) on the line provided.

1. In the rational expression $\dfrac{(x^2-81)}{x^2}$, the denominator is x^2-81. _____

2. Restrictions on the variable in a rational expression are the values of the variable that gives a numerator of zero.

3. To simplify the rational expression of $\dfrac{18}{42}$, divide the numerator and denominator by the common factor 7.

4. The rate equation written as rate equals a ratio of distance to time looks like $r=d\cdot t$. _____

Navigation Skills: Prepare, Study, Achieve

Your academic self-image is how you see yourself as a student and the level of success you see yourself achieving. Do you believe you are capable of learning any subject and succeeding in any class you take? If you believe in yourself and work hard by applying the appropriate study methods, you will succeed. If you have a poor outlook for a class, most likely your performance in that class will match that outlook. Self-doubt will negatively affect your focus. Furthermore, an inner dialogue of negative statements, such as "I'll never be able to learn this material," distract you from achieving success. Consider replacing those thoughts with three positive statements you can say when you notice your mind participating in a negative inner dialogue. Make a commitment to change your attitude for the better. Begin by thinking positively, having confidence in your abilities, and utilizing your resources. Asking for help is a sign of a successful student.

6.2 Multiplication and Division of Rational Expressions

A Multiply and divide rational expressions by factoring and then dividing out common factors.

B Convert between units using unit analysis.

Image © Katherine Heistand Shields, 2010

You may have heard of the Iditarod, the dog-sledding race in Alaska, but you probably haven't heard of the Idiotarod. An Idiotarod racing team consists of five people, four in front of and attached to a shopping cart in the "sled-dog" position and one behind the cart in the "musher" position. The race's start and finish lines and check points are mapped out through a city, but the course itself is of each team's choosing. Teams often modify their cart and wear costumes to match a certain theme. And unlike most regulated races that support a code of conduct and quality sportsmanship, bribes and sabotage are very much welcomed during the event.

Let's say that the distance between two checkpoints of an Idiotarod race is 1.5 miles. How would we determine the distance in feet between those two checkpoints? In this section, we will discuss unit analysis and introduce the concept of a conversion factor, which will help us convert miles into feet. The conversion factor is also in the form of a rational expression. But first, let's practice multiplying and dividing rational expressions.

A Multiplying and Dividing Rational Expressions

Recall that to multiply two rational numbers we simply multiply numerators and multiply denominators and then reduce to lowest terms, if possible.

$$\frac{3}{4} \cdot \frac{10}{21} = \frac{30}{84} \qquad \text{Multiply numerators.}$$
$$\text{Multiply denominators.}$$

$$= \frac{5}{14} \qquad \text{Reduce to lowest terms.}$$

Recall also that the same result can be achieved by factoring numerators and denominators first and then dividing out the factors they have in common.

$$\frac{3}{4} \cdot \frac{10}{21} = \frac{3}{2 \cdot 2} \cdot \frac{2 \cdot 5}{3 \cdot 7} \qquad \text{Factor.}$$

$$= \frac{3 \cdot 2 \cdot 5}{2 \cdot 2 \cdot 3 \cdot 7} \qquad \text{Multiply numerators.}$$
$$\text{Multiply denominators.}$$

$$= \frac{5}{14} \qquad \text{Divide out common factors.}$$

We can apply the second process to the product of two rational expressions, as the following example illustrates.

EXAMPLE 1 Multiply $\dfrac{x-2}{x+3} \cdot \dfrac{x^2-9}{2x-4}$.

Solution We begin by factoring numerators and denominators as much as possible. Then we multiply the numerators and denominators. The last step consists of dividing out all factors common to the numerator and denominator.

$$\dfrac{x-2}{x+3} \cdot \dfrac{x^2-9}{2x-4} = \dfrac{x-2}{x+3} \cdot \dfrac{(x-3)(x+3)}{2(x-2)} \quad \text{Factor completely.}$$

$$= \dfrac{(x-2)(x-3)(x+3)}{(x+3)(2)(x-2)} \quad \text{Multiply numerators and denominators.}$$

$$= \dfrac{x-3}{2} \quad \text{Divide out common factors.}$$

In Chapter 0, we defined division as the equivalent of multiplication by the reciprocal. This is how it looks with fractions:

$$\dfrac{4}{5} \div \dfrac{8}{9} = \dfrac{4}{5} \cdot \dfrac{9}{8} \quad \text{Division as multiplication by the reciprocal.}$$

$$= \dfrac{2 \cdot 2 \cdot 3 \cdot 3}{5 \cdot 2 \cdot 2 \cdot 2} \Bigg\} \quad \text{Factor and divide out common factors.}$$

$$= \dfrac{9}{10}$$

The same idea holds for division with rational expressions. The rational expression that follows the division symbol is called the *divisor;* to divide, we multiply by the reciprocal of the divisor.

EXAMPLE 2 Divide $\dfrac{3x-9}{x^2-x-20} \div \dfrac{x^2+2x-15}{x^2-25}$.

Solution We begin by taking the reciprocal of the divisor and writing the problem again in terms of multiplication. We then factor, multiply, and, finally, divide out all factors common to the numerator and denominator of the resulting expression. The complete problem looks like this:

$$\dfrac{3x-9}{x^2-x-20} \div \dfrac{x^2+2x-15}{x^2-25}$$

$$= \dfrac{3x-9}{x^2-x-20} \cdot \dfrac{x^2-25}{x^2+2x-15} \quad \text{Multiply by the reciprocal of the divisor.}$$

$$= \dfrac{3(x-3)}{(x+4)(x-5)} \cdot \dfrac{(x-5)(x+5)}{(x+5)(x-3)} \quad \text{Factor.}$$

$$= \dfrac{3(x-3)(x-5)(x+5)}{(x+4)(x-5)(x+5)(x-3)} \quad \text{Multiply.}$$

$$= \dfrac{3}{x+4} \quad \text{Divide out common factors.}$$

As you can see, factoring is the single most important tool we use in working with rational expressions. It is easier to work with rational expressions if they are in factored form. Here are more examples of multiplication and division with rational expressions:

EXAMPLE 3 Multiply $\dfrac{3a+6}{a^2} \cdot \dfrac{a}{2a+4}$.

Solution

$$\dfrac{3a+6}{a^2} \cdot \dfrac{a}{2a+4}$$

$$= \dfrac{3(a+2)}{a^2} \cdot \dfrac{a}{2(a+2)} \quad \text{Factor completely.}$$

$$= \dfrac{3(a+2)a}{a^2(2)(a+2)} \quad \text{Multiply.}$$

$$= \dfrac{3}{2a} \quad \begin{array}{l}\text{Divide numerator and denominator}\\\text{by common factors } a\,(a+2).\end{array}$$

Practice Problems

1. Multiply $\dfrac{x-5}{x+2} \cdot \dfrac{x^2-4}{3x-15}$.

2. Divide
$$\dfrac{4x+8}{x^2-x-6} \div \dfrac{x^2+7x+12}{x^2-9}.$$

3. Multiply $\dfrac{2a+10}{a^3} \cdot \dfrac{a^2}{3a+15}$.

Answers

1. $\dfrac{x-2}{3}$

2. $\dfrac{4}{x+4}$

3. $\dfrac{2}{3a}$

4. Divide
$$\frac{x^2 - x - 12}{x^2 - 16} \div \frac{x^2 + 6x + 9}{2x + 8}.$$

EXAMPLE 4 Divide $\dfrac{x^2 + 7x + 12}{x^2 - 16} \div \dfrac{x^2 + 6x + 9}{2x - 8}$.

Solution

$$\frac{x^2 + 7x + 12}{x^2 - 16} \div \frac{x^2 + 6x + 9}{2x - 8}$$

$$= \frac{x^2 + 7x + 12}{x^2 - 16} \cdot \frac{2x - 8}{x^2 + 6x + 9} \qquad \text{Division is multiplication by the reciprocal.}$$

$$= \frac{(x + 3)(x + 4)(2)(x - 4)}{(x - 4)(x + 4)(x + 3)(x + 3)} \qquad \text{Factor and multiply.}$$

$$= \frac{2}{x + 3} \qquad \text{Divide out common factors.}$$

In Example 4, we factored and multiplied the two expressions in a single step. This saves writing the problem one extra time.

5. Multiply $(x^2 - 9)\left(\dfrac{x + 2}{x + 3}\right)$.

EXAMPLE 5 Multiply $(x^2 - 49)\left(\dfrac{x + 4}{x + 7}\right)$.

Solution We can think of the polynomial $x^2 - 49$ as having a denominator of 1. Thinking of $x^2 - 49$ in this way allows us to proceed as we did in previous examples.

$$(x^2 - 49)\left(\frac{x + 4}{x + 7}\right)$$

$$= \frac{x^2 - 49}{1} \cdot \frac{x + 4}{x + 7} \qquad \text{Write } x^2 - 49 \text{ with denominator 1.}$$

$$= \frac{(x + 7)(x - 7)(x + 4)}{x + 7} \qquad \text{Factor and multiply.}$$

$$= (x - 7)(x + 4) \qquad \text{Divide out common factors.}$$

We can leave the answer in this form or multiply to get $x^2 - 3x - 28$. In this section let's agree to leave our answers in factored form.

6. Multiply
$$a(a + 2)(a - 4)\left(\frac{a + 5}{a^2 - 4a}\right).$$

EXAMPLE 6 Multiply $a(a + 5)(a - 5)\left(\dfrac{a + 4}{a^2 + 5a}\right)$.

Solution We can think of the expression $a(a + 5)(a - 5)$ as having a denominator of 1.

$$a(a + 5)(a - 5)\left(\frac{a + 4}{a^2 + 5a}\right)$$

$$= \frac{a(a + 5)(a - 5)}{1} \cdot \frac{a + 4}{a^2 + 5a}$$

$$= \frac{a(a + 5)(a - 5)(a + 4)}{a(a + 5)} \qquad \text{Factor and multiply.}$$

$$= (a - 5)(a + 4) \qquad \text{Divide out common factors.}$$

B Unit Analysis

Unit analysis is a method of converting between units of measure by multiplying by the number 1. Measuring the length of an object is done by assigning a number to its length. To let other people know what that number represents, we include with it a unit of measure. The most common units used to represent *length* in the U.S. system are inches, feet, yards, and miles. The basic unit of length is the foot. The other units are defined in terms of feet.

7. Convert 3 feet to inches.

EXAMPLE 7 If you know that 1 foot = 12 inches, convert 5 feet to inches.

Solution Because 1 foot = 12 inches, we can multiply 5 by 12 inches to get

$$5 \text{ feet} = 5 \times 12 \text{ inches}$$

$$= 60 \text{ inches}$$

Answers

4. $\dfrac{2}{x + 3}$

5. $(x - 3)(x + 2)$

6. $(a + 2)(a + 5)$

7. 36 inches

As shown in the previous example, since 1 ft = 12 in., we can say that there are 12 in. per 1 ft or 1 ft per 12 in.; that is,

$$\underbrace{\frac{12 \text{ in.}}{1 \text{ ft}}}_{} \longleftarrow \textit{Per} \qquad or \qquad \underbrace{\frac{1 \text{ ft}}{12 \text{ in.}}}_{} \longleftarrow \textit{Per}$$

We call the expressions $\frac{12 \text{ in.}}{1 \text{ ft}}$ and $\frac{1 \text{ ft}}{12 \text{ in.}}$ *conversion factors*. The fraction bar is read as "per." Both these conversion factors are really just the number 1. That is,

$$\frac{12 \text{ in.}}{1 \text{ ft}} = \frac{12 \text{ in.}}{12 \text{ in.}} = 1$$

We already know that multiplying a number by 1 leaves the number unchanged. So, to convert from one unit to the other, we can multiply by one of the conversion factors without changing value. Both the conversion factors above say the same thing about the units feet and inches. They both indicate that there are 12 inches in every foot. The one we choose to multiply by depends on what units we are starting with and what units we want to end up with. If we start with feet and we want to end up with inches, we multiply by the conversion factor

$$\frac{12 \text{ in.}}{1 \text{ ft}}$$

EXAMPLE 8 The most common ceiling height in houses is 8 feet. If you know that there are 3 feet in 1 yard, how many yards tall is the ceiling?

Solution To convert 8 feet to yards, we multiply by the conversion factor $\frac{1 \text{ yd}}{3 \text{ ft}}$ so that feet will divide out and we will be left with yards.

$$8 \text{ ft} = 8 \text{ ft} \times \frac{1 \text{ yd}}{3 \text{ ft}} \qquad \textit{Multiply by correct conversion factor.}$$

$$= \frac{8}{3} \text{ yd} \qquad\qquad 8 \times \frac{1}{3} = \frac{8}{3}$$

$$= 2\frac{2}{3} \text{ yd} \qquad\qquad \textit{Or 2.67 yd rounded to the nearest hundredth.}$$

EXAMPLE 9 A rectangular poster measures 36 inches by 24 inches. If there are 144 in^2 in 1 ft^2, how many square feet of wall space will the poster cover?

Solution One way to work this problem is to find the number of square inches the poster covers, and then convert square inches to square feet.

$$\text{Area of poster} = \text{length} \times \text{width} = 36 \text{ in.} \times 24 \text{ in.} = 864 \text{ in}^2$$

To finish the problem, we convert square inches to square feet.

$$864 \text{ in}^2 = 864 \text{ in}^2 \times \frac{1 \text{ ft}^2}{144 \text{ in}^2}$$

$$= \frac{864}{144} \text{ ft}^2$$

$$= 6 \text{ ft}^2$$

8. Suppose a house has high ceilings. Convert 12 feet to yards.

9. Suppose there is a 2 inch frame around the poster in Example 2. How many square feet of wall space will be covered?

GETTING READY FOR CLASS

After reading through the preceding section, respond in your own words and in complete sentences.

A. How do you multiply rational expressions?

B. Explain the steps used to divide rational expressions.

C. Give an example of an application that uses unit analysis.

D. Why are all conversion factors the same as the number 1?

Answers
8. 4 yards
9. 7.78 ft^2

EXERCISE SET 6.2

Vocabulary Review

Choose the correct words to fill in the blanks below.

multiply reciprocal unit analysis rational numbers

number 1 factors common conversion factor

1. To multiply two _____, multiply numerators and multiply denominators, and then reduce to lowest terms.

2. To _____ two rational expressions, first factor the numerators and denominators as much as possible, and then multiply.

3. The last step to multiply two rational expressions is to divide out all _____ common to the numerator and denominator.

4. To divide rational expressions, take the _____ of the divisor and write the problem in terms of multiplication.

5. _____ is a method of converting between units of measure by multiplying by the number 1.

6. When analyzing units, _____ units can be divided out just as common factors are.

7. A conversion factor is the _____ written in a convenient form to compare units.

8. The key to unit analysis is choosing the right _____.

Problems

A Multiply or divide as indicated. Be sure to simplify all answers. (The numerator and denominator of the answer should not have any factors in common.)

1. $\dfrac{x+y}{3} \cdot \dfrac{6}{x+y}$

2. $\dfrac{x-1}{x+1} \cdot \dfrac{5}{x-1}$

3. $\dfrac{2x+10}{x^2} \cdot \dfrac{x^3}{4x+20}$

4. $\dfrac{3x^4}{3x-6} \cdot \dfrac{x-2}{x^2}$

5. $\dfrac{9}{2a-8} \div \dfrac{3}{a-4}$

6. $\dfrac{8}{a^2-25} \div \dfrac{16}{a+5}$

7. $\dfrac{x+1}{x^2-9} \div \dfrac{2x+2}{x+3}$

8. $\dfrac{11}{x-2} \div \dfrac{22}{2x^2-8}$

9. $\dfrac{a^2+5a}{7a} \cdot \dfrac{4a^2}{a^2+4a}$

10. $\dfrac{4a^2+4a}{a^2-25} \cdot \dfrac{a^2-5a}{8a}$

11. $\dfrac{y^2-5y+6}{2y+4} \div \dfrac{2y-6}{y+2}$

12. $\dfrac{y^2-7y}{3y^2-48} \div \dfrac{y^2-9}{y^2-7y+12}$

13. $\dfrac{2x-8}{x^2-4} \cdot \dfrac{x^2+6x+8}{x-4}$

SCAN TO ACCESS

14. $\dfrac{x^2 + 5x + 1}{7x - 7} \cdot \dfrac{x - 1}{x^2 + 5x + 1}$

15. $\dfrac{x - 1}{x^2 - x - 6} \cdot \dfrac{x^2 + 5x + 6}{x^2 - 1}$

16. $\dfrac{x^2 - 3x - 10}{x^2 - 4x + 3} \cdot \dfrac{x^2 - 5x + 6}{x^2 - 3x - 10}$

17. $\dfrac{a^2 + 10a + 25}{a + 5} \div \dfrac{a^2 - 25}{a - 5}$

18. $\dfrac{a^2 + a - 2}{a^2 + 5a + 6} \div \dfrac{a - 1}{a}$

19. $\dfrac{5x - 10}{6x - 9} \div \dfrac{6 - 3x}{8x - 12}$

20. $\dfrac{10b - 15}{4b + 2} \div \dfrac{6 - 4b}{6b + 3}$

21. $\dfrac{6x^2 - 11xy + 3y^2}{2x^2 - 7xy + 6y^2} \cdot \dfrac{3x^2 + 8xy + 4y^2}{2y^2 - 3xy - 9x^2}$

22. $\dfrac{a^2 + ab - 12b^2}{3a^2 + 8ab - 16b^2} \cdot \dfrac{2a^2 + 7ab + 6b^2}{9b^2 + 3ab - 2a^2}$

23. $\dfrac{y^3 - 5y^2}{y^4 + 3y^3 + 2y^2} \div \dfrac{y^2 - 5y + 6}{y^2 - 2y - 3}$

24. $\dfrac{y^2 - 5y}{y^2 + 7y + 12} \div \dfrac{y^3 - 7y^2 + 10y}{y^2 + 9y + 18}$

25. $\dfrac{2x^2 + 17x + 21}{x^2 + 2x - 35} \cdot \dfrac{x^2 - 25}{2x^2 - 7x - 15}$

26. $\dfrac{x^2 - 13x + 42}{4x^2 - 31x + 21} \cdot \dfrac{4x^2 + 5x - 6}{x^2 - 4}$

27. $\dfrac{2x^2 + 10x + 12}{4x^2 + 24x + 32} \cdot \dfrac{2x^2 + 18x + 40}{x^2 + 8x + 15}$

28. $\dfrac{3x^2 - 3}{6x^2 + 18x + 12} \cdot \dfrac{2x^2 - 8}{x^2 - 3x + 2}$

29. $\dfrac{2a^2 + 7a + 3}{a^2 - 16} \div \dfrac{4a^2 + 8a + 3}{2a^2 - 5a - 12}$

30. $\dfrac{3a^2 + 7a - 20}{a^2 + 3a - 4} \div \dfrac{3a^2 - 2a - 5}{a^2 - 2a + 1}$

31. $\dfrac{4y^2 - 12y + 9}{y^2 - 36} \div \dfrac{2y^2 - 5y + 3}{y^2 + 5y - 6}$

32. $\dfrac{5y^2 - 6y + 1}{y^2 - 1} \div \dfrac{16y^2 - 9}{4y^2 + 7y + 3}$

33. $\dfrac{x^2 - 1}{6x^2 + 18x + 12} \cdot \dfrac{7x^2 + 17x + 6}{x + 1} \cdot \dfrac{6x + 30}{7x^2 - 11x - 6}$

34. $\dfrac{4x^2 - 1}{3x - 15} \cdot \dfrac{4x^2 - 17x - 15}{4x^2 - 9x - 9} \cdot \dfrac{3x - 3}{x^2 - 9}$

35. $\dfrac{18x^3 + 21x^2 - 60x}{21x^2 - 25x - 4} \cdot \dfrac{28x^2 - 17x - 3}{16x^3 + 28x^2 - 30x}$

36. $\dfrac{56x^3 + 54x^2 - 20x}{8x^2 - 2x - 15} \cdot \dfrac{6x^2 + 5x - 21}{63x^3 + 129x^2 - 42x}$

The next two problems are intended to give you practice reading, and paying attention to, the instructions that accompany the problems you are working. Working these problems is an excellent way to get ready for a test or quiz.

37. Work each problem according to the instructions.

 a. Reduce: $\dfrac{9-1}{27-1}$.

 b. Simplify: $\dfrac{x^2-1}{x^3-1}$.

 c. Multiply: $\dfrac{x^2-1}{x^3-1} \cdot \dfrac{x-2}{x+1}$.

 d. Divide: $\dfrac{x^2-1}{x^3-1} \div \dfrac{x-1}{x^2+x+1}$.

38. Work each problem according to the instructions.

 a. Reduce: $\dfrac{16-9}{16+24+9}$.

 b. Simplify: $\dfrac{4x^2-9}{4x^2+12x+9}$.

 c. Multiply: $\dfrac{4x^2-9}{4x^2+12x+9} \cdot \dfrac{2x+3}{2x-3}$.

 d. Divide: $\dfrac{4x^2-9}{4x^2+12x+9} \div \dfrac{2x+3}{2x-3}$.

Multiply the following expressions using the method shown in Examples 5 and 6 in this section.

39. $(x^2-9)\left(\dfrac{2}{x+3}\right)$

40. $(x^2-9)\left(\dfrac{-3}{x-3}\right)$

41. $(x^2-x-6)\left(\dfrac{x+1}{x-3}\right)$

42. $(x^2-2x-8)\left(\dfrac{x+3}{x-4}\right)$

43. $(x^2-4x-5)\left(\dfrac{-2x}{x+1}\right)$

44. $(x^2-6x+8)\left(\dfrac{4x}{x-2}\right)$

Each of the following problems involves some factoring by grouping. Remember, before you can divide out factors common to the numerators and denominators of a product, you must factor completely.

45. $\dfrac{x^2-9}{x^2-3x} \cdot \dfrac{2x+10}{xy+5x+3y+15}$

46. $\dfrac{x^2-16}{x^2-4x} \cdot \dfrac{3x+18}{xy+6x+4y+24}$

47. $\dfrac{2x^2+4x}{x^2-y^2} \cdot \dfrac{x^2+3x+xy+3y}{x^2+5x+6}$

48. $\dfrac{x^2-25}{3x^2+3xy} \cdot \dfrac{x^2+4x+xy+4y}{x^2+9x+20}$

49. $\dfrac{x^3-3x^2+4x-12}{x^4-16} \cdot \dfrac{3x^2+5x-2}{3x^2-10x+3}$

50. $\dfrac{x^3-5x^2+9x-45}{x^4-81} \cdot \dfrac{5x^2+18x+9}{5x^2-22x-15}$

Simplify each expression. Work inside parentheses first, and then divide out common factors.

51. $\left(1-\dfrac{1}{2}\right)\left(1-\dfrac{1}{3}\right)\left(1-\dfrac{1}{4}\right)\left(1-\dfrac{1}{5}\right)$

52. $\left(1+\dfrac{1}{2}\right)\left(1+\dfrac{1}{3}\right)\left(1+\dfrac{1}{4}\right)\left(1+\dfrac{1}{5}\right)$

The dots in the following problems represent factors not written that are in the same pattern as the surrounding factors. Simplify by looking for a pattern in dividing out common factors.

53. $\left(1 - \frac{1}{2}\right)\left(1 - \frac{1}{3}\right)\left(1 - \frac{1}{4}\right)\cdots\left(1 - \frac{1}{99}\right)\left(1 - \frac{1}{100}\right)$

54. $\left(1 - \frac{1}{3}\right)\left(1 - \frac{1}{4}\right)\left(1 - \frac{1}{5}\right)\cdots\left(1 - \frac{1}{98}\right)\left(1 - \frac{1}{99}\right)$

B Make the following conversions by multiplying by the appropriate conversion factor found in this section. Write your answers as whole numbers or mixed numbers.

55. 5 ft to inches

56. 9 ft to inches

57. 2 yd to feet

58. 8 yd to feet

59. 48 in. to yards

60. 56 in. to yards

61. 3 ft² to square inches

62. 5 ft² to square inches

63. 720 in² to square feet

64. 30 acres to square feet

65. 12 yd² to square feet

66. 20 yd² to square feet

Getting Ready for the Next Section

Perform the indicated operations and simplify your answer.

67. $\frac{1}{5} + \frac{3}{5}$

68. $\frac{1}{7} + \frac{5}{7}$

69. $\frac{1}{10} + \frac{3}{14}$

70. $\frac{1}{21} + \frac{4}{15}$

71. $\frac{1}{10} - \frac{3}{14}$

72. $\frac{1}{21} - \frac{4}{15}$

73. $\frac{3}{10} - \frac{4}{15}$

74. $\frac{5}{14} - \frac{2}{21}$

Simplify the following expressions.

75. $2(x - 3)$

76. $x(x + 2)$

77. $(x + 4)(x - 5)$

78. $(x + 3)(x - 4)$

79. $2(x + 1) - 3(x + 5)$

80. $5(x - 3) - 3(x + 2)$

81. $(x + 1)^2 - (x - 1)^2$

82. $(x - 2)^2 - (x + 2)^2$

Simplify each rational expression.

83. $\dfrac{x+3}{x^2-9}$

84. $\dfrac{x+7}{x^2-49}$

85. $\dfrac{x^2-x-30}{2(x+5)(x-5)}$

86. $\dfrac{x^2-x-20}{2(x+4)(x-4)}$

Simplify the following expressions.

87. $(x+4)(x-5)-10$

88. $(x+3)(x-4)-8$

89. $(-3)(10)-(x+2)(x-4)$ **90.** $-5(x-3)-(x+1)(x+2)$

Find the Mistake

Each sentence below contains a mistake. Circle the mistake and write the correct word(s) or number(s) on the line provided.

1. To multiply the rational expressions $\dfrac{(4x-6)}{x^2}$ and $\dfrac{(x^2-9)}{(x+3)}$, add the numerators and denominators. _____

2. To divide the rational expression $\dfrac{(x+7)}{(x-1)}$ by $\dfrac{(x+7)(x+5)}{4}$, multiply the divisor by the reciprocal of the first expression. _____

3. To convert 419 seconds to hours, multiply by the conversion factor 3600 seconds / 1 hour. _____

4. Using the rate equation $r=\dfrac{d}{t}$, a bicycle traveling 13,500 feet in 12 minutes is traveling at a speed of 162,000 feet per minute. _____

6.3 Addition and Subtraction of Rational Expressions

OBJECTIVES

A Add and subtract rational expressions that have the same denominators.

B Add and subtract rational expressions that have different denominators.

A Adding and Subtracting Expressions with the Same Denominator

If a human faced off against an elephant in a hot dog bun eating contest, who would win? Each July, three brave humans sign up to find out firsthand. Three elephants, Susie, Minnie, and Bunny, from Ringling Bros. and Barnum & Bailey Circus stand behind a table on which a tall pile of hot dog buns sits. A few feet away, three competitive eating contestants stand in front of their own table outfitted with stacks of buns and large cups of water in which they can dunk the buns to help their cause. When the announcer yells "Go!", the two teams begin devouring the buns. They have 6 minutes to eat as many as possible. The 2010 final tally: 41 dozen buns for the elephants, 15 dozen for the humans.

We can use this scenario to help us introduce the topic of this section: addition and subtraction of rational expressions. Suppose Susie the elephant ate x dozen hot dog buns, Minnie ate one more dozen than Susie, and Bunny ate twice as many dozen as Susie. What rational expression represents the total number of buns eaten by the three elephants? First, we can create a rational expression for each elephant by writing the amount eaten as a numerator over a denominator of 41 (dozen). Then we can add them together using our knowledge of combining fractions with the same denominator.

$$\underset{\underset{\text{Susie}}{\downarrow}}{} \quad \underset{\underset{\text{Minnie}}{\downarrow}}{} \quad \underset{\underset{\text{Bunny}}{\downarrow}}{}$$

$$\text{Add } \frac{x}{41} + \frac{x+1}{41} + \frac{2x}{41} \text{ which equals } \frac{4x+1}{41}.$$

In Chapter 0, we combined fractions having the same denominator by combining their numerators and putting the result over the common denominator. Above, we use the same process to add two rational expressions with the same denominator. Here's another example:

EXAMPLE 1 Add $\frac{5}{x} + \frac{3}{x}$.

Solution Adding numerators, we have

$$\frac{5}{x} + \frac{3}{x} = \frac{8}{x}$$

Practice Problems

1. Add $\frac{2}{x} + \frac{7}{x}$.

Answer

1. $\frac{9}{x}$

2. Add $\dfrac{x}{x^2 - 49} + \dfrac{7}{x^2 - 49}$.

EXAMPLE 2 Add $\dfrac{x}{x^2 - 9} + \dfrac{3}{x^2 - 9}$.

Solution Since both expressions have the same denominator, we add numerators and reduce to lowest terms.

$$\frac{x}{x^2 - 9} + \frac{3}{x^2 - 9} = \frac{x + 3}{x^2 - 9}$$

$$= \frac{x + 3}{(x + 3)(x - 3)}$$

$$= \frac{1}{x - 3}$$

Reduce to lowest terms by factoring the denominator and then dividing out the common factor x − 3.

B Adding and Subtracting Expressions with Different Denominators

Remember, it is the distributive property that allows us to add rational expressions by simply adding numerators. Because of this, we must begin all addition problems involving rational expressions by first making sure all the expressions have the least common denominator. Remember, the least common denominator (LCD) for a set of denominators is the simplest quantity that is exactly divisible by all the denominators.

Use the strategy for adding or subtracting any two fractions we learned previously to work the following examples.

3. Add $\dfrac{1}{21} + \dfrac{4}{15}$.

EXAMPLE 3 Add $\dfrac{1}{10} + \dfrac{3}{14}$.

Solution

Step 1: Find the LCD for 10 and 14. To do so, we factor each denominator and build the LCD from the factors.

$$\left. \begin{array}{l} 10 = 2 \cdot 5 \\ 14 = 2 \cdot 7 \end{array} \right\} \quad LCD = 2 \cdot 5 \cdot 7 = 70$$

We know the LCD is divisible by 10 because it contains the factors 2 and 5. It is also divisible by 14 because it contains the factors 2 and 7.

Step 2: Change to equivalent fractions that each have denominator 70. To accomplish this task, we multiply the numerator and denominator of each fraction by the factor of the LCD that is not also a factor of its denominator.

Original Fractions		Denominators in Factored Form		Multiply by Factor Needed to Obtain LCD		These Have the Same Value as the Original Fractions
$\dfrac{1}{10}$	=	$\dfrac{1}{2 \cdot 5}$	=	$\dfrac{1}{2 \cdot 5} \cdot \dfrac{7}{7}$	=	$\dfrac{7}{70}$
$\dfrac{3}{14}$	=	$\dfrac{3}{2 \cdot 7}$	=	$\dfrac{3}{2 \cdot 7} \cdot \dfrac{5}{5}$	=	$\dfrac{15}{70}$

The fraction $\frac{7}{70}$ has the same value as the fraction $\frac{1}{10}$. Likewise, the fractions $\frac{15}{70}$ and $\frac{3}{14}$ are equivalent; they have the same value.

Step 3: Add numerators and put the result over the LCD.

$$\frac{7}{70} + \frac{15}{70} = \frac{7 + 15}{70} = \frac{22}{70}$$

Step 4: Reduce to lowest terms.

$$\frac{22}{70} = \frac{11}{35} \qquad \textit{Divide numerator and denominator by 2.}$$

> *Note* If you have had difficulty in the past with addition and subtraction of fractions with different denominators, this is the time to get it straightened out. Go over Example 3 as many times as is necessary for you to understand the process.

Answers

2. $\dfrac{1}{x - 7}$

3. $\dfrac{11}{35}$

EXAMPLE 4 Subtract $\dfrac{3}{x} - \dfrac{1}{2}$.

Solution

Step 1: The LCD for x and 2 is $2x$. It is the smallest expression divisible by x and by 2.

Step 2: To change to equivalent expressions with the denominator $2x$, we multiply the first fraction by $\frac{2}{2}$ and the second by $\frac{x}{x}$.

$$\frac{3}{x} \cdot \frac{2}{2} = \frac{6}{2x}$$

$$\frac{1}{2} \cdot \frac{x}{x} = \frac{x}{2x}$$

Step 3: Subtracting numerators of the rational expressions in step 2, we have

$$\frac{6}{2x} - \frac{x}{2x} = \frac{6-x}{2x}$$

Step 4: Since $6 - x$ and $2x$ do not have any factors in common, we cannot reduce any further. Here is the complete problem:

$$\frac{3}{x} - \frac{1}{2} = \frac{3}{x} \cdot \frac{2}{2} - \frac{1}{2} \cdot \frac{x}{x}$$

$$= \frac{6}{2x} - \frac{x}{2x}$$

$$= \frac{6-x}{2x}$$

EXAMPLE 5 Add $\dfrac{5}{2x-6} + \dfrac{x}{x-3}$.

Solution If we factor $2x - 6$, we have $2x - 6 = 2(x - 3)$. We need only multiply the second rational expression in our problem by $\frac{2}{2}$ to have two expressions with the same denominator.

$$\frac{5}{2x-6} + \frac{x}{x-3} = \frac{5}{2(x-3)} + \frac{x}{x-3} \qquad \text{Factor the first denominator.}$$

$$= \frac{5}{2(x-3)} + \frac{2}{2}\left(\frac{x}{x-3}\right) \qquad \text{Change to equivalent rational expressions.}$$

$$= \frac{5}{2(x-3)} + \frac{2x}{2(x-3)} \qquad \text{Multiply.}$$

$$= \frac{2x+5}{2(x-3)} \qquad \text{Add numerators.}$$

EXAMPLE 6 Add $\dfrac{1}{x+4} + \dfrac{8}{x^2-16}$.

Solution After writing each denominator in factored form, we find that the least common denominator is $(x + 4)(x - 4)$. To change the first rational expression to an equivalent rational expression with the common denominator, we multiply its numerator and denominator by $x - 4$.

$$\frac{1}{x+4} + \frac{8}{x^2-16}$$

$$= \frac{1}{x+4} + \frac{8}{(x+4)(x-4)} \qquad \text{Factor each denominator.}$$

$$= \frac{1}{x+4} \cdot \frac{x-4}{x-4} + \frac{8}{(x+4)(x-4)} \qquad \text{Change to equivalent rational expressions.}$$

$$= \frac{x-4}{(x+4)(x-4)} + \frac{8}{(x+4)(x-4)} \qquad \text{Simplify.}$$

$$= \frac{x+4}{(x+4)(x-4)} \qquad \text{Add numerators.}$$

$$= \frac{1}{x-4} \qquad \text{Divide out common factor } x+4.$$

4. Subtract $\dfrac{2}{x} - \dfrac{1}{3}$.

Note The main idea in adding fractions is to write each fraction again with the LCD for a denominator. Once we have done that, we simply add numerators.

5. Add $\dfrac{7}{3x+6} + \dfrac{x}{x+2}$.

6. Add $\dfrac{1}{x-3} + \dfrac{-6}{x^2-9}$.

Answers

4. $\dfrac{6-x}{3x}$

5. $\dfrac{7+3x}{3(x+2)}$

6. $\dfrac{1}{x+3}$

7. Add $\dfrac{5}{x^2 - 7x + 12} + \dfrac{1}{x^2 - 9}$.

EXAMPLE 7 Add $\dfrac{2}{x^2 + 5x + 6} + \dfrac{x}{x^2 - 9}$.

Solution

Step 1: We factor each denominator and build the LCD from the factors.

$$\left.\begin{array}{l} x^2 + 5x + 6 = (x + 2)(x + 3) \\ x^2 - 9 = (x + 3)(x - 3) \end{array}\right\} \qquad LCD = (x + 2)(x + 3)(x - 3)$$

Step 2: Change to equivalent rational expressions.

$$\frac{2}{x^2 + 5x + 6} = \frac{2}{(x + 2)(x + 3)} \cdot \frac{(x - 3)}{(x - 3)} = \frac{2x - 6}{(x + 2)(x + 3)(x - 3)}$$

$$\frac{x}{x^2 - 9} = \frac{x}{(x + 3)(x - 3)} \cdot \frac{(x + 2)}{(x + 2)} = \frac{x^2 + 2x}{(x + 2)(x + 3)(x - 3)}$$

Step 3: Add numerators of the rational expressions produced in step 2.

$$\frac{2x - 6}{(x + 2)(x + 3)(x - 3)} + \frac{x^2 + 2x}{(x + 2)(x + 3)(x - 3)} = \frac{x^2 + 4x - 6}{(x + 2)(x + 3)(x - 3)}$$

The numerator and denominator do not have any factors in common.

8. Subtract $\dfrac{x + 3}{2x + 8} - \dfrac{4}{x^2 - 16}$.

EXAMPLE 8 Subtract $\dfrac{x + 4}{2x + 10} - \dfrac{5}{x^2 - 25}$.

Solution We begin by factoring each denominator.

$$\frac{x + 4}{2x + 10} - \frac{5}{x^2 - 25} = \frac{x + 4}{2(x + 5)} - \frac{5}{(x + 5)(x - 5)}$$

The LCD is $2(x + 5)(x - 5)$. Completing the problem, we have

$$= \frac{x + 4}{2(x + 5)} \cdot \frac{(x - 5)}{(x - 5)} + \frac{-5}{(x + 5)(x - 5)} \cdot \frac{2}{2}$$

$$= \frac{x^2 - x - 20}{2(x + 5)(x - 5)} + \frac{-10}{2(x + 5)(x - 5)}$$

$$= \frac{x^2 - x - 30}{2(x + 5)(x - 5)}$$

To see if this expression will reduce, we factor the numerator into $(x - 6)(x + 5)$.

$$= \frac{(x - 6)(x + 5)}{2(x + 5)(x - 5)}$$

$$= \frac{x - 6}{2(x - 5)}$$

Note In the last step, we reduced the rational expression to lowest terms by dividing out the common factor of $x + 5$.

Answers

7. $\dfrac{6x + 11}{(x - 4)(x - 3)(x + 3)}$

8. $\dfrac{x - 5}{2(x - 4)}$

EXAMPLE 9 Write an expression for the sum of a number and its reciprocal, and then simplify that expression.

Solution If we let $x =$ the number, then its reciprocal is $\frac{1}{x}$. To find the sum of the number and its reciprocal, we add them.

$$x + \frac{1}{x}$$

The first term x can be thought of as having a denominator of 1. Since the denominators are 1 and x, the least common denominator is x.

$$x + \frac{1}{x} = \frac{x}{1} + \frac{1}{x} \qquad \text{Write } x \text{ as } \frac{x}{1}.$$

$$= \frac{x}{1} \cdot \frac{x}{x} + \frac{1}{x} \qquad \text{The LCD is } x.$$

$$= \frac{x^2}{x} + \frac{1}{x}$$

$$= \frac{x^2 + 1}{x} \qquad \text{Add numerators.}$$

9. Write an expression for the difference of twice a number and its reciprocal. Then simplify.

GETTING READY FOR CLASS

After reading through the preceding section, respond in your own words and in complete sentences.

A. How do you add two rational expressions that have the same denominator?

B. What is the least common denominator for two fractions that include rational expressions?

C. What role does factoring play in finding a least common denominator when adding rational expressions?

D. Explain how, if possible, you would reduce the solution to an addition problem of two rational expressions.

Answer

9. $\frac{2x^2 - 1}{x}$

Vocabulary Review

Choose the correct words to fill in the blanks below.

different reduce same least common denominator

1. To add or subtract rational expressions with the _____ denominator, combine numerators and put the result over the common denominator.

2. The _____ for a set of denominators is the simplest quantity that is exactly divisible by all denominators.

3. When adding and subtracting expressions with _____ denominators, we must first make sure all the expressions have the least common denominator.

4. The last step when adding or subtracting rational expressions is to _____ to lowest terms.

Problems

A Find the following sums or differences.

1. $\dfrac{3}{x} + \dfrac{4}{x}$

2. $\dfrac{5}{x} + \dfrac{3}{x}$

3. $\dfrac{9}{a} - \dfrac{5}{a}$

4. $\dfrac{8}{a} - \dfrac{7}{a}$

5. $\dfrac{1}{x+1} + \dfrac{x}{x+1}$

6. $\dfrac{x}{x-3} - \dfrac{3}{x-3}$

7. $\dfrac{y^2}{y-1} - \dfrac{1}{y-1}$

8. $\dfrac{y^2}{y+3} - \dfrac{9}{y+3}$

9. $\dfrac{x^2}{x+2} + \dfrac{4x+4}{x+2}$

10. $\dfrac{x^2-6x}{x-3} + \dfrac{9}{x-3}$

11. $\dfrac{x^2}{x-2} - \dfrac{4x-4}{x-2}$

12. $\dfrac{x^2}{x-5} - \dfrac{10x-25}{x-5}$

13. $\dfrac{x+2}{x+6} - \dfrac{x-4}{x+6}$

14. $\dfrac{x+5}{x+2} - \dfrac{x+3}{x+2}$

15. $\dfrac{2}{x^2-4} - \dfrac{x}{x^2-4}$

16. $\dfrac{3}{x^2-9} - \dfrac{x}{x^2-9}$

B Find the following sums or differences. Remember to convert fractions to the same denominator before combining.

17. $\dfrac{y}{2} - \dfrac{2}{y}$

18. $\dfrac{3}{y} + \dfrac{y}{3}$

19. $\dfrac{1}{2} + \dfrac{a}{3}$

20. $\dfrac{2}{3} + \dfrac{2a}{5}$

21. $\dfrac{x}{x+1} + \dfrac{3}{4(x+1)}$

22. $\dfrac{x}{x-3} + \dfrac{1}{3}$

$LCD = 4(x+1)$

$\dfrac{4x}{4(x+1)} + \dfrac{3(x+1)}{4(x+1)}$

$\dfrac{4x+3x+3}{4(x+1)} = \dfrac{7x+3}{4(x+1)}$

23. $\dfrac{x+1}{x-2} - \dfrac{4x+7}{5x-10}$

24. $\dfrac{3x+1}{2x-6} - \dfrac{x+2}{x-3}$

25. $\dfrac{4x-2}{3x+12} - \dfrac{x-2}{x+4}$

26. $\dfrac{6x+5}{5x-25} - \dfrac{x+2}{x-5}$

27. $\dfrac{6}{x(x-2)} + \dfrac{3}{x}$

28. $\dfrac{10}{x(x+5)} - \dfrac{2}{x}$

29. $\dfrac{4}{a} - \dfrac{12}{a^2+3a}$

30. $\dfrac{5}{a} + \dfrac{20}{a^2-4a}$

31. $\dfrac{2}{x+5} - \dfrac{10}{x^2-25}$

32. $\dfrac{6}{x^2-1} + \dfrac{3}{x+1}$

33. $\dfrac{x-4}{x-3} + \dfrac{6}{x^2-9}$

34. $\dfrac{x+1}{x-1} - \dfrac{4}{x^2-1}$

35. $\dfrac{a-4}{a-3} + \dfrac{5}{a^2-a-6}$

36. $\dfrac{a+2}{a+1} + \dfrac{7}{a^2-5a-6}$

37. $\dfrac{8}{x^2-16} - \dfrac{7}{x^2-x-12}$

38. $\dfrac{6}{x^2-9} - \dfrac{5}{x^2-x-6}$

39. $\dfrac{4y}{y^2+6y+5} - \dfrac{3y}{y^2+5y+4}$

40. $\dfrac{3y}{y^2+7y+10} - \dfrac{2y}{y^2+6y+8}$

41. $\dfrac{4x+1}{x^2+5x+4} - \dfrac{x+3}{x^2+4x+3}$

42. $\dfrac{2x-1}{x^2+x-6} - \dfrac{x+2}{x^2+5x+6}$

43. $\dfrac{1}{x} + \dfrac{x}{3x+9} - \dfrac{3}{x^2+3x}$

44. $\dfrac{1}{x} + \dfrac{x}{2x+4} - \dfrac{2}{x^2+2x}$

45. $\dfrac{5}{2x-1} + \dfrac{6x+7}{1-4x^2}$

46. $\dfrac{x-2}{2x+3} - \dfrac{2x-11}{3-4x-4x^2}$

47. Work each problem according to the instructions.

 a. Multiply: $\dfrac{4}{9} \cdot \dfrac{1}{6}$.

 b. Divide: $\dfrac{4}{9} \div \dfrac{1}{6}$.

 c. Add: $\dfrac{4}{9} + \dfrac{1}{6}$.

 d. Multiply: $\dfrac{x+2}{x-2} \cdot \dfrac{3x+10}{x^2-4}$.

 e. Divide: $\dfrac{x+2}{x-2} \div \dfrac{3x+10}{x^2-4}$.

 f. Subtract: $\dfrac{x+2}{x-2} - \dfrac{3x+10}{x^2-4}$.

48. Work each problem according to the instructions.

 a. Multiply: $\dfrac{9}{25} \cdot \dfrac{1}{15}$.

 b. Divide: $\dfrac{9}{25} \div \dfrac{1}{15}$.

 c. Subtract: $\dfrac{9}{25} - \dfrac{1}{15}$.

 d. Multiply: $\dfrac{3x-2}{3x+2} \cdot \dfrac{15x+6}{9x^2-4}$.

 e. Divide: $\dfrac{3x-2}{3x+2} \div \dfrac{15x+6}{9x^2-4}$.

 f. Subtract: $\dfrac{3x+2}{3x-2} - \dfrac{15x+6}{9x^2-4}$.

Complete the following tables.

49.

Number x	Reciprocal $\dfrac{1}{x}$	Sum $1 + \dfrac{1}{x}$	Quotient $\dfrac{x + 1}{x}$
1			
2			
3			
4			

50.

Number x	Reciprocal $\dfrac{1}{x}$	Difference $1 - \dfrac{1}{x}$	Quotient $\dfrac{x - 1}{x}$
1			
2			
3			
4			

Add or subtract as indicated.

51. $1 + \dfrac{1}{x + 2}$

52. $1 - \dfrac{1}{x + 2}$

53. $1 - \dfrac{1}{x + 3}$

54. $1 + \dfrac{1}{x + 3}$

Applying the Concepts

55. Number Problem Write an expression for the sum of a number and twice its reciprocal. Then, simplify that expression. (If the reciprocal of a number is $\frac{1}{x}$, then twice that is $\frac{2}{x}$, not $\frac{1}{2x}$.)

56. Number Problem Write an expression for the sum of a number and 3 times its reciprocal. Then, simplify that expression.

57. Number Problem One number is twice another. Write an expression for the sum of their reciprocals. Then, simplify that expression. Hint: The numbers are x and $2x$. Their reciprocals are, respectively, $\frac{1}{x}$ and $\frac{1}{2x}$.

58. Number Problem One number is three times another. Write an expression for the sum of their reciprocals. Then, simplify that expression.

59. One number is three more than another. Write an expression for the sum of their reciprocals, then simplify the expression.

60. One number is two less than another. Write an expression for the sum of their reciprocals, then simplify the expression.

Getting Ready for the Next Section

Compute each value.

61. $6\left(\dfrac{1}{2}\right)$

62. $10\left(\dfrac{1}{5}\right)$

63. $\dfrac{0}{5}$

64. $\dfrac{0}{2}$

65. $\dfrac{5}{0}$

66. $\dfrac{2}{0}$

67. $1 - \dfrac{5}{2}$

68. $1 - \dfrac{5}{3}$

Use the distributive property to simplify each expression.

69. $6\left(\dfrac{x}{3} + \dfrac{5}{2}\right)$

70. $10\left(\dfrac{x}{2} + \dfrac{3}{5}\right)$

71. $x^2\left(1 - \dfrac{5}{x}\right)$

72. $x^2\left(1 - \dfrac{3}{x}\right)$

Solve.

73. $2x + 15 = 3$

74. $15 = 3x - 3$

75. $-2x - 9 = x - 3$

76. $a^2 - a - 22 = -2$

Find the Mistake

Each sentence below contains a mistake. Circle the mistake and write the correct word or expression on the line provided.

1. To add fraction with same denominator, place the product of the numerators over the common denominator. _____

2. When subtracting the fractions $\dfrac{x}{(x^2 - 36)}$ and $\dfrac{6}{(x^2 - 36)}$, factor the common denominator and divide out the common factor $x + 6$. _____

3. To add the fractions $\dfrac{8}{4x}$ and $\dfrac{9}{16x^2}$, first multiply the denominator of the first fraction by $16x^2$ to obtain the LCD. _____

4. The lowest common denominator for a problem that finds the difference between a number squared and three time the number's reciprocal is 3. _____

Landmark Review: Checking Your Progress

Simplify the following expressions. State any restrictions on the variable.

1. $\dfrac{3a^2 - 27}{6a - 18}$

2. $\dfrac{y - 3}{y^2 - 6y + 9}$

3. $\dfrac{3x^3 - 6x^2 - 45x}{x^2 + 5x + 6}$

4. $\dfrac{5x^2 + 22x + 8}{25x^2 - 4}$

Perform the indicated operations.

5. $\dfrac{3a^4}{6a + 18} \cdot \dfrac{a + 3}{a^3}$

6. $\dfrac{4x^2 - 12x}{x^2 - 1} \cdot \dfrac{x^2 + x}{4x}$

7. $\dfrac{y^2 - 3y}{2y^2 - 8} \div \dfrac{y^2 - 25}{y^2 + 3y - 10}$

8. $(x^2 + 8x + 15)\dfrac{2x}{x + 5}$

9. $\dfrac{8}{4x + 16} \div \dfrac{2}{x + 4}$

10. $\dfrac{4}{y + 3} + \dfrac{24}{y^2 - 9}$

11. $\dfrac{a + 5}{a + 4} + \dfrac{8}{a^2 - 16}$

12. $\dfrac{x^2}{x + 4} - \dfrac{16}{x + 4}$

13. $\dfrac{x + 2}{x - 1} - \dfrac{6}{x^2 - 1}$

6.4 Equations Involving Rational Expressions

Image © NASA JPL

For 520 days, 6 men will simulate a trip to Mars and back. The men will be locked inside a large facility that was built to look like the interior of a space shuttle. Isolated from the outside world, the men will study the effects of the mission on their own bodies. The mission includes a trip to the planet for x number of days, y number of days to perform mock experiments on the planet, and then z number of days to fly back to Earth. We can set up the following rational expressions to represent the days in each stage of the mission:

Trip to Mars
Trip to Earth

$$\frac{x}{520} + \frac{y}{520} + \frac{z}{520} = \text{Total days of mission}$$

Experiments

Then if given the value for two of the variables, we can determine the value of the third variable using a rational equation. Equations that involve rational expressions are the focus of this section.

A Solving Rational Equations

The first step in solving an equation that contains one or more rational expressions is to find the LCD for all denominators in the equation. Once the LCD has been found, we multiply both sides of the equation by it. The resulting equation should be equivalent to the original one (unless we inadvertently multiplied by zero) and free from any denominators except the number 1.

Practice Problems

1. Solve $\frac{x}{2} + \frac{3}{5} = \frac{1}{5}$.

EXAMPLE 1 Solve $\frac{x}{3} + \frac{5}{2} = \frac{1}{2}$.

Solution The LCD for 3 and 2 is 6. If we multiply both sides by 6, we have

$$6\left(\frac{x}{3} + \frac{5}{2}\right) = 6\left(\frac{1}{2}\right) \qquad \text{Multiply both sides by 6.}$$

$$6\left(\frac{x}{3}\right) + 6\left(\frac{5}{2}\right) = 6\left(\frac{1}{2}\right) \qquad \text{Distributive property}$$

$$2x + 15 = 3$$

$$2x = -12$$

$$x = -6$$

We can check our solution by replacing x with -6 in the original equation.

$$-\frac{6}{3} + \frac{5}{2} \overset{?}{=} \frac{1}{2}$$

$$\frac{1}{2} = \frac{1}{2}$$

The solution is -6.

Answer

1. $-\frac{4}{5}$

Multiplying both sides of an equation containing fractions by the LCD clears the equation of all denominators, because the LCD has the property that all denominators will divide it evenly.

EXAMPLE 2 Solve $\dfrac{3}{x-1} = \dfrac{3}{5}$.

Solution The LCD for $(x-1)$ and 5 is $5(x-1)$. Multiplying both sides by $5(x-1)$, we have

$$5(x-1) \cdot \frac{3}{x-1} = 5(x-1) \cdot \frac{3}{5}$$

$$5 \cdot 3 = (x-1) \cdot 3$$

$$15 = 3x - 3$$

$$18 = 3x$$

$$6 = x$$

If we substitute $x = 6$ into the original equation, we have

$$\frac{3}{6-1} \overset{?}{=} \frac{3}{5}$$

$$\frac{3}{5} = \frac{3}{5}$$

The solution set is 6.

2. Solve $\dfrac{4}{x-5} = \dfrac{4}{3}$.

EXAMPLE 3 Solve $1 - \dfrac{5}{x} = \dfrac{-6}{x^2}$.

Solution The LCD is x^2. Multiplying both sides by x^2, we have

$$x^2\left(1 - \frac{5}{x}\right) = x^2\left(\frac{-6}{x^2}\right) \qquad \text{Multiply both sides by } x^2.$$

$$x^2(1) - x^2\left(\frac{5}{x}\right) = x^2\left(\frac{-6}{x^2}\right) \qquad \text{Apply the distributive property to the left side.}$$

$$x^2 - 5x = -6 \qquad \text{Simplify each side.}$$

We have a quadratic equation, which we write in standard form, factor, and solve as we did in Chapter 5.

$$x^2 - 5x + 6 = 0 \qquad\qquad \text{Standard form}$$

$$(x-2)(x-3) = 0 \qquad\qquad \text{Factor.}$$

$$x - 2 = 0 \quad \text{or} \quad x - 3 = 0 \qquad \text{Set factors equal to 0.}$$

$$x = 2 \qquad\qquad x = 3$$

The two possible solutions are 2 and 3. Checking each in the original equation, we find they both give true statements. They are both solutions to the original equation.

Check $x = 2$ Check $x = 3$

$$1 - \frac{5}{2} \overset{?}{=} \frac{-6}{4} \qquad\qquad 1 - \frac{5}{3} \overset{?}{=} \frac{-6}{9}$$

$$\frac{2}{2} - \frac{5}{2} = -\frac{3}{2} \qquad\qquad \frac{3}{3} - \frac{5}{3} = -\frac{2}{3}$$

$$-\frac{3}{2} = -\frac{3}{2} \qquad\qquad -\frac{2}{3} = -\frac{2}{3}$$

3. Solve $1 - \dfrac{3}{x} = \dfrac{-2}{x^2}$.

Answers

2. 8

3. 1 or 2

4. Solve $\dfrac{x}{x^2-4} - \dfrac{2}{x-2} = \dfrac{1}{x+2}$.

EXAMPLE 4 Solve $\dfrac{x}{x^2-9} - \dfrac{3}{x-3} = \dfrac{1}{x+3}$.

Solution The factors of $x^2 - 9$ are $(x+3)(x-3)$. The LCD then, is $(x+3)(x-3)$.

$$(x+3)(x-3) \cdot \frac{x}{(x+3)(x-3)} + (x+3)(x-3) \cdot \frac{-3}{x-3}$$

$$= (x+3)(x-3) \cdot \frac{1}{x+3}$$

$$x + (x+3)(-3) = (x-3)1$$

$$x + (-3x) + (-9) = x - 3$$

$$-2x - 9 = x - 3$$

$$-3x = 6$$

$$x = -2$$

The solution is -2. It checks when replaced for x in the original equation.

5. Solve $\dfrac{x}{x-5} + \dfrac{5}{2} = \dfrac{5}{x-5}$.

EXAMPLE 5 Solve $\dfrac{x}{x-3} + \dfrac{3}{2} = \dfrac{3}{x-3}$.

Solution We begin by multiplying each term on both sides of the equation by $2(x-3)$.

$$2(x-3) \cdot \frac{x}{x-3} + 2(x-3) \cdot \frac{3}{2} = 2(x-3) \cdot \frac{3}{x-3}$$

$$2x + (x-3) \cdot 3 = 2 \cdot 3$$

$$2x + 3x - 9 = 6$$

$$5x - 9 = 6$$

$$5x = 15$$

$$x = 3$$

Our only possible solution is $x = 3$. If we substitute $x = 3$ into our original equation, we get

$$\frac{3}{3-3} + \frac{3}{2} \stackrel{?}{=} \frac{3}{3-3}$$

$$\frac{3}{0} + \frac{3}{2} = \frac{3}{0}$$

Two of the terms are undefined, so the equation is meaningless. What has happened is that we have multiplied both sides of the original equation by zero. The equation produced by doing this is not equivalent to our original equation. We must always check our solution when we multiply both sides of an equation by an expression containing the variable to make sure we have not multiplied both sides by zero.

Our original equation has no solution; that is, there is no real number x such that

$$\frac{x}{x-3} + \frac{3}{2} = \frac{3}{x-3}$$

There is no solution to the equation.

6. Solve $\dfrac{a+2}{a^2+3a} = \dfrac{-2}{a^2-9}$.

EXAMPLE 6 Solve $\dfrac{a+4}{a^2+5a} = \dfrac{-2}{a^2-25}$.

Solution Factoring each denominator, we have

$$a^2 + 5a = a(a+5)$$

$$a^2 - 25 = (a+5)(a-5)$$

The LCD is $a(a + 5)(a - 5)$. Multiplying both sides of the equation by the LCD gives us

$$a(a + 5)(a - 5) \cdot \frac{a + 4}{a(a + 5)} = \frac{-2}{(a + 5)(a - 5)} \cdot a(a + 5)(a - 5)$$

$$(a - 5)(a + 4) = -2a$$

$$a^2 - a - 20 = -2a$$

The result is a quadratic equation, which we write in standard form, factor, and solve.

$$a^2 + a - 20 = 0 \qquad \qquad \text{Add } 2a \text{ to both sides.}$$

$$(a + 5)(a - 4) = 0 \qquad \qquad \text{Factor.}$$

$$a + 5 = 0 \qquad \text{or} \quad a - 4 = 0 \qquad \text{Set each factor equal to 0.}$$

$$a = -5 \qquad \qquad a = 4$$

The two possible solutions are -5 and 4. There is no problem with the 4. It checks when substituted for a in the original equation. However, -5 is not a solution. Substituting -5 into the original equation gives

$$\frac{-5 + 4}{(-5)^2 + 5(-5)} \overset{?}{=} \frac{-2}{(-5)^2 - 25}$$

$$\frac{-1}{0} = \frac{-2}{0}$$

This indicates -5 is not a solution. The solution is 4.

GETTING READY FOR CLASS

After reading through the preceding section, respond in your own words and in complete sentences.

A. What is the first step in solving an equation that contains rational expressions?

B. Explain how to find the LCD used to clear an equation of fractions.

C. When will an equation containing rational expressions have more than one possible solution?

D. When do you check for extraneous solutions to an equation containing rational expressions?

EXERCISE SET 6.4

Vocabulary Review

Choose the correct words to fill in the blanks below.

multiply denominators least common denominator quadratic

1. The first step to solving a rational equation is to find the _____ for all denominators in the equation.
2. The second step to solving a rational equation is to _____ both sides of the equation by the LCD.
3. Multiplying a rational equation by the LCD clears the equation of all _____.
4. If we get a _____ equation once we have multiplied a rational equation by the LCD, then we write the equation in standard form, factor, and solve.

Problems

A Solve the following equations. Be sure to check each answer in the original equation if you multiply both sides by an expression that contains the variable.

1. $\dfrac{x}{3} + \dfrac{1}{2} = -\dfrac{1}{2}$

2. $\dfrac{x}{2} + \dfrac{4}{3} = -\dfrac{2}{3}$

3. $\dfrac{4}{a} = \dfrac{1}{5}$

4. $\dfrac{2}{3} = \dfrac{6}{a}$

5. $\dfrac{3}{x} + 1 = \dfrac{2}{x}$

6. $\dfrac{4}{x} + 3 = \dfrac{1}{x}$

7. $\dfrac{3}{a} - \dfrac{2}{a} = \dfrac{1}{5}$

8. $\dfrac{7}{a} + \dfrac{1}{a} = 2$

9. $\dfrac{3}{x} + 2 = \dfrac{1}{2}$

10. $\dfrac{5}{x} + 3 = \dfrac{4}{3}$

11. $\dfrac{1}{y} - \dfrac{1}{2} = -\dfrac{1}{4}$

12. $\dfrac{3}{y} - \dfrac{4}{5} = -\dfrac{1}{5}$

13. $1 - \dfrac{8}{x} = \dfrac{-15}{x^2}$

14. $1 - \dfrac{3}{x} = \dfrac{-2}{x^2}$

15. $\dfrac{x}{2} - \dfrac{4}{x} = -\dfrac{7}{2}$

16. $\dfrac{x}{2} - \dfrac{5}{x} = -\dfrac{3}{2}$

(handwritten notes) LCD = x² , x≠0

$x^2 - \dfrac{8x^2}{x} = \dfrac{-15x^2}{x^2}$

$x^2 - 8x = -15$

$x^2 - 8x + 15 = 0 = (x-3)(x-5)$

$x = 3$, $x = 5$

17. $\dfrac{x-3}{2} + \dfrac{2x}{3} = \dfrac{5}{6}$

18. $\dfrac{x-2}{3} + \dfrac{5x}{2} = 5$

19. $\dfrac{x+1}{3} + \dfrac{x-3}{4} = \dfrac{1}{6}$

20. $\dfrac{x+2}{3} + \dfrac{x-1}{5} = -\dfrac{3}{5}$

21. $\dfrac{6}{x+2} = \dfrac{3}{5}$

22. $\dfrac{4}{x+3} = \dfrac{1}{2}$

23. $\dfrac{3}{y-2} = \dfrac{2}{y-3}$

24. $\dfrac{5}{y+1} = \dfrac{4}{y+2}$

$LCD = 3(x-2)$

25. $\dfrac{x}{x-2} + \dfrac{2}{3} = \dfrac{2}{x-2}$

$\dfrac{3(x-2)x}{(x-2)} + \dfrac{3\cdot 2(x-2)}{3(x-2)} = \dfrac{2(x-2)}{(x-2)}$

$= 3x + 2x - 4 = 6$

26. $\dfrac{x}{x-5} + \dfrac{1}{5} = \dfrac{5}{x-5}$

27. $\dfrac{x}{x-2} + \dfrac{3}{2} = \dfrac{9}{2(x-2)}$

28. $\dfrac{x}{x+1} + \dfrac{4}{5} = \dfrac{-14}{5(x+1)}$

$5x - 4 = 6$

$5x = 10$

$x = 2$

No Solution

$LCD = 2(x+3)(x+2)$

29. $\dfrac{5(x+2)(x+3)}{x+2} + \dfrac{1(x+3)(x+2)}{x+3} = \dfrac{-1(x+3)(x+2)}{x^2+5x+6}$

$\dfrac{5}{x+2} + \dfrac{1}{x+3} = \dfrac{1}{(x+3)(x+2)}$

$5x+15 + x+2 = -1 = 6x+17=-1 = \dfrac{6x}{6} = \dfrac{-18}{6} \; \boxed{=-3} \;$ No Solution

30. $\dfrac{3}{x-1} + \dfrac{2}{x+3} = \dfrac{-3}{x^2+2x-3}$

31. $\dfrac{8}{x^2-4} + \dfrac{3}{x+2} = \dfrac{1}{x-2}$

32. $\dfrac{10}{x^2-25} - \dfrac{1}{x-5} = \dfrac{3}{x+5}$

33. $\dfrac{a}{2} + \dfrac{3}{a-3} = \dfrac{a}{a-3}$

34. $\dfrac{a}{2} + \dfrac{4}{a-4} = \dfrac{a}{a-4}$

35. $\dfrac{6}{y^2-4} = \dfrac{4}{y^2+2y}$

36. $\dfrac{2}{y^2-9} = \dfrac{5}{y^2+3y}$

37. $\dfrac{2}{a^2-9} = \dfrac{3}{a^2+a-12}$

38. $\dfrac{2}{a^2-1} = \dfrac{6}{a^2-2a-3}$

39. $\dfrac{3x}{x-5} - \dfrac{2x}{x+1} = \dfrac{-42}{x^2-4x-5}$

40. $\dfrac{4x}{x-4} - \dfrac{3x}{x-2} = \dfrac{-3}{x^2-6x+8}$

41. $\dfrac{2x}{x+2} = \dfrac{x}{x+3} - \dfrac{3}{x^2+5x+6}$

42. $\dfrac{3x}{x-4} = \dfrac{2x}{x-3} + \dfrac{6}{x^2-7x+12}$

43. Solve each equation.

a. $5x - 1 = 0$ **d.** $\dfrac{5}{x} - 1 = \dfrac{2}{3}$

b. $\dfrac{5}{x} - 1 = 0$ **e.** $\dfrac{5}{x^2} + 5 = \dfrac{26}{x}$

c. $\dfrac{x}{5} - 1 = \dfrac{2}{3}$

44. Solve each equation.

a. $2x - 3 = 0$ **d.** $\dfrac{3}{x} - 2 = \dfrac{1}{2}$

b. $2 - \dfrac{3}{x} = 0$ **e.** $\dfrac{1}{x} + \dfrac{3}{x^2} = 2$

c. $\dfrac{x}{3} - 2 = \dfrac{1}{2}$

45. Work each problem according to the instructions.

a. Divide: $\dfrac{7}{a^2 - 5a - 6} \div \dfrac{a + 2}{a + 1}$.

b. Add: $\dfrac{7}{a^2 - 5a - 6} + \dfrac{a + 2}{a + 1}$.

c. Solve: $\dfrac{7}{a^2 - 5a - 6} + \dfrac{a + 2}{a + 1} = 2$.

46. Work each problem according to the instructions.

a. Divide: $\dfrac{6}{x^2 - 9} \div \dfrac{x - 4}{x - 3}$.

b. Add: $\dfrac{6}{x^2 - 9} + \dfrac{x - 4}{x - 3}$.

c. Solve: $\dfrac{6}{x^2 - 9} + \dfrac{x - 4}{x - 3} = \dfrac{4}{3}$.

Getting Ready for the Next Section

Solve each equation.

47. $\dfrac{1}{x} + \dfrac{1}{2x} = \dfrac{9}{2}$

48. $\dfrac{50}{x + 5} = \dfrac{30}{x - 5}$

49. $\dfrac{1}{10} - \dfrac{1}{15} = \dfrac{1}{x}$

50. $\dfrac{15}{x} + \dfrac{15}{x + 20} = 2$

Find the value of $y = \dfrac{-6}{x}$ for the given value of x.

51. $x = -6$ **52.** $x = -3$ **53.** $x = 2$ **54.** $x = 1$

Find the Mistake

Each sentence below contains a mistake. Circle the mistake and write the correct word(s) or expression on the line provided.

1. The first step to solving an equation that contains rational expressions is to find the sum of all the numerators in the equation. _____

2. To solve the problem $\dfrac{x}{(x + 7)} + \dfrac{7}{2} = \dfrac{7}{x} + 7$, first multiply each term of the equation by 2. _____

3. The LCD for the equation $\dfrac{x + 2}{x^2 + 8x} = \dfrac{-2}{x^2 - 64}$ is $x(x + 2)(x + 8)(x - 8)$. _____

4. When checking a solution, if one or more terms in the equation are undefined, then there is no LCD that makes the equation a true statement. _____

6.5 Applications of Rational Equations

©iStockphoto.com/YuriyS

OBJECTIVES

A Solve applications whose solutions depend on solving an equation containing rational expressions.

B Convert between units using unit analysis

C Graph an equation involving a rational expression in two variables.

D Solve formulas involving rational expressions.

KEY WORDS

rational formulas

In May 2010, an American adventurer fulfilled a lifelong dream of crossing the English Channel by balloon—helium balloons, that is. Jonathon Trappe is a registered pilot for lighter-than-air systems. He strapped himself to a chair outfitted with safety equipment and hung from the bundle of large colorful balloons. Then he took to the skies from an English field. He floated over the White Cliffs of Dover, continued across the English Channel, and landed in a French cabbage patch without incident.

Imagine yourself trying to recreate Trappe's helium balloon flight across the English Channel. The flight's distance was approximately 21 miles, and the complete trip took 5 hours. If the speed that Trappe completed his flight was 5 miles per hour faster than yours, how fast was your speed? This is an example of an application that involves a rational equation. We will work similar problems in this section.

A Applications Involving Rational Expressions

In this section, we will solve some word problems whose equations involve rational expressions. Like the other word problems we have encountered, the more you work with them, the easier they become.

EXAMPLE 1 One number is twice another. The sum of their reciprocals is $\frac{9}{2}$. Find the two numbers.

Solution Let x = the smaller number. The larger then must be $2x$. Their

reciprocals are $\frac{1}{x}$ and $\frac{1}{2x}$, respectively. An equation that describes the situation is

$$\frac{1}{x} + \frac{1}{2x} = \frac{9}{2}$$

We can multiply both sides by the LCD $2x$ and then solve the resulting equation.

$$2x\left(\frac{1}{x}\right) + 2x\left(\frac{1}{2x}\right) = 2x\left(\frac{9}{2}\right)$$

$$2 + 1 = 9x$$

$$3 = 9x$$

$$x = \frac{3}{9} = \frac{1}{3}$$

The smaller number is $\frac{1}{3}$. The other number is twice as large, or $\frac{2}{3}$. If we add their reciprocals, we have

$$\frac{3}{1} + \frac{3}{2} = \frac{6}{2} + \frac{3}{2} = \frac{9}{2}$$

The solutions check with the original problem.

Practice Problems

1. One number is three times another. The sum of their reciprocals is $\frac{4}{9}$. Find the numbers.

Answers
1. 3, 9

2. A boat travels 26 miles up a river in the same amount of time it takes to travel 34 miles down the same river. If the current is 2 miles per hour, what is the speed of the boat in still water? (Hint: Begin by filling in the table that follows.)

	d	r	t
Upstream			
Downstream			

Note For the sake of simplicity we have represented our four unknowns in terms of a single variable, as opposed to four different variables, which would unnecessarily complicate our process.

Note There are two things to note about this problem. The first is that to use the rate equation we learned earlier and solve the equation $d = r \cdot t$ for t, we divide each side by r, like this:

$$\frac{d}{r} = \frac{r \cdot t}{r}$$

$$\frac{d}{r} = t$$

The second thing is this: The speed of the boat in still water is the rate at which it would be traveling if there were no current; that is, it is the speed of the boat through the water. Since the water itself is moving at 5 miles per hour, the boat is going 5 miles per hour slower when it travels against the current and 5 miles per hour faster when it travels with the current.

EXAMPLE 2 A boat travels 30 miles up a river in the same amount of time it takes to travel 50 miles down the same river. If the current is 5 miles per hour, what is the speed of the boat in still water?

Solution The easiest way to work a problem like this is with a table. The top row of the table is labeled with d for distance, r for rate, and t for time. The left column of the table is labeled with the two trips: upstream and downstream. Here is what the table looks like:

	d	r	t
Upstream			
Downstream			

The next step is to read the problem over again and fill in as much of the table as we can with the information in the problem. The distance the boat travels upstream is 30 miles and the distance downstream is 50 miles. Since we are asked for the speed of the boat in still water, we will let that be x. If the speed of the boat in still water is x, then its speed upstream (against the current) must be $x - 5$, and its speed downstream (with the current) must be $x + 5$. Putting these four quantities into the appropriate positions in the table, we have

	d	r	t
Upstream	30	$x - 5$	
Downstream	50	$x + 5$	

The last positions in the table are filled in by using an equivalent form of the rate equation $t = \dfrac{d}{r}$.

	d	r	t
Upstream	30	$x - 5$	$\dfrac{30}{x-5}$
Downstream	50	$x + 5$	$\dfrac{50}{x+5}$

Reading the problem again, we find that the time for the trip upstream is equal to the time for the trip downstream. Setting these two quantities equal to each other, we have our equation:

$$\text{Time (downstream)} = \text{Time (upstream)}$$

$$\frac{50}{x+5} = \frac{30}{x-5}$$

The LCD is $(x + 5)(x - 5)$. We multiply both sides of the equation by the LCD to clear it of all denominators. Here is the solution:

$$(x+5)(x-5) \cdot \frac{50}{x+5} = (x+5)(x-5) \cdot \frac{30}{x-5}$$

$$50x - 250 = 30x + 150$$

$$20x = 400$$

$$x = 20$$

The speed of the boat in still water is 20 miles per hour.

Answer

2. 15 miles per hour

3. Repeat Example 3 if Tina's downhill speed is 8 miles per hour faster than her uphill speed.

EXAMPLE 3 Tina is training for a triathlon. To train for the bicycle portion, she rides her bike 15 miles up hill and then 15 miles back down. The complete trip takes her 2 hours. If her downhill speed is 20 miles per hour faster than her uphill speed, how fast does she ride uphill?

Solution Again, we make a table. As in the previous example, we label the top row with distance, rate, and time. We label the left column with the two trips, uphill and downhill.

	d	r	t
Uphill			
Downhill			

Next, we fill in the table with as much information as we can from the problem. We know the distance traveled is 15 miles uphill and 15 miles downhill, which allows us to fill in the distance column. To fill in the rate column, we first note that she rides 20 miles per hour faster downhill than uphill. Therefore, if we let x equal her rate uphill, then her rate downhill is $x + 20$. Filling in the table with this information gives us

	d	r	t
Uphill	15	x	
Downhill	15	$x + 20$	

Since time is distance divided by rate, $t = \frac{d}{r}$, we can fill in the last column in the table.

	d	r	t
Uphill	15	x	$\frac{15}{x}$
Downhill	15	$x + 20$	$\frac{15}{x + 20}$

Rereading the problem, we find that the total time (the time riding uphill plus the time riding downhill) is two hours. We write our equation as follows:

$$\text{Time (uphill)} + \text{Time (downhill)} = 2$$

$$\frac{15}{x} + \frac{15}{x + 20} = 2$$

We solve this equation for x by first finding the LCD and then multiplying each term in the equation by it to clear the equation of all denominators. Our LCD is $x(x + 20)$. Here is our solution:

$$x(x + 20)\frac{15}{x} + x(x + 20)\frac{15}{x + 20} = 2 \cdot [x(x + 20)]$$

$$x(x + 20)\frac{15}{x} + x(x + 20)\frac{15}{x + 20} = 2 \cdot [x(x + 20)]$$

$$15(x + 20) + 15x = 2x(x + 20)$$

$$15x + 300 + 15x = 2x^2 + 40x$$

$$0 = 2x^2 + 10x - 300$$

$$0 = x^2 + 5x - 150 \qquad \text{Divide both sides by 2.}$$

$$x = -15 \quad \text{or} \quad x = 10$$

Since we cannot have a negative speed, our only solution is $x = 10$. Tina rides her bike at a rate of 10 miles per hour when going uphill. (Her downhill speed is $x + 20 = 30$ miles per hour.)

●

Answer

3. 12 miles/hour

4. A tub can be filled by the cold water faucet in 8 minutes. The drain empties the tub in 12 minutes. How long will it take to fill the tub if both the faucet and the drain are opened?

8 minutes to fill tub

12 minutes to empty tub

Note In solving a problem of this type, we have to assume that the thing doing the work (whether it is a pipe, a person, or a machine) is working at a constant rate; that is, as much work gets done in the first hour as is done in the last hour and any other hour in between.

EXAMPLE 4 An inlet pipe can fill a water tank in 10 hours, while an outlet pipe can empty the same tank in 15 hours. By mistake, both pipes are left open. How long will it take to fill the water tank with both pipes open?

Solution Let x = amount of time to fill the tank with both pipes open.

One method of solving this type of problem is to think in terms of how much of the job is done by a pipe in 1 hour.

1. If the inlet pipe fills the tank in 10 hours, then in 1 hour the inlet pipe fills $\frac{1}{10}$ of the tank.

2. If the outlet pipe empties the tank in 15 hours, then in 1 hour the outlet pipe empties $\frac{1}{15}$ of the tank.

3. If it takes x hours to fill the tank with both pipes open, then in 1 hour the tank is $\frac{1}{x}$ full.

Here is how we set up the equation. In 1 hour,

$$\underset{\substack{\text{Amount of water let} \\ \text{in by inlet pipe}}}{\frac{1}{10}} \quad - \quad \underset{\substack{\text{Amount of water let} \\ \text{out by outlet pipe}}}{\frac{1}{15}} \quad = \quad \underset{\substack{\text{Total amount of} \\ \text{water in tank}}}{\frac{1}{x}}$$

The LCD for our equation is $30x$. We multiply both sides by the LCD and solve.

$$30x\left(\frac{1}{10}\right) - 30x\left(\frac{1}{15}\right) = 30x\left(\frac{1}{x}\right)$$

$$3x - 2x = 30$$

$$x = 30$$

It takes 30 hours with both pipes open to fill the tank.

B Unit Analysis

Recall that unit analysis is a method of converting between units of measure by multiplying by the number 1. Here is our first illustration: Suppose you are flying in a commercial airliner and the pilot tells you the plane has reached its cruising altitude of 35,000 feet. How many miles is the plane above the ground?

If you know that 1 mile is 5,280 feet, then it is simply a matter of deciding what to do with the two numbers, 5,280 and 35,000. By using unit analysis, this decision is unnecessary.

$$35{,}000 \text{ feet} = \frac{35{,}000 \text{ feet}}{1} \cdot \frac{1 \text{ mile}}{5{,}280 \text{ feet}}$$

We treat the units common to the numerator and denominator in the same way we treat factors common to the numerator and denominator; common units can be divided out, just as common factors are. In the previous expression, we have feet common to the numerator and denominator. Dividing them out leaves us with miles only. Here is the complete problem:

$$35{,}000 \text{ feet} = \frac{35{,}000 \text{ \cancel{feet}}}{1} \cdot \frac{1 \text{ mile}}{5{,}280 \text{ \cancel{feet}}}$$

$$= \frac{35{,}000}{5{,}280} \text{ miles}$$

$$\approx 6.6 \text{ miles to the nearest tenth of a mile}$$

Answer

4. 24 minutes

EXAMPLE 5 The Mall of America in Minnesota covers 78 acres of land. If 1 square mile = 640 acres, how many square miles does the Mall of America cover? Round your answer to the nearest hundredth of a square mile.

Solution We are starting with acres and want to end up with square miles. We need to multiply by a conversion factor that will allow acres to divide out and leave us with square miles.

$$78 \text{ acres} = \frac{78 \cancel{\text{ acres}}}{1} \cdot \frac{1 \text{ square mile}}{640 \cancel{\text{ acres}}}$$

$$= \frac{78}{640} \text{ square miles}$$

$$\approx 0.12 \text{ square miles to the nearest hundredth}$$

The next example is a continuation of Example 8 from Section 6.1.

EXAMPLE 6 The Forest chair lift at the Northstar ski resort in Lake Tahoe is 5,603 feet long. If a ride on this chair lift takes 11 minutes, what is the average speed of the lift in miles per hour?

Solution First, we find the speed of the lift in feet per second, as we did in Example 8 of Section 6.1, by taking the ratio of distance to time.

$$\text{Rate} = \frac{\text{distance}}{\text{time}} = \frac{5{,}603 \text{ feet}}{11 \text{ minutes}} = \frac{5{,}603}{11} \text{ feet per minute}$$

$$= 509 \text{ feet per minute}$$

Next, we convert feet per minute to miles per hour. To do this, we need to know that

$$1 \text{ mile } = 5{,}280 \text{ feet}$$

$$1 \text{ hour } = 60 \text{ minutes}$$

$$\text{Speed } = 509 \text{ feet per minute} = \frac{509 \text{ feet}}{1 \cancel{\text{ minute}}} \cdot \frac{1 \text{ mile}}{5{,}280 \cancel{\text{ feet}}} \cdot \frac{60 \cancel{\text{ minutes}}}{1 \text{ hour}}$$

$$= \frac{509 \cdot 60}{5{,}280} \text{ miles per hour}$$

$$= 5.8 \text{ miles per hour to the nearest tenth}$$

C Graphing Rational Expressions in Two Variables

EXAMPLE 7 Graph the equation $y = \dfrac{1}{x}$.

Solution Since this is the first time we have graphed an equation of this form, we will make a table of values for x and y that satisfy the equation. Before we do, let's make some generalizations about the graph (Figure 1).

First, notice that since y is equal to 1 divided by x, y will be positive when x is positive. (The quotient of two positive numbers is a positive number.) Likewise, when x is negative, y will be negative. In other words, x and y will always have the same sign. Thus, our graph will appear in quadrants I and III only because in those quadrants x and y have the same sign.

Next, notice that the expression $\frac{1}{x}$ will be undefined when x is 0, meaning that there is no value of y corresponding to $x = 0$. Because of this, the graph will not cross the y-axis. Further, the graph will not cross the x-axis either. If we try to find the x-intercept by letting $y = 0$, we have

$$0 = \frac{1}{x}$$

But there is no value of x to divide into 1 to obtain 0. Therefore, since there is no solution to this equation, our graph will not cross the x-axis.

5. The South Coast Plaza shopping mall in Costa Mesa, California, covers an area of 2,224,750 square feet. If 1 acre = 43,560 square feet, how many acres does the South Coast Plaza shopping mall cover? Round your answer to the nearest tenth of an acre.

6. The Bear Paw chair lift at the Northstar ski resort in Lake Tahoe is 772 feet long. If a ride on this chair lift takes 2.2 minutes, what is the average speed of the lift in miles per hour?

7. Graph the equation $y = \dfrac{2}{x}$.

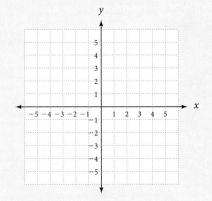

Answers

5. 51.1 acres

6. 3.99 mph

To summarize, we can expect to find the graph in quadrants I and III only, and the graph will cross neither axis.

x	y
−3	$-\frac{1}{3}$
−2	$-\frac{1}{2}$
−1	−1
$-\frac{1}{2}$	−2
$-\frac{1}{3}$	−3
0	Undefined
$\frac{1}{3}$	3
$\frac{1}{2}$	2
1	1
2	$\frac{1}{2}$
3	$\frac{1}{3}$

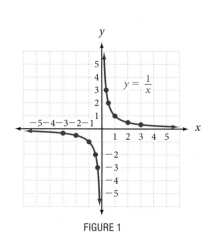

FIGURE 1

8. Graph the equation $y = \dfrac{-2}{x}$.

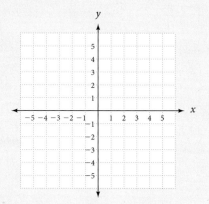

EXAMPLE 8 Graph the equation $y = \dfrac{-6}{x}$.

Solution Since y is −6 divided by x, when x is positive, y will be negative (a negative divided by a positive is negative), and when x is negative, y will be positive (a negative divided by a negative). Thus, the graph (Figure 2) will appear in quadrants II and IV only. As was the case in Example 5, the graph will not cross either axis.

x	y
−6	1
−3	2
−2	3
−1	6
0	Undefined
1	−6
2	−3
3	−2
6	−1

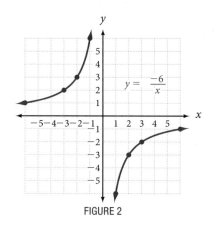

FIGURE 2

D Rational Formulas

As discussed earlier in this book, a *formula* is an equation with more than one variable that shows the relationship between the variables, or quantities. *Rational formulas* tend to be more complicated than the formulas we worked with previously as they all involve fractions.

Here are the steps for solving rational formulas:

> **HOW TO** Solve Rational Formulas
>
> **Step 1:** Identify the variable for which to solve.
>
> **Step 2:** Find the LCD and then multiply it by each term in the formula to eliminate the fractions.
>
> **Step 3:** Isolate the variable by using the addition property of equality to move terms to the opposite side of the equation.
>
> **Step 4:** Depending on the equation, factor out the variable for which you are solving the formula, or combine like terms.
>
> **Step 5:** Isolate the variable by using the multiplication property of equality to move any coefficients to the opposite side of the equation.

EXAMPLE 9 The total electrical resistance for two parallel resistors is given by the formula $\frac{1}{R} = \frac{1}{R_1} + \frac{1}{R_2}$. Solve this formula for R_2.

9. Solve the resistance formula for R.

Solution The first step is to find the LCD, which in this case is RR_1R_2. Multiply each term in the formula by the LCD.

$$RR_1R_2\left(\frac{1}{R}\right) = RR_1R_2\left(\frac{1}{R_1} + \frac{1}{R_2}\right)$$

which simplifies to

$$R_1R_2 = RR_2 + RR_1$$

Now we use the addition property of equality to isolate the terms with the variable R_2 on the left side of the equation.

$$R_1R_2 - RR_2 = RR_1$$

The next step is to factor R_2 from each term on the left side of the equation.

$$R_2(R_1 - R) = RR_1$$

Finally, to isolate R_2, we use the multiplication property of equality to divide both sides of the equation by $(R_2 - R)$.

$$R_2 = \frac{RR_1}{R_1 - R}$$

> **GETTING READY FOR CLASS**
>
> *After reading through the preceding section, respond in your own words and in complete sentences.*
>
> **A.** Why is it useful to construct a table when working an application problem involving rational expressions?
>
> **B.** Use rational expressions to show how the current of a river affects the speed of a motor boat.
>
> **C.** Write an application problem for which the solution depends on solving the equation $\frac{1}{2} + \frac{1}{3} = \frac{1}{x}$.
>
> **D.** Explain why the graph of the equation $y = \frac{-4}{x}$ would not cross the x- or the y-axis.

Answer

9. $R = \dfrac{R_1R_2}{R_2 + R_1}$

Vocabulary Review

Choose the correct words to fill in the blanks below.

| opposite | same | quadrant III | quadrant II |

1. For the equation $y = \frac{1}{x}$, x and y will always have the _____ sign.

2. The graph for the equation $y = \frac{1}{x}$ will appear in quadrant I and _____ only.

3. For the equation $y = -\frac{1}{x}$, x and y will always have _____ signs.

4. The graph for the equation $y = -\frac{1}{x}$ will appear in _____ and quadrant IV only.

A Number Problems

1. One number is 3 times as large as another. The sum of their reciprocals is $\frac{16}{3}$. Find the two numbers.

2. If $\frac{3}{5}$ is added to twice the reciprocal of a number, the result is 1. Find the number.

3. The sum of a number and its reciprocal is $\frac{13}{6}$. Find the number.

4. The sum of a number and 10 times its reciprocal is 7. Find the number.

5. If a certain number is added to both the numerator and denominator of the fraction $\frac{7}{9}$, the result is $\frac{5}{7}$. Find the number.

6. The numerator of a certain fraction is 2 more than the denominator. If $\frac{1}{3}$ is added to the fraction, the result is 2. Find the fraction.

7. The sum of the reciprocals of two consecutive even integers is $\frac{5}{12}$. Find the integers.

8. The sum of the reciprocals of two consecutive integers is $\frac{7}{12}$. Find the two integers.

Motion Problems

9. A boat travels 26 miles up the river in the same amount of time it takes to travel 38 miles down the same river. If the current is 3 miles per hour, what is the speed of the boat in still water?

	d	r	t
Upstream			
Downstream			

SCAN TO ACCESS

10. A boat can travel 9 miles up a river in the same amount of time it takes to travel 11 miles down the same river. If the current is 2 miles per hour, what is the speed of the boat in still water?

	d	r	t
Upstream			
Downstream			

11. An airplane flying against the wind travels 140 miles in the same amount of time it would take the same plane to travel 160 miles with the wind. If the wind speed is a constant 20 miles per hour, how fast would the plane travel in still air?

12. An airplane flying against the wind travels 500 miles in the same amount of time that it would take to travel 600 miles with the wind. If the speed of the wind is 50 miles per hour, what is the speed of the plane in still air?

13. One plane can travel 20 miles per hour faster than another. One of them goes 285 miles in the same time it takes the other to go 255 miles. What are their speeds?

14. One car travels 300 miles in the same amount of time it takes a second car, traveling 5 miles per hour slower than the first, to go 275 miles. What are the speeds of the cars?

15. Tina, whom we mentioned in Example 3 of this section, is training for a triathlon. To train for the running portion of the race, she runs 8 miles each day, over the same course. The first 2 miles of the course are on level ground, while the last 6 miles are downhill. She runs 3 miles per hour slower on level ground than she runs downhill. If the complete course takes 1 hour, how fast does she run on the downhill part of the course?

16. Jerri is training for the same triathlon as Tina (Example 3 and Problem 15). To train for the bicycle portion of the race, she rides 24 miles out a straight road, then turns around and rides 24 miles back. The trip out is against the wind, whereas the trip back is with the wind. If she rides 10 miles per hour faster with the wind then she does against the wind, and the complete trip out and back takes 2 hours, how fast does she ride when she rides against the wind?

17. To train for the running of a triathlon, Jerri jogs 1 hour each day over the same 9-mile course. Five miles of the course is downhill, whereas the other 4 miles is on level ground. Jerri figures that she runs 2 miles per hour faster downhill than she runs on level ground. Find the rate at which Jerri runs on level ground.

18. Travis paddles his kayak in the harbor at Morro Bay, California, where the incoming tide has caused a current in the water. From the point where he enters the water, he paddles 1 mile against the current, then turns around and paddles 1 mile back to where he started. His average speed when paddling with the current is 4 miles per hour faster than his speed against the current. If the complete trip (out and back) takes him 1.2 hours, find his average speed when he paddles against the current.

Work Problems

19. An inlet pipe can fill a pool in 12 hours, while an outlet pipe can empty it in 15 hours. If both pipes are left open, how long will it take to fill the pool?

20. A water tank can be filled in 20 hours by an inlet pipe and emptied in 25 hours by an outlet pipe. How long will it take to fill the tank if both pipes are left open?

21. A bathtub can be filled by the cold water faucet in 10 minutes and by the hot water faucet in 12 minutes. How long does it take to fill the tub if both faucets are open?

22. A water faucet can fill a sink in 12 minutes, whereas the drain can empty it in 6 minutes. If the sink is full, how long will it take to empty if both the faucet and the drain are open?

23. A sink can be filled by the cold water faucet in 3 minutes. The drain can empty a full sink in 4 minutes. If the sink is empty and both the cold water faucet and the drain are open, how long will it take the sink to fill?

24. A bathtub can be filled by the cold water faucet in 9 minutes and by the hot water faucet in 10 minutes. The drain can empty the tub in 5 minutes. Can the tub be filled if both faucets and the drain are open? If so, how long will it take?

25. One hose can fill a spa in 10 hours. Working with another hose, the spa can be filled in 5 hours. How long will it take the other hose to fill the spa if working alone?

26. One faucet can fill a sink in 4 minutes. Working with another faucet, the sink can be filled in $2\frac{2}{3}$ minutes. How long will it take the other faucet to fill the sink if working alone?

B Unit Analysis

27. Mount Whitney The top of Mount Whitney, the highest point in California, is 14,494 feet above sea level. Give this height in miles to the nearest tenth of a mile.

28. Motor Displacement The relationship between liters and cubic inches, both of which are measures of volume, is 0.0164 liters = 1 cubic inch. If a Ford Mustang has a motor with a displacement of 4.9 liters, what is the displacement in cubic inches? Round your answer to the nearest cubic inch.

29. Speed of Sound The speed of sound is 1,088 feet per second. Convert the speed of sound to miles per hour. Round your answer to the nearest whole number.

30. Average Speed A car travels 122 miles in 3 hours. Find the average speed of the car in feet per second. Round to the nearest whole number.

31. Fitness Walking The guidelines for fitness indicate that a person who walks 10,000 steps daily is physically fit. According to *The Walking Site* on the Internet, "The average person's stride length is approximately 2.5 feet long. That means it takes just over 2,000 steps to walk one mile, and 10,000 steps is close to 5 miles." Use your knowledge of unit analysis to determine if these facts are correct. (There are 5,280 feeet in a mile.)

32. Unit Analysis The photograph shows the Cone Nebula as seen by the Hubble telescope in April 2002. The distance across the photograph is about 2.5 light-years. If we assume light travels 186,000 miles in 1 second, we can find the number of miles in 1 light-year by converting 186,000 miles/second to miles/ year. Find the number of miles in 1 light-year. Write your answer in expanded form and in scientific notation.

33. Ferris Wheel A Ferris wheel called Colossus has a circumference of 518 feet. If a trip around the circumference of Colossus takes 40 seconds, find the average speed of a rider in miles per hour. Round to the nearest tenth.

34. Ferris Wheel The first Ferris wheel was built in 1893. It was a large wheel with a circumference of 785 feet. If one trip around the circumference of the wheel took 20 minutes, find the average speed of a rider in miles per hour. Round to the nearest hundredth.

C Graph each of the following equations.

35. $y = \dfrac{-4}{x}$

36. $y = \dfrac{4}{x}$

37. $y = \dfrac{8}{x}$

38. $y = \dfrac{-8}{x}$

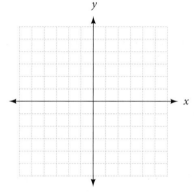

39. Graph $y = \dfrac{3}{x}$ and $x + y = 4$ on the same coordinate system. At what points do the two graphs intersect?

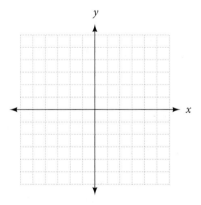

40. Graph $y = \dfrac{4}{x}$ and $x - y = 3$ on the same coordinate system. At what points do the two graphs intersect?

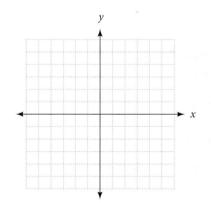

D Rational Formulas

Solve each of the following equations for the indicated variable.

41. Area of a Trapezoid $A = \dfrac{1}{2}h(a + b)$; for h

42. Area of a Triangle $A = \dfrac{1}{2}bh$; for h

43. Volume of a Cone $V = \dfrac{1}{3}\pi r^2 h$; for h

44. Volume of a Pyramid $V = \dfrac{1}{3}Bh$; for h

45. Slope $m = \dfrac{y_2 - y_1}{x_2 - x_1}$; for y_2

46. Slope $m = \dfrac{y_2 - y_1}{x_2 - x_1}$; for y_1

47. Slope $m = \dfrac{y_2 - y_1}{x_2 - x_1}$; for x_2

48. Slope $m = \dfrac{y_2 - y_1}{x_2 - x_1}$; for x_1

49. Electrical Power $R = \dfrac{V^2}{P}$; for P

50. Electrical Power $P = \dfrac{V^2}{R}$; for R

51. Electrical Resistance $\dfrac{1}{R} = \dfrac{1}{R_1} + \dfrac{1}{R_2}$ for R_1

52. Electrical Resistance $\dfrac{1}{R} = \dfrac{1}{R_1} + \dfrac{1}{R_2}$ for R

Getting Ready for the Next Section

Perform the following operations and simplify.

53. $\dfrac{1}{2} \div \dfrac{2}{3}$

54. $\dfrac{1}{3} \div \dfrac{3}{4}$

55. $1 + \dfrac{1}{2}$

56. $1 + \dfrac{2}{3}$

57. $y^5 \cdot \dfrac{2x^3}{y^2}$

58. $y^7 \cdot \dfrac{3x^5}{y^4}$

59. $\dfrac{2x^3}{y^2} \cdot \dfrac{y^5}{4x}$

60. $\dfrac{3x^5}{y^4} \cdot \dfrac{y^7}{6x^2}$

Factor the following expressions.

61. $x^2y + x$

62. $xy^2 + y$

63. $x^3 - x^2$

64. $x^3 - 4x$

Find the Mistake

Each sentence below contains a mistake. Circle the mistake and write the correct word(s) or number(s) on the line provided.

1. One number is four times as large as another. The sum of their reciprocals is $\dfrac{15}{4}$. The two numbers are $\dfrac{3}{4}$ and 3.

2. A kayak travels 3 miles upriver in the same amount of time it takes the kayak to travel 5 miles back down the same river. If the current is 5 miles per hour, what is the speed of the boat in still water? To solve this problem, we must first assign the variable x to the speed of the current. _____

3. The graph of $x = \dfrac{1}{y}$ will not cross the y-axis because the expression $\dfrac{1}{y}$ is undefined when $y = 0$. _____

4. The graph of the equation $y = \dfrac{9}{x}$ will appear in Quadrants II and IV. _____

Every week in Portland, Oregon, bicyclists called zoombombers, young and old convene on a street corner near the local zoo. Bicycles are decorated in streamers, strings of lights, and other ornate details to coincide with the often-costumed riders. Even the bikes themselves are known to cause a stir, falling into a variety of categories, such as swing bikes, mini bikes, tall bikes, and cruisers. Amidst laughter and screams of joy, the bicyclists spend the evening speeding together down the steep hills of the neighborhood.

Imagine you were a spectator on the sidewalk of a street the zoobombers were speeding down. You hear the sound of a single screaming bicyclist zooming by because of the Doppler effect, where the pitch of the sound changes as the bicyclist approaches and then passes you. The formula for the sound frequency h you hear is

$$h = \frac{f}{1 + \frac{v}{s}}$$

where f is the actual frequency of the sound being produced, s is the speed of the sound (about 740 miles per hour), and v is the velocity of the bicyclist. This formula uses a complex rational expression, which is the focus of this section.

> **DEFINITION** complex fraction
>
> A **complex fraction** is a fraction or rational expression that contains other fractions in its numerator or denominator.

Each of the following is a *complex fraction*:

$$\frac{\dfrac{1}{2}}{\dfrac{2}{3}} \qquad \frac{x + \dfrac{1}{y}}{y + \dfrac{1}{x}} \qquad \frac{\dfrac{a+1}{a^2-9}}{\dfrac{2}{a+3}}$$

A Simplifying Complex Fractions

We will begin this section by simplifying the first of these complex fractions. Before we do, though, let's agree on some vocabulary. So that we won't have to use phrases such as the numerator of the denominator, let's call the numerator of a complex fraction the *top* and the denominator of a complex fraction the *bottom*.

EXAMPLE 1 Simplify $\dfrac{\frac{1}{2}}{\frac{2}{3}}$.

Solution There are two methods we can use to solve this problem.

Method 1

We can multiply the top and bottom of this complex fraction by the LCD for both fractions. In this case, the LCD is 6.

$$\text{top}\left\{ \begin{array}{c} \dfrac{1}{2} \\ \dfrac{2}{3} \end{array} \right. \text{bottom} \quad = \quad \dfrac{6 \cdot \dfrac{1}{2}}{6 \cdot \dfrac{2}{3}} = \dfrac{3}{4}$$

Method 2

We can treat this as a division problem. Instead of dividing by $\frac{2}{3}$, we can multiply by its reciprocal $\frac{3}{2}$.

$$\dfrac{\frac{1}{2}}{\frac{2}{3}} = \dfrac{1}{2} \cdot \dfrac{3}{2} = \dfrac{3}{4}$$

Using either method, we obtain the same result.

EXAMPLE 2 Simplify $\dfrac{\frac{2x^3}{y^2}}{\frac{4x}{y^5}}$.

Solution

Method 1

The LCD for each rational expression is y^5. Multiplying the top and bottom of the complex fraction by y^5, we have

$$\dfrac{\frac{2x^3}{y^2}}{\frac{4x}{y^5}} = \dfrac{y^5 \cdot \dfrac{2x^3}{y^2}}{y^5 \cdot \dfrac{4x}{y^5}} = \dfrac{2x^3 y^3}{4x} = \dfrac{x^2 y^3}{2}$$

Method 2

Instead of dividing by $\dfrac{4x}{y^5}$ we can multiply by its reciprocal, $\dfrac{y^5}{4x}$.

$$\dfrac{\frac{2x^3}{y^2}}{\frac{4x}{y^5}} = \dfrac{2x^3}{y^2} \cdot \dfrac{y^5}{4x} = \dfrac{x^2 y^3}{2}$$

Again the result is the same, whether we use Method 1 or Method 2.

EXAMPLE 3 Simplify $\dfrac{x + \dfrac{1}{y}}{y + \dfrac{1}{x}}$.

3. Simplify $\dfrac{y - \dfrac{1}{x}}{x - \dfrac{1}{y}}$.

Solution To apply Method 2 as we did in the first two examples, we would have to simplify the top and bottom separately to obtain a single rational expression for both before we could multiply by the reciprocal. In this case, it is much easier to use Method 1 to multiply the top and bottom by the LCD xy.

$$\dfrac{x + \dfrac{1}{y}}{y + \dfrac{1}{x}} = \dfrac{xy\left(x + \dfrac{1}{y}\right)}{xy\left(y + \dfrac{1}{x}\right)} \qquad \text{Multiply top and bottom by } xy.$$

$$= \dfrac{xy \cdot x + xy \cdot \dfrac{1}{y}}{xy \cdot y + xy \cdot \dfrac{1}{x}} \qquad \text{Distributive property}$$

$$= \dfrac{x^2y + x}{xy^2 + y} \qquad \text{Simplify.}$$

We can factor an x from $x^2y + x$ and a y from $xy^2 + y$ and then reduce to lowest terms.

$$= \dfrac{x(xy + 1)}{y(xy + 1)}$$

$$= \dfrac{x}{y}$$

EXAMPLE 4 Simplify $\dfrac{1 - \dfrac{4}{x^2}}{1 - \dfrac{1}{x} - \dfrac{6}{x^2}}$.

4. Simplify $\dfrac{1 - \dfrac{9}{x^2}}{1 - \dfrac{5}{x} + \dfrac{6}{x^2}}$.

Solution The simplest way to simplify this complex fraction is to multiply the top and bottom by the LCD x^2.

$$\dfrac{1 - \dfrac{4}{x^2}}{1 - \dfrac{1}{x} - \dfrac{6}{x^2}} = \dfrac{x^2\left(1 - \dfrac{4}{x^2}\right)}{x^2\left(1 - \dfrac{1}{x} - \dfrac{6}{x^2}\right)} \qquad \begin{array}{l}\text{Multiply top and}\\\text{bottom by } x^2.\end{array}$$

$$= \dfrac{x^2 \cdot 1 - x^2 \cdot \dfrac{4}{x^2}}{x^2 \cdot 1 - x^2 \cdot \dfrac{1}{x} - x^2 \cdot \dfrac{6}{x^2}} \qquad \text{Distributive property}$$

$$= \dfrac{x^2 - 4}{x^2 - x - 6} \qquad \text{Simplify.}$$

$$= \dfrac{(x - 2)(x + 2)}{(x - 3)(x + 2)} \qquad \text{Factor.}$$

$$= \dfrac{x - 2}{x - 3} \qquad \text{Reduce.}$$

GETTING READY FOR CLASS

After reading through the preceding section, respond in your own words and in complete sentences.

A. What is a complex fraction?

B. Explain the first method of simplifying complex fractions.

C. When would you use an LCD to simplify a complex fraction rather than multiply the numerator by the denominator's reciprocal?

D. Give an example of a complex fraction problem that can be rewritten as a division problem.

Answer

3. $\dfrac{y}{x}$

4. $\dfrac{x + 3}{x - 2}$

Vocabulary Review

Choose the correct words to fill in the blanks below.

LCD reciprocal complex fraction top

1. A _____ is a fraction or rational expression that contains other fractions in its numerator or denominator.

2. We call the numerator of a complex fraction the _____ and the denominator the bottom.

3. The first method to simplify a complex fraction is to multiply the top and bottom by the _____.

4. The second method to simplify a complex fraction and multiply the top by the _____ of the bottom.

Problems

A Simplify each complex fraction.

1. $\dfrac{\frac{3}{4}}{\frac{1}{8}}$

2. $\dfrac{\frac{1}{3}}{\frac{5}{6}}$

3. $\dfrac{\frac{2}{3}}{\frac{4}{}}$

4. $\dfrac{\frac{5}{}}{\frac{1}{2}}$

5. $\dfrac{\frac{x^2}{y}}{\frac{x}{y^3}}$

6. $\dfrac{\frac{x^5}{y^3}}{\frac{x^2}{y^8}}$

7. $\dfrac{\frac{4x^3}{y^6}}{\frac{8x^2}{y^7}}$

8. $\dfrac{\frac{6x^4}{y}}{\frac{2x}{y^5}}$

9. $\dfrac{y + \frac{1}{x}}{x + \frac{1}{y}}$

10. $\dfrac{y - \frac{1}{x}}{x - \frac{1}{y}}$

11. $\dfrac{1 + \frac{1}{a}}{1 - \frac{1}{a}}$

12. $\dfrac{\frac{1}{a} - 1}{\frac{1}{a} + 1}$

13. $\dfrac{\frac{x+1}{x^2-9}}{\frac{2}{x+3}}$

14. $\dfrac{\frac{3}{x-5}}{\frac{x+1}{x^2-25}}$

15. $\dfrac{\frac{1}{a+2}}{\frac{1}{a^2-a-6}}$

16. $\dfrac{\frac{1}{a^2+5a+6}}{\frac{1}{a+3}}$

17. $\dfrac{1 - \frac{9}{y^2}}{1 - \frac{1}{y} - \frac{6}{y^2}}$

18. $\dfrac{1 - \frac{4}{y^2}}{1 - \frac{2}{y} - \frac{8}{y^2}}$

19. $\dfrac{\frac{1}{y} + \frac{1}{x}}{\frac{1}{xy}}$

20. $\dfrac{\dfrac{1}{xy}}{\dfrac{1}{y} - \dfrac{1}{x}}$

21. $\dfrac{1 - \dfrac{1}{a^2}}{1 - \dfrac{1}{a}}$

22. $\dfrac{1 + \dfrac{1}{a}}{1 - \dfrac{1}{a^2}}$

23. $\dfrac{\dfrac{1}{10x} - \dfrac{y}{10x^2}}{\dfrac{1}{10} - \dfrac{y}{10x}}$

24. $\dfrac{\dfrac{1}{2x} + \dfrac{y}{2x^2}}{\dfrac{1}{4} + \dfrac{y}{4x}}$

25. $\dfrac{\dfrac{1}{a + 1} + 2}{\dfrac{1}{a + 1} + 3}$

26. $\dfrac{\dfrac{2}{a + 1} + 3}{\dfrac{3}{a + 1} + 4}$

27. $\dfrac{\dfrac{1}{x - 2} - 3}{\dfrac{1}{x - 2} + 2}$

28. $\dfrac{\dfrac{1}{x - 1} + 4}{\dfrac{2}{x - 1} - 3}$

Although the following problems do not contain complex fractions, they do involve more than one operation. Simplify inside the parentheses first, then multiply.

29. $\left(1 - \dfrac{1}{x}\right)\left(1 - \dfrac{1}{x + 1}\right)\left(1 - \dfrac{1}{x + 2}\right)$

30. $\left(1 + \dfrac{1}{x}\right)\left(1 + \dfrac{1}{x + 1}\right)\left(1 + \dfrac{1}{x + 2}\right)$

31. $\left(1 + \dfrac{1}{x + 3}\right)\left(1 + \dfrac{1}{x + 2}\right)\left(1 + \dfrac{1}{x + 1}\right)$

32. $\left(1 - \dfrac{1}{x + 3}\right)\left(1 - \dfrac{1}{x + 2}\right)\left(1 - \dfrac{1}{x + 1}\right)$

33. Simplify each term.

$$2 + \dfrac{1}{2 + 1}, 2 + \dfrac{1}{2 + \dfrac{1}{2 + 1}}, 2 + \dfrac{1}{2 + \dfrac{1}{2 + \dfrac{1}{2 + 1}}}$$

34. Simplify each term.

$$2 + \dfrac{3}{2 + 3}, 2 + \dfrac{3}{2 + \dfrac{3}{2 + 3}}, 2 + \dfrac{3}{2 + \dfrac{3}{2 + \dfrac{3}{2 + 3}}}$$

Complete the following tables.

35.

Number	Reciprocal	Quotient	Square
x	$\dfrac{1}{x}$	$\dfrac{x}{\dfrac{1}{x}}$	x^2
1			
2			
3			
4			

36.

Number	Reciprocal	Quotient	Square
x	$\dfrac{1}{x}$	$\dfrac{\dfrac{1}{x}}{x}$	x^2
1			
2			
3			
4			

37.

Number	Reciprocal	Sum	Quotient
x	$\dfrac{1}{x}$	$1+\dfrac{1}{x}$	$\dfrac{1+\dfrac{1}{x}}{\dfrac{1}{x}}$
1			
2			
3			
4			

38.

Number	Reciprocal	Difference	Quotient
x	$\dfrac{1}{x}$	$1-\dfrac{1}{x}$	$\dfrac{1-\dfrac{1}{x}}{\dfrac{1}{x}}$
1			
2			
3			
4			

Getting Ready for the Next Section

Solve each equation.

39. $21 = 6x$

40. $72 = 2x$

41. $x^2 + x = 6$

42. $x^2 + 2x = 8$

43. $\dfrac{x}{5} = 4$

44. $\dfrac{x}{2} = -3$

45. $\dfrac{x}{5} = \dfrac{2}{3}$

46. $\dfrac{x}{4} - - \dfrac{5}{8}$

Find the Mistake

Each sentence below contains a mistake. Circle the mistake and write the correct word(s) or expression on the line provided.

1. A complex fraction contains a fraction in its numerator but not its denominator. _____

2. To simplify the complex fraction $\dfrac{\dfrac{4}{9}}{\dfrac{7}{13}}$ using the LCD for both fractions, we must multiply the top and bottom by 28.

3. One way to simplify the complex fraction $\dfrac{\dfrac{6y^2}{7x}}{\dfrac{4y}{x}}$ is to multiply $\dfrac{4y}{x}$ by $\dfrac{7x}{6y^2}$. _____

4. The easiest way to simplify the complex fraction $\dfrac{\dfrac{2}{x+3}+4}{\dfrac{2}{x}+3}$ is to multiply by the reciprocal. _____

6.7 Proportions

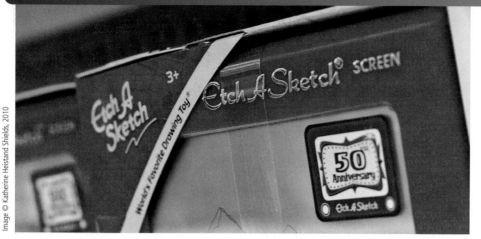

In June 2010, the Children's Museum of Indianapolis had the world's largest Etch-a-Sketch on display. The large toy measured 8 feet tall, compared to the original toy's measurements of 7 inches wide by 9 inches tall. When the screen needed to be cleared, instead of shaking it like the original, the display was designed to flip upside down and then return upright again.

We can compare the measurements of the museum's Etch-a-Sketch with the original using the concept of proportions. Let's allow x to equal the width in feet of the museum's Etch-a-Sketch. Then we can put the measurements of both toys into two rational expressions and set them equal to each other to solve for x.

$$\frac{x}{8 \text{ feet}} = \frac{7 \text{ inches}}{9 \text{ inches}}$$

After reading this section, you will be able to find the value for x. First, let's explore further the definition of proportion.

DEFINITION proportion

A **proportion** is two equal ratios; that is, if $\frac{a}{b}$ and $\frac{c}{d}$ are ratios, then

$$\frac{a}{b} = \frac{c}{d}$$

is a proportion.

Each of the four numbers in a proportion is called a *term* of the proportion. We number the terms as follows:

$$\text{First term} \to \frac{a}{b} = \frac{c}{d} \leftarrow \text{Third term}$$
$$\text{Second term} \to \qquad\qquad \leftarrow \text{Fourth term}$$

The first and fourth terms are called the *extremes*, and the second and third terms are called the *means*.

$$\text{Means} \overset{\frown}{} \frac{a}{b} = \frac{c}{d} \text{Extremes}$$

For example, in the proportion

$$\frac{3}{8} = \frac{12}{32}$$

the extremes are 3 and 32, and the means are 8 and 12.

> **PROPERTY** Means-Extremes Property
>
> If a, b, c, and d are real numbers with $b \neq 0$ and $d \neq 0$, then
>
> $$\text{if} \qquad \frac{a}{b} = \frac{c}{d}$$
>
> $$\text{then} \quad ad = bc$$
>
> *In words:* In any proportion, the product of the extremes is equal to the product of the means.

A Solving Proportions

This property of proportions comes from the multiplication property of equality. We can use it to solve for a missing term in a proportion.

Practice Problems

1. Solve the proportion $\frac{2}{9} = \frac{8}{x}$.

EXAMPLE 1 Solve the proportion $\frac{3}{x} = \frac{6}{7}$.

Solution We could solve for x by using the method developed in Section 6.4; that is, multiplying both sides by the LCD $7x$. Instead, let's use our new means-extremes property.

$$\frac{3}{x} = \frac{6}{7} \qquad \text{Extremes are 3 and 7; means are } x \text{ and 6.}$$

$$21 = 6x \qquad \text{Product of extremes} = \text{product of means.}$$

$$\frac{21}{6} = x \qquad \text{Divide both sides by 6.}$$

$$x = \frac{7}{2} \qquad \text{Reduce to lowest terms.}$$

2. Solve the proportion $\frac{x+2}{4} = \frac{2}{x}$.

EXAMPLE 2 Solve the proportion $\frac{x+1}{2} = \frac{3}{x}$.

Solution Again, we want to point out that we could solve for x by using the method we used in Section 6.4. Using the means-extremes property is simply an alternative to the method developed in Section 6.4.

$$\frac{x+1}{2} = \frac{3}{x} \qquad \text{Extremes are } x+1 \text{ and } x \text{; means are 2 and 3.}$$

$$x^2 + x = 6 \qquad \text{Product of extremes} = \text{product of means}$$

$$x^2 + x - 6 = 0 \qquad \text{Standard form for a quadratic equation}$$

$$(x+3)(x-2) = 0 \qquad \text{Factor.}$$

$$x + 3 = 0 \qquad \text{or} \qquad x - 2 = 0 \qquad \text{Set factors equal to 0.}$$

$$x = -3 \qquad\qquad x = 2$$

This time we have two solutions: -3 and 2.

B Applications Involving Proportions

3. Using the information in Example 3, determine how many out of 1,650 parts can be expected to be defective.

EXAMPLE 3 A manufacturer knows that during a production run, 8 out of every 100 parts produced by a certain machine will be defective. If the machine produces 1,450 parts, how many can be expected to be defective?

Solution The ratio of defective parts to total parts produced is $\frac{8}{100}$. If we let x represent the number of defective parts out of the total of 1,450 parts, then we can write this ratio again as $\frac{x}{1,450}$. This gives us a proportion to solve.

Answers

1. 36
2. $-4, 2$
3. 132

Defective parts in numerator
Total parts in denominator

$$\frac{x}{1{,}450} = \frac{8}{100}$$ Extremes are x and 100; means are 1,450 and 8.

$$100x = 11{,}600$$ Product of extremes = product of means

$$x = 116$$

The manufacturer can expect 116 defective parts out of the total of 1,450 parts if the machine usually produces 8 defective parts for every 100 parts it produces.

EXAMPLE 4 A woman drives her car 270 miles in 6 hours. If she continues at the same rate, how far will she travel in 10 hours?

Solution We let x represent the distance traveled in 10 hours. Using x, we translate the problem into the following proportion:

Miles \longrightarrow $\dfrac{x}{10} = \dfrac{270}{6}$ \longleftarrow Miles
Hours \longrightarrow $\phantom{\dfrac{x}{10}}$ \longleftarrow Hours

Notice that the two ratios in the proportion compare the same quantities. That is, both ratios compare miles to hours. In words, this proportion says

x miles is to 10 hours as 270 miles is to 6 hours

$$\frac{x}{10} = \frac{270}{6}$$

Next, we solve the proportion.

$$6x = 2700$$

$$x = 450 \text{ miles}$$

If the woman continues at the same rate, she will travel 450 miles in 10 hours.

EXAMPLE 5 A baseball player gets 8 hits in the first 18 games of the season. If he continues at the same rate, how many hits will he get in 45 games?

Solution We let x represent the number of hits he will get in 45 games. Then

x is to 45 as 8 is to 18

Hits \longrightarrow $\dfrac{x}{45} = \dfrac{8}{18}$ \longleftarrow Hits
Games \longrightarrow $\phantom{\dfrac{x}{45}}$ \longleftarrow Games

Notice again that the two ratios are comparing the same quantities, hits to games. We solve the proportion as follows:

$$18x = 360$$

$$x = 20$$

If he continues to hit at the rate of 8 hits in 18 games, he will get 20 hits in 45 games.

GETTING READY FOR CLASS

After reading through the preceding section, respond in your own words and in complete sentences.

A. What is a proportion?

B. What are the means and extremes of a proportion?

C. What is the relationship between the means and the extremes in a proportion?

D. Write an application problem using the means-extremes property.

4. A woman drives her car 204 miles in 3 hours. How many miles will she travel in 7 hours if she continues at the same rate?

5. A baseball pitcher gives up 4 runs in 9 innings. If he continues at the same rate, how many runs will he give up in 108 innings?

Answers
4. 476 miles
5. 48 runs

Vocabulary Review

Choose the correct words to fill in the blanks below.

extremes term proportion means-extremes property means

1. A _____ is two equal ratios given by $\frac{a}{b} = \frac{c}{d}$.

2. Each of the four numbers in a proportion is called a _____ .

3. The first and fourth terms in a proportion are called the _____ .

4. The second and third terms in a proportion are called the _____ .

5. The _____ states that the product of the extremes is equal to the product of the means in any proportion.

Problems

A Solve each of the following proportions.

1. $\dfrac{x}{2} = \dfrac{6}{12}$

2. $\dfrac{x}{4} = \dfrac{6}{8}$

3. $\dfrac{2}{5} = \dfrac{4}{x}$

4. $\dfrac{3}{8} = \dfrac{9}{x}$

5. $\dfrac{10}{20} = \dfrac{20}{x}$

6. $\dfrac{15}{60} = \dfrac{60}{x}$

7. $\dfrac{a}{3} = \dfrac{5}{12}$

8. $\dfrac{a}{2} = \dfrac{7}{20}$

9. $\dfrac{2}{x} = \dfrac{6}{7}$

10. $\dfrac{4}{x} = \dfrac{6}{7}$

11. $\dfrac{x+1}{3} = \dfrac{4}{x}$

12. $\dfrac{x+1}{6} = \dfrac{7}{x}$

13. $\dfrac{x}{2} = \dfrac{8}{x}$

14. $\dfrac{x}{9} = \dfrac{4}{x}$

15. $\dfrac{4}{a+2} = \dfrac{a}{2}$

16. $\dfrac{3}{a+2} = \dfrac{a}{5}$

17. $\dfrac{1}{x} = \dfrac{x-5}{6}$

18. $\dfrac{1}{x} = \dfrac{x-6}{7}$

19. $\dfrac{3}{x} = \dfrac{x+1}{4}$

20. $\dfrac{x}{6} = \dfrac{8}{x-2}$

21. $\dfrac{x}{3} = \dfrac{3}{x}$

22. $\dfrac{x}{4} = \dfrac{4}{x}$

23. $\dfrac{x}{11} = \dfrac{5}{2x+1}$

24. $\dfrac{x}{10} = \dfrac{4}{3x-2}$

B Applying the Concepts

25. Baseball A baseball player gets 6 hits in the first 18 games of the season. If he continues hitting at the same rate, how many hits will he get in the first 45 games?

26. Basketball A basketball player makes 8 of 12 free throws in the first game of the season. If she shoots with the same accuracy in the second game, how many of the 15 free throws she attempts will she make?

27. Mixture Problem A solution contains 12 milliliters of alcohol and 16 milliliters of water. If another solution is to have the same concentration of alcohol in water but is to contain 28 milliliters of water, how much alcohol must it contain?

28. Mixture Problem A solution contains 15 milliliters of HCl and 42 milliliters of water. If another solution is to have the same concentration of HCl in water but is to contain 140 milliliters of water, how much HCl must it contain?

29. Nutrition If 100 grams of ice cream contains 13 grams of fat, how much fat is in 350 grams of ice cream?

30. Nutrition A 6-ounce serving of grapefruit juice contains 159 grams of water. How many grams of water are in 20 ounces of grapefruit juice?

31. Map Reading A map is drawn so that every 3.5 inches on the map corresponds to an actual distance of 100 miles. If the actual distance between the two cities is 420 miles, how far apart are they on the map?

32. Map Reading The scale on a map indicates that 1 inch on the map corresponds to an actual distance of 105 miles. Two cities are 4.5 inches apart on the map. What is the actual distance between the two cities?

33. Distance A man drives his car 245 miles in 5 hours. At this rate, how far will he travel in 7 hours?

34. Distance An airplane flies 1,380 miles in 3 hours. How far will it fly in 5 hours?

Getting Ready for the Next Section

Use the formula $y = 5x$ to find y for the given value of x.

35. $x = 4$

36. $x = 3$

37. $x = -4$

38. $x = -\dfrac{2}{5}$

Use the formula $y = \dfrac{20}{x}$ to find y for the given value of x.

39. $x = 10$

40. $x = 5$

41. $x = 40$

42. $x = -6$

Use the formula $y = 2x^2$ to find x for the given value of y.

43. $y = 50$

44. $y = 72$

45. $y = 8$

46. $y = 98$

Find the Mistake

Each sentence below contains a mistake. Circle the mistake and write the correct word(s) or number(s) on the line provided.

1. A proportion is a ratio equal to a whole number. _____

2. In the proportion $\dfrac{6}{11} = \dfrac{12}{22}$, the second term is 12. _____

3. In the proportion $\dfrac{a}{b} = \dfrac{c}{d}$, the extremes are b and c. _____

4. For the proportion $\dfrac{13}{19} = \dfrac{39}{57}$, the product of the means is 507. _____

KEY WORDS

direct variation

constant of variation

inverse variation

6.8 Direct and Inverse Variation

Image © Will Parris, 2010

A comedian by the name of Gallagher is known for his stand-up act's messy grand finale. Ponchos or sheets of plastic are provided to the first few audience rows. Then Gallagher uses a large wooden mallet dubbed the "Sledge-O-Matic" to smash a variety of food objects, most notably watermelons. When purchasing fruit for his show, Gallagher must pay a price that is in direct variation to the fruit's weight. The more the fruit weighs, the more he will pay. Variation is the main focus of this section.

A Direct Variation

Two variables are said to *vary directly* if one is a constant multiple of the other. For instance, y varies directly as x if $y = kx$, where k is a constant. The constant k is called the *constant of variation*. The following table gives the relationship between direct variation statements and their equivalent algebraic equations.

Statement	Equation ($k \cdot$ variable)
y varies directly as x	$y = kx$
y varies directly as the square of x	$y = kx^2$
s varies directly as the square root of t	$s = k\sqrt{t}$
r varies directly as the cube of s	$r = ks^3$

Any time we run across a statement similar to those in the table, we immediately can write an equivalent expression involving variables and a constant of variation k.

Practice Problems

1. y varies directly as x. If y is 24 when x is 8, find y when x is 2.

EXAMPLE 1 Suppose y varies directly as x. When y is 15, x is 3. Find y when x is 4.

Solution From the first sentence we can write the relationship between x and y as
$$y = kx$$

We now use the second sentence to find the value of k. Since y is 15 when x is 3, we have
$$15 = k(3) \quad \text{or} \quad k = 5$$
Now we can rewrite the relationship between x and y more specifically as
$$y = 5x$$
To find the value of y when x is 4 we simply substitute $x = 4$ into our last equation.

$$\text{When} \rightarrow \quad x = 4$$
$$\text{the equation} \rightarrow \quad y = 5x$$
$$\text{becomes} \rightarrow \quad y = 5(4)$$
$$y = 20$$

Answer

1. 6

EXAMPLE 2 Suppose y varies directly as the square of x. When x is 4, y is 32. Find y when x is 5.

Solution The first sentence gives us

$$y = kx^2$$

Since y is 32 when x is 4, we have

$$32 = k(4)^2$$
$$32 = 16k$$
$$k = 2$$

The equation now becomes

$$y = 2x^2$$

When x is 5, we have

$$y = 2(5)^2$$
$$y = (2)(25)$$
$$y = 50$$

2. y varies directly as the square of x. When x is 5, y is 75. Find x when y is 48.

EXAMPLE 3 The cost of a certain kind of candy varies directly with the weight of the candy. If 12 ounces of the candy cost $1.68, how much will 16 ounces cost?

Solution Let $x =$ the number of ounces of candy and $y =$ the cost of the candy. Then $y = kx$. Since y is 1.68 when x is 12, we have

$$1.68 = k \cdot 12$$
$$k = \frac{1.68}{12}$$
$$= 0.14$$

The equation must be

$$y = 0.14x$$

When x is 16, we have

$$y = 0.14(16)$$
$$= 2.24$$

The cost of 16 ounces of candy is $2.24.

3. If 13 ounces of a candy cost $4.68, how much will 8 ounces cost?

B Inverse Variation

Two variables are said to vary inversely if one is a constant multiple of the reciprocal of the other. For example, y varies inversely as x if $y = \frac{k}{x}$, where k is a real number constant. Again, k is called the constant of variation. The table that follows gives some examples of inverse variation statements and their associated algebraic equations.

Statement	Equation
y varies inversely as x	$y = \frac{k}{x}$
y varies inversely as the square of x	$y = \frac{k}{x^2}$
F varies inversely as the square root of t	$F = \frac{k}{\sqrt{t}}$
r varies inversely as the cube of s	$r = \frac{k}{s^3}$

Every variation statement has an associated inverse variation equation.

Answers
2. 4
3. $2.88

4. y varies inversely as x. When y is 6, x is 12. Find y when x is 9.

EXAMPLE 4 Suppose y varies inversely as x. When y is 4, x is 5. Find y when x is 10.

Solution The first sentence gives us the relationship between x and y.

$$y = \frac{k}{x}$$

We use the second sentence to find the value of the constant k.

$$4 = \frac{k}{5}$$

or

$$k = 20$$

We can now write the relationship between x and y more specifically as

$$y = \frac{20}{x}$$

We use this equation to find the value of y when x is 10.

$$\text{When} \rightarrow \qquad x = 10$$
$$\text{the equation} \rightarrow \qquad y = \frac{20}{x}$$
$$\text{becomes} \rightarrow \qquad y = \frac{20}{10}$$
$$y = 2$$

C Applications

5. Using the information from Example 5, what is the intensity 20 feet from the source?

EXAMPLE 5 The intensity (I) of light from a source varies inversely as the square of the distance (d) from the source. Ten feet away from the source the intensity is 200 candela. What is the intensity 5 feet from the source?

Solution

$$I = \frac{k}{d^2}$$

Since $I = 200$ when $d = 10$, we have

$$200 = \frac{k}{10^2}$$
$$200 = \frac{k}{100}$$
$$k = 20{,}000$$

The equation becomes

$$I = \frac{20{,}000}{d^2}$$

When $d = 5$, we have

$$I = \frac{20{,}000}{5^2}$$
$$= \frac{20{,}000}{25}$$
$$= 800 \text{ candela}$$

> ### GETTING READY FOR CLASS
>
> *After reading through the preceding section, respond in your own words and in complete sentences.*
>
> **A.** What does "y varies directly as x" mean?
>
> **B.** Give an example of a sentence that is a direct variation statement.
>
> **C.** Translate the equation $y = \frac{k}{x}$ into words.
>
> **D.** Give an example of an everyday situation where one quantity varies inversely with another.

Answers

4. 8

5. 50 candela

Vocabulary Review

Choose the correct words to fill in the blanks below.

directly inversely constant of variation multiple reciprocal

1. Two variables vary directly if one is a constant _____ of the other given by $y = kx$.
2. The equation $s = k\sqrt{t}$ means that s varies _____ as the square root of t.
3. The constant k is called the _____.
4. The variables are said to vary inversely if one is a constant multiple of the _____ of the other, given by $y = \frac{k}{x}$.
5. The equation $r = \frac{k}{s^3}$ means that r varies _____ as the cube of s.

Problems

A For each of the following problems, y varies directly as x.

1. If $y = 10$ when $x = 5$, find y when x is 4.

2. If $y = 20$ when $x = 4$, find y when x is 11.

3. If $y = 39$ when $x = 3$, find y when x is 10.

4. If $y = -18$ when $x = 6$, find y when x is 3.

5. If $y = -24$ when $x = 4$, find x when y is -30.

6. If $y = 30$ when $x = -15$, find x when y is 8.

7. If $y = -7$ when $x = -1$, find x when y is -21.

8. If $y = 30$ when $x = 4$, find y when x is 7.

For each of the following problems, y varies directly as the square of x.

9. If $y = 75$ when $x = 5$, find y when x is 1.

10. If $y = -72$ when $x = 6$, find y when x is 3.

11. If $y = 48$ when $x = 4$, find y when x is 9.

12. If $y = 27$ when $x = 3$, find y when x is 5.

13. If $y = 100$ when $x = 2$, find y when x is 4.

14. If $y = 30$ when $x = 2$, find y when x is 6.

SCAN TO ACCESS

B For each of the following problems, y varies inversely as x.

15. If $y = 5$ when $x = 2$, find y when x is 5.

16. If $y = 2$ when $x = 10$, find y when x is 4.

17. If $y = 2$ when $x = 1$, find y when x is 4.

18. If $y = 4$ when $x = 3$, find y when x is 6.

19. If $y = 5$ when $x = 3$, find x when y is 15.

20. If $y = 12$ when $x = 10$, find x when y is 60.

21. If $y = 10$ when $x = 10$, find x when y is 20.

22. If $y = 15$ when $x = 2$, find x when y is 6.

For each of the following problems, y varies inversely as the square of x.

23. If $y = 4$ when $x = 5$, find y when x is 2.

24. If $y = 5$ when $x = 2$, y when x is 6.

25. If $y = 4$ when $x = 3$, find y when x is 2.

26. If $y = 9$ when $x = 4$, find y when x is 3.

27. If $y = 10$ when $x = 2$, find y when x is 4.

28. If $y = 30$ when $x = 2$, find y when x is 5.

Applying the Concepts

29. Tension in a Spring The tension t in a spring varies directly with the distance d the spring is stretched. If the tension is 42 pounds when the spring is stretched 2 inches, find the tension when the spring is stretched twice as far.

30. Fill Time The time t it takes to fill a bucket varies directly with the volume g of the bucket. If it takes 1 minute to fill a 4-gallon bucket, how long will it take to fill a 6-gallon bucket?

31. **Electricity** The power P in an electric circuit varies directly with the square of the current I. If $P = 30$ when $I = 2$, find P when $I = 7$.

32. **Electricity** The resistance R in an electric circuit varies directly with the voltage V. If $R = 20$ when $V = 120$, find R when $V = 240$.

33. **Wages** The amount of money M a woman makes per week varies directly with the number of hours h she works per week. If she works 20 hours and earns $157, how much does she make if she works 30 hours?

34. **Volume** The volume V of a gas varies directly as the temperature T. If $V = 3$ when $T = 150$, find V when T is 200.

35. **Weight** The weight F of a body varies inversely with the square of the distance d between the body and the center of the Earth. If a man weighs 150 pounds 4,000 miles from the center of the Earth, how much will he weigh at a distance of 5,000 miles from the center of the Earth?

36. **Light Intensity** The intensity I of a light source varies inversely with the square of the distance d from the source. Four feet from the source, the intensity is 9 footcandles. What is the intensity 3 feet from the source?

37. **Electricity** The current I in an electric circuit varies inversely with the resistance R. If a current of 30 amperes is produced by a resistance of 2 ohms, what current will be produced by a resistance of 5 ohms?

38. **Pressure** The pressure exerted by a gas on the container in which it is held varies inversely with the volume of the container. A pressure of 40 pounds per square inch is exerted on a container of volume 2 cubic feet. What is the pressure on a container whose volume is 8 cubic feet?

Maintaining Your Skills

39. Fill in each ordered pair so that it is a solution to $y = \frac{1}{2}x + 3$.

 $(-2, \quad), (0, \quad), (2, \quad)$

40. Graph the line $y = \frac{1}{2}x + 3$.

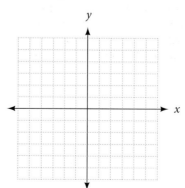

41. Graph the line $x = -2$.

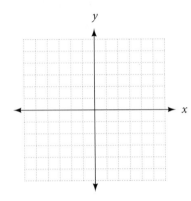

42. Graph $3x - 2y = 6$.

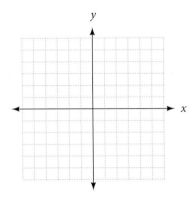

43. Find the slope of the line through $(2, 5)$ and $(0, 1)$.

44. Find the slope and y-intercept for the line $2x - 5y = 10$.

45. Find the equation of the line through $(-2, 1)$ with slope $\frac{1}{2}$.

46. Write the equation of the line with slope -2 and y-intercept $\frac{3}{2}$.

47. Find the equation of the line through $(2, 5)$ and $(0, 1)$.

48. Graph the solution set for $2x - y < 4$.

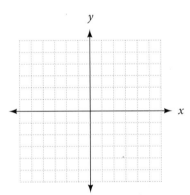

Find the Mistake

Each sentence below contains a mistake. Circle the mistake and write the correct word or variable on the line provided.

1. An algebraic equation written as an equivalent to a variation statement includes k, called the constant of multiplication. _____

2. Two variables vary directly if one is a constant divisor of the other. _____

3. The equation $y = kx$ is translated as y varies directly as k. _____

4. The equation $F = \dfrac{k}{\sqrt{t}}$ is translated as F varies directly as the square root of t. _____

A Access to internet for research

B Pencil and paper

Musical Math

Many aspects of music are related to math. For example, when two notes are a single octave apart, the ratio of their frequencies is 2:1. For this project, let's extend our knowledge of music and its relation to mathematical concepts. Begin by researching the following terms:

octave

interval

pitch

half-step

frequency

Hertz

pentatonic scale

heptatonic scale

Explain how the above terms use mathematical concepts you have learned in this course thus far. Give numerical examples, and include such concepts as reciprocals, exponents, sequences, ratios, and proportions.

Chapter 6 Summary

EXAMPLES

1. We can reduce $\frac{6}{8}$ to lowest terms by dividing the numerator and denominator by their greatest common factor 2.

$$\frac{6}{8} = \frac{2 \cdot 3}{2 \cdot 4} = \frac{3}{4}$$

Rational Numbers [6.1]

Any number that can be put in the form $\frac{a}{b}$, where a and b are integers ($b \neq 0$), is called a rational number.

Multiplying or dividing the numerator and denominator of a rational number by the same nonzero number never changes the value of the rational number.

Rational Expressions [6.1]

2. We simplify rational expressions by factoring the numerator and denominator and then dividing out any factors they have in common.

$$\frac{x-3}{x^2-9} = \frac{x-3}{(x-3)(x+3)} = \frac{1}{x+3}$$

Any expression of the form $\frac{P}{Q}$, where P and Q are polynomials ($Q \neq 0$), is a rational expression.

Multiplying or dividing the numerator and denominator of a rational expression by the same nonzero quantity always produces a rational expression equivalent to the original one.

Multiplication of Rational Expressions [6.2]

3. $\dfrac{x-1}{x^2+2x-3} \cdot \dfrac{x^2-9}{x-2}$

$$= \frac{x-1}{(x+3)(x-1)} \cdot \frac{(x-3)(x+3)}{x-2}$$

$$= \frac{x-3}{x-2}$$

To multiply two rational expressions, multiply numerators, multiply denominators, and divide out any factors common to the numerator and denominator.

For rational expressions $\dfrac{P}{Q}$ and $\dfrac{R}{S}$, $\dfrac{P}{Q} \cdot \dfrac{R}{S} = \dfrac{PR}{QS}$

Division of Rational Expressions [6.2]

4. $\dfrac{2x}{x^2-25} \div \dfrac{4}{x-5}$

$$= \frac{2x}{(x-5)(x+5)} \cdot \frac{(x-5)}{(2 \cdot 2)}$$

$$= \frac{x}{2(x+5)}$$

To divide by a rational expression, simply multiply by its reciprocal.

For rational expressions $\dfrac{P}{Q}$ and $\dfrac{R}{S}$, $\dfrac{P}{Q} \div \dfrac{R}{S} = \dfrac{P}{Q} \cdot \dfrac{S}{R}$

Addition of Rational Expressions [6.3]

5. $\dfrac{3}{x-1} + \dfrac{x}{2}$

$$= \frac{3}{x-1} \cdot \frac{2}{2} + \frac{x}{2} \cdot \frac{x-1}{x-1}$$

$$= \frac{6}{2(x-1)} + \frac{x^2-x}{2(x-1)}$$

$$= \frac{x^2-x+6}{2(x-1)}$$

To add two rational expressions, find a common denominator, change each expression to an equivalent expression having the common denominator, and then add numerators and reduce if possible.

For rational expressions $\dfrac{P}{S}$ and $\dfrac{Q}{S}$, $\dfrac{P}{S} + \dfrac{Q}{S} = \dfrac{P+Q}{S}$

Subtraction of Rational Expressions [6.3]

To subtract a rational expression, simply add its opposite.

For rational expressions $\dfrac{P}{S}$ and $\dfrac{Q}{S}$, $\dfrac{P}{S} - \dfrac{Q}{S} = \dfrac{P}{S} + \left(-\dfrac{Q}{S}\right)$

6. $\dfrac{x}{x^2 - 4} - \dfrac{2}{x^2 - 4}$

$= \dfrac{x - 2}{x^2 - 4}$

$= \dfrac{x - 2}{(x - 2)(x + 2)}$

$= \dfrac{1}{x + 2}$

Equations Involving Rational Expressions [6.4]

To solve equations involving rational expressions, first find the least common denominator (LCD) for all denominators. Then multiply both sides by the LCD and solve as usual. Check all solutions in the original equation to be sure there are no undefined terms.

7. Solve $\dfrac{1}{2} + \dfrac{3}{x} = 5$.

$2x\left(\dfrac{1}{2}\right) + 2x\left(\dfrac{3}{x}\right) = 2x(5)$

$x + 6 = 10x$

$6 = 9x$

$x = \dfrac{2}{3}$

Complex Fractions [6.6]

A rational expression that contains a fraction in its numerator or denominator is called a complex fraction. The most common method of simplifying a complex fraction is to multiply the top and bottom by the LCD for all denominators.

8. $\dfrac{1 - \dfrac{4}{x}}{x - \dfrac{16}{x}} = \dfrac{x\left(1 - \dfrac{4}{x}\right)}{x\left(x - \dfrac{16}{x}\right)}$

$= \dfrac{x - 4}{x^2 - 16}$

$= \dfrac{x - 4}{(x - 4)(x + 4)}$

$= \dfrac{1}{x + 4}$

Ratio and Proportion [6.1, 6.7]

The ratio of a to b is

$$\dfrac{a}{b}$$

Two equal ratios form a proportion. In the proportion

$$\dfrac{a}{b} = \dfrac{c}{d}$$

a and d are the *extremes*, and b and c are the *means*. In any proportion the product of the extremes is equal to the product of the means.

9. Solve $\dfrac{3}{x} = \dfrac{5}{20}$ for x.

$3 \cdot 20 = 5 \cdot x$

$60 = 5x$

$x = 12$

Direct Variation [6.8]

The variable y is said to vary directly as the variable x if $y = kx$, where k is a real number.

10. If y varies directly as the square of x, then

$$y = kx^2$$

Inverse Variation [6.8]

The variable y is said to vary inversely as the variable x if $y = \dfrac{k}{x}$, where k is a real number.

11. If y varies inversely as the cube of x, then

$$y = \dfrac{k}{x^3}$$

Reduce to lowest terms. [6.1]

1. $\dfrac{x^2 - 16}{x^2 - 7x + 12}$

2. $\dfrac{6x^2 + 15x - 36}{12x - 18}$

3. $\dfrac{xy + 6x - 4y - 24}{x^3 - 4x^2 + 2x - 8}$

Multiply or divide as indicated. [6.2]

4. $\dfrac{3x - 6}{4} \cdot \dfrac{12}{6x - 12}$

5. $\dfrac{x^2 - 25}{x - 6} \div \dfrac{x + 5}{x^2 - 36}$

6. $\dfrac{x^2 - 8x + 15}{x^2 - x - 20} \div \dfrac{2x^2 - 8x + 6}{x^2 + 2x - 8}$

7. $(x^2 - 81)\left(\dfrac{x + 3}{x - 9}\right)$

Add or subtract as indicated. [6.3]

8. $\dfrac{5}{x - 3} + \dfrac{2}{x - 3}$

9. $\dfrac{x}{x^2 - 9} - \dfrac{2}{4x + 12}$

10. $\dfrac{9}{x^2 - 3x - 18} - \dfrac{7}{x^2 - 5x - 6}$

Solve the following equations. [6.4]

11. $\dfrac{4}{7} = \dfrac{x + 5}{6}$

12. $\dfrac{16}{x - 6} = \dfrac{7}{x}$

13. $\dfrac{5}{x - 4} - \dfrac{4}{x + 1} = \dfrac{9}{x^2 - 3x - 4}$

Solve the following problems. [6.5]

14. Speed of a Boat It takes a boat 5 hours to travel upstream. It takes the same boat 4 hours to travel the same distance downstream. If the current is 1 mile per hour, what is the speed of the boat in still water?

15. Emptying a Pool An inlet pipe can fill a pool in 15 hours, the outlet pipe can empty it in 5 hours. If the pool is full and both pipes are open, how long will it take to empty?

Simplify each complex fraction. [6.6]

16. $\dfrac{1 - \dfrac{4}{x}}{2 - \dfrac{3}{x}}$

17. $\dfrac{1 - \dfrac{16}{x^2}}{1 - \dfrac{2}{x} - \dfrac{8}{x^2}}$

Solve the following problems involving ratio and proportion. [6.7]

18. Ratio A solution of alcohol and water contains 45 milliliters of alcohol and 60 milliliters of water. What is the ratio of alcohol to water and the ratio of alcohol to total volume?

19. Ratio A manufacturer knows that during a production run 3 out of every 175 parts produced by a certain machine will be defective. If the machine produces 1,575 parts, how many can he expect to be defective?

20. Direct Variation Suppose y varies directly with the square of x. If y is 81 when x is 3, find y when x is 2. [6.8]

21. Inverse Variation If y varies inversely with the square of x, and y is 4 when x is 16, find y when x is 64. [6.8]

22. Ratio Use the chart below to find the ratio of the cost of a Bugatti Veyron to the cost of a Lamborghini Reventon. Round to the nearest tenth.

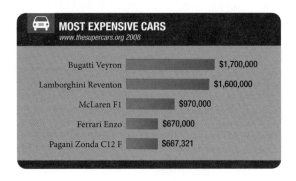

23. Unit Analysis The illustration below shows the minimum acreage of sports fields. If 1 acre equals 43,560 square feet, what is the minimum square footage of tennis, basketball and volleyball courts? Round each to the nearest tenth.

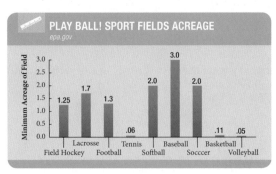

Simplify.

1. $9 - 17$

2. $-14 + 27$

3. $\dfrac{-36}{4}$

4. $\dfrac{1}{4}(8x)$

5. $6x - 3 - 2x + 7$

6. $6 - 2x + 4$

7. $\dfrac{x^{-14}}{x^{-23}}$

8. $\left(\dfrac{x^6}{x^2}\right)^{-3}$

9. $3^0 + (-5)^1 - 4^0$

10. $\dfrac{(x^{-3})^2(x^{-2})^{-4}}{x^{-4}}$

11. $6x - 11x$

12. $(3a^3 - 2a^2 - 5) - (-5a^2 - 6a - 3)$

13. $\dfrac{x^2}{x + 3} - \dfrac{2x + 15}{x + 3}$

14. $\dfrac{4}{6x - 12} + \dfrac{x}{x - 2}$

Solve each equation.

15. $6x - 1 = 4x + 5$

16. $\dfrac{2}{3}(6x - 9) = \dfrac{1}{2}(8x - 10)$

17. $24r^2 - 6 = 0$

18. $y^2 - 3y = 4y - 10$

19. $\dfrac{6}{x} - \dfrac{1}{4} = \dfrac{5}{x}$

20. $\dfrac{5}{x - 2} - \dfrac{3}{x + 4} = \dfrac{6}{x^2 + 2x - 8}$

Solve each system.

21. $-x - 2y = -7$
 $y = 3x + 14$

22. $x + 2y = 3$
 $y = -\dfrac{1}{2}x - 5$

23. $x - 5y = 17$
 $2x + 8y = -2$

24. $3x - 2y = 27$
 $-5x - y = -6$

Graph each equation on a rectangular coordinate system.

25. $y = -2x + 7$

26. $y = \dfrac{2}{3}x + 1$

Solve each inequality.

27. $-4x > 12$

28. $-\dfrac{a}{4} \geq 2$

Factor completely.

29. $xy + 7x - 5y - 35$

30. $9x^2 - 16y^2$

31. $a^2 + 2a - 15$

32. $15x^2 - 19x + 6$

33. $9x^2 - 48x + 64$

34. $5x^2 + 20$

35. For the equation $4x - 8y = 24$, find the x- and y-intercepts.

36. For the equation $4x - y = 6$, find the slope and y-intercept.

37. Find the equation for the line with slope $-\dfrac{2}{3}$ and y-intercept -5.

38. Find the equation for the line with slope $\dfrac{3}{5}$ that passes through $(5, -3)$.

39. Subtract 3 from -7.

40. Add -5 to the product of -2 and 3.

Give the property that justifies the statement.

41. $(x + 3) + 5 = (3 + x) + 5$

42. $(x + 2) + 7 = x + (2 + 7)$

43. Divide $\dfrac{x^2 + 2x - 15}{x - 3}$.

44. Multiply $\dfrac{3x - 6}{6x - 24} \cdot \dfrac{2x - 8}{3x - 9}$.

Reduce to lowest terms.

45. $\dfrac{6a - 9}{18a - 27}$

46. $\dfrac{4xy - 8x - y + 2}{2xy - 4x + 3y - 6}$

47. Direct Variation y varies directly as the square of x. If $y = -24$ when $x = 2$, find y when $x = -4$.

48. Inverse Variation y varies inversely as x. If $y = -3$ when $x = 4$, find y when $x = -2$.

Use the chart to answer the following questions. Round your answers to the nearest tenth. (*Note*: 1 mile = 1.61 kilometers)

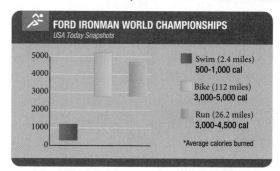

49. What is the minimum average calories burned per kilometer during the swim portion of the Ironman?

50. What is the maximum average calories burned per kilometer during the bicycling portion of the Ironman?

Reduce to lowest terms. [6.1]

1. $\dfrac{x^2 - 9}{x^2 - 6x + 9}$

2. $\dfrac{15a + 30}{5a^2 - 10a - 40}$

3. $\dfrac{xy - 2x + 3y - 6}{y^3 - 2y^2 - y + 2}$

Multiply or divide as indicated. [6.2]

4. $\dfrac{2x - 6}{3} \cdot \dfrac{9}{4x - 12}$

5. $\dfrac{x^2 - 9}{x - 4} \div \dfrac{x + 3}{x^2 - 16}$

6. $\dfrac{x^2 + x - 6}{x^2 + 4x + 3} \div \dfrac{x^2 + 2x - 8}{2x^2 - x - 3}$

7. $(x^2 - 16)\left(\dfrac{x - 1}{x - 4}\right)$

Add or subtract as indicated. [6.3]

8. $\dfrac{7}{x - 8} - \dfrac{9}{x - 8}$

9. $\dfrac{x}{x^2 - 16} + \dfrac{3}{3x - 12}$

10. $\dfrac{3}{(x - 3)(x + 3)} - \dfrac{1}{(x - 3)(x - 1)}$

Solve the following equations. [6.4]

11. $\dfrac{3}{5} = \dfrac{x + 3}{7}$

12. $\dfrac{25}{x - 3} = \dfrac{7}{x}$

13. $\dfrac{6}{x - 3} - \dfrac{5}{x + 1} = \dfrac{7}{x^2 - 2x - 3}$

Solve the following problems. [6.5]

14. **Speed of a Boat** It takes a boat 3 hours to travel upstream. It takes the same boat 2 hours to travel the same distance downstream. If the current is 3 miles per hour, what is the speed of the boat in still water?

15. **Emptying a Pool** An inlet pipe can fill a pool in 12 hours, the outlet pipe can empty it in 8 hours. If the pool is full and both pipes are open, how long will it take to empty?

Simplify each complex fraction. [6.6]

16. $\dfrac{1 + \dfrac{2}{x}}{1 - \dfrac{2}{x}}$

17. $\dfrac{1 - \dfrac{9}{x^2}}{1 - \dfrac{1}{x} - \dfrac{6}{x^2}}$

Solve the following problems involving ratio and proportion. [6.7]

18. **Ratio** A solution of alcohol and water contains 29 milliliters of alcohol and 87 milliliters of water. What is the ratio of alcohol to water and the ratio of alcohol to total volume?

19. **Ratio** A manufacturer knows that during a production run 6 out of every 150 parts produced by a certain machine will be defective. If the machine produces 2,550 parts, how many can he expect to be defective?

20. **Direct Variation** Suppose y varies directly as the cube of x. If y is 16 when x is 2, find y when x is 3. [6.8]

21. **Inverse Variation** If y varies inversely as the square of x, and y is 8 when x is 3, find y when x is 6. [6.8]

22. **Ratio** The chart shows the number of convenience stores in each state. Use the information to find the ratio of the number of convenience stores in Texas to the number of stores in Illinois. Round to the nearest tenth.

CONVENIENCE STORES AROUND THE U.S.
U.S. Census Bureau, 2002

State	Establishments
Texas	9,143
California	5,768
Florida	5,676
Ohio	3,537
Illinois	3,280
Michigan	3,402
Pennsylvania	2,996
Georgia	4,050
North Carolina	4,081

*Convenience Store Establishments per State

23. **Unit Analysis** The chart shows the number of milligrams of caffeine in different kinds of sodas. Convert each amount to grains (1 mg = 0.01543 grains) and round to the nearest hundredth.

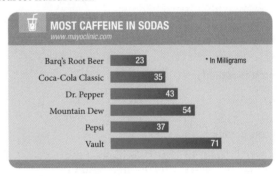

MOST CAFFEINE IN SODAS
www.mayoclinic.com

Soda	Milligrams
Barq's Root Beer	23
Coca-Cola Classic	35
Dr. Pepper	43
Mountain Dew	54
Pepsi	37
Vault	71

* In Milligrams

Systems of Linear Equations

Gray Buildings © 2008 Sanborn
Image © 2010 DigitalGlobe
Las Vegas, Nevada

I f you have ever been to Las Vegas, Nevada, you have most likely seen the Stratosphere Hotel and Casino, which contains the tallest observation deck in the United States. The observation deck stands at 869 feet above the ground, offers 360-degree views, and provides visitors access to some incredible thrill rides. The Big Shot is a roller coaster circling the tower at nearly 45 miles per hour, and is the highest coaster in the United States. The second highest thrill ride in the world, Insanity, dangles riders out over The Strip on a 64-foot arm as they spin around at three times regular g-forces. The X-Scream is a seesaw that holds patrons out over the edge of the tower more than 80 stories from the ground. The newest Stratosphere attraction, the SkyJump, allows jumpers to descend over 100 stories down to street level wearing a special suit and guided by wires. The controlled descent allows visitors the experience of a sky dive without the parachute or the airplane.

©iStockphoto.com/Kasia75

Suppose a ticket to ride Insanity costs $12, a ticket for X-Scream costs $14, a ticket for Big Shot costs $13, and a ticket for SkyJump costs $110. Let's also suppose that in one day, the thrill rides were ridden 105 times bringing in a total of $2,810. How many tickets for each ride were sold? In order to answer this question, you will need to know how to work with systems of equations, which is the topic of this chapter.

7.1 Solving Linear Systems by Graphing

Image © sxc.hu, Amyjacobs, 2007

KEY WORDS

system of linear equations

point of intersection

consistent system

inconsistent system

dependent equations

independent equations

Each year, spectators at the QuikTrip Air and Rocket Racing Show see a variety of planes, such as rocket racers, combat aircraft, biplanes, and gliders take to the skies above an airfield in Tulsa, Oklahoma. The wide-eyed ticket holders, packed tightly into the grandstands, watch planes perform dramatic acrobatic stunts. During some acts, two or more planes fly in the air at once. It it extremely important that the planes' routines are well-planned to avoid any collisions. Suppose we were to assign a linear equation, in the form of $Ax + By = C$, to the path of each plane of a two-plane act. When we consider these two equations together, we are working with a *system of linear equations*, which is the focus of this section. Then we would be able to determine if the planes were flying directly parallel to each other or if their paths will intersect.

A Solving a System of Linear Equations by Graphing

In a system of linear equations, both equations contain two variables and have graphs that are straight lines. The following are systems of linear equations:

$$x + y = 3 \qquad y = 2x + 1 \qquad 2x - y = 1$$
$$3x + 4y = 2 \qquad y = 3x + 2 \qquad 3x - 2y = 6$$

The solution set for a system of linear equations is all ordered pairs that are solutions to both equations. The intersection of the graphs are a point whose coordinates are solutions to a system; that is, if we graph both equations on a coordinate system, we can read the coordinates of the point of intersection as a solution to our system. Here is an example.

EXAMPLE 1 Solve the following system by graphing.

$$x + y = 4$$
$$x - y = -2$$

Solution Graph each equation separately. Figure 1 shows both graphs, without showing the work necessary to get them. We can see the graphs intersect at the point (1, 3). Therefore, the point (1, 3) must be a solution to our system because it is the only ordered pair that lies on both lines. Its coordinates satisfy both equations.

Practice Problems

1. Solve the system by graphing.

$$x + y = 3$$
$$x - y = 5$$

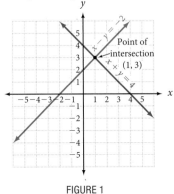

FIGURE 1

Answer

1. $(4, -1)$

We can check our results by substituting the coordinates $x = 1$, $y = 3$ into both equations to see if they work.

When →	$x = 1$	When →	$x = 1$
and →	$y = 3$	and →	$y = 3$
the equation →	$x + y = 4$	the equation →	$x - y = -2$
becomes →	$1 + 3 \overset{?}{=} 4$	becomes →	$1 - 3 \overset{?}{=} -2$
	$4 = 4$		$-2 = -2$

The point $(1, 3)$ satisfies both equations.

Here are some steps to follow when solving linear systems by graphing.

> **HOW TO** **Solving a Linear System by Graphing**
>
> **Step 1:** Graph the first equation by the methods described in Chapter 3.
>
> **Step 2:** Graph the second equation on the same set of axes used for the first equation.
>
> **Step 3:** Read the coordinates of the point of intersection of the two graphs.
>
> **Step 4:** Check the solution in both equations.

EXAMPLE 2 Solve the following system by graphing.

$$x + 2y = 8$$
$$2x - 3y = 2$$

Solution Graphing each equation on the same coordinate system, we have the lines shown in Figure 2.

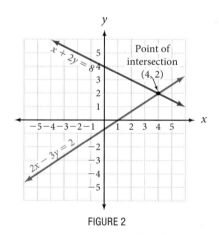

FIGURE 2

From Figure 2, we can see the solution for our system is $(4, 2)$. We check this solution as follows:

When →	$x = 4$	When →	$x = 4$
and →	$y = 2$	and →	$y = 2$
the equation →	$x + 2y = 8$	the equation →	$2x - 3y = 2$
becomes →	$4 + 2(2) \overset{?}{=} 8$	becomes →	$2(4) - 3(2) \overset{?}{=} 2$
	$4 + 4 = 8$		$8 - 6 = 2$
	$8 = 8$		$2 = 2$

The point $(4, 2)$ satisfies both equations and, therefore, must be the solution to our system.

2. Solve the system by graphing.

$$x + 2y = 6$$
$$3x - y = 4$$

Answer

2. $(2, 2)$

3. Solve by graphing.

$$y = 3x - 1$$
$$x = 2$$

4. Solve by graphing.

$$y = x + 3$$
$$y = x - 2$$

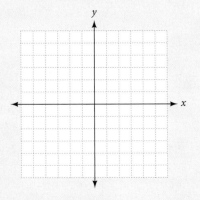

5. Graph the system.

$$6x + 2y = 12$$
$$9x + 3y = 18$$

Answers

3. $(2, 5)$
4. No solution.
5. The lines coincide.

EXAMPLE 3 Solve this system by graphing.

$$y = 2x - 3$$
$$x = 3$$

Solution Graphing both equations on the same set of axes, we have Figure 3.

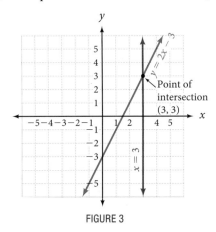

FIGURE 3

The solution to the system is the point (3, 3).

EXAMPLE 4 Solve by graphing.

$$y = x - 2$$
$$y = x + 1$$

Solution Graphing both equations produces the lines shown in Figure 4. We can see in Figure 4 that the lines are parallel and therefore do not intersect. Our system has no ordered pair as a solution because there is no ordered pair that satisfies both equations. There is no solution to the system.

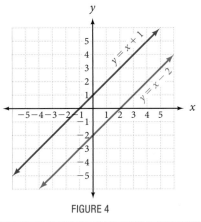

FIGURE 4

EXAMPLE 5 Graph the system.

$$2x + y = 4$$
$$4x + 2y = 8$$

Solution Both graphs are shown in Figure 5. The two graphs coincide. The reason becomes apparent when we multiply both sides of the first equation by 2.

$$2x + y = 4$$
$$2(2x + y) = 2(4) \qquad \text{Multiply both sides by 2.}$$
$$4x + 2y = 8$$

The equations have the same solution set. We say the lines coincide. Any ordered pair that is a solution to one is a solution to the system. The system has an infinite number of solutions. (Any point on the line is a solution to the system.)

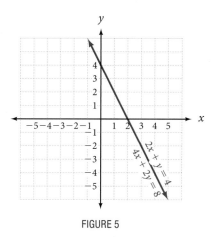

FIGURE 5

We sometimes use special vocabulary to describe the special cases shown in Examples 4 and 5. When a system of equations has no solution because the lines are parallel (as in Example 4), we say the system is *inconsistent*. When the lines coincide (as in Example 5), we say the equations are *dependent*.

The special cases illustrated in the previous two examples do not happen often. Usually, a system has a single ordered pair as a solution in which case we call the system *consistent* and the equations *independent*. Solving a system of linear equations by graphing is useful only when the ordered pair in the solution set has integers for coordinates. Two other solution methods work well in all cases. We will develop the other two methods in the next two sections.

Here is a summary of three possible types of solutions to a system of equations in two variables:

One Solution	**No Solution**	**Infinite Solutions**
Lines intersect at a single point	Lines are parallel and never cross	Lines coincide

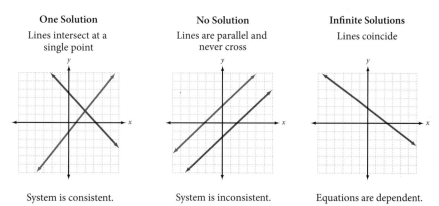

System is consistent.	System is inconsistent.	Equations are dependent.

GETTING READY FOR CLASS

After reading through the preceding section, respond in your own words and in complete sentences.

A. What is a system of linear equations?

B. What is a solution to a system of linear equations?

C. How do you solve a system of linear equations by graphing?

D. Under what conditions will a system of linear equations not have a solution?

EXERCISE SET 7.1

Vocabulary Review

Choose the correct words to fill in the blanks below.

intersection axes ordered pairs system

1. A _____ of linear equations is two linear equations considered together.
2. The solution set for a system of linear equations is all _____ that are solutions to both equations.
3. To solve a system of linear equations, first graph the two equations on the same _____.
4. Once you have graphed a system of linear equations, check the point of _____ as a solution to both equations.

Problems

A Solve the following systems of linear equations by graphing. If the system is inconsistent or dependent, state so.

1. $x + y = 3$
$x - y = 1$

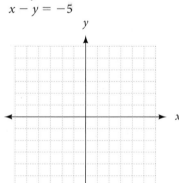

2. $x + y = 2$
$x - y = 4$

3. $x + y = 1$
$-x + y = 3$

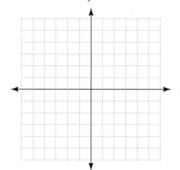

4. $x + y = 1$
$x - y = -5$

5. $x + y = 8$
$-x + y = 2$

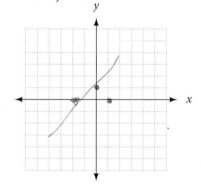

6. $x + y = 6$
$-x + y = -2$

SCAN TO ACCESS

7. $3x - 2y = 6$
 $x - y = 1$

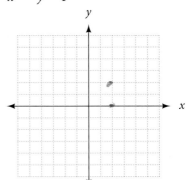

8. $5x - 2y = 10$
 $x - y = -1$

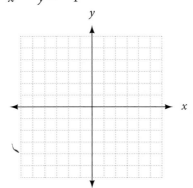

9. $6x - 2y = 12$
 $3x + y = -6$

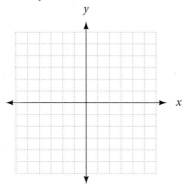

10. $4x - 2y = 8$
 $2x + y = -4$

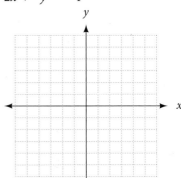

11. $4x + y = 4$
 $3x - y = 3$

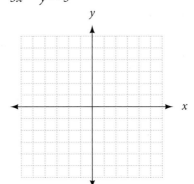

12. $5x - y = 10$
 $2x + y = 4$

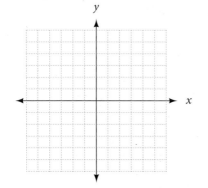

13. $x + 2y = 0$
 $2x - y = 0$

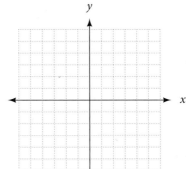

14. $3x + y = 0$
 $5x - y = 0$

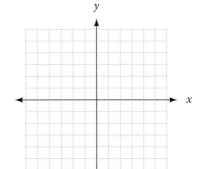

15. $3x - 5y = 15$
 $-2x + y = 4$

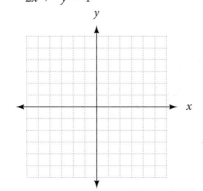

16. $2x - 4y = 8$
 $2x - y = -1$

17. $y = 2x + 1$
 $y = -2x - 3$

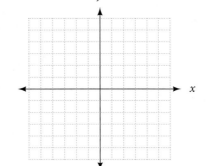

18. $y = 3x - 4$
 $y = -2x + 1$

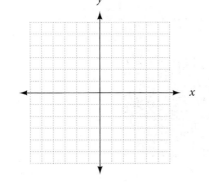

19. $x + 3y = 3$
$y = x + 5$

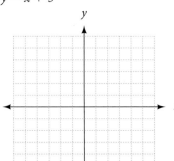

20. $2x + y = -2$
$y = x + 4$

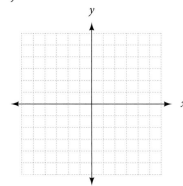

21. $x + y = 2$
$x = -3$

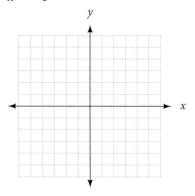

22. $x + y = 6$
$y = 2$

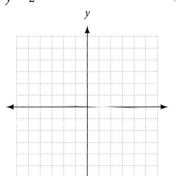

23. $x = -4$
$y = 6$

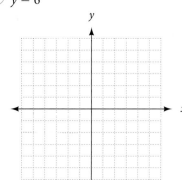

24. $x = 5$
$y = -1$

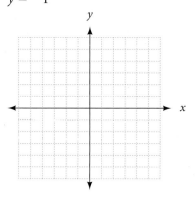

25. $x + y = 4$
$2x + 2y = -6$

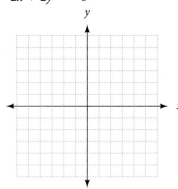

26. $x - y = 3$
$2x - 2y = 6$

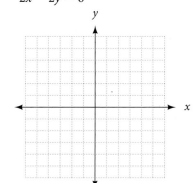

27. $4x - 2y = 8$
$2x - y = 4$

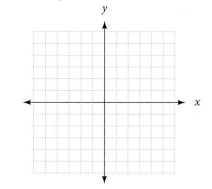

28. $3x - 6y = 6$
$x - 2y = 4$

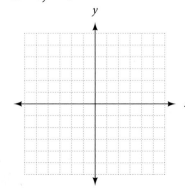

29. $y = 3x + 1$
$y = -2x + 1$

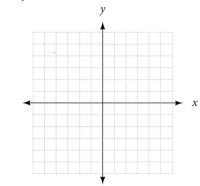

30. $y = -4x$
$y = \frac{1}{2}x$

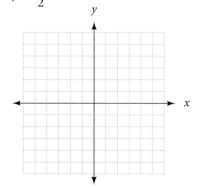

31. As you probably have guessed by now, it can be difficult to solve a system of equations by graphing if the solution to the system contains a fraction. The solution to the following system is $\left(\frac{1}{2}, 1\right)$. Solve the system by graphing.

$$y = -2x + 2$$
$$y = 4x - 1$$

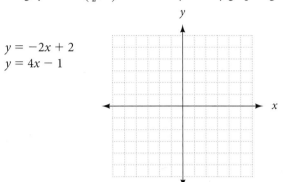

32. The solution to the following system is $\left(\frac{1}{3}, -2\right)$. Solve the system by graphing.

$$y = 3x - 3$$
$$y = -3x - 1$$

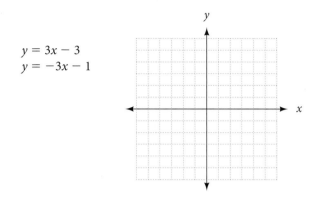

33. A second difficulty can arise in solving a system of equations by graphing if one or both of the equations is difficult to graph. The solution to the following system is $(2, 1)$. Solve the system by graphing.

$$3x - 8y = -2$$
$$x - y = 1$$

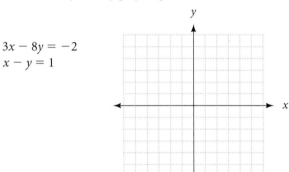

34. The solution to the following system is $(-3, 2)$. Solve the system by graphing.

$$2x + 5y = 4$$
$$x - y = -5$$

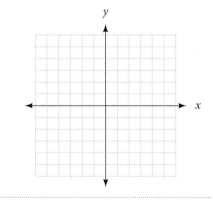

Applying the Concepts

35. Suppose you have 45 poker chips with $5 and $25 values. The total value of chips is $465. If x is the number of $5 chips and y is the number of $25 chips, then the following system of equations represents this situation:

$$x + y = 45$$
$$5x + 25y = 465$$

Solve this system by graphing each equation.

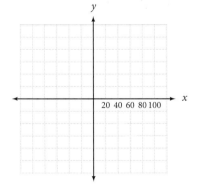

36. The stratosphere's Big Shot holds 16 passengers. Tickets can be purchased for 1 ride ($13), or for all 3 rides ($31). The total value of tickets purchased is $388. If x is the number of one-ride tickets and y is the number of three-ride tickets, then the following system of equations represents this situation:

$$x + y = 16$$
$$13x + 31y = 388$$

Solve this system by graphing each equation.

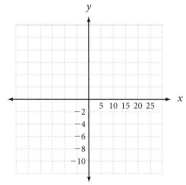

37. Job Comparison Jane is deciding between two sales positions. She can work for Marcy's and receive $8.00 per hour, or she can work for Gigi's, where she earns $6.00 per hour but also receives a $50 commission per week. The two lines in the following figure represent the money Jane will make for working at each of the jobs.

a. From the figure, how many hours would Jane have to work to earn the same amount at each of the positions?

b. If Jane expects to work less than 20 hours a week, which job should she choose?

c. If Jane expects to work more than 30 hours a week, which job should she choose?

38. Truck Rental You need to rent a moving truck for two days. Rider Moving Trucks charges $50 per day and $0.50 per mile. UMove Trucks charges $45 per day and $0.75 per mile. The following figure represents the cost of renting each of the trucks for two days.

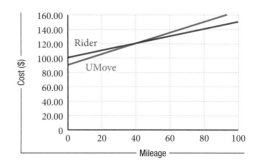

a. From the figure, after how many miles would the trucks cost the same?

b. Which company will give you a better deal if you drive less than 30 miles?

c. Which company will give you a better deal if you drive more than 60 miles?

Getting Ready for the Next Section

Solve each equation.

39. $x + (2x - 1) = 2$

40. $2x - 3(2x - 8) = 12$

41. $2(3y - 1) - 3y = 4$

42. $-2x + 4(3x + 6) = 14$

43. $4x + 2(-2x + 4) = 8$

44. $1.5x + 15 = 0.75x + 24.95$

Solve each equation for the indicated variable.

45. $x - 3y = -1$ for x

46. $-3x + y = 6$ for y

47. Let $y = 2x - 1$. If $x = 1$, find y.

48. Let $y = 2x - 8$. If $x = 5$, find y.

49. Let $x = 3y - 1$. If $y = 2$, find x.

50. Let $y = 3x + 6$. If $y = -6$, find x.

Let $y = 1.5x + 15$ for the next four problems.

51. If $x = 13$, find y.

52. If $x = 14$, find y.

53. If $y = 21$, find x.

54. If $y = 42$, find x.

Let $y = 0.75x + 24.95$ for the next four problems.

55. If $x = 12$, find y.

56. If $x = 16$, find y.

57. If $y = 39.95$, find x.

58. If $y = 41.45$, find x.

Find the Mistake

Each sentence below contains a mistake. Circle the mistake and write the correct word(s) or number(s) on the line provided.

1. The point of intersection for two straight line graphs is a solution to only one of the equations. _____

2. The point of intersection for the equations $y = 6$ and $y = 4x + 2$ is $(2, 6)$. _____

3. The point of intersection for the equations $x = 5$ and $x = 9$ is $(5, 9)$. _____

4. When lines coincide, we say that the equations are inconsistent. _____

Navigation Skills: Prepare, Study, Achieve

It is very common for students to feel burned out or feel a decrease in motivation for continuing to put in the hard work this course requires. Many factors may contribute to this burn out and vary depending on the student. It is important to recognize these factors and implement ways to combat them. If you notice yourself feeling burned out, the following suggestions may help:

• Join a study group for emotional support during tough times in this class.
• Set short-term and long-term goals.
• Take breaks during your study sessions.
• Begin study sessions with the most difficult topics, then work through the easier topics near the end of your session.
• Revisit your study calendar to see if you have overbooked your time.
• Vary your time and location for studying.
• Examine your diet and introduce healthier and more balanced meals.
• Get adequate restful sleep between studying and going to class.
• Schedule regular physical activity during your day.
• Make use of your on-campus health services for either physical or psychological needs.

7.2 The Substitution Method

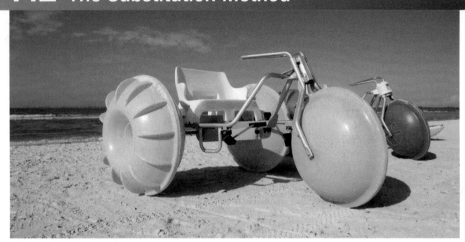

© iStockphoto.com/cstewart

Art and physical endurance enthusiasts race in an annual event called the Kinetic Sculpture Grand Championship. Teams have each engineered an amphibious work of art that can travel over any terrain. The vehicle must be able to hold its pilots, who power it over sand dunes, across a harbor, up and down steep city streets, and through a swamp. The vehicles are built and decorated as if they were elaborate parade floats, many with animated parts and their pilots dressed in costume.

Let's suppose a kinetic sculpture team races its vehicle on a river. It takes 5 minutes for the vehicle to race down the river going with a current of 3 miles per hour, and 12 minutes to travel the same distance back up the river against the current. After working through this section, you will be able to set up a linear system of equations for this problem and use an algebraic method, called the substitution method, to solve the system.

A Substitution Method

EXAMPLE 1 Solve the system by substitution.

$$x + y = 2$$
$$y = 2x - 1$$

Solution If we were to solve this system by the methods used in the previous section, we would have to graph both lines and see where they intersect. There is no need to do this, however, because the second equation tells us that y is $2x - 1$. We can replace the y variable in the first equation with the expression $2x - 1$ from the second equation; that is, we *substitute* $2x - 1$ from the second equation for y in the first equation. Here is what it looks like:

$$x + (2x - 1) = 2$$

The equation we end up with contains only the variable x. The y variable has been eliminated by substitution.

Solving the resulting equation, we have

$$x + (2x - 1) = 2$$
$$3x - 1 = 2$$
$$3x = 3$$
$$x = 1$$

This is the x-coordinate of the solution to our system. To find the y-coordinate, we substitute $x = 1$ into the second equation of our system. (We could substitute $x = 1$ into the first equation also and have the same result.)

$$y = 2(1) - 1$$
$$y = 2 - 1$$
$$y = 1$$

> *Note* Sometimes this method of solving systems of equations is confusing the first time you see it. If you are confused, you may want to read through this first example more than once and try it on your own before attempting the previous practice problem.

The solution to our system is the ordered pair $(1, 1)$. It satisfies both of the original equations. Figure 1 provides visual evidence that the substitution method yields the correct solution.

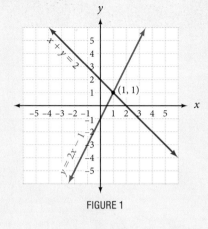

FIGURE 1

EXAMPLE 2 Solve the system by substitution.

$$2x - 3y = 12$$
$$y = 2x - 8$$

Solution Again, the second equation says y is $2x - 8$. Because we are looking for the ordered pair that satisfies both equations, the y in the first equation must also be $2x - 8$. Substituting $2x - 8$ from the second equation for y in the first equation, we have

$$2x - 3(2x - 8) = 12$$

This equation can still be read as $2x - 3y = 12$ because $2x - 8$ is the same as y. Solving the equation, we have

$$2x - 3(2x - 8) = 12$$
$$2x - 6x + 24 = 12$$
$$-4x + 24 = 12$$
$$-4x = -12$$
$$x = 3$$

To find the y-coordinate of our solution, we substitute $x = 3$ into the second equation in the original system.

When → $x = 3$
the equation → $y = 2x - 8$
becomes → $y = 2(3) - 8$
 $y = 6 - 8 = -2$

The solution to our system is $(3, -2)$.

2. Solve the system by substitution.

$$5x - 4y = -2$$
$$y = 2x + 2$$

EXAMPLE 3 Solve the following system by solving the first equation for x and then using the substitution method.

$$x - 3y = -1$$
$$2x - 3y = 4$$

Solution We solve the first equation for x by adding $3y$ to both sides to get

$$x = 3y - 1$$

Using this value of x in the second equation, we have

$$2(3y - 1) - 3y = 4$$
$$6y - 2 - 3y = 4$$
$$3y - 2 = 4$$
$$3y = 6$$
$$y = 2$$

Next, we find x.

When → $y = 2$
the equation → $x = 3y - 1$
becomes → $x = 3(2) - 1$
 $x = 6 - 1$
 $x = 5$

The solution to our system is $(5, 2)$

3. Solve the first equation for x, and then substitute the result into the second equation to solve the system by substitution.

$$x - 4y = -5$$
$$3x - 2y = 5$$

Answers

2. $(-2, -2)$

3. $(3, 2)$

Here are the steps to use in solving a system of equations by the substitution method.

> **HOW TO** Strategy for Solving a System of Linear Equations by the Substitution Method
>
> **Step 1:** Solve either one of the equations for x or y. (This step is not necessary if one of the equations is already in the correct form, as in Examples 1 and 2.)
>
> **Step 2:** Substitute the expression for the variable obtained in step 1 into the other equation and solve it.
>
> **Step 3:** Substitute the solution from step 2 into any equation in the system that contains both variables and solve it.
>
> **Step 4:** Check your results, if necessary.

4. Solve the system by substitution.

$$5x - y = 1$$
$$-2x + 3y = 10$$

EXAMPLE 4 Solve the system by substitution.

$$-2x + 4y = 14$$
$$-3x + y = 6$$

Solution We can solve either equation for either variable. If we look at the system closely, it becomes apparent that solving the second equation for y is the easiest way to go. If we add $3x$ to both sides of the second equation, we have

$$y = 3x + 6$$

Substituting the expression $3x + 6$ back into the first equation in place of y yields the following result.

$$-2x + 4(3x + 6) = 14$$
$$-2x + 12x + 24 = 14$$
$$10x + 24 = 14$$
$$10x = -10$$
$$x = -1$$

Substituting $x = -1$ into the equation $y = 3x + 6$ leaves us with

$$y = 3(-1) + 6$$
$$y = -3 + 6$$
$$y = 3$$

The solution to our system is $(-1, 3)$.

5. Solve the system by substitution.

$$6x + 3y = 1$$
$$y = -2x - 5$$

EXAMPLE 5 Solve the system by substitution.

$$4x + 2y = 8$$
$$y = -2x + 4$$

Solution Substituting the expression $-2x + 4$ for y from the second equation into the first equation, we have

$$4x + 2(-2x + 4) = 8$$
$$4x - 4x + 8 = 8$$
$$8 = 8 \qquad \text{A true statement}$$

Both variables have been eliminated, and we are left with a true statement. A true statement in this situation tells us the lines coincide; that is, the equations $4x + 2y = 8$ and $y = -2x + 4$ have exactly the same graph. Any point on that graph has coordinates that satisfy both equations and is a solution to the system. This system is dependent. If the result had been a false statement, this would tell us that the system is inconsistent, that is, the lines are parallel. If the resulting statement is false, there is no solution to the system.

Answers

4. $(1, 4)$

5. No solution. The system is inconsistent.

EXAMPLE 6 The following table shows two contract rates charged by a cellular phone company. How many text messages must a person send so that the cost will be the same regardless of which plan is chosen?

	Flat Monthly Rate	Per Text Charge
Plan 1	$40	$0.15
Plan 2	$70	$0.05

Solution If we let y = the monthly charge for x number of text messages, then the equations for each plan are

Plan 1: $y = 0.15x + 40$

Plan 2: $y = 0.05x + 70$

We can solve this system using the substitution method by replacing the variable y in Plan 2 with the expression $0.15x + 40$ from Plan 1. If we do so, we have

$$0.15x + 40 = 0.05x + 70$$

$$0.10x + 40 = 70$$

$$0.10x = 30$$

$$x = 300$$

The monthly bill is based on the number of text messages you send. We can use this value of x to gather more information by plugging it into either plan equation to find y.

Plan 1: $y = 0.15x + 40$	or	Plan 2: $y = 0.05x + 70$
$y = 0.15(300) + 40$		$y = 0.05(300) + 70$
$y = 45 + 40$		$y = 15 + 70$
$y = 85$		$y = 85$

Therefore, when you send 300 text messages in a month, the total cost for that month will be $85 regardless of which plan you used.

GETTING READY FOR CLASS

After reading through the preceding section, respond in your own words and in complete sentences.

A. What is the first step in solving a system of linear equations by substitution?

B. When would substitution be more efficient than graphing in solving two linear equations?

C. What does it mean when you solve a system of linear equations by the substitution method and you end up with the statement $8 = 8$?

D. How would you begin solving the following system using the substitution method?

$$x + y = 2$$
$$y = 2x - 1$$

6. The rates for two garbage companies are given in the following table. How many bags of trash must be picked up in one month for the two companies to charge the same amount?

	Flat Rate	Per Bag Charge
Co. 1	$13.00	$1.50
Co. 2	$20.00	$1.00

Vocabulary Review

The following is a list of steps for solving a system of linear equations by the substitution method. Choose the correct words to fill in the blanks below.

substitute variables check equations

Step 1: Solve either one of the _____ for x or y.

Step 2: _____ the expression for one variable into the other equation and solve it.

Step 3: Substitute the solution from Step 2 into any equation in the system that contains both _____ and solve it.

Step 4: _____ your results.

Problems

A Solve the following systems by substitution. Substitute the expression in the second equation into the first equation and solve.

1. $x + y = 11$
$y = 2x - 1$

2. $x - y = -3$
$y = 3x + 5$

3. $x + y = 20$
$y = 5x + 2$

4. $3x - y = -1$
$x = 2y - 7$

5. $-2x + y = -1$
$y = -4x + 8$

6. $4x - y = 5$
$y = -4x + 1$

7. $3x - 2y = -2$
$x = -y + 6$

8. $2x - 3y = 17$
$x = -y + 6$

9. $5x - 4y = -16$
$y = 4$

10. $6x + 2y = 18$
$x = 3$

11. $5x + 4y = 7$
$y = -3x$

12. $10x + 2y = -6$
$y = -5x$

Solve the following systems by solving one of the equations for x or y and then using the substitution method.

13. $x + 3y = 4$
$x - 2y = -1$

14. $x - y = 5$
$x + 2y = -1$

15. $2x + y = 1$
$x - 5y = 17$

16. $2x - 2y = 2$
$x - 3y = -7$

17. $3x + 5y = -3$
$x - 5y = -5$

18. $2x - 4y = -4$
$x + 2y = 8$

19. $5x + 3y = 0$
$x - 3y = -18$

SCAN TO ACCESS

20. $x - 3y = -5$
$x - 2y = 0$

21. $-3x - 9y = 7$
$x + 3y = 12$

$-362 7$

22. $2x + 6y = -18$
$x + 3y = -9$

23. $3x + 2y = -4$
$6x + 4y = -8$

24. $y = 4x + 7$
$y = 4x - 3$

Solve the following systems using the substitution method.

25. $5x - 8y = 7$
$y = 2x - 5$

26. $3x + 4y = 10$
$y = 8x - 15$

27. $7x - 6y = -1$
$x = 2y - 1$

28. $4x + 2y = 3$
$x = 4y - 3$

29. $-3x + 2y = 6$
$y = 3x$

30. $-2x - y = -3$
$y = -3x$

31. $5x - 6y = -4$
$x = y$

32. $2x - 4y = 0$
$y = x$

33. $3x + 3y = 9$
$y = 2x - 12$

34. $7x + 6y = -9$
$y = -2x + 1$

35. $7x - 11y = 16$
$y = 10$

36. $9x - 7y = -14$
$x = 7$

37. $-4x + 4y = -8$
$y = x - 2$

38. $-4x + 2y = -10$
$y = 2x - 5$

39. $-6x + 3y = 4$
$y = 2x + 1$

40. $5x - 10y = 4$
$y = \dfrac{1}{2}x + \dfrac{2}{5}$

Solve each system by substitution. You can eliminate the decimals if you like, but you don't have to. The solution will be the same in either case.

41. $0.05x + 0.10y = 1.70$
$y = 22 - x$

42. $0.20x + 0.50y = 3.60$
$y = 12 - x$

Applying the Concepts

43. In the introduction to the section, it took 5 minutes to race down river with a current of 3 mph, and it took 12 minutes to travel back upriver against the current. Use the following system of equations to find the rate r of the kinetic sculpture team.

$$5(r + 3) = d$$
$$12(r - 3) = d$$

44. San Francisco Bay has a current of approximately 4 mph. It takes 20 minutes to power a boat with the current, but 35 minutes against the current. Use the following system of equations to find the rate r of the boat.

4 mph

$$20(r + 4) = d$$
$$35(r - 4) = d$$

45. Gas Mileage Daniel is trying to decide whether to buy a car or a truck. The truck he is considering will cost him $150 a month in loan payments, and it gets 20 miles per gallon in gas mileage. The car will cost $180 a month in loan payments, but it gets 35 miles per gallon in gas mileage. Daniel estimates that he will pay $2.50 per gallon for gas. This means that the monthly cost to drive the truck x miles will be $y = \frac{2.50}{20}x + 150$. The total monthly cost to drive the car x miles will be $y = \frac{2.50}{35}x + 180$. The following figure shows the graph of each equation.

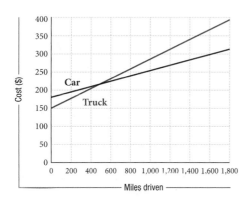

a. At how many miles do the car and the truck cost the same to operate?

b. If Daniel drives more than 1,200 miles, which will be cheaper?

c. If Daniel drives fewer than 300 miles, which will be cheaper?

d. Why do the graphs appear in the first quadrant only?

46. Video Production Pat runs a small company that duplicates DVDs. The daily cost and daily revenue for a company duplicating DVDs are shown in the following figure. The daily cost for duplicating x DVDs is $y = \frac{6}{5}x + 20$; the daily revenue (the amount of money he brings in each day) for duplicating x DVDs is $y = 1.7x$. The graphs of the two lines are shown in the following figure.

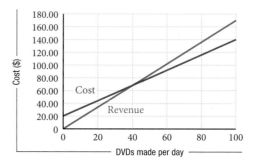

a. Pat will "break even" when his cost and his revenue are equal. How many DVDs does he need to duplicate to break even?

b. Pat will incur a loss when his revenue is less than his cost. If he duplicates 30 DVDs in one day, will he incur a loss?

c. Pat will make a profit when his revenue is larger than his costs. For what values of x will Pat make a profit?

d. Why does the graph appear in the first quadrant only?

Getting Ready For the Next Section

Simplify each of the following.

47. $(x + y) + (x - y)$

48. $(x + 2y) + (-x + y)$

49. $3(2x - y) + (x + 3y)$

50. $3(2x + 3y) - 2(3x + 5y)$

51. $-4(3x + 5y) + 5(5x + 4y)$

52. $(3x + 8y) - (3x - 2y)$

53. $6\left(\frac{1}{2}x - \frac{1}{3}y\right)$

54. $12\left(\frac{1}{4}x + \frac{2}{3}y\right)$

55. $-4(3x - 2y)$

56. $-8(2x - 3y)$

57. $4(3x - 2y) + 2(5x + 4y)$

58. $-3(4x - 3y) + 4(3x - 5y)$

59. Let $x + y = 4$. If $x = 3$, find y.

60. Let $x + 2y = 4$. If $x = 3$, find y.

61. Let $x + 3y = 3$. If $x = 3$, find y.

62. Let $2x + 3y = -1$. If $y = -1$, find x.

63. Let $3x + 5y = -7$. If $x = 6$, find y.

64. Let $3x - 2y = 12$. If $y = 6$, find x.

Find the Mistake

The table below shows systems and the first step taken to solve each system by the substitution method. Circle the mistake(s) in the first step and write the correct way to begin on the line provided.

System	First Step
1. $x = 6y - 5$ $2x + 4y = 1$	Substitute the value for x from the first equation for y in the second equation. _____
2. $x - 3y = 9$ $7y - 3x = -2$	Subtract x from both sides of the first equation to isolate y. _____
3. $y = 8x + 4$ $3x = 7y$	Divide both sides of the second equation by 3 to isolate x. _____
4. $x = \frac{1}{2}y + \frac{3}{4}$ $4y - 3x = 12$	Multiply all terms in the first equation by the LCD 4. _____

OBJECTIVES

A Use the elimination method to solve a system of linear equations in two variables.

KEY WORDS

elimination

coefficients

Practice Problems

1. Solve the following system using the elimination method.

$$x + y = 3$$
$$x - y = 5$$

Note The graphs with our first three examples are not part of their solutions. The graphs are there simply to show you that the results we obtain by the elimination method are consistent with the results we would obtain by graphing.

Answer

1. $(4, -1)$

7.3 The Elimination Method

©iStockphoto.com/Vipre77

Remember the exciting QuikTrip Air and Rocket Racing Show we discussed earlier in the chapter. Let's suppose 1720 general admission tickets were sold at the main gate for a total of $24,590. If it cost $15 for adults and $10 for children, how many of each kind of ticket were sold? In this section, we will learn another method of solving systems of linear equations in two variables, such as the one needed to answer the above question. This method is called the elimination method and makes use of the addition property.

The addition property states that if equal quantities are added to both sides of an equation, the solution set is unchanged. In the past, we have used this property to help solve equations in one variable. We will now use it to solve systems of linear equations. Here is another way to state the addition property of equality.

Let A, B, C, and D represent algebraic expressions.

$$\begin{aligned} \text{If} \rightarrow \qquad A &= B \\ \text{and} \rightarrow \qquad C &= D \\ \text{then} \rightarrow \quad A + C &= B + D \end{aligned}$$

Because C and D are equal (that is, they represent the same number), what we have done is added the same amount to both sides of the equation $A = B$. Let's see how we can use this form of the addition property of equality to solve a system of linear equations by way of the elimination, or addition, method.

A The Elimination Method

EXAMPLE 1 Solve the following system using the elimination method.

$$x + y = 4$$
$$x - y = 2$$

Solution The system is written in the form of the addition property of equality as shown above. It looks like this:

$$A = B$$
$$C = D$$

where A is $x + y$, B is 4, C is $x - y$, and D is 2.

We use the addition property of equality to add the left sides together and the right sides together.

$$\begin{aligned} x + y &= 4 \\ \underline{x - y} &= \underline{2} \\ 2x + 0 &= 6 \end{aligned}$$

Chapter 7 Systems of Linear Equations

We now solve the resulting equation for x.

$$2x + 0 = 6$$

$$2x = 6$$

$$x = 3$$

The value we get for x is the value of the x-coordinate of the point of intersection of the two lines $x + y = 4$ and $x - y = 2$. To find the y-coordinate, we simply substitute $x = 3$ into either of the two original equations. Using the first equation, we get

$$3 + y = 4$$

$$y = 1$$

The solution to our system is the ordered pair $(3, 1)$. It satisfies both equations.

When →	$x = 3$	When →	$x = 3$
and →	$y = 1$	and →	$y = 1$
the equation →	$x + y = 4$	the equation →	$x - y = 2$
becomes →	$3 + 1 \stackrel{?}{=} 4$	becomes →	$3 - 1 \stackrel{?}{=} 2$
	$4 = 4$		$2 = 2$

Figure 1 is visual evidence that the solution to our system is $(3, 1)$.

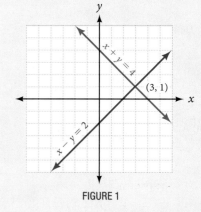

FIGURE 1

The most important part of this method of solving linear systems is eliminating one of the variables when we add the left and right sides together. In our first example, the equations were written so that the y variable was eliminated when we added the left and right sides together. If the equations are not set up this way to begin with, we have to work on one or both of them separately before we can add them together to eliminate one variable.

EXAMPLE 2 Solve the following system using the elimination method.

$$x + 2y = 4$$
$$x - y = -5$$

Solution Notice that if we were to add the equations together as they are, the resulting equation would have terms in both x and y. Let's eliminate the variable x by multiplying both sides of the second equation by -1 before we add the equations together. (As you will see, we can choose to eliminate either the x or the y variable.) Multiplying both sides of the second equation by -1 will not change its solution, so we do not need to be concerned that we have altered the system.

$$x + 2y = 4 \quad \xrightarrow{\text{No change}} \quad x + 2y = 4$$

$$x - y = -5 \xrightarrow[\text{Multiply by } -1.]{} -x + y = 5$$

$$0 + 3y = 9 \qquad \text{Add left and right sides.}$$

$$3y = 9$$

$$y = 3 \qquad \left\{ \begin{array}{l} y\text{-coordinate of the} \\ \text{point of intersection} \end{array} \right.$$

Substituting $y = 3$ into either of the two original equations, we get $x = -2$. The solution to the system is $(-2, 3)$. It satisfies both equations. Figure 2 shows the solution to the system as the point where the two lines cross.

2. Solve the following system using the elimination method.

$$x + 3y = 5$$
$$x - y = 1$$

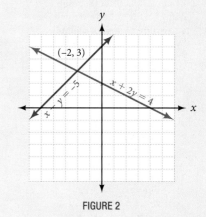

FIGURE 2

3. Solve the following system using the elimination method.

$$3x - y = 7$$
$$x + 2y = 7$$

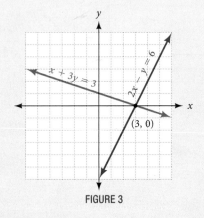

FIGURE 3

4. Solve the system using the elimination method.

$$3x + 2y = 3$$
$$2x + 5y = 13$$

Note If you are having trouble understanding this method of solution, it is probably because you can't see why we chose to multiply by 3 and −2 in the first step of Example 4. Look at the result of doing so: the $6x$ and $-6x$ will add to 0. We chose to multiply by 3 and −2 because they produce $6x$ and $-6x$, which will add to 0.

EXAMPLE 3 Solve the following system using the elimination method.

$$2x - y = 6$$
$$x + 3y = 3$$

Solution Let's eliminate the y variable from the two equations. We can do this by multiplying the first equation by 3 and leaving the second equation unchanged.

$$
\begin{array}{lcl}
2x - y = 6 & \xrightarrow{\text{Multiply by 3.}} & 6x - 3y = 18 \\
x + 3y = 3 & \xrightarrow{\text{No change}} & x + 3y = 3
\end{array}
$$

The important thing about our system now is that the *coefficients* (the numbers in front) of the y variables are opposites. When we add the terms on each side of the equal sign, then the terms in y will add to zero and be eliminated.

$$
\begin{array}{l}
6x - 3y = 18 \\
\underline{x + 3y = 3} \\
\quad 7x = 21 \qquad \text{Add corresponding terms.}
\end{array}
$$

This gives us $x = 3$. Using this value of x in the second equation of our original system, we have

$$3 + 3y = 3$$
$$3y = 0$$
$$y = 0$$

We could substitute $x = 3$ into any of the equations with both x and y variables and also get $y = 0$. The solution to our system is the ordered pair $(3, 0)$. Figure 3 is a picture of the system of equations showing the solution $(3, 0)$.

EXAMPLE 4 Solve the system using the elimination method.

$$2x + 3y = -1$$
$$3x + 5y = -2$$

Solution Let's eliminate x from the two equations. If we multiply the first equation by 3 and the second by −2, the coefficients of x will be 6 and −6, respectively. The x terms in the two equations will then add to zero.

$$
\begin{array}{lcl}
2x + 3y = -1 & \xrightarrow{\text{Multiply by 3.}} & 6x + 9y = -3 \\
3x + 5y = -2 & \xrightarrow{\text{Multiply by } -2.} & -6x - 10y = 4
\end{array}
$$

We now add the left and right sides of our new system together.

$$
\begin{array}{l}
6x + 9y = -3 \\
\underline{-6x - 10y = \ \ 4} \\
\quad -y = \ \ 1 \\
\quad \ \ y = -1
\end{array}
$$

Substituting $y = -1$ into the first equation in our original system, we have

$$2x + 3(-1) = -1$$
$$2x - 3 = -1$$
$$2x = \ \ 2$$
$$x = \ \ 1$$

The solution to our system is $(1, -1)$. It is the only ordered pair that satisfies both equations.

EXAMPLE 5 Solve the system using the elimination method.

$$3x + 5y = -7$$
$$5x + 4y = 10$$

Solution Let's eliminate y by multiplying the first equation by -4 and the second equation by 5.

$$3x + 5y = -7 \xrightarrow{\text{Multiply by } -4.} -12x - 20y = 28$$

$$5x + 4y = 10 \xrightarrow[\text{Multiply by 5.}]{} \underline{25x + 20y = 50}$$

$$13x \qquad = 78$$

$$x = 6$$

Substitute $x = 6$ into either equation in our original system, and the result will be $y = -5$. Therefore, the solution is $(6, -5)$.

EXAMPLE 6 Solve the system using the elimination method.

$$\frac{1}{2}x - \frac{1}{3}y = 2$$

$$\frac{1}{4}x + \frac{2}{3}y = 6$$

Solution Although we could solve this system without clearing the equations of fractions, there is probably less chance for error if we have only integer coefficients to work with. So let's begin by multiplying both sides of the top equation by 6 and both sides of the bottom equation by 12, to clear each equation of fractions.

$$\frac{1}{2}x - \frac{1}{3}y = 2 \xrightarrow{\text{Multiply by 6.}} 3x - 2y = 12$$

$$\frac{1}{4}x + \frac{2}{3}y = 6 \xrightarrow[\text{Multiply by 12.}]{} 3x + 8y = 72$$

Now we can eliminate x by multiplying the top equation by -1 and leaving the bottom equation unchanged.

$$3x - 2y = 12 \xrightarrow{\text{Multiply by } -1.} -3x + 2y = -12$$

$$3x + 8y = 72 \xrightarrow[\text{No change}]{} \underline{3x + 8y = \quad 72}$$

$$10y = \quad 60$$

$$y = \quad 6$$

We can substitute $y = 6$ into any equation that contains both x and y. Let's use $3x - 2y = 12$.

$$3x - 2(6) = 12$$

$$3x - 12 = 12$$

$$3x = 24$$

$$x = 8$$

The solution to the system is $(8, 6)$.

Our next two examples will remind us what happens when we try to solve a system of equations consisting of parallel lines and a system in which the lines coincide.

5. Solve the system using the elimination method.

$$5x + 4y = -6$$
$$2x + 3y = -8$$

6. Solve the system using the elimination method.

$$\frac{1}{3}x + \frac{1}{2}y = 1$$

$$x + \frac{3}{4}y = 0$$

Answers

5. $(2, -4)$

6. $(-3, 4)$

7. Solve the system using the elimination method.

$$x - 3y = 2$$
$$-3x + 9y = 2$$

EXAMPLE 7 Solve the system using the elimination method.

$$2x - y = 2$$
$$4x - 2y = 12$$

Solution Let us choose to eliminate y from the system. We can do this by multiplying the first equation by -2 and leaving the second equation unchanged.

Multiply by -2.

$$2x - \quad y = \ 2 \xrightarrow{\hspace{2cm}} -4x + 2y = -4$$
$$4x - 2y = 12 \xrightarrow[\text{No change}]{\hspace{2cm}} \quad 4x - 2y = \ 12$$

If we add both sides of the resulting system, we have

$$-4x + 2y = -4$$
$$\underline{\quad 4x - 2y = \ 12 \quad}$$
$$0 + 0 = \quad 8$$
$$\quad 0 = \quad 8 \qquad \text{A false statement.}$$

Both variables have been eliminated and we end up with the false statement $0 = 8$. We have tried to solve a system that consists of two parallel lines. There is no solution and the system is inconsistent. Figure 4 is a visual representation of the situation and is conclusive evidence that there is no solution to our system.

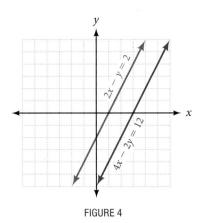

FIGURE 4

8. Solve the system using the elimination method.

$$5x - y = 1$$
$$10x - 2y = 2$$

EXAMPLE 8 Solve the system using the elimination method.

$$4x - 3y = 2$$
$$8x - 6y = 4$$

Solution Multiplying the top equation by -2 and adding, we can eliminate the variable x.

Multiply by -2.

$$4x - 3y = 2 \xrightarrow{\hspace{2cm}} -8x + 6y = -4$$
$$8x - 6y = 4 \xrightarrow[\text{No change}]{\hspace{2cm}} \quad \underline{8x - 6y = \quad 4}$$
$$0 = \quad 0$$

Both variables have been eliminated, and the resulting statement $0 = 0$ is true. In this case the lines coincide because the equations are equivalent. The solution set consists of all ordered pairs that satisfy either equation, and the system is dependent.

Answers

7. No solution (inconsistent)

8. Lines coincide (dependent)

The preceding two examples remind us of the the two special cases in which the graphs of the equations in the system either coincide or are parallel.

Here is a summary:

Both variables are eliminated and the resulting statement is false.	↔	The lines are parallel and there is no solution to the system.
Both variables are eliminated and the resulting statement is true.	↔	The lines coincide and there is an infinite number of solutions to the system.

The main idea in solving a system of linear equations by the elimination method is to use the multiplication property of equality on one or both of the original equations, if necessary, to make the coefficients of either variable opposites. The following box shows some steps to follow when solving a system of linear equations by the elimination method.

HOW TO Solving a System of Linear Equations by the Elimination Method

Step 1: Decide which variable to eliminate. (In some cases one variable will be easier to eliminate than the other. With some practice you will notice which one it is.)

Step 2: Use the multiplication property of equality on each equation separately to make the coefficients of the variable that is to be eliminated opposites.

Step 3: Add the respective left and right sides of the system together.

Step 4: Solve for the variable remaining.

Step 5: Substitute the value of the variable from step 4 into an equation containing both variables and solve for the other variable.

Step 6: Check your solution in both equations, if necessary.

GETTING READY FOR CLASS

After reading through the preceding section, respond in your own words and in complete sentences.

A. How do you use the addition property of equality in the elimination method of solving a system of linear equations?

B. What happens when you use the elimination method to solve a system of linear equations consisting of two parallel lines?

C. How would you use the multiplication property of equality to solve a system of linear equations?

D. What is the first step in solving a system of linear equations that contains fractions?

Vocabulary Review

Choose the correct words to fill in the blanks below.

substitute solution add variable solve multiplication property of equality

Solving a System of Linear Equations by the Elimination Method

Step 1: Decide which _____ to eliminate.

Step 2: Use the _____ on each equation separately to make the coefficients of the
variable that is to be eliminated opposites.

Step 3: _____ the respective left and right sides of the system together.

Step 4: _____ for the variable remaining.

Step 5: _____ the value of the variable from step 4 into an equation containing both
variable and solve for the other variable.

Step 6: Check your _____ in both equations.

Problems

A Solve the following systems of linear equations by elimination.

1. $x + y = 3$
$x - y = 1$

2. $x + y = -2$
$x - y = 6$

3. $x + y = 10$
$-x + y = 4$

4. $x - y = 1$
$-x - y = -7$

5. $x - y = 7$
$-x - y = 3$

6. $x - y = 4$
$2x + y = 8$

7. $x + y = -1$
$3x - y = -3$

8. $2x - y = -2$
$-2x - y = 2$

9. $3x + 2y = 1$
$-3x - 2y = -1$

10. $-2x - 4y = 1$
$2x + 4y = -1$

11. $3x - 2y = 4$
$-3x + 2y = 6$

12. $x - 2y = 5$
$-x + 2y = 5$

SCAN TO ACCESS

Solve each of the following systems by eliminating the *y* variable.

13. $3x - y = 4$
$2x + 2y = 24$

14. $2x + y = 3$
$3x + 2y = 1$

15. $5x - 3y = -2$
$10x - y = 1$

16. $4x - y = -1$
$2x + 4y = 13$

17. $11x - 4y = 11$
$5x + y = 5$

18. $3x - y = 7$
$10x - 5y = 25$

19. $2x - y = 6$
$x + 4y = 3$

20. $2x + y = 5$
$-2x + 2y = 4$

Solve each of the following systems by eliminating the *x* variable.

21. $3x - 5y = 7$
$-x + y = -1$

22. $4x + 2y = 32$
$x + y = -2$

23. $-x - 8y = -1$
$-2x + 4y = 13$

24. $-x + 10y = 1$
$-5x + 15y = -9$

25. $-3x - y = 7$
$6x + 7y = 11$

26. $-5x + 2y = -6$
$10x + 7y = 34$

27. $-3x + 2y = 6$
$6x - y = -9$

28. $-4x + 5y = 20$
$8x - 6y = -12$

Solve each of the following systems of linear equations by the elimination method.

29. $6x - y = -8$
$2x + y = -16$

30. $5x - 3y = -3$
$3x + 3y = -21$

31. $x + 3y = 9$
$2x - y = 4$

32. $x + 2y = 0$
$2x - y = 0$

33. $x - 6y = 3$
$4x + 3y = 21$

34. $8x + y = -1$
$4x - 5y = 16$

35. $2x + 9y = 2$
$5x + 3y = -8$

36. $5x + 2y = 11$
$7x + 8y = 7$

$2 \quad \overline{4x + 3y = 21} \quad 4 \cdot 2$
$\overline{}$

$9x = 45$

$x = 5$

$5 - 6y = 3$

$6y = 5 - 3$

$6y \quad 2$

$y = \frac{2}{6} \to \frac{1}{3}$

LCD=12 LCD=6

4.2 + 3y = 14
8 + 6y = 14 → 3y = 14 ·
y = 2

37. $\frac{1}{3}x + \frac{1}{4}y = \frac{7}{6}$ · 2

38. $\frac{7}{12}x - \frac{1}{2}y = \frac{1}{6}$

39. $3x + 2y = -1$
$6x + 4y = 0$

40. $8x - 2y = 2$
$4x - y = 2$

$\frac{3}{2}x - \frac{1}{3}y = \frac{7}{3}$ · 3 · 2

$\frac{2}{5}x - \frac{1}{3}y = \frac{11}{15}$

2(4x + 3y = 14) 8x + 6y = 28
3(9x - 2y = 14) 27x - 6y = 42
35x = 70 → x = 2

41. $11x + 6y = 17$
$5x - 4y = 1$

42. $3x - 8y = 7$
$10x - 5y = 45$

43. $\frac{1}{2}x + \frac{1}{6}y = \frac{1}{3}$
$-x - \frac{1}{3}y = -\frac{1}{6}$

44. $-\frac{1}{3}x - \frac{1}{2}y = -\frac{2}{3}$
$-\frac{2}{3}x - y = -\frac{4}{3}$

45. Multiply both sides of the second equation in the following system by 100, and then solve as usual.

$x + y = 22$

$0.05x + 0.10y = 1.70$

46. Multiply both sides of the second equation in the following system by 100, and then solve as usual.

$x + y = 15,000$

$0.06x + 0.07y = 980$

Applying the Concepts

47. Recall at the beginning of this section that adult tickets x for the QuikTrip Air and Rocket Racing Show cost $15 and children's tickets y cost $10 Suppose 1,720 tickets are sold at the gate for a total of $19,750.

$x + y = 1,720$
$15x - 10y = 19,750$

Solve the above system of equations to find the number of adult and children tickets sold.

48. A local community theater sells adult tickets x for $10 and children's tickets y for $5. For one evening performance of a popular musical, the theater sells out 375 seats and collects a revenue of $2,875. The system of equations for this situation is

$x + y = 375$
$10x + 5y = 2,875$

Solve this system to find the number of adult and children tickets sold.

Getting Ready for the Next Section

49. One number is eight more than five times another; their sum is 26. Find the numbers.

50. One number is three less than four times another; their sum is 27. Find the numbers.

51. The difference of two positive numbers is nine. The larger number is six less than twice the smaller number. Find the numbers.

52. The difference of two positive numbers is 17. The larger number is one more than twice the smaller number. Find the numbers.

53. The length of a rectangle is five inches more than three times the width. The perimeter is 58 inches. Find the length and width.

54. The length of a rectangle is three inches less than twice the width. The perimeter is 36 inches. Find the length and width.

55. John has $1.70 in nickels and dimes in his pocket. He has four more nickels than he does dimes. How many of each does he have?

56. Jamie has $2.65 in dimes and quarters in her pocket. She has two more dimes than she does quarters. How many of each does she have?

Find the Mistake

The table below shows systems and the first step taken to solve each system by the elimination method. Circle the mistake(s) in the first step and write the correct way to begin on the line provided.

System	First Step
1. $4x - y = -8$ $-4x - y = 8$	Multiply both sides of the equation by -1 to eliminate the variable y.
2. $x - 7y = 11$ $3x + 2y = 15$	Multiply both sides of the first equation by 3 to eliminate x.
3. $5x + y = 9$ $-5x + 3y = 21$	Multiply both sides of the first equation by 3 to eliminate y.

Circle the mistake in the sentence below and write the correct word or phrase on the line provided.

4. When solving a system by the elimination method, if both variables are eliminated and the resulting statement is true, then the lines are parallel. _____

Landmark Review: Checking Your Progress

Solve the following systems.

1. $y = 2x + 5$
$y = -3x - 5$

2. $2x - y = -2$
$2x - 3y = 6$

3. $2x - y = 1$
$y = -2x + 5$

4. $2x - 5y = 0$
$x - 5y = -9$

5. $-3x + y = 8$
$x = -y + 4$

6. $-6x + 2y = -9$
$y = 3x + 4$

7. $x - y = 11$
$x + y = -1$

8. $-3x + y = -15$
$-2x + 3y = -17$

9. $2x + 5y = 17$
$x - 4y = -24$

10. $x + 4y = 5$
$3x + 8y = 14$

7.4 Applications of Systems of Equations

OBJECTIVES

A Apply the Blueprint for Problem Solving to a variety of application problems involving systems of equations.

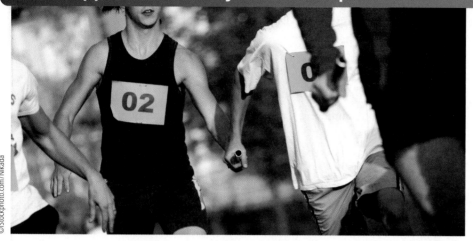

The Mad Hatter Relay Race in Boise, Idaho is the epitome of a "fun run." Two-person teams from across the state participate in 4-mile combined races, 2 miles for each team member. All participants must wear silly or theme hats while running, and are strongly encouraged to wear costumes as well. For the 2010 event, pre-registration for each team cost $28, with an additional $10 for race-day registration. Let's suppose x teams pre-registered for the race and y teams registered on race day for a total of 29 teams. Let's also suppose that the total registration fees paid for this race were $922. Can you use the information to create a linear system and solve for its variables? Read through this section and practice the application problems. You will be asked to solve this relay race problem in the problem set.

I often have heard students remark about the word problems in introductory algebra: "What does this have to do with real life?" Many of the word problems we will encounter in this section don't have much to do with "real life." We are actually just practicing. Ultimately, all problems requiring the use of algebra are word problems; that is, they are stated in words first, then translated to symbols. The problem then is solved by some system of mathematics, like algebra. If word problems frustrate you, take a deep breath and tackle them one step at a time. You can do this.

A The Blueprint for Problem Solving

The word problems in this section have two unknown quantities. We will write two equations in two variables (each of which represents one of the unknown quantities), which of course is a system of equations. We then solve the system by one of the methods developed previously. Here are the steps to follow in solving these word problems.

> **BLUEPRINT FOR PROBLEM SOLVING** Using a System of Equations
>
> **Step 1:** *Read* the problem, and then mentally *list* the items that are known and the items that are unknown.
>
> **Step 2:** *Assign variables* to each of the unknown items; that is, let $x =$ one of the unknown items and $y =$ the other unknown item. Then *translate* the other *information* in the problem to expressions involving the two variables.
>
> **Step 3:** *Reread* the problem, and then *write a system of equations,* using the items and variables listed in steps 1 and 2, that describes the situation. Drawing a sketch or creating a table may help you organize your data.
>
> **Step 4:** *Solve the system* found in step 3.
>
> **Step 5:** *Write* your *answers* using complete sentences.
>
> **Step 6:** *Reread* the problem, and *check* your solution with the original words in the problem.

Remember, the more problems you work, the more problems you will be able to work. If you have trouble getting started on the problem set, come back to the examples and work through them yourself. The examples are similar to the problems found in the problem set.

Number Problem

EXAMPLE 1 One number is 2 more than 5 times another number. Their sum is 20. Find the two numbers.

Solution Applying the steps in our blueprint, we have

Step 1: *Read and list.*
> We know that the two numbers have a sum of 20 and that one of them is 2 more than 5 times the other. We don't know what the numbers themselves are.

Step 2: *Assign variables and translate information.*
> Let x represent one of the numbers and y represent the other. "One number is 2 more than 5 times another" translates to

$$y = 5x + 2$$

> "Their sum is 20" translates to

$$x + y = 20$$

Step 3: *Write a system of equations.*
> The system that describes the situation must be

$$x + y = 20$$
$$y = 5x + 2$$

Step 4: *Solve the system.*
> We can solve this system by substituting the expression $5x + 2$ in the second equation for y in the first equation.

$$x + (5x + 2) = 20$$
$$6x + 2 = 20$$
$$6x = 18$$
$$x = 3$$

> Using $x = 3$ in either of the first two equations and then solving for y, we get $y = 17$.

Step 5: *Write answers.*
> So 17 and 3 are the numbers we are looking for.

Step 6: *Read and check.*
> The number 17 is 2 more than 5 times 3, and the sum of 17 and 3 is 20.

Age Problem

EXAMPLE 2 Kaci is 3 more than 10 times the age of her daughter, Amelie. The sum of their ages is 36 years. Find their ages.

Solution Begin by mentally listing what we know and don't know.

Step 1: *Read and list.*
> We know that the two ages have a sum of 36, and that one of them is 3 more than 10 times the other.

Step 2: *Assign variables and translate information.*

Let x represent Amelie's age and y represent Kaci's age. "Kaci is 3 more than 10 times the age of Amelie" translates to

$$y = 10x + 3$$

"The sum of their ages is 36 years" translates to

$$x + y = 36$$

Step 3: *Write a system of equations.*

The system for this problem is

$$y = 10x + 3$$
$$x + y = 36$$

Step 4: *Solve the system.*

We can solve the system by substituting the expression $10x + 3$ in the first equation for y in the second equation.

$$x + (10x + 3) = 36$$
$$11x + 3 = 36$$
$$11x = 33$$
$$x = 3$$

Step 5: *Write answers.*

Amelie is 3 years old, and Kaci is 33 years old.

Step 6: *Read and check.*

Kaci's age of 33 is 3 more than 10 times Amelie's age of 3, and the sum of 33 and 3 is 36.

Interest Problem

EXAMPLE 3 Mr. Hicks had $15,000 to invest. He invested some at 6% and the rest at 7%. If he earns $980 in interest, how much did he invest at each rate?

Solution Remember, step 1 is done mentally.

Step 1: *Read and list.*

We do not know the specific amounts invested in the two accounts. We do know that their sum is $15,000 and that the interest rates on the two accounts are 6% and 7%.

Step 2: *Assign variables and translate information.*

Let x = the amount invested at 6% and y = the amount invested at 7%. Because Mr. Hicks invested a total of $15,000, we have

$$x + y = 15,000$$

The interest he earns comes from 6% of the amount invested at 6% and 7% of the amount invested at 7%. To find 6% of x, we multiply x by 0.06, which gives us $0.06x$. To find 7% of y, we multiply 0.07 times y and get $0.07y$.

$$\begin{array}{ccc} \text{Interest} & \text{Interest} & \text{Total} \\ \text{at 6%} & + \quad \text{at 7%} & = \quad \text{interest} \\ 0.06x & + \quad 0.07y & = \quad 980 \end{array}$$

Step 3: *Write a system of equations.*

The system is

$$x + y = 15,000$$
$$0.06x + 0.07y = 980$$

3. Amy has $10,000 to invest. She invests part at 6% and the rest at 7%. If she earns $630 in interest for the year, how much does she have invested at each rate?

Step 4: *Solve the system.*

We multiply the first equation by -6 and the second by 100 to eliminate x.

$$x + y = 15{,}000 \xrightarrow{\text{Multiply by } -6.} -6x - 6y = -90{,}000$$

$$0.06x + 0.07y = 980 \xrightarrow[\text{Multiply by 100.}]{} \quad \underline{6x + 7y = 98{,}000}$$

$$y = 8{,}000$$

Substituting $y = 8{,}000$ into the first equation and solving for x, we get $x = 7{,}000$.

Step 5: *Write answers.*

He invested $7,000 at 6% and $8,000 at 7%.

Step 6: *Read and check.*

Checking our solutions in the original problem, we have the following:
The sum of $7,000 and $8,000 is $15,000, the total amount he invested. To complete our check, we find the total interest earned from the two accounts.

$$\text{The interest on \$7,000 at 6\% is } 0.06(7{,}000) = 420$$
$$\underline{\text{The interest on \$8,000 at 7\% is } 0.07(8{,}000) = 560}$$
$$\text{The total interest is \$980.}$$

Coin Problem

4. Patrick has $1.85 in dimes and quarters. He has a total of 14 coins. How many of each kind does he have?

EXAMPLE 4 John has $1.70 in dimes and nickels. He has a total of 22 coins. How many of each kind does he have?

Solution

Step 1: *Read and list.*

We know that John has 22 coins that are dimes and nickels. We know that a dime is worth 10 cents and a nickel is worth 5 cents. We do not know the specific number of dimes and nickels he has.

Step 2: *Assign variables and translate information.*

Let $x =$ the number of nickels and $y =$ the number of dimes. The total number of coins is 22, so

$$x + y = 22$$

The total amount of money he has is $1.70, which comes from nickels and dimes.

$$\begin{array}{c} \text{Amount of money} \\ \text{in nickels} \end{array} + \begin{array}{c} \text{Amount of money} \\ \text{in dimes} \end{array} = \begin{array}{c} \text{Total amount} \\ \text{of money} \end{array}$$

$$0.05x \quad + \quad 0.10y \quad = \quad 1.70$$

Step 3: *Write a system of equations.*

The system that represents the situation is

$$x + y = 22 \qquad \text{The number of coins}$$
$$0.05x + 0.10y = 1.70 \qquad \text{The value of the coins}$$

Step 4: *Solve the system.*

We multiply the first equation by -5 and the second by 100 to eliminate the variable x.

$$x + y = 22 \xrightarrow{\text{Multiply by } -5.} -5x - 5y = -110$$

$$0.05x + 0.10y = 1.70 \xrightarrow[\text{Multiply by 100.}]{} \quad \underline{5x + 10y = 170}$$

$$5y = 60$$

$$y = 12$$

Substituting $y = 12$ into our first equation, we get $x = 10$.

Step 5: *Write answers.*
John has 12 dimes and 10 nickels.

Step 6: *Read and check.*
Twelve dimes and 10 nickels total 22 coins.

$$12 \text{ dimes are worth } 12(0.10) = 1.20$$
$$\underline{10 \text{ nickels are worth } 10(0.05) = 0.50}$$
$$\text{The total value is } \$1.70.$$

Mixture Problem

EXAMPLE 5 How much of a 20% alcohol solution and 50% alcohol solution must be mixed to get 12 gallons of 30% alcohol solution?

Solution To solve this problem we must first understand that a 20% alcohol solution is 20% alcohol and 80% water.

Step 1: *Read and list.*
We know there are two solutions that together must total 12 gallons. 20% of one of the solutions is alcohol and the rest is water, whereas the other solution is 50% alcohol and 50% water. We do not know how many gallons of each individual solution we need.

Step 2: *Assign variables and translate information.*
Let x = the number of gallons of 20% alcohol solution needed and y = the number of gallons of 50% alcohol solution needed. Because the total number of gallons we will end up with is 12, and this 12 gallons must come from the two solutions we are mixing, our first equation is

$$x + y = 12$$

To obtain our second equation, we look at the amount of alcohol in our two original solutions and our final solution. The amount of alcohol in the x gallons of 20% solution is $0.20x$, and the amount of alcohol in y gallons of 50% solution is $0.50y$. The amount of alcohol in the 12 gallons of 30% solution is $0.30(12)$. Because the amount of alcohol we start with must equal the amount of alcohol we end up with, our second equation is

$$0.20x + 0.50y = 0.30(12)$$

The information we have so far can also be summarized with a table. Sometimes by looking at a table like the one that follows, it is easier to see where the equations come from.

	20% Solution	50% Solution	Final Solution
Number of Gallons	x	y	12
Gallons of Alcohol	$0.20x$	$0.50y$	$0.30(12)$

Step 3: *Write a system of equations.*
Our system of equations is

$$x + y = 12$$
$$0.20x + 0.50y = 0.30(12)$$

5. How much 30% alcohol solution and 60% alcohol solution must be mixed to get 25 gallons of 48% solution?

Answers
5. 10 gallons of 30%, 15 gallons of 60%

Step 4: *Solve the system.*

We can solve this system by substitution. Solving the first equation for y and substituting the result into the second equation, we have

$$0.20x + 0.50(12 - x) = 0.30(12)$$

Multiplying each side by 10 gives us an equivalent equation that is a little easier to work with.

$$2x + 5(12 - x) = 3(12)$$
$$2x + 60 - 5x = 36$$
$$-3x + 60 = 36$$
$$-3x = -24$$
$$x = 8$$

If x is 8, then y must be 4 because $x + y = 12$.

Step 5: *Write answers.*

It takes 8 gallons of 20% alcohol solution and 4 gallons of 50% alcohol solution.

Step 6: *Read and check.*

Eight gallons of 20% alcohol solution plus four gallons of 50% alcohol solution will equal twelve gallons 30% alcohol solution.

$$0.20(8) + 0.50(4) = 0.30(12)$$
$$1.6 + 2 = 3.6$$
$$3.6 = 3.6 \qquad \text{True statement}$$

GETTING READY FOR CLASS

After reading through the preceding section, respond in your own words and in complete sentences.

A. What is the first step of the Blueprint for Problem Solving when using a system of equations?

B. Is the Blueprint for Problem Solving when using a system of equations any different than the Blueprint you learned in Chapter 2? Explain.

C. Why do you have to understand systems of linear equations to work the application problems in this section?

D. Write an application problem for which the solution depends on solving a system of equations.

Vocabulary Review

The following is a list of steps for the Blueprint for Problem Solving for systems. Choose the correct words to fill in the blanks below.

system	expressions	equations	sentences
known	problem	unknown	solution

Step 1: Read the problem and mentally list the items that are _____ and unknown.

Step 2: Assign a variable to each of the _____ items, and then translate the other information in the problem to _____ involving the variables.

Step 3: Reread the problem and write a system of _____.

Step 4: Solve the _____.

Step 5: Write your answers using complete _____.

Step 6: Reread the _____ and check your _____.

Problems

A Solve the following word problems using the Blueprint for Problem Solving. Be sure to show the equations used.

Number Problems

1. Two numbers have a sum of 25. One number is 5 more than the other. Find the numbers.

2. The difference of two numbers is 6. Their sum is 30. Find the two numbers.

3. The sum of two numbers is 15. One number is 4 times the other. Find the numbers.

4. The difference of two positive numbers is 28. One number is 3 times the other. Find the two numbers.

5. Two positive numbers have a difference of 5. The larger number is one more than twice the smaller. Find the two numbers.

6. One number is 2 more than 3 times another. Their sum is 26. Find the two numbers.

7. One number is 5 more than 4 times another. Their sum is 35. Find the two numbers.

8. The difference of two positive numbers is 8. The larger is twice the smaller decreased by 7. Find the two numbers.

SCAN TO ACCESS

Age Problems

9. Linda is 8 more than 2 times the age of her daughter, Audra. The sum of their ages is 101 years. Find their ages.

10. A mathematics professor is 3 less than 3 times the average age of his students. The difference between the professor's age and the average student is 35 years. Find the ages.

11. Kate took some time off between earning her bachelor's degree and beginning her master's program. On the first day of Kate's master's program, she was 5 less than twice the age she was when she graduated with her bachelor's. The difference between the two ages was 16 years.

12. Michael is scheduled to receive an inheritance when he is 6 more than a third the age of his grandfather. The difference between their ages will be 38 years. Find their ages at the time of inheritance.

Interest Problems

13. Mr. Wilson invested money in two accounts. His total investment was $20,000. If one account pays 6% interest and the other pays 8% interest, how much does he have in each account if he earned a total of $1,380 in interest in 1 year?

14. A total of $11,000 was invested. Part of the $11,000 was invested at 4%, and the rest was invested at 7%. If the investments earn $680 per year, how much was invested at each rate?

15. A woman invested 4 times as much at 5% as she did at 6%. The total amount of interest she earns in 1 year from both accounts is $520. How much did she invest at each rate?

16. Ms. Hagan invested twice as much money in an account that pays 7% interest as she did in an account that pays 6% interest. Her total investment pays her $1,000 a year in interest. How much did she invest at each rate?

Coin Problems

17. Ron has 14 coins with a total value of $2.30. The coins are nickels and quarters. How many of each coin does he have?

18. Diane has $0.95 in dimes and nickels. She has a total of 11 coins. How many of each kind does she have?

19. Suppose Tom has 21 coins totaling $3.45. If he has only dimes and quarters, how many of each type does he have?

20. A coin collector has 31 dimes and nickels with a total face value of $2.40. (They are actually worth a lot more.) How many of each coin does she have?

Mixture Problems

21. How many liters of 50% alcohol solution and 20% alcohol solution must be mixed to obtain 18 liters of 30% alcohol solution?

	50% Solution	20% Solution	Final Solution
Number of Liters	x	y	
Liters of Alcohol			

22. How many liters of 10% alcohol solution and 5% alcohol solution must be mixed to obtain 40 liters of 8% alcohol solution?

	10% Solution	5% Solution	Final Solution
Number of Liters	x	y	
Liters of Alcohol			

23. A mixture of 8% disinfectant solution is to be made from 10% and 7% disinfectant solutions. How much of each solution should be used if 30 gallons of 8% solution are needed?

24. How much 50% antifreeze solution and 40% antifreeze solution should be combined to give 50 gallons of 46% antifreeze solution?

Miscellaneous Problems

25. For a Saturday matinee, adult tickets cost $5.50 and kids under 12 pay only $4.00. If 70 tickets are sold for a total of $310, how many of the tickets were adult tickets and how many were sold to kids under 12?

26. The Bishop's Peak 4-H club is having its annual fundraising dinner. Adults pay $15 each and children pay $10 each. If the number of adult tickets sold is twice the number of children's tickets sold, and the total income for the dinner is $1,600, how many of each kind of ticket did the 4-H club sell?

27. A farmer has 96 feet of fence with which to make a corral. If he arranges it into a rectangle that is twice as long as it is wide, what are the dimensions?

$2x$

x

28. If a 22-inch rope is to be cut into two pieces so that one piece is 3 inches longer than twice the other, how long is each piece?

x $2x + 3$

22 inches

29. Suppose you finish a session of blackjack with $5 chips and $25 chips. You have 45 chips in all, with a total value of $465, how many of each kind of chip do you have?

30. Tyler has been saving his winning lottery tickets. He has 23 tickets that are worth a total of $175. If each ticket is worth either $5 or $10, how many of each does he have?

31. Mary Jo spends $2,550 to buy stock in two companies. She pays $11 a share to one of the companies and $20 a share to the other. If she ends up with a total of 150 shares, how many shares did she buy at $11 a share and how many did she buy at $20 a share?

32. Kelly sells 62 shares of stock she owns for a total of $433. If the stock was in two different companies, one selling at $6.50 a share and the other at $7.25 a share, how many of each did she sell?

33. Recall the relay race problem introduced in the beginning of this section. Registration costs were $28 for pre-registration and $38 for race day registration. Suppose x teams are pre-registered and y teams registered on race day, and 29 teams register for a total fee of $922. Set up a system of equations and solve it to find how many teams pre-registered and how many teams registered the day of the race.

34. Pre-event tickets for a local theater fundraiser cost $30 and $40 for at-the-door tickets. Organizers sell a total of 200 tickets and generate a total revenue of $6,650. How many pre-event and at-the-door tickets were sold?

Maintaining Your Skills

Solve each system of equations by the elimination method.

35. $2x + y = 3$
$3x - y = 7$

36. $3x - y = -6$
$4x + y = -8$

37. $4x - 5y = 1$
$x - 2y = -2$

38. $6x - 4y = 2$
$2x + y = 10$

Solve each system of equations by the substitution method.

39. $5x + 2y = 7$
$y = 3x - 2$

40. $-7x - 5y = -1$
$y = x + 5$

41. $2x - 3y = 4$
$x = 2y + 1$

42. $4x - 5y = 2$
$x = 2y - 1$

Find the Mistake

Each sentence below contains a mistake. Circle the mistake and write the correct word(s) or number(s) on the line provided.

1. The first step in the Blueprint for Problem Solving is to assign variables to each of the unknown items.

2. If there are two unknown quantities, we write one equation to get our answer. _____

3. If we are solving a problem involving nickels and dimes and we know we have a total of 15 coins, we can use $0.05n + 0.10d = 15$ as one of our equations. _____

4. One number is 3 times another and their sum is 8. One of our equations will be $3x = 8$. _____

A Access to internet for research

B Pencil and paper

Mathematical Bees

In this course, we have discussed the great mathematician Fibonacci. One of Fibonacci's many mathematical discoveries, the Fibonacci sequence, can be used to explain a honey bee's family tree. How are the Fibonacci sequence and honey bee ancestry related?

Another interesting fact about honey bees is that they perform choreographed dances to communicate the location of a food source, such as a flower. The specific movements in the dance depend on the location of the sun, the hive, and the flower. The communicated information includes angles and distance measurements needed to find the food source from the hive's origin. Research this amazing process. What math concepts do the bees use to make a dance successful? Now on graph paper, create a diagram of a chosen food source in relation to the sun and the hive. Give coordinates for the three locations, and determine a system of linear equations that would represent lines drawn through the following coordinates:

 1. The hive and the sun

 2. The hive and the food source

 3. The sun and the food source

Definitions [7.1]

EXAMPLES

1. The solution to the system
$$x + 2y = 4$$
$$x - y = 1$$
is the ordered pair (2, 1). It is the only ordered pair that satisfies both equations.

1. A *system of linear equations*, as the term is used in this book, is two linear equations that each contain the same two variables.

2. The *solution set* for a system of equations is the set of all ordered pairs that satisfy *both* equations. The solution set to a system of linear equations will contain:

Case I One ordered pair when the graphs of the two equations intersect at only one point (this is the most common situation)

Case II No ordered pairs when the graphs of the two equations are parallel lines

Case III An infinite number of ordered pairs when the graphs of the two equations coincide (are the same line)

Strategy for Solving a System by Graphing [7.1]

2. Solving the system in Example 1 by graphing looks like

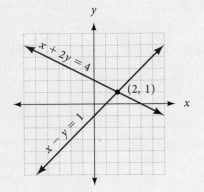

Step 1: Graph the first equation.

Step 2: Graph the second equation on the same set of axes.

Step 3: Read the coordinates of the point where the graphs cross each other (the coordinates of the point of intersection.)

Step 4: Check the solution to see that it satisfies *both* equations.

Strategy for Solving a System by the Substitution Method [7.2]

3. We can apply the substitution method to the system in Example 1 by first solving the second equation for x to get $x = y + 1$. Substituting this expression for x into the first equation, we have
$$(y + 1) + 2y = 4$$
$$3y + 1 = 4$$
$$3y = 3$$
$$y = 1$$
Using $y = 1$ in either of the original equations gives $x = 2$.

Step 1: Solve either of the equations for one of the variables (this step is not necessary if one of the equations has the correct form already.)

Step 2: Substitute the results of step 1 into the other equation, and solve.

Step 3: Substitute the results of step 2 into an equation with both x and y variables, and solve. (The equation produced in step 1 is usually a good one to use.)

Step 4: Check your solution, if necessary.

Strategy for Solving a System by the Elimination Method [7.3]

Step 1: Look the system over to decide which variable will be easier to eliminate.

Step 2: Use the multiplication property of equality on each equation separately to ensure that the coefficients of the variable to be eliminated are opposites.

Step 3: Add the left and right sides of the system produced in step 2, and solve the resulting equation.

Step 4: Substitute the solution from step 3 back into any equation with both x and y variables, and solve.

Step 5: Check your solution in both equations, if necessary.

4. We can eliminate the y variable from the system in Example 1 by multiplying both sides of the second equation by 2 and adding the result to the first equation

$$
\begin{array}{ll}
& \text{No Change} \\
x + 2y = 4 & \xrightarrow{\hspace{1cm}} x + 2y = 4 \\
x - y = 1 & \xrightarrow{\hspace{1cm}} 2x - 2y = 2 \\
& \text{Multiply by 2} \\
\hline
& 3x \quad\quad = 6 \\
& x \quad\quad = 2
\end{array}
$$

Substituting $x = 2$ into either of the original two equations gives $y = 1$. The solution is $(2, 1)$.

Special Cases [7.1, 7.2, 7.3]

In some cases, using the elimination or substitution method eliminates both variables. The situation is interpreted as follows.

1. If the resulting statement is *false*, then the lines are parallel and there is no solution to the system. This is called an inconsistent system.

2. If the resulting statement is *true*, then the equations represent the same line (the lines coincide). In this case any ordered pair that satisfies either equation is a solution to the system. This is called a dependent system.

Blueprint for Problem Solving: Using a System of Equations [7.4]

Step 1: *Read* the problem, and then mentally *list* the items that are known and the items that are unknown.

Step 2: *Assign variables* to each of the unknown items; that is, let $x =$ one of the unknown items and $y =$ the other unknown item. Then *translate* the other *information* in the problem to expressions involving the two variables.

Step 3: *Reread* the problem, and then *write a system of equations,* using the items and variables listed in steps 1 and 2, that describes the situation.

Step 4: *Solve the system* found in step 3.

Step 5: *Write* your *answers* using complete sentences.

Step 6: *Reread* the problem, and *check* your solution with the original words in the problem.

COMMON MISTAKE

The most common mistake encountered in solving linear systems is the failure to complete the problem. Here is an example:

$$
\begin{array}{l}
x + y = 8 \\
x - y = 4 \\
\hline
2x = 12 \\
x = 6
\end{array}
$$

This is only half the solution. To find the other half, we must substitute the 6 back into one of the original equations and then solve for y.

Remember, solutions to systems of linear equations always consist of ordered pairs. We need an x-coordinate and a y-coordinate; $x = 6$ can never be a solution to a system of linear equations.

Solve the following systems by graphing. [7.1]

1. $x + y = 3$
$x - y = -5$

2. $x + y = 8$
$-x + y = 2$

3. $2x - 3y = -14$
$2x - y = -2$

4. $3x + 2y = 0$
$4x + y = 5$

5. $y = 3x + 3$
$y = 5x + 5$

6. $y = -2x - 7$
$y = x + 5$

Solve the following systems by substitution. [7.2]

7. $x + y = 8$
$y = 2x - 1$

8. $x - y = 3$
$x = 2y + 9$

9. $3x - y = -7$
$y - 2x + 6$

10. $4x - 2y = 12$
$y = 3x - 10$

11. $3x - y = -9$
$x - y = -5$

12. $4x - 3y = 5$
$x = 2y + 5$

13. $12x - 3y = -6$
$x = y + 4$

14. $4y + 8x = 6$
$-2y - 4x = -3$

Solve the following systems by the elimination method. [7.3]

15. $x - y = 3$
$2x - 5y = -3$

16. $2x + 2y = 2$
$-3x - y = 5$

17. $3x - 4y = 3$
$5x + 2y = 5$

18. $2x - 3y = -1$
$-6x + 9y = 3$

19. $-2x + y = 8$
$x - 3y = -14$

20. $2x + 5y = -9$
$-4x + 6y = 2$

21. $5x - 2y = -10$
$7x + 3y = 15$

22. $6x - 3y = 7$
$-2x + y = 5$

Solve the following word problems. Be sure to show the equations used. [7.4]

23. Number Problem The sum of two numbers is 10. If one more than twice the smaller number is the larger, find the two numbers.

24. Number Problem The difference of two positive numbers is 3. One number is $\frac{3}{4}$ of the other number. Find the two numbers.

25. Investing A total of $9,000 was invested. Part of the $9,000 was invested at 8%, and the rest was invested at 5%. If the interest for one year is $510, how much was invested at each rate?

26. Investing A total of $14,000 was invested. Part of the $14,000 was invested at 9%, and the rest was invested at 11%. If the interest for one year is $1,440, how much was invested at each rate?

27. Coin Problem Nate has $3.00 in dimes and quarters. He has a total of 21 coins. How many of each does he have?

28. Coin Problem Maria has $2.95 in dimes and quarters. She has a total of 19 coins. How many of each does she have?

29. Mixture Problem How many liters of 20% alcohol solution and 80% alcohol solution must be mixed to obtain 11 liters of a 58% alcohol solution? Round to the nearest liter.

30. Mixture Problem How many liters of 35% alcohol solution and 10% alcohol solution must be mixed to obtain 30 liters of a 20% alcohol solution? Round to the nearest liter.

Simplify.

1. $6 \cdot (-2) + 7$

2. $9 - 3(6 - 2)$

3. $5 \cdot 2^3 + 2(4 - 10)$

4. $9[-2 + 7] + 2(-8 + 2)$

5. $\dfrac{4 - 13}{3 - 3}$

6. $\dfrac{4(2 - 9) - 2(12 - 8)}{7 - 3 - 8}$

7. $\dfrac{9}{48} + \dfrac{3}{72}$

8. $\dfrac{5}{4} + \dfrac{1}{2} - \dfrac{3}{8}$

9. $3(x - 2) + 7$

10. $4 - 3(4a - 6) + 4$

Solve each equation.

11. $-4 - 3 = -3y + 5 + y$

12. $-5x = 0$

13. $4(x + 3) = -4$

14. $8 + 3(y - 6) = 11$

Solve each inequality, and graph the solution.

15. $0.2x - 0.8 \geq -3$

16. $7x - 18 \geq 9x + 2$

Graph on a rectangular coordinate system.

17. $y = \dfrac{1}{3}x - 4$

18. $y = -\dfrac{3}{5}x$

Solve each system by graphing.

19. $4x - 2y = 16$
$3x + 4y = 1$

20. $x - y = 4$
$3x - 5y = 10$

Solve each system.

21. $x + y = 6$
$3x + 3y = 18$

22. $4x + y = -4$
$4x - 2y = 2$

23. $5x + 3y = -21$
$x - y = -9$

24. $2x + 5y = -6$
$3x - y = -9$

25. $8x - 3y = -5$
$12x - 5y = -11$

26. $3x - 2y = 0$
$2x + 4y = -32$

27. $2x - y = 21$
$x + 2y = 3$

28. $5x + 3y = 11$
$6x - y = -19$

29. $5x + 2y = -1$
$y = -6x + 10$

30. $5x - 7y = -17$
$x = 9y - 11$

31. What is the quotient of -32 and 4?

32. Subtract 12 from -7.

Factor into the product of primes.

33. 270

34. 420

Find the value of each expression when x is 4.

35. $-3x + 7 + 5x$

36. $-x - 4x - 2x$

37. Given $2x + 5y = -1$, complete the ordered pair $(-3, \ \)$.

38. Given $y = -\dfrac{2}{3}x + 1$, complete the ordered pairs $(\ \ , 3)$ and $(2, \ \)$.

39. Find the x- and y-intercepts for the line $5x - 6y = 30$.

40. Find the slope of the line $x = 4y - 8$.

Find the slope of the line through the given pair of points.

41. $(3, -4), (5, -9)$

42. $\left(8, \dfrac{1}{3}\right), \left(4, \dfrac{7}{6}\right)$

43. Find the equation of the line with slope $\dfrac{5}{4}$ and y-intercept -2.

44. Find the equation of the line with slope $\dfrac{1}{3}$ if it passes through $(3, 9)$.

45. Find the equation of the line through $(3, -4)$ and $(5, 2)$.

46. Find the x- and y-intercepts for the line that passes through the points $(-6, 2)$ and $(3, 5)$.

47. Coin Problem Dave has 12 coins with a value of $2.25. The coins are quarters and dimes. How many of each does he have?

48. Investing I have invested money in two accounts. One account pays 7% annual interest, and the other pays 11%. I have $300 more in the 11% account than I have in the 7% account. If the total amount of interest was $141, how much do I have in each account?

The chart shows the results of a survey of iPhones buyers. The percentages show what they reported as their previous phone. Use the information to answer the following questions.

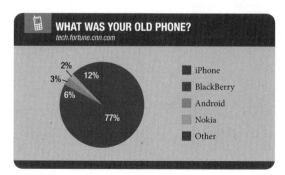

WHAT WAS YOUR OLD PHONE?
tech.fortune.cnn.com

- iPhone
- BlackBerry
- Android
- Nokia
- Other

2%
3%
12%
6%
77%

49. If 26,000 people were surveyed, how many reported replacing an Android phone with their new iPhone?

50. If 24 people say they replaced a Blackberry with their iPhone, how many people were surveyed?

1. Write the solution to the system which is graphed below. [7.1]

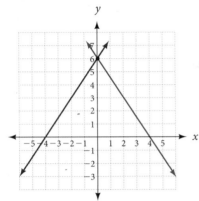

Solve each system by graphing. [7.1]

2. $4x - 2y = 8$
$\quad y = \dfrac{2}{3}x$

3. $3x - 2y = 13$
$\quad y = 4$

4. $2x - 2y = -12$
$\quad -3x - y = 2$

Solve each system by the substitution method. [7.2]

5. $3x - y = 12$
$\quad y = 2x - 8$

6. $3x + 6y = 3$
$\quad x = 4y - 17$

7. $2x - 3y = -18$
$\quad 3x + y = -5$

8. $2x - 3y = 13$
$\quad x - 4y = -1$

Solve each system by the elimination method. [7.3]

9. $x - y = -9$
$\quad 2x + 3y = 7$

10. $3x - y = 1$
$\quad 5x - y = 3$

11. $2x + 3y = -3$
$\quad x + 6y = 12$

12. $2x + 3y = 4$
$\quad 4x + 6y = 8$

Solve the following word problems. In each case, be sure to show the system of equations that describes the situation. [7.4]

13. Number Problem The sum of two numbers is 15. Their difference is 11. Find the numbers.

14. Number Problem The sum of two numbers is 18. One number is 2 more than 3 times the other. Find the two numbers.

15. Investing Dave has $2,000 to invest. He would like to earn $135.20 per year in interest. How much should he invest at 6% if the rest is to be invested at 7%?

16. Coin Problem Maria has 19 coins that total $1.35. If the coins are all nickels and dimes, how many of each type does she have?

17. Perimeter Problem A rancher wants to build a rectangular corral using 198 feet of fence. If the length of the corral is to be 15 feet longer than twice the width, find the dimensions of the corral.

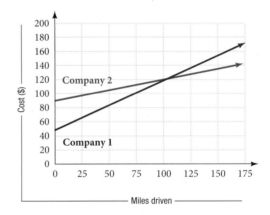

18. Moving Truck Rental One moving truck company charges $50 per day plus $0.75 per mile to rent a 28-foot truck for local use. Another company charges $90 per day plus $0.35 per mile. The total costs to use the trucks for a day and drive x miles are

$$\text{Company 1: } y = 0.75x + 50$$

$$\text{Company 2: } y = 0.35x + 90$$

The graphs of these equations are shown below.

a. At how many miles do the two companies charge the same amount for the truck rental?

b. If the truck is driven less than 100 miles, which company would charge less?

c. For which values of x will Company 2 charge less than Company 1?

Radical Expressions and Equations

Disney World
Orlando, Florida
© 2010 Google

If you have ever visited Epcot Center inside Disney World, you have no doubt seen the attraction called Spaceship Earth. The original storyline for the attraction was created by science fiction writer Ray Bradbury, who also helped with the design. His concept was to take visitors on a trip through time showing how each human generation had changed its world, and how those changes have laid the groundwork for discoveries by future generations. Recent renovations have expanded the scope of the attraction to look more toward the future. New scenes and effects have been added to the existing story, and interactive touch screens have been installed to allow visitors to personalize their experience. The ride now ends at the very top of Spaceship Earth, where guests arrive at the present. They are invited to visit Project Tomorrow, which includes games and displays showcasing new technologies that focus on medicine, transportation, and energy management.

The shape of Spaceship Earth is called a sphere. Knowing this allows us to make several calculations if we are given the attraction's dimensions. If the volume V of Spaceship Earth is 2,200,000 cubic feet, you can calculate the radius of the structure r by using the formula

$$r = \sqrt[3]{\frac{3V}{4\pi}}$$

Volume

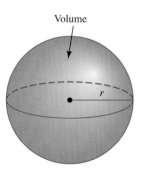

As you work through this chapter, you will become familiar with radical expressions such as the one used in the above formula.

OBJECTIVES

A Evaluate radical expressions involving numbers using the definition for roots.

B Simplify radical expressions involving variables.

KEY WORDS

positive square root

cube root

radical expression

radicand

index

golden ratio

8.1 Radical Expressions

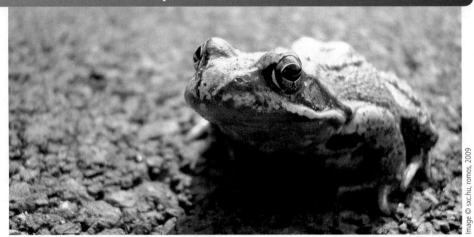

Image © sxc.hu, romos, 2009

In Greece, millions of frogs decided to migrate all at once across the busy Egnatia highway. Scientists believe the frogs left a nearby lake in search of food, but needless to say, the amphibian influx caused a major traffic jam. Greek officials shut down the highway for two hours out of concern for driver safety. Previously, three motorists had skidded off the road trying to avoid the frogs.

Let's assume the closed stretch of road was 30 yards long by 30 yards wide. Imagine a single frog hopping the length of the closed road's diagonal from one corner to its opposite. In order to find the square of the length of this diagonal, we need to apply the Pythagorean theorem. Now, let's simplify this concept by examining a square in which each of the four sides is only 1 inch long.

$$c^2 = 1^2 + 1^2$$
$$c^2 = 2$$

1 inch

1 inch

FIGURE 1

Because we know that c is positive and that its square is 2, we call c the *positive square root* of 2, and we write $c = \sqrt{2}$. Associating numbers, such as $\sqrt{2}$, with the diagonal of a square or rectangle allows us to analyze some interesting items from geometry. One particularly interesting geometric object that we will study in this section is shown in Figure 2. It is constructed from a right triangle, and the length of the diagonal is found from the Pythagorean theorem. We will come back to this figure at the end of this section.

The Golden Rectangle

A　　　　B　　F

2

1　　1

C　O　D　E

FIGURE 2

A Radical Expressions Involving Numbers

In Chapter 0, we developed notation (exponents) to give us the square, cube, or any other power of a number. For instance, if we wanted the square of 3, we wrote $3^2 = 9$. If we wanted the cube of 3, we wrote $3^3 = 27$. In this section, we will develop notation that will take us in the reverse direction, that is, from the square of a number, say 25, back to the original number, 5.

DEFINITION positive square root

If x is a nonnegative real number, then the expression \sqrt{x} is called the *positive square root*, or the principal square root, of x and is such that

$$(\sqrt{x})^2 = x$$

In words: \sqrt{x} is the positive number we square to get x.

The negative square root of x, $-\sqrt{x}$, is defined in a similar manner.

EXAMPLE 1 The positive square root of 64 is 8 because 8 is the positive number with the property $8^2 = 64$. The negative square root of 64 is -8 because -8 is the negative number whose square is 64. We can summarize these facts by saying

$$8^2 = 64 \quad \text{and} \quad \sqrt{64} = 8, \text{ whereas } (-8)^2 = 64 \quad \text{and} \quad -\sqrt{64} = -8$$

The higher roots, cube roots, fourth roots, and so on, are defined by definitions similar to that of square roots.

DEFINITION

If x is a real number and n is a positive integer, then

Positive square root of x, \sqrt{x}, is such that $(\sqrt{x})^2 = x$ $x \geq 0$

Cube root of x, $\sqrt[3]{x}$, is such that $(\sqrt[3]{x})^3 = x$

Positive fourth root of x, $\sqrt[4]{x}$, is such that $(\sqrt[4]{x})^4 = x$ $x \geq 0$

Fifth root of x, $\sqrt[5]{x}$, is such that $(\sqrt[5]{x})^5 = x$

$$\begin{array}{cc} \cdot & \cdot \\ \cdot & \cdot \\ \cdot & \cdot \end{array}$$

The nth root of x, $\sqrt[n]{x}$, is such that $(\sqrt[n]{x})^n = x$ $x \geq 0$ if n is even

The following is a table of the most common positive roots used in this book. Any of the roots that are unfamiliar should be memorized.

Square Roots		Cube Roots	Fourth Roots
$\sqrt{0} = 0$	$\sqrt{49} = 7$	$\sqrt[3]{0} = 0$	$\sqrt[4]{0} = 0$
$\sqrt{1} = 1$	$\sqrt{64} = 8$	$\sqrt[3]{1} = 1$	$\sqrt[4]{1} = 1$
$\sqrt{4} = 2$	$\sqrt{81} = 9$	$\sqrt[3]{8} = 2$	$\sqrt[4]{16} = 2$
$\sqrt{9} = 3$	$\sqrt{100} = 10$	$\sqrt[3]{27} = 3$	$\sqrt[4]{81} = 3$
$\sqrt{16} = 4$	$\sqrt{121} = 11$	$\sqrt[3]{64} = 4$	
$\sqrt{25} = 5$	$\sqrt{144} = 12$	$\sqrt[3]{125} = 5$	
$\sqrt{36} = 6$	$\sqrt{169} = 13$		

Notation An expression like $\sqrt[3]{8}$ that involves a root is called a *radical expression*. In the expression $\sqrt[3]{8}$, the 3 is called the *index*, the $\sqrt{}$ is the *radical sign*, and 8 is called the *radicand*. The index of a radical must be a positive integer greater than 1. If no index is written, it is assumed to be 2.

Note It is a common mistake to assume that an expression like $\sqrt{25}$ indicates both square roots, 5 and -5. The expression $\sqrt{25}$ indicates only the positive square root of 25, which is 5. If we want the negative square root, we must use a negative sign: $-\sqrt{25} = -5$.

Practice Problems

1. Give the two square roots of 36.

Note We have restricted the even roots in this definition to nonnegative numbers. Even roots of negative numbers exist, but are not represented by real numbers. That is, $\sqrt{-4}$ is not a real number because there is no real number whose square is -4.

Answers

1. 6, -6

Roots and Negative Numbers

When dealing with negative numbers and radicals, the only restriction concerns negative numbers under even roots. We can have negative signs in front of radicals and negative numbers under odd roots and still obtain real numbers. Here are some examples to help clarify this.

EXAMPLE 2 Simplify each expression, if possible.

a. $\sqrt[3]{-8} = -2$ because $(-2)^3 = -8$.

b. $\sqrt{-4}$ is not a real number because there is no real number whose square is -4.

c. $-\sqrt{25} = -5$, because -5 is the negative square root of 25.

d. $\sqrt[5]{-32} = -2$ because $(-2)^5 = -32$.

e. $\sqrt[4]{-81}$ is not a real number because there is no real number we can raise to the fourth power and obtain -81.

B Radical Expressions Involving Variables

From the preceding examples, it is clear that we must be careful that we do not try to take an even root of a negative number. For this reason, we will begin our work with radical expressions involving variables assuming that all variables appearing under a radical sign represent nonnegative numbers.

EXAMPLE 3 Assume all variables represent nonnegative numbers, and simplify each expression as much as possible.

a. $\sqrt{25a^4b^6} = 5a^2b^3$ because $(5a^2b^3)^2 = 25a^4b^6$.

b. $\sqrt[3]{x^6y^{12}} = x^2y^4$ because $(x^2y^4)^3 = x^6y^{12}$.

c. $\sqrt[4]{81r^8s^{20}} = 3r^2s^5$ because $(3r^2s^5)^4 = 81r^8s^{20}$.

FACTS FROM GEOMETRY The Pythagorean Theorem

Now that we have had some experience working with square roots, we can rewrite the Pythagorean theorem using a square root. If triangle ABC is a right triangle with $C = 90°$, then the length of the longest side is the **positive square root** of the sum of the squares of the other two sides (see Figure 3).

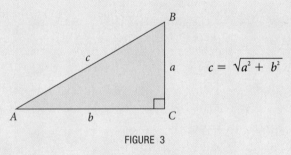

FIGURE 3

You may have heard of the golden rectangle. Its origins can be traced back over 2,000 years to the Greek civilization that produced Pythagoras, Socrates, Plato, Aristotle, and Euclid. The most important mathematical work to come from that Greek civilization was Euclid's *Elements,* an elegantly written summary of all that was known about geometry at that time in history. Euclid's *Elements,* according to Howard Eves, an authority on the history of mathematics, exercised a greater influence on scientific thinking than any other work. Here is how we construct a golden rectangle from a square of side 2, using the same method that Euclid used in his *Elements.*

2. Simplify, if possible.

a. $\sqrt[3]{-64}$

b. $\sqrt{-25}$

c. $-\sqrt{4}$

d. $\sqrt[5]{-1}$

e. $\sqrt[4]{-16}$

Note To extend our discussion in Example 2 part e, it is important to mention the opposite of the nth root of the opposite of a is not the nth root of a. In other words,
$$-\sqrt[n]{a} \neq \sqrt[n]{-a}$$

3. Simplify.

a. $\sqrt{81a^4b^8}$

b. $\sqrt[3]{8x^3y^9}$

c. $\sqrt[4]{81a^4b^8}$

Answers

2. a. -4 **b.** Not a real number
 c. -2 **d.** -1
 e. Not a real number
3. a. $9a^2b^4$ **b.** $2xy^3$ **c.** $3ab^2$

Constructing a Golden Rectangle From a Square of Side 2

Step 1: Draw a square with a side of length 2. Connect the midpoint of side *CD* to corner *B*. (Note that we have labeled the midpoint of segment *CD* with the letter *O*.)

Step 2: Drop the diagonal from step 1 down so it aligns with side *CD*.

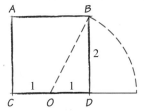

Step 3: Form rectangle *ACEF*. This is a golden rectangle.

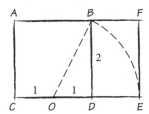

All golden rectangles are constructed from squares. Every golden rectangle, no matter how large or small it is, will have the same shape. To associate a number with the shape of the golden rectangle, we use the ratio of its length to its width. This ratio is called the *golden ratio*. To calculate the golden ratio, we must first find the length of the diagonal we used to construct the golden rectangle. Figure 4 shows the golden rectangle we constructed from a square of side 2. The length of the diagonal *OB* is found by applying the Pythagorean theorem to triangle *OBD*.

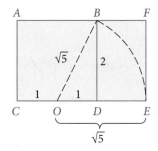

FIGURE 4

The length of segment *OE* is equal to the length of diagonal *OB*; both are $\sqrt{5}$. Because the distance from *C* to *O* is 1, the length *CE* of the golden rectangle is $1 + \sqrt{5}$. Now we can find the golden ratio.

$$\text{Golden ratio} = \frac{\text{Length}}{\text{Width}} = \frac{CE}{EF} = \frac{1 + \sqrt{5}}{2}$$

GETTING READY FOR CLASS

After reading through the preceding section, respond in your own words and in complete sentences.

A. Every real number has two square roots. Explain the notation you use to tell them apart. Use the square roots of 3 for examples.

B. Explain why a square root of −4 is not a real number.

C. We use the notation $\sqrt{2}$ to represent the positive square root of 2. Explain why there isn't a simpler way to express the positive square root of 2.

D. What is a golden rectangle?

EXERCISE SET 8.1

Vocabulary Review

Choose the correct words to fill in the blanks below.

 radical sign cube index golden ratio square radicand radical

1. If x is a nonnegative real number, then the expression \sqrt{x} is called the positive _____ root of x.

2. If x is a real number, then the _____ root of x, $\sqrt[3]{x}$ is such that $(\sqrt[3]{x})^3 = x$.

3. An expression that involves a root is called a _____ expression.

4. In the expression $\sqrt[n]{x}$, the n is called the _____, the $\sqrt{}$ is called the _____, and the x is called the _____.

5. The _____ is given by the expression $\frac{1 + \sqrt{5}}{2}$.

Problems

A Simplify each of the following radicals, if possible.

1. $\sqrt{144}$

2. $-\sqrt{144}$

3. $\sqrt{-144}$

4. $\sqrt{-49}$

5. $-\sqrt{49}$

6. $\sqrt{49}$

7. $\sqrt[3]{-27}$

8. $-\sqrt[3]{27}$

9. $\sqrt[4]{16}$

10. $-\sqrt[4]{16}$

11. $\sqrt[4]{-16}$

12. $-\sqrt[4]{-16}$

13. $\sqrt{0.04}$

14. $\sqrt{0.81}$

15. $\sqrt[3]{0.008}$

16. $\sqrt[3]{0.125}$

17. $\sqrt[3]{125}$

18. $\sqrt[3]{-125}$

19. $-\sqrt[3]{125}$

20. $-\sqrt[3]{-125}$

21. $\sqrt[4]{-81}$

22. $-\sqrt[4]{81}$

23. $\sqrt[5]{-32}$

24. $-\sqrt[5]{-32}$

25. $-\sqrt[3]{-8}$

26. $-\sqrt[3]{8}$

27. $\sqrt[3]{-8}$

28. $-\sqrt[3]{-8}$

SCAN TO ACCESS

B Simplify each expression. Assume all variables represent nonnegative numbers.

29. $\sqrt{36a^2}$

30. $\sqrt{25x^2}$

31. $\sqrt{81x^2}$

32. $\sqrt{49y^2}$

33. $\sqrt{25x^4}$

34. $\sqrt{49y^4}$

35. $\sqrt{49x^6}$

36. $\sqrt{64y^8}$

37. $\sqrt{9x^2y^4}$

38. $\sqrt{9x^4y^6}$

39. $\sqrt{121a^6b^2}$

40. $\sqrt{121a^4b^8}$

41. $\sqrt{25x^{12}y^{10}}$

42. $\sqrt{49x^{14}y^8}$

43. $\sqrt{144a^8b^{16}}$

44. $\sqrt{64a^{12}b^{18}}$

45. $\sqrt[3]{27a^{12}}$

46. $\sqrt[3]{8a^{15}}$

47. $\sqrt[3]{x^3y^6}$

48. $\sqrt[3]{x^6y^3}$

49. $\sqrt[5]{32x^{10}y^5}$

50. $\sqrt[5]{32x^5y^{10}}$

51. $\sqrt[4]{16a^{12}b^{20}}$

52. $\sqrt[4]{81a^{24}b^8}$

53. $\sqrt[3]{-27x^9y^6}$

54. $\sqrt[3]{-8x^{12}y^{15}}$

55. $\sqrt[3]{-8a^6b^{15}}$

56. $\sqrt[3]{-64a^{15}b^{21}}$

57. $\sqrt[4]{16x^{12}y^{20}}$

58. $\sqrt[4]{81x^{20}y^{32}}$

59. $\sqrt[5]{-32a^{10}b^{30}}$

60. $\sqrt[5]{-32a^{60}b^{100}}$

Applying the Concepts

61. Maximum Speed The maximum speed v that an automobile can travel around a curve of radius r without skidding is given by the equation

$$v = \sqrt{\frac{5r}{2}}$$

where v is in miles per hour and r is measured in feet. What is the maximum speed a car can travel around a curve with a radius of 250 feet without skidding?

62. Relativity The equation

$$L = \sqrt{1 - \frac{v^2}{c^2}}$$

gives the relativistic length of a 1-foot ruler traveling with velocity v. Find L if

$$\frac{v}{c} = \frac{3}{5}$$

63. Golden Ratio The golden ratio is the ratio of the length to the width in any golden rectangle. The exact value of this number is $\frac{1+\sqrt{5}}{2}$. Use a calculator to find a decimal approximation of this number and round it to the nearest thousandth.

64. Golden Ratio The reciprocal of the golden ratio is $\frac{2}{1+\sqrt{5}}$. Find a decimal approximation of this number that is accurate to the nearest thousandth.

65. Chemistry Figure 5 shows part of a model of a magnesium oxide (MgO) crystal. Each corner of the square is at the center of one oxygen ion (O^{2-}), and the center of the middle ion is at the center of the square. The radius for each oxygen ion is 60 picometers (pm), and the radius for each magnesium ion (Mg^{2+}) is 150 picometers.

FIGURE 5

a. Find the length of the side of the square. Write your answer in picometers.

b. Find the length of the diagonal of the square. Write your answer in picometers.

c. If 1 meter is 10^{12} picometers, give the length of the diagonal of the square in meters.

66. Geometry The length of each side of the cube shown in Figure 6 is 1 inch.

FIGURE 6

a. Find the length of the diagonal CH.

b. Find the length of the diagonal CF.

67. In the introduction to this chapter, the radius of a sphere is given by the formula

$$r = \sqrt[3]{\frac{3V}{4\pi}}$$

If the volume of Spaceship Earth is 2,200,000 cubic feet, find its radius.

68. With an unknown origin, the store spheres of Costa Rica are an archaeological mystery. Over 300 spheres have been discovered, and the largest has a volume of approximately 183.6 cubic feet. Using the formula given in the previous exercise, find the diameter of this sphere.

69. If A represents the surface area of a sphere, the radius is given by the formula $r = \sqrt{\frac{A}{4\pi}}$. The Montreal Biosphere is a large geodesic dome with a total surface area of 196,350 square feet. Find the diameter of the sphere.

70. The famous dome at the U.S. capital has an area of 15,708 square feet. Use the formula given in the previous exercise to find its diameter. Hint: Since the capital is a half-dome, you will need to double the area before using the formula.

Getting Ready for the Next Section

Simplify. Assume all variable are positive real numbers.

71. $\sqrt{25}$

72. $\sqrt{4}$

73. $\sqrt{6^2}$

74. $\sqrt{3^2}$

75. $\sqrt{16x^4y^2}$

76. $\sqrt{4x^6y^8}$

77. $\sqrt{(5y)^2}$

78. $\sqrt{(8x^3)^2}$

79. $\sqrt[3]{27}$

80. $\sqrt[3]{-8}$

81. $\sqrt[3]{2^3}$

82. $\sqrt[3]{(-5)^3}$

83. $\sqrt[3]{8a^3b^3}$

84. $\sqrt[3]{64a^6b^3}$

85. $\sqrt[4]{625x^8y^4}$

86. $\sqrt[5]{(-x)^5}$

Find the Mistake

Each sentence below contains a mistake. Circle the mistake and write the correct word(s) or expression on the line provided.

1. The n^{th} root of the x is such that $\sqrt[n]{x} = x$. _____

2. Simplifying $\sqrt[6]{-729}$ equals 3 because $3^6 = 729$. _____

3. Simplifying $\sqrt[4]{256x^{16}y^{12}z^8}$ equals $4x^{12}y^8z^4$. _____

4. The golden ratio is the ratio of a rectangle's perimeter to width, which equals $\frac{1 + \sqrt{5}}{2}$. _____

Navigation Skills: Prepare, Study, Achieve

The chapters in this course are organized such that each chapter builds on the previous chapters. So you already have learned the tools to master the final topics in this book and be successful on the final exam. However, studying for the final exam may still seem like an overwhelming task. To ease that anxiety, begin now to lay out a study plan. Dedicate time in your day to do the following:

- Stay calm and maintain a positive attitude.
- Scan each section of the book to review headers, graphics, definitions, rules, properties, formulas, italicized words, and margin notes. Take new notes as you scan.
- Review your notes and homework.
- Rework problems in each section and from your difficult problems list.
- Review chapter summaries, and make your own outlines for each chapter.
- Make and review flashcards of definitions, properties, or formulas.
- Explain concepts to a friend or out loud to yourself.
- Schedule time to meet with a study partner or group.

Another way to prepare is to visualize yourself being successful on the exam. Close your eyes and picture yourself arriving on exam day, receiving the test, staying calm, and working through difficult problems successfully. Now picture yourself receiving a high score on the exam. Complete this visualization multiple times before the day of the exam. If you picture yourself achieving success, you are more likely to be successful.

OBJECTIVES

A Write radical expressions in simplified form.

B Rationalize a denominator that contains only one term.

KEY WORDS

product property for radicals

quotient property for radicals

simplified form

rationalizing the denominator

absolute value property for radicals

spiral of roots

In the previous section, we showed how the Pythagorean theorem can be used to construct a golden rectangle. In a similar manner, the Pythagorean theorem can be used to construct the attractive spiral shown here.

This spiral is called the Spiral of Roots because each of the diagonals is the positive square root of one of the positive integers. At the end of this section, we will use the Pythagorean theorem and some of the material in this section to construct this spiral.

We will begin by stating two properties of radicals. Following this, we will give a definition for simplified form for radical expressions. The examples in this section show how we use the properties of radicals to write radical expressions in simplified form.

A Simplified Form

Here are the first two properties of radicals.

> *Note* Restrictions are an important part of any mathematical definition. If asked to define a property, don't forget to include the restrictions.

> **PROPERTY** Product Property for Radicals
>
> If a and b are nonnegative real numbers whenever n is an even number, then
> $$\sqrt[n]{ab} = \sqrt[n]{a}\sqrt[n]{b}$$
>
> *In words:* The nth root of a product is the product of the nth roots.

> *Note* There is not a property for radicals that says the nth root of a sum is the sum of the nth roots. That is,
>
> $$\sqrt[n]{a+b} \neq \sqrt[n]{a} + \sqrt[n]{b}$$

> **PROPERTY** Quotient Property for Radicals
>
> If a and b are nonnegative real numbers whenever n is an even number, then
> $$\sqrt[n]{\frac{a}{b}} = \frac{\sqrt[n]{a}}{\sqrt[n]{b}} = \quad (b \neq 0)$$
>
> *In words:* The nth root of a quotient is the quotient of the nth roots.

These two properties of radicals allow us to change the form of and simplify radical expressions without changing their value.

RULE Simplified Form for Radical Expressions

A radical expression is in *simplified form* if

1. None of the factors of the radicand (the quantity under the radical sign) can be written as powers greater than or equal to the index—that is, no perfect squares can be factors of the quantity under a square root sign, no perfect cubes can be factors of what is under a cube root sign, and so forth.

2. There are no fractions under the radical sign.

3. There are no radicals in the denominator.

Satisfying the first condition for simplified form actually amounts to taking as much out from under the radical sign as possible. The following examples illustrate the first condition for simplified form.

EXAMPLE 1 Write $\sqrt{50}$ in simplified form.

Solution The largest perfect square that divides 50 is 25. We write 50 as $25 \cdot 2$ and apply the product property for radicals.

$$\sqrt{50} = \sqrt{25 \cdot 2} \qquad 50 = 25 \cdot 2$$
$$= \sqrt{25}\sqrt{2} \qquad \text{Product property for radicals}$$
$$= 5\sqrt{2} \qquad \sqrt{25} = 5$$

We have taken as much as possible out from under the radical sign—in this case, factoring 25 from 50 and then writing $\sqrt{25}$ as 5.

EXAMPLE 2 Write $\sqrt{48x^4y^3}$, where $x, y \geq 0$ in simplified form.

Solution The largest perfect square that is a factor of the radicand is $16x^4y^2$. Applying the product property again, we have

$$\sqrt{48x^4y^3} = \sqrt{16x^4y^2 \cdot 3y}$$
$$= \sqrt{16x^4y^2}\sqrt{3y}$$
$$= 4x^2y\sqrt{3y}$$

EXAMPLE 3 Write $\sqrt[3]{40a^5b^4}$ in simplified form.

Solution We now want to factor the largest perfect cube from the radicand. We write $40a^5b^4$ as $8a^3b^3 \cdot 5a^2b$ and proceed as we did in Examples 1 and 2.

$$\sqrt[3]{40a^5b^4} = \sqrt[3]{8a^3b^3 \cdot 5a^2b}$$
$$= \sqrt[3]{8a^3b^3}\sqrt[3]{5a^2b}$$
$$= 2ab\sqrt[3]{5a^2b}$$

Here are some further examples concerning the first condition for simplified form.

EXAMPLE 4 Write each expression in simplified form.

a. $\sqrt{12x^7y^6} = \sqrt{4x^6y^6 \cdot 3x}$
$$= \sqrt{4x^6y^6}\sqrt{3x}$$
$$= 2x^3y^3\sqrt{3x}$$

b. $\sqrt[3]{54a^6b^2c^4} = \sqrt[3]{27a^6c^3 \cdot 2b^2c}$
$$= \sqrt[3]{27a^6c^3}\sqrt[3]{2b^2c}$$
$$= 3a^2c\sqrt[3]{2b^2c}$$

Practice Problems

1. Write $\sqrt{18}$ in simplified form.

2. Write $\sqrt{50x^2y^3}$ in simplified form. Assume $x, y \geq 0$.

3. Write $\sqrt[3]{54a^4b^3}$ in simplified form.

4. Write each expression in simplified form.

 a. $\sqrt{75x^5y^8}$

 b. $\sqrt[4]{48a^8b^5c^4}$

Answers

1. $3\sqrt{2}$
2. $5xy\sqrt{2y}$
3. $3ab\sqrt[3]{2a}$
4. **a.** $5x^2y^4\sqrt{3x}$ **b.** $2a^2bc\sqrt[4]{3b}$

5. Simplify $\sqrt{\dfrac{5}{9}}$.

B Rationalizing the Denominator

The quotient property for radicals is used to simplify a radical that contains a fraction.

EXAMPLE 5 Simplify $\sqrt{\dfrac{3}{4}}$.

Solution Applying the quotient property for radicals, we have

$$\sqrt{\dfrac{3}{4}} = \dfrac{\sqrt{3}}{\sqrt{4}} \qquad \textit{Quotient property for radicals}$$

$$= \dfrac{\sqrt{3}}{2} \qquad \sqrt{4} = 2$$

The last expression is in simplified form because it satisfies all three conditions for simplified form.

6. Write $\sqrt{\dfrac{2}{3}}$ in simplified form.

EXAMPLE 6 Write $\sqrt{\dfrac{5}{6}}$ in simplified form.

Solution Proceeding as in Example 5, we have

$$\sqrt{\dfrac{5}{6}} = \dfrac{\sqrt{5}}{\sqrt{6}}$$

The resulting expression satisfies the second condition for simplified form because neither radical contains a fraction. It does, however, violate Condition 3 because it has a radical in the denominator. Getting rid of the radical in the denominator is called *rationalizing the denominator* and is accomplished, in this case, by multiplying the numerator and denominator by $\sqrt{6}$.

$$\dfrac{\sqrt{5}}{\sqrt{6}} = \dfrac{\sqrt{5}}{\sqrt{6}} \cdot \dfrac{\sqrt{6}}{\sqrt{6}}$$

$$= \dfrac{\sqrt{30}}{\sqrt{6^2}}$$

$$= \dfrac{\sqrt{30}}{6}$$

7. Rationalize the denominator.

a. $\dfrac{5}{\sqrt{2}}$

b. $\dfrac{3\sqrt{5x}}{\sqrt{2y}}$

EXAMPLE 7 Rationalize the denominator.

a. $\dfrac{4}{\sqrt{3}} = \dfrac{4}{\sqrt{3}} \cdot \dfrac{\sqrt{3}}{\sqrt{3}}$

$$= \dfrac{4\sqrt{3}}{\sqrt{3^2}}$$

$$= \dfrac{4\sqrt{3}}{3}$$

b. $\dfrac{2\sqrt{3x}}{\sqrt{5y}} = \dfrac{2\sqrt{3x}}{\sqrt{5y}} \cdot \dfrac{\sqrt{5y}}{\sqrt{5y}}$

$$= \dfrac{2\sqrt{15xy}}{\sqrt{(5y)^2}}$$

$$= \dfrac{2\sqrt{15xy}}{5y}$$

When the denominator involves a cube root, we must multiply by a radical that will produce a perfect cube under the cube root sign in the denominator, as Example 8 on the next page illustrates.

Answers

5. $\dfrac{\sqrt{5}}{3}$

6. $\dfrac{\sqrt{6}}{3}$

7. a. $\dfrac{5\sqrt{2}}{2}$ **b.** $\dfrac{3\sqrt{10xy}}{2y}$

EXAMPLE 8 Rationalize the denominator in $\dfrac{7}{\sqrt[3]{4}}$.

Solution Because $4 = 2^2$, we can multiply both numerator and denominator by $\sqrt[3]{2}$ and obtain $\sqrt[3]{2^3}$ in the denominator.

$$\frac{7}{\sqrt[3]{4}} = \frac{7}{\sqrt[3]{2^2}}$$

$$= \frac{7}{\sqrt[3]{2^2}} \cdot \frac{\sqrt[3]{2}}{\sqrt[3]{2}}$$

$$= \frac{7\sqrt[3]{2}}{\sqrt[3]{2^3}}$$

$$= \frac{7\sqrt[3]{2}}{2}$$

EXAMPLE 9 Simplify $\sqrt{\dfrac{12x^5y^3}{5z}}$.

Solution We use the quotient property for radicals to write the numerator and denominator as two separate radicals.

$$\sqrt{\frac{12x^5y^3}{5z}} = \frac{\sqrt{12x^5y^3}}{\sqrt{5z}}$$

Simplifying the numerator, we have

$$\frac{\sqrt{12x^5y^3}}{\sqrt{5z}} = \frac{\sqrt{4x^4y^2}\sqrt{3xy}}{\sqrt{5z}}$$

$$= \frac{2x^2y\sqrt{3xy}}{\sqrt{5z}}$$

To rationalize the denominator, we multiply the numerator and denominator by $\sqrt{5z}$.

$$\frac{2x^2y\sqrt{3xy}}{\sqrt{5z}} \cdot \frac{\sqrt{5z}}{\sqrt{5z}} = \frac{2x^2y\sqrt{15xyz}}{\sqrt{(5z)^2}}$$

$$= \frac{2x^2y\sqrt{15xyz}}{5z}$$

Square Root of a Perfect Square

So far in this chapter, we have assumed that all our variables are nonnegative when they appear under a square root symbol. There are times, however, when this is not the case.

Consider the following two statements:

$$\sqrt{3^2} = \sqrt{9} = 3 \qquad \text{and} \qquad \sqrt{(-3)^2} = \sqrt{9} = 3$$

Whether we operate on 3 or -3, the result is the same: Both expressions simplify to 3. The other operation we have worked with in the past that produces the same result is absolute value. That is,

$$|3| = 3 \qquad \text{and} \qquad |-3| = 3$$

This leads us to the next property of radicals.

> **PROPERTY** Absolute Value Property for Radicals
>
> If a is a real number, then $\sqrt{a^2} = |a|$.

The result of this discussion and the absolute value property for radicals is simply this:

If we know a is non-negative, then $\sqrt{a^2} = a$.

If we know a is negative, then $\sqrt{a^2} = |a|$.

If we don't know if a is positive or negative, then $\sqrt{a^2} = |a|$.

8. Rationalize the denominator.
$$\frac{5}{\sqrt[3]{9}}$$

9. Simplify $\sqrt{\dfrac{48x^3y^4}{7z}}$.

Answers

8. $\frac{5\sqrt[3]{3}}{3}$

9. $\frac{3\sqrt{10xy}}{2y}$

10. Simplify each expression. Do not assume the variables represent non-negative numbers.

a. $\sqrt{16x^2}$

b. $\sqrt{25x^3}$

c. $\sqrt{x^2 + 10x + 25}$

d. $\sqrt{2x^3 + 7x^2}$

11. Simplify each expression.

a. $\sqrt[3]{(-3)^3}$

b. $\sqrt[3]{(-1)^3}$

c. $\sqrt[3]{(-3x)^3}$

EXAMPLE 10 Simplify each expression. Do *not* assume the variables represent positive numbers.

a. $\sqrt{9x^2} = 3|x|$

b. $\sqrt{x^3} = |x|\sqrt{x}$

c. $\sqrt{x^2 - 6x + 9} = \sqrt{(x-3)^2} = |x-3|$

d. $\sqrt{x^3 - 5x^2} = \sqrt{x^2(x-5)} = |x|\sqrt{x-5}$

As you can see, we must use absolute value symbols when we take a square root of a perfect square, unless we know the base of the perfect square is a non-negative number. The same idea holds for higher even roots, but not for odd roots. With odd roots, no absolute value symbols are necessary.

EXAMPLE 11 Simplify each expression.

a. $\sqrt[3]{(-2)^3} = \sqrt[3]{-8} = -2$

b. $\sqrt[3]{(-5)^3} = \sqrt[3]{-125} = -5$

c. $\sqrt[3]{(-2n)^3} = \sqrt[3]{-8n^3} = -2n$

We can extend this discussion to all roots as follows:

> **PROPERTY** Extending the Absolute Value Property for Radicals
>
> If a is a real number, then
>
> $$\sqrt[n]{a^n} = |a| \qquad \text{if} \qquad n \text{ is even}$$
> $$\sqrt[n]{a^n} = a \qquad \text{if} \qquad n \text{ is odd}$$

The Spiral of Roots

To visualize the square roots of the positive integers, we can construct the spiral of roots that we mentioned in the introduction to this section. To begin, we draw two line segments, each of length 1, at right angles to each other. Then we use the Pythagorean theorem to find the length of the diagonal. Figure 1 illustrates this procedure.

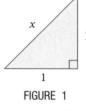

FIGURE 1

$$x = \sqrt{1^2 + 1^2}$$
$$= \sqrt{2}$$

Next, we construct a second triangle by connecting a line segment of length 1 to the end of the first diagonal so that the angle formed is a right angle. We find the length of the second diagonal using the Pythagorean theorem. Figure 2 illustrates this procedure. Continuing to draw new triangles by connecting line segments of length 1 to the end of each new diagonal, so that the angle formed is a right angle, the spiral of roots begins to appear (Figure 3).

FIGURE 2 FIGURE 3

$$x = \sqrt{(\sqrt{2})^2 + 1^2} \qquad x = \sqrt{(\sqrt{3})^2 + 1^2}$$
$$= \sqrt{2 + 1} \qquad\qquad = \sqrt{3 + 1}$$
$$= \sqrt{3} \qquad\qquad\quad = \sqrt{4}$$

Answers

10. a. $4|x|$ **b.** $5|x|\sqrt{x}$ **c.** $|x+5|$
 d. $|x|\sqrt{2x+7}$

11. a. -3 **b.** -1 **c.** $-3x$

The Spiral of Roots and Function Notation

Looking over the diagrams and calculations in the preceding discussion, we see that each diagonal in the spiral of roots is found by using the length of the previous diagonal.

First diagonal: $\sqrt{1^2 + 1^2} = \sqrt{2}$

Second diagonal: $\sqrt{(\sqrt{2})^2 + 1^2} = \sqrt{3}$

Third diagonal: $\sqrt{(\sqrt{3})^2 + 1^2} = \sqrt{4}$

Fourth diagonal: $\sqrt{(\sqrt{4})^2 + 1^2} = \sqrt{5}$

A process like this one, in which the answer to one calculation is used to find the answer to the next calculation, is called a *recursive* process. In this particular case, we can use function notation to model the process. If we let x represent the length of any diagonal, then the length of the next diagonal is given by

$$f(x) = \sqrt{x^2 + 1}$$

To begin the process of finding the diagonals, we let $x = 1$.

$$f(1) = \sqrt{1^2 + 1} = \sqrt{2}$$

To find the next diagonal, we substitute $\sqrt{2}$ for x to obtain

$$f(f(1)) = f(\sqrt{2}) = \sqrt{(\sqrt{2})^2 + 1} = \sqrt{3}$$
$$f(f(f(1))) = f(\sqrt{3}) = \sqrt{(\sqrt{3})^2 + 1} = \sqrt{4}$$

We will explore functions and function notation more in the next chapter.

First Input
x

Function Machine

Output
$f(x)$

USING TECHNOLOGY

As our preceding discussion indicates, the length of each diagonal in the spiral of roots is used to calculate the length of the next diagonal. The ANS key on a graphing calculator can be used effectively in a situation like this. To begin, we store the number 1 in the variable ANS. Next, we key in the formula used to produce each diagonal using ANS for the variable. After that, it is simply a matter of pressing ENTER, as many times as we like, to produce the lengths of as many diagonals as we like. Here is a summary of what we do:

Enter This	Display Shows
1 ENTER	1.000
$\sqrt{\ }$ (ANS² + 1) ENTER	1.414
ENTER	1.732
ENTER	2.000
ENTER	2.236

If you continue to press the ENTER key, you will produce decimal approximations for as many of the diagonals in the spiral of roots as you like.

GETTING READY FOR CLASS

After reading through the preceding section, respond in your own words and in complete sentences.

A. Describe the product property for radicals.

B. Describe the quotient property for radicals.

C. What does it mean to rationalize the denominator in an expression?

D. What is useful about the spiral of roots?

EXERCISE SET 8.2

Vocabulary Review

Choose the correct words to fill in the blanks below.

rationalizing even quotient radicand

fractions odd product denominator

simplified form index absolute value spiral of roots

1. The _____ property for radicals states that the n^{th} root of a product is the product of the n^{th} roots.

2. The _____ property for radicals states that the n^{th} root of a quotient is the quotient of the n^{th} roots.

3. A radical expression is in _____ if (1) none of the factors of the _____ can be written as powers greater than or equal to the _____, (2) there are no _____ under the radical sign, and (3) there are no radicals in the _____.

4. Getting rid of the radical in a denominator is called _____ the denominator.

5. The _____ property for radicals states that $\sqrt{a^2} = |a|$ if a is a real number.

6. To extend the absolute value property for radicals, we say that $\sqrt[n]{a^n} = |a|$ if n is _____ and $\sqrt[n]{a^n} = a$ if n is _____.

7. The _____ helps us visualize the square roots of positive integers.

Problems

Substitute the given numbers into the expression $\sqrt{b^2 - 4ac}$, and then simplify. You will encounter this expression again in the next chapter.

1. $a = 2, b = -6, c = 3$ 2. $a = 6, b = 7, c = -5$ 3. $a = 1, b = 2, c = 6$

4. $a = 2, b = 5, c = 3$ 5. $a = \dfrac{1}{2}, b = -\dfrac{1}{2}, c = -\dfrac{5}{4}$ 6. $a = \dfrac{7}{4}, b = -\dfrac{3}{4}, c = -2$

SCAN TO ACCESS

Use the product property for radicals to write each of the following expressions in simplified form. Assume all variables are nonnegative.

7. $\sqrt{8}$

8. $\sqrt{32}$

9. $\sqrt{98}$

10. $\sqrt{75}$

11. $\sqrt{288}$

12. $\sqrt{128}$

13. $\sqrt{80}$

14. $\sqrt{200}$

15. $\sqrt{48}$

16. $\sqrt{27}$

17. $\sqrt{675}$

18. $\sqrt{972}$

19. $\sqrt[3]{54}$

20. $\sqrt[3]{24}$

21. $\sqrt[3]{128}$

22. $\sqrt[3]{162}$

23. $\sqrt[3]{432}$

24. $\sqrt[3]{1,536}$

25. $\sqrt[5]{64}$

26. $\sqrt[4]{48}$

27. $\sqrt{18x^3}$

28. $\sqrt{27x^5}$

29. $\sqrt[4]{32y^7}$

30. $\sqrt[5]{32y^7}$

31. $\sqrt[3]{40x^4y^7}$

32. $\sqrt[3]{128x^6y^2}$

33. $\sqrt{48a^2b^3c^4}$

34. $\sqrt{72a^4b^3c^2}$

35. $\sqrt[3]{48a^2b^3c^4}$

36. $\sqrt[3]{72a^4b^3c^2}$

37. $\sqrt[5]{64x^8y^{12}}$

38. $\sqrt[4]{32x^9y^{10}}$

39. $\sqrt[5]{243x^7y^{10}z^5}$

40. $\sqrt[5]{64x^8y^4z^{11}}$

41. $\sqrt{128x^7y^5}$

42. $\sqrt{50x^8y^{15}}$

$\sqrt{128x^6 \cdot x \, y^4 \, y}$

$2 \cdot 64x^6 \cdot x \, y^4 \cdot y$

$\sqrt{2 \cdot 8^2 \, x^6 \cdot y^4 \, y}$

$8x^3y^2\sqrt{2xy}$

Use the quotient property for radicals to write each of the following expressions in simplified form. Assume all variables are nonnegative.

43. $\sqrt{\dfrac{49}{4}}$

44. $\sqrt{\dfrac{5}{121}}$

45. $\sqrt[3]{\dfrac{-125}{64}}$

46. $\sqrt{\dfrac{63x^4}{4}}$

47. $\sqrt{\dfrac{72x^6}{y^2}}$

48. $\sqrt[4]{\dfrac{a^4d^{10}}{b^4c^8}}$

49. $\sqrt[5]{\dfrac{1,024}{32a^5}}$

50. $\sqrt[6]{\dfrac{729}{a^6b^{12}c^{18}}}$

B Rationalize the denominator in each of the following expressions. Assume all variables represent positive numbers.

51. $\dfrac{2}{\sqrt{3}}$

52. $\dfrac{3}{\sqrt{2}}$

53. $\dfrac{5}{\sqrt{6}}$

54. $\dfrac{7}{\sqrt{5}}$

55. $\sqrt{\dfrac{1}{2}}$

56. $\sqrt{\dfrac{1}{3}}$

57. $\sqrt{\dfrac{1}{5}}$

58. $\sqrt{\dfrac{1}{6}}$

59. $\dfrac{4}{\sqrt[3]{2}}$

60. $\dfrac{5}{\sqrt[3]{3}}$

61. $\dfrac{2}{\sqrt[3]{9}}$

62. $\dfrac{3}{\sqrt[3]{4}}$

63. $\sqrt[4]{\dfrac{3}{2x^2}}$

64. $\sqrt[4]{\dfrac{5}{3x^2}}$

65. $\sqrt[4]{\dfrac{8}{y}}$

66. $\sqrt[4]{\dfrac{27}{y}}$

67. $\sqrt[3]{\dfrac{4x}{3y}}$

68. $\sqrt[3]{\dfrac{7x}{6y}}$

69. $\sqrt[3]{\dfrac{2x}{9y}}$

70. $\sqrt[3]{\dfrac{5x}{4y}}$

71. $\sqrt[4]{\dfrac{1}{8x^3}}$

72. $\sqrt[4]{\dfrac{8}{9x^3}}$

73. $\sqrt[3]{\dfrac{4}{9x^2}}$

74. $\sqrt[3]{\dfrac{40}{9xy^2}}$

Write each of the following in simplified form. Assume all variables represent positive numbers.

75. $\sqrt{\dfrac{27x^3}{5y}}$

76. $\sqrt{\dfrac{12x^5}{7y}}$

77. $\sqrt{\dfrac{75x^3y^2}{2z}}$

78. $\sqrt{\dfrac{50x^2y^3}{3z}}$

79. $\sqrt[3]{\dfrac{16a^4b^3}{9c}}$

80. $\sqrt[3]{\dfrac{54a^5b^4}{25c^2}}$

81. $\sqrt[3]{\dfrac{8x^3y^6}{9z}}$

82. $\sqrt[3]{\dfrac{27x^6y^3}{2z^2}}$

Simplify each expression. Do *not* assume the variables represent positive numbers.

83. $\sqrt{25x^2}$

84. $\sqrt{49x^2}$

85. $\sqrt{27x^3y^2}$

86. $\sqrt{40x^3y^2}$

87. $\sqrt{x^2 - 10x + 25}$

88. $\sqrt{x^2 - 16x + 64}$

89. $\sqrt{4x^2 + 12x + 9}$

90. $\sqrt{16x^2 + 40x + 25}$

91. $\sqrt{4a^4 + 16a^3 + 16a^2}$

92. $\sqrt{9a^4 + 18a^3 + 9a^2}$

93. $\sqrt{4x^3 - 8x^2}$

94. $\sqrt{18x^3 - 9x^2}$

95. Show that the statement $\sqrt{a + b} = \sqrt{a} + \sqrt{b}$ is not true by replacing a with 9 and b with 16 and simplifying both sides.

96. Find a pair of values for a and b that will make the statement $\sqrt{a + b} = \sqrt{a} + \sqrt{b}$ true.

Applying the Concepts

97. Diagonal Distance The distance d between opposite corners of a rectangular room with length l and width w is given by

$$d = \sqrt{l^2 + w^2}$$

How far is it between opposite corners of a living room that measures 10 by 15 feet?

98. Radius of a Sphere As discussed in the chapter opener, the radius r of a sphere with volume V can be found by using the formula

$$r = \sqrt[3]{\dfrac{3V}{4\pi}}$$

Find the radius of a sphere with volume 9 cubic feet. Write your answer in simplified form.

(Use $\dfrac{22}{7}$ for π.)

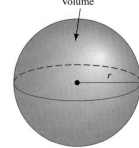
Volume
r

99. Distance to the Horizon If you are at a point k miles above the surface of the Earth, the distance you can see, in miles, is approximated by the equation $d = \sqrt{8000k + k^2}$.

 a. How far can you see from a point that is 1 mile above the surface of Earth?

 b. How far can you see from a point that is 2 miles above the surface of Earth?

 c. How far can you see from a point that is 3 miles above the surface of Earth?

100. Investing If you invest P dollars and you want the investment to grow to A dollars in t years, the interest rate that must be earned if interest is compounded annually is given by the formula

$$r = \sqrt[t]{\frac{A}{P}} - 1$$

If you invest \$4,000 and want to have \$7,000 in 8 years, what interest rate must be earned?

101. The hypotenuse h of a right triangle is given by $h = \sqrt{a^2 + b^2}$, where a and b are the length of the sides. If the two sides are 6 inches and 4 inches, find and simplify the hypotenuse.

102. Repeat exercise 101, if the two sides are 6 meters and 9 meters.

103. Repeat exercise 101, if the two sides are x and $2x$.

104. Repeat exercise 101, if the two sides are $6x$ and $12x$.

105. Spiral of Roots Construct your own spiral of roots by using a ruler. Draw the first triangle by using two 1-inch lines. The first diagonal will have a length of $\sqrt{2}$ inches. Each new triangle will be formed by drawing a 1-inch line segment at the end of the previous diagonal so the angle formed is 90°.

106. Spiral of Roots Construct a spiral of roots by using line segments of length 2 inches. The length of the first diagonal will be $2\sqrt{2}$ inches. The length of the second diagonal will be $2\sqrt{3}$ inches.

Getting Ready for the Next Section

Simplify the following.

107. $5x - 4x + 6x$

108. $12x + 8x - 7x$

109. $35xy^2 - 8xy^2$

110. $20a^2b + 33a^2b$

111. $\frac{1}{2}x + \frac{1}{3}x$

112. $\frac{2}{3}x + \frac{5}{8}x$

113. $5(x + 1)^2 - 12(x + 1)^2$

114. $-7(a + 2)^3 + 4(a + 2)^3$

Write in simplified form for radicals.

115. $\sqrt{18}$

116. $\sqrt{8}$

117. $\sqrt{75xy^3}$

118. $\sqrt{12xy}$

119. $\sqrt[3]{8a^4b^2}$

120. $\sqrt[3]{27ab^2}$

121. $\sqrt[4]{32x^8y^5}$

122. $\sqrt[4]{81a^6b^{11}}$

Find the Mistake

Each sentence below contains a mistake. Circle the mistake and write the correct word or expression on the line provided.

1. Product property for radicals states that the n^{th} root of a quotient is the product of the n^{th} roots. _____

2. Using the quotient property for radicals, the expression $\sqrt[5]{\dfrac{3}{7}}$ is equal to $\dfrac{\sqrt[5]{3}}{7}$. _____

3. The process of rationalizing the denominator is used to get rid of a radical in the numerator. _____

4. According to the extension of the absolute value property for radicals, $\sqrt[n]{a^n} = a$ if n is an even number. _____

Landmark Review: Checking Your Progress

Write each radical in simplified form.

1. $\sqrt{25a^2b^2}$

2. $\sqrt[3]{8x^3}$

3. $\sqrt{64x^2y^4}$

4. $\sqrt[3]{125x^6y^9}$

5. $\sqrt{45}$

6. $\sqrt{32}$

7. $x\sqrt[3]{8x^4}$

8. $\dfrac{4}{5}\sqrt{150x^3}$

9. $\sqrt[3]{\dfrac{125x^3}{8y^3}}$

10. $\sqrt{\dfrac{25}{64x^4}}$

11. $\sqrt{\dfrac{1}{10}}$

12. $\sqrt{\dfrac{5}{7}}$

13. $\dfrac{y\sqrt{6}}{\sqrt{2y}}$

14. $\dfrac{2\sqrt{3}}{\sqrt{2a}}$

KEY WORDS

similar radicals

Note Similar radicals may also
be referred to as *like radicals*.

Practice Problems

1. Combine $3\sqrt{5} - 2\sqrt{5} + 4\sqrt{5}$.

Answer

1. $5\sqrt{5}$

8.3 Addition and Subtraction of Radicals

If it were possible to drop a penny off the top of Mount Everest and have it free fall all the way to the bottom, it would take $\frac{1}{4}\sqrt{29{,}035}$ seconds to reach the ground. Now consider dropping a penny off a slightly shorter mountain called Kangchenjunga. This second penny takes $\frac{1}{4}\sqrt{28{,}208}$ seconds to reach the ground. In this section, we will discover how to add and subtract radical expressions, such as those for the pennies. Then you can come back and figure out how much longer the penny takes to fall from Mt. Everest than from Kangchenjunga.

In Chapter 1, we found we could add similar terms when combining polynomials. The same idea applies to addition and subtraction of radical expressions.

A Combining Similar Terms

> **DEFINITION** similar radicals
>
> Two radicals are said to be **similar radicals** if they have the same index and the same radicand.

The expressions $5\sqrt[3]{7}$ and $-8\sqrt[3]{7}$ are similar since the index is 3 in both cases and the radicands are 7. The expressions $3\sqrt[4]{5}$ and $7\sqrt[3]{5}$ are not similar because they have different indices, and the expressions $2\sqrt[5]{8}$ and $3\sqrt[5]{9}$ are not similar because the radicands are not the same.

We add and subtract radical expressions in the same way we add and subtract polynomials—by combining similar terms under the distributive property.

EXAMPLE 1 Combine $5\sqrt{3} - 4\sqrt{3} + 6\sqrt{3}$.

Solution All three radicals are similar. We apply the distributive property to get

$$5\sqrt{3} - 4\sqrt{3} + 6\sqrt{3} = (5 - 4 + 6)\sqrt{3}$$
$$= 7\sqrt{3}$$

EXAMPLE 2 Combine $3\sqrt{8} + 5\sqrt{18}$.

Solution The two radicals do not seem to be similar. We must write each in simplified form before applying the distributive property.

$$3\sqrt{8} + 5\sqrt{18} = 3\sqrt{4 \cdot 2} + 5\sqrt{9 \cdot 2}$$
$$= 3\sqrt{4}\sqrt{2} + 5\sqrt{9}\sqrt{2}$$
$$= 3 \cdot 2\sqrt{2} + 5 \cdot 3\sqrt{2}$$
$$= 6\sqrt{2} + 15\sqrt{2}$$
$$= (6 + 15)\sqrt{2}$$
$$= 21\sqrt{2}$$

The result of Example 2 can be generalized to the following rule for sums and differences of radical expressions.

> **RULE** Addition or Subtraction of Radical Expressions
>
> To add or subtract radical expressions, put each in simplified form and apply the distributive property, if possible. We can add only similar radicals. We must write each expression in simplified form for radicals before we can tell if the radicals are similar.

EXAMPLE 3 Combine $7\sqrt{75xy^3} - 4y\sqrt{12xy}$, where $x, y \geq 0$.

Solution We write each expression in simplified form and combine similar radicals.

$$7\sqrt{75xy^3} - 4y\sqrt{12xy} = 7\sqrt{25y^2}\sqrt{3xy} - 4y\sqrt{4}\sqrt{3xy}$$
$$= 35y\sqrt{3xy} - 8y\sqrt{3xy}$$
$$= (35y - 8y)\sqrt{3xy}$$
$$= 27y\sqrt{3xy}$$

EXAMPLE 4 Combine $10\sqrt[3]{8a^4b^2} + 11a\sqrt[3]{27ab^2}$.

Solution Writing each radical in simplified form and combining similar terms, we have

$$10\sqrt[3]{8a^4b^2} + 11a\sqrt[3]{27ab^2} = 10\sqrt[3]{8a^3}\sqrt[3]{ab^2} + 11a\sqrt[3]{27}\sqrt[3]{ab^2}$$
$$= 20a\sqrt[3]{ab^2} + 33a\sqrt[3]{ab^2}$$
$$= 53a\sqrt[3]{ab^2}$$

2. Combine $4\sqrt{50} + 3\sqrt{8}$.

3. Assume $x, y \geq 0$ and combine.
$4\sqrt{18x^2y} - 3x\sqrt{50y}$.

4. Combine $2\sqrt[3]{27a^2b^4} + 3b\sqrt[3]{125a^2b}$.

Answers
2. $26\sqrt{2}$
3. $-3x\sqrt{2y}$
4. $21b\sqrt[3]{a^2b}$

5. Combine $\dfrac{\sqrt{5}}{3} + \dfrac{1}{\sqrt{5}}$.

EXAMPLE 5 Combine $\dfrac{\sqrt{3}}{2} + \dfrac{1}{\sqrt{3}}$.

Solution We begin by writing the second term in simplified form.

$$\frac{\sqrt{3}}{2} + \frac{1}{\sqrt{3}} = \frac{\sqrt{3}}{2} + \frac{1}{\sqrt{3}} \cdot \frac{\sqrt{3}}{\sqrt{3}}$$

$$= \frac{\sqrt{3}}{2} + \frac{\sqrt{3}}{3}$$

$$= \frac{1}{2}\sqrt{3} + \frac{1}{3}\sqrt{3}$$

$$= \left(\frac{1}{2} + \frac{1}{3} \right)\sqrt{3}$$

$$= \frac{5}{6}\sqrt{3} = \frac{5\sqrt{3}}{6}$$

B The Golden Rectangle

6. Construct a golden rectangle from a square of side 6. Then show that the ratio of the length to the width is the golden ratio.

EXAMPLE 6 Construct a golden rectangle from a square of side 4. Then show that the ratio of the length to the width is the golden ratio $\frac{1 + \sqrt{5}}{2}$.

Solution Figure 1 shows the golden rectangle constructed from a square of side 4. The length of the diagonal *OB* is found from the Pythagorean theorem.

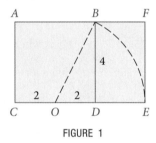

FIGURE 1

$$OB = \sqrt{2^2 + 4^2} = \sqrt{4 + 16} = \sqrt{20} = 2\sqrt{5}$$

The ratio of the length to the width for the rectangle is the golden ratio.

$$\text{Golden ratio} = \frac{CE}{EF} = \frac{2 + 2\sqrt{5}}{4} = \frac{2(1 + \sqrt{5})}{2 \cdot 2} = \frac{1 + \sqrt{5}}{2}$$

As you can see, showing that the ratio of length to width in this rectangle is the golden ratio depends on our ability to write $\sqrt{20}$ as $2\sqrt{5}$ and our ability to reduce to lowest terms by factoring and then dividing out the common factor 2 from the numerator and denominator.

> **GETTING READY FOR CLASS**
>
> *After reading through the preceding section, respond in your own words and in complete sentences.*
>
> **A.** How is adding radical expressions similar to adding polynomial terms?
>
> **B.** Give an example of using the distributive property to add two radical expressions.
>
> **C.** What is the first step when adding or subtracting expressions containing radicals?
>
> **D.** Explain why it is not possible to add $\sqrt{2}$ and $\sqrt{3}$.

Answer

5. $\dfrac{8\sqrt{5}}{15}$

Vocabulary Review

Choose the correct words to fill in the blanks below.

simplified form	index	similar
distributive property	polynomials	radicand

1. Two radicals are said to be similar radicals if they have the same _____ and the same _____ .

2. We add and subtract radical expressions in the same way we add and subtract _____ .

3. To add radical expressions, first put each in _____ and apply the _____ .

4. Only _____ radicals can be added or subtracted.

Problems

A Combine the following expressions. (Assume any variables under an even root are nonnegative.)

1. $3\sqrt{5} + 4\sqrt{5}$

2. $6\sqrt{3} - 5\sqrt{3}$

3. $3x\sqrt{7} - 4x\sqrt{7}$

4. $6y\sqrt{a} + 7y\sqrt{a}$

5. $5\sqrt[3]{10} - 4\sqrt[3]{10}$

6. $6\sqrt[4]{2} + 9\sqrt[4]{2}$

7. $8\sqrt[5]{6} - 2\sqrt[5]{6} + 3\sqrt[5]{6}$

8. $7\sqrt[6]{7} - \sqrt[6]{7} + 4\sqrt[6]{7}$

9. $3x\sqrt{2} - 4x\sqrt{2} + x\sqrt{2}$

10. $5x\sqrt{6} - 3x\sqrt{6} - 2x\sqrt{6}$

11. $\sqrt{20} - \sqrt{80} + \sqrt{45}$

12. $\sqrt{8} - \sqrt{32} - \sqrt{18}$

13. $4\sqrt{8} - 2\sqrt{50} - 5\sqrt{72}$

14. $\sqrt{48} - 3\sqrt{27} + 2\sqrt{75}$

15. $5x\sqrt{8} + 3\sqrt{32x^2} - 5\sqrt{50x^2}$

4. $2\sqrt{5}_2 - 2\cdot 3\sqrt{5}_2 - 5\cdot 6\sqrt{5}_2$

$8\sqrt{5}_2 - 6\sqrt{5}_2 - 30\sqrt{5}_2 = -32\sqrt{5}_2$

16. $2\sqrt{50x^2} - 8x\sqrt{18} - 3\sqrt{72x^2}$

17. $5\sqrt[3]{16} - 4\sqrt[3]{54}$

18. $\sqrt[3]{81} + 3\sqrt[3]{24}$

19. $\sqrt[3]{x^4 y^2} + 7x\sqrt[3]{xy^2}$

20. $2\sqrt[3]{x^8y^6} - 3y^2\sqrt[3]{8x^8}$

21. $5a^2\sqrt{27ab^3} - 6b\sqrt{12a^5b}$

22. $9a\sqrt{20a^3b^2} + 7b\sqrt{45a^5}$

23. $b\sqrt[3]{24a^5b} + 3a\sqrt[3]{81a^2b^4}$

24. $7\sqrt[3]{a^4b^3c^2} - 6ab\sqrt[3]{ac^2}$

25. $5x\sqrt[4]{3y^5} + y\sqrt[4]{243x^4y} + \sqrt[4]{48x^4y^5}$

26. $x\sqrt[4]{5xy^8} + y\sqrt[4]{405x^5y^4} + y^2\sqrt[4]{80x^5}$

27. $\dfrac{\sqrt{2}}{2} + \dfrac{1}{\sqrt{2}}$

28. $\dfrac{\sqrt{3}}{3} + \dfrac{1}{\sqrt{3}}$

29. $\dfrac{\sqrt{5}}{3} + \dfrac{1}{\sqrt{5}}$

30. $\dfrac{\sqrt{6}}{2} + \dfrac{1}{\sqrt{6}}$

31. $\sqrt{x} - \dfrac{1}{\sqrt{x}}$

32. $\sqrt{x} + \dfrac{1}{\sqrt{x}}$

33. $\dfrac{\sqrt{18}}{6} + \sqrt{\dfrac{1}{2}} + \dfrac{\sqrt{2}}{2}$

34. $\dfrac{\sqrt{12}}{6} + \sqrt{\dfrac{1}{3}} + \dfrac{\sqrt{3}}{3}$

35. $\sqrt{6} - \sqrt{\dfrac{2}{3}} + \sqrt{\dfrac{1}{6}}$

36. $\sqrt{15} - \sqrt{\dfrac{3}{5}} + \sqrt{\dfrac{5}{3}}$

37. $\sqrt[3]{25} + \dfrac{3}{\sqrt[3]{5}}$

38. $\sqrt[4]{8} + \dfrac{1}{\sqrt[4]{2}}$

39. Use a calculator to find a decimal approximation for $\sqrt{12}$ and for $2\sqrt{3}$.

40. Use a calculator to find decimal approximations for $\sqrt{50}$ and $5\sqrt{2}$.

41. Use a calculator to find a decimal approximation for $\sqrt{8} + \sqrt{18}$. Is it equal to the decimal approximation for $\sqrt{26}$ or $\sqrt{50}$?

42. Use a calculator to find a decimal approximation for $\sqrt{3} + \sqrt{12}$. Is it equal to the decimal approximation for $\sqrt{15}$ or $\sqrt{27}$?

Each of the following statements is false. Correct the right side of each one to make the statement true.

43. $3\sqrt{2x} + 5\sqrt{2x} = 8\sqrt{4x}$

44. $5\sqrt{3} - 7\sqrt{3} = -2\sqrt{9}$

45. $\sqrt{9 + 16} = 3 + 4$

46. $\sqrt{36 + 64} = 6 + 8$

Applying the Concepts

47. Golden Rectangle Construct a golden rectangle from a square of side 8. Then show that the ratio of the length to the width is the golden ratio $\frac{1 + \sqrt{5}}{2}$.

48. Golden Rectangle Construct a golden rectangle from a square of side 10. Then show that the ratio of the length to the width is the golden ratio $\frac{1 + \sqrt{5}}{2}$.

49. Golden Rectangle Use a ruler to construct a golden rectangle from a square of side 1 inch. Then show that the ratio of the length to the width is the golden ratio.

50. Golden Rectangle Use a ruler to construct a golden rectangle from a square of side $\frac{2}{3}$ inch. Then show that the ratio of the length to the width is the golden ratio.

51. Golden Rectangle To show that all golden rectangles have the same ratio of length to width, construct a golden rectangle from a square of side $2x$. Then show that the ratio of the length to the width is the golden ratio.

52. Golden Rectangle To show that all golden rectangles have the same ratio of length to width, construct a golden rectangle from a square of side x. Then show that the ratio of the length to the width is the golden ratio.

53. Isosceles Right Triangles A triangle is isosceles if it has two equal sides, and a triangle is a right triangle if it has a right angle in it. Sketch an isosceles right triangle, and find the ratio of the hypotenuse to a leg.

54. Equilateral Triangles A triangle is equilateral if it has three equal sides. The triangle in the figure is equilateral with each side of length $2x$. Find the ratio of the height to a side.

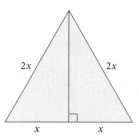

55. Pyramids The following solid is called a regular square pyramid because its base is a square and all eight edges are the same length, 5. It is also true that the vertex, V, is directly above the center of the base.

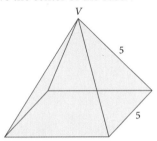

Find the ratio of a diagonal of the base to the length of a side.

56. Pyramids Refer to the diagram of a square pyramid below. Find the ratio of the height h of the pyramid to the altitude a.

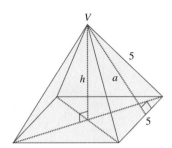

57. In the introduction to this section, you were introduced to the formula $t = \frac{1}{4}\sqrt{h}$ to represent the time (in seconds) it takes an object to free fall h feet, neglecting air resistance. Compute the times it would take a coin to fall from Mount Everest (29,035 feet) and Kangchenjanga (28,208 feet).

58. Two of the tallest buildings in the world are the Burj Khalifa in Dubai (2,717 feet) and the Taipei 101 in Taipei (1,670 feet). Using the formula from exercise 57, find the time it would take for a coin to hit the ground in free fall if dropped from the top of each building.

Getting Ready for the Next Section

Simplify the following.

59. $3 \cdot 2$

60. $5 \cdot 7$

61. $(x + y)(4x - y)$

62. $(2x + y)(x - y)$

63. $(x + 3)^2$

64. $(3x - 2y)^2$

65. $(x - 2)(x + 2)$

66. $(2x + 5)(2x - 5)$

Simplify the following expressions.

67. $2\sqrt{18}$

68. $5\sqrt{36}$

69. $(\sqrt{6})^2$

70. $(\sqrt{2})^2$

71. $(3\sqrt{x})^2$

72. $(2\sqrt{y})^2$

73. $(4\sqrt{3x})^2$

74. $(-5\sqrt{2x})^2$

Find the Mistake

Each sentence below contains a mistake. Circle the mistake and write the correct word(s) or expression on the line provided.

1. Two radicals are similar if they have the same index and different radicands. _____

2. Combining $2\sqrt{6} + 2\sqrt{6} + 2\sqrt{6}$ gives us $2^3 \cdot \sqrt{6}$. _____

3. We can combine $\sqrt[3]{x^2}$ and $\sqrt[2]{x^3}$ to get $2\sqrt[6]{x}$. _____

4. The first step to combining $\frac{\sqrt{4}}{6}$ and $\frac{9}{\sqrt{4}}$ is to factor 3 from both expressions. _____

8.4 Multiplication and Division of Radicals

We have worked with the golden rectangle more than once in this chapter. The following is one such golden rectangle.

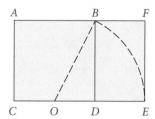

By now you know that, in any golden rectangle constructed from a square (of any size), the ratio of the length to the width will be

$$\frac{1 + \sqrt{5}}{2}$$

which we call the golden ratio. What is interesting is that the smaller rectangle on the right, *BFED*, is also a golden rectangle. We will use the mathematics developed in this section to confirm this fact.

Some people believe Dutch painter Piet Mondrian used the golden ratio in several of his works. He is said to believe that using ratios and straight lines enabled his paintings to achieve harmony. Look at the illustration at the beginning of this section, which was inspired by Mondrian's *Composition with yellow, blue, and red, 1937–42.* Can you see how he used golden rectangles to create this work of art?

In this section, we will expand our mathematical work with the golden ratio to include multiplication and division of radical expressions.

A Multiplication of Radical Expressions

EXAMPLE 1 Multiply $(3\sqrt{5})(2\sqrt{7})$.

Solution We can rearrange the order and grouping of the numbers in this product by applying the commutative and associative properties. Following this, we apply product property for radicals and multiply.

$$(3\sqrt{5})(2\sqrt{7}) = (3 \cdot 2)(\sqrt{5}\sqrt{7}) \qquad \text{Communicative and associative properties}$$

$$= (3 \cdot 2)(\sqrt{5 \cdot 7}) \qquad \text{Product proeprty for radicals}$$

$$= 6\sqrt{35} \qquad \text{Multiply.}$$

Practice Problems

1. Multiply $(7\sqrt{3})(5\sqrt{11})$.

Answer

1. $35\sqrt{33}$

2. Multiply $\sqrt{2}(3\sqrt{5} - 4\sqrt{2})$.

In practice, it is not necessary to show the first two steps.

EXAMPLE 2 Multiply $\sqrt{3}(2\sqrt{6} - 5\sqrt{12})$.

Solution Applying the distributive property, we have

$$\sqrt{3}(2\sqrt{6} - 5\sqrt{12}) = \sqrt{3} \cdot 2\sqrt{6} - \sqrt{3} \cdot 5\sqrt{12}$$
$$= 2\sqrt{18} - 5\sqrt{36}$$

Writing each radical in simplified form gives

$$2\sqrt{18} - 5\sqrt{36} = 2\sqrt{9}\sqrt{2} - 5\sqrt{36}$$
$$= 6\sqrt{2} - 30$$

3. Multiply $(\sqrt{2} + \sqrt{7})(\sqrt{2} - 3\sqrt{7})$.

EXAMPLE 3 Multiply $(\sqrt{3} + \sqrt{5})(4\sqrt{3} - \sqrt{5})$.

Solution The same principle that applies when multiplying two binomials applies to this product. We must multiply each term in the first expression by each term in the second one. Any convenient method can be used. Let's use the FOIL method.

$$
\begin{array}{cccc}
\text{F} & \text{O} & \text{I} & \text{L}
\end{array}
$$
$$(\sqrt{3} + \sqrt{5})(4\sqrt{3} - \sqrt{5}) = \sqrt{3} \cdot 4\sqrt{3} - \sqrt{3}\sqrt{5} + \sqrt{5} \cdot 4\sqrt{3} - \sqrt{5}\sqrt{5}$$
$$= 4 \cdot 3 - \sqrt{15} + 4\sqrt{15} - 5$$
$$= 12 + 3\sqrt{15} - 5$$
$$= 7 + 3\sqrt{15}$$

4. Expand and simplify $(\sqrt{x} + 5)^2$.

EXAMPLE 4 Expand and simplify $(\sqrt{x} + 3)^2$.

Solution 1 We can write this problem as a multiplication problem and proceed as we did in Example 3.

$$(\sqrt{x} + 3)^2 = (\sqrt{x} + 3)(\sqrt{x} + 3)$$

$$
\begin{array}{cccc}
\text{F} & \text{O} & \text{I} & \text{L}
\end{array}
$$
$$= \sqrt{x} \cdot \sqrt{x} + 3\sqrt{x} + 3\sqrt{x} + 3 \cdot 3$$
$$= x + 3\sqrt{x} + 3\sqrt{x} + 9$$
$$= x + 6\sqrt{x} + 9$$

Solution 2 We can obtain the same result by applying the formula for the square of a sum $(a + b)^2 = a^2 + 2ab + b^2$.

$$(\sqrt{x} + 3)^2 = (\sqrt{x})^2 + 2(\sqrt{x})(3) + 3^2$$
$$= x + 6\sqrt{x} + 9$$

5. Expand $(5\sqrt{a} - 3\sqrt{b})^2$ and simplify the result.

EXAMPLE 5 Expand $(3\sqrt{x} - 2\sqrt{y})^2$ and simplify the result.

Solution Let's apply the formula for the square of a difference, $(a - b)^2 = a^2 - 2ab + b^2$.

$$(3\sqrt{x} - 2\sqrt{y})^2 = (3\sqrt{x})^2 - 2(3\sqrt{x})(2\sqrt{y}) + (2\sqrt{y})^2$$
$$= 9x - 12\sqrt{xy} + 4y$$

6. Expand and simplify $(\sqrt{x + 3} - 1)^2$.

EXAMPLE 6 Expand and simplify $(\sqrt{x + 2} - 1)^2$.

Solution Applying the formula $(a - b)^2 = a^2 - 2ab + b^2$, we have

$$(\sqrt{x + 2} - 1)^2 = (\sqrt{x + 2})^2 - 2\sqrt{x + 2}(1) + 1^2$$
$$= x + 2 - 2\sqrt{x + 2} + 1$$
$$= x + 3 - 2\sqrt{x + 2}$$

Answers
2. $3\sqrt{10} - 8$
3. $-19 - 2\sqrt{14}$
4. $x + 10\sqrt{x} + 25$
5. $25a - 30\sqrt{ab} + 9b$
6. $x + 4 - 2\sqrt{x + 3}$

EXAMPLE 7 Multiply $(\sqrt{6} + \sqrt{2})(\sqrt{6} - \sqrt{2})$.

Solution We notice the product is of the form $(a + b)(a - b)$, which always gives the difference of two squares, $a^2 - b^2$.

$$(\sqrt{6} + \sqrt{2})(\sqrt{6} - \sqrt{2}) = (\sqrt{6})^2 - (\sqrt{2})^2$$
$$= 6 - 2$$
$$= 4$$

In Example 7, the two expressions $(\sqrt{6} + \sqrt{2})$ and $(\sqrt{6} - \sqrt{2})$ are called *conjugates*. In general, the conjugate of $(\sqrt{a} + \sqrt{b})$ is $(\sqrt{a} - \sqrt{b})$. If a and b are integers, multiplying conjugates of this form always produces a rational number. That is, if a and b are positive integers, then

$$(\sqrt{a} + \sqrt{b})(\sqrt{a} - \sqrt{b}) = \sqrt{a}\sqrt{a} - \sqrt{a}\sqrt{b} + \sqrt{a}\sqrt{b} - \sqrt{b}\sqrt{b}$$
$$= a - \sqrt{ab} + \sqrt{ab} - b$$
$$= a - b$$

which is rational if a and b are rational.

B Division of Radical Expressions

Division with radical expressions is the same as rationalizing the denominator. In Section 8.2, we were able to see how we can divide $\sqrt{3}$ by $\sqrt{2}$ by rationalizing the denominator.

$$\frac{\sqrt{3}}{\sqrt{2}} = \frac{\sqrt{3}}{\sqrt{2}} \cdot \frac{\sqrt{2}}{\sqrt{2}} = \frac{\sqrt{6}}{2}$$

We can accomplish the same result with expressions such as

$$\frac{6}{\sqrt{5} - \sqrt{3}}$$

by multiplying the numerator and denominator by the conjugate of the denominator.

EXAMPLE 8 Divide $\dfrac{6}{\sqrt{5} - \sqrt{3}}$. (Rationalize the denominator.)

Solution Because the product of two conjugates is a rational number, we multiply the numerator and denominator by the conjugate of the denominator.

$$\frac{6}{\sqrt{5} - \sqrt{3}} = \frac{6}{\sqrt{5} - \sqrt{3}} \cdot \frac{(\sqrt{5} + \sqrt{3})}{(\sqrt{5} + \sqrt{3})}$$

$$= \frac{6\sqrt{5} + 6\sqrt{3}}{(\sqrt{5})^2 - (\sqrt{3})^2}$$

$$= \frac{6\sqrt{5} + 6\sqrt{3}}{5 - 3}$$

$$= \frac{6\sqrt{5} + 6\sqrt{3}}{2}$$

The numerator and denominator of this last expression have a factor of 2 in common. We can reduce to lowest terms by factoring 2 from the numerator and then dividing both the numerator and denominator by 2.

$$= \frac{2(3\sqrt{5} + 3\sqrt{3})}{2}$$

$$= 3\sqrt{5} + 3\sqrt{3}$$

7. Multiply $(\sqrt{5} + \sqrt{3})(\sqrt{5} - \sqrt{3})$.

8. Divide $\dfrac{3}{\sqrt{7} - \sqrt{3}}$.

Answers

7. 2

8. $\dfrac{3\sqrt{7} + 3\sqrt{3}}{4}$

9. Rationalize the denominator.

$$\frac{\sqrt{10} - 3}{\sqrt{10} + 3}$$

EXAMPLE 9 Rationalize the denominator $\dfrac{\sqrt{5} - 2}{\sqrt{5} + 2}$.

Solution To rationalize the denominator, we multiply the numerator and denominator by the conjugate of the denominator.

$$\frac{\sqrt{5} - 2}{\sqrt{5} + 2} = \frac{\sqrt{5} - 2}{\sqrt{5} + 2} \cdot \frac{(\sqrt{5} - 2)}{(\sqrt{5} - 2)}$$

$$= \frac{5 - 2\sqrt{5} - 2\sqrt{5} + 4}{(\sqrt{5})^2 - 2^2}$$

$$= \frac{9 - 4\sqrt{5}}{5 - 4}$$

$$= \frac{9 - 4\sqrt{5}}{1}$$

$$= 9 - 4\sqrt{5}$$

10. If side *AC* in Figure 1 were 4 instead of 2, show that the smaller rectangle *BDEF* is also a golden rectangle by finding the ratio of its length to its width.

EXAMPLE 10 A golden rectangle constructed from a square of side 2 is shown in Figure 1. Show that the smaller rectangle *BDEF* is also a golden rectangle by finding the ratio of its length to its width.

Solution First, find expressions for the length and width of the smaller rectangle.

$$\text{Length} = EF = 2$$

$$\text{Width} = DE = \sqrt{5} - 1$$

Next, we find the ratio of length to width.

$$\text{Ratio of length to width} = \frac{EF}{DE}$$

$$= \frac{2}{\sqrt{5} - 1}$$

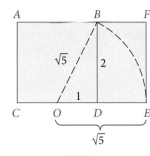

FIGURE 1

To show that the small rectangle is a golden rectangle, we must show that the ratio of length to width is the golden ratio. We do so by rationalizing the denominator.

$$\frac{2}{\sqrt{5} - 1} = \frac{2}{\sqrt{5} - 1} \cdot \frac{\sqrt{5} + 1}{\sqrt{5} + 1}$$

$$= \frac{2(\sqrt{5} + 1)}{5 - 1}$$

$$= \frac{2(\sqrt{5} + 1)}{4}$$

$$= \frac{\sqrt{5} + 1}{2} \qquad \text{Divide out common factor 2.}$$

Because addition is commutative, this last expression is the golden ratio. Therefore, the small rectangle in Figure 1 is a golden rectangle.

GETTING READY FOR CLASS

After reading through the preceding section, respond in your own words and in complete sentences.

A. Describe how you would use the commutative and associative properties to multiply $3\sqrt{5}$ and $2\sqrt{7}$.

B. Explain why $(\sqrt{5} + \sqrt{2})^2 \neq 5 + 2$.

C. Explain in words how you would rationalize the denominator in the expression $\dfrac{\sqrt{3}}{\sqrt{5} - \sqrt{2}}$.

D. What are conjugates?

Answers

9. $19 - 6\sqrt{10}$

Vocabulary Review

Choose the correct words to fill in the blanks below.

| rational | division | conjugate | multiply |

1. The _____ of $\sqrt{a} + \sqrt{b}$ is $\sqrt{a} - \sqrt{b}$.

2. Multiplying conjugates of the form $\sqrt{a} + \sqrt{b}$ and $\sqrt{a} - \sqrt{b}$ will always produce a _____ number.

3. _____ with radical expressions is the same as rationalizing the denominator.

4. To rationalize a denominator that contains a radical expression, _____ the numerator and denominator by the conjugate of the denominator.

Problems

A Multiply. (Assume all expressions appearing under a square root symbol represent nonnegative numbers throughout this problem set.)

1. $\sqrt{6}\sqrt{3}$

2. $\sqrt{6}\sqrt{2}$

3. $(2\sqrt{3})(5\sqrt{7})$

4. $(5\sqrt{5})(2\sqrt{2})$

5. $(3\sqrt{5})(2\sqrt{7})$

6. $(4\sqrt{35})(2\sqrt{21})(5\sqrt{15})$

7. $(3\sqrt[3]{3})(6\sqrt[3]{9})$

8. $(2\sqrt[3]{2})(6\sqrt[3]{4})$

9. $\sqrt{3}(\sqrt{2} - 3\sqrt{3})$

10. $\sqrt{2}(5\sqrt{3} + 4\sqrt{2})$

11. $6\sqrt[3]{4}(2\sqrt[3]{2} + 1)$

12. $7\sqrt[3]{5}(3\sqrt[3]{25} - 2)$

13. $(\sqrt{3} + \sqrt{2})(3\sqrt{3} - \sqrt{2})$

14. $(\sqrt{5} - \sqrt{2})(3\sqrt{5} + 2\sqrt{2})$

15. $(\sqrt{x} + 5)(\sqrt{x} - 3)$

16. $(\sqrt{x} + 4)(\sqrt{x} + 2)$

17. $(3\sqrt{6} + 4\sqrt{2})(\sqrt{6} + 2\sqrt{2})$

18. $(\sqrt{7} - 3\sqrt{3})(2\sqrt{7} - 4\sqrt{3})$

19. $(\sqrt{3} + 4)^2$ **20.** $(\sqrt{5} - 2)^2$ **21.** $(\sqrt{x} - 3)^2$

22. $(\sqrt{x} + 4)^2$ **23.** $(2\sqrt{a} - 3\sqrt{b})^2$ **24.** $(5\sqrt{a} - 2\sqrt{b})^2$

25. $(\sqrt{x-4} + 2)^2$ **26.** $(\sqrt{x-3} + 2)^2$ **27.** $(\sqrt{x-5} - 3)^2$

28. $(\sqrt{x-3} - 4)^2$ **29.** $(\sqrt{3} - \sqrt{2})(\sqrt{3} + \sqrt{2})$ **30.** $(\sqrt{5} - \sqrt{2})(\sqrt{5} + \sqrt{2})$

31. $(\sqrt{a} + 7)(\sqrt{a} - 7)$ **32.** $(\sqrt{a} + 5)(\sqrt{a} - 5)$ **33.** $(5 - \sqrt{x})(5 + \sqrt{x})$

34. $(3 - \sqrt{x})(3 + \sqrt{x})$ **35.** $(\sqrt{x-4} + 2)(\sqrt{x-4} - 2)$ **36.** $(\sqrt{x+3} + 5)(\sqrt{x+3} - 5)$

37. $(\sqrt{3} + 1)^3$ **38.** $(\sqrt{5} - 2)^3$ **39.** $(\sqrt{3} - 2)^3$

40. $(\sqrt{5} - 4)^2$ **41.** $(\sqrt{x+2} - \sqrt{x})^2$ **42.** $(\sqrt{n+1} - \sqrt{n})^2$

B Rationalize the denominator in each of the following.

43. $\dfrac{\sqrt{2}}{\sqrt{6} - \sqrt{2}}$ **44.** $\dfrac{\sqrt{5}}{\sqrt{5} + \sqrt{3}}$ **45.** $\dfrac{\sqrt{5}}{\sqrt{5} + 1}$ **46.** $\dfrac{\sqrt{7}}{\sqrt{7} - 1}$

47. $\dfrac{\sqrt{x}}{\sqrt{x} - 3}$ **48.** $\dfrac{\sqrt{x}}{\sqrt{x} + 2}$ **49.** $\dfrac{\sqrt{5}}{2\sqrt{5} - 3}$ **50.** $\dfrac{\sqrt{7}}{3\sqrt{7} - 2}$

51. $\dfrac{3}{\sqrt{x} - \sqrt{y}}$

52. $\dfrac{2}{\sqrt{x} + \sqrt{y}}$

53. $\dfrac{\sqrt{6} + \sqrt{2}}{\sqrt{6} - \sqrt{2}}$

54. $\dfrac{\sqrt{5} - \sqrt{3}}{\sqrt{5} + \sqrt{3}}$

55. $\dfrac{\sqrt{7} - 2}{\sqrt{7} + 2}$

56. $\dfrac{\sqrt{11} + 3}{\sqrt{11} - 3}$

57. $\dfrac{\sqrt{a} + \sqrt{b}}{\sqrt{a} - \sqrt{b}}$

58. $\dfrac{\sqrt{a} - \sqrt{b}}{\sqrt{a} + \sqrt{b}}$

59. $\dfrac{\sqrt{x} + 2}{\sqrt{x} - 2}$

60. $\dfrac{\sqrt{x} - 3}{\sqrt{x} + 3}$

61. $\dfrac{2\sqrt{3} - \sqrt{7}}{3\sqrt{3} + \sqrt{7}}$

62. $\dfrac{5\sqrt{6} + 2\sqrt{2}}{\sqrt{6} - \sqrt{2}}$

63. $\dfrac{3\sqrt{x} + 2}{1 + \sqrt{x}}$

64. $\dfrac{5\sqrt{x} - 1}{2 + \sqrt{x}}$

65. $\dfrac{2\sqrt{x} - 1}{1 - 3\sqrt{x}}$

66. $\dfrac{4\sqrt{x} - 3}{2\sqrt{x} + 3}$

67. Show that the product below is 5.

$$(\sqrt[3]{2} + \sqrt[3]{3})(\sqrt[3]{4} - \sqrt[3]{6} + \sqrt[3]{9})$$

68. Show that the product below is $x + 8$.

$$(\sqrt[3]{x} + 2)(\sqrt[3]{x^2} - 2\sqrt[3]{x} + 4)$$

Each of the following statements below is false. Correct the right side of each one to make it true.

69. $5(2\sqrt{3}) = 10\sqrt{15}$

70. $3(2\sqrt{x}) = 6\sqrt{3x}$

71. $(\sqrt{x} + 3)^2 = x + 9$

72. $(\sqrt{x} - 7)^2 = x - 49$

73. $(5\sqrt{3})^2 = 15$

74. $(3\sqrt{5})^2 = 15$

75. $(x + \sqrt{3})^2 = x^2 + 3$

76. $(\sqrt{x} - \sqrt{3})^2 = x - 3$

Applying the Concepts

77. Gravity If an object is dropped from the top of a 100-foot building, the amount of time t (in seconds) that it takes for the object to be h feet from the ground is given by the formula

$$t = \frac{\sqrt{100 - h}}{4}$$

How long does it take before the object is 50 feet from the ground? How long does it take to reach the ground? (When it is on the ground, h is 0.)

78. Gravity Use the formula given in Problem 77 to determine h if t is 1.25 seconds.

79. Golden Rectangle Rectangle *ACEF* in Figure 2 is a golden rectangle. If side *AC* is 6 inches, show that the smaller rectangle *BDEF* is also a golden rectangle.

80. Golden Rectangle Rectangle *ACEF* in Figure 2 is a golden rectangle. If side *AC* is 1 inch, show that the smaller rectangle *BDEF* is also a golden rectangle.

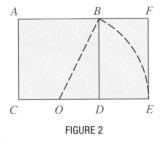

FIGURE 2

81. Golden Rectangle If side *AC* in Figure 2 is 2*x*, show that rectangle *BDEF* is a golden rectangle.

82. Golden Rectangle If side *AC* in Figure 2 is *x*, show that rectangle *BDEF* is a golden rectangle.

83. The Mona Lisa Considered by some the most famous painting in the world, Leonardo da Vinci's Mona Lisa has a height of 77 cm and a width of 53 cm. It is hypothesized that da Vinci used the golden ratio to design the painting . Compare the height-to-width ratio with the golden ratio $\frac{1 + \sqrt{5}}{2}$.

84. Famous Paintings Listed below are famous paintings by expressionists and modern artists. Compare the length: width ratios of each and compare these ratios with the golden ratio $\frac{1+\sqrt{5}}{2}$.

Painting	Artist	Location	Length	Width
a. The Persistence of Memory	Salvador Dali	Museum of Modern Art, New York	33 cm	24 cm
b. The Old Guitarist	Pablo Picasso	Art Institute of Chicago, Chicago	122.9 cm	82.6 cm
c. The Starry Night	Vincent van Gogh	Museum of Modern Art, New York	92.1 cm	73.7 cm
d. The Scream	Edvard Munch	National Gallery, Oslo	91 cm	73.5 cm

Getting Ready for the Next Section

Simplify.

85. $(t + 5)^2$

86. $(x - 4)^2$

87. $\sqrt{x} \cdot \sqrt{x}$

88. $\sqrt{3x} \cdot \sqrt{3x}$

Solve.

89. $3x + 4 = 5^2$

90. $4x - 7 = 3^2$

91. $t^2 + 7t + 12 = 0$

92. $x^2 - 3x - 10 = 0$

93. $t^2 + 10t + 25 = t + 7$

94. $x^2 - 4x + 4 = x - 2$

95. $(x + 4)^2 = x + 6$

96. $(x - 6)^2 = x - 4$

97. Is 7 a solution to $\sqrt{3x + 4} = 5$?

98. Is 4 a solution to $\sqrt{4x - 7} = -3$?

99. Is -6 a solution to $t + 5 = \sqrt{t + 7}$?

100. Is -3 a solution to $t + 5 = \sqrt{t + 7}$?

Find the Mistake

Each sentence below contains a mistake. Circle the mistake and write the correct word(s) or expression on the line provided.

1. When finding the product of $5\sqrt{6}$ and $7\sqrt{2}$, we can set $(5\sqrt{6})(7\sqrt{2})$ equal to $(\sqrt{5 \cdot 7})(\sqrt{6 \cdot 2})$. _____

2. Multiplying conjugates of the form $\sqrt{a} + \sqrt{b}$ and $\sqrt{a} - \sqrt{b}$ will always produce an irrational number.

3. The conjugate of the expression $2\sqrt{11} + \sqrt{6}$ is $2\sqrt{66}$. _____

4. To rationalize the denominator of the expression $\dfrac{\sqrt{7}}{(\sqrt{5} - \sqrt{3})}$, you must multiply the numerator and the denominator by $\sqrt{7}$. _____

A Solve equations containing radicals.

B Graph simple square root and cube root equations in two variables.

8.5 Radical Equations and Graphs

Image © sxc.hu, bretwalda, 2003

Imagine you are standing 220 feet above San Francisco Bay on the Golden Gate Bridge. If you accidentally drop your car keys over the rail, how long will it take them to reach 10 feet above the water? How long before they hit the surface of the water? If we assign the surface of the water a value of 0, both answers can be found by solving the following equation:

$$t = \frac{\sqrt{220 - h}}{4}$$

where t is the time of the keys' fall and h is their height at any given time. Solving radical equations like this is our focus in this section. We will work with equations that involve one or more radicals. The first step in solving an equation that contains a radical is to eliminate the radical from the equation. To do so, we need an additional property.

A Solving Equations with Radicals

> **PROPERTY** Squaring Property of Equality
>
> If both sides of an equation are squared, the solutions to the original equation are solutions to the resulting equation.

We will never lose solutions to our equations by squaring both sides. We may, however, introduce *extraneous solutions*. Extraneous solutions satisfy the equation obtained by squaring both sides of the original equation, but do not satisfy the original equation.

We know that if two real numbers a and b are equal, then so are their squares.

$$\text{If} \quad a = b$$
$$\text{then} \quad a^2 = b^2$$

On the other hand, extraneous solutions are introduced when we square opposites. That is, even though opposites are not equal, their squares are. For example,

$$5 = -5 \qquad \text{A false statement}$$
$$(5)^2 = (-5)^2 \qquad \text{Square both sides.}$$
$$25 = 25 \qquad \text{A true statement}$$

We are free to square both sides of an equation any time it is convenient. We must be aware, however, that doing so may introduce extraneous solutions. We must, therefore, check all our solutions in the original equation if at any time we square both sides of the original equation.

EXAMPLE 1 Solve $\sqrt{3x + 4} = 5$.

1. Solve $\sqrt{2x + 4} = 4$.

Solution We square both sides and proceed as usual.

$$\sqrt{3x + 4} = 5$$
$$(\sqrt{3x + 4})^2 = 5^2$$
$$3x + 4 = 25$$
$$3x = 21$$
$$x = 7$$

Checking $x = 7$ in the original equation, we have

$$\sqrt{3(7) + 4} \stackrel{?}{=} 5$$
$$\sqrt{21 + 4} \stackrel{?}{=} 5$$
$$\sqrt{25} \stackrel{?}{=} 5$$
$$5 = 5$$

The solution 7 satisfies the original equation.

> *Note* Just as a reminder, be careful when writing your radicals, paying special attention to the length of your radical signs. In other words, mistakenly writing $\sqrt{3x + 4}$ as $\sqrt{3x} + 4$ will give a wrong answer, so be sure to complete your work as neatly as possible.

EXAMPLE 2 Solve $\sqrt{4x - 7} = -3$.

2. Solve $\sqrt{7x - 3} = -5$.

Solution Squaring both sides, we have

$$\sqrt{4x - 7} = -3$$
$$(\sqrt{4x - 7})^2 = (-3)^2$$
$$4x - 7 = 9$$
$$4x = 16$$
$$x = 4$$

Checking $x = 4$ in the original equation gives

$$\sqrt{4(4) - 7} \stackrel{?}{=} -3$$
$$\sqrt{16 - 7} \stackrel{?}{=} -3$$
$$\sqrt{9} \stackrel{?}{=} -3$$
$$3 = -3$$

The value 4 produces a false statement when checked in the original equation. Because $x = 4$ was the only possible solution, there is no solution to the original equation. The possible solution $x = 4$ is an extraneous solution. It satisfies the equation obtained by squaring both sides of the original equation, but does not satisfy the original equation.

> *Note* The fact that there is no solution to the equation in Example 2 was obvious to begin with. Notice that the left side of the equation is the positive square root of $4x - 7$, which must be a positive number or 0. The right side of the equation is -3. Because we cannot have a number that is either positive or zero equal to a negative number, there is no solution to the equation.

Answers

1. 6

2. No Solution

3. Solve $\sqrt{4x + 5} + 2 = 7$.

EXAMPLE 3 Solve $\sqrt{5x - 1} + 3 = 7$.

Solution We must isolate the radical on the left side of the equation. If we attempt to square both sides without doing so, the resulting equation will also contain a radical. Adding -3 to both sides, we have

$$\sqrt{5x - 1} + 3 = 7$$

$$\sqrt{5x - 1} = 4$$

We can now square both sides and proceed as usual.

$$(\sqrt{5x - 1})^2 = 4^2$$

$$5x - 1 = 16$$

$$5x = 17$$

$$x = \frac{17}{5}$$

Checking $x = \frac{17}{5}$, we have

$$\sqrt{5\left(\frac{17}{5}\right) - 1} + 3 \stackrel{?}{=} 7$$

$$\sqrt{17 - 1} + 3 \stackrel{?}{=} 7$$

$$\sqrt{16} + 3 \stackrel{?}{=} 7$$

$$4 + 3 \stackrel{?}{=} 7$$

$$7 = 7$$

The solution is $\frac{17}{5}$.

4. Solve $t - 6 = \sqrt{t - 4}$.

EXAMPLE 4 Solve $t + 5 = \sqrt{t + 7}$.

Solution This time, squaring both sides of the equation results in a quadratic equation.

$$(t + 5)^2 = (\sqrt{t + 7})^2 \qquad \text{Square both sides.}$$

$$t^2 + 10t + 25 = t + 7$$

$$t^2 + 9t + 18 = 0 \qquad \text{Standard form}$$

$$(t + 3)(t + 6) = 0 \qquad \text{Factor the left side.}$$

$$t + 3 = 0 \quad \text{or} \quad t + 6 = 0 \qquad \text{Set factors equal to 0.}$$

$$t = -3 \qquad\qquad t = -6$$

We must check each solution in the original equation.

Check $t = -3$ Check $t = -6$

$-3 + 5 \stackrel{?}{=} \sqrt{-3 + 7}$ $-6 + 5 \stackrel{?}{=} \sqrt{-6 + 7}$

$2 \stackrel{?}{=} \sqrt{4}$ $-1 \stackrel{?}{=} \sqrt{1}$

$2 = 2$ A true statement $-1 = 1$ A false statement

Because $t = -6$ does not check, our only solution is -3.

5. Solve $\sqrt{x - 9} = \sqrt{x} - 3$.

EXAMPLE 5 Solve $\sqrt{x - 3} = \sqrt{x} - 3$.

Solution We begin by squaring both sides. Note carefully what happens when we square the right side of the equation, and compare the square of the right side with the square of the left side. You must convince yourself that these results are correct. (The note in the margin will help if you are having trouble convincing yourself that what is written below is true.)

$$(\sqrt{x - 3})^2 = (\sqrt{x} - 3)^2$$

$$x - 3 = x - 6\sqrt{x} + 9$$

Now we still have a radical in our equation, so we will have to square both sides again. Before we do, though, let's isolate the remaining radical.

$$x - 3 = x - 6\sqrt{x} + 9$$

$$-3 = -6\sqrt{x} + 9 \qquad \text{Add } -x \text{ to each side.}$$

$$-12 = -6\sqrt{x} \qquad \text{Add } -9 \text{ to each side.}$$

$$2 = \sqrt{x} \qquad \text{Divide each side by } -6.$$

$$4 = x \qquad \text{Square each side}$$

Our only possible solution is $x = 4$, which we check in our original equation as follows:

$$\sqrt{4 - 3} \stackrel{?}{=} \sqrt{4} - 3$$

$$\sqrt{1} \stackrel{?}{=} 2 - 3$$

$$1 = -1 \qquad \text{A false statement}$$

Substituting 4 for x in the original equation yields a false statement. Because 4 was our only possible solution, there is no solution to our equation.

Here is another example of an equation for which we must apply our squaring property twice before all radicals are eliminated.

EXAMPLE 6 Solve $\sqrt{x + 1} = 1 - \sqrt{2x}$.

Solution This equation has two separate terms involving radical signs.

Squaring both sides gives

$$x + 1 = 1 - 2\sqrt{2x} + 2x$$

$$-x = -2\sqrt{2x} \qquad \text{Add } -2x \text{ and } -1 \text{ to both sides.}$$

$$x^2 = 4(2x) \qquad \text{Square both sides.}$$

$$x^2 - 8x = 0 \qquad \text{Standard form}$$

Our equation is a quadratic equation in standard form. To solve for x, we factor the left side and set each factor equal to 0.

$$x(x - 8) = 0 \qquad \text{Factor left side.}$$

$$x = 0 \quad \text{or} \quad x - 8 = 0$$

$$x = 8 \qquad \text{Set factors equal to 0.}$$

Because we squared both sides of our equation, we have the possibility that one or both of the solutions are extraneous. We must check each one in the original equation.

Check $x = 8$ Check $x = 0$

$$\sqrt{8 + 1} \stackrel{?}{=} 1 - \sqrt{2 \cdot 8} \qquad \sqrt{0 + 1} \stackrel{?}{=} 1 - \sqrt{2 \cdot 0}$$

$$\sqrt{9} \stackrel{?}{=} 1 - \sqrt{16} \qquad \sqrt{1} \stackrel{?}{=} 1 - \sqrt{0}$$

$$3 \stackrel{?}{=} 1 - 4 \qquad 1 \stackrel{?}{=} 1 - 0$$

$$3 = -3 \quad \text{A false statement} \qquad 1 = 1 \quad \text{A true statement}$$

Because 8 does not check, it is an extraneous solution. Our only solution is 0.

> **Note** It is very important that you realize that the square of $(\sqrt{x} - 3)$ is not $x + 9$. Remember, when we square a difference with two terms, we use the formula
>
> $$(a - b)^2 = a^2 - 2ab + b^2$$
>
> Applying this formula to $(\sqrt{x} - 3)^2$, we have
>
> $$(\sqrt{x} - 3)^2 =$$
> $$(\sqrt{x})^2 - 2(\sqrt{x})(3) + 3^2$$
> $$= x - 6\sqrt{x} + 9$$

6. Solve $\sqrt{x + 4} = 2 - \sqrt{3x}$.

Answer

6. 0

7. Solve $\sqrt{x + 2} = \sqrt{x + 3} - 1$.

EXAMPLE 7 Solve $\sqrt{x + 1} = \sqrt{x + 2} - 1$.

Solution Squaring both sides we have

$$(\sqrt{x + 1})^2 = (\sqrt{x + 2} - 1)^2$$

$$x + 1 = x + 2 - 2\sqrt{x + 2} + 1$$

Once again, we are left with a radical in our equation. Before we square each side again, we must isolate the radical on the right side of the equation.

$x + 1 = x + 3 - 2\sqrt{x + 2}$	Simplify the right side.
$1 = 3 - 2\sqrt{x + 2}$	Add $-x$ to each side.
$-2 = -2\sqrt{x + 2}$	Add -3 to each side.
$1 = \sqrt{x + 2}$	Divide each side by -2.
$1 = x + 2$	Square both sides.
$-1 = x$	Add -2 to each side.

Checking our only possible solution, -1, in our original equation, we have

$$\sqrt{-1 + 1} \stackrel{?}{=} \sqrt{-1 + 2} - 1$$

$$\sqrt{0} \stackrel{?}{=} \sqrt{1} - 1$$

$$0 \stackrel{?}{=} 1 - 1$$

$$0 = 0 \qquad \text{A true statement}$$

Our solution checks.

It is also possible to raise both sides of an equation to powers greater than 2. We only need to check for extraneous solutions when we raise both sides of an equation to an even power. Raising both sides of an equation to an odd power will not produce extraneous solutions.

8. Solve $\sqrt[3]{3x - 7} = 2$.

EXAMPLE 8 Solve $\sqrt[3]{4x + 5} = 3$.

Solution Cubing both sides, we have

$$(\sqrt[3]{4x + 5})^3 = 3^3$$

$$4x + 5 = 27$$

$$4x = 22$$

$$x = \frac{22}{4}$$

$$x = \frac{11}{2}$$

We do not need to check $x = \frac{11}{2}$ because we raised both sides to an odd power.

Answers

7. -2

8. 5

B Graphing Radical Equations

We end this section by looking at graphs of some equations that contain radicals.

EXAMPLE 9 Graph $y = \sqrt{x}$ and $y = \sqrt[3]{x}$.

Solution The graphs are shown in Figures 1 and 2. Notice that the graph of $y = \sqrt{x}$ begins at the origin and then continues in the first quadrant only, because in the equation $y = \sqrt{x}$, x and y cannot be negative.

The graph of $y = \sqrt[3]{x}$ passes through the origin and continues through Quadrants 1 and 3 because the cube root of a positive number is also a positive number, and the cube root of a negative number is a negative number. That is, when x is positive, y will be positive, and when x is negative, y will be negative.

The graphs of both equations will contain the origin, because $y = 0$ when $x = 0$ in both equations. It is important to note that the origin exists on its own and does not appear in any quadrant.

x	y
-4	Undefined
-1	Undefined
0	0
1	1
4	2
9	3
16	4

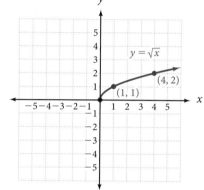

FIGURE 1

x	y
-27	-3
-8	-2
-1	-1
0	0
1	1
8	2
27	3

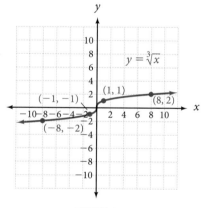

FIGURE 2

9. Graph each equation.

$$y = \sqrt{x} + 3$$
$$y = \sqrt{x + 3}$$

GETTING READY FOR CLASS

After reading through the preceding section, respond in your own words and in complete sentences.

A. What is the squaring property of equality?

B. Under what conditions could you obtain extraneous solutions to equations that contain radical expressions?

C. Why are there no real solutions to the equation $\sqrt{2x - 9} = -9$?

D. Why does the graph of $y = \sqrt{x}$ only appear in quadrant 1?

Vocabulary Review

Choose the correct words to fill in the blanks below.

extraneous solutions resulting quadrant 1

squaring property of equality original quadrant 3

1. The_____ states that the solutions to the _____ equation are solutions to the _____ equation if both sides of a equation are squared.

2. _____ satisfy an equation obtained by squaring both sides of the original equation, but do not satisfy the original equation.

3. The graph of the radical equation $y = \sqrt{x}$ begins at the origin and continues through _____ only.

4. The graph of the radical equation $y = \sqrt[3]{x}$ passes through the origin appears in quadrant 1 and _____.

Problems

A Solve each of the following equations.

1. $\sqrt{2x + 1} = 3$

2. $\sqrt{3x + 1} = 4$

3. $\sqrt{4x + 1} = -5$

4. $\sqrt{6x + 1} = -5$

5. $\sqrt{2y - 1} = 3$

6. $\sqrt{3y - 1} = 2$

7. $\sqrt{5x - 7} = -1$

8. $\sqrt{8x + 3} = -6$

9. $\sqrt{2x - 3} - 2 = 4$

10. $\sqrt{3x + 1} - 4 = 1$

11. $\sqrt{4a + 1} + 3 = 2$

12. $\sqrt{5a - 3} + 6 = 2$

13. $\sqrt[4]{3x + 1} = 2$

14. $\sqrt[4]{4x + 1} = 3$

15. $\sqrt[3]{2x - 5} = 1$

16. $\sqrt[3]{5x + 7} = 2$

17. $\sqrt[3]{3a + 5} = -3$

18. $\sqrt[3]{2a + 7} = -2$

SCAN TO ACCESS

19. $\sqrt{y - 3} = y - 3$

20. $\sqrt{y + 3} = y - 3$

21. $\sqrt{a + 2} = a + 2$

22. $\sqrt{a + 10} = a - 2$

23. $\sqrt{2x + 4} = \sqrt{1 - x}$

24. $\sqrt{3x + 4} = -\sqrt{2x + 3}$

25. $\sqrt{4a + 7} = -\sqrt{a + 2}$

26. $\sqrt{7a - 1} = \sqrt{2a + 4}$

27. $\sqrt[4]{5x - 8} = \sqrt[4]{4x - 1}$

28. $\sqrt[4]{6x + 7} = \sqrt[4]{x + 2}$

29. $x + 1 = \sqrt{5x + 1}$

30. $x - 1 = \sqrt{6x + 1}$

31. $t + 5 = \sqrt{2t + 9}$

32. $t + 7 = \sqrt{2t + 13}$

33. $\sqrt{y - 8} = \sqrt{8 - y}$

34. $\sqrt{2y + 5} = \sqrt{5y + 2}$

35. $\sqrt[3]{3x + 5} = \sqrt[3]{5 - 2x}$

36. $\sqrt[3]{4x + 9} = \sqrt[3]{3 - 2x}$

The following equations will require that you square both sides twice before all the radicals are eliminated. Solve each equation using the methods shown in Examples 5, 6, and 7.

37. $\sqrt{x - 8} = \sqrt{x} - 2$

38. $\sqrt{x + 3} = \sqrt{x} - 3$

39. $\sqrt{x + 1} = \sqrt{x} + 1$

40. $\sqrt{x - 1} = \sqrt{x} - 1$

41. $\sqrt{x + 8} = \sqrt{x - 4} + 2$

42. $\sqrt{x + 5} = \sqrt{x - 3} + 2$

43. $\sqrt{x - 5} - 3 = \sqrt{x - 8}$

44. $\sqrt{x - 3} - 4 = \sqrt{x - 3}$

45. $\sqrt{x + 4} = 2 - \sqrt{2x}$

46. $\sqrt{5x + 1} = 1 + \sqrt{5x}$

47. $\sqrt{2x + 4} = \sqrt{x + 3} + 1$

48. $\sqrt{2x - 1} = \sqrt{x - 4} + 2$

B Graph each equation.

49. $y = 2\sqrt{x}$

50. $y = -2\sqrt{x}$

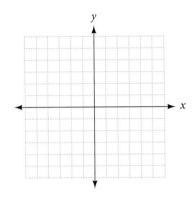

51. $y = \sqrt{x} - 2$

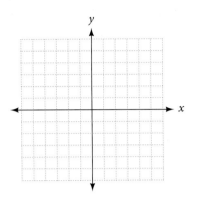

52. $y = \sqrt{x} + 2$

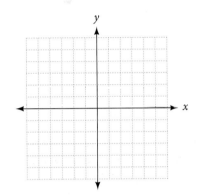

53. $y = \sqrt{x - 2}$

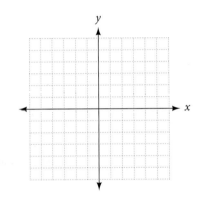

54. $y = \sqrt{x + 2}$

55. $y = 3\sqrt[3]{x}$

56. $y = -3\sqrt[3]{x}$

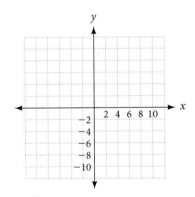

57. $y = \sqrt[3]{x} + 3$

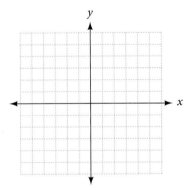

58. $y = \sqrt[3]{x} - 3$

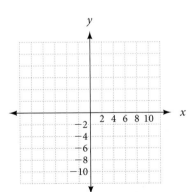

59. $y = \sqrt[3]{x + 3}$

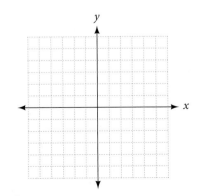

60. $y = \sqrt[3]{x - 3}$

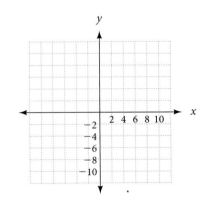

Applying the Concepts

61. Solving a Formula Solve the following formula for h.

$$t = \frac{\sqrt{100 - h}}{4}$$

62. Solving a Formula Solve the following formula for h.

$$t = \sqrt{\frac{2h - 40t}{g}}$$

63. Pendulum Clock The length of time T in seconds it takes the pendulum of a clock to swing through one complete cycle is given by the formula

$$T = 2\pi\sqrt{\frac{L}{32}}$$

where L is the length, in feet, of the pendulum, and π is approximately $\frac{22}{7}$. How long must the pendulum be if one complete cycle takes 2 seconds?

64. Pendulum Clock Solve the formula in Problem 63 for L.

65. Similar Rectangles Two rectangles are similar if their vertices lie along the same diagonal, as shown in the following diagram.

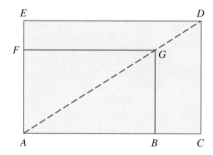

If two rectangles are similar, then corresponding sides are in proportion, which means that $\frac{ED}{DC} = \frac{FG}{GB}$. Use these facts in the following diagram to express the length of the larger rectangle l in terms of x.

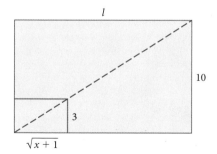

66. Volume Recall that the volume of a box V can be found from the formula $V = (\text{length})(\text{width})(\text{height})$. Find the volume of the following box in terms of x.

$$\left(\sqrt[4]{x + 1}\right)^5$$

$$\sqrt[3]{x + 1}$$

$$\sqrt{x + 1}$$

Pollution A long straight river, 100 meters wide, is flowing at 1 meter per second. A pollutant is entering the river at a constant rate from one of its banks. As the pollutant disperses in the water, it forms a plume that is modeled by the equation $y = \sqrt{x}$. Use this information to answer the following questions.

Pollution plume

67. How wide is the plume 25 meters down river from the source of the pollution?

68. How wide is the plume 100 meters down river from the source of the pollution?

69. How far down river from the source of the pollution does the plume reach halfway across the river?

70. How far down the river from the source of the pollution does the plume reach the other side of the river?

71. For the situation described in the instructions and modeled by the equation $y = \sqrt{x}$, what is the range of values that y can assume?

72. If the river was moving at 2 meters per second, would the plume be larger or smaller 100 meters downstream from the source?

73. Recall from the introduction to this section the equation $t = \frac{\sqrt{220 - h}}{4}$ gives the time (in seconds) for an object dropped off the Golden Gate Bridge to reach a height of h feet above water. Find the time required for your car keys to reach the following heights:

 a. 200 feet

 b. 100 feet

 c. 0 feet (the water)

74. The Burj Khalifa tower in Dubai is one of the tallest buildings in the world at 2,717 feet. The equation $t = \frac{\sqrt{2,717 - h}}{4}$ gives the time (in seconds) for an object dropped from the top to reach a height of h feet above the ground. If you dropped your wallet from the roof, find the time required for it to reach the following heights:

 a. 2,000 feet

 b. 1,000 feet

 c. 0 feet (the ground)

Maintaining Your Skills

Simplify.

75. $\sqrt{25}$

76. $\sqrt{49}$

77. $\sqrt{12}$

78. $\sqrt{50}$

79. $(-1)^{15}$

80. $(-1)^{20}$

81. $(-1)^{50}$

82. $(-1)^{5}$

Solve each equation.

83. $3x = 12$

84. $4 = 8y$

85. $4x - 3 = 5$

86. $7 = 2y - 1$

Perform the indicated operation.

87. $(3 + 4x) + (7 - 6x)$

88. $(2 - 5x) + (-1 + 7x)$

89. $(7 + 3x) - (5 + 6x)$

90. $(5 - 2x) - (9 - 4x)$

91. $(3 - 4x)(2 + 5x)$

92. $(8 + x)(7 - 3x)$

93. $2x(4 - 6x)$

94. $3x(7 + 2x)$

95. $(2 + 3x)^2$

96. $(3 + 5x)^2$

97. $(2 - 3x)(2 + 3x)$

98. $(4 - 5x)(4 + 5x)$

Find the Mistake

Each sentence below contains a mistake. Circle the mistake and write the correct word(s) or equation on the line provided.

1. The squaring property of equality states that the solutions to the original equation are solutions to the resulting equation when both sides of the original equation are factored. _____

2. Squaring both sides of the equation $\sqrt{7x - 5} = -3$ gives us $(7x - 5)^2 = -9$. _____

3. An extraneous solution satisfies the original equation. _____

4. The graph of $y = \sqrt[5]{x}$ appears in Quadrant I only because x and y cannot be negative. _____

8.6 Rational Exponents

A Simplify rational exponents.

B Use properties of exponents to simplify expressions with rational exponents.

C Use the negative exponent property to simplify expressions with negative exponents.

KEY WORDS

rational exponent

Suppose you were asked to build a bridge at the base of a waterfall. You were given the following equation that represents the shape of the first half of the bridge:

$$y = x^{1/2}$$

This equation uses a fraction as an exponent. Expressions of this type are the focus of this section. Let's look at the graph of this equation:

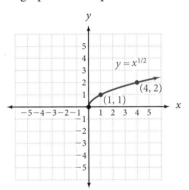

The above graph is the same as the one we encountered in the previous section for $y = \sqrt{x}$. Therefore, $y = x^{1/2}$ is equivalent to $y = \sqrt{x}$; in other words x raised to the fractional exponent of $\frac{1}{2}$ is equal to the positive square root of x.

Consider the expression $3^{1/2}$ and the expression $\sqrt{3}$. The second expression is the positive square root of 3. We have never encountered the first expression before. The properties of exponents apply to rational exponents in the same way they apply to integer exponents if bases or radicands are non-negative. Based on this assumption, we can look at what happens when we square both of the preceding quantities:

$$(3^{1/2})^2 = 3^1 \qquad (\sqrt{3}) = \sqrt{3} \cdot \sqrt{3}$$
$$= 3 \qquad\qquad = \sqrt{9}$$
$$= 3$$

The result is the same in both cases. The result gives us the following:

If x represents a nonnegative real number, then

$$x^{1/2} = \sqrt{x}$$

$$x^{1/3} = \sqrt[3]{x}$$

$$x^{1/4} = \sqrt[4]{x}$$

and in general,
$$x^{1/n} = \sqrt[n]{x}$$

Note The discussion at the left also holds if x is negative as long as n is an odd number. That is, the base x can be negative as long as the exponent gives an odd root.

A Simplifying Rational Exponents

We can use our discussion on the previous page to change any expression involving radicals to an expression that involves only exponents.

EXAMPLE 1 Write each of the following as a radical and then simplify.

a. $25^{1/2} = \sqrt{25} = 5$ The exponent $\frac{1}{2}$ indicates that we are to find the square root of 25.

b. $9^{1/2} = \sqrt{9} = 3$ Again, the exponent $\frac{1}{2}$ indicates that we are to find the square root of 9.

c. $8^{1/3} = \sqrt[3]{8} = 2$ The exponent $\frac{1}{3}$ indicates that we are to find the cube root of 8.

d. $81^{1/4} = \sqrt[4]{81} = 3$ The exponent $\frac{1}{4}$ indicates that we are to find the fourth root of 81.

Consider the expression $9^{3/2}$. If we assume that all our properties of exponents hold for rational exponents, we can rewrite this expression as a power raised to another power.

$$9^{3/2} = (9^{1/2})^3$$

Since $9^{1/2}$ is equivalent to $\sqrt{9}$, we can simplify further.

$$9^{3/2} = (9^{1/2})^3 = (\sqrt{9})^3 = 3^3 = 27$$

We can summarize this discussion with the following:

> If x represents a nonnegative real number, and n and m are integers, then
>
> $$x^{m/n} = \sqrt[n]{x^m} = (\sqrt[n]{x})^m$$
>
> We can think of the exponent $\frac{m}{n}$ as consisting of two parts. The numerator m is the power, and the denominator n is the root.

EXAMPLE 2 Simplify $8^{2/3}$.

Solution Using the preceding definition, we have

$$
\begin{aligned}
8^{2/3} &= (8^{1/3})^2 && \text{Separate exponents.}\\
&= (\sqrt[3]{8})^2 && \text{Write as cube root.}\\
&= 2^2 && \sqrt[3]{8} = 2\\
&= 4 && 2^2 = 4
\end{aligned}
$$

EXAMPLE 3 Simplify $27^{4/3}$.

Solution
$$
\begin{aligned}
27^{4/3} &= (27^{1/3})^4 && \text{Separate exponents.}\\
&= (\sqrt[3]{27})^4 && \text{Write as cube root.}\\
&= 3^4 && \sqrt[3]{27} = 3\\
&= 81 && 3^4 = 81
\end{aligned}
$$

B Simplifying Expressions

We can apply the properties of exponents to expressions that involve rational exponents. Use these properties to simplify each expression. Assume that all variables represent positive numbers.

EXAMPLE 4 Multiply $x^{1/3} \cdot x^{2/3}$.

Solution To multiply with the same base, we add exponents. The property we use is $a^r a^s = a^{r+s}$.

$$
\begin{aligned}
x^{1/3} \cdot x^{2/3} &= x^{1/3 + 2/3}\\
&= x^{3/3}\\
&= x
\end{aligned}
$$

Practice Problems

1. Simplify $125^{1/3}$.

> *Note* The exponent $\frac{m}{n}$ in this summary must be in simplified form for the terms of the identity to be equal.

2. Simplify $27^{2/3}$.

3. Simplify $9^{5/2}$.

4. Multiply $a^{3/4} \cdot a^{5/4}$.

Answers
1. 5
2. 9
3. 243
4. a^2

5. Simplify $(a^{3/2})^2$.

EXAMPLE 5 Simplify $(x^{1/4})^8$.

Solution To raise a power to another power, we multiply exponents. The property we use is $(a^r)^s = a^{r \cdot s}$.

$$(x^{1/4})^8 = x^{(1/4)8}$$
$$= x^2$$

6. Simplify $\dfrac{x^{7/8}}{x^{5/8}}$.

EXAMPLE 6 Simplify $\dfrac{a^{5/6}}{a^{4/6}}$.

Solution To divide with the same base, we subtract the exponent in the denominator from the exponent in the numerator.

$$\frac{a^{5/6}}{a^{4/6}} = a^{5/6 - 4/6}$$
$$= a^{1/6}$$

7. Simplify $(4a^4)^{3/2}$.

EXAMPLE 7 Simplify $(8y^{12})^{1/3}$.

Solution Distributing the exponent $\frac{1}{3}$ across the product, we have

$$(8y^{12})^{1/3} = 8^{1/3}(y^{12})^{1/3}$$
$$= 2y^4$$

When simplifying expressions that involve negative bases, it is important to note whether the base is surrounded by parentheses. An absence of parentheses will change your answer. To understand this point, compare the following problems:

$$-64^{1/2} = -\sqrt{64} = -\sqrt{8^2} = -8$$
$$(-64)^{1/2} = -\sqrt{-64} = \text{undefined}$$

C Negative Rational Exponents

Recall that negative exponents indicate reciprocals. That is, $a^{-r} = \dfrac{1}{a^r}$. We can apply this property to our work with rational exponents when the exponent appears as a negative fraction.

$$a^{-1/r} = \frac{1}{\sqrt[r]{a}}$$

8. Simplify $27^{-1/3}$.

EXAMPLE 8 Simplify $36^{-1/2}$.

Solution The negative exponent indicates a reciprocal. Therefore, 36 raised to a $-\frac{1}{2}$ power equals the reciprocal of the positive square root of 36.

$$36^{-1/2} = \frac{1}{\sqrt{36}}$$
$$= \frac{1}{6}$$

GETTING READY FOR CLASS

After reading through the preceding section, respond in your own words and in complete sentences.

A. Explain how rational exponents and radicals relate to each other.

B. Use symbols to show how you would write $x^{m/n}$ as a radical.

C. How you would apply the distributive property for exponents to an expression involving rational exponents?

D. How do negative rational exponents and reciprocals relate to each other?

Answers

5. a^3

6. $x^{1/4}$

7. $8a^6$

8. $\frac{1}{3}$

Vocabulary Review

Choose the correct words to fill in the blanks below.

 power nonnegative cube integer

1. If x represents a _____ real number, then $\sqrt[n]{x} = x^{1/n}$.

2. The properties of exponents apply to fractional exponents in the same way they apply to _____ exponents.

3. The expression $64^{1/3}$ indicates that we are to find the _____ root of 64.

4. If $\frac{m}{n}$ appears as an exponent, we can think of the numerator m as the _____ and the denominator n as the root.

Problems

A Change each of the following to an expression involving roots and then simplify.

1. $4^{1/2}$ **2.** $9^{1/2}$ **3.** $16^{1/2}$ **4.** $25^{1/2}$

5. $27^{1/3}$ **6.** $8^{1/3}$ **7.** $125^{1/3}$ **8.** $16^{1/4}$

9. $81^{1/4}$ **10.** $36^{1/2}$ **11.** $81^{1/2}$ **12.** $144^{1/2}$

13. $8^{2/3}$ **14.** $25^{3/2}$ **15.** $125^{2/3}$ **16.** $36^{3/2}$

17. $16^{3/4}$ **18.** $9^{3/2}$ **19.** $16^{3/2}$ **20.** $8^{5/3}$

21. $4^{3/2}$ **22.** $4^{5/2}$ **23.** $(-8)^{2/3}$

24. $(-27)^{2/3}$ **25.** $(-32)^{1/5}$ **26.** $(-32)^{3/5}$

27. $4^{1/2} + 9^{1/2}$ **28.** $16^{1/2} + 25^{1/2}$ **29.** $16^{3/4} + 27^{2/3}$

30. $49^{1/2} + 64^{1/2}$ **31.** $4^{1/2} + 27^{1/3}$ **32.** $16^{1/2} + 25^{1/2}$

B Use the properties of exponents to simplify each of the following expressions. Assume that all variables represent positive numbers.

33. $x^{1/4} \cdot x^{3/4}$

34. $x^{1/8} \cdot x^{3/8}$

35. $(x^{2/3})^3$

36. $(x^{1/4})^{12}$

37. $\dfrac{a^{3/5}}{a^{1/5}}$

38. $\dfrac{a^{5/7}}{a^{3/7}}$

39. $(27y^6)^{1/3}$

40. $(81y^8)^{1/4}$

41. $(9a^4b^8)^{1/2}$

42. $(25a^8b^4)^{1/2}$

43. $\dfrac{x^{3/5} \cdot x^{4/5}}{x^{2/5}}$

44. $\dfrac{x^{4/7} \cdot x^{6/7}}{x^{3/7}}$

C Use the negative exponent property to simplify the following expressions.

45. $25^{-1/2}$

46. $8^{-1/3}$

47. $8^{-2/3}$

48. $81^{-3/4}$

49. $27^{-2/3}$

50. $16^{-3/4}$

Find the Mistake

Each sentence below contains a mistake. Circle the mistake and write the correct word(s) or expression(s) on the line provided.

1. If $\sqrt[n]{x} = x^{1/n}$, then $\sqrt[8]{x} = \frac{x}{8}$. _____

2. For the expression $243^{1/5}$, the exponent $\frac{1}{5}$ indicates that we are to find the fifth root of 243; that is, $5 \cdot \sqrt{423}$.

3. The expression $(b^{2/5})^{9/11}$ is equivalent to $b^{2/5 + 9/11}$. _____

4. To simplify $(64x^{18})^{1/3}$, distribute the exponent 18 across the product. _____

Pythagorean Proof

As we have discussed, the Pythagorean theorem, $a^2 + b^2 = c^2$, is named after Pythagoras, a Greek mathematician and philosopher from the 6th century BC. Pythagoras proved the theorem using the following illustration:

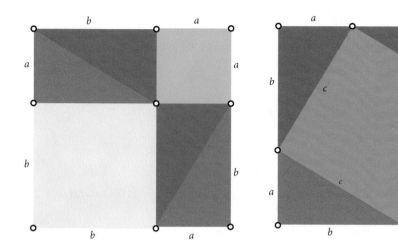

If we set the area of the square on the left $(4(12ab) + a^2 + b^2)$ equal to the area of the square on the right $(4(12ab) + c^2)$, and use the addition property of equality to eliminate $4(12ab)$ from both sides, our result is the Pythagorean theorem.

$$4(12ab) + a^2 + b^2 = 4(12ab) + c^2$$

$$a^2 + b^2 = c^2$$

Historical evidence suggests that other mathematicians knew of and used this theorem long before Pythagoras. Research the history of the Pythagorean theorem and give examples of how other ancient cultures and mathematicians illustrated the theorem.

Chapter **8** Summary

The numbers in brackets refer to the section(s) in which the topic can be found.

EXAMPLES

1. The number 49 has two square roots, 7 and -7. They are written like this:

$$\sqrt{49} = 7 \qquad -\sqrt{49} = -7$$

Square Roots [8.1]

Every positive real number x has two square roots. The *positive square root* of x is written \sqrt{x}, and the *negative square root* of x is written $-\sqrt{x}$. Both the positive and the negative square roots of x are numbers we square to get x; that is,

$$\left.\begin{array}{c} (\sqrt{x})^2 = x \\ \text{and} \qquad (-\sqrt{x})^2 = x \end{array}\right\} \text{ for } x \geq 0$$

2. $\sqrt[3]{8} = 2$

$\sqrt[3]{-27} = -3$

General Roots [8.1]

In the expression $\sqrt[n]{a}$, n is the *index*, a is the *radicand*, and $\sqrt{}$ is the *radical sign*. The expression $\sqrt[n]{a}$ is such that

$$(\sqrt[n]{a})^n = a \qquad a \geq 0 \text{ when } n \text{ is even}$$

3. $\sqrt{4 \cdot 5} = \sqrt{4}\,\sqrt{5} = 2\sqrt{5}$

$\sqrt{\dfrac{7}{9}} = \dfrac{\sqrt{7}}{\sqrt{9}} = \dfrac{\sqrt{7}}{3}$

Properties of Radicals [8.2]

If a and b are nonnegative real numbers whenever n is even, then

1. $\sqrt[n]{ab} = \sqrt[n]{a}\,\sqrt[n]{b}$ *Product property for radicals*

2. $\sqrt[n]{\dfrac{a}{b}} = \dfrac{\sqrt[n]{a}}{\sqrt[n]{b}} \qquad (b \neq 0)$ *Quotient property for radicals*

3. $\sqrt{a^2} = |a|$ *Absolute value property for radicals*

4. $\sqrt{\dfrac{4}{5}} = \dfrac{\sqrt{4}}{\sqrt{5}}$

$= \dfrac{2}{\sqrt{5}} \cdot \dfrac{\sqrt{5}}{\sqrt{5}}$

$= \dfrac{2\sqrt{5}}{5}$

Simplifying Radicals [8.2]

A radical expression is said to be in *simplified form* if all of the following are true:

1. There is no factor of the radicand that can be written as a power greater than or equal to the index;

2. There are no fractions under the radical sign.

3. There are no radicals in the denominator.

5. $5\sqrt{3} - 7\sqrt{3} = (5 - 7)\sqrt{3}$

$= -2\sqrt{3}$

$\sqrt{20} + \sqrt{45} = 2\sqrt{5} + 3\sqrt{5}$

$= (2 + 3)\sqrt{5}$

$= 5\sqrt{5}$

Addition and Subtraction of Radicals [8.3]

We add and subtract radical expressions by using the distributive property to combine similar radicals. Similar radicals are radicals with the same index and the same radicand.

Multiplication of Radicals [8.4]

We multiply radical expressions in the same way that we multiply polynomials. We can use the distributive property and the FOIL method.

6. $(\sqrt{x} + 2)(\sqrt{x} + 3)$

$\quad = \sqrt{x}\,\sqrt{x} + 3\sqrt{x} + 2\sqrt{x} + 2 \cdot 3$

$\quad = x + 5\sqrt{x} + 6$

Rationalizing the Denominator [8.2, 8.4]

When a fraction contains a square root in the denominator, we rationalize the denominator by multiplying the numerator and denominator by

1. The square root itself if there is only one term in the denominator, or

2. The conjugate of the denominator if there are two terms in the denominator.

Rationalizing the denominator is also called division of radical expressions.

7. $\dfrac{3}{\sqrt{2}} = \dfrac{3}{\sqrt{2}} \cdot \dfrac{\sqrt{2}}{\sqrt{2}} = \dfrac{3\sqrt{2}}{2}$

$\dfrac{3}{\sqrt{5} - \sqrt{3}} = \dfrac{3}{\sqrt{5} - \sqrt{3}} \cdot \dfrac{\sqrt{5} + \sqrt{3}}{\sqrt{5} + \sqrt{3}}$

$\qquad = \dfrac{3\sqrt{5} + 3\sqrt{3}}{5 - 3}$

$\qquad = \dfrac{3\sqrt{5} + 3\sqrt{3}}{2}$

Squaring Property of Equality [8.5]

We may square both sides of an equation any time it is convenient to do so, as long as we check all resulting solutions in the original equation.

8. $\sqrt{2x + 1} = 3$

$\quad (\sqrt{2x + 1})^2 = 3^2$

$\qquad 2x + 1 = 9$

$\qquad\qquad x = 4$

Rational Exponents [8.6]

If x represents a nonnegative real number, then
$$x^{1/2} = \sqrt{x}$$
$$x^{1/3} = \sqrt[3]{x}$$
$$x^{1/4} = \sqrt[4]{x}$$
$$x^{1/n} = \sqrt[n]{x} \quad \text{if } n \text{ is even}$$

9. $36^{1/2} = \sqrt{36}$

$\qquad = 6$

Find the following roots. [8.1]

1. $\sqrt{49}$ **2.** $-\sqrt{16}$

3. The square root of 169

4. $\sqrt[3]{125}$ **5.** $\sqrt[3]{-8}$

6. $-\sqrt[4]{81}$

Put the following expressions into simplified form. [8.2]

7. $\sqrt{112}$ **8.** $\sqrt{98}$

9. $\sqrt{\dfrac{5}{7}}$ **10.** $\dfrac{5}{\sqrt[3]{4}}$

11. $4\sqrt{18y^4}$ **12.** $\sqrt{\dfrac{24x^3y^4}{3}}$

Combine. [8.3]

13. $4\sqrt{12} - \sqrt{48}$ **14.** $2x\sqrt{45} + 3\sqrt{5x^2}$

Multiply. [8.4]

15. $\sqrt{3}(\sqrt{5} + 2)$ **16.** $(\sqrt{2} - 7)(\sqrt{2} + 4)$

17. $(\sqrt{x} - 6)(\sqrt{x} + 6)$ **18.** $(\sqrt{3} + \sqrt{2})^2$

Divide. (Rationalize the denominator.) [8.4]

19. $\dfrac{\sqrt{3} - \sqrt{5}}{\sqrt{3} + \sqrt{5}}$ **20.** $\dfrac{\sqrt{x}}{\sqrt{x} + 4}$

Solve the following equations. [8.5]

21. $\sqrt{3x + 6} - 7 = 5$ **22.** $\sqrt{3x - 5} = x - 3$

23. $\sqrt{5x + 7} + 9 = 2$

24. The difference of a number and 8 is equal to twice the positive square root of the number. Find the number.

25. Find x in the following right triangle. [8.1]

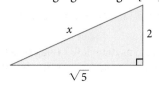

For each equation, complete the given table and then use the table values to match the equation with its graph. [8.5]

26. $y = \sqrt[3]{x} - 5$ **27.** $y = \sqrt{x} + 5$

x	y
-8	
-1	
0	
1	
8	

x	y
0	
1	
4	
9	

28. $y = \sqrt[3]{x + 4}$ **29.** $y = \sqrt{x + 3}$

x	y
-4	
-3	
4	
23	

x	y
-3	
-2	
1	
6	

a.

b.

c.

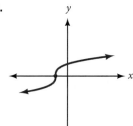

d.

Simplify the following expressions. [8.6]

30. $144^{1/2}$ **31.** $-243^{1/5}$

32. $16^{3/4} + 27^{1/3}$ **33.** $125^{-1/3}$

Simplify.

1. $\left(\dfrac{2}{5}\right)\left(\dfrac{5}{8}\right)$

2. $-(-5)$

3. $\left|-\dfrac{3}{5}\right|$

4. $-35 + (-2)$

5. $13 - 8 + 2$

6. $-2(3)(-5)$

7. $\dfrac{a^{10}}{a^3}$

8. $\dfrac{(b^2)^3(b^5)^2}{(b^3)^7}$

9. $\dfrac{30a^7}{6a^3} + \dfrac{28a^{12}}{7a^8}$

10. $\dfrac{12x^2y^{13}}{2xy^7} + \dfrac{32x^5y^9}{8x^4y^3}$

11. $\left(\dfrac{1}{3}y - 4\right)\left(\dfrac{1}{3}y + 4\right)$

12. $3x(9x - 5)$

13. $\dfrac{\dfrac{1}{a+2} + 6}{\dfrac{1}{a+2} + 3}$

14. $\dfrac{\dfrac{x+4}{x^2 + 2x - 15}}{\dfrac{3}{x+5}}$

15. $\dfrac{4a}{a^2 - 5a - 24} + \dfrac{2}{a + 3}$

16. $\dfrac{2x}{2x - 5} - \dfrac{5}{2x - 5}$

Simplify. Assume all variables represent positive numbers.

17. $\sqrt{147x^4y^5}$

18. $\sqrt{\dfrac{180b^2}{196}}$

Solve.

19. $4(3x - 2) = 2(5x - 1) + 16$

20. $3x - 5 = 9 - x$

21. $2x(3x - 15)(x + 4) = 0$ **22.** $1 - \dfrac{2}{x} = \dfrac{15}{x^2}$

23. $\sqrt{x - 5} = 2$

24. $\sqrt{x - 3} + 7 = 0$

Solve each system.

25. $x + y = -2$
$y = -2x - 5$

26. $3x + 5y = 14$
$3x + 2y = 2$

27. $2x - 5y = 19$
$x - y = 8$

28. $2x - y = 14$
$y = 3x - 17$

Graph on a rectangular coordinate system.

29. $x = -6$

30. $6x - 4y = 12$

31. $y \geq -3$

32. $y < -2x + 6$

33. For the equation $y = 2x - 5$, find the x- and y-intercepts.

34. For the equation $y = x - 3$, find the slope and y-intercept.

35. Find the equation for the line with slope $-\frac{1}{3}$ that passes through $(6, 2)$.

36. Find y if the line through $(-2, 2)$ and $(1, y)$ has slope $-\frac{1}{3}$.

Factor completely.

37. $y^4 - 16$

38. $t^2 - 9t + 18$

39. $x^4 + 4x^3 - 12x^2$

40. $48x^2y - 27y^3$

Simplify and write your answer in scientific notation.

41. $(6 \times 10^6)(3.6 \times 10^4)$

42. $\dfrac{(5 \times 10^6)(3 \times 10^{-3})}{1.2 \times 10^{-4}}$

Rationalize the denominator.

43. $\dfrac{3}{\sqrt{5}}$

44. $\dfrac{\sqrt{6} + \sqrt{2}}{\sqrt{6} - \sqrt{2}}$

45. Find the value of the expression $3x - 7$ when $x = 5$.

46. Solve $V = \pi r^2 h$ for r.

47. Multiply $\dfrac{6x - 3}{x^2 + 2x - 15} \cdot \dfrac{x^2 - x - 6}{2x^2 - 3x + 1}$.

48. Divide $\dfrac{3x^2 - 7x - 6}{x^2 - x - 6} \div \dfrac{3x^2 + 20x + 12}{x^2 - 7x + 10}$.

49. Geometry A rectangle has a perimeter of 22 feet. If the length is 5 less than 3 times the width, find the dimensions.

50. Number Problem One number is 3 more than four times another number. The sum of the numbers is 38. Find the numbers.

The graph shows the number of Facebook fans for some of its most popular pages. Use the information to answer the following questions.

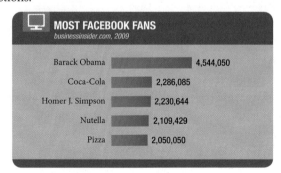

MOST FACEBOOK FANS
businessinsider.com, 2009

Barack Obama — 4,544,050
Coca-Cola — 2,286,085
Homer J. Simpson — 2,230,644
Nutella — 2,109,429
Pizza — 2,050,050

51. How many more fans does Coca Cola have than Pizza?

52. What is the ratio of Barack Obama's fans to Homer J. Simpson? Round to the nearest hundredth.

Find the following roots. [8.1]

1. $\sqrt{36}$ **2.** $-\sqrt{81}$

3. The square root of 121

4. $\sqrt[3]{64}$ **5.** $\sqrt[3]{27}$

6. $-\sqrt[4]{625}$

Put the following expressions into simplified form. [8.2]

7. $\sqrt{72}$ **8.** $\sqrt{96}$

9. $\sqrt{\dfrac{3}{5}}$ **10.** $\dfrac{6}{\sqrt[3]{9}}$

11. $4\sqrt{12y^2}$ **12.** $\sqrt{\dfrac{18x^3y^5}{3}}$

Combine. [8.3]

13. $3\sqrt{20} - \sqrt{45}$ **14.** $3m\sqrt{12} + 2\sqrt{3m^2}$

Multiply. [8.4]

15. $\sqrt{2}(\sqrt{3} - 3)$ **16.** $(\sqrt{3} - 5)(\sqrt{3} + 2)$

17. $(\sqrt{x} - 9)(\sqrt{x} + 9)$ **18.** $(\sqrt{7} - \sqrt{2})^2$

Divide. (Rationalize the denominator.) [8.4]

19. $\dfrac{\sqrt{5} - \sqrt{7}}{\sqrt{5} + \sqrt{7}}$ **20.** $\dfrac{\sqrt{x}}{\sqrt{x} - 3}$

Solve the following equations. [8.5]

21. $\sqrt{4x - 3} - 3 = 4$ **22.** $\sqrt{2x + 8} = x + 4$

23. $\sqrt{4x - 3} + 8 = 3$

24. The difference of a number and 3 is equal to twice the positive square root of the number. Find the number.

25. Find x in the following right triangle. [8.1]

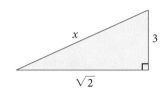

For each equation, complete the given table and then use the table values to match the equation with its graph. [8.5]

26. $y = \sqrt{x} + 2$

x	y
0	
1	
4	
9	

27. $y = \sqrt{x + 2}$

x	y
-2	
-1	
2	
7	

28. $y = \sqrt[3]{x} - 2$

x	y
-8	
-1	
0	
1	
8	

29. $y = \sqrt[3]{x} + 2$

x	y
-10	
-3	
-2	
-1	
6	

a.

b.

c.

d.

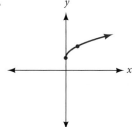

Simplify the following expressions. [8.6]

30. $27^{1/3}$ **31.** $-64^{1/3}$

32. $125^{2/3} - 16^{3/4}$ **33.** $256^{-1/4}$

Quadratic Equations and Functions

Saint Louis, Missouri
Gray Buildings © 2008 Sanborn

One of the most recognizable structures in the United States is the St. Louis Gateway Arch. Known as the "Gateway to the West," its construction was completed on October 28, 1965, and cost a total of $13 million. As the tallest monument in the country, the arch stands 630 feet tall and spans 630 feet across its base. The cross-sections of its legs are hollow isosceles triangles, which contain a unique tram system that brings visitors to an observation deck at the top of the arch. Nearly a million visitors ride the trams to the top each year. The trams have been in operation for over 30 years, traveling a total of 250,000 miles and carrying over 25 million passengers.

While the architect used what is called a catenary curve in its design, the shape of the arch can be closely modeled by the quadratic equation $y = -\frac{1}{150}x^2 + \frac{21}{5}x$. The equation can be graphed on a rectangular coordinate system and the resulting shape is called a parabola, shown below.

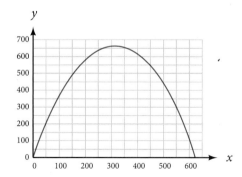

Parabolas are one of the topics covered in this chapter, and you will be asked to sketch graphs like this one in some of the problem sets.

A Solve quadratic equations by factoring.

B Solve quadratic equations by taking square roots.

C Use quadratic equations to solve for missing parts of right triangles.

KEY WORDS

square root property for equations

right triangle

9.1 Factoring and the Square Root Property

The table below is taken from the trail map given to skiers at the Northstar at Tahoe Ski Resort in Lake Tahoe, California. The table gives the length of each chair lift at Northstar, along with the change in elevation from the beginning of the lift to the end of the lift.

Right triangles are good mathematical models for chair lifts. In this section, we will use our knowledge of right triangles, along with the new material developed here, to solve problems involving chair lifts and a variety of other examples.

From the Trail Map for the Northstar at Tahoe Ski Resort		
Lift Information		
Lift	Vertical Rise (feet)	Length (feet)
Big Springs Gondola	480	4,100
Bear Paw Double	120	790
Echo Triple	710	4,890
Aspen Express Quad	900	5,100
Forest Double	1,170	5,750
Lookout Double	960	4,330
Comstock Express Quad	1,250	5,900
Rendezvous Triple	650	2,900
Schaffer Camp Triple	1,860	6,150
Chipmunk Tow Lift	28	280
Bear Cub Tow Lift	120	750

A Solving Quadratic Equations by Factoring

Recall from an earlier chapter that quadratic equations, which are equations that can appear in the form $ax^2 + bx + c = 0$, can be solved by factoring the left-hand side and setting each factor equal to 0. The first two examples in this section will review this technique for solving quadratic equations.

Practice Problems

1. Solve $x^2 - 3x = 28$.

EXAMPLE 1 Solve $x^2 + 2x = 15$.

Solution First write the equation in the form $ax^2 + bx + c = 0$, then factor the lefthand side.

$$x^2 + 2x = 15$$
$$x^2 + 2x - 15 = 0 \qquad \text{Subtract 15 from each side.}$$
$$(x + 5)(x - 3) = 0 \qquad \text{Factor.}$$
$$x + 5 = 0 \ \text{ or } \ x - 3 = 0 \qquad \text{Set factors equal to 0.}$$
$$x = -5 \ \text{ or } \ x = 3 \qquad \text{Solve for x.}$$

The solutions are $-5, 3$.

2. Solve $x(x - 1) = x + 24$.

EXAMPLE 2 Solve $x(x + 1) = 2(x + 10)$

Solution First write the equation in the form $ax^2 + bx + c = 0$, then factor the lefthand side.

$$x(x + 1) = 2(x + 10)$$
$$x^2 + x = 2x + 20 \qquad \text{Distributive property}$$
$$x^2 - x - 20 = 0 \qquad \text{Subtract 2x and 20 from each side.}$$
$$(x - 5)(x + 4) = 0 \qquad \text{Factor.}$$
$$x - 5 = 0 \ \text{or} \ x + 4 = 0 \qquad \text{Set factors equal to 0.}$$
$$x = 5 \ \text{or} \ x = -4 \qquad \text{Solve for x.}$$

The solutions are $-4, 5$.

B Solving Quadratic Equations by Taking Square Roots

In the next section, we will develop the first of our new methods of solving quadratic equations. The new method is called *completing the square*. Completing the square on a quadratic equation allows us to obtain solutions, regardless of whether the equation can be factored. Before we solve equations by completing the square, we need to learn how to solve equations by taking square roots of both sides.

Consider the equation

$$x^2 = 16$$

We could solve it by writing it in standard form, factoring the left side, and proceeding as we did in Chapter 5. We can shorten our work considerably, however, if we simply notice that x must be either the positive square root of 16 or the negative square root of 16. That is,

$$\text{if} \qquad x^2 = 16$$
$$\text{then} \qquad x = \sqrt{16} \ \text{ or } \qquad x = -\sqrt{16}$$
$$x = 4 \qquad\qquad x = -4$$

We can generalize this result as follows:

PROPERTY Square Root Property for Equations

If $a^2 = b$, where b is a real number, then $a = \sqrt{b}$ or $a = -\sqrt{b}$.

Notation The expression $a = b$ or $a = -b$ can be written in shorthand form as $a = \pm b$. The symbol \pm is read "plus or minus."

We can apply the square root property for equations to some fairly complicated quadratic equations.

Answers

1. $-4, 7$
2. $-4, 6$

3. Solve $(3x + 2)^2 = 16$.

EXAMPLE 3 Solve $(2x - 3)^2 = 25$.

Solution

$$(2x - 3)^2 = 25$$

$$2x - 3 = \pm\sqrt{25} \qquad \textit{Square root property for equations}$$

$$2x - 3 = \pm 5 \qquad \sqrt{25} = 5$$

$$2x = 3 \pm 5 \qquad \textit{Add 3 to both sides.}$$

$$x = \frac{3 \pm 5}{2} \qquad \textit{Divide both sides by 2.}$$

The last equation can be written as two separate statements.

$$x = \frac{3 + 5}{2} \quad \text{or} \quad x = \frac{3 - 5}{2}$$

$$= \frac{8}{2} \qquad\qquad = \frac{-2}{2}$$

$$= 4 \qquad\qquad = -1$$

The solutions are $4, -1$.

Notice that we could have solved the equation in Example 3 by expanding the left side, writing the resulting equation in standard form, and then factoring. The problem would look like this:

$$(2x - 3)^2 = 25 \qquad \textit{Original equation}$$

$$4x^2 - 12x + 9 = 25 \qquad \textit{Expand the left side.}$$

$$4x^2 - 12x - 16 = 0 \qquad \textit{Add } -25 \textit{ to each side.}$$

$$4(x^2 - 3x - 4) = 0 \qquad \textit{Begin factoring.}$$

$$4(x - 4)(x + 1) = 0 \qquad \textit{Factor completely.}$$

$$x - 4 = 0 \quad \text{or} \quad x + 1 = 0 \qquad \textit{Set variable factors equal to 0.}$$

$$x = 4 \qquad\qquad x = -1$$

Notice that solving the equation by factoring leads to the same two solutions.

4. Solve $(4x - 3)^2 = 50$.

EXAMPLE 4 Solve $(3x - 1)^2 = 12$ for x.

Solution

$$(3x - 1)^2 = 12$$

$$3x - 1 = \pm\sqrt{12} \qquad \textit{Square root property for equations}$$

$$3x - 1 = \pm 2\sqrt{3} \qquad \sqrt{12} = 2\sqrt{3}$$

$$3x = 1 \pm 2\sqrt{3} \qquad \textit{Add 1 to both sides.}$$

$$x = \frac{1 \pm 2\sqrt{3}}{3} \qquad \textit{Divide both sides by 3.}$$

The solutions are $\dfrac{1 + 2\sqrt{3}}{3}, \dfrac{1 - 2\sqrt{3}}{3}$. We can combine these as $\dfrac{1 \pm 2\sqrt{3}}{3}$.

Here is a check of the first solution:

When → $\qquad\qquad\qquad\qquad x = \dfrac{1 + 2\sqrt{3}}{3}$

the equation → $\qquad\qquad\qquad (3x - 1)^2 = 12$

becomes → $\qquad\qquad \left(3 \cdot \dfrac{1 + 2\sqrt{3}}{3} - 1\right)^2 \overset{?}{=} 12$

$$(1 + 2\sqrt{3} - 1)^2 \overset{?}{=} 12$$

$$(2\sqrt{3})^2 \overset{?}{=} 12$$

$$4 \cdot 3 \overset{?}{=} 12$$

$$12 = 12$$

Answers

3. $\frac{2}{3}, -2$

4. $\frac{3 \pm 5\sqrt{2}}{4}$

EXAMPLE 5 Solve $x^2 + 6x + 9 = 12$.

Solution We can solve this equation as we have the equations in Examples 3 and 4 if we first write the left side as $(x + 3)^2$.

$$x^2 + 6x + 9 = 12 \qquad \textit{Original equation}$$
$$(x + 3)^2 = 12 \qquad \textit{Write } x^2 + 6x + 9 \textit{ as } (x + 3)^2.$$
$$x + 3 = \pm 2\sqrt{3} \qquad \textit{Square root property for equations}$$
$$x = -3 \pm 2\sqrt{3} \qquad \textit{Add } -3 \textit{ to each side.}$$

We have two solutions: $-3 + 2\sqrt{3}$ and $-3 - 2\sqrt{3}$. What is important about this problem, however, is the fact that the equation was easy to solve because the left side was a perfect square trinomial.

C Quadratic Equations and Right Triangles

FACTS FROM GEOMETRY More Special Triangles

The triangles shown in Figures 1 and 2 occur frequently in mathematics.

FIGURE 1

FIGURE 2

Note that both of the triangles are right triangles. We refer to the triangle in Figure 1 as a 30°– 60°– 90° triangle, and the triangle in Figure 2 as a 45°– 45°– 90° triangle.

EXAMPLE 6 If the shortest side in a 30°–60°–90° triangle is 1 inch, find the lengths of the other two sides.

Solution In Figure 3, triangle ABC is a 30° – 60° – 90° triangle in which the shortest side AC is 1 inch long. Triangle DBC is also a 30° – 60° – 90° triangle in which the shortest side DC is 1 inch long.

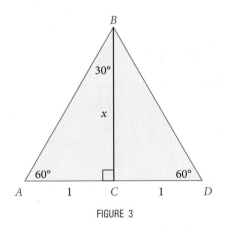

FIGURE 3

5. Solve $x^2 + 10x + 25 = 20$.

6. If the shortest side in a 30°–60°–90° triangle is 2 inches long, find the lengths of the other two sides.

Answers

5. $-5 \pm 2\sqrt{5}$

6. 4 inches and $2\sqrt{3}$ inches

Notice that the large triangle ABD is an equilateral triangle because each of its interior angles is 60°. Each side of triangle ABD is 2 inches long. Side AB in triangle ABC is therefore 2 inches. To find the length of side BC, we use the Pythagorean theorem.

$$BC^2 + AC^2 = AB^2$$
$$x^2 + 1^2 = 2^2$$
$$x^2 + 1 = 4$$
$$x^2 = 3$$
$$x = \sqrt{3} \text{ inches}$$

Note that we write only the positive square root because x is the length of a side in a triangle and is therefore a positive number.

7. The table in the introduction to this section gives the vertical rise of the Lookout Double chair lift as 960 feet and the length of the chair lift as 4,330 feet. To the nearest foot, find the horizontal distance covered by a person riding the lift.

EXAMPLE 7 Table 1 in the introduction to this section gives the vertical rise of the Forest Double chair lift as 1,170 feet and the length of the chair lift as 5,750 feet. To the nearest foot, find the horizontal distance covered by a person riding this lift.

Solution Figure 4 is a model of the Forest Double chair lift. A rider gets on the lift at point A and exits at point B. The length of the lift is AB.

FIGURE 4

To find the horizontal distance covered by a person riding the chair lift, we use the Pythagorean theorem.

$5{,}750^2 = x^2 + 1{,}170^2$	Pythagorean theorem
$33{,}062{,}500 = x^2 + 1{,}368{,}900$	Simplify the squares.
$x^2 = 33{,}062{,}500 - 1{,}368{,}900$	Solve for x^2.
$x^2 = 31{,}693{,}600$	Simplify the right side.
$x = \sqrt{31{,}693{,}600}$	Square root property for equations
$= 5{,}630 \text{ feet}$	Round to the nearest foot.

A rider getting on the lift at point A and riding to point B will cover a horizontal distance of approximately 5,630 feet.

GETTING READY FOR CLASS

After reading through the preceding section, respond in your own words and in complete sentences.

A. In what form is an equation that can be solved by factoring?

B. What is the square root property for equations?

C. What is the relationship between the shortest side and the longest side in a 30°–60°–90° triangle?

D. How would we use the Pythagorean theorem and the square root property for equations in the same problem?

Answer

7. 4,222 feet

Vocabulary Review

Choose the correct words to fill in the blanks below.

square root property completing the square shorthand factored

1. _____ on a quadratic equation allows us to obtain solutions, regardless of whether the equation can be _____ .

2. The _____ for equations says that $a = b$ or $a = -b$ if $a^2 = b$.

3. The _____ form for the expression $a = b$ or $a = -b$ is $a = \pm b$, read "plus or minus."

Problems

A Solve the following equations by factoring.

1. $x^2 - 4x = 21$

2. $x^2 - 4x = 45$

3. $x^2 - 8x = -15$

4. $x^2 = -6x - 5$

5. $x(x + 1) = 4(2x - 3)$

6. $x(x + 2) = 3(2x + 7)$

7. $2x^2 = 4x$

8. $9x^2 = 27x$

9. $x(x + 4) = 3(2x + 1)$

10. $2x(x + 2) = 3(3x + 1)$

11. $3x(x + 3) = 2(2x + 1)$

12. $10x(x + 2) = 3(3x + 2)$

13. $6x(x + 3) = 5x - 6$

14. $6(x^2 + 1) = 2(x + 3)$

15. $4x(2x + 3) = 3(2x + 3)$

16. $4(x - 1)^2 = 5 - 8x$

B Solve the following equations using the square root property.

17. $x^2 = 25$

18. $x^2 = 16$

19. $a^2 = 9$

20. $a^2 = 49$

21. $y^2 = \dfrac{3}{4}$

22. $y^2 = \dfrac{5}{9}$

23. $x^2 - 12 = 0$

24. $x^2 - 8 = 0$

25. $4a^2 - 45 = 0$

SCAN TO ACCESS

26. $9a^2 - 20 = 0$ **27.** $(2y - 1)^2 = 25$ **28.** $(3y + 7)^2 = 1$ **29.** $(2a + 3)^2 = 9$

$$2a + 3 = \pm 3$$

$$2a + 3 = 3 \qquad 2a + 3 = -3$$
$$\quad -3 \; -3 \qquad \qquad -3 \; -3$$
$$2a = 0 \qquad \qquad 2a = -6$$
$$a = \frac{0}{2} \qquad \qquad a = -3$$
$$a = 0$$

30. $(3a - 5)^2 = 49$ **31.** $(5x + 2)^2 = 8$ **32.** $(6x - 7)^2 = 75$

Solve the following equations by taking square roots. Note that the left-hand side is a perfect square.

33. $x^2 + 8x + 16 = 27$ **34.** $x^2 - 12x + 36 = 8$ **35.** $4a^2 - 12a + 9 = 4$ **36.** $9a^2 - 12a + 4 = 9$

37. $x^2 + 4x + 4 = 8$ **38.** $x^2 - 6x + 9 = 15$ **39.** $4x^2 + 4x + 1 = 20$ **40.** $9x^2 - 6x + 1 = 32$

41. Solve each equation.

 a. $5x - 7 = 0$

 b. $5x - 7 = 8$

 c. $(5x - 7)^2 = 8$

 d. $\sqrt{5x - 7} = 8$

 e. $\dfrac{5}{2} - \dfrac{7}{2x} = \dfrac{4}{x}$

42. Solve each equation.

 a. $5x + 11 = 0$

 b. $5x + 11 = 9$

 c. $(5x + 11)^2 = 9$

 d. $\sqrt{5x + 11} = 9$

 e. $\dfrac{5}{3} - \dfrac{11}{3x} = \dfrac{3}{x}$

Applying the Concepts

43. Geometry If the shortest side in a 30°–60°–90° triangle is $\frac{1}{2}$ inch long, find the lengths of the other two sides.

44. Geometry If the length of the longest side of a 30°–60°–90° triangle is x, find the lengths of the other two sides in terms of x.

45. Geometry If the length of the shorter sides of a $45° - 45° - 90°$ triangle is 1 inch, find the length of the hypotenuse.

46. Geometry If the length of the shorter sides of a $45° - 45° - 90°$ triangle is x, find the length of the hypotenuse, in terms of x.

47. Chair Lift Use Table 1 from the introduction to this section to find the horizontal distance covered by a person riding the Bear Paw Double chair lift. Round your answer to the nearest foot.

48. Interest Rate Suppose a deposit of $3,000 in a savings account that paid an annual interest rate r (compounded yearly) is worth $3,456 after 2 years. Using the formula $A = P(1 + r)^t$, we have

$$3,456 = 3,000(1 + r)^2$$

Solve for r to find the annual interest rate.

49. Special Triangles In Figure 5, triangle ABC has angles $45°$ and $30°$, and height x. Find the lengths of sides AB, BC, and AC, in terms of x.

FIGURE 5

50. Dimensions of a Tent A two-person tent has a height at the center of 4 feet. If the sides of the tent meet the ground at an angle of $60°$ and the tent is 6 feet in length, how many square feet of material will be needed to make the tent? (Figure 7; assume that the tent has a floor and is closed at both ends.) Give your answer to the nearest square foot.

FIGURE 7

51. Length of an Escalator The longest single escalator in the United States is at the Wheaton Station of the Washington D.C. Metro system. Its length is 230 feet and it has a vertical rise of 115 feet. What is the horizontal distance used for this escalator?

52. Length of an Escalator One of the longest escalators in the world is at the Kreschatik station of the Kiev Metro system. Its length is 432 feet and it has a vertical rise of 216 feet. What is the horizontal distance used for this escalator?

Getting Ready for the Next Section

Simplify the following.

53. $\left(\dfrac{1}{2} \cdot 6\right)^2$

54. $\left(\dfrac{1}{2} \cdot 12\right)^2$

55. $\left[\dfrac{1}{2} \cdot (-6)\right]^2$

56. $\left[\dfrac{1}{2} \cdot (-8)\right]^2$

57. $\left(\dfrac{1}{2} \cdot 7\right)^2$

58. $\left(\dfrac{1}{2} \cdot 9\right)^2$

59. $\left[\dfrac{1}{2} \cdot (-7)\right]^2$

60. $\left[\dfrac{1}{2} \cdot (-5)\right]^2$

Factor the following as a perfect square trinomial.

61. $x^2 + 4x + 4$

62. $x^2 - 8x + 16$

63. $x^2 - 10x + 25$

64. $x^2 + 12x + 36$

65. $4x^2 - 12x + 9$

66. $9x^2 - 12x + 4$

67. $9x^2 - 24x + 16$

68. $25x^2 - 30x + 9$

69. $x^2 - \dfrac{2}{3}x + \dfrac{1}{9}$

70. $x^2 + \dfrac{4}{3}x + \dfrac{4}{9}$

71. $x^2 - \dfrac{3}{2}x + \dfrac{9}{16}$

72. $x^2 - \dfrac{8}{5}x + \dfrac{16}{25}$

Find the Mistake

Each sentence below contains a mistake. Circle the mistake and write the correct word(s) or expression(s) on the line provided.

1. The first step to solving the equation $x^2 - 5x = 14$ is to factor the left-hand side.

2. Using the square root property for equations, the solution to $x^2 = 100$ is 10. _____

3. Solving $x^2 + 16x + 64 = 20$ gives the solution $x = -8 + 2\sqrt{5}$. _____

4. If the shortest side of a 30°– 60°– 90° triangle is 3 feet and the hypotenuse is 3 feet, then the other side measures 6 feet.

Navigation Skills: Prepare, Study, Achieve

It is important to reward yourself for working hard in this class. Recognizing your achievement will help foster success through the end of this course, and enable you to carry over the skills you've learned to future classes. What are some things you can do to reward and celebrate yourself? Also, for future classes, you could decide on these rewards at the beginning of the course. Schedule smaller short-term rewards to work towards as you complete each chapter. And then have a long-term reward that you achieve at the end of the course to celebrate your overall success. You deserve it. Congratulations!

9.2 Completing the Square

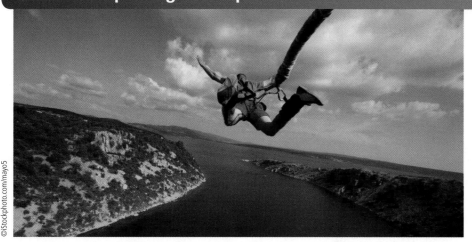

Suppose a bungee jumper's ankles are bound to a 100-foot cord. He leaps off a bridge with an initial velocity of 10 feet per second. The relationship between the distance fallen d and time t is given by $d = -16t^2 - 10t + 100$. How long does it take for the jumper to fall 100 feet, after which the cord begins to stretch? As we previously mentioned, we will begin to develop a new method of solving quadratic equations, called completing the square. In this section, we will work with equations similar to the one above using this method.

A Method of Completing the Square

The method of completing the square is simply a way of transforming any quadratic equation into an equation that can be solved using the square root property.

The key to understanding the method of completing the square lies in recognizing the relationship between the last two terms of any perfect square trinomial whose leading coefficient is 1.

Consider the following list of perfect square trinomials and their corresponding binomial squares:

$$x^2 - 6x + 9 = (x - 3)^2$$

$$x^2 + 8x + 16 = (x + 4)^2$$

$$x^2 - 10x + 25 = (x - 5)^2$$

$$x^2 + 12x + 36 = (x + 6)^2$$

In each case, the leading coefficient is 1. A more important observation comes from noticing the relationship between the linear and constant terms (middle and last terms) in each trinomial. Observe that the constant term in each case is the square of half the coefficient of x in the middle term. For example, in the last expression, the constant term 36 is the square of half of 12, where 12 is the coefficient of x in the middle term. (Notice also that the second terms in all the binomials on the right side are half the coefficients of the middle terms of the trinomials on the left side.) We can use these observations to build our own perfect square trinomials and, in doing so, solve some quadratic equations.

1. Fill in the blanks to complete the square.

 a. $x^2 + 8x +$ _____ $= (x +$ _____ $)^2$

 b. $x^2 - 10x +$ _____ $= (x -$ _____ $)^2$

EXAMPLE 1 Fill in the blanks to complete the square.

 a. $x^2 + 4x +$ _____ $= (x +$ _____ $)^2$

 b. $x^2 - 6x +$ _____ $= (x -$ _____ $)^2$

Solution

 a. Following the patterns established above, we find the square of $\frac{1}{2}$ of the middle term coefficient.

$$\left[\frac{1}{2}(4)\right]^2 = [2]^2 = 4$$

Therefore: $x^2 + 4x + 4 = (x + 2)^2$

 b. We find the square of $\frac{1}{2}$ of the middle term coefficient.

$$\left[\frac{1}{2}(-6)\right]^2 = [-3]^2 = 9$$

Therefore: $x^2 - 6x + 9 = (x - 3)^2$.

B Solving Quadratic Equations by Completing the Square

We can use the process of completing the square to solve quadratic equations. Since completing the square converts a quadratic expression to a square, we can then take square roots to solve the equation. The next three examples illustrate this method.

2. Solve $x^2 + 3x - 4 = 0$ by completing the square.

EXAMPLE 2 Solve $x^2 - 6x + 5 = 0$ by completing the square.

Solution We begin by adding -5 to both sides of the equation. We want just $x^2 - 6x$ on the left side so that we can add our own final term to get a perfect square trinomial.

$$x^2 - 6x + 5 = 0$$
$$x^2 - 6x \qquad = -5 \qquad \text{Add } -5 \text{ to both sides.}$$

Now we can add 9 to both sides and the left side will be a perfect square.

$$x^2 - 6x + 9 = -5 + 9$$
$$(x - 3)^2 = 4$$

The final line is in the form of the equations we solved previously.

$$x - 3 = \pm 2$$
$$x = 3 \pm 2 \qquad \text{Add 3 to both sides.}$$
$$x = 3 + 2 \quad \text{or} \quad x = 3 - 2$$
$$x = 5 \qquad\qquad x = 1$$

The two solutions are 5 and 1.

Note The equation in Example 2 can be solved quickly by factoring:

$$x^2 - 6x + 5 = 0$$
$$(x - 5)(x - 1) = 0$$
$$x - 5 = 0 \quad \text{or} \quad x - 1 = 0$$
$$x = 5 \qquad\qquad x = 1$$

The reason we didn't solve it by factoring is we want to practice completing the square on some simple equations.

3. Solve $x^2 - 5x + 2 = 0$ by completing the square.

EXAMPLE 3 Solve $x^2 + 5x - 2 = 0$ by completing the square.

Solution We must begin by adding 2 to both sides. (The left side of the equation, as it is, is not a perfect square, because it does not have the correct constant term. We will simply "move" that term to the other side and use our own constant term.)

$$x^2 + 5x = 2 \qquad \text{Add 2 to each side.}$$

We complete the square by adding the square of half the coefficient of the linear term to both sides.

$$x^2 + 5x + \frac{25}{4} = 2 + \frac{25}{4} \qquad \text{Half of 5 is } \frac{5}{2}, \text{ the square of which is } \left(\frac{5}{2}\right)^2 = \frac{25}{4}.$$

$$\left(x + \frac{5}{2}\right)^2 = \frac{33}{4} \qquad\qquad 2 + \frac{25}{4} = \frac{8}{4} + \frac{25}{4} = \frac{33}{4}$$

$$x + \frac{5}{2} = \pm\sqrt{\frac{33}{4}} \qquad \textit{Square root property for equations}$$

$$x + \frac{5}{2} = \pm\frac{\sqrt{33}}{2} \qquad \textit{Simplify the radical.}$$

$$x = -\frac{5}{2} \pm \frac{\sqrt{33}}{2} \qquad \textit{Add } -\frac{5}{2} \textit{ to both sides.}$$

$$x = \frac{-5 \pm \sqrt{33}}{2}$$

The solutions are $\dfrac{-5 \pm \sqrt{33}}{2}$.

We can use a calculator to get decimal approximations to these solutions. If $\sqrt{33} \approx 5.74$, then

$$\frac{-5 + 5.74}{2} = 0.37$$

$$\frac{-5 - 5.74}{2} = -5.37$$

EXAMPLE 4 Solve $3x^2 - 8x + 2 = 0$.

Solution

$$3x^2 - 8x + 2 = 0$$

$$3x^2 - 8x = -2 \qquad \textit{Add } -2 \textit{ to both sides.}$$

We cannot complete the square on the left side because the leading coefficient is not 1. We take an extra step and divide both sides by 3.

$$\frac{3x^2}{3} - \frac{8x}{3} = -\frac{2}{3}$$

$$x^2 - \frac{8}{3}x = -\frac{2}{3}$$

Half of $\frac{8}{3}$ is $\frac{4}{3}$, the square of which is $\left(\frac{4}{3}\right)^2 = \frac{16}{9}$.

$$x^2 - \frac{8}{3}x + \frac{16}{9} = -\frac{2}{3} + \frac{16}{9} \qquad \textit{Add } \frac{16}{9} \textit{ to both sides.}$$

$$\left(x - \frac{4}{3}\right)^2 = \frac{10}{9} \qquad \textit{Simplify the right side.}$$

$$x - \frac{4}{3} = \pm\sqrt{\frac{10}{9}} \qquad \textit{Square root property for equations}$$

$$x - \frac{4}{3} = \pm\frac{\sqrt{10}}{3} \qquad \sqrt{\frac{10}{9}} = \frac{\sqrt{10}}{3}$$

$$x = \frac{4}{3} \pm \frac{\sqrt{10}}{3} \qquad \textit{Add } \frac{4}{3} \textit{ to both sides.}$$

$$x = \frac{4 \pm \sqrt{10}}{3}$$

The solutions are $\dfrac{4 \pm \sqrt{10}}{3}$.

4. Solve: $5x^2 - 3x - 2 = 0$.

Answers

4. $-\frac{2}{5}, 1$

HOW TO Solve a Quadratic Equation by Completing the Square

To summarize the method used in the preceding two examples, we list the following steps:

Step 1: Write the equation in the form $ax^2 + bx = c$.

Step 2: If the leading coefficient is not 1, divide both sides by the coefficient so that the resulting equation has a leading coefficient of 1. That is, if $a \neq 1$, then divide both sides by a.

Step 3: Add the square of half the coefficient of the linear term to both sides of the equation.

Step 4: Write the left side of the equation as the square of a binomial, and simplify the right side if possible.

Step 5: Apply the square root property for equations, and solve as usual.

GETTING READY FOR CLASS

After reading through the preceding section, respond in your own words and in complete sentences.

A. What kind of equation do you solve using the method of completing the square?

B. What are the steps necessary to complete the square on a trinomial?

C. Explain in words how you would complete the square on $x^2 - 6x = 5$.

D. Why do you use the square root property for equations as the last step of completing the square?

Vocabulary Review

The following is a list of steps for solving a quadratic equation by completing the square. Choose the correct words to fill in the blanks below.

 linear term square leading coefficient

 equation binomial square root property

Step 1: Write the _____ in the form $ax^2 + bx = c$.

Step 2: If the _____ is not 1, divide both sides by the coefficient so that the resulting equation has a leading coefficient of 1.

Step 3: Add the _____ of half the coefficient of the _____ to both sides of the equation.

Step 4: Write the left side of the equation as the square of a _____, and simplify the right side if possible.

Step 5: Apply the _____ for equations and solve as usual.

Problems

A Copy each of the following, and fill in the blanks so the left side of each is a perfect square trinomial. That is, complete the square.

1. $x^2 + 12x +$ ___ $= (x +$ ___$)^2$

2. $x^2 + 6x +$ ___ $= (x +$ ___$)^2$

3. $x^2 - 4x +$ ___ $= (x -$ ___$)^2$

4. $x^2 - 2x +$ ___ $= (x -$ ___$)^2$

5. $a^2 - 10a +$ ___ $= (a -$ ___$)^2$

6. $a^2 - 8a +$ ___ $= (a -$ ___$)^2$

7. $x^2 + 5x +$ ___ $= (x +$ ___$)^2$

8. $x^2 + 3x +$ ___ $= (x +$ ___$)^2$

9. $y^2 - 7y +$ ___ $= (y -$ ___$)^2$

10. $y^2 - y +$ ___ $= (y -$ ___$)^2$

11. $x^2 + \frac{1}{2}x +$ ___ $= (x +$ ___$)^2$

12. $x^2 - \frac{3}{4}x +$ ___ $= (x -$ ___$)^2$

13. $x^2 + \frac{2}{3}x +$ ⅓ $= (x +$ ⅓$)^2$

14. $x^2 - \frac{4}{5}x +$ ___ $= (x -$ ___$)^2$

B Solve each of the following quadratic equations by completing the square.

15. $x^2 + 4x = 12$

16. $x^2 - 2x = 8$

17. $x^2 + 12x = -27$

18. $x^2 - 6x = 16$

19. $a^2 + 2a - 5 = 0$

20. $a^2 + 10a + 22 = 0$

21. $y^2 - 8y + 1 = 0$

22. $y^2 + 6y - 1 = 0$

23. $x^2 - 5x - 3 = 0$

24. $x^2 - 5x - 2 = 0$

25. $2x^2 - 4x - 8 = 0$

26. $3x^2 - 9x - 12 = 0$

27. $3t^2 - 8t + 1 = 0$

28. $5t^2 + 12t - 1 = 0$

29. $4x^2 - 3x - 5 = 0$

30. $7x^2 - 5x - 2 = 0$

31. $3x^2 + 4x - 1 = 0$

32. $2x^2 + 6x - 1 = 0$

33. $2x^2 - 10x = 11$

34. $25x^2 - 20x = 1$

35. $4x^2 - 10x - 11 = 0$

36. $4x^2 - 6x + 1 = 0$

37. $3x^2 - 5x - 1 = 0$

38. $4x^2 - 5x - 3 = 0$

39. $\frac{1}{2}b^2 = \frac{9}{10}b - \frac{1}{5}$

40. $\frac{1}{4}x^2 - \frac{19}{20}x + \frac{4}{5} = 0$

41. $\frac{2}{3}x^2 - \frac{1}{5} = \frac{4}{15}x$

42. $\frac{1}{3}x^2 - \frac{5}{15}x = \frac{2}{5}$

43. $0.03x^2 - 0.01x - 0.01 = 0$ **44.** $0.02b^2 + 0.03b = 0.01$

45. $0.02y^2 - 0.03 = 0.04y$

46. $0.03x^2 = 0.06x + 0.07$

47. Consider the equation $x^2 = 9$.

 a. Can it be solved by factoring?

 b. Solve it.

48. Consider the equation $x^2 - 10x + 18 = 0$.

 a. Can it be solved by factoring?

 b. Solve it.

$$x^2 - 10x + 5^2 - 18 + \underline{\quad} \; 25$$
$$(x - 5)^2 = \pm\sqrt{7}$$
$$x - 5 = \pm\sqrt{7} \quad x = 5 \pm \sqrt{7}$$

49. Solve the equation $x^2 - 6x = 0$

 a. by factoring.

 b. by completing the square.

$$x^2 - 6x + 3^2 = 0 + 9 \quad x - 3 = \pm 3$$
$$(x - 3)^2 = \pm\sqrt{9} \quad x = 0$$
$$x = 6$$

50. Solve the equation $x^2 + ax = 0$

 a. by factoring.

 b. by completing the square.

51. Solve the equation $x^2 + 2x = 35$

 a. by factoring.

 b. by completing the square.

52. Solve the equation $8x^2 - 10x - 25 = 0$

 a. by factoring.

 b. by completing the square.

53. Is $x = -3 + \sqrt{2}$ a solution to $x^2 - 6x = 7$?

54. Is $x = 2 - \sqrt{5}$ a solution to $x^2 - 4x = 1$?

55. A quadratic equation has solutions of $\pm\sqrt{5}$. What was the original equation?

Hint: Let $x = \pm\sqrt{5}$, then square each side.

56. A quadratic equation has solutions of $\pm 2\sqrt{3}$. What was the original equation?

57. A quadratic equation has solutions of $1 \pm \sqrt{3}$. What was the original equation?

Hint: Let $x = 1 \pm \sqrt{3}$, then isolate the radical before squaring.

58. A quadratic equation has solutions of $-2 \pm \sqrt{5}$. What was the original equation?

59. A quadratic equation has solutions of $\frac{1 \pm \sqrt{5}}{3}$. What was the original equation?

60. A quadratic equation has solutions of $\frac{-2 \pm \sqrt{3}}{4}$. What was the original equation?

Getting Ready for the Next Section

Simplify.

61. $49 - 4(6)(-5)$

62. $49 - 4(6)(2)$

63. $(-27)^2 - 4(0.1)(1,700)$

64. $25 - 4(4)(-10)$

65. $\dfrac{-7 + \sqrt{169}}{12}$

66. $\dfrac{-7 - \sqrt{169}}{12}$

67. $\dfrac{-4 + \sqrt{36}}{2}$

68. $\dfrac{-5 - \sqrt{81}}{4}$

Factor.

69. $27t^3 - 8$

70. $125t^3 + 1$

71. $2x^3 + 54$

72. $8y^3 - 27$

Find the Mistake

Each sentence below contains a mistake. Circle the mistake and write the correct word(s) or number(s) on the line provided.

1. A perfect square trinomial has a constant term that is the square of half the leading coefficient. _____

2. To solve the equation $x^2 - 9x + 11 = 0$ by completing the square, add -11 to both sides of the equation and then add 81 to both sides. _____

3. The first step to solving the equation $4x^2 - 6x + 5 = 0$ is to add half the coefficient of the linear term to the constant term. _____

4. The square of half the coefficient of the linear term in the above equation is $\frac{25}{64}$. _____

9.3 The Quadratic Formula

If you go on to take a business course or an economics course, you will find yourself spending lots of time with the three expressions that form the mathematical foundation of business: profit, revenue, and cost. Many times these expressions are given as polynomials. The relationship between the three equations is known as the profit equation.

$$\text{Profit} = \text{Revenue} - \text{Cost}$$

$$P(x) = R(x) - C(x)$$

The table and graphs below were produced on a graphing calculator. They give numerical and graphical descriptions of revenue, profit, and cost for a company that manufactures and sells prerecorded DVDs according to the following equations:

$$R(x) = 11.5x - 0.05x^2 \quad \text{and} \quad C(x) = 200 + 2x$$

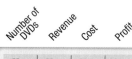

X	Y_1	Y_2	Y_3
0	0	200	−200
50	450	300	150
100	650	400	250
150	600	500	100
200	300	600	−300

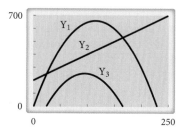

FIGURE 1

By studying the material in this section, you will get a more thorough look at the equations and relationships that are emphasized in business and economics.

In this section, we will use the method of completing the square from the preceding section to derive the quadratic formula. The *quadratic formula* is a very useful tool in mathematics. It allows us to solve all types of quadratic equations.

A The Quadratic Formula

THEOREM **The Quadratic Formula**

For any quadratic equation in the form $ax^2 + bx + c = 0$, $a \neq 0$, the two solutions are

$$x = \frac{-b + \sqrt{b^2 - 4ac}}{2a} \quad \text{and} \quad x = \frac{-b - \sqrt{b^2 - 4ac}}{2a}$$

KEY WORDS

profit

revenue

cost

quadratic formula

Note $P(x)$, $R(x)$ and $C(x)$ are examples of function notation. We will discuss functions and function notation in more detail later in the chapter.

Note These solutions may also be written together as

$$x = \frac{-b \pm \sqrt{b^4 - 4ac}}{2a}$$

Proof We will prove the quadratic formula by completing the square on $ax^2 + bx + c = 0$.

$$ax^2 + bx + c = 0$$

$$ax^2 + bx = -c \qquad \text{Add } -c \text{ to both sides.}$$

$$x^2 + \frac{b}{a}x = -\frac{c}{a} \qquad \text{Divide both sides by } a.$$

To complete the square on the left side, we add the square of $\frac{1}{2}$ of $\frac{b}{a}$ to both sides $\left(\frac{1}{2} \text{ of } \frac{b}{a} \text{ is } \frac{b}{2a}\right)$.

$$x^2 + \frac{b}{a}x + \left(\frac{b}{2a}\right)^2 = -\frac{c}{a} + \left(\frac{b}{2a}\right)^2$$

We now simplify the right side as a separate step. We combine the two terms by writing each with the least common denominator $4a^2$.

$$-\frac{c}{a} + \left(\frac{b}{2a}\right)^2 = -\frac{c}{a} + \frac{b^2}{4a^2} = \frac{4a}{4a}\left(\frac{-c}{a}\right) + \frac{b^2}{4a^2} = \frac{-4ac + b^2}{4a^2}$$

It is convenient to write this last expression as

$$\frac{b^2 - 4ac}{4a^2}$$

Continuing with the proof, we have

$$x^2 + \frac{b}{a}x + \left(\frac{b}{2a}\right)^2 = \frac{b^2 - 4ac}{4a^2}$$

$$\left(x + \frac{b}{2a}\right)^2 = \frac{b^2 - 4ac}{4a^2} \qquad \text{Write the left side as a binomial square.}$$

$$x + \frac{b}{2a} = \pm\frac{\sqrt{b^2 - 4ac}}{2a} \qquad \text{Square root property for equations}$$

$$x = -\frac{b}{2a} \pm \frac{\sqrt{b^2 - 4ac}}{2a} \qquad \text{Add } -\frac{b}{2a} \text{ to both sides.}$$

$$= \frac{-b \pm \sqrt{b^2 - 4ac}}{2a}$$

Our proof is now complete. What we have is this: If our equation is in the form $ax^2 + bx + c = 0$ (standard form), where $a \neq 0$, the two solutions are always given by the formula

$$x = \frac{-b \pm \sqrt{b^2 - 4ac}}{2a}$$

This formula is known as the *quadratic formula*. If we substitute the coefficients a, b, and c of any quadratic equation in standard form into the formula, we need only perform some basic arithmetic to arrive at the solution set.

EXAMPLE 1 Solve $x^2 - 5x - 6 = 0$ using the quadratic formula.

Solution To use the quadratic formula, we must make sure the equation is in standard form; identify a, b, and c; substitute them into the formula; and work out the arithmetic.

For the equation $x^2 - 5x - 6 = 0$, $a = 1$, $b = -5$, and $c = -6$.

$$x = \frac{-b \pm \sqrt{b^2 - 4ac}}{2a}$$

$$= \frac{-(-5) \pm \sqrt{(-5)^2 - 4(1)(-6)}}{2(1)}$$

$$= \frac{5 \pm \sqrt{49}}{2}$$

$$= \frac{5 \pm 7}{2}$$

Practice Problems

1. Solve $x^2 - 9x + 18 = 0$ using the quadratic formula.

Note Whenever the solutions to our quadratic equations turn out to be rational numbers, as in Example 1, it means the original equation could have been solved by factoring. (We didn't solve the equation in Example 1 by factoring because we were trying to get some practice with the quadratic formula.)

Answers

1. $3, 6$

$$x = \frac{5+7}{2} \quad \text{or} \quad x = \frac{5-7}{2}$$

$$x = \frac{12}{2} \qquad\qquad x = -\frac{2}{2}$$

$$x = 6 \qquad\qquad x = -1$$

The two solutions are 6 and -1.

EXAMPLE 2 Solve $2x^2 = -4x + 3$ using the quadratic formula.

Solution Before we can identify a, b, and c, we must write the equation in standard form. To do so, we add $4x$ and -3 to each side of the equation.

$$2x^2 = -4x + 3$$

$$2x^2 + 4x - 3 = 0 \qquad \textit{Add } 4x \textit{ and } -3 \textit{ to each side.}$$

Now that the equation is in standard form, we see that $a = 2$, $b = 4$, and $c = -3$. Using the quadratic formula we have

$$x = \frac{-b \pm \sqrt{b^2 - 4ac}}{2a}$$

$$= \frac{-4 \pm \sqrt{4^2 - 4(2)(-3)}}{2(2)}$$

$$= \frac{-4 \pm \sqrt{40}}{4}$$

$$= \frac{-4 \pm 2\sqrt{10}}{4}$$

We can reduce the final expression in the preceding equation to lowest terms by factoring 2 from the numerator and denominator and then dividing it out.

$$x = \frac{2(-2 \pm \sqrt{10})}{2 \cdot 2}$$

$$= \frac{-2 \pm \sqrt{10}}{2}$$

Our two solutions are $\dfrac{-2 + \sqrt{10}}{2}$ and $\dfrac{-2 - \sqrt{10}}{2}$.

EXAMPLE 3 Solve $x^2 - 6x = -7$ using the quadratic formula.

Solution We begin by writing the equation in standard form.

$$x^2 - 6x = -7$$

$$x^2 - 6x + 7 = 0 \qquad \textit{Add 7 to each side.}$$

Using $a = 1$, $b = -6$, and $c = 7$ in the quadratic formula

$$x = \frac{-b \pm \sqrt{b^2 - 4ac}}{2a}$$

we have

$$x = \frac{-(-6) \pm \sqrt{(-6)^2 - 4(1)(7)}}{2(1)}$$

$$= \frac{6 \pm \sqrt{36 - 28}}{2}$$

$$= \frac{6 \pm \sqrt{8}}{2}$$

$$= \frac{6 \pm 2\sqrt{2}}{2}$$

2. Solve $6x^2 + 2 = -7x$ using the quadratic formula.

3. Solve $x^2 - 5x = 4$ using the quadratic formula.

Answers

2. $-\frac{1}{2}, -\frac{2}{3}$

3. $5 \pm \dfrac{\sqrt{41}}{2}$

The two terms in the numerator have a 2 in common. We reduce to lowest terms by factoring the 2 from the numerator and then dividing numerator and denominator by 2.

$$= \frac{2(3 \pm \sqrt{2})}{2}$$

$$= 3 \pm \sqrt{2}$$

The two solutions are $3 + \sqrt{2}$ and $3 - \sqrt{2}$. For practice, let's check our solutions in the original equation $x^2 - 6x = -7$.

Checking $x = 3 + \sqrt{2}$, we have

$$(3 + \sqrt{2})^2 - 6(3 + \sqrt{2}) \overset{?}{=} -7$$

$9 + 6\sqrt{2} + 2 - 18 - 6\sqrt{2} \overset{?}{=} -7$	Multiply.
$11 - 18 + 6\sqrt{2} - 6\sqrt{2} \overset{?}{=} -7$	Add 9 and 2.
$-7 + 0 \overset{?}{=} -7$	Subtract.
$-7 = -7$	A true statement

Checking $x = 3 - \sqrt{2}$, we have

$$(3 - \sqrt{2})^2 - 6(3 - \sqrt{2}) \overset{?}{=} -7$$

$9 - 6\sqrt{2} + 2 - 18 + 6\sqrt{2} \overset{?}{=} -7$	Multiply.
$11 - 18 - 6\sqrt{2} + 6\sqrt{2} \overset{?}{=} -7$	Add 9 and 2.
$-7 + 0 \overset{?}{=} -7$	Subtract.
$-7 = -7$	A true statement

As you can see, both solutions yield true statements when used in place of the variable in the original equation.

4. Solve $\frac{x^2}{2} + x = \frac{1}{3}$ using the quadratic formula.

EXAMPLE 4 Solve: $\frac{1}{10}x^2 - \frac{1}{5}x = \frac{1}{2}$ using the quadratic formula.

Solution It will be easier to apply the quadratic formula if we clear the equation of fractions. Multiplying both sides of the equation by the LCD 10 gives us

$$x^2 - 2x = 5$$

Next, we add -5 to both sides to put the equation into standard form.

$$x^2 - 2x - 5 = 0 \qquad \text{Add } -5 \text{ to both sides.}$$

Applying the quadratic formula with $a = 1$, $b = -2$, and $c = -5$, we have

$$x = \frac{-(-2) \pm \sqrt{(-2)^2 - 4(1)(-5)}}{2(1)} = \frac{2 \pm \sqrt{24}}{2} = \frac{2 \pm 2\sqrt{6}}{2}$$

Dividing the numerator and denominator by 2, we have the two solutions.

$$x = 1 \pm \sqrt{6}$$

The two solutions are $1 + \sqrt{6}$ and $1 - \sqrt{6}$.

5. Solve $(3x - 2)(x - 1) = 3$ using the quadratic formula.

EXAMPLE 5 Solve $(2x - 3)(2x - 1) = 4$ using the quadratic formula.

Solution Even though the left side of this equation begins as two binomials, we cannot use the quadratic formula until we set one side equal to zero. Therefore, we multiply the binomials on the left side and then add -4 to each side to write the equation in standard form. From there we identify a, b, and c and apply the quadratic formula.

$$(2x - 3)(2x - 1) = 4$$

$4x^2 - 8x + 3 = 4$	Multiply binomials on left side.
$4x^2 - 8x - 1 = 0$	Add -4 to each side.

Answers

4. $\dfrac{-3 \pm \sqrt{15}}{3}$

5. $\dfrac{5 \pm \sqrt{37}}{6}$

Placing $a = 4$, $b = -8$, and $c = -1$ in the quadratic formula we have

$$x = \frac{-(-8) \pm \sqrt{(-8)^2 - 4(4)(-1)}}{2(4)}$$

$$= \frac{8 \pm \sqrt{64 + 16}}{8}$$

$$= \frac{8 \pm \sqrt{80}}{8}$$

$$= \frac{8 \pm 4\sqrt{5}}{8} \qquad \sqrt{80} = 4\sqrt{5}$$

To reduce this final expression to lowest terms, we factor a 4 from the numerator and denominator and then divide the numerator and denominator by 4.

$$= \frac{4(2 \pm \sqrt{5})}{4 \cdot 2}$$

$$= \frac{2 \pm \sqrt{5}}{2}$$

B Applications

EXAMPLE 6 If an object is thrown downward with an initial velocity of 20 feet per second, the distance d, in feet, it travels in t seconds is given by the equation $d = 20t + 16t^2$. How long does it take the object to fall 40 feet?

Solution We let $d = 40$, and solve for t.

When →	$d = 40$
the equation →	$d = 20t + 16t^2$
becomes →	$40 = 20t + 16t^2$

$$16t^2 + 20t - 40 = 0$$

$$4t^2 + 5t - 10 = 0 \qquad \text{Divide by 4.}$$

Using the quadratic formula, we have

$$t = \frac{-5 \pm \sqrt{25 - 4(4)(-10)}}{2(4)}$$

$$= \frac{-5 \pm \sqrt{185}}{8}$$

$$t = \frac{-5 + \sqrt{185}}{8} \quad \text{or} \quad t = \frac{-5 - \sqrt{185}}{8}$$

The second solution is impossible because it is a negative number and time t must be positive. It takes

$$t = \frac{-5 + \sqrt{185}}{8} \quad \text{or approximately} \quad \frac{-5 + 13.60}{8} \approx 1.08 \text{ seconds}$$

for the object to fall 40 feet.

As discussed in the beginning of the section, the relationship between profit, revenue, and cost is given by the formula

$$P(x) = R(x) - C(x)$$

where $P(x)$ is the profit, $R(x)$ is the total revenue, and $C(x)$ is the total cost of producing and selling x items.

Note It would be a mistake to try to reduce this final expression further. Sometimes first-year algebra students will try to divide the 2 in the denominator into the 2 in the numerator, which is a mistake. Remember, when we reduce to lowest terms, we do so by dividing the numerator and denominator by any factors they have in common. In this case 2 is not a factor of the numerator. This expression is in lowest terms.

6. An object thrown upward with an initial velocity of 32 feet per second rises and falls according to the equation $s = 32t - 16t^2$ where s is the height of the object above the ground at any time t. At what times t will the object be 12 feet above the ground?

Answers

6. $\frac{3}{2}$ seconds, $\frac{1}{2}$ second

7. Use the information in Example 7 to find the number of programs the company must sell each week for its weekly profit to be $1,320.

EXAMPLE 7 A company produces and sells copies of an accounting program for home computers. The total weekly cost (in dollars) to produce x copies of the program is $C(x) = 8x + 500$, and the weekly revenue for selling all x copies of the program is $R(x) = 35x - 0.1x^2$. How many programs must be sold each week for the weekly profit to be $1,200?

Solution Substituting the given expressions for $R(x)$ and $C(x)$ in the equation $P(x) = R(x) - C(x)$, we have a polynomial in x that represents the weekly profit $P(x)$.

$$P(x) = R(x) - C(x)$$
$$= 35x - 0.1x^2 - (8x + 500)$$
$$= 35x - 0.1x^2 - 8x - 500$$
$$= -500 + 27x - 0.1x^2$$

Setting this expression equal to 1,200, we have a quadratic equation to solve that gives us the number of programs x that need to be sold each week to bring in a profit of $1,200.

$$1,200 = -500 + 27x - 0.1x^2$$

We can write this equation in standard form by adding the opposite of each term on the right side of the equation to both sides of the equation. Doing so produces the following equation.

$$0.1x^2 - 27x + 1,700 = 0$$

Applying the quadratic formula to this equation with $a = 0.1$, $b = -27$, and $c = 1,700$, we have

$$x = \frac{27 \pm \sqrt{(-27)^2 - 4(0.1)(1,700)}}{2(0.1)}$$

$$= \frac{27 \pm \sqrt{729 - 680}}{0.2}$$

$$= \frac{27 \pm \sqrt{49}}{0.2}$$

$$= \frac{27 \pm 7}{0.2}$$

Writing this last expression as two separate expressions, we have our two solutions.

$$x = \frac{27 + 7}{0.2} \quad \text{or} \quad x = \frac{27 - 7}{0.2}$$

$$= \frac{34}{0.2} \qquad\qquad = \frac{20}{0.2}$$

$$= 170 \qquad\qquad = 100$$

The weekly profit will be $1,200 if the company produces and sells 100 programs or 170 programs.

GETTING READY FOR CLASS

After reading through the preceding section, respond in your own words and in complete sentences.

A. What is the quadratic formula?

B. Under what circumstances should the quadratic formula be applied?

C. When would the quadratic formula result in rational solutions?

D. When will the quadratic formula result in only one solution?

Answers

7. 130 or 140 programs

Vocabulary Review

Choose the correct words to fill in the blanks below.

solutions revenue rational standard form

1. _____ for a quadratic equation is $ax^2 + bx + c = 0$.

2. If a quadratic equation is in standard form, the two _____ are given by the formula $x = \dfrac{-b \pm \sqrt{b^2 - 4ac}}{2a}$.

3. If solutions found using the quadratic formula are _____ numbers, then the original equation could have been solved by factoring.

4. The relationship between profit, _____, and cost is given by the formula $P(x) = R(x) - C(x)$.

Problems

A Solve each equation. Use factoring or the quadratic formula, whichever is appropriate. (Try factoring first. If you have any difficulty factoring, then try the quadratic formula.)

1. $x^2 + 5x + 6 = 0$

2. $x^2 + 5x - 6 = 0$

3. $a^2 - 4a + 1 = 0$

4. $a^2 + 4a + 1 = 0$

5. $\dfrac{1}{6}x^2 - \dfrac{1}{2}x + \dfrac{1}{3} = 0$

6. $\dfrac{1}{6}x^2 + \dfrac{1}{2}x + \dfrac{1}{3} = 0$

7. $\dfrac{x^2}{2} + 1 = \dfrac{3x}{2}$

8. $\dfrac{x^2}{4} = \dfrac{x}{3} + \dfrac{1}{3}$

9. $y^2 - 5y = 0$

10. $2y^2 + 10y = 0$

11. $30x^2 + 40x = 0$

12. $50x^2 - 20x = 0$

13. $\dfrac{2t^2}{3} - t = -\dfrac{1}{6}$

14. $\dfrac{x^2}{3} - \dfrac{x}{2} = \dfrac{1}{2}$

15. $0.01x^2 + 0.06x - 0.08 = 0$

16. $0.03x^2 - 0.03x - 0.03 = 0$

17. $-3x^2 = 4x - 2$

18. $4x^2 - 4x = 44$

SCAN TO ACCESS

19. $100x^2 - 200x + 100 = 0$

20. $100x^2 - 600x + 900 = 0$

21. $\frac{1}{3}r^2 = \frac{r}{6} + \frac{3}{2}$

22. $\frac{1}{4}r^2 = \frac{2}{5}r + \frac{1}{10}$

23. $(x - 3)(x - 5) = 1$

24. $(x - 3)(x + 1) = 6$

25. $(x + 3)^2 + (x - 8)(x - 1) = 16$

26. $(x - 4)^2 + (x - 2)(x + 4) = 12$

27. $\frac{x^2}{3} - \frac{5x}{6} = \frac{1}{2}$

28. $\frac{x^2}{6} - \frac{5}{6} = -\frac{x}{3}$

Multiply both sides of each equation by its LCD. Then solve the resulting equation.

29. $\frac{1}{x + 1} + \frac{1}{x} = \frac{1}{2}$

30. $\frac{1}{x + 1} + \frac{1}{x} = \frac{1}{3}$

31. $\frac{1}{y - 1} + \frac{1}{y + 1} = 1$

32. $\frac{2}{y + 2} + \frac{3}{y - 2} = 1$

33. $\frac{1}{x + 2} + \frac{1}{x + 3} = 1$

34. $\frac{1}{x + 3} + \frac{1}{x + 4} = 1$

35. $\frac{6}{r^2 - 1} - \frac{1}{2} = \frac{1}{r + 1}$

36. $2 + \frac{5}{r - 1} = \frac{12}{(r - 1)^2}$

37. $\frac{1}{x^2 - 4} + \frac{x}{x - 2} = 2$

38. $\frac{1}{x^2} - \frac{1}{x} = 2$

39. $2 - \frac{3}{x} = \frac{1}{x^2}$

40. $4 - \frac{1}{x} = \frac{2}{x^2}$

Each of the following equations has three solutions. Factor the greatest common factor and then use the quadratic formula to find all solutions.

41. $2x^3 + 2x^2 - 3x = 0$

42. $6x^3 - 4x^2 - 6x = 0$

43. $3y^4 = 6y^3 + 6y^2$

44. $4y^4 = 16y^3 + 24y^2$

45. $8t^5 + 4t^4 = 2t^3$

46. $8t^5 + 2t^4 = 12t^3$

47. Which two of the expressions below are equivalent?

 a. $\dfrac{6 + 2\sqrt{3}}{4}$

 b. $\dfrac{3 + \sqrt{3}}{2}$

 c. $6 + \dfrac{\sqrt{3}}{2}$

48. Which two of the expressions below are equivalent?

 a. $\dfrac{8 - 4\sqrt{2}}{4}$

 b. $2 - 4\sqrt{3}$

 c. $2 - \sqrt{2}$

49. Solve $3x^2 - 5x = 0$

 a. by factoring.

 b. using the quadratic formula.

50. Solve $3x^2 + 23x - 70 = 0$

 a. by factoring.

 b. using the quadratic formula.

51. Can the equation $x^2 - 4x - 7 = 0$ be solved by factoring? Solve it.

52. Can the equation $x^2 = 5$ be solved by factoring? Solve it.

53. Is $x = -1 + \sqrt{3}$ a solution to $x^2 + 2x = 2$?

54. Is $x = 2 - \sqrt{6}$ a solution to $(x - 2)^2 = 6$?

Applying the Concepts

55. Falling Object An object is thrown downward with an initial velocity of 5 feet per second. The relationship between the distance s it travels and time t is given by $s = 5t + 16t^2$. How long does it take the object to fall 74 feet?

56. Coin Toss A coin is tossed upward with an initial velocity of 32 feet per second from a height of 16 feet above the ground. The equation giving the object's height h at any time t is $h = 16 + 32t - 16t^2$. Does the object ever reach a height of 32 feet?

57. Profit The total cost (in dollars) for a company to manufacture and sell x items per week is $C = 60x + 300$, whereas the revenue brought in by selling all x items is $R = 100x - 0.5x^2$. How many items must be sold to obtain a weekly profit of $300?

58. Profit Suppose a company manufactures and sells x picture frames each month with a total cost of $C = 1{,}200 + 3.5x$ dollars. If the revenue obtained by selling x frames is $R = 9x - 0.002x^2$, find the number of frames it must sell each month if its monthly profit is to be $2,300.

59. Profit In the introduction to this section, the revenue and cost equations for selling x DVDs was given by

$$R(x) = 11.5x - 0.05x^2 \text{ and } C(x) = 200 + 2x$$

a. Find the profit equation.
b. Find the profit from selling 100 DVDs and 200 DVDs.
c. Use the quadratic formula to find when profit is 0.

60. Profit The revenue and cost equations for selling x DVDs is given by

$$R(x) = 150x - 0.01x^2 \text{ and } C(x) = 1{,}000 + 50x$$

a. Find an expression for the profit equation.
b. Find the profit from selling 100 DVDs and 200 DVDs.
c. Use the quadratic formula to find when profit is 0.

61. Photograph Cropping Suppose a photographic image is printed on a piece of paper that is 10.5 centimeters by 8.2 centimeters. The overall area of the paper photo is to be reduced to 80% of its original area by cutting off (cropping) equal strips on all four sides. What is the width of the strip that is cut from each side?

62. Area of a Garden A garden measures 20.3 meters by 16.4 meters. To double the area of the garden, strips of equal width are added to all four sides.

a. Draw a diagram that illustrates these conditions.

b. What are the new overall dimensions of the garden?

63. Area and Perimeter A rectangle has a perimeter of 20 yards and an area of 15 square yards.

 a. Write two equations that state these facts in terms of the rectangle's length, l, and its width, w.

 b. Solve the two equations from part a to determine the actual length and width of the rectangle.

 c. Explain why two answers are possible to part b.

64. Population Size Writing in 1829, former President James Madison made some predictions about the growth of the population of the United States. The populations he predicted fit the equation

$$P(x) = 0.029x^2 - 1.39x + 42$$

where $P(x)$ is the population in millions of people x years from 1829.

 a. Use the equation to determine the approximate year President Madison would have predicted that the U.S. population would reach 100,000,000.

 b. If the U.S. population in 2010 was approximately 308,000,000, were President Madison's predictions accurate in the long term? Explain why or why not.

Getting Ready for the Next Section

65. Evaluate $y = 3x^2 - 6x + 1$ for $x = 1$.

66. Evaluate $y = -2x^2 + 6x - 5$ for $x = \frac{3}{2}$.

67. Let $P = -0.1x^2 + 27x - 500$ and find P when $x = 135$.

68. Let $P = -0.1x^2 + 12x - 400$ and find P when $x = 600$.

Solve.

69. $0 = a(80)^2 + 70$

70. $0 = a(80)^2 + 90$

71. $x^2 - 6x + 5 = 0$

72. $x^2 - 3x - 4 = 0$

73. $-x^2 - 2x + 3 = 0$

74. $-x^2 + 4x + 12 = 0$

75. $x^2 - 6x - 5 = 0$

76. $x^2 - 4x - 6 = 0$

Fill in the blanks to complete the square.

77. $x^2 - 6x + \boxed{} = (x - \boxed{})^2$

78. $x^2 - 10x + \boxed{} = (x - \boxed{})^2$

79. $y^2 + 2y + \boxed{} = (y + \boxed{})^2$

80. $y^2 - 12y + \boxed{} = (y - \boxed{})^2$

Find the Mistake

Each sentence below contains a mistake. Circle the mistake and write the correct number(s) or expression(s) on the line provided.

1. The solutions for any quadratic equation in the form of $ax^2 + bx + c = 0$ are $x = \dfrac{-b \pm \sqrt{a^2 - 4bc}}{2b}$.

2. Using the quadratic formula to solve the equation $2x^2 - 7x = -11$, $a = 2$, $b = 7$, and $c = -11$. _____

3. Solving $x^2 = -8x + 12$ using the quadratic formula gives us $x = -6$ and $x = -2$. _____

4. To solve $3x^2 - 6x - 4 = 0$, apply the quadratic formula to get $x = \dfrac{-6 \pm \sqrt{6^2 - 4(3)(-4)}}{2(3)}$ _____

Landmark Review: Checking Your Progress

Solve the following equations.

1. $4y^2 = 80$

2. $4x^2 = 108$

3. $(x - 3)^2 = 81$

4. $(2a - 4)^2 = 25$

5. $x^2 + 8x = 9$

6. $y^2 + 10y = -24$

7. $2x^2 - 4x - 4 = 0$

8. $3y^2 - 30y + 42 = 0$

9. $x^2 - 5x - 5 = 0$

10. $4x^2 + 12x + 3 = 0$

11. $3x^2 - 3x - 18 = 0$

12. $2x^2 - 4x = 4$

9.4 Graphing Quadratic Equations

OBJECTIVES

A Graph a quadratic equation.

B Solve application problems using information from a graph.

C Find an equation from its graph.

KEY WORDS

parabola

vertex

concave

Every year on the first weekend after Halloween, Delaware hosts the World Championship Punkin' Chunkin. During the competition, people use slingshots, catapults, and other contraptions to hurl pumpkins as far as possible. The current world record for farthest pumpkin chucked is 1351 meters! Suppose the path of a chucked pumpkin is given by the equation $y = 1600 - (x - 400)^2$. We are now in a position to put our knowledge of quadratic equations to work to graph a variety of equations, such as this one.

Let's begin with finding the solution set for the following equation:

$$y = x^2 - 3$$

The solution set will consist of ordered pairs, allowing us to easily graph the equation. One method of graphing the solution set is to find a number of ordered pairs that satisfy the equation and to graph them. We can obtain some ordered pairs that are solutions to $y = x^2 - 3$ by use of a table as follows:

x	$y = x^2 - 3$	y	Solutions
-3	$y = (-3)^2 - 3 = 9 - 3 = 6$	6	$(-3, 6)$
-2	$y = (-2)^2 - 3 = 4 - 3 = 1$	1	$(-2, 1)$
-1	$y = (-1)^2 - 3 = 1 - 3 = -2$	-2	$(-1, -2)$
0	$y = 0^2 - 3 = 0 - 3 = -3$	-3	$(0, -3)$
1	$y = 1^2 - 3 = 1 - 3 = -2$	-2	$(1, -2)$
2	$y = 2^2 - 3 = 4 - 3 = 1$	1	$(2, 1)$
3	$y = 3^2 - 3 = 9 - 3 = 6$	6	$(3, 6)$

Graphing these solutions and then connecting them with a smooth curve, we have the graph of $y = x^2 - 3$. (See Figure 1.)

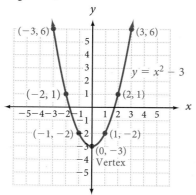

FIGURE 1

This graph is an example of a *parabola*. All equations of the form $y = ax^2 + bx + c$, $a \neq 0$, have vertical parabolas for graphs.

Although it is always possible to graph parabolas by making a table of values of x and y that satisfy the equation, there are other methods that are faster and, in some cases, more accurate.

The important points associated with the graph of a parabola are the highest (or lowest) point on the graph and the x-intercepts. The y-intercept can also be useful.

A Intercepts for Vertical Parabolas

The graph of the function $y = ax^2 + bx + c$ crosses the y-axis at $y = c$, because when $x = 0$, $y = a(0)^2 + b(0) + c = c$.

Because the graph crosses the x-axis when $y = 0$, the x-intercepts are those values of x that are solutions to the quadratic equation $0 = ax^2 + bx + c$.

The Vertex of a Parabola

The highest or lowest point on a parabola is called the *vertex*. The vertex for the graph of $y = ax^2 + bx + c$ will always occur when

$$x = \frac{-b}{2a}$$

To see this, we must transform the right side of $y = ax^2 + bx + c$ into an expression that contains x in just one of its terms. This is accomplished by completing the square on the first two terms. Here is what it looks like:

$$y = ax^2 + bx + c$$
$$= a\left(x^2 + \frac{b}{a}x\right) + c$$
$$= a\left[x^2 + \frac{b}{a}x + \left(\frac{b}{2a}\right)^2\right] + c - a\left(\frac{b}{2a}\right)^2$$
$$= a\left(x + \frac{b}{2a}\right)^2 + \frac{4ac - b^2}{4a}$$

It may not look like it, but this last line indicates that the vertex of the graph of $y = ax^2 + bx + c$ has an x-coordinate of $\frac{-b}{2a}$. Because a, b, and c are constants, the only quantity that is varying in the last expression is the x in $\left(x + \frac{b}{2a}\right)^2$. Because the quantity $\left(x + \frac{b}{2a}\right)^2$ is the square of $x + \frac{b}{2a}$, the smallest it will ever be is 0, and that will happen when $x = \frac{-b}{2a}$.

We can use the vertex point along with the x- and y-intercepts to sketch the graph of any equation of the form $y = ax^2 + bx + c$. Here is a summary of the preceding information.

Graphing Vertical Parabolas I

The graph of $y = ax^2 + bx + c$, $a \neq 0$, will be a parabola with
1. a y-intercept at $y = c$
2. x-intercepts (if they exist) at

$$x = \frac{-b \pm \sqrt{b^2 - 4ac}}{2a}$$

3. a vertex when $x = \frac{-b}{2a}$

EXAMPLE 1 Sketch the graph of $y = x^2 - 6x + 5$.

Solution To find the x-intercepts, we let $y = 0$ and solve for x.

$$0 = x^2 - 6x + 5$$

$$0 = (x - 5)(x - 1)$$

$$x = 5 \quad \text{or} \quad x = 1$$

To find the coordinates of the vertex, we first find

$$x = \frac{-b}{2a} = \frac{-(-6)}{2(1)} = 3$$

The x-coordinate of the vertex is 3. To find the y-coordinate, we substitute 3 for x in our original equation.

$$y = 3^2 - 6(3) + 5 = 9 - 18 + 5 = -4$$

The graph crosses the x-axis at 1 and 5 and has its vertex at $(3, -4)$. Plotting these points and connecting them with a smooth curve, we have the graph shown in Figure 2.

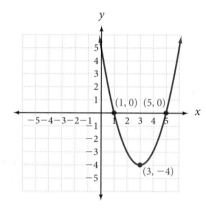

FIGURE 2

The graph is a parabola that opens up, so we say the graph is *concave up*. The vertex is the lowest point on the graph. (Note that the graph crosses the y-axis at 5, which is the value of y we obtain when we let $x = 0$.)

Finding the Vertex by Completing the Square

Another way to locate the vertex of the parabola in Example 1 is by completing the square on the first two terms on the right side of the equation $y = x^2 - 6x + 5$. In this case, we would do so by adding 9 to and subtracting 9 from the right side of the equation. This amounts to adding 0 to the equation, so we know we haven't changed its solutions. This is what it looks like:

$$y = (x^2 - 6x \quad) + 5$$

$$= (x^2 - 6x + 9) + 5 - 9$$

$$= (x - 3)^2 - 4$$

You may have to look at this last equation awhile to see this, but when $x = 3$, then $y = (3 - 3)^2 - 4 = 0^2 - 4 = -4$ is the smallest y will ever be. That is why the vertex is at $(3, -4)$. As a matter of fact, this is the same kind of reasoning we used when we derived the formula $x = \frac{-b}{2a}$ for the x-coordinate of the vertex.

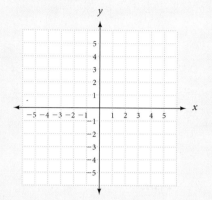

2. Graph $y = -x^2 + 2x + 8$.

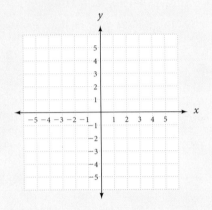

EXAMPLE 2 Graph $y = -x^2 - 2x + 3$.

Solution To find the x-intercepts, we let $y = 0$.

$$0 = -x^2 - 2x + 3$$

$$0 = x^2 + 2x - 3 \qquad \text{\textit{Multiply each side by} } -1.$$

$$0 = (x + 3)(x - 1)$$

$$x = -3 \quad \text{or} \quad x = 1$$

The x-coordinate of the vertex is given by

$$x = \frac{-b}{2a} = \frac{-(-2)}{2(-1)} = \frac{2}{-2} = -1$$

To find the y-coordinate of the vertex, we substitute -1 for x in our original equation to get

$$y = -(-1)^2 - 2(-1) + 3 = -1 + 2 + 3 = 4$$

Our parabola has x-intercepts at -3 and 1, and a vertex at $(-1, 4)$. Figure 3 shows the graph.

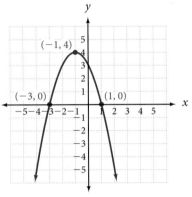

FIGURE 3

We say the graph is *concave down* because it opens downward. Again, we could have obtained the coordinates of the vertex by completing the square on the first two terms on the right side of our equation. To do so, we must first factor -1 from the first two terms. (Remember, the leading coefficient must be 1 to complete the square.) When we complete the square, we add 1 inside the parentheses, which actually decreases the right side of the equation by -1 because everything in the parentheses is multiplied by -1. To make up for it, we add 1 outside the parentheses.

$$y = -1(x^2 + 2x \qquad) + 3$$

$$= -1(x^2 + 2x + 1) + 3 + 1$$

$$= -1(x + 1)^2 + 4$$

The last line tells us that the *largest* value of y will be 4, and that will occur when $x = -1$.

Recall from Chapter 6 that rational numbers are any numbers that can be put in the form $\frac{a}{b}$ where b is nonzero. Irrational numbers are numbers that cannot be put in this form. They are nonrepeating, nonterminating decimals. Some examples of irrational numbers are π, $\sqrt{2}$, and $\sqrt{5}$. This information will be useful in the next examples.

EXAMPLE 3 Graph $y = 3x^2 - 6x + 1$.

Solution To find the x-intercepts, we let $y = 0$ and solve for x.

$$0 = 3x^2 - 6x + 1$$

It is impossible to factor the right side of this equation, and the discriminant is not a perfect square. Therefore, when solving for x, we find that the x-intercepts are irrational numbers and will have to be approximated with decimals using the quadratic formula. Rather than use the quadratic formula, we will find some other points on the graph, but first let's find the vertex.

Here are both methods of finding the vertex:

Using the formula that gives us the x-coordinate of the vertex, we have $$x = \frac{-b}{2a} = \frac{-(-6)}{2(3)} = 1$$ Substituting 1 for x in the equation gives us the y-coordinate of the vertex. $$y = 3 \cdot 1^2 - 6 \cdot 1 + 1 = -2$$	To complete the square on the right side of the equation, we factor 3 from the first two terms, add 1 inside the parentheses, and add -3 outside the parentheses (this amounts to adding 0 to the right side). $$y = 3(x^2 - 2x \quad) + 1$$ $$= 3(x^2 - 2x + 1) + 1 - 3$$ $$= 3(x - 1)^2 - 2$$

In either case, the vertex is $(1, -2)$.

If we can find two points, one on each side of the vertex, we can sketch the graph. Let's let $x = 0$ and $x = 2$, because each of these numbers is the same distance from $x = 1$, and $x = 0$ will give us the y-intercept.

When $x = 0$	When $x = 2$
$y = 3(0)^2 - 6(0) + 1$	$y = 3(2)^2 - 6(2) + 1$
$= 0 - 0 + 1$	$= 12 - 12 + 1$
$= 1$	$= 1$

The two points just found are $(0, 1)$ and $(2, 1)$. Plotting these two points along with the vertex $(1, -2)$, we have the graph shown in Figure 4.

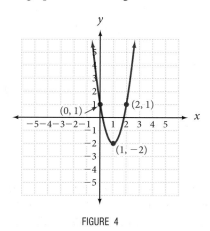

FIGURE 4

EXAMPLE 4 Graph $y = -2x^2 + 6x - 5$.

Solution Letting $y = 0$, we have

$$0 = -2x^2 + 6x - 5$$

Solving for x we find that there are no real numbers that solve the equation. This means that our original equation does not have x-intercepts. The graph does not cross the x-axis.

3. Graph $y = 2x^2 - 4x + 1$.

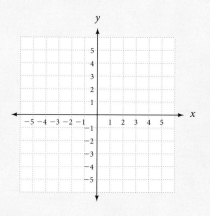

4. Graph $y = -x^2 + 4x - 5$.

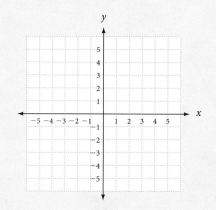

Let's find the vertex.

Using our formula for the x-coordinate of the vertex, we have

$$x = \frac{-b}{2a} = \frac{-6}{2(-2)} = \frac{6}{4} = \frac{3}{2}$$

To find the y-coordinate, we let $x = \frac{3}{2}$.

$$y = -2\left(\frac{3}{2}\right)^2 + 6\left(\frac{3}{2}\right) - 5$$

$$= \frac{-18}{4} + \frac{18}{2} - 5$$

$$= \frac{-18 + 36 - 20}{4}$$

$$= -\frac{1}{2}$$

Finding the vertex by completing the square is a more complicated matter. To make the coefficient of x^2 a 1, we must factor -2 from the first two terms. To complete the square inside the parentheses, we add $\frac{9}{4}$. Since each term inside the parentheses is multiplied by -2, we add $\frac{9}{2}$ outside the parentheses so that the net result is the same as adding 0 to the right side.

$$y = -2(x^2 - 3x \quad\quad) - 5$$

$$= -2\left(x^2 - 3x + \frac{9}{4}\right) - 5 + \frac{9}{2}$$

$$= -2\left(x - \frac{3}{2}\right)^2 - \frac{1}{2}$$

The vertex is $\left(\frac{3}{2}, -\frac{1}{2}\right)$. Because this is the only point we have so far, we must find two others. Let's let $x = 3$ and $x = 0$, because each point is the same distance from $x = \frac{3}{2}$ and on either side.

When $x = 3$

$$y = -2(3)^2 + 6(3) - 5$$

$$= -18 + 18 - 5$$

$$= -5$$

When $x = 0$

$$y = -2(0)^2 + 6(0) - 5$$

$$= 0 + 0 - 5$$

$$= -5$$

The two additional points on the graph are $(3, -5)$ and $(0, -5)$. Figure 5 shows the graph.

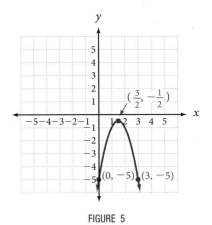

FIGURE 5

The graph is concave down. The vertex is the highest point on the graph.

By looking at the equations and graphs in Examples 1 through 4, we can conclude that the graph of $y = ax^2 + bx + c$ will be concave up when a is positive, and concave down when a is negative. Taking this even further, if $a > 0$, then the vertex is the *lowest* point on the graph, and if $a < 0$, the vertex is the *highest* point on the graph. Finally, if we complete the square on x in the equation $y = ax^2 + bx + c$, $a \neq 0$, we can rewrite the equation of our parabola as $y = a(x - h)^2 + k$. When the equation is in this form, the vertex is at the point (h, k). Here is a summary:

Graphing Vertical Parabolas II

The graph of

$$y = a(x - h)^2 + k, a \neq 0$$

will be a parabola with a vertex at $(h, k,)$. The vertex will be the highest point on the graph when $a < 0$, and the lowest point on the graph when $a > 0$.

Note Pay special attention to the sign of h. As is, the sign is positive. If the equation takes the form $y = a(x + h)^2 + k$, the sign of h is negative, because $y = a(x - (-h))^2 + k$.

B Applications

EXAMPLE 5 A company selling copies of an accounting program for home computers finds that it will make a weekly profit of P dollars from selling x copies of the program, according to the equation

$$P = -0.1x^2 + 27x - 500$$

How many copies of the program should it sell to make the largest possible profit, and what is the largest possible profit?

Solution Because the coefficient of x^2 is negative, we know the graph of this parabola will be concave down, meaning that the vertex is the highest point of the curve. We find the vertex by first finding its x-coordinate.

$$x = \frac{-b}{2a} = \frac{-27}{2(-0.1)} = \frac{27}{0.2} = 135$$

This represents the number of programs the company needs to sell each week to make a maximum profit. To find the maximum profit, we substitute 135 for x in the original equation. (A calculator is helpful for these kinds of calculations.)

$$P = -0.1(135)^2 + 27(135) - 500$$
$$= -0.1(18,225) + 3,645 - 500$$
$$= -1,822.5 + 3,645 - 500$$
$$= 1,322.5$$

The maximum weekly profit is $1,322.50 and is obtained by selling 135 programs a week.

5. Find the largest value of p if $p = -0.01x^2 + 12x - 400$.

EXAMPLE 6 The owners of an art supply store find that they can sell x sketch pads each week at p dollars each, according to the equation $x = 900 - 300p$. Graph the revenue equation $R = xp$. Then use the graph to find the price p that will bring in the maximum revenue. Finally, find the maximum revenue.

Solution As it stands, the revenue equation contains three variables. Because we are asked to find the value of p that gives us the maximum value of R, we rewrite the equation using just the variables R and p. Because $x = 900 - 300p$, we have

$$R = xp = (900 - 300p)p = -300p^2 + 900p$$

The graph of this equation is shown in Figure 6. The graph appears in the first quadrant only, because R and p are both positive quantities.

6. Repeat Example 6 if the number of sketch pads they can sell each week is related to the price by the equation $x = 800 - 200p$.

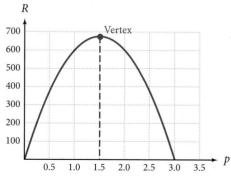

FIGURE 6

From the graph, we see that the maximum value of R occurs when $p = \$1.50$. We can calculate the maximum value of R from the equation.

$$\text{When} \rightarrow \qquad\qquad p = 1.5$$
$$\text{the equation} \rightarrow \qquad R = (900 - 300p)p$$
$$\text{becomes} \rightarrow \qquad R = (900 - 300 \cdot 1.5)1.5$$
$$= (900 - 450)1.5$$
$$= 450 \cdot 1.5$$
$$= 675$$

The maximum revenue is $\$675$. It is obtained by setting the price of each sketch pad at $p = \$1.50$.

C Finding the Equation from the Graph

7. How would the equation in Example 7 look if the maximum height were 80 feet? (Assume the maximum distance is still 160 feet).

EXAMPLE 7 At the 1997 Washington County Fair in Oregon, David Smith, Jr., The Bullet, was shot from a cannon. As a human cannonball, he reached a height of 70 feet before landing in a net 160 feet from the cannon. Sketch the graph of his path, and then find the equation of the graph.

Solution We assume that the path taken by the human cannonball is a parabola. If the origin of the coordinate system is at the opening of the cannon, then the net that catches him will be at 160 on the x-axis. Figure 7 shows the graph.

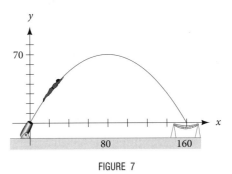

FIGURE 7

Because the curve is a parabola, we know the equation will have the form

$$y = a(x - h)^2 + k$$

Because the vertex of the parabola is at $(80, 70)$, we can fill in two of the three constants in our equation, giving us

$$y = a(x - 80)^2 + 70$$

To find a, we note that the landing point will be $(160, 0)$. Substituting the coordinates of this point into the equation, we solve for a.

$$y = a(160 - 80)^2 + 70$$
$$0 = a(80)^2 + 70$$
$$0 = 6{,}400a + 70$$
$$a = -\frac{70}{6{,}400} = -\frac{7}{640}$$

The equation that describes the path of the human cannonball is

$$y = -\frac{7}{640}(x - 80)^2 + 70 \quad \text{for} \quad 0 \le x \le 160$$

Answer

3. $y = -\frac{1}{80}(x - 80)^2 + 80$

Vocabulary Review

Choose the correct words to fill in the blanks below.

 y-intercept concave graph ordered pairs vertex parabola

1. One method of graphing the solution set to a quadratic equation is to find a number of _____ that satisfy the equation and graph them.

2. An equation of the form $y = ax^2 + bx + c$ will have a _____ for a graph.

3. The graph of the function $y = ax^2 + bx + c$ has a _____ at $y = c$.

4. The highest or lowest point on a vertical parabola is called the _____.

5. The graph of a vertical parabola that opens up is said to be _____ up, whereas one that opens down is said to be concave down.

6. The _____ of $y = a(x - h)^2 + k$ will be a vertical parabola with a vertex at (h, k).

Problems

A For each of the following equations, give the x-intercepts and the coordinates of the vertex, and sketch the graph.

1. $y = x^2 + 2x - 3$ **2.** $y = x^2 - 2x - 3$ **3.** $y = -x^2 - 4x + 5$

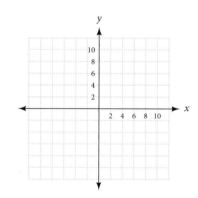

4. $y = x^2 + 4x - 5$ **5.** $y = x^2 - 1$

6. $y = x^2 - 4$

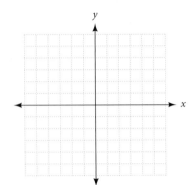

7. $y = -x^2 + 9$

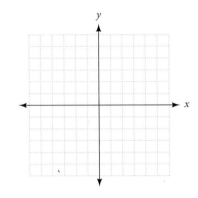

8. $y = -x^2 + 1$

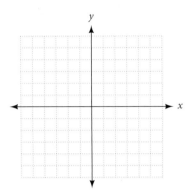

9. $y = 2x^2 - 4x - 6$

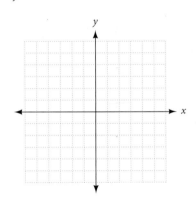

10. $y = 2x^2 + 4x - 6$

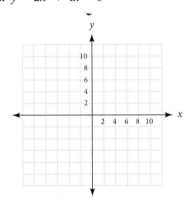

11. $y = x^2 - 2x - 4$

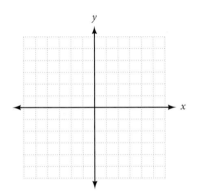

Graph each parabola. Label the vertex and any intercepts that exist.

12. $y = x^2 - 2x - 2$

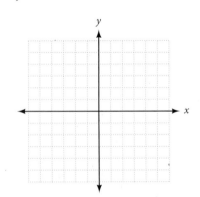

13. $y = 2(x - 1)^2 + 3$

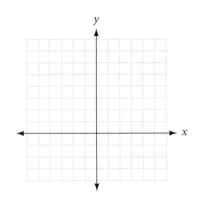

14. $y = 2(x + 1)^2 - 3$

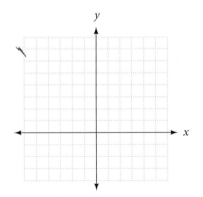

15. $y = -(x + 2)^2 + 4$

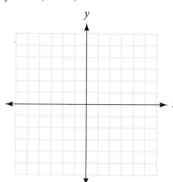

16. $y = -(x - 3)^2 + 1$

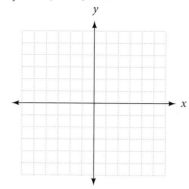

17. $y = \frac{1}{2}(x - 2)^2 - 4$

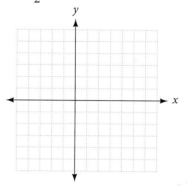

18. $y = \frac{1}{3}(x - 3)^2 - 3$

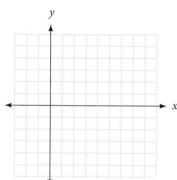

19. $y = -2(x - 4)^2 - 1$

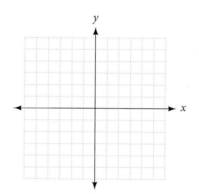

20. $y = -4(x - 1)^2 + 4$

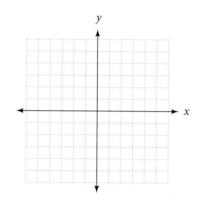

Find the vertex and any two convenient points to sketch the graphs of the following equations.

21. $y = x^2 - 4x - 4$

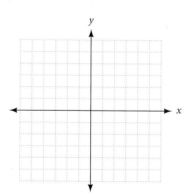

22. $y = x^2 - 2x + 3$

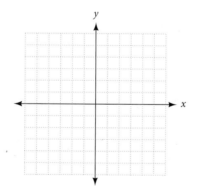

23. $y = -x^2 + 2x - 5$

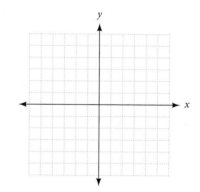

24. $y = -x^2 + 4x - 2$

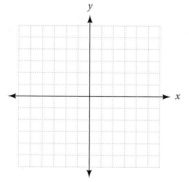

25. $y = x^2 + 1$

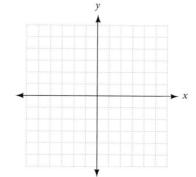

26. $y = x^2 + 4$

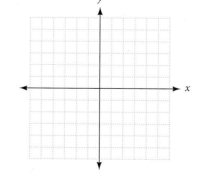

27. $y = -x^2 - 3$

28. $y = -x^2 - 2$

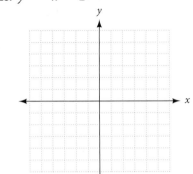

29. $y = 3x^2 + 4x + 1$

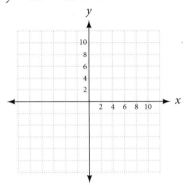

30. $y = 2x^2 + 4x + 3$

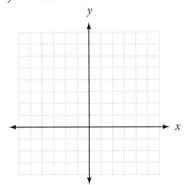

31. $y = -2x^2 + 8x - 3$

32. $y = -3x^2 + 6x$

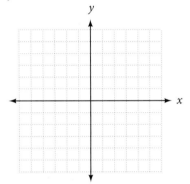

For each of the following equations, find the coordinates of the vertex, and indicate whether the vertex is the highest point on the graph or the lowest point on the graph. (Do not graph.)

33. $y = x^2 - 6x + 5$

34. $y = -x^2 + 6x - 5$

35. $y = -x^2 + 2x + 8$

36. $y = x^2 - 2x - 8$

37. $y = 12 + 4x - x^2$

38. $y = -12 - 4x + x^2$

39. $y = -x^2 - 8x$

40. $y = x^2 + 8x$

Applying the Concepts

41. Maximum Profit A company finds that it can make a profit of P dollars each month by selling x sewing patterns, according to the formula $P = -0.002x^2 + 3.5x - 800$. How many patterns must it sell each month to have a maximum profit? What is the maximum profit?

42. Maximum Profit A company selling picture frames finds that it can make a profit of P dollars each month by selling x frames, according to the formula $P = -0.002x^2 + 5.5x - 1,200$. How many frames must it sell each month to have a maximum profit? What is the maximum profit?

43. Maximum Height Kristen is tossing a softball into the air with an underhand motion. The distance of the ball above her hand at any time is given by the equation

$$h = 32t - 16t^2 \text{ for } 0 \le t \le 2$$

where h is the height of the ball (in feet) and t is the time (in seconds). Find the times at which the ball is in her hand, and the maximum height of the ball.

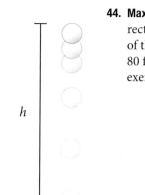

44. Maximum Area Justin wants to fence three sides of a rectangular exercise yard for his dog. The fourth side of the exercise yard will be a side of the house. He has 80 feet of fencing available. Find the dimensions of the exercise yard that will enclose the maximum area.

45. Maximum Revenue A company that manufactures hair ribbons knows that the number of ribbons x it can sell each week is related to the price p of each ribbon by the equation $x = 1{,}200 - 100p$. Graph the revenue equation $R = xp$. Then use the graph to find the price p that will bring in the maximum revenue. Finally, find the maximum revenue.

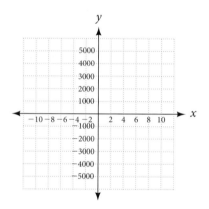

46. Maximum Revenue A company that manufactures CDs for home computers finds that it can sell x CDs each day at p dollars per CD, according to the equation $x = 800 - 100p$. Graph the revenue equation $R = xp$. Then use the graph to find the price p that will bring in the maximum revenue. Finally, find the maximum revenue.

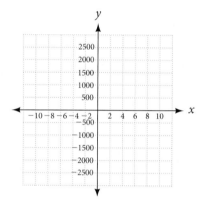

47. Maximum Revenue The relationship between the number of calculators x a company sells each day and the price p of each calculator is given by the equation $x = 1{,}700 - 100p$. Graph the revenue equation $R = xp$, and use the graph to find the price p that will bring in the maximum revenue. Then find the maximum revenue.

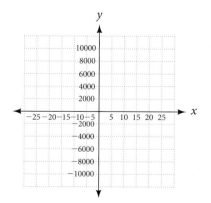

48. Maximum Revenue The relationship between the number x of pencil sharpeners a company sells each week and the price p of each sharpener is given by the equation $x = 1{,}800 - 100p$. Graph the revenue equation $R = xp$, and use the graph to find the price p that will bring in the maximum revenue. Then find the maximum revenue.

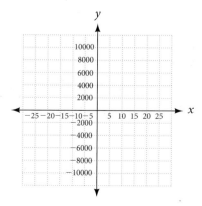

49. Human Cannonball　A human cannonball is shot from a cannon at the county fair. He reaches a height of 60 feet before landing in a net 180 feet from the cannon. Sketch the graph of his path, and then find the equation of the graph.

50. Interpreting Graphs　The graph below shows the different paths taken by the human cannonball when his velocity out of the cannon is 50 miles/hour, and his cannon is inclined at varying angles.

a. If his landing net is placed 104 feet from the cannon, at what angle should the cannon be inclined so that he lands in the net?

b. Approximately where do you think he would land if the cannon was inclined at 45°?

c. If the cannon was inclined at 45°, approximately what height do you think he would attain?

d. Do you think there is another angle for which he would travel the same distance he travels at 80°? Give an estimate of that angle.

e. The fact that every landing point can come from two different paths makes us think that the equations that give us the landing points must be what type of equations?

Find the Mistake

Each sentence below contains a mistake. Circle the mistake and write the correct word(s) or expression on the line provided.

1. All equations of the form $ax^3 = 0$ have parabolas for graphs. _____

2. The graph of the function $y = ax^2 + bx + c$ crosses the x-axis where $y = 0$. _____

3. The vertex for the graph $y = 4x^2 + 6x + 7$ occurs when $y = -\dfrac{4}{12} = -\dfrac{1}{3}$. _____

4. The graph of $y = ax^2 + bx + c$ will be concave down when a is positive. _____

9.5 Introduction to Functions

OBJECTIVES

A Construct a table or a graph from a function rule.

B Identify the domain and range of a function.

C Determine whether a relation is also a function.

D Use the vertical line test to determine whether a graph is a function.

KEY WORDS

relation

function

domain

range

function map

vertical line test

Before the harvest each fall, a Nebraskan farmer climbs onto his riding lawn mower and rides through acres of corn. He uses the mower to carve an intricate maze in his cornfield. One of his most impressive designs took the shape of a Star Wars starship. He is known to fly in a plane over his freshly cut maze to ensure its perfection. Thousands flock to Grandpa John's Amazing Maze; each visitor paying $5.00 to walk through it.

In this section, we will introduce you to functions. Using Grandpa John's Amazing Maze as an example, the amount of money the maze earns for the farmer each season depends on the number of people who visit it. In mathematics, we say that his seasonal earnings are a *function* of how many people visit the maze.

Relations

Before we begin our discussion of functions, let's examine what happens when a coordinate from a set of inputs is paired with one or more elements from a set of outputs.

Table 1 shows the prices of used Ford Mustangs that were listed in the local newspaper. The diagram in Figure 1 is called a *scatter diagram*. It gives a visual representation of the data in Table 1.

Table 1

Used Mustang Prices

Year x	Price ($) y
2009	17,995
2009	16,945
2009	25,985
2008	15,995
2008	14,867
2007	16,960
2006	14,999
2006	19,150
2005	11,352

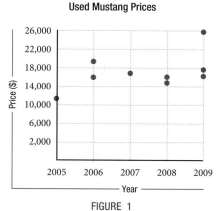

FIGURE 1

Ordered Pairs

(2009, 17,995)

(2009, 16,945)

(2009, 25,985)

(2008, 15,995)

(2008, 14,867)

(2007, 16,960)

(2006, 14,999)

(2006, 19,150)

(2005, 11,352)

Looking at the graph, you can see that the year 2009 is paired with three different prices: $17,995, $16,945, and $25,985. This data is considered a *relation*. Furthermore, all sets of paired data are considered relations where each number in the domain is paired with one or more numbers in the range. The following are two formal definitions of a relation.

A Constructing Tables and Graphs

A special case occurs when each number in a domain is paired with exactly one number in a range. This is called a *function*.

Suppose you have a job that pays $9.50 per hour and that requires you to work anywhere from 0 to 40 hours per week. If we let the variable x represent hours and the variable y represent the money you make, then the relationship between x and y can be written as

$$y = 9.5x \qquad \text{for} \qquad 0 \le x \le 40$$

EXAMPLE 1 Construct a table and graph for the function

$$y = 9.5x \qquad \text{for} \qquad 0 \le x \le 40$$

Solution Table 2 gives some of the paired data that satisfy the equation $y = 9.5x$. Figure 2 is the graph of the equation with the restriction $0 \le x \le 40$.

Table 2	Weekly Wages	
Hours Worked	Rule	Pay
x	$y = 9.5x$	y
0	$y = 9.5(0)$	0
10	$y = 9.5(10)$	95
20	$y = 9.5(20)$	190
30	$y = 9.5(30)$	285
40	$y = 9.5(40)$	380

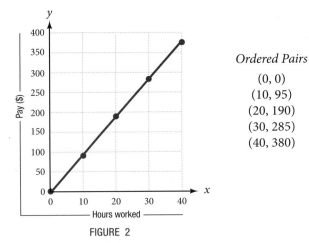

Ordered Pairs

(0, 0)
(10, 95)
(20, 190)
(30, 285)
(40, 380)

FIGURE 2

The equation $y = 9.5x$ with the restriction $0 \le x \le 40$, Table 2, and Figure 2 are three ways to describe the same relationship between the number of hours you work in one week and your gross pay for that week. In all three, we *input* values of x, and then use the function rule to *output* values of y.

B Domain and Range of a Function

We began this discussion by saying that the number of hours worked during the week was from 0 to 40, so these are the values that x can assume. From the line graph in Figure 2, we see that the values of y range from 0 to 380. We call the complete set of values that x can assume the *domain* of the function. The values that are assigned to y are called the *range* of the function.

Domain:
The set of all inputs → The Function Rule → Range:
The set of all outputs

EXAMPLE 2 State the domain and range for the function

$$y = 9.5x, \quad 0 \le x \le 40$$

Solution From the previous discussion, we have

$$\text{Domain} = \{x \mid 0 \le x \le 40\} = [0, 40]$$
$$\text{Range} = \{y \mid 0 \le y \le 380\} = [0, 380]$$

We are familiar with writing domain and range using set-builder notation and interval notation, shown in Example 2. We can also write the domain and range using *roster notation*. This notation lists all the elements in a set one by one. In the case of Example 1, our set consists of the coordinates

$$(0, 0), (10, 95), (20, 190), (30, 285), (40, 380)$$

Therefore, using roster notation, we give the domain as $\{0, 10, 20, 30, 40\}$, and the range as $\{0, 95, 190, 285, 380\}$. Here is a summary of the three types of notation using the values from our first example:

	Set-Builder Notation	Roster Notation	Interval Notation
Domain	$\{x \mid 0 \le x \le 40\}$	$\{0, 10, 20, 30, 40\}$	$[0, 40]$
Range	$\{y \mid 0 \le y \le 380\}$	$\{0, 95, 190, 285, 380\}$	$[0, 380]$

Function Maps

Another way to visualize the relationship between x and y is with the diagram in Figure 2, which we call a *function map*.

Domain Rule: Multiply by 9.5. Range

10 → 95
20 → 190
x → $9.5x$

FIGURE 3

Although the diagram in Figure 3 does not show all the values that x and y can assume, it does give us a visual description of how x and y are related. It shows that values of y in the range come from values of x in the domain according to a specific rule (multiply by 9.5 each time).

We are now ready for the formal definition of a *function*.

> **DEFINITION** function
>
> A *function* is a rule that pairs each element in one set, called the **domain**, with exactly one element from a second set, called the **range**. In other words, a function is a rule for which each input is paired with exactly one output.

2. State the domain and range for $y = 8x, 0 \le x \le 40$.

3. Determine the domain and range for the following relations.

a. $\{(3, 3), (1, -1), (0, -3), (a, b)\}$

b.

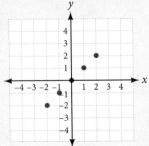

c.

Input		Output
5		0
7		1
3		4
-2		2

EXAMPLE 3 Determine the domain and range of the following relations.

a. $\{(5, 7), (0, -3), (2, 1), (x, y)\}$

b.

c.

Input	Rule: Population in millions	Output
Ohio		25
Florida		19
Michigan		11
Texas		10

Solution

a. For the set of ordered pairs, the first coordinate is the domain and the second coordinate is the range.

$$\text{Domain} = \{5, 0, 2, x\}$$
$$\text{Range} = \{7, -3, 1, y\}$$

b. The points on the graph represent ordered pairs. The set of x-coordinates are the domain and the set of y-coordinates are the range.

The relation is $\{(-3, 2), (-2, -1), (-1, 0), (1, 1), (2, -1)\}$.

$$\text{Domain} = \{-3, -2, -1, 1, 2\}$$
$$\text{Range} = \{2, -1, 0, 1\}$$

c. The domain is the set of states, and the range is the set of numbers that correspond to each state.

$$\text{Domain} = \{\text{Ohio, Florida, Michigan, Texas}\}$$
$$\text{Range} = \{10, 11, 19, 25\}$$

EXAMPLE 4 Kendra tosses a softball into the air with an underhand motion. The distance of the ball above her hand is given by the function

$$h = 32t - 16t^2 \qquad \text{for} \qquad 0 \leq t \leq 2$$

where h is the height of the ball in feet and t is the time in seconds. Construct a table that gives the height of the ball at quarter-second intervals, starting with $t = 0$ and ending with $t = 2$, then graph the function.

Solution We construct Table 3 using the following values of t: $0, \frac{1}{4}, \frac{1}{2}, \frac{3}{4}, 1, \frac{5}{4}, \frac{3}{2}, \frac{7}{4}, 2$. Then we construct the graph in Figure 4 from the table. The graph appears only in the first quadrant because neither t nor h can be negative.

Table 3 Tossing a Softball into the Air

Input		Output
Time (sec) t	Function Rule $h = 32t - 16t^2$	Distance (ft) h
0	$h = 32(0) - 16(0)^2 = 0 - 0 = 0$	0
$\frac{1}{4}$	$h = 32\left(\frac{1}{4}\right) - 16\left(\frac{1}{4}\right)^2 = 8 - 1 = 7$	7
$\frac{1}{2}$	$h = 32\left(\frac{1}{2}\right) - 16\left(\frac{1}{2}\right)^2 = 16 - 4 = 12$	12
$\frac{3}{4}$	$h = 32\left(\frac{3}{4}\right) - 16\left(\frac{3}{4}\right)^2 = 24 - 9 = 15$	15
1	$h = 32(1) - 16(1)^2 = 32 - 16 = 16$	16
$\frac{5}{4}$	$h = 32\left(\frac{5}{4}\right) - 16\left(\frac{5}{4}\right)^2 = 40 - 25 = 15$	15
$\frac{3}{2}$	$h = 32\left(\frac{3}{2}\right) - 16\left(\frac{3}{2}\right)^2 = 48 - 36 = 12$	12
$\frac{7}{4}$	$h = 32\left(\frac{7}{4}\right) - 16\left(\frac{7}{4}\right)^2 = 56 - 49 = 7$	7
2	$h = 32(2) - 16(2)^2 = 64 - 64 = 0$	0

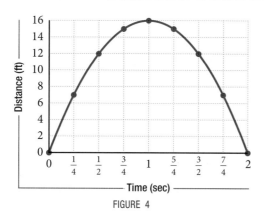

FIGURE 4

4. Kendra is tossing a softball into the air so that the distance h the ball is above her hand t seconds after she begins the toss is given by $h = 48t - 16t^2$, $0 \leq t \leq 3$. Construct a table and line graph for this function.

t	h
0	
$\frac{1}{2}$	
1	
$\frac{3}{2}$	
2	
$\frac{5}{2}$	
3	

Here is a summary of what we know about functions as it applies to this example: We input values of t and output values of h according to the function rule

$$h = 32t - 16t^2 \qquad \text{for} \qquad 0 \leq t \leq 2$$

The domain is given by the inequality that follows the equation; it is

$$\text{Domain} = \{t \mid 0 \leq t \leq 2\} = [0, 2]$$

The range is the set of all outputs that are possible by substituting the values of t from the domain into the equation. From our table and graph, it seems that the range is

$$\text{Range} = \{h \mid 0 \leq h \leq 16\} = [0, 16]$$

As you can see from the examples we have done to this point, the function rule produces ordered pairs of numbers. We use this result to write an alternative definition for a function.

ALTERNATE DEFINITION function

A *function* is a set of ordered pairs in which no two different ordered pairs have the same first coordinate. The set of all first coordinates is called the *domain* of the function. The set of all second coordinates is called the *range* of the function. The restriction on first coordinates keeps us from assigning a number in the domain to more than one number in the range.

C Determining Relations and Functions

Here are some facts that will help clarify the distinction between relations and functions:

1. Any rule that assigns numbers from one set to numbers in another set is a relation. If that rule makes the assignment so no input has more than one output, then it is also a function.
2. Any set of ordered pairs is a relation. If none of the first coordinates of those ordered pairs is repeated, the set of ordered pairs is also a function.
3. Every function is a relation.
4. Not every relation is a function.

5. Use the equation $x = y^2 - 4$ to fill in the following table. Then use the table to sketch the graph.

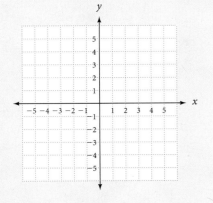

EXAMPLE 5 Sketch the graph of $x = y^2$.

Solution Without going into much detail, we graph the equation $x = y^2$ by finding a number of ordered pairs that satisfy the equation, plotting these points, then drawing a smooth curve that connects them. A table of values for x and y that satisfy the equation follows, along with the graph of $x = y^2$ shown in Figure 5.

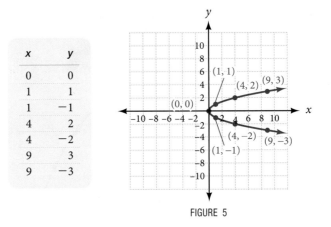

FIGURE 5

As you can see from looking at the table and the graph in Figure 5, several ordered pairs whose graphs lie on the curve have repeated first coordinates, for instance (1, 1) and (1, −1), (4, 2) and (4, −2), as well as (9, 3) and (9, −3). Therefore, the graph is not the graph of a function.

D Vertical Line Test

Look back at the scatter diagram for used Mustang prices shown in Figure 1. Notice that some of the points on the diagram lie above and below each other along vertical lines. This is an indication that the data do not constitute a function. Two data points that lie on the same vertical line must have come from two ordered pairs with the same first coordinates.

Now, look at the graph shown in Figure 1. The reason this graph is the graph of a relation, but not of a function, is that some points on the graph have the same first coordinates, for example, the points (4, 2) and (4, −2). Furthermore, any time two points on a graph have the same first coordinates, those points must lie on a vertical line. [To convince yourself, connect the points (4, 2) and (4, −2) with a straight line. You will see that it must be a vertical line.] This allows us to write the following test that uses the graph to determine whether a relation is also a function.

RULE Vertical Line Test

If a vertical line crosses the graph of a relation in more than one place, the relation cannot be a function. If no vertical line can be found that crosses a graph in more than one place, then the graph represents a function.

If we look back to the graph of $h = 32t - 16t^2$ as shown in Figure 4, we see that no vertical line can be found that crosses this graph in more than one place. The graph shown in Figure 4 is therefore the graph of a function.

EXAMPLE 6 Match each relation with its graph, then indicate which relations are functions.

a. $y = |x| - 4$ **b.** $y = x^2 - 4$ **c.** $y = 2x + 2$

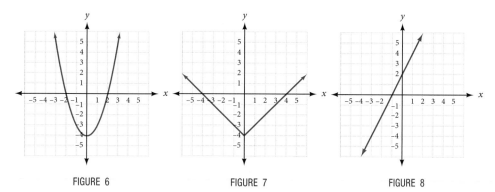

FIGURE 6 FIGURE 7 FIGURE 8

Solution Using the basic graphs for a guide along with our knowledge of translations, we have the following:

a. Figure 7 **b.** Figure 6 **c.** Figure 8

And, since all graphs pass the vertical line test, all are functions.

6. Graph $y = |x| - 3$.

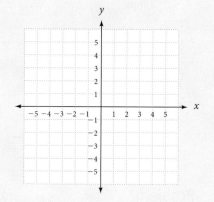

GETTING READY FOR CLASS

After reading through the preceding section, respond in your own words and in complete sentences.

A. What is a relation?

B. What is a function?

C. Which variable is usually associated with the domain of a function? Which variable is associated with the range?

D. What is the vertical line test?

EXERCISE SET 9.5

Vocabulary Review

Choose the correct words to fill in the blanks below.

function	domain	right	left
range	relation	vertical line test	

1. The _____ of a function is the complete set of inputs, whereas the _____ is the set of all outputs.

2. On a function map, the values for the domain of the function appear on the _____ and the values for the range appear on the _____.

3. A _____ is a rule that pairs each element in one set with exactly one element from a second set.

4. A _____ is a rule that pairs each element in one set with one or more elements from a second set.

5. The _____ can be used to determine whether the graph of a relation is a function.

Problems

A For each of the following functions, construct a table and graph.

1. $y = 4x$ for $-2 \leq x \leq 5$

x	y
-2	
0	
2	
5	

2. $y = -3x$ for $-3 \leq x \leq 2$

x	y
-3	
-1	
0	
1	
2	

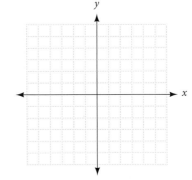

3. $y = 100 - 4x$ for $0 \leq x \leq 25$

x	y
0	
5	
10	
15	
20	
25	

4. $y = 250 - 10x$ for $0 \leq x \leq 25$

x	y
0	
5	
10	
15	
20	
25	

SCAN TO ACCESS

5. $y = 1600 - 10x$ for $0 \le x \le 160$

x	y
0	
40	
80	
120	
160	

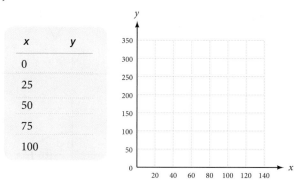

6. $y = 1200 - 40x$ for $0 \le x \le 30$

x	y
0	
10	
20	
30	

7. $y = 40 + 2.5x$ for $0 \le x \le 100$

x	y
0	
25	
50	
75	
100	

8. $y = 100 + 0.5x$ for $0 \le x \le 40$

x	y
0	
10	
20	
30	
40	

B C For each of the following relations, give the domain and range using roster notation, and indicate which are also functions.

9. $\{(1, 2), (3, 4), (5, 6), (7, 8)\}$

10. $\{(2, 1), (4, 3), (6, 5), (8, 7)\}$

11. $\{(2, 5), (3, 4), (1, 4), (0, 6)\}$

12. $\{(0, 4), (1, 6), (2, 4), (1, 5)\}$

13. $\{(a, 3), (b, 4), (c, 3), (d, 5)\}$

14. $\{(a, 5), (b, 5), (c, 4), (d, 5)\}$

15. $\{(a, 1), (a, 2), (a, 3), (a, 4)\}$

16. $\{(a, 1), (b, 1), (c, 1), (d, 1)\}$

Give the domain and range for the following function maps.

17.

Stock Symbol	Rule	Price

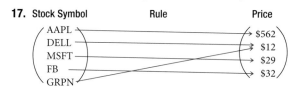

18.

Stock Symbol	Rule	Price

19.

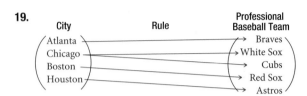

City	Rule	Professional Baseball Team
Atlanta		Braves
Chicago		White Sox
Boston		Cubs
Houston		Red Sox
		Astros

20.

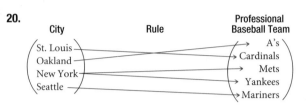

City	Rule	Professional Baseball Team
St. Louis		A's
Oakland		Cardinals
New York		Mets
Seattle		Yankees
		Mariners

C D State whether each of the following graphs represents the graph of a function.

21.

22.

23.

24.

25.

26.

27.

28.

29.

30.

31.

32.

Determine the domain and range of the following functions. Assume the *entire* function is shown. Write your answers in set-builder notation and interval notation.

33.

34.

35.

36.

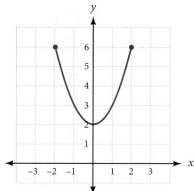

Graph each of the following relations. In each case, use the graph to find the domain and range, and indicate whether the graph is the graph of a function.

37. $y = x^2 - 1$

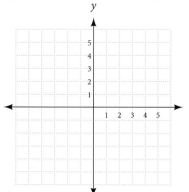

38. $y = x^2 + 1$

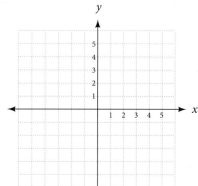

39. $y = x^2 + 4$

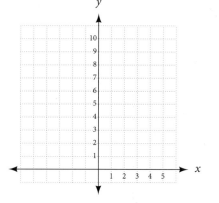

40. $y = x^2 - 9$

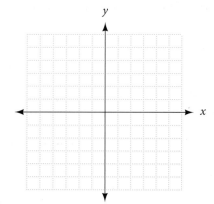

41. $x = y^2 - 1$

42. $x = y^2 + 1$

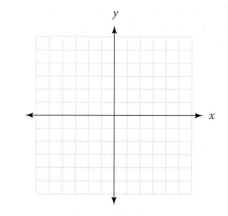

43. $y = (x + 2)^2$

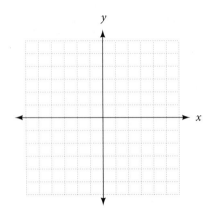

44. $y = (x - 3)^2$

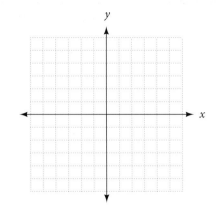

45. $x = (y + 1)^2$

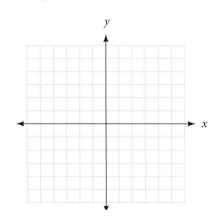

46. $x = 3 - y^2$

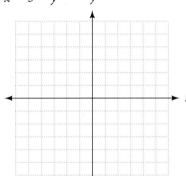

47. $y = |x| + 2$

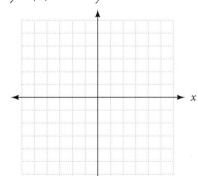

48. $x = |y| - 3$

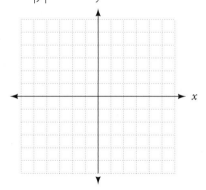

Applying the Concepts

49. iPod Memory The chart shows the memory capacity for different generations of iPods from 2001 to 2009. Using the chart, list all the values in the domain and range for the memory capacity.

50. Red Box Kiosks The chart shows the number of Redbox DVD rental kiosks installed from 2005 to 2010. Use the chart to state the domain and range of the function for Redbox kiosk installation.

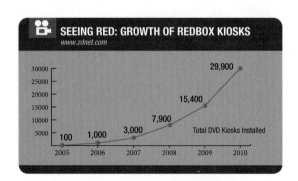

51. Weekly Wages Suppose you have a job that pays $8.50 per hour and you work anywhere from 10 to 40 hours per week.

a. Write an equation, with a restriction on the variable x, that gives the amount of money, y, you will earn for working x hours in one week.

b. Use the function rule you have written in part **a.** to complete Table 4.

Table 4 Weekly Wages		
Hours Worked	Function Rule	Gross Pay ($)
x		y
10		
20		
30		
40		

c. Construct a line graph from the information in Table 4.

Weekly Wages

d. State the domain and range of this function.

e. What is the minimum amount you can earn in a week with this job? What is the maximum amount?

52. Weekly Wages The ad shown here was in the local newspaper. Suppose you are hired for the job described in the ad.

a. If x is the number of hours you work per week and y is your weekly gross pay, write the equation for y. (Be sure to include any restrictions on the variable x that are given in the ad.)

b. Use the function rule you have written in part a to complete Table 5.

Table 5 Weekly Wages		
Hours Worked	Function Rule	Gross Pay ($)
x		y
15		
20		
25		
30		

c. Construct a line graph from the information in Table 5.

Weekly Wages

d. State the domain and range of this function.

e. What is the minimum amount you can earn in a week with this job? What is the maximum amount?

53. Global Solar Energy Demand The chart shows the global solar power demand from 2008 to 2012. Using the chart:

a. List the values for the domain and for the range.

b. Did demand increase more in 2011 or 2012 (projected)?

c. In what year did demand decrease?

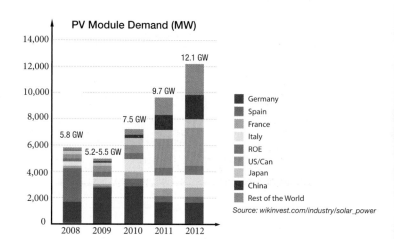

Source: wikinvest.com/industry/solar_power

54. Global Wind Power Capacity The chart shows the global wind power capacity from 1996 to 2010:

Global Wind Power Cumulative Capacity (Data: GWEC)

Source: gwec.net

Use the information from the chart to answer the following questions.

a. List the values for the domain and for the range.

b. Did capacity increase more in 2006 or 2007?

c. In which year did capacity increase the most?

55. Profits Match each of the following statements to the appropriate graph indicated by labels I-IV.

a. Sarah works 25 hours to earn $250.

b. Justin works 35 hours to earn $560.

c. Rosemary works 30 hours to earn $360.

d. Marcus works 40 hours to earn $320.

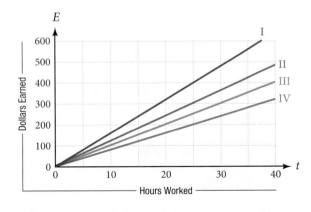

FIGURE 9

56. Find an equation for each of the functions shown in the Figure 9. Show dollars earned, E, as a function of hours worked, t. Then indicate the domain and range of each function.

a. Graph I: $E =$

 Domain $= \{ t \mid$ $\}$

 Range $= \{ E \mid$ $\}$

b. Graph II: $E =$

 Domain $= \{ t \mid$ $\}$

 Range $= \{ E \mid$ $\}$

c. Graph III: $E =$

 Domain $= \{ t \mid$ $\}$

 Range $= \{ E \mid$ $\}$

d. Graph IV: $E =$

 Domain $= \{ t \mid$ $\}$

 Range $= \{ E \mid$ $\}$

Getting Ready for the Next Section

57. If $s = \dfrac{60}{t}$, find s when

 a. $t = 10$

 b. $t = 8$

58. If $y = 3x^2 + 2x - 1$, find y when

 a. $x = 0$

 b. $x = -2$

59. Find the value of $x^2 + 2$ for

 a. $x = 5$

 b. $x = -2$

60. Find the value of $125 \cdot 2^t$ for

 a. $t = 0$

 b. $t = 1$

For the equation $y = x^2 - 3$:

61. Find y if x is 2.

62. Find y if x is -2.

63. Find y if x is 0.

64. Find y if x is -4.

65. For which value of x is $y = -3$?

66. For which values of x is $y = 6$? (There are two answers)

The problems that follow review some of the more important skills you have learned in previous sections and chapters.

67. If $x - 2y = 4$, and $x = \dfrac{8}{5}$ find y.

68. If $\dfrac{x^2}{25} + \dfrac{y^2}{9} = 1$, find y when x is -4.

69. Let $x = 0$ and $y = 0$ in $y = a(x - 8)^2 + 70$ and solve for a.

70. Find R if $p = 2.5$ and $R = (900 - 300p)p$.

Find the Mistake

Each sentence below contains a mistake. Circle the mistake and write the correct word or phrase on the line provided.

1. The set of all inputs for a function is called the range. _____

2. For the relation $(1, 2), (3, 4), (5, 6), (7, 8)$, the domain is $\{2, 4, 6, 8\}$. _____

3. Graphing the relation $y = x^2 - 3$ will show that the domain is $\{ x \mid x \geq -3 \}$. _____

4. If a horizontal line crosses the graph of a relation in more than one place, the relation cannot be a function.

A Use function notation to find the value of a function for a given value of the variable.

B Use graphs to visualize the relationship between a function and a variable.

C Use function notation in formulas.

dependent variable

independent variable

function notation

inputs

outputs

Note An important thing to remember is that f is the function name, it is not a variable and cannot be replaced with a number. Furthermore, $f(x)$ is read "f of x" but does not imply multiplication.

Practice Problems

1. If $f(x) = 8x$, find
 a. $f(0)$
 b. $f(5)$
 c. $f(10.5)$

Answers
1. a. 0 **b.** 40 **c.** 84

©iStockphoto.com/tap10

A Evaluate a Function at a Point

Pogopalooza is an annual world championship for stunt pogo stick athletes. Pogo jumpers compete in various events, such as the most or the least jumps in a minute, the highest jump, and numerous exhibitions of acrobatic stunts. Imagine a group of local stunt pogo enthusiasts are holding a pogo stick competition in your city. They have gathered a variety a sponsors, and their budget allows them to hire you to market and promote the event. If the job pays $8.50 per hour for working from 0 to 40 hours a week, then the amount of money y earned in one week is a function of the number of hours worked x. The exact relationship between x and y is written

$$y = 8.5x \quad \text{for} \quad 0 \le x \le 40$$

Because the amount of money earned y depends on the number of hours worked x, we call y the *dependent variable* and x the *independent variable*. Furthermore, if we let f represent all the ordered pairs produced by the equation, then we can write

$$f = \{(x, y) \,|\, y = 8.5x \quad \text{and} \quad 0 \le x \le 40\}$$

Once we have named a function with a letter, we can use an alternative notation to represent the dependent variable y. The alternative notation for y is $f(x)$. It is read "f of x" and can be used instead of the variable y when working with functions. The notation y and the notation $f(x)$ are equivalent. That is,

$$y = 8.5x \Leftrightarrow f(x) = 8.5x$$

When we use the notation $f(x)$ we are using *function notation*. The benefit of using function notation is that we can write more information with fewer symbols than we can by using just the variable y. For example, asking how much money a person will make for working 20 hours is simply a matter of asking for $f(20)$. Without function notation, we would have to say, "Find the value of y that corresponds to a value of $x = 20$." To illustrate further, using the variable y, we can say "y is 170 when x is 20." Using the notation $f(x)$, we simply say "$f(20) = 170$." Each expression indicates that you will earn $170 for working 20 hours.

EXAMPLE 1 If $f(x) = 7.5x$, find $f(0)$, $f(10)$, and $f(20)$.

Solution To find $f(0)$, we substitute 0 for x in the expression $7.5x$ and simplify. We find $f(10)$ and $f(20)$ in a similar manner — by substitution.

$$\text{If} \qquad f(x) = 7.5x$$
$$\text{then} \qquad f(0) = 7.5(0) = 0$$
$$f(10) = 7.5(10) = 75$$
$$f(20) = 7.5(20) = 150$$

If we changed the example in the discussion that opened this section so the hourly wage was $6.50 per hour, we would have a new equation to work with, namely,

$$y = 6.5x \quad \text{for} \quad 0 \le x \le 40$$

Suppose we name this new function with the letter g. Then

$$g = \{(x, y) \mid y = 6.5x \quad \text{and} \quad 0 \le x \le 40\}$$

and

$$g(x) = 6.5x$$

If we want to talk about both functions in the same discussion, having two different letters, f and g, makes it easy to distinguish between them. For example, since $f(x) = 7.5x$ and $g(x) = 6.5x$, asking how much money a person makes for working 20 hours is simply a matter of asking for $f(20)$ or $g(20)$, avoiding any confusion over which hourly wage we are talking about.

The diagrams shown in Figure 1 further illustrate the similarities and differences between the two functions we have been discussing.

$x \in$ Domain and $f(x) \in$ Range $x \in$ Domain and $g(x) \in$ Range

FIGURE 1

B Function Notation and Graphs

We can visualize the relationship between x and $f(x)$ on the graph of the function. Figure 2 shows the graph of $f(x) = 7.5x$ along with two additional line segments. The horizontal line segment corresponds to $x = 20$, and the vertical line segment corresponds to $f(20)$. (Note that the domain is restricted to $0 \le x \le 40$.)

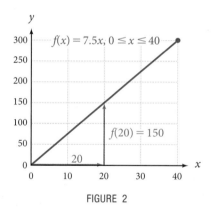

FIGURE 2

We can use functions and function notation to talk about numbers in the following chart showing the number of millionaire households in the U.S. Let's let x represent one of the years in the chart.

Note Some students like to think of functions as machines. Values of x are put into the machine, which transforms them into values of $f(x)$, which are then output by the machine.

If the function *f* pairs each year in the chart with the number of millionaire households for that year, then each statement below is true:

$$f(2006) = 9.0$$

The domain of $f = \{2000, 2002, 2004, 2006, 2008, 2010\}$

In general, when we refer to the function *f* we are referring to the domain, the range, and the rule that takes elements in the domain and outputs elements in the range. When we talk about $f(x)$ we are talking about the rule itself, or an element in the range, or the variable *y*.

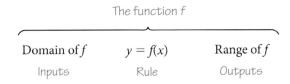

The function *f*

Domain of *f*	$y = f(x)$	Range of *f*
Inputs	Rule	Outputs

C Using Function Notation

The remaining examples in this section show a variety of ways to use and interpret function notation.

EXAMPLE 2 If Lorena takes *t* minutes to run a mile, then her average speed *s*, in miles per hour, is given by the formula

$$s(t) = \frac{60}{t} \qquad \text{for} \qquad t > 0$$

Find $s(10)$ and $s(8)$, and then explain what they mean.

Solution To find $s(10)$, we substitute 10 for *t* in the equation and simplify.

$$s(10) = \frac{60}{10} = 6$$

In words: When Lorena runs a mile in 10 minutes, her average speed is 6 miles per hour. We calculate $s(8)$ by substituting 8 for *t* in the equation. Doing so gives us

$$s(8) = \frac{60}{8} = 7.5$$

In words: Running a mile in 8 minutes is running at a rate of 7.5 miles per hour.

2. When Lorena runs a mile in *t* minutes, then her average speed in feet per second is given by

$$s(t) = \frac{88}{t}, t > 0$$

 a. Find $s(8)$ and explain what it means.

 b. Find $s(11)$ and explain what it means.

3. A medication has a half-life of 5 days. If the concentration of the medication in a patient's system is 80 ng/mL, and the patient stops taking it, then *t* days later the concentration will be

$$C(t) = 80\left(\frac{1}{2}\right)^{t/5}$$

Find each of the following, and explain what they mean.

 a. $C(5)$

 b. $C(10)$

EXAMPLE 3 A painting is purchased as an investment for $125. If its value increases continuously so that it doubles every 5 years, then its value is given by the function

$$V(t) = 125 \cdot 2^{t/5} \qquad \text{for} \qquad t \geq 0$$

where *t* is the number of years since the painting was purchased, and *V* is its value (in dollars) at time *t*. Find $V(5)$ and $V(10)$, and explain what they mean.

©Marissa McKeague, 2011

Solution The expression $V(5)$ is the value of the painting when $t = 5$ (5 years after it is purchased). We calculate $V(5)$ by substituting 5 for *t* in the equation $V(t) = 125 \cdot 2^{t/5}$. Here is our work:

$$V(5) = 125 \cdot 2^{5/5} = 125 \cdot 2^1 = 125 \cdot 2 = 250$$

In words: After 5 years, the painting is worth $250.

Note Remember, as stated in the order of operations, work the exponents before any multiplication the problem requires.

Answers

2. a. $s(8) = 11$; runs a mile in 8 minutes, average speed is 11 feet per second.

 b. $s(11) = 8$; runs a mile in 11 minutes, average speed is 8 feet per second.

3. a. $C(5) = 40$ ng/mL; after 5 days the concentration is 40 ng per mL.

 b. $C(10) = 20$ ng/mL; after 10 days the concentration is 20 ng per mL.

The expression $V(10)$ is the value of the painting after 10 years. To find this number, we substitute 10 for t in the equation:

$$V(10) = 125 \cdot 2^{10/5} = 125 \cdot 2^2 = 125 \cdot 4 = 500$$

In words: The value of the painting 10 years after it is purchased is $500.

EXAMPLE 4 A balloon has the shape of a sphere with a radius of 3 inches. Use the following formulas to find the volume and surface area of the balloon.

$$V(r) = \frac{4}{3}\pi r^3 \qquad S(r) = 4\pi r^2$$

Solution As you can see, we have used function notation to write the formulas for volume and surface area, because each quantity is a function of the radius. To find these quantities when the radius is 3 inches, we evaluate $V(3)$ and $S(3)$.

$$V(3) = \frac{4}{3}\pi \cdot 3^3 = \frac{4}{3}\pi \cdot 27$$

$$= 36\pi \text{ cubic inches, or } 113 \text{ cubic inches}$$
$$\text{(to the nearest whole number)}$$

$$S(3) = 4\pi \cdot 3^2$$

$$= 36\pi \text{ square inches, or } 113 \text{ square inches}$$
$$\text{(to the nearest whole number)}$$

The fact that $V(3) = 36\pi$ means that the ordered pair $(3, 36\pi)$ belongs to the function V. Likewise, the fact that $S(3) = 36\pi$ tells us that the ordered pair $(3, 36\pi)$ is a member of function S.

We can generalize the discussion at the end of Example 4 this way:

$$(a, b) \in f \qquad \text{if and only if} \qquad f(a) = b$$

USING TECHNOLOGY More About Example 4

If we look at Example 4, we see that when the radius of a sphere is 3, the numerical values of the volume and surface area are equal. How unusual is this? Are there other values of r for which $V(r)$ and $S(r)$ are equal? We can answer this question by looking at the graphs of both V and S.

To graph the function $V(r) = \frac{4}{3}\pi r^3$, set $Y_1 = 4\pi X^3/3$. To graph $S(r) = 4\pi r^2$, set $Y_2 = 4\pi X^2$. Graph the two functions in each of the following windows:

Window 1: X from -4 to 4, Y from -2 to 10

Window 2: X from 0 to 4, Y from 0 to 50

Window 3: X from 0 to 4, Y from 0 to 150

Then use the Trace and Zoom features of your calculator to locate the point in the first quadrant where the two graphs intersect. How do the coordinates of this point compare with the results in Example 4?

EXAMPLE 5 If $f(x) = 3x^2 + 2x - 1$, find $f(0), f(3)$, and $f(-2)$.

Solution Since $f(x) = 3x^2 + 2x - 1$, we have

$$f(0) = 3(0)^2 + 2(0) - 1 = 0 + 0 - 1 = -1$$

$$f(3) = 3(3)^2 + 2(3) - 1 = 27 + 6 - 1 = 32$$

$$f(-2) = 3(-2)^2 + 2(-2) - 1 = 12 - 4 - 1 = 7$$

4. The following formulas give the circumference and area of a circle with radius r. Use the formulas to find the circumference and area of a circular plate if the radius is 5 inches.
 a. $C(r) = 2\pi r$
 b. $A(r) = \pi r^2$

5. If $f(x) = 4x^2 - 3$, find
 a. $f(0)$
 b. $f(3)$
 c. $f(-2)$

Answers
 4. $C(5) = 10\pi$
 $A(5) = 25\pi$
 5. a. -3 **b.** 33 **c.** 13

In Example 5, the function f is defined by the equation $f(x) = 3x^2 + 2x - 1$. We could just as easily have said $y = 3x^2 + 2x - 1$. That is, $y = f(x)$. Saying $f(-2) = 7$ is exactly the same as saying y is 7 when x is -2.

6. If $f(x) = 2x + 1$ and $g(x) = x^2 - 3$, find

 a. $f(5)$

 b. $g(5)$

 c. $f(-2)$

 d. $g(-2)$

 e. $f(a)$

 f. $g(a)$

EXAMPLE 6 If $f(x) = 4x - 1$ and $g(x) = x^2 + 2$, find the value of each for $x = 5, -2, 0, z, a,$ and $a + 3$.

Solution $f(5) = 4(5) - 1 = 19$ and $g(5) = 5^2 + 2 = 27$

$f(-2) = 4(-2) - 1 = -9$ and $g(-2) = (-2)^2 + 2 = 6$

$f(0) = 4(0) - 1 = -1$ and $g(0) = 0^2 + 2 = 2$

$f(z) = 4z - 1$ and $g(z) = z^2 + 2$

$f(a) = 4a - 1$ and $g(a) = a^2 + 2$

$f(a + 3) = 4(a + 3) - 1$ $g(a + 3) = (a + 3)^2 + 2$

$= 4a + 12 - 1$ $= (a^2 + 6a + 9) + 2$

$= 4a + 11$ $= a^2 + 6a + 11$

USING TECHNOLOGY **More About Example 6**

Most graphing calculators can use tables to evaluate functions. To work Example 6 using a graphing calculator table, set Y_1 equal to $4X - 1$ and Y_2 equal to $X^2 + 2$. Then set the independent variable in the table to Ask instead of Auto. Go to your table and input 5, -2, and 0. Under Y_1 in the table, you will find $f(5), f(-2)$, and $f(0)$. Under Y_2, you will find $g(5), g(-2)$, and $g(0)$.

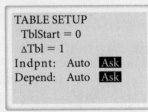

The table will look like this:

X	Y_1	Y_2
5	19	27
-2	-9	6
0	-1	2

Although the calculator asks us for a table increment, the increment doesn't matter because we are inputting the X values ourselves.

Answers

6. a. 11 **b.** 22 **c.** -3

 d. 1 **e.** $2a + 1$ **f.** $a^2 - 3$

EXAMPLE 7 If the function f is given by $f = \{(-2, 0), (3, -1), (2, 4), (7, 5)\}$, find $f(-2), f(3), f(2),$ and $f(7)$.

Solution $f(-2) = 0, f(3) = -1, f(2) = 4,$ and $f(7) = 5.$

EXAMPLE 8 If $f(x) = 2x^2$ and $g(x) = 3x - 1$, find
a. $f(g(2))$ **b.** $g(f(2))$
Solution The expression $f(g(2))$ is read "f of g of 2."
a. Because $g(2) = 3(2) - 1 = 5,$

$$f(g(2)) = f(5) = 2(5)^2 = 50$$

b. Because $f(2) = 2(2)^2 = 8,$

$$g(f(2)) = g(8) = 3(8) - 1 = 23$$

EXAMPLE 9 Let $f(x) = x^2 + 2x$ and $g(x) = 3x - 5$. Evaluate the following:
a. $f(3) + g(-3)$
b. $2 f(x) + 4 g(x)$
c. $g(x^2)$
d. $[f(-4)]^2$
Solution
a. Because $f(3) = (3)^2 + 2(3)$ and $g(-3) = 3(-3) - 5$ then

$$f(3) + g(-3) = (3)^2 + 2(3) + 3(-3) - 5 = 9 + 6 - 9 - 5 = 1$$

b. Because $2 f(x) = 2(x^2 + 2x)$ and $4 g(x) = 4(3x - 5)$ then

$$2 f(x) + 4 g(x) = 2(x^2 + 2x) + 4(3x - 5)$$

$$= 2x^2 + 4x + 12x - 20$$

$$= 2x^2 + 16x - 20$$

c. $g(x^2) = 3x^2 - 5$
d. $[f(-4)]^2 = [(-4)^2 + 2(-4)]^2 = (16 - 8)^2 = 64$

GETTING READY FOR CLASS

After reading through the preceding section, respond in your own words and in complete sentences.

A. Explain what you are calculating when you find $f(2)$ for a given function f.

B. If $s(t) = \frac{60}{t}$ how do you find $s(10)$?

C. If $f(2) = 3$ for a function f, what is the relationship between the numbers 2 and 3 and the graph of f?

D. If $f(6) = 0$ for a particular function f, then you can immediately graph one of the intercepts. Explain.

7. If $f = \{(-4, 1), (2, -3), (7, 9)\}$, find
 a. $f(-4)$
 b. $f(2)$
 c. $f(7)$

8. If $f(x) = 3x^2$ and $g(x) = 4x + 1$, find
 a. $f(g(2))$
 b. $g(f(2))$

9. Let $f(x) = 2x^2 + 3x$ and $g(x) = 5x - 6$. Evaluate the following:
 a. $f(-2) - g(1)$
 b. $3 f(x) - 2 g(x)$
 c. $f(x^2)$
 d. $[g(-3)]^2$

Answers
7. a. 1 **b.** −3 **c.** 9
8. a. 243 **b.** 49
9. a. 3 **b.** $6x^2 - x + 12$
 c. $2x^4 + 3x^2$ **d.** 441

Vocabulary Review

Choose the correct words to fill in the blanks below.

dependent independent $f(x)$ function f

1. The variable that represents the input of a function is called the _____ variable.
2. The variable that represents the output of a function is called the _____ variable.
3. When we use the notation _____, we are using function notation.
4. The _____ refers to the domain, the range, and the rule that takes elements in the domain and outputs elements in the range.

Problems

A Let $f(x) = 2x - 5$ and $g(x) = x^2 + 3x + 4$. Evaluate the following.

1. $f(2)$

2. $f(3)$

3. $f(-3)$

4. $g(-2)$

5. $g(-1)$

6. $f(-4)$

7. $g(-3)$

8. $g(2)$

9. $g(a)$

10. $f(a)$

11. $f(a + 6)$

12. $g(a + 6)$

Let $f(x) = 3x^2 - 4x + 1$ and $g(x) = 2x - 1$. Evaluate the following.

13. $f(0)$

14. $g(0)$

15. $g(-4)$

16. $f(1)$

17. $f(-1)$

18. $g(-1)$

19. $g\left(\dfrac{1}{2}\right)$

20. $g\left(\dfrac{1}{4}\right)$

21. $f(a)$

22. $g(a)$

23. $f(a + 2)$

24. $g(a + 2)$

SCAN TO ACCESS

If $f = \{(1, 4), (-2, 0), \left(3, \frac{1}{2}\right), (\pi, 0)\}$ and $g = \{(1, 1),(-2, 2), \left(\frac{1}{2}, 0\right)\}$, find each of the following values of f and g.

25. $f(1)$

26. $g(1)$

27. $g\left(\dfrac{1}{2}\right)$

28. $f(3)$

29. $g(-2)$

30. $f(\pi)$

31. $f(g(1))$

32. $f(-2) - g(-2)$

33. For which value of x is $g(x) = f(\pi)$?

34. Is $f(1) = g(1)$?

Let $f(x) = x^2 - 2x$ and $g(x) = 5x - 4$. Evaluate the following.

35. $f(-4)$

36. $g(-3)$

37. $f(-2) + g(-1)$

38. $f(-1) + g(-2)$

39. $2f(x) - 3g(x)$

40. $f(x) - g(x^2)$

41. $f(g(3))$

42. $g(f(3))$

43. $f(g(2)) - g(f(2))$

44. $g(f(1)) - f(g(1))$

45. $[f(-3)]^2$

46. $[g(3)]^2$

Let $f(x) = \dfrac{1}{x + 3}$ and $g(x) = \dfrac{1}{x} + 1$. Evaluate the following.

47. $f\left(\dfrac{1}{3}\right)$

48. $g\left(\dfrac{1}{3}\right)$

49. $f\left(-\dfrac{1}{2}\right)$

50. $g\left(-\dfrac{1}{2}\right)$

51. $f(-3)$

52. $g(0)$

53. $g(-2)$

54. $f(-1)$

55. For the function $f(x) = x^2 - 4$, evaluate each of the following expressions.

a. $f(a) - 3$

b. $f(a - 3)$

c. $f(x) + 2$

d. $f(x + 2)$

e. $f(a + b)$

f. $f(x + h)$

56. For the function $f(x) = 3x^2$, evaluate each of the following expressions.

a. $f(a) - 2$

b. $f(a - 2)$

c. $f(x) + 5$

d. $f(x + 5)$

e. $f(a + b)$

f. $f(x + h)$

B

57. Graph the function $f(x) = \frac{1}{2}x + 2$. Then draw and label the line segments that represent $x = 4$ and $f(4)$.

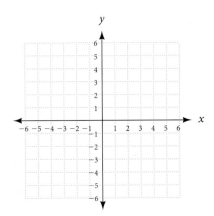

58. Graph the function $f(x) = -\frac{1}{2}x + 6$. Then draw and label the line segments that represent $x = 4$ and $f(4)$.

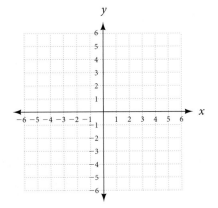

59. For the function $f(x) = \frac{1}{2}x + 2$, find the value of x for which $f(x) = x$.

60. For the function $f(x) = -\frac{1}{2}x + 6$, find the value of x for which $f(x) = x$.

61. Graph the function $f(x) = x^2$. Then draw and label the line segments that represent $x = 1$ and $f(1)$, $x = 2$ and $f(2)$ and, finally, $x = 3$ and $f(3)$.

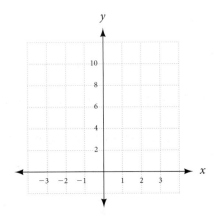

62. Graph the function $f(x) = x^2 - 2$. Then draw and label the line segments that represent $x = 2$ and $f(2)$ and the line segments corresponding to $x = 3$ and $f(3)$.

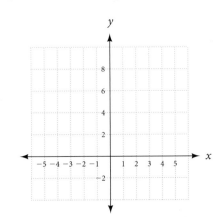

Applying the Concepts

63. Investing in Art Charlie purchased a painting as an investment for $150. If its value increases continuously so that it doubles every 3 years, then its value is given by the function

$$V(t) = 150 \cdot 2^{t/3} \quad \text{for} \quad t \geq 0$$

where t is the number of years since the painting was purchased, and $V(t)$ is its value (in dollars) at time t. Find $V(3)$ and $V(6)$, and then explain what they mean.

64. Average Speed If Minke takes t minutes to run a mile, then her average speed $s(t)$, in miles per hour, is given by the formula

$$s(t) = \frac{60}{t} \quad \text{for} \quad t > 0$$

Find $s(4)$ and $s(5)$, and then explain what they mean.

65. iPhone Sales The chart below shows the percentage of respondents who purchased the iPhone on each of its first day of sales, shown by their current carrier. Suppose x represents one of the carriers in the chart. We have three functions $f, g,$ and h that do the following:

f pairs each year with the percentage of respondents who were AT&T customers for that year.

g pairs each year with the percentage of respondents who were Verizon customers for that year.

h pairs each year with the percentage of respondents who were T-Mobile customers for that year.

For each statement below, indicate whether the statement is true or false.

a. The domain of g is {2007, 2008, 2009, 2010}

b. The domain of f is $\{x \mid 2008 \leq x \leq 2009\}$

c. $f(2008) > g(2008)$

d. $h(2009) > 30\%$

e. $h(2008) < g(2008) < f(2010)$

66. Newspaper Sales Suppose x represents one of the publications in the chart. We have three functions $f, g,$ and h that do the following:

f pairs each publication with the number of paper subscribers for that publication.

g pairs each publication with the number of online viewers for that publication.

h is such that $h(x) = f(x) + g(x)$.

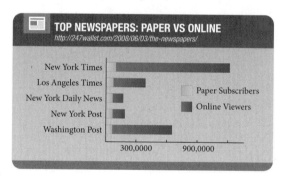

For each statement below, indicate whether the statement is true or false.

a. The domain of f is {New York Times, Los Angeles Times, New York Daily News, New York Post, Washington Post}

b. h(Los Angeles Times) = 600,000

c. g(New York Times) $> f$(New York Times)

d. g(New York Daily News) $< g$(Washington Post)

e. h(New York Times) $> h$(Los Angeles Times) $> h$(New York Post)

Straight-Line Depreciation Straight-line depreciation is an accounting method used to help spread the cost of new equipment over a number of years. It takes into account both the cost when new and the salvage value, which is the value of the equipment at the time it gets replaced.

67. Value of a Copy Machine The function
$V(t) = -3,300t + 18,000$, where V is value and t is time in years, can be used to find the value of a large copy machine during the first 5 years of use.

a. What is the value of the copier after 3 years and 9 months?

b. What is the salvage value of this copier if it is replaced after 5 years?

c. State the domain of this function.

d. Sketch the graph of this function.

Value of a Copy Machine

e. What is the range of this function?

f. After how many years will the copier be worth only $10,000?

68. Step Function Figure 3 shows the graph of the step function C that was used to calculate the first-class postage on a large envelope weighing x ounces in 2010. Use this graph to answer the following questions.

FIGURE 3

a. Fill in the following table:

Weight (ounces)	0.6	1.0	1.1	2.5	3.0	4.8	5.0	5.3
Cost (cents)								

b. If a letter cost 122 cents to mail, how much does it weigh? State your answer in words and as an inequality.

c. If the entire function is shown in Figure 3, state the domain.

d. State the range of the function shown in Figure 3.

69. Wind Turbine The annual output from a wind turbine with an average wind velocity of 10mph is given by the function $f(d) = 13.28d^2$, where d is the diameter (in feet) of the blade and f is kWh/year.

a. Complete the table for various diameter blades.

d (feet)	f(d) (kWh/year)
0	
5	
10	
15	
20	

Source: builditsolar.com/Projects/Wind/Wind.htm

b. What is the range of this function for $0 \le d \le 20$?

70. Wind Turbine The annual output from a wind turbine with a 10-foot-diameter blade is given by the function $f(v) = 1.328v^3$, where v is the average velocity of the wind (in mph) and f is kWh/year.

a. Complete the table for various diameter blades.

v (mph)	f(v) (kWh/year)
0	
10	
20	
30	
40	—

Source: builditsolar.com/Projects/Wind/Wind.htm

b. What is the range of this function for $0 \le v \le 40$?

Getting Ready for the Next Section

Multiply.

71. $x(35 - 0.1x)$
72. $0.6(M - 70)$
73. $(4x - 3)(x - 1)$
74. $(4x - 3)(4x^2 - 7x + 3)$

Simplify.

75. $(35x - 0.1x^2) - (8x + 500)$
76. $(4x - 3) + (4x^2 - 7x + 3)$

77. $(4x^2 + 3x + 2) - (2x^2 - 5x - 6)$
78. $(4x^2 + 3x + 2) + (2x^2 - 5x - 6)$

79. $4(2)^2 - 3(2)$
80. $4(-1)^2 - 7(-1)$

Simplify.

81. $2(x - 3)^2 - 5(x - 3)$
82. $5(3x - 7) - 4$

83. $-8(-3x + 1) + 2$
84. $4(x + 1)^2 - 6(x + 1)$

Find the Mistake

Each sentence below contains a mistake. Circle the mistake and write the correct word or phrase on the line provided.

1. To evaluate the function $f(x) = 10x + 3$ for $f(1)$ we begin by setting $10x + 3$ equal to 1. _____

2. For the average speed function $s(t) = \frac{60}{t}$, finding $s(6)$ shows an average speed of 6 miles per hour. _____

3. If the volume of a ball is given by $V(r) = \frac{4}{3}\pi r^3$ where r is the radius, we see that the volume decreases as the radius of the ball increases. _____

4. If $f(x) = x^2$ and $g(x) = 2x + 4$, evaluating $f(g(2))$ is the same as multiplying $f(2) \cdot g(2)$. _____

A Resources for research

B Pencil and paper

Jet-Powered Parabolas

Picture a dolphin leaping out of the water. The path of the leap is similar in shape to a parabola, and therefore, can be represented by a quadratic equation. Now imagine a human leaping out of the water in the same manner as a dolphin. Sound impossible? Not so if you have $6600 to purchase a machine called the Flyboard. Invented by French jet ski racer Franky Zapata, the Flyboard combines a jet ski motor, jet boots, hand stabilizers, and a few hoses, and allows a person wearing the invention to explode out of a body of water upwards of thirty feet, just like a dolphin.

1. Based on what you have learned in this chapter, create an application problem that revolves around the Flyboard and involves an accurate quadratic function.

2. Research another method of movement, either human-powered or machine-assisted, that can be modelled using a quadratic equation. Create a second application problem that involves this quadratic equation.

The Square Root Property for Equations [9.1]

If $a^2 = b$, where b is a real number, then

$$a = \sqrt{b} \quad \text{or} \quad a = -\sqrt{b}$$

which can be written as $a = \pm\sqrt{b}$.

1. If $(x - 3)^2 = 25$
then $x - 3 = \pm 5$
$x = 3 \pm 5$
$x = 8$ or $x = -2$

To Solve a Quadratic Equation by Completing the Square [9.2]

Step 1: Write the equation in the form $ax^2 + bx = c$.
Step 2: If $a \neq 1$, divide through by the constant a so the coefficient of x^2 is 1.
Step 3: Complete the square on the left side by adding the square of $\frac{1}{2}$ the coefficient of x to both sides.
Step 4: Write the left side of the equation as the square of a binomial. Simplify the right side if possible.
Step 5: Apply the square root property for equations, and solve as usual.

2. Solve $x^2 - 6x - 6 = 0$
$x^2 - 6x = 6$
$x^2 - 6x + 9 = 6 + 9$
$(x - 3)^2 = 15$
$x - 3 = \pm\sqrt{15}$
$x = 3 \pm \sqrt{15}$

The Quadratic Formula [9.3]

For any quadratic equation in the form $ax^2 + bx + c = 0$, $a \neq 0$, the two solutions are

$$x = \frac{-b \pm \sqrt{b^2 - 4ac}}{2a}$$

This equation is known as the *quadratic formula*.

3. If $2x^2 + 3x - 4 = 0$, then

$$x = \frac{-3 \pm \sqrt{9 - 4(2)(-4)}}{2(2)}$$

$$= \frac{-3 \pm \sqrt{41}}{4}$$

Graphing Vertical Parabolas [9.4]

The graph of any equation of the form $y = ax^2 + bx + c$ where $a \neq 0$ is a vertical *parabola*. The graph is *concave up* if $a > 0$ and *concave down* if $a < 0$. The highest or lowest point on the graph is called the *vertex* and always has an x-coordinate of $x = \dfrac{-b}{2a}$.

4. The graph of $y = x^2 - 4$ will be a parabola. It will cross the x-axis at 2 and -2, and the vertex will be $(0, -4)$.

Relations and Functions [9.5]

A *relation* is any set of ordered pairs. The set of all first coordinates is called the *domain* of the relation, and the set of all second coordinates is the *range* of the relation. A *function* is a rule that pairs each element in one set, called the *domain,* with exactly one element from a second set, called the *range*. A function is a relation in which no two different ordered pairs have the same first coordinates. Also, if a vertical line crosses the graph of a relation in more than one place, the relation cannot be a function. If no vertical line can be found that crosses the graph in more than one place, the relation must be a function.

5. The relation $\{(8, 1), (6, 1), (-3, 0)\}$ is also a function because no ordered pairs have the same first coordinates. The domain is $\{8, 6, -3\}$ and the range is $\{1, 0\}$.
 The graph of $x = y^2$ shown in Figure 5 in Section 9.5 fails the vertical line test. It is not the graph of a function.

Function Notation [9.6]

The alternative notation for y is $f(x)$. It is read "f of x" and can be used instead of the variable y when working with functions. The notation y and the notation $f(x)$ are equivalent; that is, $y = f(x)$.

6. If $f(x) = 5x - 3$, then
$f(0) = 5(0) - 3 = -3$
$f(1) = 5(1) - 3 = 2$
$f(-2) = 5(-2) - 3 = -13$
$f(a) = 5a - 3$

Solve the following quadratic equations. [9.1, 9.2, 9.3]

1. $x^2 + 2x - 15 = 0$

2. $(x - 4)^2 = 24$

3. $\left(x - \dfrac{3}{5}\right)^2 = \dfrac{32}{25}$

4. $\dfrac{x^2}{6} = \dfrac{3}{2} + \dfrac{3x}{4}$

5. $3x^2 + x - 10 = 0$

6. $(x + 3)(x + 2) = 2$

7. $9x^2 - 24x + 16 = 0$

8. Find the height of the isosceles triangle. [9.1]

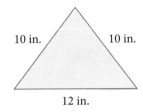

10 in. 10 in.

12 in.

Solve by completing the square. [9.2]

9. $x^2 - 8x + 4 = 0$

10. $4n^2 + 7n - 3 = 0$

Solve using the quadratic formula. [9.3]

11. $3x^2 - 3x = 33$

12. $(x - 5)(x + 2) = 4$

Graph the following equations. [9.4]

13. $y = x^2 + 5$

14. $y = (x - 4)^2$

15. $y = (x + 3)^2 + 1$

16. $y = x^2 + 2x - 3$

Match each equation with its graph. [9.4]

17. $y = (x - 4)^2$

18. $y = x^2 - 4$

19. $y = (x + 4)^2$

20. $y = x^2 + 4$

a.

b.

c.

d.

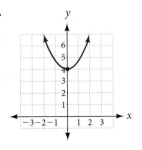

Match each equation with its graph. [9.4]

21. $y = (x - 3)^2 + 1$

22. $y = (x + 1)^2 + 3$

23. $y = (x + 3)^2 - 1$

24. $y = (x - 1)^2 - 3$

a.

b.

c.

d.

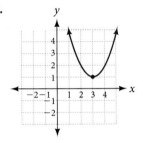

For each of the following relations, give the domain and range, and indicate which are also functions. [9.5]

25. $(a, 4), (5, a), (x, 8), (12, x)$

26. $(n, 4), (1, 3), (n, 7), (n, 8)$

Let $f(x) = 5x^2 - 3x + 1$ and $g(x) = 4x - 2$. Evaluate the following: [9.6]

27. $f(4)$

28. $g\left(\dfrac{1}{2}\right)$

29. $f(2) - g(4)$

30. $f(g(3))$

Simplify.

1. $\dfrac{3(-7) - 4(6)}{9(-1)}$

2. $\dfrac{5}{9} \div \left(-\dfrac{10}{21}\right)$

3. $\dfrac{x^{-6}}{x^4}$

4. $\dfrac{(2x^3)(9x^4)}{6x^{12}}$

5. $\dfrac{\frac{4x^7}{3y^2}}{\frac{16x^3}{9y^8}}$

6. $\dfrac{1 - \frac{9}{x^2}}{1 + \frac{1}{x} - \frac{12}{x^2}}$

7. $\sqrt{49}$

8. $-\sqrt{256}$

9. $\dfrac{10}{\sqrt{6}}$

10. $\dfrac{3}{\sqrt{11} - \sqrt{2}}$

11. $(a - 3)(a^2 + 4a - 6)$

12. $\dfrac{x^2 + 11x + 28}{x + 7}$

Simplify. Assume all variables represent positive numbers.

13. $\sqrt{196x^5y^4}$

14. $\sqrt{\dfrac{432b^2}{16}}$

Solve each equation.

15. $a - \dfrac{1}{2} = \dfrac{3}{2}a$

16. $x + \dfrac{5}{6} = \dfrac{1}{3}$

17. $5 - 2(7x - 5) = -4x$

18. $5x^3 = 13x^2 + 6x$

19. $3x = -18x^2$

20. $(3x - 5)^2 = 16$

21. $\dfrac{x}{x + 3} = \dfrac{8}{5}$

22. $\dfrac{x}{3} = \dfrac{4}{x + 4}$

Graph the solution set on the number line.

23. $-5 \le 3x - 2$

24. $x < -2$

Graph on a rectangular coordinate system.

25. $3x - y = 7$

26. $y = \dfrac{2}{3}x - 4$

27. $y = x^2 + 4$

28. $y = (x + 1)^2 - 6$

Solve each system.

29. $x - y = 6$
 $4x + 3y = -4$

30. $x = 3 - 5y$
 $2x - 6y = 22$

Factor completely.

31. $x^2 - 2x - 35$

32. $6x^2 + 11x - 35$

33. $81 - a^2$

34. $25x^2 + 20x + 4$

35. Find the slope of the line that passes through the points $(5, -3)$ and $(2, -1)$.

36. Find the slope and y-intercept for the equation $15x - 9y = 18$.

37. Divide $18a^2b^4 - 24ab^3 + 15a^3b$ by $3ab$.

38. Divide $\dfrac{6x^3 - 5x^2 - 7x - 8}{2x - 1}$.

Subtract.

39. $\dfrac{a - 2}{a - 6} - \dfrac{a + 7}{a - 6}$

40. $\dfrac{2}{x^2 + 6x + 8} - \dfrac{1}{x^2 + 7x + 12}$

Combine the expressions. Assume all variables represent positive numbers.

41. $2\sqrt{108} + 2\sqrt{48}$

42. $3x\sqrt{96} - \sqrt{294x^2}$

43. **Investing** Randy invested $25,000 in two accounts. One paid 6% simple interest, and the other paid 7%. He earned $460 more in interest on the 6% account than he did on the 7% account. How much did he invest in each account?

44. **Investing** Ms. Jones invested $17,000 in two accounts. One account pays 5% simple interest, and the other pays 9%. Her total interest for the year was $1,010. How much did she have in each account?

45. **Direct Variation** y varies directly as the cube root of x. If $y = 9$ when $x = 27$, find y when $x = 64$.

46. **Inverse Variation** y varies inversely as the square of x. If $y = 27$ when $x = 1$, find y when $x = 3$.

The illustration shows the percentage of Netflix customers choosing each of their plans. Use the information to answer the following questions.

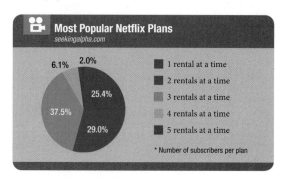

47. If 350,000 Netflix customers were surveyed, how many had signed up for the 3 rental at a time plan?

48. If 451,375 people reported choosing the 1 rental at a time plan, how many people were surveyed? Round to the nearest person

Solve the following quadratic equations. [9.1, 9.2, 9.3]

1. $x^2 - 8x - 9 = 0$

2. $(x - 2)^2 = 18$

3. $\left(x - \dfrac{7}{3}\right)^2 = \dfrac{50}{9}$

4. $\dfrac{1}{3}x^2 = \dfrac{5}{6}x - \dfrac{1}{2}$

5. $4x^2 + 7x - 2 = 0$

6. $(x - 4)(x + 1) = 6$

7. $4x^2 - 20x + 25 = 0$

8. Find the height of the isosceles triangle. [9.1]

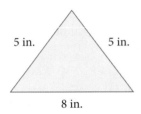

5 in. 5 in.

8 in.

Solve by completing the square. [9.2]

9. $x^2 - 3x - 7 = 0$

10. $2y^2 + 5y - 2 = 0$

Solve using the quadratic formula. [9.3]

11. $5x^2 - 7x = 22$

12. $(x - 3)(x + 4) = 8$

Graph the following equations. [9.4]

13. $y = x^2 + 3$

14. $y = (x + 2)^2$

15. $y = (x - 4)^2 + 2$

16. $y = x^2 - 6x + 8$

Match each equation with its graph. [9.4]

17. $y = x^2 - 3$

18. $y = (x + 3)^2$

19. $y = x^2 + 3$

20. $y = (x - 3)^2$

a.

b.

c.

d.

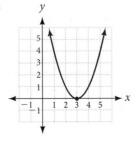

Match each equation with its graph. [9.4]

21. $y = (x + 2)^2 - 4$

22. $y = (x + 2)^2 + 4$

23. $y = (x - 4)^2 - 2$

24. $y = (x + 4)^2 - 2$

a.

b.

c.

d.

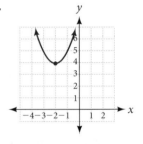

For each of the following relations, give the domain and range, and indicate which are also functions. [9.5]

25. $(x, 3), (7, x), (y, 8), (10, y)$

26. $(a, 8), (7, 4), (a, 9), (a, 12)$

Let $f(x) = 4x^2 - 7x + 1$ and $g(x) = 2x - 3$. Evaluate the following. [9.6]

27. $f(2)$

28. $g\left(\dfrac{1}{2}\right)$

29. $f(2) - g\left(\dfrac{1}{2}\right)$

30. $f(g(4))$

Answers to Odd-Numbered Problems

Chapter 0

Exercise Set 0.1

Vocabulary Review **1.** prime number **2.** composite number
3. numerator **4.** denominator **5.** equivalent **6.** lowest terms
Problems **1.** Prime **3.** Composite; 3, 5, and 7 are factors
5. Composite; 3 is a factor **7.** Prime **9.** $2^2 \cdot 3$
11. 3^4 **13.** $5 \cdot 43$ **15.** $3 \cdot 5$ **17.** $\frac{1}{2}$ **19.** $\frac{2}{3}$ **21.** $\frac{4}{5}$ **23.** $\frac{9}{5}$
25. $\frac{7}{11}$ **27.** $\frac{3}{5}$ **29.** $\frac{1}{7}$ **31.** $\frac{7}{9}$ **33.** $\frac{7}{5}$ **35.** $\frac{1}{5}$ **37.** $\frac{11}{7}$
39. $\frac{5}{3}$ **41.** $\frac{8}{9}$ **43.** $\frac{42}{55}$ **45.** $\frac{17}{19}$ **47.** $\frac{14}{33}$
49. a. $\frac{2}{17}$ **b.** $\frac{3}{36}$ **c.** $\frac{1}{9}$ **d.** $\frac{3}{28}$ **e.** $\frac{2}{19}$
51. a. $\frac{1}{45}$ **b.** $\frac{1}{30}$ **c.** $\frac{1}{18}$ **d.** $\frac{1}{15}$ **e.** $\frac{1}{10}$
53. a. $\frac{1}{3}$ **b.** $\frac{5}{6}$ **c.** $\frac{1}{5}$ **55.** $\frac{9}{16}$
57. $\frac{1}{2} = \frac{2}{4} = \frac{4}{8} = \frac{8}{16}$ **59.** $\frac{5}{4} = \frac{10}{8} = \frac{20}{16}$

61. $\frac{1}{4}$ **63. a.** More than half **b.** Less than one-third
65. $\frac{2}{3}$ **67.** $\frac{7}{15}$ **69.** $\frac{1}{4}$ **71.** 0
73.

Duration	Total Cost	Cost per Day
1 day	$80	$80
2 days	$158	$79
3 days	$199	$66.33
4 days	$219	$54.75
5 days	$231	$46.20

Source: disneyland.disney.go.com

75. a. 2.47 times **b.** 7.25 times **77.** a and b **79.** c
81. 1,482 **83.** 39,968 **85.** 9 **87.** 7
Find the Mistake **1.** A number is a divisor of another number if it
divides it without a remainder. **2.** 31 is a prime number because
the only numbers that divide it without a remainder are 31 and 1.
3. Factoring 112 into the product of primes gives us $2^4 \cdot 7$.
4. To get the fraction $\frac{24}{36}$ in lowest terms, we divide the numerator
and the denominator by $2 \cdot 2 \cdot 3$.

Exercise Set 0.2

Vocabulary Review **1.** multiplication **2.** division **3.** reciprocal
4. least common denominator **5.** factors
Problems **1.** $\frac{8}{15}$ **3.** $\frac{7}{8}$ **5.** 1 **7.** $\frac{27}{4}$ **9.** 1 **11.** $\frac{1}{24}$
13. $\frac{24}{125}$ **15.** $\frac{105}{8}$ **17.** $\frac{8}{27}$ **19.** $\frac{3}{5}$ **21.** 9 **23.** 1
25. 8 **27.** $\frac{1}{15}$ **29.** $\frac{15}{4}$ **31.** $\frac{4}{3}$ **33.** 9 **35.** 200
37. $\frac{3}{8}$ **39.** 1 **41.** $\frac{49}{64}$ **43.** $\frac{3}{4}$ **45.** $\frac{15}{16}$ **47.** $\frac{1}{6}$
49. 6 **51.** $\frac{5}{18}$ **53.** $\frac{9}{2}$ **55.** $\frac{2}{9}$ **57.** $\frac{2}{3}$ **59.** $\frac{1}{4}$

61. $\frac{1}{2}$ **63.** $\frac{1}{3}$ **65.** $\frac{3}{2}$ **67.** $\frac{x+6}{2}$ **69.** $\frac{4}{5}$
71. $\frac{10}{3}$ **73.** $\frac{7}{9}$ **75.** $\frac{7}{3}$ **77.** $\frac{7}{4}$ **79.** $\frac{7}{6}$ **81.** $\frac{9}{20}$
83. $\frac{7}{10}$ **85.** $\frac{19}{24}$ **87.** $\frac{13}{60}$ **89.** $\frac{31}{100}$ **91.** $\frac{67}{144}$
Find the Mistake **1.** To multiply the fractions $\frac{2}{8}$ and $\frac{3}{4}$, we put
the product of 2 and 3 in the numerator over the product of 8 and
4 in the denominator. **2.** The reciprocal of $\frac{5}{20}$ is $\frac{20}{5}$.
3. To divide $\frac{7}{8}$ by $\frac{2}{5}$, multiply $\frac{7}{8}$ by $\frac{5}{2}$. **4.** The least common
denominator for the fractions $\frac{3}{7}$ and $\frac{5}{6}$ is 42.

Exercise Set 0.3

Vocabulary Review **1.** multiply **2.** divide **3.** decimal point
4. tenths **5.** hundredths
Problems **1.** Three tenths **3.** Fifteen thousandths
5. Three and four tenths **7.** Fifty-two and seven tenths
9. $405\frac{36}{100}$ **11.** $9\frac{9}{1,000}$ **13.** $1\frac{234}{1,000}$ **15.** $\frac{305}{100,000}$
17. Tens **19.** Tenths **21.** Hundred-thousandths
23. Ones **25.** Hundreds **27.** Hundred-thousandths
29. 0.55 **31.** 6.9 **33.** 11.11 **35.** 100.02 **37.** 3,000.003

Round to the Nearest				
	Whole Number	Tenth	Hundredth	Thousandth
39.	48	47.5	47.55	47.548
41.	1	0.8	0.82	0.818
43.	0	0.2	0.16	0.156
45.	2,789	2,789.3	2,789.32	2,789.324
47.	100	100.0	100.00	100.000

49. 0.002 0.005 0.02 0.025 0.05 0.052 **51.** 7.451 and 7.54
53. $\frac{1}{4}$ **55.** $\frac{7}{20}$ **57.** $\frac{1}{8}$ **59.** $\frac{5}{8}$ **61.** $\frac{7}{8}$ **63.** $\frac{3}{250}$

65. Building	Height in Words
Burj Khalifa	Two thousand, seven hundred sixteen
Taipei 101	One thousand, six hundred sixty-seven
World Financial Center	One thousand, six hundred fourteen
Petronus Tower 1	One thousand, four hundred eighty-three
Petronus Tower 2	One thousand, four hundred eighty-three

67. 9.99 **69.** 10.05 **71.** 0.05 **73.** 0.01
Find the Mistake **1.** The value of each column to the left of the
decimal point is 10 times the value of the column on its right.
2. The fourth column to the right of the decimal point is called
the ten thousandths column. **3.** 2,568 rounded to the nearest
hundred is 2,600. **4.** 2.568 rounded to the nearest tenth is 2.6.

Exercise Set 0.4

Vocabulary Review 1. add **2.** multiplying, right
3. divide, above
Problems 1. 6.19 **3.** 1.13 **5.** 6.29 **7.** 9.042 **9.** 8.021
11. 11.7843 **13.** 24.342 **15.** 24.111 **17.** 258.8414
19. 666.66 **21.** 38.0916 **23.** 11.11 **25.** 3.57 **27.** 4.22
29. 120.41 **31.** 44.933 **33.** 7.673 **35.** 530.865
37. 8.058 **39.** 27.89 **41.** 35.64 **43.** 411.438
45. 6 **47.** 1 **49.** 3.1 **51.** 5.9 **53.** 3.272
55. 4.001 **57.** 0.28 **59.** 0.028 **61.** 0.0027 **63.** 0.78
65. 0.792 **67.** 0.0156 **69.** 24.29821 **71.** 0.03
73. 187.85 **75.** 0.002 **77.** 27.96 **79.** 0.43 **81.** 49,940
83. 9,876,540 **85.** 1 **87.** 1.89 **89.** 0.0025 **91** 5.1106
93. 7.3485 **95.** 4.4 **97.** 2.074 **99.** 3.58 **101.** 187.4
103. 116.64 **105.** 20.75 **107.** 19.7 **109.** 6.2 **111.** 5.2
113. 11.04 **115.** 4.8 **117.** 9.7 **119.** 2.63 **121.** 4.24
123. 2.55 **125.** 1.35 **127.** 6.5 **129.** 9.9 **131.** 0.05
133. 89 **135.** 2.2 **137.** 12 **139.** 1.35 **141.** 16.97
143. 0.25 **145.** 2.71 **147.** 7.5 miles **149.** \$7.55/hour
151. 22.4 miles **153.** 3 hours **155.** 7 minutes
157. Federal budget: $-\$1.102$ trillion; Congressional plan: $-\$.999$
 trillion. The alternative plan is better.

Find the Mistake 1. To subtract 2.35 from 16.854, line up the
decimals and subtract as usual. **2.** The product of 7.19 and 189.4
will have 3 digits to the right of the decimal. **3.** To estimate the
answer to the problem 6.79×28.6, we round the numbers to 7 and
30 and multiply to get 210. **4.** When dividing with decimal
numbers, the decimal point in the answer is placed directly above
the decimal in the dividend.

Landmark Review 1. Composite, 5^2 **2.** Prime **3.** Prime
4. Composite, $3 \cdot 7$ **5.** $\frac{1}{3}$ **6.** $\frac{16}{15}$ **7.** 1 **8.** $\frac{1}{2}$ **9.** $15\frac{7}{10}$
10. $\frac{46}{10,000}$ **11.** 1.4 **12.** 11.11 **13.** 6.084 **14.** 120.12
15. 0.69 **16.** 6.098 **17.** 35.667 **18.** 88.667

Exercise Set 0.5

Vocabulary Review 1. percent **2.** denominator **3.** left
4. right **5.** fraction **6.** decimal
Problems 1. $\frac{20}{100}$ **3.** $\frac{60}{100}$ **5.** $\frac{24}{100}$ **7.** $\frac{65}{100}$ **9.** $\frac{48}{100}$
11. $\frac{10}{100}$ **13.** 0.23 **15.** 0.92 **17.** 0.09 **19.** 0.034
21. 0.0634 **23.** 0.009 **25.** 23% **27.** 92% **29.** 45%
31. 3% **33.** 60% **35.** 80% **37.** 27% **39.** 123% **41.** 70%
43. $\frac{3}{5}$ **45.** $\frac{3}{4}$ **47.** $\frac{1}{25}$ **49.** $\frac{53}{200}$ **51.** $\frac{7,187}{10,000}$ **53.** $\frac{3}{400}$
55. $\frac{1}{16}$ **57.** $\frac{1}{3}$ **59.** $\frac{1}{40}$ **61.** 50% **63.** 75% **65.** $33\frac{1}{3}$%
67. 80% **69.** 87.5% **71.** 14% **73.** 325% **75.** 150%
77. 250% **79.** 48.8% **81.** 78.4% **83.** 11.8%
85. 72.2% **87.** 71.0% **89.** 0.50; 0.75
91. a. $\frac{49}{100}, \frac{7}{25}, \frac{7}{50}, \frac{1}{25}, \frac{3}{100}, \frac{1}{50}, \frac{1}{100}$
 b. 0.49, 0.28, 0.14, 0.04, 0.03, 0.02, 0.01 **93.** 20%
95. $\frac{4}{25} = 16\%, \frac{3}{25} = 12\%, \frac{11}{50} = 22\%, \frac{1}{10} = 10\%, \frac{3}{20} = 15\%, \frac{1}{4} = 25\%$
97. $\frac{1}{17} \approx 5.9\%; \frac{1}{19} \approx 5.3\%$ **99.** 48%, 73% **101.** 38%

Find the Mistake 1. The fraction $\frac{36}{100}$ given as a percent is 36%.
2. Changing the percent 57.4% to a decimal gives us 0.574.
3. Changing the decimal 0.1465 to a percent gives us 14.65%.
4. To change a fraction to a percent, first change the fraction to a
 decimal and then change to a percent.

Exercise Set 0.6

Vocabulary Review 1. base, exponent **2.** factor **3.** first
4. zero **5.** sum **6.** difference **7.** product
8. quotient **List of steps:** 3, 2, 1, 4
Problems 1. Base 4, Exponent 5 **3.** Base 3, Exponent 6
5. Base 8, Exponent 2 **7.** Base 9, Exponent 1
9. Base 4, Exponent 0 **11.** Base 4, Exponent 8
13. 36 **15.** 8 **17.** 1 **19.** 1 **21.** 81 **23.** 10
25. 12 **27.** 1 **29.** 27 **31.** 12 **33.** 100 **35.** 4
37. 43 **39.** 16 **41.** 84 **43.** 14 **45.** 74 **47.** 12,768
49. 104 **51.** 416 **53.** 66 **55.** 21 **57.** 7 **59.** 16
61. 84 **63.** 40 **65.** 41 **67.** 18 **69.** 405 **71.** 124
73. 11 **75.** 91 **77.** 7 **79.** $\frac{1}{2}$ **81.** $\frac{9}{100}$ **83.** 3
85. 40 **87.** $\frac{7}{10}$ **89.** 13 **91.** 12 **93.** 186 **95.** 646
97. $8(4 + 2) = 48$ **99.** $2(10 + 3) = 26$ **101.** $3(3 + 4) + 4 = 25$
103. $(20 \div 2) - 9 = 1$ **105.** $(8 \cdot 5) + (5 \cdot 4) = 60$

Find the Mistake 1. The expression 8^5 is read "8 to the 5th power."
2. In the expression 16^3, the 16 is used as a factor three times.
3. To simplify $9^2 + 4 - 6 \cdot 7$ using the order of operations, we first
work the exponent. **4.** To work the problem $5 + 6 - 2 + 18$
using the order of operations, we add and subtract left to right.

Chapter 0 Review

1. $3^2 \cdot 5^2$ **2.** $2 \cdot 3^3 \cdot 5$ **3.** $2^2 \cdot 7 \cdot 11$ **4.** $3^3 \cdot 7$ **5.** $\frac{7}{9}$
6. $\frac{4}{13}$ **7.** $\frac{5}{12}$ **8.** $\frac{8}{17}$ **9.** $\frac{6}{7}$ **10.** $\frac{5}{9}$ **11.** $\frac{3}{7}$ **12.** $\frac{6}{7}$
13. $\frac{29}{24}$ **14.** $\frac{5}{24}$ **15.** $\frac{1}{18}$ **16.** $\frac{91}{96}$ **17.** $\frac{3}{10}$ **18.** 3 **19.** $\frac{3}{5}$
20. $\frac{49}{200}$ **21.** $\frac{88}{125}$ **22.** $\frac{17}{50}$ **23.** 22.001 **24.** 17.035
25. 23.129 **26.** 48.13 **27.** 24.39 **28.** 12.8115 **29.** 19.95
30. 17.0238 **31.** 4.375 **32.** 0.7 **33.** 0.78, $\frac{39}{50}$ **34.** 0.90, $\frac{9}{10}$
35. 88%, $\frac{22}{25}$ **36.** 125%, $\frac{5}{4}$ **37.** 52%, 0.52 **38.** 35%, 0.35
39. 1 **40.** 256 **41.** 125 **42.** 9 **43.** 7 **44.** 16
45. $\frac{1}{6}$ **46.** 36 **47.** 18 **48.** 37 **49.** 0 **50.** 6
51. 0.2012, $\frac{503}{2500}$ **52.** 5.11%

Chapter 0 Test

1. $2 \cdot 3^2 \cdot 5$ **2.** $2^2 \cdot 3 \cdot 7$ **3.** $2 \cdot 3^2 \cdot 5 \cdot 7$ **4.** $2^3 \cdot 3 \cdot 7$
5. $\frac{9}{16}$ **6.** $\frac{3}{5}$ **7.** $\frac{4}{7}$ **8.** $\frac{11}{13}$ **9.** $\frac{5}{9}$ **10.** $\frac{4}{7}$ **11.** 14
12. $\frac{12}{7}$ **13.** $\frac{1}{2}$ **14.** $\frac{53}{36}$ **15.** $\frac{1}{4}$ **16.** $\frac{47}{48}$ **17.** $\frac{2}{7}$
18. $\frac{5}{18}$ **19.** $\frac{12}{25}$ **20.** $\frac{5}{16}$ **21.** $\frac{7}{16}$ **22.** $\frac{91}{200}$
23. 12.02 **24.** 28.076 **25.** 22.273 **26.** 21.783
27. 13.59 **28.** 12.59 **29.** 25.28 **30.** 52.0312 **31.** 5.25
32. 3.5 **33.** 0.48, $\frac{12}{25}$ **34.** 0.36, $\frac{9}{25}$ **35.** 55%, $\frac{11}{20}$
36. 180%, $\frac{9}{5}$ **37.** 62.5%, 0.625 **38.** 18.75%, 0.1875 **39.** 64
40. 1 **41.** 25 **42.** 81 **43.** 1 **44.** 10 **45.** $\frac{19}{6}$ **46.** 8
47. 15 **48.** 13 **49.** 38 **50.** 5 **51.** 0.42, $\frac{21}{50}$ **52.** 36%

Chapter 1

Exercise Set 1.1

Vocabulary Review 1. real number line **2.** coordinate
3. real **4.** equivalent **5.** nonzero **6.** absolute value
7. opposites **8.** algebraic expression, variables
Problems 1-7.

9. $\frac{18}{24}$ **11.** $\frac{12}{24}$ **13.** $\frac{15}{24}$ **15.** $\frac{36}{60}$ **17.** $\frac{22}{60}$ **19.** $\frac{50}{60}$
21. 6 **23.** 22 **25.** 3 **27.** 7 **29.** 3 **31.** 17
33. $-10, 10$ **35.** $-\frac{3}{4}, \frac{3}{4}$ **37.** $-\frac{11}{2}, \frac{11}{2}$ **39.** 3, 3
41. $\frac{2}{5}, \frac{2}{5}$ **43.** $-x, |x|$ **45.** $<$ **47.** $>$ **49.** $>$
51. $>$ **53.** $<$ **55.** $<$ **57.** $x + 5 = 14$
59. $5y < 30$ **61.** $5y \geq y - 16$ **63.** $\frac{x}{3} = x + 2$
65. 4 inches; 1 square inch **67.** 4.5 inches; 1.125 square inches
69. 10.25 centimeters; 5 square centimeters **71.** $-8, -2$
73. $-64°\,F; -54°\,F$ **75.** $-15°\,F$ **77.** -100 feet; -105 feet
79. 93.5 square inches, 39 inches **81.** 1,387 calories
83. 654 more calories **85. a.** \$263,800,000 **b.** False **c.** True
Find the Mistake **1.** Writing the fraction $\frac{4}{11}$ as an equivalent fraction with a denominator of 33 gives us $\frac{12}{33}$. **2.** A real number and its opposite will have the same absolute value.
3. In the algebraic expression $2x - y$, the x and the y are the variables. **4.** The perimeter of a rectangle with a length of 32 and a width of 8 is 80.

Exercise Set 1.2

Vocabulary Review **1.** same **2.** different **3.** opposite
4. complementary **5.** supplementary
Problems **1.** 3 **3.** -7 **5.** -14 **7.** -3 **9.** -25
11. -12 **13.** -19 **15.** -25 **17.** -3 **19.** -6 **21.** 0
23. -10 **25.** -16 **27.** -12 **29.** -7 **31.** 35
33. 0 **35.** 9 **37.** -8 **39.** -4 **41.** -28 **43.** 6
45. 6 **47.** 8 **49.** 18 **51.** 10 **53.** 17 **55.** 13
57. 3 **59.** 15 **61.** 1 **63.** 1 **65.** 38 **67.** 68
69. $5 + 9 = 14$ **71.** $[-7 + (-5)] + 4 = -8$
73. $-7 - 4 = -11$ **75.** $[4 + (-5)] - 17 = -18$
77. $8 - 5 = 3$ **79.** $8 - (-5) = 13$ **81.** 3 **83.** 10
85. $35°$ **87.** $60°$ **89.** $73 + 10 - 8; 75°\,F$ **91.** 188 degrees
93. 265 degrees **95.** \$2,000 **97.** 2006; \$500
99. a. 5.7 million **b.** 8.5 million **c.** 41.3 million
Find the Mistake **1.** The difference of -22 and -17 is -5.
2. To subtract one real number from another, add its opposite.
3. To work the problem $4^3 - (-8 + 4) - 2$ using the order of operations, we work the parentheses, then simplify the exponential expression, and then subtract the 2. **4.** The supplement of an angle $88°$ is 92.

Exercise Set 1.3

Vocabulary Review **1.** absolute values **2.** positive
3. negative **4.** zero
Problems **1.** -42 **3.** -16 **5.** 3 **7.** 121 **9.** 6 **11.** -60
13. 24 **15.** 49 **17.** -27 **19.** 6 **21.** 10 **23.** 9
25. 45 **27.** 14 **29.** -2 **31.** 216 **33.** -2 **35.** -18
37. 29 **39.** 38 **41.** -5 **43.** 37 **45.** 80 **47.** 17
49. 40 **51.** $-\frac{10}{21}$ **53.** -4 **55.** 1 **57.** $\frac{9}{16}$ **59.** 0 **61.** 0
63. 0 **65.** 0 **67.** 4 **69.** 24 **71.** 81 **73.** $\frac{9}{4}$ **75.** $1°\,F$
77. 275 calories **79.** \$264.60 **81.** \$848.20 was gained
Find the Mistake **1.** Multiplying a positive number by a negative number will give a negative answer. **2.** Multiplying two numbers with the same sign will give a positive answer.
3. Simplifying the expression $(-15)(-3)(6)(-2)$ gives us -540.
4. The number -12 raised to the third power is the opposite of the number 12 raised to the third power.

Exercise Set 1.4

Vocabulary Review **1.** reciprocal **2.** positive **3.** negative
4. lowest terms **5.** zero
Problems **1.** -2 **3.** -3 **5.** $-\frac{1}{3}$ **7.** 3 **9.** $\frac{1}{7}$ **11.** 0
13. 9 **15.** -15 **17.** -36 **19.** $-\frac{1}{4}$ **21.** $\frac{16}{15}$ **23.** $\frac{4}{3}$
25. $-\frac{8}{13}$ **27.** -1 **29.** 1 **31.** $\frac{5}{9}$ **33.** $\frac{3}{5}$ **35.** $-\frac{5}{3}$
37. -2 **39.** -3 **41.** Undefined **43.** 0 **45.** 5
47. $-\frac{7}{3}$ **49.** -1 **51.** -7 **53.** $\frac{15}{17}$ **55.** $-\frac{32}{17}$ **57.** $\frac{1}{3}$
59. 1 **61.** 1 **63.** -2 **65.** $\frac{9}{7}$ **67.** $\frac{16}{11}$ **69.** -1
71. -1 **73. a.** 25 **b.** -25 **c.** -25 **d.** -25 **e.** 25
75. a. 10 **b.** 0 **c.** -100 **d.** -20 **77.** 3 **79.** -10
81. -3 **83.** -8 **85.** \$350 **87.** Drops $3.5°\,F$ each hour
89. 14 blankets **91.** 48 bags **93.** 6 **95.** 28
97. a. \$20,000 **b.** \$50,000 **c.** Yes, the projected revenue for 5,000 e-mail addresses is \$10,000, which is \$5,000 more than the list costs. **99.** $24.5°\,F$/minute, $9.4°\,F$/minute
Find the Mistake **1.** The division problem $88 \div 4$ is the same as $88 \cdot \left(\frac{1}{4}\right)$. **2.** Dividing two numbers with unlike signs gives a negative answer. **3.** Dividing -55 by -5 will give us 11.
4. Simplifying $\frac{-3(6)}{9}$ gives us -2.

Landmark Review **1.** $-\frac{3}{4}, \frac{3}{4}$ **2.** 27, 27 **3.** 11 **4.** 9
5. $y + 15 = 27$ **6.** $3x < 24$ **7.** -39 **8.** -18
9. -3 **10.** -30 **11.** b **12.** a **13.** 25 **14.** 50
15. -3 **16.** Undefined **17.** -3 **18.** -4

Exercise Set 1.5

Vocabulary Review **1.** e **2.** b **3.** d **4.** h **5.** c
6. f **7.** j **8.** i **9.** a **10.** g
Problems **1.** $(4 + 2) + x = 6 + x$ **3.** $x + (2 + 7) = x + 9$
5. $(3 \cdot 5)x = 15x$ **7.** $(9 \cdot 6)y = 54y$ **9.** $\left(\frac{1}{2} \cdot 3\right)a = \frac{3}{2}a$
11. $\left(\frac{1}{3} \cdot 3\right)x = x$ **13.** $\left(\frac{1}{2} \cdot 2\right)y = y$ **15.** $\left(\frac{3}{4} \cdot \frac{4}{3}\right)x = x$
17. $\left(\frac{6}{5} \cdot \frac{5}{6}\right)a = a$ **19.** $(3 \cdot 5)xy = 15xy$ **21.** $(15 \cdot 2)xy = 30xy$
23. $8x + 16$ **25.** $8x - 16$ **27.** $4y + 4$ **29.** $18x + 15$
31. $6a + 14$ **33.** $54y - 72$ **35.** $x + 2$ **37.** $-3x + 12y$
39. $12x + 18y$ **41.** $12a - 8b$ **43.** $3x + 2y$ **45.** $4a + 25$
47. $-4a + 11$ **49.** $6x + 12$ **51.** $14x + 38$ **53.** $8x - 17$
55. $2x + 1$ **57.** $6x - 3$ **59.** $-2x + 4$ **61.** $5x + 10$
63. $6x + 5$ **65.** $-3x - 8$ **67.** $5x + 6$ **69.** $6m - 5$
71. $7 + 3x$ **73.** $3x - 2y$ **75.** $0.09x + 180$ **77.** $0.12x + 60$
79. $a + 1$ **81.** $1 - a$ **83.** Commutative
85. Multiplicative inverse **87.** Commutative
89. Distributive **91.** Commutative, associative
93. Commutative, associative **95.** Commutative
97. Symmetric **99.** Commutative **101.** Additive inverse
103. Multiplicative inverse **105.** Symmetric **107.** $3x + 6$
109. $9a + 9b$ **111.** 0 **113.** 0 **115.** 10 **117.** 1
119. No **121.** $8 \div 4 \neq 4 \div 8$
123. $4(2 + 3) = 20$ $(4 \cdot 2) + (4 \cdot 3) = 20$

Find the Mistake **1.** To simplify the problem $8\left(\frac{4}{7}x\right)$, we used the associative property of multiplication. **2.** The distributive property states that multiplication distributes over addition. **3.** The multiplicative identity property states that $a(1) = a$ and $1(a) = a$.
4. The expression $35\left(\frac{2}{3}y + 18z\right)$ is equal to $35\left(\frac{2}{3}y\right) + 35(18z)$ because of the distributive property.

Exercise Set 1.6

Vocabulary Review **1.** expression **2.** term **3.** similar
4. value, variable **5.** parentheses
Problems **1.** $-3x$ **3.** $-a$ **5.** $10x + 2y$ **7.** $8a - 2b$
9. $\frac{9}{2}x - 3$ **11.** $a + 5$ **13.** $5x - 5$ **15.** $4a + 2$
17. $-9x - 2$ **19.** $0.75b - 11$ **21.** $0.17x$ **23.** $-0.16x$
25. $10x - 1$ **27** $21y + 6$ **29.** $-6x + 8$ **31.** $3a - 2$
33. $-4x + 26$ **35.** $y - 10$ **37.** $-11x + 4$ **39.** $2x - 12$
41. $10a + 33$ **43.** $1.1x - 3.2$ **45.** $6y - 46$ **47.** $-\frac{5}{2}x - \frac{7}{2}$
49. $2x$ **51.** $-2x - 9$ **53.** $7x$ **55.** $10y$ **57.** $0.17x + 180$
59. $0.02x + 400$ **61.** 5 **63.** -9 **65.** 4 **67.** 4
69. -37 **71.** -41 **73.** 64 **75.** 64 **77.** 144
79. 144 **81.** 49 **83.** 40 **85.** 3 **87.** 0 **89.** 15
91. 6 **93. a.** $42°F$ **b.** $28°F$ **c.** $-14°F$ **d.** $-32°F$
95. a. $\$52$ **b.** $\$55$ **c.** $\$60$ **97. a.** 20.7 lb/in^2 **b.** 24.7 lb/in^2
c. 34.7 lb/in^2 **99.** $0.29H$; 2.32 hours **101.** $\frac{21}{2}a - 5$
Find the Mistake **1.** Simplifying the expression $12z + 9z - 3z$
by combining similar terms gives us $18z$. **2.** The value of
the expression $8y + \frac{1}{3}y$ when $y = 3$ is 25. **3.** To simplify the
expression $-3(5x - 4) - 8x$, first apply distributive property and
then combine similar terms. **4.** The value of the expression
$(2x - 4)^2$ when $x = \frac{1}{2}$ is different from the value when $x = 4$.

Chapter 1 Review

1. $25 - x = 9$ **2.** $7a = 63$ **3.** $5 + n > 10$ **4.** $\frac{10}{y} \le 2$
5. 1 **6.** 0 **7.** $4, 4$ **8.** $-\frac{7}{3}, \frac{7}{3}$ **9.** -6 **10.** 9 **11.** -19
12. 13 **13.** -30 **14.** 60 **15.** -8 **16.** $\frac{9}{16}$ **17.** -17
18. -13 **19.** 2 **20.** -9 **21.** 13 **22.** -4 **23.** e
24. c **25.** d **26.** b **27.** $15 - 2x$ **28.** $-36x$
29. $-12x + 15$ **30.** $6x - 4$ **31.** $-3b + 12$ **32.** $n - 13m$
33. $-2x - y$ **34.** $0.08x + 300$ **35.** $\$596,252,200$
36. $\$1,246,427,700$ **37.** $\$228,095,700$

Chapter 1 Cumulative Review

1. $2^4 \cdot 3 \cdot 5$ **2.** $2 \cdot 3^2 \cdot 13$ **3.** $2 \cdot 3^2 \cdot 17$ **4.** $3^2 \cdot 5 \cdot 11$
5. $\frac{3}{2}$ **6.** $\frac{1}{2}$ **7.** $\frac{5}{24}$ **8.** $\frac{7}{48}$ **9.** 7.245 **10.** 1.167
11. 20.22621 **12.** 2.5 **13.** 0 **14.** 0.06 **15.** 16
16. $\frac{98}{3}$ **17.** 18 **18.** 17 **19.** 9 **20.** 64 **21.** 17
22. 9 **23.** -5 **24.** -9 **25.** 3 **26.** -13 **27.** 60
28. -60 **29.** 6 **30.** -4 **31.** $6x - 12$ **32.** $-3x - 21$
33. $30x - 5$ **34.** $6x - 9$ **35.** $4a + 10$ **36.** $-4a - 2$
37. $8x + 21$ **38.** $-x + 20$ **39.** $6x - 5y$ **40.** $4a - 13$
41. $2x - 3$ **42.** $-x - 2y - 3$ **43.** -9 **44.** -5
45. 3 **46.** 16 **47.** -26 **48.** -3

Chapter 1 Test

1. $15 - x = 12$ **2.** $6a = 30$ **3.** $7 + a > 15$ **4.** $\frac{12}{n} \le 5$
5. -1 **6.** -11 **7.** $6, 6$ **8.** $-\frac{5}{3}, \frac{5}{3}$ **9.** -3 **10.** 12
11. -11 **12.** -7 **13.** -28 **14.** 24 **15.** -3
16. $-\frac{27}{8}$ **17.** -13 **18.** -10 **19.** -7 **20.** 62 **21.** 2
22. -6 **23.** d **24.** e **25.** a **26.** c **27.** $12 + 3x$
28. $-15y$ **29.** $-10x + 15$ **30.** $-2x - 4$ **31.** $-4n + 10$
32. $-2a - 16b$ **33.** $x + y$ **34.** $0.11n + 240$
35. $123,760$ dogs **36.** $82,446$ dogs **37.** $4,771$ dogs

Chapter 2

Exercise Set 2.1

Vocabulary Review **1.** solution set **2.** equivalent
3. addition **4.** subtraction
Problems **1.** Yes **3.** Yes **5.** No **7.** Yes **9.** No
11. Yes **13.** 11 **15.** 4 **17.** $-\frac{3}{4}$ **19.** -5.8 **21.** -17
23. $-\frac{1}{8}$ **25.** -4 **27.** -3.6 **29.** 1 **31.** $-\frac{7}{45}$ **33.** 3
35. $\frac{11}{8}$ **37.** 21 **39.** 7 **41.** 3.5 **43.** 22 **45.** -2 **47.** -16
49. -3 **51.** 10 **53.** -12 **55.** 4 **57.** 2 **59.** -5
61. -1 **63.** -3 **65.** 8 **67.** -8 **69.** 2 **71.** 11
73. a. 280 tickets **b.** $\$2,100$ **75.** $67°$ **77. a.** 225 **b.** $\$11,125$
79.

n	t
32	72
50	90
44	84
96	136

81. $x + 55 + 55 = 180$; $70°$ **83.** y
85. x **87.** 6 **89.** 6 **91.** -9
93. $-\frac{15}{8}$ **95.** 8 **97.** $-\frac{5}{4}$
99. $3x$ **101.** $-9x$

Find the Mistake **1.** The number -2 is a solution to the equation
$3x - 9 = 8x + 1$. **2.** Saying that the number 12 is a solution to
the equation $\frac{1}{6}y + 3 = 5$ is a true statement. **3.** The addition
property of equality states that $A + C = B + C$ if $A = B$.
4. To solve the equation $\frac{5}{6}b + 6 = 11$ for b, begin by adding -6 to
both sides.

Exercise Set 2.2

Vocabulary Review **1.** nonzero **2.** multiplication
3. division **4.** reciprocal
Problems **1.** 2 **3.** 4 **5.** $-\frac{1}{2}$ **7.** -2 **9.** 3 **11.** 4
13. 0 **15.** 0 **17.** 6 **19.** -50 **21.** $\frac{3}{2}$ **23.** 12
25. -3 **27.** 32 **29.** -8 **31.** $\frac{1}{2}$ **33.** 4 **35.** 8
37. -12 **39.** -3 **41.** -4 **43.** 4 **45.** -15 **47.** $-\frac{1}{2}$
49. 3 **51.** 1 **53.** $\frac{1}{4}$ **55.** -3 **57.** 3 **59.** 2
61. $-\frac{3}{2}$ **63.** $-\frac{3}{2}$ **65.** 1 **67.** 1 **69.** -2 **71.** -2
73. 200 tickets **75.** $\$1,390.85$ per month **77.** 31
79. 21 **81.** 20 songs **83. a.** 20 ft. **b.** 35 ft. **c.** 75 ft.
85. 55 hours **87.** 2 **89.** 6 **91.** $3,000$ **93.** 650
95. $3x - 11$ **97.** $0.09x + 180$ **99.** $-6y + 4$ **101.** $4x - 11$
Find the Mistake **1.** The multiplication property of equality states
that multiplying both sides of an equation by the same nonzero
number will not change the solution set. **2.** To solve $-\frac{2b}{9} = 3$
for b, we multiply both sides of the equation by the reciprocal of $-\frac{2}{9}$.
3. To begin solving the equation $\frac{5}{6}x + \frac{1}{2} = -\frac{1}{4}$, we can either add
$-\frac{1}{2}$ to each side of the equation or multiply both sides by the LCD 12.
4. To solve the equation $-\frac{8}{9}y = -3$, we begin by multiplying both
sides of the equation by 9.

Exercise Set 2.3

Vocabulary Review 1. a. distributive **b.** LCD, decimals **c.** similar
2. addition, constant **3.** multiplication **4.** solution
Problems **1.** 3 **3.** -2 **5.** -1 **7.** 2 **9.** -4 **11.** -2
13. 0 **15.** 1 **17.** $\frac{1}{2}$ **19.** 7 **21.** 8 **23.** All real numbers
25. $\frac{3}{4}$ **27.** 75 **29.** 2 **31.** 6 **33.** 8 **35.** 0 **37.** $\frac{3}{7}$
39. No solution **41.** 1 **43.** -1 **45.** 6 **47.** $\frac{3}{4}$
49. 3 **51.** No solution **53.** 8 **55.** 6 **57.** -2
59. -2 **61.** 2 **63.** All real numbers **65.** 2 **67.** 20
69. $4,000$ **71.** 700 **73.** 11 **75.** 7

77. a. $\frac{5}{4} = 1.25$ **b.** $\frac{15}{2} = 7.5$ **c.** $6x + 20$ **d.** 15 **e.** $4x - 20$ **f.** $\frac{45}{2} = 22.5$
79. $x = 3$ or 3 robot events were entered **81.** $4,200
83. 14 **85.** -3 **87.** $\frac{1}{4}$ **89.** $\frac{1}{3}$ **91.** $-\frac{3}{2}x + 3$
93. $\frac{5}{4}x - 5$ **95.** $4x - 3$ **97.** $8x + 3$
Find the Mistake **1.** A linear equation in one variable is any equation that can be put in the form of $Ax + B = C$.
2. $12 + 6y - 4 = 3y$ is a linear equation simplified to $3y + 8 = 0$.
3. To solve the linear equation $3(2x + 7) = 2(x - 1)$, begin by applying the distributive property. **4.** To solve a linear equation, make sure to combine similar terms on each side.

Exercise Set 2.4

Vocabulary Review 1. variable **2.** formula **3.** isolate
4. complementary **5.** supplement
Problems 1. 100 feet **3.** 0 **5.** 2 **7.** 15 **9.** 10
11. -2 **13.** 1 **15. a.** 2 **b.** 4 **17. a.** 5 **b.** 18
19. $l = \frac{A}{w}$ **21.** $h = \frac{V}{lw}$ **23.** $a = P - b - c$
25. $x = 3y - 1$ **27.** $y = 3x + 6$ **29.** $y = -\frac{2}{3}x + 2$
31. $y = -2x - 5$ **33.** $y = -\frac{2}{3}x + 1$ **35.** $w = \frac{P - 2l}{2}$
37. $v = \frac{h - 16t^2}{t}$ **39.** $h = \frac{A - \pi r^2}{2\pi r}$ **41.** $t = \frac{I}{pr}$
43. a. $y = \frac{3}{5}x + 1$ **b.** $y = \frac{1}{2}x + 2$ **c.** $y = 4x + 3$
45. a. $y = -\frac{1}{2}x - 2$ **b.** $y = -3x + 2$ **c.** $y = -\frac{1}{3}x - 1$
47. $y = \frac{3}{7}x - 3$ **49.** $y = 2x + 8$ **51.** $60°; 150°$
53. $45°; 135°$ **55.** 10 **57.** 240 **59.** 25% **61.** 35%
63. 64 **65.** 2,000 **67.** $100°C$; yes **69.** $20°C$; yes
71. $C = \frac{5}{9}(F - 32)$ **73.** 46.3% **75.** 5.3% **77.** 7 meters
79. $\frac{3}{2}$ or 1.5 inches **81.** 132 feet **83.** $\frac{2}{9}$ centimeters
85.

Pitcher, Team	Rolaids Points
Heath Bell, San Diego	144
Brian Wilson, San Francisco	135
Carlos Marmol, Chicago	107
Billy Wagner, Atlanta	128
Francisco Cordero, Cincinnati	130

87. The sum of 4 and 1 is 5. **89.** The difference of 6 and 2 is 4.
91. The difference of a number and 5 is -12.
93. The sum of a number and 3 is four times the difference of that number and 3. **95.** $2(6 + 3) = 18$ **97.** $2(5) + 3 = 13$
99. $x + 5 = 13$ **101.** $5(x + 7) = 30$
Find the Mistake 1. A formula contains more than one variable.
2. To solve for a variable in a formula, make sure the variable is isolated on one side of the equation. **3.** To solve the formula $2ab = a + 8$ for b, divide each side of the equation by $2a$.
4. The question "What percent of 108 is 9?" can be translated into the equation $x \cdot 108 = 9$.

Landmark Review 1. 12 **2.** 25 **3.** 5.2 **4.** $\frac{13}{15}$ **5.** 4
6. 10 **7.** 4 **8.** 3 **9.** 3 **10.** 4 **11.** $\frac{A}{w} = l$
12. $P - a - b = c$ **13.** $18 - 3y = x$ **14.** $-\frac{4}{3}x + 3 = y$

Exercise Set 2.5

Vocabulary Review 1. known **2.** unknown **3.** variable
4. equation **5.** sentence **6.** problem
Problems 1. 8 **3.** 5 **5.** -1 **7.** 3 and 5 **9.** 6 and 14
11. Shelly is 39; Michele is 36 **13.** Evan is 11; Cody is 22
15. Barney is 27; Fred is 31 **17.** Lacy is 16; Jack is 32
19. Patrick is 18; Pat is 38 **21.** $s = 9$ inches **23.** $s = 15$ feet
25. 11 feet, 18 feet, 33 feet **27.** 26 feet, 13 feet, 14 feet

29. $l = 11$ inches; $w = 6$ inches **31.** $l = 25$ meters; $w = 9$ meters
33. $l = 15$ feet; $w = 3$ feet **35.** 9 dimes; 14 quarters
37. 12 quarters; 27 nickels **39.** 8 nickels; 17 dimes
41. 7 nickels; 10 dimes, 12 quarters **43.** 3 nickels; 9 dimes, 6 quarters
45. $5x$ **47.** $1.075x$ **49.** $0.09x + 180$
Find the Mistake 1. False, $7 + \frac{m}{n} = n + 9$ **2.** True
3. False, $2(3x + 6) + 2x = 42$ **4.** False, $10(2x - 4) + 5x = 310$

Exercise Set 2.6

Vocabulary Review
1. consecutive **2.** even **3.** integers **4.** sum
Problems 1. 5 and 6 **3.** -4 and -5 **5.** 13 and 15 **7.** 52 and 54
9. -14 and -16 **11.** 17, 19, and 21 **13.** 42, 44, and 46
15. $4,000 at 8%, $6,000 at 9% **17.** $700 at 10%; $1,200 at 12%
19. $500 at 8%, $1,000 at 9%, $1,500 at 10% **21.** $45°, 45°, 90°$
23. $22.5°, 45°, 112.5°$ **25.** $53°, 90°$ **27.** $80°, 60°, 40°$
29. 16 adult and 22 children's tickets **31.** 16 minutes
33. 39 hours **35.** They are in offices 7329 and 7331.
37. Kendra is 8 years old and Marissa is 10 years old. **39.** Jeff
41. $10.38 **43.** $l = 12$ meters; $w = 10$ meters **45.** $59°, 60°, 61°$
47. 18 cm, 24 cm, 36 cm **49.** $54.00 **51.** Yes **53.** 5
55. a. 9 **b.** 3 **c.** -9 **d.** -3 **57. a.** -8 **b.** 8 **c.** 8 **d.** -8
59. -2.3125 **61.** $\frac{10}{3}$ **63.** $y = \frac{2}{3}x - 2$ **65.** $x = \frac{2}{3}y + \frac{5}{3}$
Find the Mistake 1. True **2.** False, $0.07(4x) + 0.08x = 396$
3. True **4.** True

Exercise Set 2.7

Vocabulary Review 1. inequalities **2.** positive **3.** negative
4. graph **5.** compound
Problems
1. $x < 12$

3. $a \le 12$

5. $x > 13$

7. $y \ge 4$

9. $x > 9$

11. $x < -6$

13. $x < 2$

15. $a \le 5$

17. $x > 15$

19. $x < -3$

```
◄————●————|————►
     -3    0
```

21. $x \le 6$

```
◄————|————●————►
     0    6
```

23. $x \ge -50$

```
◄————●————|————►
    -50    0
```

25. $y < -6$

```
◄————●————|————►
    -6    0
```

27. $x < -4$

```
◄————●————|————►
    -4    0
```

29. $x > -12$

```
◄————●————|————►
    -12    0
```

31. $x < 6$

```
◄————|————●————►
     0    6
```

33. $y \ge -5$

```
◄————●————|————►
    -5    0
```

35. $x < 3$

```
◄————|————●————►
     0    3
```

37. $x \le 18$

```
◄————|————●————►
     0    18
```

39. $a < -20$

```
◄————●————|————►
    -20    0
```

41. $y < 25$

```
◄————|————●————►
     0    25
```

43. $a \le 3$

```
◄————|————●————►
     0    3
```

45. $x \ge \frac{15}{2}$

```
◄————|————●————►
     0    15
          ──
          2
```

47. $x < -1$

```
◄————●————|————►
    -1    0
```

49. $y \ge -2$

```
◄————●————|————►
    -2    0
```

51. $x < -1$

```
◄————●————|————►
    -1    0
```

53. $m \le -6$

```
◄————●————|————►
    -6    0
```

55. $x \le -5$

```
◄————●————|————►
    -5    0
```

57. $x \le 3$

```
◄————|————●————►
     0    3
```

59. $x \le \frac{25}{9}$

```
◄————|————●————►
     0    25
          ──
          9
```

61. $y < -\frac{3}{2}x + 3$ **63.** $y < \frac{2}{5}x - 2$ **65.** $y \le \frac{3}{7}x + 3$
67. $y \le \frac{1}{2}x + 1$ **69. a.** 3 **b.** 2 **c.** No **d.** $x > 2$
71. $x < 3$ **73.** $x \le 3$

75. $x \le -7$ or $x \ge -3$

```
◄————●————————●————►
    -7         -3
```

77. $x \le -1$ or $x \ge \frac{3}{5}$

```
◄————●————————●————►
    -1          3
                ──
                5
```

79. $x < -10$ or $x > 6$

```
◄————●————————●————►
   -10         6
```

81. $x < -3$ or $x \ge 2$

```
◄————●————————●————►
    -3         2
```

83. $3 \le m \ge 7$

```
◄————————●————●————►
         3    7
```

85. $-3 < a < 0$

```
◄————●————————●————►
    -3          0
```

87. $4 \le a \le 6$

```
◄————————●————●————►
         4    6
```

89. $-4 < x < 2$

```
◄————●————————●————►
    -4         2
```

91. $-3 < x < 3$

```
◄————●————————●————►
    -3         3
```

93. $0 \le x < 6$

```
◄————————●————●————►
         0    6
```

95. $-\frac{2}{3} \le x \le 0$

```
◄————————●————●————►
         -2    0
         ──
         3
```

97. $-\frac{3}{2} \le x \le \frac{17}{2}$

```
◄————————●————●————►
         -3    17
         ──    ──
         2     2
```

99. $\frac{3}{8} \le x \le \frac{3}{2}$

```
◄————————●————●————►
         3    3
         ──   ──
         8    2
```

101. At least 291 **103.** $x < 2$ **105.** $x > -\frac{8}{3}$
107. $x \ge 6$; the width is at least 6 meters. **109.** $x > 6$; the shortest side is even and greater than 6 inches. **111.** $t \ge 100$
113. Lose money if they sell less than 200 tickets. Make a profit if they sell more than 200 tickets.
115. 8 **117.** 24 **119.** 25% **121.** 10% **123.** 80 **125.** 400
127. -5 **129.** 5 **131.** 7 **133.** 9 **135.** 6 **137.** $2x - 3$
Find the Mistake 1. If the solution set to an inequality is all the real numbers greater than 9, the solution is $x > 9$. **2.** To include the number 14 in the solution to the inequality $x \ge 14$, we mark the number on the graph with a closed circle. **3.** The last step in solving an inequality is to graph the solution set.
4. Upon solving the inequality $-5y < 4x + 9$ for y, the inequality symbol is reversed.

Chapter 2 Review

1. No **2.** Yes **3.** Yes **4.** No **5.** 15 **6.** 11
7. 16 **8.** $\frac{1}{2}$ **9.** 6 **10.** $-\frac{3}{4}$ **11.** 5 **12.** 4 **13.** 372
14. 3 **15.** -2 **16.** 7 **17.** 10 **18.** $-\frac{3}{4}$ **19.** 0
20. $y = \frac{3}{2}x - 3$ **21.** $T_1 = \frac{9}{k}x + T_2$ **22.** Brian is 31, Michelle is 27
23. $l = 15$ in., $w = 7$ in. **24.** 3 quarters, 4 dimes
25. \$6,000 at 7% and \$7,500 at 15% **26.** $x \le -12$ **27.** $y < -6$
28. $x > -4$ **29.** $n \ge 2$ **30.** $x < 4$ **31.** $x \le 7$
32. $x \ge -5$ **33.** $x \le 7$ **34.** $x > -3$ **35.** 23.8% **36.** 7.4%

Chapter 2 Cumulative Review

1. -5 **2.** 11 **3.** -3 **4.** -12 **5.** -13 **6.** 12 **7.** $-\frac{27}{8}$
8. 25 **9.** $-\frac{1}{3}$ **10.** $\frac{4}{5}$ **11.** $-\frac{3}{2}$ **12.** $\frac{6}{35}$ **13.** $\frac{3}{4}$
14. $-\frac{7}{4}$ **15.** $\frac{13}{15}$ **16.** $-\frac{7}{32}$ **17.** $6x$ **18.** $-32x$
19. $6x - 8$ **20.** $-3x + 1$ **21.** -3 **22.** 4 **23.** -25
24. 9 **25.** 8 **26.** 13 **27.** 2 **28.** 2 **29.** $\frac{14}{5}$
30. $-\frac{1}{3}$ **31.** $w = \frac{P - 2l}{2}$ **32.** $y = -\frac{8}{9}x + \frac{8}{3}$ **33.** $x \ge 5$
34. $x < -6$ **35.** $x \ge 7$ **36.** $x > -1$ **37.** $y - 7 = 17$
38. $x + 3 = -5$ **39.** -5 **40.** -16 **41.** 28 **42.** 36%
43. 21 **44.** -3 **45.** $53°, 90°$ **46.** $41°$ **47.** 18 hours
48. 5 oz. **49.** 13,400,000 viewers **50.** 323,000 viewers

Chapter 2 Test

1. No **2.** Yes **3.** Yes **4.** No **5.** 17 **6.** 5
7. 5.2 **8.** 3 **9.** -5 **10.** 1 **11.** 4 **12.** 1 **13.** 55
14. -3 **15.** $\frac{5}{2}$ **16.** 2 **17.** 5 **18.** -1 **19.** 8
20. $y = -\frac{1}{3}x + 2$ **21.** $a = \frac{x^2 - v^2}{2d}$ **22.** Paul is 36, Becca is 18
23. $l = 55$ cm, $w = 20$ cm **24.** 14 dimes, 6 nickels
25. \$700 at 6% and \$1,200 at 12% **26.** $x > 10$ **27.** $y \ge -4$
28. $x > -4$ **29.** $n \le -2$ **30.** $x < 1$ **31.** $x \ge 2$ **32.** $x > -3$
33. $x \ge -1$ **34.** $x < -2$ **35.** 27.9% **36.** 20.4%

Chapter 3

Exercise Set 3.1

Vocabulary Review **1.** scatter diagram/plot **2.** line graph
3. ordered pair **4.** x–axis, y–axis **5.** quadrants **6.** origin
Problems
1.

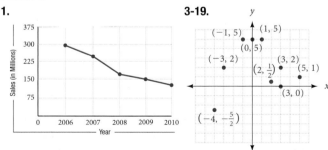

3-19.

21. $(-4, 4)$; QII **23.** $(-4, 2)$; QII **25.** $(-3, 0)$; no quadrant, x-axis
27. $(2, -2)$; QIV **29.** $(-5, -5)$; QIII

31. Yes **33.** No **35.** Yes **37.** No

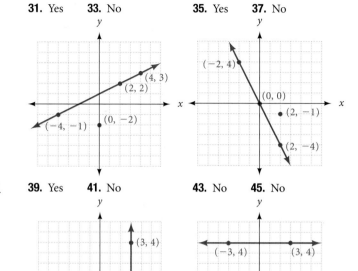

39. Yes **41.** No **43.** No **45.** No

47. a. $(5, 40), (10, 80), (20, 160)$, Answers may vary **b.** \$320
 c. 30 hours **d.** No, if she works 35 hours, she should be paid \$280.
49. $(1999, 70), (2001, 73), (2003, 68), (2005, 75), (2007, 80), (2009, 73)$
51. $A = (1, 2), B = (6, 7)$ **53.** $A = (2, 2), B = (2, 5), C = (7, 5)$
55. a. -3 **b.** 6 **c.** 0 **4.** -4 **57. a.** 4 **b.** 2 **c.** -1 **d.** 9
59. -3 **61.** 2 **63.** 0 **65.** $y = -5x + 4$ **67.** $y = \frac{3}{2}x - 3$
Find the Mistake **1.** A scatter diagram uses dots to show data in
the form of order pairs. **2.** In the ordered pair $(-3, 6)$, the
x–coordinate is -3. **3.** To graph the ordered pair (7, 3), start the
origin, move 7 units to the right, and 3 units up. **4.** The ordered
pair $(-2, 5)$ appears in the second quadrant.

Exercise Set 3.2

Vocabulary Review **1.** first **2.** second **3.** solution
4. graphing **5.** linear **6.** ordered pairs **7.** graph
8. straight line **9.** horizontal, vertical
Problems **1.** $(0, 6), (3, 0), (6, -6)$ **3.** $(0, 3), (4, 0), (-4, 6)$
5. $(1, 1), (\frac{3}{4}, 0), (5, 17)$ **7.** $(2, 13), (1, 6), (0, -1)$
9. $(-5, 4), (-5, -3), (-5, 0)$

11.

x	y
1	3
-3	-9
4	12
6	18

13.

x	y
0	0
$-\frac{1}{2}$	-2
-3	-12
3	12

15.

x	y
2	3
3	2
5	0
9	-4

17.

x	y
2	0
3	2
1	-2
-3	-10

19.

x	y
0	-1
-1	-7
-3	-19
$\frac{3}{2}$	8

21.

x	y
0	3
-2	7
2	-1
5	-7

23. $(0, -2)$ **25.** $(1, 5), (0, -2),$ and $(-2, -16)$

27. $(1, 6)$ and $(0, 0)$ **29.** $(2, -2)$ **31.** $(3, 0)$ and $(3, -3)$

33. $(0, 4), (2, 2), (4, 0)$

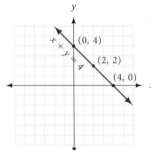

35. $(0, 3), (2, 1), (4, -1)$

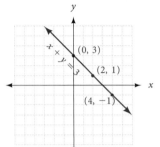

37. $(0, 0), (-2, -4), (2, 4)$

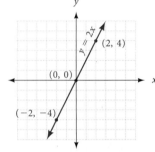

39. $(-3, -1), (0, 0), (3, 1)$

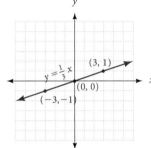

41. $(0, 1), (-1, -1), (1, 3)$

...

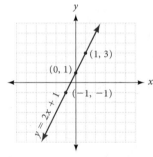

43. $(0, 4), (-1, 4), (2, 4)$

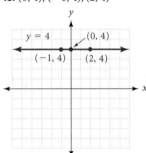

45. $(-2, 2), (0, 3), (2, 4)$

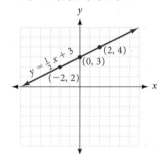

47. $(-3, 3), (0, 1), (3, -1)$

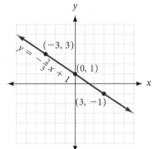

49. $(-1, 5), (0, 3), (1, 1)$

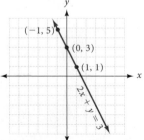

51. $(0, 3), (2, 0), (4, -3)$

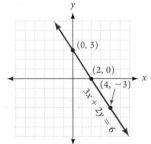

53. $(-2, 2), (0, 3), (2, 4)$

55.

57.

59.

61.

63.

65.

67.

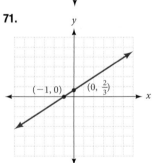

69.

71.

73.

Equation	H, V, and/or O
$x = 3$	V
$y = 3$	H
$y = 3x$	O
$y = 0$	O, H

75.

Equation	H, V, and/or O
$x = -\frac{3}{5}$	V
$y = -\frac{3}{5}$	H
$y = -\frac{3}{5}x$	O
$x = 0$	O, V

77.

x	y
−4	−3
−2	−2
0	−1
2	0
6	2

79. 12 inches
81. a. Yes **b.** No, she should earn $108 for working 9 hours. **c.** No, she should earn $84 for working 7 hours. **d.** Yes
83. a. $375,000 **b.** At the end of 6 years. **c.** No, the crane will be worth $195,000 after 9 years. **d.** $600,000
85. a. 2 **b.** 3 **87. a.** −4 **b.** 2
89. a. 6 **b.** 2

Find the Mistake **1.** A solution to a linear equation in two variable will be an ordered pair that makes the equation a true statement. **2.** A solution to the equation $3y = 6x + 9$ is (2, 7). **3.** To graph a linear equation in two variables, you must find two or more solutions. **4.** The line $y = \frac{1}{2}x$ intersects both axes, or passes through the origin.

Exercise Set 3.3

Vocabulary Review 1. x-intercept **2.** y-intercept
3. x-axis **4.** y-axis
Problems

1.

Equation	x-intercept	y-intercept
$3x + 4y = 12$	4	3
$3x + 4y = 4$	$\frac{4}{3}$	1
$3x + 4y = 3$	1	$\frac{3}{4}$
$3x + 4y = 2$	$\frac{2}{3}$	$\frac{1}{2}$

3.

Equation	x-intercept	y-intercept
$x - 3y = 2$	2	$-\frac{2}{3}$
$y = \frac{1}{3}x - \frac{2}{3}$	2	$-\frac{2}{3}$
$x - 3y = 0$	0	0
$y = \frac{1}{3}x$	0	0

5.

7.

9.

11.

13.

15.

17.

19.

21.

23.

25.

27.

29.

31.

33.

35. a. 0 **b.** $-\frac{3}{2}$ **c.** 1

d.

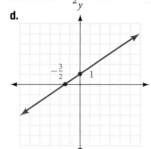

e. $y = \frac{2}{3}x + 1$

37. x-intercept $= 3$; y-intercept $= 5$
39. x-intercept $= -1$; y-intercept $= -3$

41.

43.

45.

47.

49.

51.

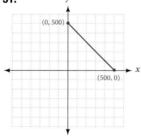

53. a. $10x + 12y = 480$
b. x-intercept $= 48$;
y-intercept $= 40$
c.

d. 10 hours **e.** 18 hours

55. a. $50x + 60y = 450$
b. x-intercept $= 9$;
y-intercept $= 7.5$
c.

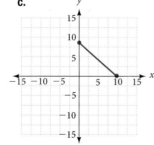

d. 5 hours **e.** 4.17 hours

57. a. $\frac{3}{2}$ **b.** $\frac{3}{2}$ **59. a.** $\frac{3}{2}$ **b.** $\frac{3}{2}$

Find the Mistake **1.** The y–intercept is where the graph of a line crosses the y–axis at the point $(0, y)$. **2.** The x-intercept for the linear equation $9x = \frac{1}{3}y + 5$ is $\left(\frac{5}{9}, 0\right)$. **3.** The y–intercept for the linear equation $2x + 7y = 3$ is $\left(0, \frac{3}{7}\right)$. **4.** The intercepts for the linear equation $15 = 3y - 5x$ are $(-3, 0)$ and $(0, 5)$.

Exercise Set 3.4

Vocabulary Review **1.** ratio **2.** vertical **3.** horizontal
4. slope **5.** zero **6.** undefined
Problems **1.**

3.

5.

7.

9.

11.

13.

15.

17.

19.

21.

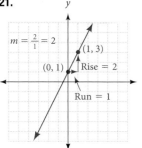

$m = \frac{2}{1} = 2$ (1, 3) (0, 1) Rise = 2 Run = 1

23.

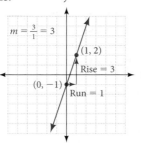

$m = \frac{3}{1} = 3$ (1, 2) (0, −1) Rise = 3 Run = 1

25.

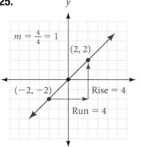

$m = \frac{4}{4} = 1$ (2, 2) (−2, −2) Rise = 4 Run = 4

27. Slope = 3; y-intercept = 2
29. Slope = 2; y-intercept = −2
31.

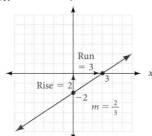

Run = 3 Rise = 2 $m = \frac{2}{3}$

33.

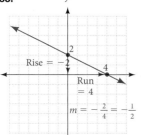

Rise = −2 Run = 4 $m = -\frac{2}{4} = -\frac{1}{2}$

35.

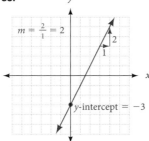

$m = \frac{2}{1} = 2$ y-intercept = −3

37.

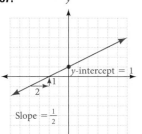

y-intercept = 1 Slope = $\frac{1}{2}$

39. 6 **41.** All real numbers

43.

Equation	Slope
$x = 3$	Undefined
$y = 3$	0
$y = 3x$	3

45.

Equation	Slope
$y = -\frac{2}{3}$	0
$x = -\frac{2}{3}$	Undefined
$y = -\frac{2}{3}x$	$-\frac{2}{3}$

47. Slopes: A, 3.3; B, 3.1, C, 5.3, D, 1.9
49. Slopes: A, −50; B, −75, C, −25

51. a.

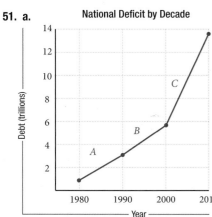

National Deficit by Decade

Debt (trillions) / Year

b. 0.23, 0.25, 0.79
c. The rate of change is increasing

53. 878 finishers. The number of finishers increased by 878 in one year.
55. a. 15,800 feet **b.** $-\frac{7}{100}$
57. 0.64% per year. Between 2003 and 2010 the percentage of computer users who use Macs has increased an average of 0.64% per year.
59. $y = 4x - 2$ **61.** $y = -\frac{3}{2}x + 3$ **63.** $y = \frac{1}{3}x - 2$
65. $y = -\frac{3}{4}x + 3$ **67.** $y = -\frac{2}{5}x - 2$ **69.** $y = -2x + 1$
71. $y = \frac{1}{2}x + 1$ **73.** $y = -\frac{3}{2}x - 1$

Find the Mistake **1.** A horizontal line has zero slope.
2. A line that rises from the left to the right has a positive slope.
3 The slope of the line that passes through the points (−1, 6) and (4, −9) is −3. **4.** Use the point (2, −1) to graph a line with a y-intercept −6 and a slope $\frac{5}{2}$.

Landmark Review **1.** The origin **2.** 4 **3.** No
4. $(0, -3), \left(\frac{3}{4}, 0\right), (3, 9)$ **5.** $(-3, 3), \left(1, \frac{21}{2}\right), \left(6, \frac{33}{2}\right)$
6. x-intercept = 1; y-intercept = 1
7. x-intercept = 6; y-intercept = −3 **8.** $\frac{1}{4}$ **9.** −2
10. No slope **11.** Rise **12.** There isn't one

Exercise Set 3.5

Vocabulary Review **1.** slope-intercept **2.** point-slope **3.** substitute
4. y-intercept **5.** slope **6.** parallel **7.** perpendicular
Problems **1.** $y = \frac{2}{3}x + 1$ **3.** $y = \frac{3}{2}x - 1$ **5.** $y = -\frac{2}{3}x + 3$
7. $y = 2x - 4$

9. $m = 2; b = 4$

$y = 2x + 4$

11. $m = -3; b = 3$

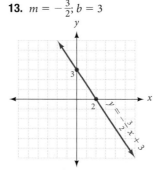

$y = -3x + 3$

13. $m = -\frac{3}{2}; b = 3$

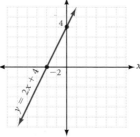

$y = -\frac{3}{2}x + 3$

15. $m = \frac{4}{5}; b = -4$

$y = \frac{4}{5}x - 4$

17. $m = -\frac{2}{5}; b = -2$

$y = -\frac{2}{5}x - 2$

19. $m = \frac{3}{2}; b = -3$

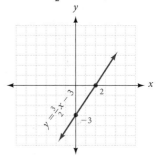

$y = \frac{3}{2}x - 3$

21. $y = 2x - 1$ **23.** $y = -\frac{1}{2}x - 1$ **25.** $y = \frac{3}{2}x - 6$

27. $y = -3x + 1$ **29.** $y = -4$ **31.** $x = -4$ **33.** $y = x - 2$

35. $y = 2x - 3$ **37.** $y = \frac{4}{3}x + 2$ **39.** $y = -\frac{2}{3}x - 3$

41. $y = 2$ **43.** $x = -5$ **45.** $m = 3, b = 3, y = 3x + 3$

47. $m = \frac{1}{4}, b = -1, y = \frac{1}{4}x - 1$

49. a. $-\frac{5}{2}$ **b.** $y = 2x + 6$ **51.** $y = -\frac{2}{3}x + 2$

 c. 6 **d.** 2 **53.** $y = -\frac{5}{2}x - 5$ **55.** $x = 3$

 e. **57.** Any line given by $y = 4x + b$

59. $m = -1$ **61.** $m = -2$

63. a. \$6,000

 b. 3 years

 c. Slope $= -3,000$

 d. \$3,000

 e. $V = -3,000t + 21,000$

65. a. \$3,000 **b.** 500 shirts **c.** Slope $= 2$ **d.** \$2 **e.** $y = 2x + 1,000$

Find the Mistake **1.** The equation $y = -\frac{4}{7}x + 2$ is in slope–intercept form. **2.** For the equation $-8x + 2y = -4$, the slope is 4 and the y–intercept is -2. **3.** The line represented by the equation $(y - 2) = 6(x - 4)$ has a slope of 6. **4.** The lines represented by the equation $y = \frac{7}{2}x + \frac{1}{3}$ and $y = -\frac{2}{7}x - 5$ are perpendicular.

Exercise Set 3.6

Vocabulary Review **1.** linear inequality **2.** solution set
3. boundary **4.** solid **5.** broken

Problems

1.

3.

5.

7.

9.

11.

13.

15.

17.

19.

21.

23.

25. a. $y < \frac{8}{3}$ **b.** $y > -\frac{8}{3}$
 c. $y = -\frac{4}{3}x + 4$
 d.

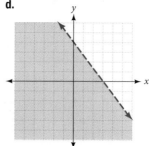

27. a. $y = \frac{2}{5}x + 2$
 b. $y < \frac{2}{5}x + 2$
 c. $y > \frac{2}{5}x + 2$

29. $-6x + 11$ **31.** -8 **33.** -4 **35.** $w = \dfrac{P - 2l}{2}$

37.

39.

41. $y \geq \dfrac{3}{2}x - 6$ **43.** width = 2 inches; length = 11 inches

Find the Mistake **1.** The solution set for the linear inequality $y < 6x - \dfrac{3}{5}$ does not contain the points through which the boundary line passes. **2.** The boundary for the linear inequality $2 - 8y \geq x + 6$ is represented by a solid line. **3.** If a test point substituted into an original inequality results in a false statement, the solution set for the inequality lies on the opposite. **4.** The solution set for the linear inequality $x < -8$, includes all points to the left of the line.

Chapter 3 Review

1-4.

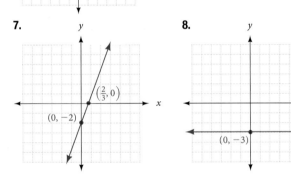

5. $\left(3, -\dfrac{4}{5}\right), \left(-2, -\dfrac{14}{5}\right)$ $\left(\dfrac{20}{3}, \dfrac{2}{3}\right), \left(\dfrac{1}{2}, -\dfrac{9}{5}\right)$

6. $(10, 0), (8, -1.6)$

7. **8.**

9. x-intercept = 6; y-intercept = -1

10. x-intercept = $\dfrac{20}{3}$; y-intercept = -2

11. x-intercept = -2; y-intercept = 2

12. x-intercept = -2; y-intercept = -2

13. $\dfrac{1}{3}$ **14.** 2 **15.** -2 **16.** 0 **17.** Undefined slope

18. 2 **19.** $y = 2x - 7$ **20.** $y = -3x + 6$ **21.** $y = \dfrac{1}{5}x + \dfrac{8}{5}$

22. $y = -\dfrac{6}{5}x + \dfrac{18}{5}$ **23.** $y = -\dfrac{1}{2}x + 2$ **24.** $m = 4, b = -3$

25. $m = \dfrac{4}{5}, b = -2$ **26.** $y = 3x - 4$ **27.** $y = \dfrac{3}{2}x + 8$

28. **29.**

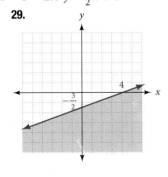

Chapter 3 Cumulative Review

1. -1 **2.** -1 **3.** 2 **4.** -8 **5.** 32 **6.** -40

7. Undefined **8.** $\dfrac{1}{3}$ **9.** $\dfrac{5}{7}$ **10.** 0 **11.** $36x$

12. $3a - 6$ **13.** -2 **14.** -4 **15.** 1 **16.** 4 **17.** $-\dfrac{3}{2}$

18. -98 **19.** $x > -6$ **20.** $x \geq 7$

21. **22**

23. **24.**

25. **26.**

27. No, yes **28.** y-intercept = 6

29. x-intercept = 4; y-intercept = 6

30.

31. $m = -1$ **32.** $y = -2x + 5$

33. $m = 3$ **34.** $y = \dfrac{2}{3}x$

35. $y = \dfrac{3}{2}x - 8$

36. $y = \dfrac{3}{5}x - 3$

37. $(5, 1), (0, -3)$

38. $(2,3), (6, 1)$

39. $24 - 2(6) = 12$

40. -20 **41.** $\dfrac{3}{5}, -\dfrac{5}{3}, \dfrac{3}{5}$

42. $-4, \dfrac{1}{4}, 4$ **43.** -9 **44.** 1

45. length = 13 in.; width = 7 in. **46.** 7 quarters, 13 dimes

47. 0.6, The attendance from 1955 to 1970 was increasing at an average rate of 0.6 million people every year **48.** 0.1475, The attendance from 1970 to 2010 was increasing at an average rate of 0.1475 million people every year

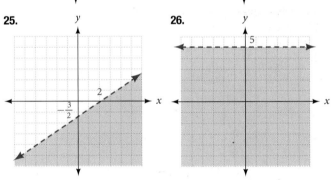

Chapter 3 Test

1-4.

5. $(0, -3), (2, 0), (4, 3), (-2, -6)$
6. $(0, 7), (4, -5)$

7.

8.

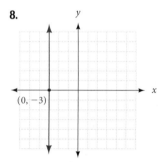

9. x-intercept = 2; y-intercept = -4
10. x-intercept = -4; y-intercept = 6
11. x-intercept = -2; y-intercept = 4
12. $-\frac{1}{2}$ **13.** $-\frac{8}{7}$ **14.** -3 **15.** Undefined slope **16.** 0
17. $y = -\frac{1}{2}x + 3$ **18.** $y = 3x - 5$ **19.** $y = -\frac{2}{3}x - 2$
20. $y = -3x - 6$ **21.** $y = \frac{3}{2}x + 6$ **22.** $y = -3x - 17$
23. $y = \frac{3}{2}x + 4$ **24.** $y = \frac{3}{2}x + 7$ **25.** $y = 2x - 6$

26.

27.

Chapter 4

Exercise Set 4.1

Vocabulary Review **1.** base, exponent **2.** product
3. power **4.** distributive **5.** scientific notation
Problems **1.** 16 **3.** 0.09 **5.** 64 **7.** 25 **9.** -8
11. 81 **13.** $\frac{4}{9}$ **15.** $\frac{1}{16}$ **17.** $\frac{9}{16}$ **19.** $-\frac{1}{36}$
21. a.

Number x	1	2	3	4	5	6	7
Square x^2	1	4	9	16	25	36	49

b. Larger

23. x^9 **25.** y^{30} **27.** 2^{12} **29.** x^{28} **31.** x^{10} **33.** 5^{12}
35. y^9 **37.** 2^{50} **39.** a^{3x} **41.** b^{xy} **43.** $16x^2$ **45.** $32y^5$
47. $81x^4$ **49.** $0.25a^2b^2$ **51.** $64x^3y^3z^3$ **53.** $-64x^3y^3$ **55.** $8x^{12}$
57. $16a^6$ **59.** x^{14} **61.** a^{11} **63.** $128x^7$ **65.** $432x^{10}$
67. $16x^4y^6$ **69.** $\frac{8}{27}a^{12}b^{15}$

71.

Number x	-3	-2	-1	0	1	2	3
Square x^2	9	4	1	0	1	4	9

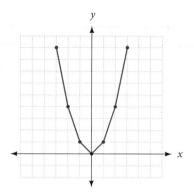

73.

Number x	-2.5	-1.5	-0.5	0	0.5	1.5	2.5
Square x^2	6.25	2.25	0.25	0	0.25	2.25	6.25

75. 4.32×10^4 **77.** 5.7×10^2 **79.** 2.38×10^5 **81.** 1.0×10^{10}
83. 2,490 **85.** 352 **87.** 28,000 **89.** $-310,000$
91.

Name	Location	Area (sq. mi.)	Scientific Notation
Caspian Sea	5 countries	152,239	1.52×10^5
Superior	US, Canada	31,820	3.18×10^4
Victoria	Tanzania, Uganda	26,828	2.68×10^4
Huron	US, Canada	23,010	2.30×10^4
Michigan	US	22,400	2.24×10^4

93. 27 inches³ **95.** 15.6 inches³ **97.** 36 inches³
99. Answers will vary **101.** 6.5×10^8 seconds **103.** $740,000
105. $180,000 **107.** $30,957.48 **109.** 219 inches³
111. 182 inches³ **113.** -3 **115.** 11 **117.** -5
119. 5 **121.** 2 **123.** 6 **125.** 4 **127.** 3
Find the Mistake **1.** Simplifying the expression $\left(-\frac{12}{32}\right)^2$ will give a negative answer. **2.** You would use the product property for exponents to simplify the expression $x^8 \cdot x^4 \cdot x^{10}$. **3.** To simplify the expression $(x^3 \cdot y)^4 (y \cdot z^2)^8$, we must first use the distributive property of exponents. **4.** 5.83×10^5 in an example of a number written in scientific notation.

Exercise Set 4.2

Vocabulary Review **1.** negative **2.** quotient
3. expanded **4.** exponents
Problems **1.** $\frac{1}{9}$ **3.** $\frac{1}{36}$ **5.** $\frac{1}{64}$ **7.** $\frac{1}{125}$ **9.** $\frac{2}{x^3}$
11. $\frac{1}{8x^3}$ **13.** $\frac{1}{25y^2}$ **15.** $\frac{1}{100}$
17.

Number x	Square x^2	Power of 2 2^x
-3	9	$\frac{1}{8}$
-2	4	$\frac{1}{4}$
-1	1	$\frac{1}{2}$
0	0	1
1	1	2
2	4	4
3	9	8

19. $\frac{1}{25}$ **21.** x^6 **23.** 64 **25.** $8x^3$ **27.** 6^{10} **29.** $\frac{1}{6^{10}}$ **31.** $\frac{1}{2^8}$
33. 2^8 **35.** $27x^3$ **37.** $16x^2$ **39.** $\frac{x^3}{27}$ **41.** $\frac{16}{y^2}$ **43.** $-\frac{125}{x^3}$

45. $\dfrac{64}{y^7}$ **47.** 1 **49.** $-2y$ **51.** $-3ab^2$ **53.** 1 **55.** $81x^4y^4$
57. 1 **59.** $2a^2b$ **61.** $\dfrac{1}{49y^6}$ **63.** $\dfrac{1}{x^8}$ **65.** $\dfrac{1}{y^3}$ **67.** x^2
69. a^6 **71.** $\dfrac{1}{y^9}$ **73.** y^{40} **75.** $\dfrac{1}{x}$ **77.** x^9 **79.** $\dfrac{4b^8}{a^4}$
81. a^{16} **83.** $\dfrac{1}{a^4}$ **85.** $\dfrac{27}{16x^2y^4}$

87.

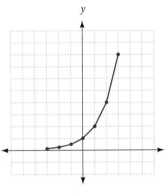

Number	Power of x
x	2^x
-3	$\dfrac{1}{8}$
-2	$\dfrac{1}{4}$
-1	$\dfrac{1}{2}$
0	1
1	2
2	4
3	8

89. 4.8×10^{-3} **91.** 2.5×10^1 **93.** 9×10^{-6} **95.** -8.7×10^{-5}
97. 0.00423 **99.** 0.00008 **101.** 4.2 **103.** -0.0036
105.

Standard Form	Scientific Notation $n \times 10^r$
0.000357	3.57×10^{-4}
0.00357	3.57×10^{-3}
0.0357	3.57×10^{-2}
0.357	3.57×10^{-1}
3.57	3.57×10^{0}
35.7	3.57×10^{1}
357	3.57×10^{2}
3,570	3.57×10^{3}
35,700	3.57×10^{4}

107. 0.002 **109.** $4 \times 10^{-2}, 1.1 \times 10^{-1}, 1.4 \times 10^{-1}, 6.7 \times 10^{-1}$
111. 2.5×10^4 **113.** 2.35×10^5 **115.** 8.2×10^{-4}
117. 100 inches2, 400 inches2; 4 **119.** x^2; $4x^2$; 4
121. 216 inches3; 1,728 inches3; 8 **123.** x^3; $8x^3$; 8 **125.** 1,407
127. 13.5 **129.** 8 **131.** 26.52 **133.** 12 **135.** x^8
137. x **139.** $\dfrac{1}{y^2}$ **141.** 1 **143.** 340 **145.** 0.000035

Find the Mistake 1. The negative exponent property says that $x^{-y} = \left(\dfrac{1}{x}\right)^y$. **2.** When using the quotient property of exponents to simplify the expression $\dfrac{y^{12}}{y^8}$, we get y^4. **3.** When using the expanded distributive property of exponents to simplify the expression $\left(\dfrac{x}{3}\right)^4$, we get $\dfrac{x^4}{81}$. **4.** The number 0.00238 written in scientific notation is 2.38×10^{-3}.

Exercise Set 4.3
Vocabulary Review 1. constant **2.** terms **3.** coefficient
4. variable **5.** subtraction
Problems 1. $12x^7$ **3.** $-16y^{11}$ **5.** $32x^2$ **7.** $200a^6$
9. $-24a^3b^3$ **11.** $24x^6y^8$ **13.** $3x$ **15.** $\dfrac{6}{y^3}$ **17.** $\dfrac{1}{2a}$
19. $-\dfrac{3a}{b^2}$ **21.** $\dfrac{x^2}{9z^2}$ **23.** $-\dfrac{y^2}{4x^3z}$
25.

a	b	ab	$\dfrac{a}{b}$	$\dfrac{b}{a}$
10	$5x$	$50x$	$\dfrac{2}{x}$	$\dfrac{x}{2}$
$20x^3$	$6x^2$	$120x^5$	$\dfrac{10x}{3}$	$\dfrac{3}{10x}$
$25x^5$	$5x^4$	$125x^9$	$5x$	$\dfrac{1}{5x}$
$3x^{-2}$	$3x^2$	9	$\dfrac{1}{x^4}$	x^4
$-2y^4$	$8y^7$	$-16y^{11}$	$-\dfrac{1}{4y^3}$	$-4y^3$

27. $4x^3$ **29.** $\dfrac{1}{b^2}$ **31.** $\dfrac{6y^{10}}{x^4}$ **33.** $-\dfrac{4x^2}{3y}$ **35.** 6×10^8
37. 1.75×10^{-1} **39.** 1.21×10^{-6} **41.** 4.2×10^3
43. 3×10^{10} **45.** 5×10^{-3} **47.** -5×10^{-4} **49.** 2×10^6
51. 1×10^1 **53.** 4.2×10^{-6} **55.** $8x^2$ **57.** $-11x^5$ **59.** 0
61. $4x^3$ **63.** $31ab^2$ **65.** $-11x^2y$
67.

a	b	ab	$a + b$
$5x$	$3x$	$15x^2$	$8x$
$4x^2$	$2x^2$	$8x^4$	$6x^2$
$3x^3$	$6x^3$	$18x^6$	$9x^3$
$2x^4$	$-3x^4$	$-6x^8$	$-x^4$
x^5	$7x^5$	$7x^{10}$	$8x^5$

69. $x^2y + x$ **71.** $x + y$ **73.** $x^2 - 4$ **75.** $x^2 - x - 6$
77. $x^2 - 5x$ **79.** $x^2 - 8x$ **81.** $9x^3$ **83.** $-20a^2$
85. $6x^5y^2$ **87.** -5 **89.** 6 **91.** 76 **93.** $6x^2$ **95.** $2x$
97. $-2x - 9$ **99.** 11 **101.** 36
Find the Mistake 1. The coefficient of the monomial $12x^4y^2z$ is 12.
2. Multiplying the monomials $-15x^3y$ and $2x^2$ gives us $-30x^5y$.
3. When dividing $28x^6y^2$ by $4x^2y$, write the problem as the separate fractions $\dfrac{28}{4} \cdot \dfrac{x^6}{x^2} \cdot \dfrac{y^2}{y}$. **4.** Monomials with similar terms have the same variable parts, but coefficients may differ.
Landmark Review 1. x^8 **2.** y^{27} **3.** x^6 **4.** a^{xy} **5.** $27a^3$
6. $256y^5$ **7.** 25 **8.** $27x^3$ **9.** $\dfrac{1}{4}$ **10.** $\dfrac{1}{a^2}$ **11.** $4xy$ **12.** 1
13. 3×10^2 **14.** 2×10^3 **15.** $18x^5y^4$ **16.** $\dfrac{a^2b}{2c^3}$

Exercise Set 4.4
Vocabulary Review 1. polynomial **2.** degree **3.** similar
4. opposite
Problems 1. Trinomial, 3 **3.** Trinomial, 3 **5.** Binomial, 1
7. Binomial, 2 **9.** Monomial, 2 **11.** Monomial, 0
13. $5x^2 + 5x + 9$ **15.** $5a^2 - 9a + 7$ **17.** $x^2 + 6x + 8$
19. $6x^2 - 13x + 5$ **21.** $x^2 - 9$ **23.** $3y^2 - 11y + 10$
25. $6x^3 + 5x^2 - 4x + 3$ **27.** $2x^2 - x + 1$ **29.** $2a^2 - 2a - 2$
31. $-\dfrac{1}{9}x^3 - \dfrac{2}{3}x^2 - \dfrac{5}{2}x + \dfrac{7}{4}$ **33.** $-4y^2 + 15y - 22$
35. $x^2 - 33x + 63$ **37.** $8y^2 + 4y + 26$ **39.** $75x^2 - 150x - 75$
41. $12x + 2$ **43.** 4 **45.** 18π inches3 **47.** 5 **49.** -6
51. $-20x^2$ **53.** $-21x$ **55.** $2x$ **57.** $-8x$ **59.** $6x - 18$
61. $-3xy + 12$
Find the Mistake 1. The degree of the polynomial $x^2 + 6 + 4x - x^7$ is 7. **2.** Adding $x^3 + 4x^2 - 8 + x^2$ gives us $x^3 + 5x^2 - 8$.
3. The first step when subtracting $(x - 5)^2$ from $(x + 5)^2$ is to apply the exponents to expand each expression. **4.** The value of the polynomial $7a^2 + 5a - 10$ when a is 4 is 122.

Exercise Set 4.5
Vocabulary Review 1. distributive **2.** term **3.** binomials
4. products **5.** column
Problems 1. $6x^2 + 2x$ **3.** $6x^4 - 4x^3 + 2x^2$ **5.** $2a^3b - 2a^2b^2 + 2ab$
7. $3y^4 + 9y^3 + 12y^2$ **9.** $8x^5y^2 + 12x^4y^3 + 32x^2y^4$
11. $x^2 + 7x + 12$ **13.** $x^2 + 7x + 6$ **15.** $x^2 + 2x + \dfrac{3}{4}$
17. $a^2 + 2a - 15$ **19.** $xy + bx - ay - ab$ **21.** $y^2 - \dfrac{16}{49}$
23. $3x^2 - 11x + 10$ **25.** $3a^2 - 16a - 12$ **27.** $6x^2 + 9x - 6$
29. $2ax - 8x - 3a + 12$ **31.** $36x^2 - 25$ **33.** $4x^2 + \dfrac{1}{2}x - \dfrac{3}{4}$
35. $3 - 7a - 6a^2$ **37.** $(x + 4)(x + 5) = x^2 + 4x + 5x + 20 = x^2 + 9x + 20$
39. $(2x + 1)(2x + 2) = 4x^2 + 6x + 2$ **41.** $a^3 + 7a^2 + 13a + 15$
43. $x^3 + 27$ **45.** $3x^3 - 23x^2 + 38x - 16$ **47.** $2x^4 - 7x^3 + 3x^2 - x + 3$
49. $20x^6 - 24x^3 - 32$ **51.** $28a^7 - 42a^4 - 32a^3 + 48$
53. $x^3 - 12x^2 + 47x - 60$ **55.** $6x^2 - 11x - 14$

57. $15x^2 + 19x - 4$ **59.** $x^2 - x - 20$ **61.** $6x + 19$
63. $x^2 + 4x$ **65.** $2x^2 - 8x + 6$ **67.** $a^2 - 4a + 8$
69. $A = x(3x + 2) = 3x^2 + 2x$ **71.** $A = x(x + 2) = x^2 + 2x$
73. a. $A = \pi(r + 2)^2 = \pi r^2 + 4\pi r + 4\pi$ **b.** $A = 4\pi(r + 1) = 4\pi r + 4\pi r$
75. $R = 1{,}200p - 100p^2$ **77.** 169 **79.** $-10x$ **81.** 0 **83.** 0
85. $-12x + 16$ **87.** $x^2 + x - 2$ **89.** $x^2 + 6x + 9$ **91.** $25x^2 - 9$
Find the Mistake 1. To find the product of two binomials, multiply each term in the first by each term in the second.
2. When using the FOIL method to multiply the binomials $(9x - 2)$ and $(5x + 3)$, the product of the two outside terms is $27x$.
3. For the problem $(x^2 - 4)(3x^2 + 6)$, adding the products of the two outside terms with the two inside terms gives us $-6x^2$.
4. For the problem $4x^2(3x^2 + 4x + 9)$, the degree of the answer is 4.

Exercise Set 4.6

Vocabulary Review 1. sum **2.** square of a binomial
3. subtract **4.** difference of two squares
Problems 1. $x^2 - 4x + 4$ **3.** $a^2 + 6a + 9$ **5.** $x^2 - 10x + 25$
7. $a^2 - a + \frac{1}{4}$ **9.** $x^2 + 20x + 100$ **11.** $a^2 + 1.6a + 0.64$
13. $4x^2 - 4x + 1$ **15.** $16a^2 + 40a + 25$ **17.** $9x^2 - 12x + 4$
19. $9a^2 + 30ab + 25b^2$ **21.** $16x^2 - 40xy + 25y^2$
23. $49m^2 + 28mn + 4n^2$ **25.** $36x^2 - 120xy + 100y^2$
27. $x^4 + 10x^2 + 25$ **29.** $a^4 + 2a^2 + 1$ **31.** $x^6 - 8x^3 + 16$
33.

x	$(x + 3)^2$	$x^2 + 9$	$x^2 + 6x + 9$
1	16	10	16
2	25	13	25
3	36	18	36
4	49	25	49

35.

a	1	3	3	4
b	1	5	4	5
$(a + b)^2$	4	64	49	81
$a^2 + b^2$	2	34	25	41
$a^2 + ab + b^2$	3	49	37	61
$a^2 + 2ab + b^2$	4	64	49	81

37. $a^2 - 25$ **39.** $y^2 - 1$ **41.** $81 - x^2$ **43.** $4x^2 - 25$
45. $16x^2 - \frac{1}{9}$ **47.** $4a^2 - 49$ **49.** $36 - 49x^2$ **51.** $x^4 - 9$
53. $a^4 - 16$ **55.** $25y^8 - 64$ **57.** $2x^2 - 34$
59. $-12x^2 + 20x + 8$ **61.** $a^2 + 4a + 6$ **63.** $8x^3 + 36x^2 + 54x + 27$
65. $20x$ **67.** $-4xy$ **69.** $3x^2 + 2$ **71.** $8a^3 - 12a^2 + 6a - 1$
73. $(50 - 1)(50 + 1) = 2{,}500 - 1 = 2{,}499$ **75.** Both equal 25.
77. $x^2 + (x + 1)^2 = 2x^2 + 2x + 1$
79. $x^2 + (x + 1)^2 + (x + 2)^2 = 3x^2 + 6x + 5$
81. $a^2 + ab + ba + b^2 = a^2 + 2ab + b^2$
83. $A = (x - 4)^2 = x^2 - 8x + 16$ **85.** $2x^2$ **87.** x^2 **89.** $3x$ **91.** $3xy$
Find the Mistake 1. To find the square of the binomial $(5a^2 - 9a)$, we multiply $(5a^2 - 9a)$ by itself. **2.** The second term in a binomial square is twice the product of the two terms.
3. Simplifying $(7x - 3)^2$ gives us $49x^2 - 42x + 9$.
4. The problem $(4a - 6)(4a + 6) = 16a^2 - 36$ is an example of a difference of two squares.

Exercise Set 4.7

Vocabulary Review 1. division, distributive **2.** polynomial, term
Problems 1. $x - 2$ **3.** $3 - 2x^2$ **5.** $5xy - 2y$
7. $7x^4 - 6x^3 + 5x^2$ **9.** $10x^4 - 5x^2 + 1$ **11.** $-4x^2 + 3x - 1$
13. $-4a + 2$ **15.** $-8a^4 - 12a^3$ **17.** $-4b - 5a$
19. $-6a^2b + 3ab^2 - 7b^3$ **21.** $-\frac{a}{2} - b - \frac{b^2}{2a}$
23. $\frac{3}{2}a^3 + 4a^2 - 6a$ **25.** $3x + 4y$ **27.** $-y + 3$ **29.** $5y - 4$
31. $xy - x^2y^2$ **33.** $-1 + xy$ **35.** $-a + 1$ **37.** $x^2 - 3xy + y^2$

39. $2 - 3b + 5b$ **41.** $-2xy + 1$ **43.** $xy - \frac{1}{2}$ **45.** $\frac{1}{4x} - \frac{1}{2a} + \frac{3}{4}$
47. $\frac{4x^2}{3} + \frac{2}{3x} + \frac{1}{x^2}$ **49.** $3a^{3m} - 9a^m$ **51.** $2x^{4m} - 5x^{2m} + 7$
53. $3x^2 - x + 6$ **55.** 4 **57.** $x + 5$ **59.** Both equal 7.
61. $\frac{3(10) + 8}{2} = 19; 3(10) + 4 = 34$ **63.** $146\frac{20}{27}$ **65.** $2x + 5$
67. $x^2 - 3x$ **69.** $2x^3 - 10x^2$ **71.** $2x^3 - 3x^2$ **73.** $-3x + 21$
75. $-2x$ **77.** 2 **79.** 5 **81.** $-5x$
Find the Mistake 1. To divide a polynomial by a monomial, divide each term of the polynomial by the monomial.
2. Dividing $15x^3 - 6x$ by $-3x$ gives us $-5x^2 + 2$. **3.** The quotient of $12x^2y + 4x^4y^2 - x^3y^2$ and x^2y gives us $4x^2y - xy + 12$.
4. If you divide $9x^2 - 12x$ by $-3x$, the answer is $3x + 4$.

Exercise Set 4.8

Vocabulary Review 1. long divison **2.** estimate
3. opposite **4.** variable **5.** powers
Problems 1. $x - 2$ **3.** $a + 4$ **5.** $x - 3$ **7.** $x + 3$
9. $a - 5$ **11.** $x + 2 + \frac{2}{x + 3}$ **13.** $a - 2 + \frac{12}{a + 5}$
15. $x + 4 + \frac{9}{x - 2}$ **17.** $x + 4 + \frac{-10}{x + 1}$ **19.** $a + 1 + \frac{-1}{a + 2}$
21. $x - 3 + \frac{17}{2x + 4}$ **23.** $3a - 2 + \frac{7}{2a + 3}$ **25.** $2a^2 - a - 3$
27. $4x^2 + 3x - \frac{1}{2} + \frac{-\frac{11}{2}}{2x - 1}$ **29.** $x - 3$ **31.** $x^2 - x + 3$
33. $x^2 + 2x + 1$ **35.** $x^2 - x + 5$ **37.** $x^2 + x + 1$
39. $x^2 + 2x + 4$ **41.** $x^2 - 4x + 16$ **43.** 6.7 goals
45. 11.8 goals **47.** 47 **49.** 14 **51.** 70 **53.** 35 **55.** 5
57. 6,540 **59.** 1,760 **61.** 20 **63.** 63
Find the Mistake 1. For a division of polynomials problem, you have found the answer when there is nothing left to bring down.
2. To begin the problem $\frac{x^2 - 8x + 4}{x - 4}$, we first estimate $x^2 \div x = x$.
3. The quotient of $x^3 - 9x + 2$ and $x - 1$ has a remainder of -6.
4. When dividing $9x^3 - 7x - 4$ by $3x - 2$, we first write the problem as $3x - 2 \overline{)9x^3 + 0x^2 - 7x - 4}$.

Chapter 4 Review

1. $-\frac{1}{4}$ **2.** $\frac{8}{27}$ **3.** $108x^{12}$ **4.** 34,700 **5.** $-\frac{1}{27}$ **6.** 1
7. $\frac{1}{a^2}$ **8.** x **9.** 3.478×10^{-2} **10.** $\frac{5xz}{2}$ **11.** $\frac{18a^3}{b^2}$
12. $5x^3$ **13.** 1.2×10^{-4} **14.** $8x^2 - x - 2$ **15.** $2x^2 - 7x + 11$
16. $x - 14$ **17.** 24 **18.** $8a^4 - 10a^3 + 17a^2$ **19.** $x^2 + x - 6$
20. $12x^2 - 24x + 12$ **21.** $x^3 + 6x^2 + 18x + 27$
22. $x^3 + 3x^2 + 3x + 1$ **23.** $9a^2 - 12a + 4$ **24.** $4x^2 - b^2$
25. $b^4 - 4$ **26.** $-2x^2 + 3x + 3$ **27.** $4x - 8$ **28.** $3x + 1$
29. $V = 3.375$ mi^3 **30.** $V = 2w^3$
31. Bugatti: 1.7×10^6, Lamborghini: 1.6×10^6, McLaren: 9.7×10^5, Ferrari: 6.7×10^5, Pagani: 6.7×10^5
32. Rebensburg/Maze: 4×10^{-2}, Lee/Wolf: 5×10^{-2}, Defago/Svindal: 8×10^{-2}, Razzoli/Kostelic: 1.6×10^{-1}

Chapter 4 Cumulative Review

1. $\frac{1}{16}$ **2.** -8 **3.** 26 **4.** 72 **5.** -25 **6.** 12 **7.** 8
8. -30 **9.** $-\frac{x}{2} + y$ **10.** 6 **11.** y^7 **12.** $27y^3$ **13.** $72a^8b^{17}$
14. $\frac{y^2}{4x^2}$ **15.** 2.45 **16.** 3×10^{-2} **17.** $-2a^2b + 10ab$
18. $-2x^2 - 10x - 3$ **19.** $25x^2 + 10x + 1$ **20.** $x^3 - 8$ **21.** -2
22. $\frac{5}{3}$ **23.** -2 **24.** -7 **25.** $a > 12$ **26.** $t \le -3$
27. $6 < x < 7$ **28.** $-2 \le x \le 4$ **29.** $2x + 3$ **30.** $x^3 + 2x^2$

31.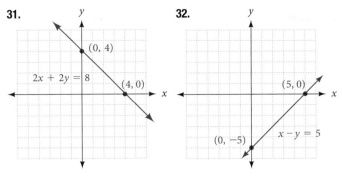

32.

$2x + 2y = 8$ $(0, 4)$ $(4, 0)$

$(5, 0)$ $(0, -5)$ $x - y = 5$

33. -11 **34.** 4 **35.** $h = \frac{2A}{b}$ **36.** $y = \frac{-2x + 1}{2}$

37. 2 **38.** $-\frac{\pi}{2}, -\sqrt{2}$ **39.** Additive inverse property

40. Commutative property of addition

41. x-intercept $= \frac{2}{3}$; y-intercept $= -2$ **42.** $m = \frac{1}{2}$

43. $80{,}185$ more **44.** $8{,}862$ fewer

Chapter 4 Test

1. -243 **2.** $\frac{4}{25}$ **3.** $72x^{10}$ **4.** $247{,}000$ **5.** $-\frac{1}{9}$

6. $7a^2b^2$ **7.** $\frac{1}{a^3}$ **8.** $\frac{1}{x^3}$ **9.** 2.78×10^1 **10.** $\frac{5x^2y^2}{z^2}$

11. $9a^3$ **12.** $5x$ **13.** 1.6×10^4 **14.** $8x^2 - x - 12$

15. $x^2 + 7x$ **16.** $-6x - 10$ **17.** 11 **18.** $6a^4 + 10a^3 - 8a^2$

19. $x^2 - x - \frac{3}{4}$ **20.** $8x^2 - 2x - 6$ **21.** $x^2 - 9$

22. $x^4 - 18x^2 + 81$ **23.** $4a^2 - 4ab + b^2$ **24.** $3x^2 - 8y^2$

25. $a^4 - 3$ **26.** $3x + 1$ **27.** $2x - \frac{1}{3x - 1}$ **28.** 0.125 ft^3

29. $V = w^3$ **30.** 7.83×10^8 **31.** 5.55×10^8

32. 2×10^{-2} **33.** 4×10^{-2}

Chapter 5

Exercise Set 5.1

Vocabulary Review **1.** monomial **2.** coefficient

3. grouping, factors

Problems **1.** $5(3x + 5)$ **3.** $3(2a + 3)$ **5.** $4(x - 2y)$

7. $3(x^2 - 2x + 3)$ **9.** $3(a^2 - a + 20)$ **11.** $4(6y^2 - 13y + 9)$

13. $x^2(9 - 8x)$ **15.** $13a^2(1 - 2a)$ **17.** $7xy(3x - 4y)$

19. $11ab^2(2a - 1)$ **21.** $7x(x^2 + 3x + 4)$ **23.** $11(11y^4 + x^4)$

25. $25x^2(4x^2 - 2x + 1)$ **27.** $8(a^2 + 2b^2 + 4c^2)$

29. $4ab(a - 4b + 8ab)$ **31.** $6a^2b(ab + a - b)$

33. $11a^2b^2(11a - 2b + 3ab)$ **35.** $12x^2y^3(1 - 6x^3 - 3x^2y)$

37. $(x + 3)(y + 5)$ **39.** $(x + 2)(y + 6)$ **41.** $(a - 3)(b + 7)$

43. $(a - b)(x + y)$ **45.** $(2x - 5)(a + 3)$ **47.** $(b - 2)(3x - 4)$

49. $(x + 2)(x + a)$ **51.** $(x - b)(x - a)$ **53.** $(x + y)(a + b + c)$

55. $(3x + 2)(2x + 3)$ **57.** $(10x - 1)(2x + 5)$

59. $(4x + 5)(5x + 1)$ **61.** $(x + 2)(x^2 + 3)$ **63.** $(3x - 2)(2x^2 + 5)$

65. $(3x + 2)(x^2 - 5)$ **67.** 6 **69.** $3(4x^2 + 2x + 1)$

71. $A = 1{,}000(1 + r)$; $\$1{,}120.00$ **73. a.** $A = 1{,}000{,}000(1 + r)$

b. $1{,}300{,}000$ **75.** $x^2 - 5x - 14$ **77.** $x^2 - x - 6$

79. $x^3 + 27$ **81.** $2x^3 + 9x^2 - 2x - 3$ **83.** $18x^7 - 12x^6 + 6x^5$

85. $x^2 + x + \frac{2}{9}$ **87.** $12x^2 - 10xy - 12y^2$ **89.** $81a^2 - 1$

91. $x^2 - 18x + 81$ **93.** $x^3 + 8$

Find the Mistake **1.** The greatest common factor for a polynomial is the monomial with the greatest coefficient and highest power of the variable that divides each term of the polynomial.

2. The greatest common factor of the polynomial $12x^4 + 3x^3 - 6x$ is $3x$. **3.** Using the process of factoring by grouping, we can factor the polynomial $8xy + 2x - 12y - 3$ to get $2x(4y + 1) - 3(4y + 1) = (2x - 3)(4y + 1)$. **4.** To factor the polynomial $9xy - 27x - 2y + 6$ by grouping, factor $9x$ from the first two terms, and factor 2 from the last two terms to get

$9x(y - 3) - 2(y - 3) = (9x - 2)(y - 3)$.

Exercise Set 5.2

Vocabulary Review coefficient, sum, product, leading, greatest common factor, trinomial

Problems **1.** $(x + 3)(x + 4)$ **3.** $(x + 1)(x + 2)$

5. $(a + 3)(a + 7)$ **7.** $(x - 2)(x - 5)$ **9.** $(y - 3)(y - 7)$

11. $(x - 4)(x + 3)$ **13.** $(y + 4)(y - 3)$ **15.** $(x + 7)(x - 2)$

17. $(r - 9)(r + 1)$ **19.** $(x - 6)(x + 5)$ **21.** $(a + 7)(a + 8)$

23. $(y + 6)(y - 7)$ **25.** $(x + 6)(x + 7)$ **27.** $(a - 9)(a + 7)$

29. $2(x + 1)(x + 2)$ **31.** $3(a + 4)(a - 5)$ **33.** $100(x - 2)(x - 3)$

35. $100(p - 5)(p - 8)$ **37.** $x^2(x + 3)(x - 4)$ **39.** $2r(r + 5)(r - 3)$

41. $2y^2(y + 1)(y - 4)$ **43.** $x^3(x + 2)(x + 2)$

45. $3y^2(y + 1)(y - 5)$ **47.** $4x^2(x - 4)(x - 9)$

49. $-2x(x - 3)(x - 2)$ **51.** $-7x^2(x + 2)(x + 1)$

53. $(x + 2y)(x + 3y)$ **55.** $(x - 4y)(x - 5y)$ **57.** $(a + 4b)(a - 2b)$

59. $(a - 5b)(a - 5b)$ **61.** $(a + 5b)(a + 5b)$ **63.** $(x - 6a)(x + 8a)$

65. $(x + 4b)(x - 9b)$ **67.** $2x(x - 3y)(x + y)$

69. $xy(x - 3y)(x - 4y)$ **71.** $-2a(a + 4b)(a - 2b)$

73. $(x^2 - 3)(x^2 - 2)$ **75.** $(x - 100)(x + 20)$

77. $\left(x - \frac{1}{2}\right)\left(x - \frac{1}{2}\right)$ **79.** $(x + 0.2)(x + 0.4)$ **81.** $x + 16$

83. $4x^2 - x - 3$ **85.** -2 **87.** -15

89. a. 5 GB **b.** 70 GB **c.** 240 GB **91.** $6a^2 + 13a + 2$

93. $6a^2 + 7a + 2$ **95.** $6a^2 + 8a + 2$ **97.** $4a^3 + 10a^2 - 24a$

Find the Mistake **1.** The leading coefficient of the polynomial $x^2 - 2x - 35$ is 1. **2.** Factoring the polynomial in the above sentence gives us $(x - 7)(x + 5)$. **3.** For the polynomial $8x^3 - 30x^2 - 108x$, the greatest common factor is $2x$. **4.** Factoring the polynomial in the above sentences gives us $2x(x - 6)(4x + 9)$.

Exercise Set 5.3

Vocabulary Review leading coefficient, trial and error, factors, products, multiply, polynomial

Problems **1.** $(2x + 1)(x + 3)$ **3.** $(2a - 3)(a + 1)$

5. $(3x + 5)(x - 1)$ **7.** $(3y + 1)(y - 5)$ **9.** $(2x + 3)(3x + 2)$

11. $(2x - 3y)(2x - 3y)$ **13.** $(4y + 1)(y - 3)$ **15.** $(4x - 5)(5x - 4)$

17. $(10a - b)(2a + 5b)$ **19.** $(4x - 5)(5x + 1)$ **21.** $(6m - 1)(2m + 3)$

23. $(4x + 5)(5x + 3)$ **25.** $(3a - 4b)(4a - 3b)$ **27.** $(3x - 7y)(x + 2y)$

29. $(2x + 5)(7x - 3)$ **31.** $(3x - 5)(2x - 11)$ **33.** $(5t - 19)(3t - 2)$

35. $(7t - 9)(4t + 3)$ **37.** $2(2x + 3)(x - 1)$ **39.** $2(4a - 3)(3a - 4)$

41. $x(5x - 4)(2x - 3)$ **43.** $x^2(3x + 2)(2x - 5)$ **45.** $2a(5a + 2)(a - 1)$

47. $3x(5x + 1)(x - 7)$ **49.** $5y(7y + 2)(y - 2)$ **51.** $a^2(5a + 1)(3a - 1)$

53. $3y(2x - 3)(4x + 5)$ **55.** $2y(2x - y)(3x - 7y)$

57. Both equal 25. **59.** $4x^2 - 9$ **61.** $x^4 - 81$ **63.** 1 **65.** -20

67.

Time t (seconds)	0	1	2	3	4
Height h (feet)	8	54	68	50	0

$h = 2(4 - t)(1 + 8t)$

69. a. $V = x \cdot (11 - 2x)(9 - 2x)$ **b.** 11 inches \times 9 inches

71. $x^2 - 9$ **73.** $x^2 - 25$ **75.** $x^2 - 49$ **77.** $x^2 - 81$

79. $4x^2 - 9y^2$ **81.** $x^4 - 16$ **83.** $x^2 + 6x + 9$

85. $x^2 + 10x + 25$ **87.** $x^2 + 14x + 49$ **89.** $x^2 + 18x + 81$

91. $4x^2 + 12x + 9$ **93.** $16x^2 - 16xy + 4y^2$

Find the Mistake **1.** $(2x + 1)$ or $(3x + 2)$ is a factor of the polynomial $6x^2 + 7x + 12$. **2.** The first step to factoring $8a^3 - 4a^2 + 6a$ is to factor out the greatest common factor. **3.** All of the factors for $8a^3 - 4a^2 - 6a$ are $2a(4a - 1)(a + 3)$. **4.** The factors of the polynomial $40x^3z - 16x^2z - 12xz$ are $4xz(5x + 3)(2x - 2)$.

Landmark Review **1.** A greatest common factor. **2.** $2a(2a^2 + 4a - 5)$
3. $8xy(5x + 3y + 2xy)$ **4.** $(x - 1)(x - 8)$ **5.** $(x - 2)(y + 6)$
6. $(x + 2)(x - 2)(x - 7)$ **7.** $(x^2 + 6)(x - 8)$ **8.** $(x + 3)(x + 7)$
9. $(x + 4)(x - 7)$ **10.** $(x - 4)(x - 3)$ **11.** $(x - 4)(x + 6)$
12. $(3x + 2y)(2x - 3y)$ **13.** $(3a + 5)(a + 2)$ **14.** $(2y - 1)(3y - 2)$
15. $(5x - 6y)(4x - 5y)$

Exercise Set 5.4

Vocabulary Review **1.** perfect squares, difference of two squares
2. sum **3.** trinomial **4.** square
Problems **1.** $(x + 3)(x - 3)$ **3.** $(a + 6)(a - 6)$
5. $(x + 7)(x - 7)$ **7.** $4(a + 2)(a - 2)$ **9.** Cannot be factored.
11. $(5x + 13)(5x - 13)$ **13.** $(3a + 4b)(3a - 4b)$
15. $(3 + m)(3 - m)$ **17.** $(5 + 2x)(5 - 2x)$ **19.** $2(x + 3)(x - 3)$
21. $32(a + 2)(a - 2)$ **23.** $2y(2x + 3)(2x - 3)$
25. $(a^2 + b^2)(a + b)(a - b)$ **27.** $(4m^2 + 9)(2m + 3)(2m - 3)$
29. $3xy(x + 5y)(x - 5y)$ **31.** $-3x(x + 4)(x - 4)$ **33.** $(x - 1)^2$
35. $(x + 1)^2$ **37.** $(a - 5)^2$ **39.** $(y + 2)^2$ **41.** $(x - 2)^2$
43. $(m - 6)^2$ **45.** $(2a + 3)^2$ **47.** $(7x - 1)^2$ **49.** $(3y - 5)^2$
51. $(x + 5y)^2$ **53.** $(3a + b)^2$ **55.** $(2x - 3y)^2$ **57.** $3(a + 3)^2$
59. $2(x + 5y)^2$ **61.** $5x(x + 3y)^2$ **63.** $-4x(x - 3)^2$
65. $(x + 3 + y)(x + 3 - y)$ **67.** $(x + y + 3)(x + y - 3)$
69. 14 **71.** -10 **73.** 25 **75.** 16 **77.** 9
79. a. $x^2 - 16$ **b.** $(x + 4)(x - 4)$ **81.** $a^2 - b^2 = (a + b)(a - b)$
83. a. 1 **b.** 8 **c.** 27 **d.** 64 **e.** 125
85. a. $x^3 - x^2 + x$ **b.** $x^2 - x + 1$ **c.** $x^3 + 1$
87. a. $x^3 - 2x^2 + 4x$ **b.** $2x^2 - 4x + 8$ **c.** $x^3 + 8$
89. a. $x^3 - 3x^2 + 9x$ **b.** $3x^2 - 9x + 27$ **c.** $x^3 + 27$
91. a. $x^3 - 4x^2 + 16x$ **b.** $4x^2 - 16x + 64$ **c.** $x^3 + 64$
93. a. $x^3 - 5x^2 + 25x$ **b.** $5x^2 - 25x + 125$ **c.** $x^3 + 125$
Find the Mistake **1.** The difference of two squares says that
$a^2 - b^2 = (a + b)(a - b)$. **2.** Factoring $81y^6 - 196$ gives us
$(9y^3 + 14)(9y^3 - 14)$. **3.** The polynomial $18x^2 - 84x + 98$ is an
example of a perfect square trinomial. **4.** The polynomial in the
above sentence factors to $2(3x - 7)^2$.

Exercise Set 5.5

Vocabulary Review **1.** cubes **2.** difference **3.** factored **4.** sum
Problems **1.** $(x - y)(x^2 + xy + y^2)$ **3.** $(a + 2)(a^2 - 2a + 4)$
5. $(3 + x)(9 - 3x + x^2)$ **7.** $(y - 1)(y^2 + y + 1)$
9. $(y - 4)(y^2 + 4y + 16)$ **11.** $(5h - t)(25h^2 + 5ht + t^2)$
13. $(x - 6)(x^2 + 6x + 36)$ **15.** $2(y - 3)(y^2 + 3y + 9)$
17. $2(a - 4b)(a^2 + 4ab + 16b^2)$ **19.** $2(x + 6y)(x^2 - 6xy + 36y^2)$
21. $10(a - 4b)(a^2 + 4ab + 16b^2)$ **23.** $10(r - 5)(r^2 + 5r + 25)$
25. $(4 + 3a)(16 - 12a + 9a^2)$ **27.** $(2x - 3y)(4x^2 + 6xy + 9y^2)$
29. $\left(t + \frac{1}{3}\right)\left(t^2 - \frac{1}{3}t + \frac{1}{9}\right)$ **31.** $\left(3x - \frac{1}{3}\right)\left(9x^2 + x + \frac{1}{9}\right)$
33. $(4a + 5b)(16a^2 - 20ab + 25b^2)$ **35.** $\left(\frac{1}{2}x - \frac{1}{3}y\right)\left(\frac{1}{4}x^2 + \frac{1}{6}xy + \frac{1}{9}y^2\right)$
37. $(a - b)(a^2 + ab + b^2)(a + b)(a^2 - ab + b^2)$
39. $(2x - y)(4x^2 + 2xy + y^2)(2x + y)(4x^2 - 2xy + y^2)$
41. $(x - 5y)(x^2 + 5xy + 25y^2)(x + 5y)(x^2 - 5xy + 25y^2)$
43. $(x^9 - 3)(x^{18} + 3x^9 + 9)$ **45.** 2 **47.** 125 **49.** $2x^5 - 8x^3$
51. $3x^4 - 18x^3 + 27x^2$ **53.** $y^3 + 25y$ **55.** $15a^2 - a - 2$
57. $4x^4 - 12x^3 - 40x^2$ **59.** $2ab^5 - 8ab^4 + 2ab^3$
Find the Mistake **1.** The difference of two cubes says that
$a^3 - b^3 = (a - b)(a^2 + ab + b^2)$. **2.** The expression $a^3 + 8b^3$ is
an example of a sum of two cubes. **3.** The factors of the sum of
two cubes in the above sentence are $(a + 8b)$ and $(a^2 - 8ab + 64b^2)$.
4. Factoring $x^{12} - 343$ gives us $(x^4 - 7)(x^8 + 7x^4 + 49)$.

Exercise Set 5.6

Vocabulary Review **1.** greatest common factor **2.** squares,
cubes **3.** perfect square trinomial, trial and error
4. grouping **5.** factors
Problems **1.** $(x + 9)(x - 9)$ **3.** $(x + 5)(x - 3)$ **5.** $(x + 3)^2$
7. $(y - 5)^2$ **9.** $2ab(a^2 + 3a + 1)$ **11.** Cannot be factored.
13. $3(2a + 5)(2a - 5)$ **15.** $(3x - 2y)^2$ **17.** $4x(x^2 + 4y^2)$
19. $2y(y + 5)^2$ **21.** $a^4(a^2 + 4b^2)$ **23.** $(x + 4)(y + 3)$
25. $(x^2 + 4)(x + 2)(x - 2)$ **27.** $(x + 2)(y - 5)$
29. $5(a + b)^2$ **31.** Cannot be factored. **33.** $3(x + 2y)(x + 3y)$
35. $(2x + 19)(x - 2)$ **37.** $100(x - 2)(x - 1)$ **39.** $(x + 8)(x - 8)$
41. $(x + a)(x + 3)$ **43.** $a^5(7a + 3)(7a - 3)$
45. Cannot be factored. **47.** $a(5a + 1)(5a + 3)$ **49.** $(x + y)(a - b)$
51. $3a^2b(4a + 1)(4a - 1)$ **53.** $5x^2(2x + 3)(2x - 3)$
55. $(3x + 41y)(x - 2y)$ **57.** $2x^3(2x - 3)(4x - 5)$
59. $(2x + 3)(x + a)$ **61.** $(y^2 + 1)(y + 1)(y - 1)$
63. $3x^2y^2(2x + 3y)^2$ **65.** $-4x^5(x + 2)(x^2 - 2x + 4)$
67. $(x + 4)^2(x - 1)$ **69.** $(x + 2)^2(x^2 + 4x + 9)$
71. $4x(x + 1)(2x + 1)$ **73.** $(2x - 3)(4x^2 - 12x + 7)$

75. (2004, 1), (2005, 5.5), (2006, 12), (2007, 20), (2008, 100),
(2009, 175), (2010, 400) **77.** 5 **79.** $-\frac{3}{2}$ **81.** $-\frac{3}{4}$
Find the Mistake **1.** If a binomial is the sum of two squares, it
will not factor. **2.** To factor $14y^4 - 126y^2$, we first factor out
the greatest common factor $14y^2$. **3.** Once you factored out
the greatest common factor from the polynomial in the previous
sentence, you are left with a difference of two squares.
4. To factor the polynomial $9x^2 + 3xy - 4y + 16x^2$, group the first
two terms and the last two terms.

Exercise Set 5.7

Vocabulary Review **1.** quadratic equation **2.** standard form
3. constant **4.** zero-factor property
List of Steps. 3, 4, 2, 5, 1
Problems **1.** $-2, 1$ **3.** 4, 5 **5.** $0, -1, 3$ **7.** $-\frac{2}{3}, -\frac{3}{2}$
9. $0, -\frac{4}{3}, \frac{4}{3}$ **11.** $0, -\frac{1}{3}, -\frac{3}{5}$ **13.** $-1, -2$ **15.** 4, 5 **17.** $6, -4$
19. 2, 3 **21.** -3 **23.** $4, -4$ **25.** $\frac{3}{2}, -4$ **27.** $-\frac{2}{3}$ **29.** 5
31. $4, -\frac{5}{2}$ **33.** $\frac{5}{3}, -4$ **35.** $\frac{7}{2}, -\frac{7}{2}$ **37.** $0, -6$ **39.** 0, 3
41. 0, 4 **43.** 0, 5 **45.** 2, 5 **47.** $\frac{1}{2}, -\frac{4}{3}$ **49.** $4, -\frac{5}{2}$ **51.** $8, -10$
53. 5, 8 **55.** 6, 8 **57.** -4 **59.** 5, 8 **61.** $6, -8$ **63.** $0, -\frac{3}{2}, -4$
65. $0, 3, -\frac{5}{2}$ **67.** $0, \frac{1}{2}, -\frac{5}{2}$ **69.** $0, \frac{3}{5}, -\frac{3}{2}$ **71.** $\frac{1}{2}, \frac{3}{2}$
73. $-5, 4$ **75.** $-7, -6$ **77.** $-3, -1$ **79.** 2, 3 **81.** $-15, 10$
83. $-5, 3$ **85.** $-3, -2, 2$ **87.** $-4, -1, 4$ **89.** $x(x + 1) = 72$
91. $x(x + 2) = 99$ **93.** $x(x + 2) = 5[x + (x + 2)] - 10$
95. Bicycle $75, suit $15 **97.** House $2,400, lot $600
Find the Mistake **1.** Standard form for a quadratic equation is
$a^2 + bx + c = 0$. **2.** The zero-factor property says that if
$(a + 8)(b - 3) = 0$, then $a + 8 = 0$ or $b - 3 = 0$. **3.** To solve the
equation $x^2 = -3x + 108$, we must first put the equation in standard
form. **4.** The solutions to the equation $3x^2 + 77 = 40x$ are $x = \frac{7}{3}$
and $x = 11$.

Exercise Set 5.8

Vocabulary Review **1.** consecutive **2.** standard form
3. hypotenuse **4.** Pythagorean theorem
Problems **1.** 8 and 10, or -10 and -8 **3.** 9 and 11, or -11 and -9
5. 8 and 10, or 0 and 2 **7.** 8 and 6 **9.** 2 and 12, or $-\frac{12}{5}$ and -10
11. 5 and 20, or 0 and 0 **13.** Width $= 3$ inches, length $= 4$ inches
15. 3 inches **17.** 6 inches and 8 inches **19.** 12 meters
21. 200 or 500 items **23.** 200 or 800 DVDs **25.** $4 or $8

27. \$7 or \$10 **29. a.** 5 feet **b.** 12 feet **31.** 6 inches by 12 inches

33. a. 25 seconds later

b.

t	h
0	100
5	1,680
10	2,460
15	2,440
20	1,620
25	0

35. $200x^{24}$ **37.** x^7
39. 8×10^1 **41.** $10ab^2$
43. $6x^4 + 6x^3 - 2x^2$
45. $9y^2 - 30y + 25$ **47.** $4a^2 - 49$

Find the Mistake **1.** To find the two integers, you must solve for x in the equation $x^2 - 34x = -288$ or $x^2 - 34x + 288 = 0$.
2. The longest side of the rectangle measures 71 inches.
3. The Pythagorean theorem states that the square of the hypotenuse is equal to the sum of the squares of the legs. **4.** To find the lengths of the two legs, we must solve for x in the equation $x^2 + (2x + 2)^2 = 169$.

Chapter 5 Review

1. $4(x - 6)$ **2.** $7ab(2a - 4 + 3b)$ **3.** $(2x - b)(x + 4a)$
4. $(3y - 4)(2x - 1)$ **5.** $(x - 4)(x + 7)$ **6.** $(x - 2)(x - 3)$
7. $(x - 7)(x + 7)$ **8.** $(x^2 + 9)(x - 3)(x + 3)$
9. Does not factor **10.** $3(2x - 3y)(2x + 3y)$ **11.** $(x^2 + 4)(x - 5)$
12. $(x + 4a)(2x - b)$ **13.** $3(2x - 3)(x + 4)$ **14.** $(3x + 5)(x - 2)$
15. $(4x - 1)(4x + 3)$ **16.** $2x(3x - 2)(2x - 3)$
17. $(x - 6)(x^2 + 6x + 36)$ **18.** $3(3x - 2)(9x^2 + 6x + 4)$
19. $-6, 3$ **20.** $-5, -2$ **21.** $-6, 6$ **22.** $-7, 5$ **23.** $-9, 5$
24. $-4, 0, 4$ **25.** $-2, \frac{4}{3}$ **26.** $-1, 0, \frac{3}{4}$ **27.** 7 and 13
28. 8 and 10 **29.** 9 ft by 6 ft **30.** 15 ft, 8 ft **31.** 600 items
32. \$3 or \$6 **33. a.** 8 ft **b.** 15 ft **34. a.** $t = \frac{1}{2}$ sec **b.** $t = \frac{5}{4}$ sec

Chapter 5 Cumulative Review

1. 4 **2.** 37 **3.** -8 **4.** -7 **5.** -32 **6.** $\frac{1}{27}$ **7.** 1
8. -3 **9.** $2a + 2$ **10.** $-15a - 4$ **11.** x^{21} **12.** 1
13. $6x^2 - 22x + 20$ **14.** $x^4 - 9$ **15.** -7 **16.** 12 **17.** -15
18. 2 **19.** $-\frac{4}{3}, 0, 3$ **20.** $-3, 3$ **21.** $x < -5$ **22.** $x > 4$
23.

24.

25.

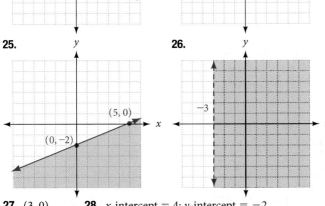

26.

27. $(3, 0)$ **28.** x-intercept $= 4$; y-intercept $= -2$
29. $m = 2$ **30.** $y = \frac{1}{4}x - 6$ **31.** $y = -\frac{3}{7}x + \frac{1}{3}$
32. slope $= -\frac{2}{5}$; y-intercept $= -3$ **33.** $(x - 7)(x - 4)$
34. $(3x - 4y)(5x + 2y)$ **35.** $(6 - x)(6 + x)$ **36.** $(6x - 1)^2$

37. $2x(x - 2y)(2x + 3y)$ **38.** $4y(x - 2)(4x + 9)$
39. $(3x - 1)(5y - 4)$ **40.** $(x - 3)(x^2 + 3x + 9)$ **41.** $-\frac{3}{2}, \frac{2}{3}, \frac{3}{2}$
42. $5, -\frac{1}{5}, 5$ **43.** $5x + 7 + \frac{19}{x - 4}$ **44.** $x^2 - 4x + 16$
45. 14 inches, 22 inches **46.** 3 burgers, 5 fries **47.** 595 apps
48. 2,905 apps

Chapter 5 Test

1. $6(x + 3)$ **2.** $4ab(3a - 6 + 2b)$ **3.** $(x - 2b)(x + 3a)$
4. $(3 - x)(5y - 4)$ **5.** $(x - 3)(x + 4)$ **6.** $(x - 7)(x + 3)$
7. $(x - 5)(x + 5)$ **8.** $(x - 2)(x + 2)(x^2 + 4)$ **9.** Does not factor
10. $2(3x - 4y)(3x + 4y)$ **11.** $(x^2 - 3)(x + 4)$ **12.** $(x - 3)(x + b)$
13. $2(2x - 5)(x + 1)$ **14.** $(4n - 3)(n + 4)$ **15.** $(3c - 2)(4c + 3)$
16. $3x(2x - 1)(2x + 3)$ **17.** $(x + 5y)(x^2 - 5xy + 25y^2)$
18. $2(3b - 4)(9b^2 + 12b + 16)$ **19.** $-3, 5$ **20.** $3, 4$ **21.** $-5, 5$
22. $-2, 7$ **23.** $-6, 5$ **24.** $-3, 0, 3$ **25.** $-4, \frac{3}{2}$ **26.** $-\frac{5}{3}, 6, 0$
27. 6 and 12 **28.** $-4, -2$ and $4, 6$ **29.** 4 ft by 13 ft
30. 6 ft, 8 ft **31.** 200 or 300 **32.** \$3 or \$5 **33. a.** 5 ft **b.** 12 ft
34. a. $t = 1, \frac{5}{8}$ sec **b.** $t = 2$ sec

Chapter 6

Exercise Set 6.1

Vocabulary Review **1.** rational expression
2. numerator, denominator **3.** undefined **4.** nonzero
5. simplify **6.** ratio **7.** rate equation **8.** distance

Problems **1.** $\frac{1}{x - 2}, x \neq 2$ **3.** $\frac{1}{a + 3}, a \neq -3, 3$ **5.** $\frac{1}{x - 5}, x \neq -5, 5$
7. $\frac{(x + 2)(x - 2)}{2}$ **9.** $\frac{2(x - 5)}{3(x - 2)}, x \neq 2$ **11.** 2 **13.** $\frac{5(x - 1)}{4}$
15. $\frac{1}{x - 3}$ **17.** $\frac{1}{x + 3}$ **19.** $\frac{1}{a - 5}$ **21.** $\frac{1}{3x + 2}$ **23.** $\frac{x + 5}{x + 2}$
25. $\frac{2m(m + 2)}{m - 2}$ **27.** $\frac{x - 1}{x - 4}$ **29.** $\frac{2(2x - 3)}{2x + 3}$ **31.** $\frac{2x - 3}{2x + 3}$
33. $\frac{1}{(x^2 + 9)(x - 3)}$ **35.** $\frac{3x - 5}{(x^2 + 4)(x - 2)}$ **37.** $\frac{2x(7x + 6)}{2x + 1}$
39. $\frac{x^2 + xy + y^2}{x + y}$ **41.** $\frac{x^2 - 2x + 4}{x - 2}$ **43.** $\frac{x^2 - 2x + 4}{x - 1}$
45. $-\frac{a + 5x}{3a + 5x}$ **47.** $-\frac{3a - 5b}{3a + 4b}$ **49.** $\frac{x + 2}{x + 5}$ **51.** $\frac{x + a}{x + b}$
53. a. $x^2 - 16$ **b.** $x^2 - 8x + 16$ **c.** $4x^3 - 32x^2 + 64x$ **d.** $\frac{x}{4}$
55. $\frac{4}{3}$ **57.** $\frac{4}{5}$ **59.** $\frac{8}{1}$ **61.** $\frac{1}{5}$

63.

Checks Written	Total Cost	Cost per Check
x	2.00 + 0.15x	$\frac{2.00 + 0.15x}{x}$
0	2.00	undefined
5	2.75	0.55
10	3.50	0.35
15	4.25	0.28
20	5	0.25

65. 40.7 miles/hour **67.** US 4.31 gallons; Mexico 2.78 gallons; Norway 2.68 gallons; Ireland 2.67 gallons; Canada 2.58 gallons
69. 0.125 miles per minute

71.

Stock	Price	Earnings per Share	P/E
Yahoo	37.80	1.07	35.33
Google	381.24	4.51	84.53
Disney	24.96	1.34	18.63
Nike	85.46	4.88	17.51
Ebay	40.96	0.73	56.11

Disney and Nike

73. $\frac{5}{14}$ **75.** $\frac{9}{10}$ **77.** $(a-5)(a+4)$ **79.** $(x+3)(x-3)$

81. $3(x-3)$ **83.** $(x-5)(x+4)$ **85.** $a(a+5)$ **87.** 11.8

89. 0 **91.** $16a^2b^2$ **93.** $19x^4 + 21x^2 - 42$

95. $7a^4b^4 + 9b^3 - 11a^3$

Find the Mistake **1.** In the rational expression $\frac{x^2-81}{x^2}$, the denominator is x^2. **2.** Restrictions on the variable in a rational expression are the values of the variable that gives a denominator of zero. **3.** To simplify the rational expression of $\frac{18}{42}$, divide the numerator and denominator by the common factors 2 and 3. **4.** The rate equation written as rate equals a ratio of distance to time looks like $r = \frac{d}{t}$.

Exercise Set 6.2

Vocabulary Review **1.** rational numbers **2.** multiply **3.** factors **4.** reciprocal **5.** unit analysis **6.** common **7.** number 1 **8.** conversion factor

Problems **1.** 2 **3.** $\frac{x}{2}$ **5.** $\frac{3}{2}$ **7.** $\frac{1}{2(x-3)}$ **9.** $\frac{4a(a+5)}{7(a+4)}$ **11.** $\frac{y-2}{4}$

13. $\frac{2(x+4)}{x-2}$ **15.** $\frac{x+3}{(x-3)(x+1)}$ **17.** 1 **19.** $-\frac{20}{9}$ **21.** $-\frac{x+2y}{x-2y}$

23. $\frac{y-5}{(y+2)(y-2)}$ **25.** $\frac{x+5}{x-5}$ **27.** 1 **29.** $\frac{a+3}{a+4}$ **31.** $\frac{2y-3}{y-6}$

33. $\frac{(x-1)(x+5)}{(x+1)(x-2)}$ **35.** $\frac{3}{2}$ **37. a.** $\frac{4}{13}$ **b.** $\frac{x+1}{x^2+x+1}$ **c.** $\frac{x-2}{x^2+x+1}$ **d.** $\frac{x+1}{x-1}$

39. $2(x-3)$ **41.** $(x+2)(x+1)$ **43.** $-2x(x-5)$ **45.** $\frac{2(x+5)}{x(y+5)}$

47. $\frac{2x}{x-y}$ **49.** $\frac{1}{x-2}$ **51.** $\frac{1}{5}$ **53.** $\frac{1}{100}$ **55.** 60 inches

57. 6 feet **59.** $1\frac{1}{3}$ yards **61.** 432 in² **63.** 5 ft² **65.** 108 ft²

67. $\frac{4}{5}$ **69.** $\frac{11}{35}$ **71.** $-\frac{4}{35}$ **73.** $\frac{1}{30}$ **75.** $2x-6$ **77.** x^2-x-20

79. $-x-13$ **81.** $4x$ **83.** $\frac{1}{x-3}$ **85.** $\frac{x-6}{2(x-5)}$

87. x^2-x-30 **89.** $-x^2+2x-22$

Find the Mistake **1.** To multiply the rational expressions $\frac{4x-6}{x^2}$ and $\frac{x^2-9}{x+3}$, multiply the numerators and denominators. **2.** To divide the rational expression $\frac{x+7}{x-1}$ by $\frac{(x+7)(x+5)}{4}$, multiply the first expression by the reciprocal of divisor. **3.** To convert 419 seconds to hours, multiply by the conversion factor $\frac{1\ \text{hour}}{3600\ \text{seconds}}$. **4.** Using the rate equation $r = \frac{d}{t}$, a bicycle traveling 13,500 feet in 12 minutes is traveling at a speed of 1,125 feet per minute.

Exercise Set 6.3

Vocabulary Review **1.** same **2.** least common denominator **3.** different **4.** reduce

Problems **1.** $\frac{7}{x}$ **3.** $\frac{4}{a}$ **5.** 1 **7.** $y+1$ **9.** $x+2$

11. $x-2$ **13.** $\frac{6}{x+6}$ **15.** $-\frac{1}{x+2}$ **17.** $\frac{(y+2)(y-2)}{2y}$

19. $\frac{3+2a}{6}$ **21.** $\frac{7x+3}{4(x+1)}$ **23.** $\frac{1}{5}$ **25.** $\frac{1}{3}$ **27.** $\frac{3}{x-2}$

29. $\frac{4}{a+3}$ **31.** $\frac{2(x-10)}{(x+5)(x-5)}$ **33.** $\frac{x+2}{x+3}$ **35.** $\frac{a+1}{a+2}$

37. $\frac{1}{(x+3)(x+4)}$ **39.** $\frac{y}{(y+5)(y+4)}$ **41.** $\frac{3(x-1)}{(x+4)(x+1)}$ **43.** $\frac{1}{3}$

45. $\frac{2}{2x+1}$ **47. a.** $\frac{2}{27}$ **b.** $\frac{8}{3}$ **c.** $\frac{11}{18}$ **d.** $\frac{3x+10}{(x-2)^2}$ **e.** $\frac{(x+2)^2}{3x+10}$ **f.** $\frac{x+3}{x+2}$

49.

Number	Reciprocal	Sum	Quotient
x	$\frac{1}{x}$	$1+\frac{1}{x}$	$\frac{x+1}{x}$
1	1	2	2
2	$\frac{1}{2}$	$\frac{3}{2}$	$\frac{3}{2}$
3	$\frac{1}{3}$	$\frac{4}{3}$	$\frac{4}{3}$
4	$\frac{1}{4}$	$\frac{5}{4}$	$\frac{5}{4}$

51. $\frac{x+3}{x+2}$ **53.** $\frac{x+2}{x+3}$ **55.** $x+\frac{2}{x} = \frac{x^2+2}{x}$ **57.** $\frac{1}{x}+\frac{1}{2x} = \frac{3}{2x}$

59. $\frac{1}{x}+\frac{1}{x+3} = \frac{2x+3}{x(x+3)}$ **61.** 3 **63.** 0 **65.** Undefined

67. $-\frac{3}{2}$ **69.** $2x+15$ **71.** x^2-5x **73.** -6 **75.** -2

Find the Mistake **1.** To add fractions with same denominator, place the sum of the numerators over the common denominator. **2.** When subtracting the fractions $\frac{x}{x^2-36}$ and $\frac{6}{x^2-36}$, factor the common denominator and divide out the common factor $x-6$. **3.** To add the fractions $\frac{8}{4x}$ and $\frac{9}{16x^2}$, first multiply the denominator of the first fraction by $4x$ to obtain the LCD. **4.** The lowest common denominator for a problem that finds the difference between a number squared and three times the number's reciprocal is x or any variable the student chooses to set equal to the number.

Landmark Review **1.** $\frac{a+3}{2}, a \neq 3$ **2.** $\frac{1}{y-3}, y \neq 3$

3. $\frac{3x(x-5)}{x+2}, x \neq -3, -2$ **4.** $\frac{x+4}{5x-2}, x \neq \pm\frac{2}{5}$ **5.** $\frac{a}{2}$

6. $\frac{x(x-3)}{x-1}$ **7.** $\frac{y(y-3)}{2(y+2)(y-5)}$ **8.** $2x^2+6x$ **9.** 1

10. $\frac{4}{y-3}$ **11.** $\frac{a-3}{a-4}$ **12.** $x-4$ **13.** $\frac{x+4}{x+1}$

Exercise Set 6.4

Vocabulary Review **1.** least common denominator **2.** multiply **3.** denominators **4.** quadratic

Problems **1.** -3 **3.** 20 **5.** -1 **7.** 5 **9.** -2 **11.** 4

13. 3, 5 **15.** $-8, 1$ **17.** 2 **19.** 1 **21.** 8 **23.** 5

25. No solution **27.** 3 **29.** No solution **31.** 0 **33.** 2 **35.** -4

37. -1 **39.** $-6, -7$ **41.** -1 **43. a.** $\frac{1}{5}$ **b.** 5 **c.** $\frac{25}{3}$ **d.** 3 **e.** $\frac{1}{5}$, 5

45. a. $\frac{7}{(a-6)(a+2)}$ **b.** $\frac{a-5}{a-6}$ **c.** 7 **47.** $\frac{1}{3}$ **49.** 30 **51.** 1 **53.** -3

Find the Mistake **1.** The first step to solving an equation that contains rational expressions is to find the LCD for all the denominators in the equation. **2.** To solve the problem $\frac{x}{x+7}+\frac{7}{2} = \frac{7}{x+7}$, first multiply each term of the equation by $2(x+7)$. **3.** The LCD for the equation $\frac{x+2}{x^2+8x} = \frac{-2}{x^2-64}$ is $x(x+8)(x-8)$. **4.** When checking a solution, if one or more terms in the equation are undefined, then there is no real number that makes the equation a true statement.

Exercise Set 6.5

Vocabulary Review **1.** same **2.** quadrant III **3.** opposite **4.** quadrant II

Problems **1.** $\frac{1}{4}, \frac{3}{4}$ **3.** $\frac{2}{3}$ and $\frac{3}{2}$ **5.** -2 **7.** 4, 6

9. 16 miles per hour **11.** 300 miles per hour

13. 170 miles per hour; 190 miles per hour **15.** 9 miles per hour

17. 8 miles per hour **19.** 60 hours **21.** $5\frac{5}{11}$ minutes

23. 12 minutes **25.** 10 hours **27.** 2.7 miles

29. 742 miles per hour **31.** The facts are correct: 2,000 steps ≈ 0.95 mile; 5,000 steps ≈ 4.7 miles **33.** 8.8 miles per hour

35. **37.**

39.

41. $\frac{2A}{a+b} - h$ **43.** $\frac{3V}{\pi r^2} = h$

45. $y_2 = m(x_2 - x_1) - y_1$

47. $x_2 = \frac{y_2 - y_1}{m} + x_1$ **49.** $P = \frac{V^2}{R}$

51. $R_1 = \frac{RR_2}{R_2 - R}$ **53.** $\frac{3}{4}$

55. $\frac{3}{2}$ **57.** $2x^3y^3$ **59.** $\frac{x^2y^3}{2}$

61. $x(xy + 1)$ **63.** $x^2(x - 1)$

Find the Mistake 1. One number is four times as large as another. The sum of their reciprocals is $\frac{15}{4}$. The two numbers are $\frac{1}{3}$ and $\frac{4}{3}$. **2.** A kayak travels 3 miles upriver in the same amount of time it takes the kayak to travel 5 miles back down the same river. If the current is 5 miles per hour, what is the speed of the boat in still water? To solve this problem, we must first assign the variable x to the speed of the boat in still water. **3.** The graph of $x = \frac{1}{y}$ will not cross the x–axis because the expression $\frac{1}{y}$ is undefined when $y = 0$. **4.** The graph of the equation $y = \frac{9}{x}$ will appear in quadrants I and III.

Exercise Set 6.6

Vocabulary Review 1. complex fraction **2.** top
3. LCD **4.** reciprocal
Problems 1. 6 **3.** $\frac{1}{6}$ **5.** xy^2 **7.** $\frac{xy}{2}$ **9.** $\frac{y}{x}$ **11.** $\frac{a+1}{a-1}$

13. $\frac{x+1}{2(x-3)}$ **15.** $a - 3$ **17.** $\frac{y+3}{y+2}$ **19.** $x + y$ **21.** $\frac{a+1}{a}$ **23.** $\frac{1}{x}$

25. $\frac{2a+3}{3a+4}$ **27.** $\frac{-3x+7}{2x-3}$ **29.** $\frac{x-1}{x+2}$ **31.** $\frac{x+4}{x+1}$ **33.** $\frac{7}{3}, \frac{17}{7}, \frac{41}{17}$

35.

Number x	Reciprocal $\frac{1}{x}$	Quotient $\frac{x}{\frac{1}{x}}$	Square x^2
1	1	1	1
2	$\frac{1}{2}$	4	4
3	$\frac{1}{3}$	9	9
4	$\frac{1}{4}$	16	16

37.

Number x	Reciprocal $\frac{1}{x}$	Sum $1 + \frac{1}{x}$	Quotient $\frac{1 + \frac{1}{x}}{\frac{1}{x}}$
1	1	2	2
2	$\frac{1}{2}$	$\frac{3}{2}$	3
3	$\frac{1}{3}$	$\frac{4}{3}$	4
4	$\frac{1}{4}$	$\frac{5}{4}$	5

39. $\frac{7}{2}$ **41.** $-3, 2$ **43.** 20 **45.** $\frac{10}{3}$
Find the Mistake 1. A complex fraction contains fractions in its numerator or its denominator. **2.** To simplify the complex fraction $\frac{\frac{4}{9}}{\frac{7}{13}}$ using the LCD for both fractions, we must multiply the top and bottom by 117. **3.** One way to simplify the complex fraction $\frac{\frac{6y^2}{7x}}{\frac{4y}{x}}$ is to multiply $\frac{6y^2}{7x}$ by $\frac{x}{4y}$.

4. The easiest way to simplify the complex fraction $\frac{\frac{2}{x+3} + 4}{\frac{2}{x} + 3}$ is to multiply the top and bottom by the LCD $x(x + 3)$.

Exercise Set 6.7

Vocabulary Review 1. proportion **2.** term **3.** extremes
4. means **5.** means-extremes property
Problems 1. 1 **3.** 10 **5.** 40 **7.** $\frac{5}{4}$ **9.** $\frac{7}{3}$ **11.** 3, -4
13. 4, -4 **15.** 2, -4 **17.** 6, -1 **19.** $-4, 3$ **21.** $-3, 3$
23. $-\frac{11}{2}, 5$ **25.** 15 hits **27.** 21 millimeters **29.** 45.5 grams
31. 14.7 inches **33.** 343 miles **35.** $y = 20$ **37.** $y = -20$
39. $y = 2$ **41.** $y = \frac{1}{2}$ **43.** $x = \pm 5$ **45.** $x = \pm 2$
Find the Mistake 1. A proportion is a ratio equal to a ratio.
2. In the proportion $\frac{6}{11} = \frac{12}{22}$, the second term is 11.
3. In the proportion $\frac{a}{b} = \frac{c}{d}$, the extremes are a and d.
4. For the proportion $\frac{13}{19} = \frac{39}{57}$, the product of the means is 741.

Exercise Set 6.8

Vocabulary Review 1. multiple **2.** directly
3. constant of variation **4.** reciprocal **5.** inversely
Problems 1. 8 **3.** 130 **5.** 5 **7.** -3 **9.** 3 **11.** 243
13. 400 **15.** 2 **17.** $\frac{1}{2}$ **19.** 1 **21.** 5 **23.** 25 **25.** 9
27. $\frac{5}{2}$ **29.** 84 pounds **31.** $\frac{735}{2}$ or 367.5 **33.** \$235.50
35. 96 lbs **37.** 12 amperes **39.** $(-2, 2), (0, 3), (2, 4)$
41.

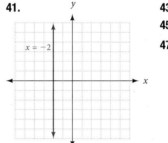

43. 2
45. $y = \frac{1}{2}x + 2$
47. $y = 2x + 1$

Find the Mistake 1. An algebraic equation written as an equivalent to a variation statement includes k, called the constant of variation. **2.** Two variables vary directly if one is a constant multiple of the other. **3.** The equation $y = kx$ is translated as y varies directly as x.
4. The equation $F = \frac{k}{\sqrt{t}}$ is translated as F varies inversely as the square root of t.

Chapter 6 Review

1. $\frac{x+4}{x-3}$ **2.** $\frac{(x+4)}{2}$ **3.** $\frac{y+6}{x^2+2}$ **4.** $\frac{3}{2}$ **5.** $(x - 5)(x + 6)$
6. $\frac{x-2}{2x-2}$ **7.** $(x + 3)(x + 9)$ **8.** $\frac{7}{x-3}$ **9.** $\frac{1}{2(x-3)}$
10. $\frac{2}{(x+1)(x+3)}$ **11.** $-\frac{11}{7}$ **12.** $-\frac{14}{3}$ **13.** -12
14. 9 miles/hour **15.** 7.5 hours **16.** $\frac{x-4}{2x-3}$ **17.** $\frac{x+4}{x+2}$
18. $\frac{3}{4}, \frac{3}{7}$ **19.** 27 parts **20.** 36 **21.** $\frac{1}{4}$ **22.** 1.1
23. Tennis: 2,613.6 ft²; Basketball: 4,791.6 ft²; Volleyball: 2,178 ft²

Chapter 6 Cumulative Review

1. -8 **2.** 13 **3.** -9 **4.** $2x$ **5.** $4x + 4$ **6.** $-2x + 10$
7. x^9 **8.** $\frac{1}{x^{12}}$ **9.** -5 **10.** $\frac{1}{x^2}$ **11.** $-5x$
12. $3a^3 + 3a^2 + 6a - 2$ **13.** $x - 5$ **14.** $\frac{3x+2}{3(x-2)}$ **15.** 3
16. \varnothing **17.** $-\frac{1}{2}, \frac{1}{2}$ **18.** 2, 5 **19.** 4 **20.** -10

21. $(-3, 5)$ **22.** lines parallel \varnothing **23.** $(7, -2)$ **24.** $(3, -9)$

25.

$y = -2x + 7$

26.

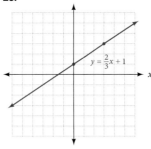

$y = \frac{2}{3}x + 1$

27. $x < -3$ **28.** $a \le -8$ **29.** $(x - 5)(y + 7)$ **30.** $(3x - 4y)(3x + 4y)$

31. $(a - 3)(a + 5)$ **32.** $(3x - 2)(5x - 3)$ **33.** $(3x - 8)^2$

34. $5(x^2 + 4)$ **35.** x-intercept $= 6$; y-intercept $= -3$

36. $m = 4$; y-intercept $= -6$ **37.** $y = -\frac{2}{3}x - 5$ **38.** $y = \frac{3}{5}x - 6$

39. -10 **40.** -11 **41.** Communtative property of addition

42. Associative property of addition **43.** $x + 5$ **44.** $\frac{x - 2}{3(x - 3)}$

45. $\frac{1}{3}$ **46.** $\frac{4x - 1}{2x + 3}$ **47.** -96 **48.** 6 **49.** 129.4 cal/km

50. 27.7 cal/km

Chapter 6 Test

1. $\frac{x + 3}{x - 3}$ **2.** $\frac{3}{a - 4}$ **3.** $\frac{x + 3}{(y - 1)(y + 1)}$ **4.** $\frac{3}{2}$ **5.** $(x - 3)(x + 4)$

6. $\frac{2x - 3}{x + 4}$ **7.** $(x + 4)(x - 1)$ **8.** $-\frac{2}{x - 8}$ **9.** $\frac{2x + 4}{(x - 4)(x + 4)}$

10. $\frac{2}{(x + 3)(x - 1)}$ **11.** $\frac{6}{5}$ **12.** $-\frac{7}{6}$ **13.** -14

14. 15 miles/hour **15.** 24 hours **16.** $\frac{x + 2}{x - 2}$ **17.** $\frac{x + 3}{x + 2}$ **18.** $\frac{1}{3}, \frac{1}{4}$

19. 102 parts **20.** 54 **21.** 2 **22.** 2.8

23. Barq's Root Beer: 0.35 grains; Coca Cola Classic: 0.54 grains;
Dr. Pepper: 0.66 grains; Mountain Dew: 0.83 grains; Pepsi: 0.57
grains; Vault: 1.10 grains

Chapter 7

Exercise Set 7.1

Vocabulary Review **1.** system **2.** ordered pairs **3.** axes

4. intersection

Problems

1.

3.

5.

7.

9.

11.

13.

15.

17.

19.

21.

23.

25.

No solution (inconsistent)

27.

Lines coincide (dependent)

29.

31.

33.

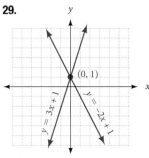

35. 33 $5 chips, 12 $25 chips

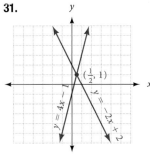

37. a. 25 hours **b.** Gigi's **c.** Marcy's
39. 1 **41.** 2 **43.** All real numbers **45.** $x = 3y - 1$
47. 1 **49.** 5 **51.** 34.5 **53.** 4 **55.** 33.95 **57.** 20
Find the Mistake 1. The point of intersection for two straight line graphs is a solution to both equations. **2.** The point of intersection for the equations $y = 6$ and $y = 4x + 2$ is (1, 6).
3. The point of intersection for the equations $x = 5$ and $x = 9$ is not (5, 9) because there is no point of intersection; the lines are parallel.
4. When lines coincide, we say that the equations are dependent.

Exercise Set 7.2

Vocabulary Review 1. equations **2.** substitute **3.** variables
4. check
Problems 1. (4, 7) **3.** (3, 17) **5.** $\left(\frac{3}{2}, 2\right)$ **7.** (2, 4) **9.** (0, 4)
11. (−1, 3) **13.** (1, 1) **15.** (2, −3) **17.** $\left(-2, \frac{3}{5}\right)$ **19.** (−3, 5)
21. No solution (inconsistent) **23.** Lines coincide (dependent)
25. (3, 1) **27.** $\left(\frac{1}{2}, \frac{3}{4}\right)$ **29.** (2, 6) **31.** (4, 4) **33.** (5, −2)
35. (18, 10) **37.** Lines coincide (dependent)
39. No solution (inconsistent) **41.** (10, 12) **43.** 7.29 miles per hour
45. a. 560 miles **b.** Car **c.** Truck **d.** We are only working with positive numbers
47. $2x$ **49.** $7x$ **51.** $13x$ **53.** $3x - 2y$ **55.** $-12x + 8y$
57. $22x$ **59.** 1 **61.** 0 **63.** −5
Find the Mistake 1. Substitute the value for x from the first equation for x in the second equation. **2.** Add $3y$ to both sides of the first equation to isolate x. **3.** Substitute the value for y from the first equation for y in the second equation. **4.** Substitute the value for x from the first equation for x in the second equation.

Exercise Set 7.3

Vocabulary Review 1. variable **2.** multiplication property of equality
3. add **4.** solve **5.** substitute **6.** solution
Problems 1. (2, 1) **3.** (3, 7) **5.** (2, −5) **7.** (−1, 0)
9. Lines coincide (dependent) **11.** No solution (inconsistent)
13. (4, 8) **15.** $\left(\frac{1}{5}, 1\right)$ **17.** (1, 0) **19.** (3, 0) **21.** (−1, −2)
23. $\left(-5, \frac{3}{4}\right)$ **25.** (−4, 5) **27.** $\left(-\frac{4}{3}, 1\right)$ **29.** (−3, −10)
31. (3, 2) **33.** $\left(5, \frac{1}{3}\right)$ **35.** $\left(-2, \frac{2}{3}\right)$ **37.** (2, 2)

39. No solution (inconsistent) **41.** (1, 1)
43. No solution (inconsistent) **45.** (10, 12)
47. 1,478 adult tickets, 242 children tickets **49.** 3 and 23
51. 15 and 24 **53.** Length = 23 in.; width = 6 in.
55. 14 nickels and 10 dimes
Find the Mistake 1. Add the original equations to eliminate x and give a false statement, making the system inconsistent.
2. Multiply both sides of the first equation by −3 to eliminate x.
3. Add the original equations to eliminate x and solve for y.
4. When solving a system by the elimination method, if both variables are eliminated and the resulting statement is true, then the lines coincide.
Landmark Review 1. (−2, 1) **2.** (−3, −4) **3.** $\left(\frac{3}{2}, 2\right)$
4. $\left(9, \frac{18}{5}\right)$ **5.** (−1, 5) **6.** No solution (inconsistent)
7. (5, −6) **8.** (4, −3) **9.** (−4, 5) **10.** $\left(4, \frac{1}{4}\right)$

Exercise Set 7.4

Vocabulary Review 1. known **2.** unknown, expressions
3. equations **4.** system **5.** sentences **6.** problem, solution
Problems 1. 10 and 15 **3.** 3 and 12 **5.** 4 and 9 **7.** 6 and 29
9. Linda is 70 years old, and Audra is 31 years old.
11. Kate was 21 years old when she received her bachelor's, and 37 years old when she began her master's program.
13. $9,000 at 8%, $11,000 at 6% **15.** $2,000 at 6%, $8,000 at 5%
17. 6 nickels, 8 quarters **19.** 12 dimes, 9 quarters
21. 6 liters of 50% solution, 12 liters of 20% solutions
23. 10 gallons of 10% solution, 20 gallons of 7% solution
25. 20 adults, 50 kids **27.** 16 feet wide, 32 feet long
29. 33 $5 chips, 12 $25 chips **31.** 50 at $11, 100 at $20
33. 18 pre-registration, 11 race day registration
35. (2, −1) **37.** (4, 3) **39.** (1, 1) **41.** (5, 2)
Find the Mistake 1. The second step in the Blueprint for Problem Solving is to assign variables to each of the unknown items.
2. If there are two unknown quantities, we write two equations to get our answer. **3.** If we are solving a problem involving nickels and dimes and we know we have a total of 15 coins, we can use $n + d = 15$ as one of our equations. **4.** One of our equations will be $y = 3x$ or $x + y = 8$.

Chapter 7 Review

Problems 1. (−1, 4) **2.** (3, 5) **3.** (2, 6) **4.** (2, −3)
5. (−1, 0) **6.** (−4, 1) **7.** (3, 5) **8.** (−3, −6) **9.** (−1, 4)
10. (4, 2) **11.** (−2, 3) **12.** (−1, −3) **13.** (−2, −6)
14. Lines coincide **15.** (6, 3) **16.** (−3, 4) **17.** (1, 0)
18. Lines coincide **19.** (−2, 4) **20.** (−2, −1) **21.** (0, 5)
22. No solution (inconsistent) **23.** 7, 3 **24.** 12, 9
25. $2,000 at 8%, $7,000 at 5% **26.** $5,000 at 9%, $9,000 at 11%
27. 15 dimes, 6 quarters **28.** 12 dimes, 7 quarters
29. 4 liters of 20% solution, 7 liters of 80% solution
30. 12 liters of 35% solution, 18 liters of 10% solution

Chapter 7 Cumulative Review

Problems 1. −5 **2.** −3 **3.** 28 **4.** 33 **5.** Undefined
6. 9 **7.** $\frac{11}{48}$ **8.** $\frac{11}{8}$ **9.** $3x + 1$ **10.** $-12a + 26$ **11.** 6
12. 0 **13.** −4 **14.** 7 **15.** $x \geq -11$ **16.** $x \leq -10$

17.

18.

19. $(3, -2)$ **20.** $(5, 1)$ **21.** Lines coincide **22.** $\left(-\frac{1}{2}, -2\right)$
23. $(-6, 3)$ **24.** $(-3, 0)$ **25.** $(2, 7)$ **26.** $(-4, -6)$
27. $(9, -3)$ **28.** $(-2, 7)$ **29.** $(3, -8)$ **30.** $(-2, 1)$
31. -8 **32.** -19 **33.** $2 \cdot 3^3 \cdot 5$ **34.** $2^2 \cdot 3 \cdot 5 \cdot 7$
35. 15 **36.** -28 **37.** $(-3, 1)$ **38.** $(-3, 3)$ and $\left(2, -\frac{1}{3}\right)$
39. x-intercept $= 6$; y-intercept $= -5$ **40.** $m = \frac{1}{4}$ **41.** $m = -\frac{5}{2}$
42. $m = -\frac{5}{24}$ **43.** $y = \frac{5}{4}x - 2$ **44.** $y = \frac{1}{3}x + 8$ **45.** $y = 3x - 13$
46. x-intercept $= -12$; y-intercept $= 4$ **47.** 5 dimes, 7 quarters
48. $600 at 7%, $900 at 11% **49.** 780 people **50.** 400 people

Chapter 7 Test

Problems 1. $(0, 6)$ **2.** $(3, 2)$ **3.** $(7, 4)$ **4.** $(-2, 4)$
5. $(4, 0)$ **6.** $(-5, 3)$ **7.** $(-3, 4)$ **8.** $(11, 3)$ **9.** $(-4, 5)$
10. $(1, 2)$ **11.** $(-6, 3)$ **12.** Lines coincide **13.** 2, 13
14. 4, 14 **15.** $480 at 6% **16.** 8 dimes, 11 nickels
17. 28 ft by 71ft **18. a.** 100 miles **b.** Company 1 **c.** $x > 100$

Chapter 8

Exercise Set 8.1

Vocabulary Review 1. square **2.** cube **3.** radical
4. index, radical sign, radicand **5.** golden ratio
Problems 1. 12 **3.** Not a real number **5.** -7 **7.** -3
9. 2 **11.** Not a real number **13.** 0.2 **15.** 0.2 **17.** 5
19. -5 **21.** Not a real number **23.** -2 **25.** 2 **27.** -2
29. $6a$ **31.** $9x$ **33.** $5x^2$ **35.** $7x^3$ **37.** $3xy^2$ **39.** $11a^3b$
41. $5x^6y^5$ **43.** $12a^4b^8$ **45.** $3a^4$ **47.** xy^2 **49.** $2x^3y$ **51.** $2a^3b^5$
53. $-3x^3y^2$ **55.** $-2a^2b^5$ **57.** $2x^3y^5$ **59.** $-2a^2b^6$ **61.** 25 mph
63. 1.618 **65. a.** 420 picometers **b.** 594 picometers
c. 5.94×10^{-10} meters **67.** 80.7 ft **69.** 250 ft **71.** 5
73. 6 **75.** $4x^2y$ **77.** $5y$ **79.** 3 **81.** 2 **83.** $2ab$
85. $5x^2y$
Find the Mistake 1. The n^{th} root of the x is such that $\left(\sqrt[n]{x}\right)^n = x$.
2. Simplifying $\sqrt[6]{-729}$ doesn't equal a real number.
3. Simplifying $\sqrt[4]{256x^{16}y^{12}z^8}$ equals $4x^4y^3z^2$. **4.** The golden ratio
is the ratio of a rectangle's length to width, which equals $\frac{1 + \sqrt{5}}{2}$.

Exercise Set 8.2

Vocabulary Review 1. product **2.** quotient
3. simplified form, radicand, index, fractions, deonominator
4. rationalizing **5.** absolute value **6.** even, odd **7.** spiral of roots
Problems 1. $2\sqrt{3}$ **3.** $\sqrt{-20}$, not a real number **5.** $\frac{\sqrt{11}}{2}$
7. $2\sqrt{2}$ **9.** $7\sqrt{2}$ **11.** $12\sqrt{2}$ **13.** $4\sqrt{5}$ **15.** $4\sqrt{3}$
17. $15\sqrt{3}$ **19.** $3\sqrt[3]{2}$ **21.** $4\sqrt[3]{2}$ **23.** $6\sqrt[3]{2}$ **25.** $2\sqrt[5]{2}$
27. $3x\sqrt{2x}$ **29.** $2y\sqrt[4]{2y^3}$ **31.** $2xy^2\sqrt[3]{5xy}$ **33.** $4abc^2\sqrt{3b}$
35. $2bc\sqrt[3]{6a^2c}$ **37.** $2xy^2\sqrt[5]{2x^3y^2}$ **39.** $3xy^2z\sqrt[5]{x^2}$

41. $8x^3y^2\sqrt{2xy}$ **43.** $\frac{7}{2}$ **45.** $\frac{-5}{4}$ **47.** $\frac{6x^3\sqrt{2}}{y}$ **49.** $\frac{2}{a}$
51. $\frac{2\sqrt{3}}{3}$ **53.** $\frac{5\sqrt{6}}{6}$ **55.** $\frac{\sqrt{2}}{2}$ **57.** $\frac{\sqrt{5}}{5}$ **59.** $2\sqrt[3]{4}$
61. $\frac{2\sqrt{3}}{3}$ **63.** $\frac{\sqrt[4]{24x^2}}{2x}$ **65.** $\frac{\sqrt[4]{8y^3}}{y}$ **67.** $\frac{\sqrt[3]{36xy^2}}{3y}$ **69.** $\frac{\sqrt[3]{6xy^2}}{3y}$
71. $\frac{\sqrt[4]{2x}}{2x}$ **73.** $\frac{\sqrt[3]{12x}}{3x}$ **75.** $\frac{3x\sqrt{15xy}}{5y}$ **77.** $\frac{5xy\sqrt{6xz}}{2z}$
79. $\frac{2ab\sqrt[3]{6ac^2}}{3c}$ **81.** $\frac{2xy^2\sqrt[3]{3z^2}}{3z}$ **83.** $5|x|$ **85.** $7|xy|\sqrt{3x}$
87. $|x - 5|$ **89.** $|2x + 3|$ **91.** $2|a(a + 2)|$ **93.** $2|x|\sqrt{x - 2}$
95. $\sqrt{9 + 16} \overset{?}{=} \sqrt{9} + \sqrt{16}$; $\sqrt{25} \overset{?}{=} 3 + 4$; $5 \neq 7$ **97.** $5\sqrt{13}$ feet
99. a. 89.4 miles **b.** 126.5 miles **c.** 154.9 miles **101.** $2\sqrt{13}$ in.
103. $x\sqrt{5}$ **107.** $7x$ **109.** $27xy^2$ **111.** $\frac{5}{6}x$
113. $-7(x + 1)^2$ **115.** $3\sqrt{2}$ **117.** $5y\sqrt{3xy}$ **119.** $2a\sqrt[3]{ab^2}$
121. $2x^2y\sqrt[4]{2y}$

Find the Mistake 1. Product property for radicals states that the n^{th}
root of a product is the product of the n^{th} roots. **2.** Using the
quotient property for radicals, the expression $\sqrt[5]{\frac{3}{7}}$ is equal to $\frac{\sqrt[5]{3}}{\sqrt[5]{7}}$.
3. The process of rationalizing the denominator is used to
get rid of a radical in the denominator.
4. According to the extension of the absolute value property for
radicals, $\sqrt[n]{a^n} = a$ if n is an odd number.

Landmark Review 1. $5ab$ **2.** $2x$ **3.** $8xy^2$ **4.** $5x^2y^3$
5. $3\sqrt{5}$ **6.** $4\sqrt{2}$ **7.** $2x^2\sqrt[3]{x}$ **8.** $4x\sqrt{6x}$ **9.** $\frac{5x}{2y}$
10. $\frac{5}{8x^2}$ **11.** $\frac{\sqrt{10}}{10}$ **12.** $\frac{\sqrt{35}}{7}$ **13.** $\sqrt{3y}$ **14.** $\frac{\sqrt{6a}}{a}$

Exercise Set 8.3

Vocabulary Review 1. index, radicand **2.** polynomials
3. simplified form, distributive property **4.** similar
Problems 1. $7\sqrt{5}$ **3.** $-x\sqrt{7}$ **5.** $\sqrt[3]{10}$ **7.** $9\sqrt[5]{6}$ **9.** 0
11. $\sqrt[3]{5}$ **13.** $-32\sqrt{2}$ **15.** $-3x\sqrt{2}$ **17.** $-2x\sqrt[3]{2}$
19. $8x\sqrt[3]{xy^2}$ **21.** $3a^2b\sqrt{3ab}$ **23.** $11ab\sqrt[3]{3a^2b}$ **25.** $10xy\sqrt[4]{3y}$
27. $\sqrt{2}$ **29.** $\frac{8\sqrt{5}}{15}$ **31.** $\frac{(x - 1)\sqrt{x}}{x}$ **33.** $\frac{3\sqrt{2}}{2}$ **35.** $\frac{5\sqrt{6}}{6}$
37. $\frac{8\sqrt[3]{25}}{5}$ **39.** $\sqrt{12} \approx 3.464$; $2\sqrt{3} \approx 2(1.732) = 3.464$
41. $\sqrt{8} + \sqrt{18} \approx 2.828 + 4.243 = 7.071$; $\sqrt{50} \approx 7.071$; $\sqrt{26} \approx 5.099$
43. $8\sqrt{2x}$ **45.** 5 **53.** $\sqrt{2}$ **55.** $\frac{\sqrt{2}}{1} \leq \frac{1.414}{1}$
57. Mt. Everest: 42.6 seconds; Kangchenjanga: 42.0 seconds **59.** 6
61. $4x^2 + 3xy - y^2$ **63.** $x^2 + 6x + 9$ **65.** $x^2 - 4$ **67.** $6\sqrt{2}$
69. 6 **71.** $9x$ **73.** $48x$

Find the Mistake 1. Two radicals are similar if they have the same
index and the same radicand. **2.** Combining $2\sqrt{6} + 2\sqrt{6} + 2\sqrt{6}$
gives us $(2 + 2 + 2)\sqrt{6} = 6\sqrt{6}$. **3.** We can not combine
$\sqrt[3]{x^2}$ and $\sqrt[3]{x^3}$ to get $2\sqrt[6]{x}$ because the radicals are not similar,
cannot be combined. **4.** The first step to combining $\frac{\sqrt{4}}{6}$ and $\frac{9}{\sqrt{4}}$
is to rationalize the denominator of the second radical expression.

Exercise Set 8.4

Vocabulary Review 1. conjugate **2.** rational **3.** division
4. multiply
Problems 1. $3\sqrt{2}$ **3.** $10\sqrt{21}$ **5.** $6\sqrt{35}$ **7.** 54 **9.** $\sqrt{6} - 9$
11. $24 + 6\sqrt[3]{4}$ **13.** $7 + 2\sqrt{6}$ **15.** $x + 2\sqrt{x} - 15$
17. $34 + 20\sqrt{3}$ **19.** $19 + 8\sqrt{3}$ **21.** $x - 6\sqrt{x} + 9$
23. $4a - 12\sqrt{ab} + 9b$ **25.** $x + 4\sqrt{x - 4}$ **27.** $x - 6\sqrt{x - 5} + 4$
29. 1 **31.** $a - 49$ **33.** $25 - x$ **35.** $x - 8$ **37.** $10 + 6\sqrt{3}$
39. $15\sqrt{3} - 26$ **41.** $2x + 2 - 2\sqrt{x^2 + 2x}$ **43.** $\frac{\sqrt{3} + 1}{2}$

45. $\frac{5\quad\sqrt{5}}{4}$ **47.** $\frac{x+3\sqrt{x}}{x-9}$ **49.** $\frac{10+3\sqrt{5}}{11}$ **51.** $\frac{3\sqrt{x}+3\sqrt{y}}{x-y}$

53. $2+\sqrt{3}$ **55.** $\frac{11-4\sqrt{7}}{3}$ **57.** $\frac{a+2\sqrt{ab}+b}{a-b}$ **59.** $\frac{x+4\sqrt{x}+4}{x-4}$

61. $\frac{5-\sqrt{21}}{4}$ **63.** $\frac{\sqrt{x}-3x+2}{1-x}$ **65.** $\frac{6x-\sqrt{x}-1}{1-9x}$ **69.** $10\sqrt{3}$

71. $x+6\sqrt{x}+9$ **73.** 75 **75.** $x^2+2x\sqrt{3}+3$

77. $\frac{5\sqrt{2}}{4}$ second; $\frac{5}{2}$ second **83.** $\frac{\text{height}}{\text{width}}\approx 1.45$, golden ratio ≈ 1.62

85. $t^2+10t+25$ **87.** x **89.** 7 **91.** $-4,-3$ **93.** $-6,-3$

95. $-5,-2$ **97.** Yes **99.** No

Find the Mistake **1.** When finding the product of $5\sqrt{6}$ and $7\sqrt{2}$, we can set $(5\sqrt{6})(7\sqrt{2})$ equal to $(5\cdot7)(\sqrt{6\cdot2})$. **2.** Multiplying conjugates of the form $\sqrt{a}+\sqrt{b}$ and $\sqrt{a}-\sqrt{b}$ will always produce a rational number. **3.** The conjugate of the expression $2\sqrt{11}+\sqrt{6}$ is $2\sqrt{11}-\sqrt{6}$. **4.** To rationalize the denominator of the expression $\frac{\sqrt{7}}{(\sqrt{5}-\sqrt{3})}$, you must multiply the numerator and the denominator by $\sqrt{5}+\sqrt{3}$.

Exercise Set 8.5

Vocabulary Review 1. squaring property of equality, original, resulting **2.** extraneous solution **3.** quadrant 1 **4.** quadrant 3

Problems 1. 4 **3.** No solution **5.** 5 **7.** No solution

9. $\frac{39}{2}$ **11.** No solution **13.** 5 **15.** 3 **17.** $-\frac{32}{3}$ **19.** 3, 4

21. $-1,-2$ **23.** -1 **25.** No solution **27.** 7 **29.** 0, 3

31. -4 **33.** 8 **35.** 0 **37.** 9 **39.** 0 **41.** 8

43. No solution **45.** 0 **47.** 6

49. **51.**

53. **55.**

57. **59.**

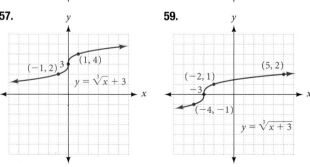

61. $h=100-16t^2$ **63.** $\frac{392}{121}\approx 3.24$ feet **65.** $l=\frac{10}{3}(\sqrt{x+1})$

67. 5 meters **69.** 2,500 meters **71.** $0\le y\le 100$

73. a. 1.12 seconds **b.** 2.74 seconds **c.** 3.71 seconds **75.** 5

77. $2\sqrt{3}$ **79.** -1 **81.** 1 **83.** 4 **85.** 2 **87.** $10-2x$

89. $2-3x$ **91.** $6+7x+20x^2$ **93.** $8x-12x^2$

95. $4+12x+9x^2$ **97.** $4-9x^2$

Find the Mistake **1.** The squaring property of equality states that the solutions to the original equation are solutions to the resulting equation when both sides of the original equation are squared. **2.** Squaring both sides of the equation $\sqrt{7x-5}=-3$ gives us $7x-5=9$. **3.** An extraneous solution satifies an equation obtained by squaring both sides of the original equation. **4.** The graph of $y=\sqrt[3]{x}$ appears in quadrant I and III because when x is positive, y is positive and when x is negative, y is negative.

Exercise Set 8.6

Vocabulary Review 1. nonnegative **2.** integer **3.** cube **4.** power
Problems 1. 2 **3.** 4 **5.** 3 **7.** 5 **9.** 3 **11.** 9 **13.** 4

15. 25 **17.** 8 **19.** 64 **21.** 8 **23.** 4 **25.** -2 **27.** 5

29. 17 **31.** 5 **33.** x **35.** x^2 **37.** $x^{2/5}$ **39.** $3y^2$

41. $3a^2b^4$ **43.** x **45.** $\frac{1}{5}$ **47.** $\frac{1}{4}$ **49.** $\frac{1}{9}$

Find the Mistake **1.** If $\sqrt[n]{x}=x^{1/n}$, then $\sqrt[8]{x}=x^{1/8}$. **2.** For the expression $243^{1/5}$, the exponent $\frac{1}{5}$ indicates that we are to find the fifth root of 243; that is, $\sqrt[5]{243}$. **3.** The expression $(b^{2/5})^{9/11}$ is equivalent to $b^{(2/5)(9/11)}$. **4.** To simplify $(64x^{18})^{1/3}$, distribute the exponent $\frac{1}{3}$ across the product.

Chapter 8 Review

Problems 1. 7 **2.** -4 **3.** 13 **4.** 5 **5.** -2 **6.** -3

7. $4\sqrt{7}$ **8.** $7\sqrt{2}$ **9.** $\frac{\sqrt{35}}{7}$ **10.** $\frac{5\sqrt[3]{2}}{2}$ **11.** $12y^2\sqrt{2}$

12. $2xy^2\sqrt{2x}$ **13.** $4\sqrt{3}$ **14.** $9x\sqrt{5}$ **15.** $\sqrt{15}+2\sqrt{3}$

16. $-3\sqrt{2}-26$ **17.** $x-36$ **18.** $5+2\sqrt{6}$ **19.** $-4+\sqrt{15}$

20. $\frac{x-4\sqrt{x}}{x-16}$ **21.** 46 **22.** 7, 2 does not check

23. No solution **24.** 16 **25.** 3

26. a

x	y
-8	-7
-1	-6
0	-5
1	-4
8	-3

27. d

x	y
0	5
1	6
4	7
9	8

28. c

x	y
-4	0
-3	1
4	2
23	3

29. b

x	y
-3	0
-2	1
1	2
6	3

30. 12 **31.** -3 **32.** 11 **33.** $\frac{1}{5}$

Chapter 8 Cumulative Review

Problems 1. $\frac{1}{4}$ **2.** 5 **3.** $\frac{3}{5}$ **4.** -37 **5.** 7 **6.** 30 **7.** a^7

8. $\frac{1}{b^5}$ **9.** $9a^4$ **10.** $10xy^6$ **11.** $\frac{1}{9}y^2-16$ **12.** $27x^2-15x$

13. $\frac{6a+13}{3a+7}$ **14.** $\frac{x+4}{3(x-3)}$ **15.** $\frac{2(3a-8)}{(a-8)(a+3)}$ **16.** 1

17. $7x^2y^2\sqrt{3y}$ **18.** $\frac{3b\sqrt{5}}{7}$ **19.** 11 **20.** $\frac{7}{2}$ **21.** $-4, 0, 5$

22. $-3, 5$ **23.** 9 **24.** \varnothing **25.** $(-3, 1)$ **26.** $(-2, 4)$
27. $(7, -1)$ **28.** $(3, -8)$
29.

30.

31.

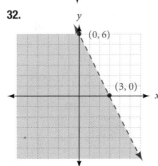

32.

33. x-intercept $= \frac{5}{2}$; y-intercept $= -5$ **34.** $m = 1$; y-intercept $= -3$
35. $y = -\frac{1}{3}x + 4$ **36.** 1 **37.** $(y-2)(y+2)(y^2+4)$ **38.** $(t-3)(t-6)$
39. $x^2(x+6)(x-2)$ **40.** $3y(4x-3y)(4x+3y)$ **41.** 2.16×10^{11}
42. 1.25×10^8 **43.** $\frac{3\sqrt{5}}{5}$ **44.** $2 + \sqrt{3}$ **45.** 8 **46.** $r = \sqrt{\frac{V}{\pi h}}$
47. $\frac{3(x+2)}{(x+5)(x-1)}$ **48.** $\frac{(x-5)(x-2)}{(x+6)(x+2)}$
49. width $= 4$ ft; length $= 7$ ft **50.** 7 and 31
51. $236{,}035$ more fans **52.** 2.04

Chapter 8 Test

Problems 1. 6 **2.** -9 **3.** 11 **4.** 4 **5.** 3 **6.** -5
7. $6\sqrt{2}$ **8.** $4\sqrt{6}$ **9.** $\frac{\sqrt{15}}{5}$ **10.** $2\sqrt[3]{3}$ **11.** $8y\sqrt{3}$
12. $xy^2\sqrt{6xy}$ **13.** $3\sqrt{5}$ **14.** $8m\sqrt{3}$ **15.** $\sqrt{6} - 3\sqrt{2}$
16. $-7 - 3\sqrt{3}$ **17.** $x - 81$ **18.** $9 - 2\sqrt{14}$
19. $-6 + \sqrt{35}$ **20.** $\frac{x + 3\sqrt{x}}{x - 9}$ **21.** 13 **22.** $-2, -4$
23. No solution **24.** 9 **25.** $\sqrt{11}$

26. d

x	y
0	2
1	3
4	4
9	5

27. b

x	y
-2	0
-1	1
2	2
7	3

28. a

x	y
-8	-4
-1	-3
0	-2
1	-1
8	0

29. c

x	y
-10	-2
-3	-1
-2	0
-1	1
6	2

30. 3 **31.** -4 **32.** 17 **33.** $\frac{1}{4}$

Chapter 9

Exercise Set 9.1

Vocabulary Review 1. completing the square, factored
2. square root property **3.** shorthand
Problems 1. $-3, 7$ **3.** $3, 5$ **5.** $3, 4$ **7.** $0, 2$ **9.** $-1, 3$
11. $-2, \frac{1}{3}$ **13.** $-\frac{3}{2}, -\frac{2}{3}$ **15.** $-\frac{3}{2}, \frac{3}{4}$ **17.** ± 5 **19.** ± 3
21. $\pm \frac{\sqrt{3}}{2}$ **23.** $\pm 2\sqrt{3}$ **25.** $\pm \frac{3\sqrt{5}}{2}$ **27.** $-2, 3$ **29.** $-3, 0$
31. $\frac{-2 \pm 2\sqrt{2}}{5}$ **33.** $-4 \pm 3\sqrt{3}$ **35.** $\frac{1}{2}, \frac{5}{2}$ **37.** $-2 \pm 2\sqrt{2}$
39. $\frac{-1 \pm 2\sqrt{5}}{2}$ **41. a.** $\frac{7}{5}$ **b.** 3 **c.** $\frac{7 \pm 2\sqrt{2}}{5}$ **d.** $\frac{71}{5}$ **e.** 3
43. $\frac{\sqrt{3}}{2}$ inch, 1 inch **45.** $\sqrt{2}$ inches **47.** 781 feet
49. $AB = x\sqrt{2}$; $BC = 2x$; $AC = (1 + \sqrt{3})x$ **51.** $115\sqrt{3} \approx 199.19$ feet
53. 9 **55.** 9 **57.** $\frac{49}{4}$ **59.** $\frac{49}{4}$ **61.** $(x+2)^2$ **63.** $(x-5)^2$
65. $(2x-3)^2$ **67.** $(3x-4)^2$ **69.** $\left(x - \frac{1}{3}\right)^2$ **71.** $\left(x - \frac{3}{4}\right)^2$
Find the Mistake 1. The first step to solving the equation $x^2 - 5x = 14$
is to write the equation in the form $ax^2 + bx + c = 0$. **2.** Using
the square root property for equations, the solution $x^2 = 100$ is ± 10.
3. Solving $x^2 + 16x + 64 = 20$ gives the solutions $x = -8 + 2\sqrt{5}$
and $x = -8 - 2\sqrt{5}$. **4.** If the shortest side of a $30°$–$60°$–$90°$
triangle is 3 feet and the hypotenuse is 6 feet, then the other side
measures $3\sqrt{3}$ feet.

Exercise Set 9.2

Vocabulary Review 1. equation **2.** leading coefficient
3. square, linear term **4.** binomial **5.** square root property
Problems 1. $36, 6$ **3.** $4, 2$ **5.** $25, 5$ **7.** $\frac{25}{4}, \frac{5}{2}$
9. $\frac{49}{4}, \frac{7}{2}$ **11.** $\frac{1}{16}, \frac{1}{4}$ **13.** $\frac{1}{9}, \frac{1}{3}$ **15.** $-6, 2$ **17.** $-3, -9$
19. $-1 \pm \sqrt{6}$ **21.** $4 \pm \sqrt{15}$ **23.** $\frac{5 \pm \sqrt{37}}{2}$ **25.** $1 \pm \sqrt{5}$
27. $\frac{4 \pm \sqrt{13}}{3}$ **29.** $\frac{3 \pm \sqrt{89}}{8}$ **31.** $\frac{-2 \pm \sqrt{7}}{3}$ **33.** $\frac{5 \pm \sqrt{47}}{2}$
35. $\frac{5 \pm \sqrt{69}}{4}$ **37.** $\frac{5 \pm \sqrt{37}}{6}$ **39.** $\frac{9 \pm \sqrt{41}}{10}$ **41.** $\frac{2 \pm \sqrt{34}}{10}$
43. $\frac{1 \pm \sqrt{13}}{6}$ **45.** $\frac{2 \pm \sqrt{10}}{2}$ **47. a.** Yes **b.** ± 3
49. a. $0, 6$ **b.** $0, 6$ **51. a.** $-7, 5$ **b.** $-7, 5$ **53.** No **55.** $x^2 = 5$
57. $x^2 - 2x - 2 = 0$ **59.** $9x^2 - 6x - 4 = 0$ **61.** 169 **63.** 49
65. $\frac{1}{2}$ **67.** 1 **69.** $(3t-2)(9t^2 + 6t + 4)$ **71.** $2(x+3)(x^2 - 3x + 9)$
Find the Mistake 1. A perfect square trinomial has a constant term
that is the square of half the linear term. **2.** To solve the equation
$x^2 - 9x + 11 = 0$ by completing the square, add -11 to both sides
of the equation and then add $\frac{81}{4}$ to both sides. **3.** The first step to
solving the equation $4x^2 - 6x + 5 = 0$ is to add -5 to both sides and
then divide both sides by 4 because the leading coefficient is not 1.
4. The square of half the coefficient of the linear term in the above
equation is $\frac{9}{16}$.

Exercise Set 9.3

Vocabulary Review 1. standard form **2.** solutions
3. rational **4.** revenue
Problems 1. $-3, -2$ **3.** $2 \pm \sqrt{3}$ **5.** $1, 2$ **7.** $1, 2$ **9.** $0, 5$
11. $0, -\frac{4}{3}$ **13.** $\frac{3 \pm \sqrt{5}}{4}$ **15.** $-3 \pm \sqrt{17}$ **17.** $\frac{-2 \pm \sqrt{10}}{3}$
19. 1 **21.** $\frac{1 \pm \sqrt{73}}{4}$ **23.** $4 \pm \sqrt{2}$ **25.** $\frac{1}{2}, 1$ **27.** $-\frac{1}{2}, 3$
29. $\frac{3 \pm \sqrt{17}}{2}$ **31.** $1 \pm \sqrt{2}$ **33.** $\frac{-3 \pm \sqrt{5}}{2}$ **35.** $3, -5$
37. $1 \pm \sqrt{10}$ **39.** $\frac{3 \pm \sqrt{17}}{4}$ **41.** $0, \frac{-1 \pm \sqrt{7}}{2}$ **43.** $0, 1 \pm \sqrt{3}$
45. $0, \frac{-1 \pm \sqrt{5}}{4}$ **47.** a and b **49. a.** $\frac{5}{3}, 0$ **b.** $\frac{5}{3}, 0$

51. No, $2 \pm \sqrt{11}$ **53.** Yes **55.** 2 seconds **57.** 20 or 60 items

59. a. $P(x) = 9.5x - 0.05x^2 - 200$
 b. $P(100) = 250$, $P(200) = -300$ **c.** $x = 24.11, 165.89$

61. 0.49 centimeter (8.86 cm is impossible)

63. a. $l + w = 10$, $lw = 15$ **b.** 8.16 yards, 1.84 yards
 c. Two answers are possible because either dimension (long or short) may be considered the length.

65. -2 **67.** 1,322.5 **69.** $-\dfrac{7}{640}$ **71.** 1, 5 **73.** $-3, 1$

75. $3 \pm \sqrt{14}$ **77.** 9, 3 **79.** 1, 1

Find the Mistake **1.** The solutions for any quadratic equation in the form of $ax^2 + bx + c = 0$ are $x = \dfrac{-b \pm \sqrt{b^2 - 4ac}}{2a}$. **2.** Using the quadratic formula to solve the equation $2x^2 - 7x = -11$, $a = 2$, $b = -7$, $c = 11$. **3.** Solving $x^2 = -8x + 12$ using the quadratic formula gives us $x = -4 \pm 2\sqrt{7}$. **4.** To solve $3x^2 - 6x - 4 = 0$, apply the quadratic formula to get $x = \dfrac{-(-6) \pm \sqrt{(-6)^2 - 4(3)(-4)}}{2(3)}$.

Landmark Review **1.** $y = \pm 2\sqrt{5}$ **2.** $x = \pm 3\sqrt{3}$

3. $x = -6, 12$ **4.** $a = -\dfrac{1}{2}, \dfrac{9}{2}$ **5.** $y = -9, 1$ **6.** $y = -6, -4$

7. $x = 1 \pm \sqrt{3}$ **8.** $y = 5 \pm \sqrt{11}$ **9.** $\dfrac{5 \pm 3\sqrt{5}}{2}$

10. $x = \dfrac{-3 \pm \sqrt{6}}{2}$ **11.** $x = -2, 3$ **12.** $1 \pm \sqrt{3}$

Exercise Set 9.4

Vocabulary Review **1.** ordered pairs **2.** parabola **3.** y-intercept
4. vertex **5.** concave **6.** graph

Problems

1.

x-intercepts $= -3, 1$;
vertex $= (-1, -4)$

3.
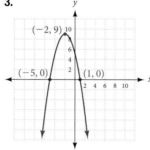
x-intercepts $= -5, 1$;
vertex $= (-2, 9)$

5.
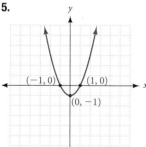
x-intercepts $= -1, 1$;
vertex $= (0, -1)$

7.
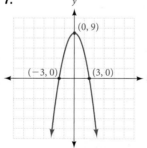
x-intercepts $= -3, 3$;
vertex $= (0, 9)$

9.
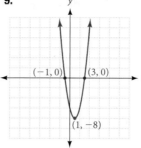
x-intercepts $= -1, 3$;
vertex $= (1, -8)$

11.

x-intercepts $= 1 - \sqrt{5}, 1 + \sqrt{5}$;
vertex $= (1, -5)$

13.

15.

17.

19.

21.
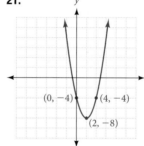
Vertex $= (2, -8)$

23.
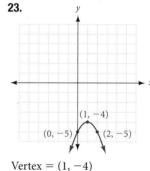
Vertex $= (1, -4)$

25.
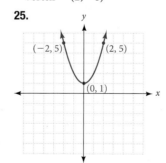
Vertex $= (0, 1)$

27.
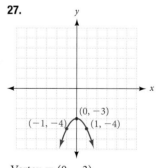
Vertex $= (0, -3)$

29.

Vertex $= \left(-\frac{2}{3}, -\frac{1}{3}\right)$

31.

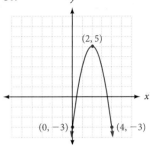

Vertex $= (2, 5)$

Problems

1.

x	y
−2	−8
0	0
2	8
5	20

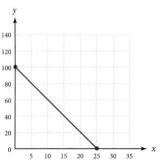

33. $(3, -4)$, lowest **35.** $(1, 9)$, highest **37.** $(2, 16)$, highest
39. $(-4, 16)$, highest **41.** 875 patterns; maximum profit $731.25
43. The ball is in her hand when $h = 0$, which means $t = 0$ or $t = 2$ seconds. Maximum height is 16 feet.

45.

Maximum $R = \$3,600$ when $p = \$6.00$

47.

Maximum $R = \$7,225$ when $p = \$8.50$

3.

x	y
0	100
5	80
10	60
15	40
20	20
25	0

5.

x	y
0	1600
40	1200
80	800
120	400
160	0

49.

$y = -\frac{1}{135}(x - 90)^2 + 60$

Find the Mistake 1. All equations of the form $ax^2 + bx + c = 0$ have parabolas for graphs. **2.** The graph of the function $y = ax^2 + bx + c$ crosses the y–axis where $y = 0$. **3.** The vertex for the graph $y = 4x^2 + 6x + 7$ occurs when $y = -\frac{6}{8} = -\frac{3}{4}$.
4. The graph of $y = ax^2 + bx + c$ will be concave down when a is negative.

Exercise Set 9.5
Vocabulary Review 1. domain, range **2.** left, right
3. function **4.** relation **5.** vertical line test

7.

x	y
0	40.0
25	102.5
50	165.0
75	227.5
100	290.0

9. Domain = {1, 3, 5, 7}; Range = {2, 4, 6, 8}; A function
11. Domain = {0, 1, 2, 3}; Range = {4, 5, 6}; A function
13. Domain = {a, b, c, d }; Range = {3, 4, 5}; A function
15. Domain = {a}; Range = {1, 2, 3, 4}; Not a function
17. Domain = {AAPL, DELL, MSFT, FB, GRPN};
Range = {$562, $12, $29, $32}
19. Domain = {Atlanta, Chicago, Boston, Houston};
Range = {Braves, White Sox, Cubs, Red Sox, Astros}
21. Yes **23.** No **25.** No **27.** Yes **29.** Yes **31.** Yes
33. Domain = $\{x \mid -5 \le x \le 5\}$, $[-5, 5]$;
Range = $\{y \mid 0 \le y \le 5\}$, $[0, 5]$
35. Domain = $\{x \mid -5 \le x \le 3\}$, $[-5, 3]$;
Range = $\{y \mid y = 3\}$, $[3, 3]$

37.

$y = x^2 - 1$

Domain = All real numbers;
Range = $\{y \mid y \ge -1\}$; A function

39.

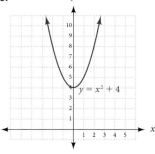

$y = x^2 + 4$

Domain = All real numbers;
Range = $\{y \mid y \ge 4\}$; A function

c.

d. Domain = $\{x \mid 10 \le x \le 40\}$; Range = $\{y \mid 85 \le y \le 340\}$
e. Minimum = \$85; Maximum = \$340

53. a. Domain = {2008, 2009, 2010, 2011, 2012};
Range = {5.8, 5.2-5.5, 7.5, 9.7, 12.1} **b.** 2012 **c.** 2009

55. a. III **b.** I **c.** II **d.** IV **57. a.** 6 **b.** 7.5 **59. a.** 27 **b.** 6

61. 1 **63.** -3 **65.** $x = 0$ **67.** $-\frac{6}{5}$ **69.** $-\frac{35}{32}$

Find the Mistake 1. The set of all inputs for a function is called the domain. **2.** For the relation (1, 2), (3, 4), (5, 6), (7, 8), the range is {2, 4, 6, 8}. **3.** Graphing the relation $y = x^2 - 3$ will show that the domain is all real numbers. **4.** If a vertical line crosses the graph of a relation in more than one place, the relation cannot be a function.

41.

$x = y^2 - 1$

Domain = $\{x \mid x \ge -1\}$;
Range = All real numbers;
Not a function

43.

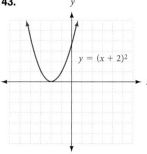

$y = (x + 2)^2$

Domain = All real numbers;
Range = $\{y \mid y \ge 0\}$; A function

Exercise Set 9.6

Vocabulary Review 1. independent **2.** dependent **3.** $f(x)$
4. function f

Problems 1. -1 **3.** -11 **5.** 2 **7.** 4 **9.** $a^2 + 3a + 4$
11. $2a + 7$ **13.** 1 **15.** -9 **17.** 8 **19.** 0 **21.** $3a^2 - 4a + 1$
23. $3a^2 + 8a + 5$ **25.** 4 **27.** 0 **29.** 2 **31.** 4
33. $x = \frac{1}{2}$ **35.** 24 **37.** -1 **39.** $2x^2 - 19x + 12$ **41.** 99
43. 28 **45.** 225 **47.** $\frac{3}{10}$ **49.** $\frac{2}{5}$ **51.** Undefined **53.** $\frac{1}{2}$
55. a. $a^2 - 7$ **b.** $a^2 - 6a + 5$ **c.** $x^2 - 2$ **d.** $x^2 + 4x$
e. $a^2 + 2ab + b^2 - 4$ **f.** $x^2 + 2xh + h^2 - 4$

45.

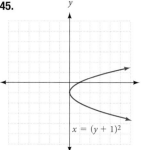

$x = (y + 1)^2$

Domain = $\{x \mid x \ge 0\}$;
Range = All real numbers;
Not a function

47.

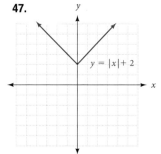

$y = |x| + 2$

Domain = All real numbers;
Range = $\{y \mid y \ge 2\}$; A function

49. Domain = {2001, 2003, 2005, 2007, 2009};
Range = {5, 10, 40, 80, 160}

51. a. $y = 8.5x$ for $10 \le x \le 40$
b.

Table 4 Weekly Wages		
Hours Worked	**Function Rule**	**Gross Pay ($)**
x	$y = 8.5x$	y
10	$y = 8.5(10) = 85$	85
20	$y = 8.5(20) = 170$	170
30	$y = 8.5(30) = 255$	255
40	$y = 8.5(40) = 340$	340

57.

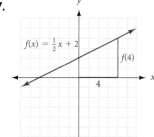

$f(x) = \frac{1}{2}x + 2$

$f(4)$

59. $x = 4$

61.

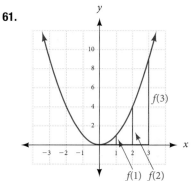

$f(3)$

$f(1)$ $f(2)$

63. $V(3) = 300$, the painting is worth \$300 in 3 years; $V(6) = 600$, the painting is worth \$600 in 6 years.

65. a. True **b.** False **c.** True **d.** False **e.** True

67. a. $5,625 **b.** $1,500 **c.** $\{t \mid 0 \le t \le 5\}$

d.
e.
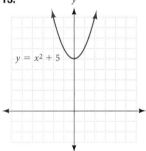

Value of a Copy Machine

$\{V(t) \mid 1,500 \le V(t) \le 18,000\}$ **f.** About 2.42 years

69. a.

d (feet)	f(d) (KWH/year)
0	0
5	332
10	1328
15	2988
20	5312

Source: builditsolar.com/Projects/Wind/Wind.htm

b. $0 \le f(d) \le 5312$

71. $-0.1x^2 + 35x$ **73.** $4x^2 - 7x + 3$ **75.** $-0.1x^2 + 27x - 500$

77. $2x^2 + 8x + 8$ **79.** 10 **81.** $2x^2 - 17x + 33$ **83.** $24x - 6$

Find the Mistake 1. To evaluate the function $f(x) = 10x + 3$ for $f(1)$ we begin by replacing x with 1. **2.** For the average speed function $s(t) = \dfrac{60}{t}$, finding $s(6)$ shows an average speed of 10 miles per hour. **3.** If the volume of a ball is given by $V(r) = \dfrac{4}{3}\pi r^3$ where r is the radius, we see that the volume increases as the radius of the ball increases. **4.** If $f(x) = x^2$ and $g(x) = 2x + 4$, evaluating $f(g(2))$ is the same as finding $g(2)$ and using that value in $f(x) = x^2$ for x.

Chapter 9 Review

1. $-5, 3$ **2.** $4 \pm 2\sqrt{6}$ **3.** $\dfrac{3 \pm 4\sqrt{2}}{5}$ **4.** $-\dfrac{3}{2}, 6$ **5.** $\dfrac{5}{3}, -2$

6. $-4, -1$ **7.** $\dfrac{4}{3}$ **8.** 8 in. **9.** $4 \pm 2\sqrt{3}$ **10.** $\dfrac{-7 \pm \sqrt{97}}{8}$

11. $\dfrac{1 \pm 3\sqrt{5}}{2}$ **12.** $\dfrac{3 \pm \sqrt{65}}{2}$

13.

$y = x^2 + 5$

14.
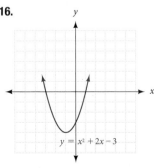
$y = (x - 4)^2$

15.
$y = (x + 3)^2 + 1$

16.
$y = x^2 + 2x - 3$

17. b **18.** a **19.** c **20.** d **21.** d **22.** b **23.** a **24.** c

25. Domain = $\{a, 5, x, 12\}$; Range = $\{4, a, 8, x\}$; a function

26. Domain = $\{n, 1\}$; Range = $\{4, 3, 7, 8\}$; not a function

27. 69 **28.** 0 **29.** 1 **30.** 471

Chapter 9 Cumulative Review

1. 5 **2.** $-\dfrac{7}{6}$ **3.** $\dfrac{1}{x^{10}}$ **4.** $\dfrac{3}{x^5}$ **5.** $\dfrac{3x^4y^6}{4}$ **6.** $\dfrac{x + 3}{x + 4}$

7. 7 **8.** -16 **9.** $\dfrac{5\sqrt{6}}{3}$ **10.** $\dfrac{\sqrt{11} + \sqrt{2}}{3}$

11. $a^3 + a^2 - 18a + 18$ **12.** $x + 4$ **13.** $14x^2y^2\sqrt{x}$

14. $3b\sqrt{3}$ **15.** -1 **16.** $-\dfrac{1}{2}$ **17.** $\dfrac{3}{2}$ **18.** $-\dfrac{2}{5}, 0, 3$

19. $-\dfrac{1}{6}, 0$ **20.** $\dfrac{1}{3}, 3$ **21.** -8 **22.** $-6, 2$

23.

24.

25.

26.

27.
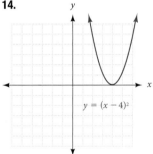
$y = x^2 + 4$

28.
$y = (x + 1)^2 - 6$

29. $(2, -4)$ **30.** $(8, -1)$ **31.** $(x - 7)(x + 5)$

32. $(3x - 5)(2x + 7)$ **33.** $(9 - a)(9 + a)$ **34.** $(5x + 2)^2$

35. $m = -\dfrac{2}{3}$ **36.** $m = \dfrac{5}{3}$; y-intercept $= -2$ **37.** $6ab^3 - 8b^2 + 5a^2$

38. $3x^2 - x - 4 - \dfrac{12}{2x - 1}$ **39.** $-\dfrac{9}{a - 6}$ **40.** $\dfrac{1}{(x + 2)(x + 3)}$

41. $20\sqrt{3}$ **42.** $5x\sqrt{6}$ **43.** $17,000 at 6%, $8,000 at 7%

44. $13,000 at 5%, $4,000 at 9% **45.** 12 **46.** 3

47. 131,250 people **48.** 1,777,067 people

Chapter 9 Test

1. $-1, 9$ **2.** $2 \pm 3\sqrt{2}$ **3.** $\dfrac{7 \pm 5\sqrt{2}}{3}$ **4.** $\dfrac{3}{2}, 1$ **5.** $\dfrac{1}{4}, -2$

6. $-2, 5$ **7.** $\dfrac{5}{2}$ **8.** 3 in. **9.** $\dfrac{3 \pm \sqrt{37}}{2}$ **10.** $\dfrac{-5 \pm \sqrt{41}}{4}$

11. $\dfrac{7 \pm \sqrt{489}}{10}$ **12.** $-5, 4$

13.

14.

15.

16.

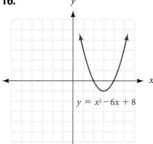

17. c **18.** b **19.** a **20.** d **21.** c **22.** d **23.** a **24.** b

25. Domain = $\{x, 7, y, 10\}$; Range = $\{3, x, 8, y\}$; a function

26. Domain = $\{a, 7\}$; Range = $\{8, 4, 9, 12\}$; not a function

27. 3 **28.** -2 **29.** 5 **30.** 66

Index